AF334177

APPLICATIONS OF MICRODIALYSIS IN PHARMACEUTICAL SCIENCE

APPLICATIONS OF MICRODIALYSIS IN PHARMACEUTICAL SCIENCE

Edited by

TUNG-HU TSAI
National Yang-Ming University
Taipei, Taiwan

 **WILEY**

A JOHN WILEY & SONS, INC., PUBLICATION

Library of Congress Cataloging-in-Publication Data:

Applications of microdialysis in pharmaceutical science / [edited by] Tung-Hu Tsai.
    p. ; cm.
  Includes bibliographical references and index.
  ISBN 978-0-470-40928-2 (cloth : alk. paper)
  1. Pharmaceutical chemistry. 2. Drug development. 3. Brain macrodialysis.
I. Tsai, Tung-Hu.
  [DNLM: 1. Chemistry, Pharmaceutical–methods. 2. Microdialysis–methods. QV 744]
  RM301.25.A67 2011
  615'.19–dc22

                        2011010963

Printed in Singapore

oBook ISBN: 9781118011294
ePDF ISBN: 9781118011270
ePub ISBN: 9781118011287

10 9 8 7 6 5 4 3 2 1

CONTENTS

CONTRIBUTORS

Ellen E. Blaak, Maastricht University Medical Centre, Maastricht, The Netherlands

Martin Brunner, Medical University of Vienna, Vienna, Austria

Anna Ciechanowska, Polish Academy of Sciences, Warsaw, Poland

Ralph Clinckers, Vrije Universiteit Brussels, Brussels, Belgium

Luc Denoroy, Université de Lyon and Lyon Neuroscience Research Center, BioRaN Team, Lyon, France; Université Lyon 1, Villeurbanne, France

Hartmut Derendorf, University of Florida, Gainesville, Florida

Giuseppe Di Giovanni, University of Malta, Msida, Malta; Cardiff University, Cardiff, UK

Vincenzo Di Matteo, Istituto di Richerche Farmacologiche Consorzio Mario Negri Sud, Santa Maria Imbaro, Italy

Thomas Feurstein, Medical University of Vienna, Vienna, Austria

Wolfgang Frenzel, Technical University of Berlin, Berlin, Germany

James M. Gallo, Mount Sinai School of Medicine, New York, New York

Gijs H. Goossens, Maastricht University Medical Centre, Maastricht, The Netherlands

Christian Höcht, Universidad de Buenos Aires, Buenos Aires, Argentina

Rie Ikeda, Nagasaki University, Nagasaki, Japan

Janusz Krzymien, Medical University of Warsaw, Warsaw, Poland

Courtney D. Kuhnline, University of Kansas, Lawrence, Kansas

Piotr Ladyzynski, Polish Academy of Sciences, Warsaw, Poland

Wen-Chuan Lee, National Yang-Ming University, Taipei, Taiwan

Susan M. Lunte, University of Kansas, Lawrence, Kansas

Iwona Maruniak-Chudek, Medical University of Silesia, Katowice, Poland

Yvette Michotte, Vrije Universiteit Brussels, Brussels, Belgium

Manuel Miró, University of the Balearic Islands, Palma de Mallorca, Illes Balears, Spain

Runa Naik, University of Florida, Gainesville, Florida

Kenichiro Nakashima, Nagasaki University, Nagasaki, Japan

Pradyot Nandi, University of Kansas, Lawrence, Kansas

Greg Nowak, Karolinska Institute, Karolinska University Hospital Huddinge, Stockholm, Sweden

Sandrine Parrot, Université de Lyon and Lyon Neuroscience Research Center, NeuroChem, Lyon, France; Université Lyon 1, Villeurbanne, France

Massimo Pierucci, University of Malta, Msida, Malta

Bernard Renaud, Université de Lyon and Lyon Neuroscience Research Center, NeuroChem, Lyon, France; Université Lyon 1, Villeurbanne, France

Martina Sahre, University of Florida, Gainesville, Florida

Wim H. M. Saris, Maastricht University Medical Centre, Maastricht, The Netherlands

Ilse Smolders, Vrije Universiteit Brussels, Brussels, Belgium

Tung-Hu Tsai, National Yang-Ming University and Taipei City Hospital, Taipei, Taiwan

Mitsuhiro Wada, Nagasaki University, Nagasaki, Japan

Jan M. Wojcicki, Polish Academy of Sciences, Warsaw, Poland

Yu-Tse Wu, National Yang-Ming University, Taipei, Taiwan

Markus Zeitlinger, Medical University of Vienna, Vienna, Austria

Qingyu Zhou, Mount Sinai School of Medicine, New York, New York

Luc Zimmer, Université de Lyon and Lyon Neuroscience Research Center, BioRaN Team, Lyon, France; Université Lyon 1, Villeurbanne, France

1

INTRODUCTION TO APPLICATIONS OF MICRODIALYSIS IN PHARMACEUTICAL SCIENCE

Tung-Hu Tsai

Institute of Traditional Medicine, National Yang-Ming University, and Taipei City Hospital, Taipei, Taiwan

Microdialysis is a very useful sampling tool that can be used in vivo to acquire concentration variations of protein-unbound molecules located in interstitial or extracellular spaces. This technique relies on the passive diffusion of substances across a dialysis membrane driven by a concentration gradient. After a microdialysis probe has been implanted in the target site for sampling, generally a blood vessel or tissue, a perfused solution consisting of physiological buffer solution flows slowly across the dialysis membrane, carrying away small molecules that come from the extracellular space on the other side of the dialysis membrane. The resulting dialysis solution can be analyzed to determine drug or target molecules in microdialysis samples by liquid chromatography or other suitable analytical techniques. In addition, it can be applied to introduce a substance into the extracellular space by the microdialysis probe, a technique referred to as *reverse microdialysis*. In this way, regional drug administration and simultaneous sampling of endogenous compounds in the extracellular compartments can be performed at the same time.

Initially, miniaturized microdialysis equipment was developed to monitor neurotransmitters continuously [1], and over the decades its use has extended to different fields, especially for drug discovery and clinical medicine. The main objectives in the early stages of drug development are to choose promising candidates and to determine optimally safe and effective dosages. Pharmacokinetic (PK) simulation is concerned with the time course of drug concentration in the body, and pharmacodynamic (PD) simulation deals with the relationship of drug effect versus concentration. The method of PK–PD modeling can be used to determine the clinically relevant relationship between time and therapeutic effect. It also expedites drug development and helps make critical decisions, such as selecting the optimal dosage regimen and planning the costly clinical trials that are critical in determining the fate of a new compound [2–4]. The conventional concept for PK–PD evaluation of medicines is to measure total drug concentrations (including bound- and free-form drug molecules) in the blood circulation. However, only free-form drug molecules can reach specific tissues for therapeutic effect, and thus determining drug levels at the site of action is a more effective method of obtaining accurate PK–PD relationships of drugs.

The case of antibiotics serves as a good example to elucidate this concept. Most infections occur in peripheral tissues (extracellular fluid) but not in plasma, and the distribution of antibiotics to the target sites is a main determinant of clinical outcome [5]. Hence, the non-protein-bound (free-form) drug concentration at the infection site should be a better indicator for therapeutic efficacy of antibiotics than indices such as the time above the minimum inhibitory concentration (MIC), the maximum concentration of drug in serum (C_{max})/MIC, or the area under the curve over 24 h (AUC_{24})/MIC derived from the total plasma concentration [6]. Recently, regulatory authorities, including the U.S. Food and Drug Administration, have also emphasized the value of human-tissue drug concentration data and support the use of clinical microdialysis to obtain this type of pharmacokinetic information [7], further indicating the significance of this technique.

This book focuses on the utilization of microdialysis in various organs and tissues for PK and PD studies, covering the range of current clinical uses for microdialysis. Topics include applications of this device for drug discovery, analytical consideration of samples, central neurological disease investigations, sampling at different organs, diabetes evaluations, tumor response estimations, and comparison of microdialysis with other image techniques. Special applications of microdialysis such as in vitro sampling for cell media, drug–drug interaction studies, and environmental monitoring are also included. Drug discovery and the role of microdialysis in drug development are described in Chapter 2. Due to the cost and time required for drug development, a more complete understanding of the pharmacokinetic, pharmacodynamic, and toxicological properties of leading drug candidates during the early stages of their development is fundamental to prevent failure. The use of microdialysis in early drug development involves the estimation of plasma protein binding, in vivo pharmacodynamic models, in vivo pharmacokinetics, and PK–PD relationships.

Chapter 3 presents general considerations for microdialysis sampling and microdialysis sample analysis. The homogeneity or heterogeneity of a sampling site must be considered initially, and selecting the appropriate microdialysis probe and sampling parameters helps improve the spatial resolution within a specific region. Moreover, optimization of testing parameters, such as perfusion flow rate and modification of perfusion solutions, increases the extraction efficiency for more reproducible results. In addition, the advancement of analytical methodology supports a wider use of microdialysis, because highly sensitive detection instruments are capable of detecting trace analytes contained in the very small volume samples.

Microdialysis applications for several nervous system diseases, such as dopamine-related disorders, glutamate- and r-aminobutyric acid (GABA)-linked neurobiological events, as well as the neurobiological mechanisms of seizures and antiepileptic drug action, are discussed in detail in Chapters 4 to 6. Dopamine is a neurotransmitter with multiple functions, and abnormal concentrations in the body have been known to lead to movement, cognitive, motivational, and learning deficits [8,9]. In the central nervous system, glutamic acid and aspartic acid are the chief excitatory amino acid neurotransmitters, while GABA and glycine are the main inhibitory transmitters. One of the chronic neurological diseases associated with these neurotransmitters is epilepsy, so GABA neurotransmission is a target for the design and development of drugs to treat epilepsy. In addition, cerebral microdialysis can help clarify the mechanisms of action of psychostimulants, addictive drugs, and analgesics, as well as contributing to studies on the control of amino acid–related neurons by receptors. A combination of microdialysis with brain imaging and immunological detection methods can further confirm and correct the results from those investigations. Microdialysis allows experiments to be performed in animals while conscious and with minimal movement restrictions, so that seizure-related behavioral changes can be both determined more accurately and correlated more closely with the fluctuation of neurotransmitters observed. As mentioned above, microdialysis is the method of choice for pharmacokinetic evaluations, because it samples the pharmacodynamically active free-form drug molecules. Microdialysis also permits the disposition and transport across the blood–brain barrier of antiepileptic drugs to be assessed. In short, microdialysis is an indispensable tool for the evaluation of neurotransmitters and thereby contributes to understanding the pathophysiology of neurological illnesses.

The range of current applications of microdialysis for clinical evaluation and basic research on different organs is presented in Chapters 7 to 14. Chapter 7 cover microdialysis in the lung for monitoring exogenous and endogenous compounds. Implanting a microdialysis probe in interstitial lung tissue is much more complex than is implanting probe in other peripheral tissues (e.g., skin, muscle, or adipose), because the lung has a protected anatomical position and is a highly vulnerable organ. Clinically, thoracotomy is generally required to avoid the risk from the abnormal presence of air in the pleural cavity, which results in collapse of the lung in clinical studies, thus limiting lung microdialysis experiments in patients with elective thoracic surgery. Due to the clinical significance of infections in the lower respiratory tract, studies have focused on the pharmacokinetics of antimicrobial agents in lung tissue and the epithelial lining fluid to understand the amount of drugs that penetrate to the infection site. Another vital organ, the liver, is not only responsible for many metabolic processes but also produces bile, which contains surfactant-like components that facilitate digestive processes. Chapter 8 demonstrates how microdialysis offers an alternative way to monitor drug metabolism in the rat liver. By using microdialysis to investigate drug metabolism, the integrity and physiological conditions of the animal can be maintained, and more of the actual metabolic processes of xenobiotic compounds can be observed than with heptocyte culture systems and in vitro enzymatic reactions. In the field of organ transplants, microdialysis combined with an enzymatic analyzer has been employed successfully to determine glucose, pyruvate, lactate, and glycerol to monitor tissue metabolism after liver transplants in humans, as discussed in Chapter 9.

The ability of microdialysis to measure free drug concentrations at the site of drug action makes it an excellent tool for bioavailability and bioequivalence assessment. Therefore, it has been used to determine bioequivalence of topical dermatological products according to industry and regulatory recommendations [10]. Chapter 10 reviews microdialysis applications to skin and soft tissues and their impact on clinical drug development. White adipose tissue is generally considered to be the main site for lipid storage in the human body. However, it is now also viewed as an active and important organ involved in various metabolic processes by secreting several hormones and a variety of substances called *adipokines*. Practical considerations and applications of microdialysis on adipose tissue in humans are detailed further in Chapter 11. Microdialysis has been used to observe the regulation of lipolysis in human adipose tissue by determining the extracellular concentrations of glycerol as an indicator. Disturbances of adipose tissue metabolism may lead to illness, and obesity has been determined as a major risk factor for hyperlipidemia, cardiovascular diseases, and type 2 diabetes [11]. Diabetes is a metabolic disorder in which the body produces insufficient insulin (type 1 diabetes) or where there is insulin resistance (type 2 diabetes). Long-term metabolic control in diabetic patients is crucial, and the microdialysis system is a suitable technique for continuous measurement of glucose concentrations. Chapter 12 describes the application of microdialysis to diabetes-related events in patients, including the diabetic patient's metabolic state and the monitoring of antibiotic therapies for the feet of diabetics.

Cancer affects people worldwide and is the leading cause of death in modern societies, and chemotherapy research is pursuing more specific antineoplastic agents to reduce adverse drug effects in patients. Chapter 13 focuses on the PK–PD evaluation of anticancer drugs by microdialysis and describes its recent employment to evaluate drug disposition and response in solid tumors. In addition to microdialysis, advanced imaging techniques such as positron-emission tomography and magnetic resonance spectroscopy have also become available to assess drug distribution, and Chapter 14 compares microdialysis with imaging approaches for evaluating in vivo drug distribution. Their advantages and drawbacks are reviewed, and their values as translational tools for clinical decisions and drug development are discussed.

Chapters 15 to 17 introduce special applications of microdialysis in studies of cell culture assays, drug–drug interactions, and environmental monitoring. Cell-based assays are essential in the preclinical phase of drug development, because these in vitro systems can speed up the processes of screening lead compounds, assessing metabolic stability, and evaluating permeation across membranes such as the gastrointestinal tract and the blood–brain barrier. Microdialysis sampling of cell culture systems, enzyme kinetics, and protein-binding assays are discussed in Chapter 15. Drug interaction is an important topic for clinical pharmacy, especially since the incidence of drug interactions is expected to increase with the increasing number of new drugs brought to the market. Exploring the relevance and mechanisms of drug interactions will assist clinicians in avoiding these often serious events. Herbal products, dietary supplements, and foods can also induce drug interactions. The reduced concentration of a free-form drug can cause treatment failure, while side effects or toxicity may occur when the drug level increases. In Chapter 16, the use of microdialysis as a tool to evaluate drug–drug or food–drug interactions is described. Recent pharmacokinetic and pharmacodynamic reports of drug–drug interactions are reviewed. Chapter 17 illustrates microdialysis as an in situ sample system by providing to the experimenter simultaneous sampling, cleanup, and real-time monitoring of targeted analytes for monitoring aqueous or solid environmental compartments or plant tissues. Although the designs of microdialysis probes for in vivo sampling are similar, modifications for monitoring particular environments can be made to enhance extraction efficiency by manipulating membrane materials, effective length of dialysis membrane, and perfusate composition. Several practical examples for environmental monitoring are also presented.

Compared with other methods of sampling intact tissue or body fluids, microdialysis offers several advantages for the experimenter. It provides the free fraction of drug molecules, which is the bioactive portion, so that more accurate PK–PD relationships can be constructed to help drug development and clinical therapeutic regimens. In addition, temporal resolution of data is improved dramatically by its continuous sampling, which can be used to observe, almost in real time, in vivo and in vitro enzymatic processes and reactions. Furthermore, the in situ measurement and sample preparation characteristics of microdialysis provide relatively clear dialysate that is ready for analysis; and sample contamination and dilution can be avoided when further treatments and extraction are performed. In sum, a broad range of studies applying microdialysis have been realized, as shown by the various topics presented in this book, making microdialysis an indispensable tool for pharmaceutical studies.

REFERENCES

[1] Ungerstedt, U., Pycock, C. (1974). Functional correlates of dopamine neurotransmission. *Bulletin der Schweizerischen Akademie der Medizinischen Wissenschaften, 30*, 44–55.

[2] Miller, R., Ewy, W., Corrigan, B.W., Ouellet, D., Hermann, D., Kowalski, K.G., Lockwood, P., Koup, J.R., Donevan, S., El-Kattan, A., Li, C.S., et al. (2005). How modeling and simulation have enhanced decision making in new drug development. *Journal of Pharmacokinetics and Pharmacodynamics, 32*, 185–197.

[3] Lalonde, R.L., Kowalski, K.G., Hutmacher, M.M., Ewy, W., Nichols, D.J., Milligan, P.A., Corrigan, B.W., Lockwood, P.A., Marshall, S.A., Benincosa, L.J., et al. (2007). Model-based drug development. *Clinical Pharmacology & Therapeutics, 82*, 21–32.

[4] Schmidt, S., Barbour, A., Sahre, M., Rand, K.H., Derendorf, H. (2008). PK/PD: new insights for antibacterial and antiviral applications. *Current Opinion in Pharmacology, 8*, 549–556.

[5] Liu, P., Müller, M., Derendorf, H. (2002). Rational dosing of antibiotics: the use of plasma concentrations versus tissue concentrations. *International Journal of Antimicrobial Agents, 19*, 285–290.

[6] Brunner, M., Derendorf, H., Müller, M. (2005). Microdialysis for in vivo pharmacokinetic/pharmacodynamic characterization of anti-infective drugs. *Current Opinion in Pharmacology, 5*, 495–499.

[7] Chaurasia, C.S., Müller, M., Bashaw, E.D., Benfeldt, E., Bolinder, J., Bullock, R., Bungay, P.M., DeLange, E.C., Derendorf, H., Elmquist, W.F., et al. (2007). AAPS–FDA Workshop White Paper: Microdialysis Principles, Application, and Regulatory Perspectives. *Journal of Clinical Pharmacology, 47*, 589–603.

[8] Bjorklund, A., Dunnett, S.B. (2007). Fifty years of dopamine research. *Trends in Neurosciences, 30*, 185–187.

[9] Schultz, W. (2007). Multiple dopamine functions at different time courses. *Annual Review of Neuroscience, 30*, 259–288.

[10] Schmidt, S., Banks, R., Kumar, V., Rand, K.H., Derendorf, H. (2008). Clinical microdialysis in skin and soft tissues: an update. *Journal of Clinical Pharmacology, 48*, 351–364.

[11] Alberti, K.G., Eckel, R.H., Grundy, S.M., Zimmet, P.Z., Cleeman, J.I., Donato, K.A., Fruchart, J.C., James, W.P., Loria, C.M., Smith, S.C., Jr. (2009). Harmonizing the metabolic syndrome: a joint interim statement of the International Diabetes Federation Task Force on Epidemiology and Prevention; National Heart, Lung, and Blood Institute; American Heart Association; World Heart Federation; International Atherosclerosis Society; and International Association for the Study of Obesity. *Circulation, 120*, 1640–1645.

2

MICRODIALYSIS IN DRUG DISCOVERY

CHRISTIAN HÖCHT

Instituto de Fisiopatología y Bioquímica Clínica, Universidad de Buenos Aires, Buenos Aires, Argentina

1. INTRODUCTION

Drug development is a highly cost- and time-demanding science with a high risk of drug failure in the late clinical phases or during commercialization of the drug [1]. The cost of developing new chemical entities is also increasing, with some estimates now exceeding $802 million. Therefore, there is a need to improve efficiency in drug development by means of a better drug candidate selection in the early-phases of drug development, especially during preclinical research. Even a small improvement could have a considerable impact, in light of the fact that preventing 5% of phase III failures could reduce costs by 5.5 to 7.1% [2].

Attrition during drug development is mostly a consequence of inadequate bioavailability at the target site, inadequate clinical efficacy, and an inadequate safety profile of the new chemical entity [1,3]. Strategies to predict late-phase safety and efficacy based on preclinical and early-phase clinical data with sufficient accuracy are highly encouraging in facilitating early termination of eventual failures. Therefore, pharmacokinetic, pharmacodynamic, and toxicological properties of new chemical entities must be fully characterized during preclinical drug development and early clinical phases (I and IIa). In recent

Applications of Microdialysis in Pharmaceutical Science, First Edition. Edited by Tung-Hu Tsai.
© 2011 John Wiley & Sons, Inc. Published 2011 by John Wiley & Sons, Inc.

years, a great number of different modern techniques have been included in drug development, including in silico approaches [4], and in vivo imaging techniques and microdialysis [5], which enhance knowledge of drug–receptor interactions and drug distribution at the target site, allowing better characterization of pharmacological properties of new chemical entities. In addition, development of mechanism-based pharmacokinetic–pharmacodynamic models and the discovery of new biomarkers have also improved the efficacy of drug development [6,7]. With regard to these points, the aim of the present chapter is to describe modern drug development, emphasizing the role of microdialysis in preclinical and clinical phases of drug development.

2. PHASES OF DRUG DEVELOPMENT

Efficient drug development is based on the learn-and-confirm paradigm of consecutive phases as described in Table 1. Preclinical studies are designed to first learn the pharmacological and safety properties of new chemical entities, allowing the identification of lead candidates to follow clinical drug development [8]. To achieve these objectives, it is necessary to demonstrate biological activity in experimental animal models of disease and to accrue toxicology data to support initial dosing in humans [8].

Inadequate pharmacokinetic properties explain most compounds' failure during drug development, and therefore complete pharmacokinetic profiles of new chemical entities must be a part of early drug development. In silico approaches, in vitro systems, and in vivo experiments are combined for satisfactory descriptions of the absorption, distribution, metabolism, and excretion of new chemical entities [9,10]. Most commonly used in vitro systems include assessment of metabolic stability and enzymology, and permeation across membranes such as the gastrointestinal tract and the blood–brain barrier (BBB) [10].

However, an important issue in preclinical drug development is to establish if sufficiently high concentrations of lead compounds can be attained and maintained at the target site in order to exert the desirable effect. Different modern sampling techniques, including imaging techniques and microdialysis, have been introduced in drug development for the estimation of target-site concentrations of new chemical entities in animal models of efficacy [5].

During preclinical studies it is also necessary to establish if the lead compound interacts with the target receptor to exert the pharmacological response. In vivo drug–receptor interactions can be characterized by means of imaging techniques, including positron-emission tomography (PET) [11]. In addition, to completely understand the biological activity of lead compounds, an estimation of the effects of new chemical entities on biomarkers can help to determine a relationship between the molecular actions of investigational compounds and the clinical efficacy proposed.

TABLE 1 Aspects of Various Phases of Drug Development and the Utility of Microdialysis Sampling

Phase of Drug Development	Main Objectives	Attrition Rate (Number of Compounds Tested)	Number of Experimental Subjects	Duration (years)	Costs ($ millions)	Applicability of Microdialysis[a]
Drug discovery	Design of compounds with optimal in vitro pharmacokinetic and pharmacodynamic properties	10,000		2–3	335	N.A.
Preclinical	Demonstration of pharmacological activity in experimental animal models of disease Accrual of toxicology data to support initial dosing in humans Identification of lead candidates	50	2000	1–1.5		++++
Phase I	Assess dosing interval Assess pharmacokinetic and pharmacodynamic characteristics	10	20–80	0.5–1	467	++
Phase II	Demonstration of efficacy in the intended population Optimal use in target population	6.8	100–300	1–2		N.A.
Phase III	Demonstration of safety and efficacy for clinical use	3.6	1000–3000	2–3		N.A.

[a]N.A., not applicable due to low throughput of microdialysis sampling.

Another important objective of preclinical drug development is the establishment of the dosing interval of lead compounds to be used in early clinical trials. At this point, development of mechanism-based models has improved knowledge of the interaction between the PK–PD properties of drugs and the clinical response (for a review, see [6,7]). Mechanism-based PK–PD modeling integrates parameters for describing drug-specific characteristics with biological system-specific properties, and therefore establishes the causal pathway between drug exposure and drug response [12]. By estimating drug target-site distribution, target binding, and activation and transduction process, mechanism-based PK–PD models make it possible to translate doses used in animal models of efficacy to human beings [12].

After assessment of efficacy and safety of new chemical entities in animal models, lead compounds are first tested mostly in volunteers, with the aim of understanding their safety and pharmacokinetics in human beings. Phase I studies include the evaluation in 20 to 80 subjects of the maximum tolerable dose, pharmacokinetic properties, and pharmacodynamic effect of new chemical entities [8]. In addition, the inclusion of PK–PD modeling and the evaluation of the effects on biomarkers could greatly improve knowledge of the pharmacological and toxicological properties of new chemical entities in this early clinical phase [13]. For example, PK–PD modeling allows the selection of intended dosing regimens in the target population by means of simulation of the relationship between exposure and response, also allowing quantification of intersubject variability [13].

After the initial phase I studies, randomized and controlled clinical phase IIa studies are designed with the aim of confirming the pharmacological properties of new chemical entities in the target population (10 to 20 patients) [8]. In this phase of drug development, use of mechanism-based PK–PD models could help us to understand the time course of disease progression and dose–response relationship to drug intervention [6]. If new chemical entities confirm efficacy in this phase, compounds are evaluated further in phase IIb clinical trials, which are designed to establish the optimal use of investigational compounds in the target population. These randomized and controlled clinical studies are used to assess the efficacy, safety, and dose ranging of a drug or drug combination in larger groups of patients (hundreds of patients) [8].

Finally, efficacy and safety shown in phase II studies must be confirmed during drug development by large, randomized controlled phase III trials involving thousands of patients. In this phase of drug development, it is important to establish if the intended dose exerts the desired safety and efficacy in the target population and if special population of patients (with comorbidities) will require changes in dose requirements [8]. Considering that costs and number of patients are increasing as drug development moves forward, it is highly desirable to detect inappropriate drug candidates early during preclinical drug development and phase I and IIa clinical trials.

The availability of several modern techniques, and new concepts in drug development can greatly reduce attrition during drug approval. Use of high-throughput in silico approaches in drug discovery, imaging techniques, and microdialysis during preclinical and early clinical phases with the estimation of drug effects on validate biomarkers by means of PK–PD modeling may improve knowledge of pharmacokinetic, pharmacodynamic, and toxicological properties of new chemical entities in the initial steps of drug development.

3. ROLE OF BIOMARKERS IN DRUG DEVELOPMENT

A *biomarker*, as defined by a U.S. National Institutes of Health (NIH) working group, is an indicator of normal biological or pathogenic processes or pharmacological responses that is measured objectively in patients or experimental subjects [14]. Although biomarkers in clinical practice are still physiological measures, such as blood pressure or plasma glucose level, in drug development, different types of biomarkers, including genotype patterns, perturbation of gene expression, and changes in protein and metabolite levels, could help to define the efficacy and safety profile of a new chemical entity early in the process [15] (Table 2).

Different classifications of biomarkers have been proposed. Biomarkers can be classified into target, mechanism, or outcome categories. *Target biomarkers* assess a direct pharmacological effect as a result of an interaction with the target receptor, enzyme, or transport protein (e.g., elevation of substrate levels with enzyme inhibition). A *mechanism biomarker* is one that is able to directly relate a measured pharmacological effect to the mechanism of action expected from a drug (e.g., vasodilatation due to α-receptor blockade) [16]. Finally, *outcome biomarkers* might substitute clinical efficacy or safety outcome and are clearly associated with clinical benefits (e.g., blood pressure reduction

TABLE 2 Biomarkers and Role of Microdialysis Sampling

Classification of Biomarkers	Utility of Microdialysis
Genotype or phenotype	Not applicable
Concentration of drug and/or metabolite	Estimation of complete time profile of unbound extracellular levels of lead compounds and their metabolites at the target site
Target occupancy	Not applicable
Target activation	Assessment of changes in endogenous compounds as a consequence of receptor activation
Physiological measures or laboratory tests	Estimation of drug effects on different endogenous compounds, including neurotransmitters and their metabolites, peptides, and hormones, among others
Disease processes	Not applicable
Clinical scales	Not applicable

in hypertension) [16]. Ideally, a biomarker should be linked to the disease process and to the efficacy and safety of drug treatment, in order to predict clinical outcome. If biomarker changes are shown to correlate with a disease state or treatment effect, these markers, called *surrogate markers*, can substitute clinical outcomes to establish the benefits and safety of a drug treatment [16]. These biomarkers are highly attractive when measurement of clinical outcome (e.g., survival) is delayed relative to predictive biochemical changes or the clinical effects of the new molecular entity. Nevertheless, surrogate biomarkers should be used in drug development only if they have a rational theoretical basis, are proven in preclinical or clinical experience, and are measured using validated methods [16].

Introduction of new techniques, such as imaging techniques, microdialysis, polymerase chain reaction (PCR) approaches, and mass spectrometry (MS), have expanded the number of possible biomarkers available to characterize pharmacological and toxicological properties of new chemical entities during drug development [15]. Therefore, Danhof et al. [17] have recently proposed a new classification of biomarkers based on a mechanistic point of view. As shown in Table 2, effects of new chemical entities could be described by means of biomarkers at different levels, such as genotype or phenotype, target site concentration of drug and/or metabolite, receptor occupancy and/or activation, physiological or biochemical response induced by drug–receptor interaction, interference in disease processes, and finally, drug effects on clinical scales [17]. The role of microdialysis in the assessment of biomarkers is described in Table 2.

Microdialysis is a powerful technique for continuous monitoring of biomarkers, especially in the preclinical phase of drug development. According to the biomarker classification of Danhof et al. [17], by introducing a microdialysis probe into target tissue, microdialysis sampling allows the continuous estimation of unbound concentration of drug and/or metabolite. In addition, as microdialysis also recovers endogenous compounds, this technique monitors the effect of target activation on endogenous compounds, such as metabolites, neurotransmitters, or endogenous peptides. Therefore, microdialysis allows not only the evaluation of target-site distribution of new chemical entities, but also the assessment of their effects on physiological variables and disease processes.

4. ROLE OF PHARMACOKINETIC–PHARMACODYNAMIC MODELING IN DRUG DEVELOPMENT

PK–PD modeling describes the relationship between the pharmacokinetics and pharmacodynamics of a drug, allowing an estimation of PK–PD parameters and a prediction of these derived clinically relevant parameters [18]. PK–PD modeling has several advantages over classical dose–response studies. PK–PD modeling allows not only better pharmacodynamic characterization of drugs, but also permits screening and dosage–regimen selection [19]. As shown in Table 3, introduction of PK–PD modeling during preclinical and clinical drug development could greatly improve knowledge of pharmacological properties of new chemical entities, thereby reducing costs and attrition of drug development [13,20].

PK–PD modeling offers great value in preclinical drug development, as it improves the selection of lead compounds because of a better description of the efficacy and safety of new chemical entities in animal models [8,13]. In addition, the introduction of pathological processes in mechanism-based PK–PD models also allows the prediction of clinical potency and the dose range to be tested in early clinical trials [6]. However, a limitation of PK–PD modeling is the necessity of simultaneous measurement of drug tissue levels and its corresponding pharmacological effect at multiple time points in order to design accurate PK–PD models [20]. To obtain the greatest precision in estimating PK–PD parameters, the number of measurements of drug tissue levels and their corresponding effect must be as large as possible [21].

Traditional sampling techniques such as blood sampling and biopsies, which have traditionally been used for this purpose, have the disadvantages that the removal of samples by themselves may interfere with pharmacokinetic and pharmacodynamic drug behavior, especially in preclinical studies with small animals, or allow us to obtain only a single time point in each experiment [22]. Furthermore, traditional sampling techniques allow the measurement of plasma concentrations of pharmacological agents rather than levels of drugs in the target tissue.

Conversely, microdialysis samples the bioactive concentration of drugs at the target site continuously without fluid loss or need of tissue biopsy. In addition, microdialysis allows endogenous compound sampling and an estimation of the effects of new chemical entities on biochemical markers, including neurotransmitters, metabolites, hormones, glucose, lactate, and peptides [23]. Therefore, this technique not only makes possible the study of drug tissue concentrations but also the effect of the compounds on physiological functions. Use of microdialysis for PK–PD modeling during preclinical drug development is supported by the fact that this technique allows the simultaneous determination of drug concentrations in one or more tissues and its effect on biochemical and clinical markers in the same animal and with high temporal resolution. Microdialysis has been used for the study of PK–PD models of various therapeutic drugs and new chemical entities in animal models [20].

PK–PD modeling also improves knowledge of pharmacological and safety properties of new chemical entities in clinical phases of drug development (Table 3). PK–PD simulations help to fully understand the dose–concentration–pharmacological effects and dose–concentration–toxicity relationship in healthy volunteers for determining optimal dosing regimens for phase II studies [8,13]. In phase II clinical trials, PK–PD modeling confirms and explores the relationship between dose–concentration–effect in patients, also examining a variety of therapeutic endpoints with the aim to select the most adequate

TABLE 3   Role of PK–PD Modeling in Drug Development and Rationale of Microdialysis

Stage of Drug Development	Benefits of PK–PD Modeling	Rationale of Microdialysis Sampling
Preclinical	Precise definition of the dose–concentration–pharmacological effects and dose–concentration–toxicity relationship. Determination of the appropriate dosing regimen for phase I studies. Identification of biomarkers and animal models for efficacy and toxicity. Exploration of any dissociation between plasma concentration and duration and onset of pharmacological effect. Providing information on drug effects that would be difficult to obtain in human subjects. Reducing the cost of preclinical phase by a reduction in the number of animals used.	Microdialysis allows continuous and simultaneous monitoring of target site concentrations of lead compounds and their effect on endogenous compounds. Implantation of multiple microdialysis probe is feasible, allowing evaluation of multiple PK-PD relationships. Microdialysis permits study of mechanisms involved in delay in drug response. Microdialysis allows the study of possible link between changes in endogenous compounds and physiological responses.
Phase I	Understanding the dose–concentration–pharmacological effects and dose–concentration–toxicity relationship in healthy volunteers. Characterization of PK and PD in a special population. Study of tolerance development. Determination of the dosing regimens for phase II studies.	Microdialysis is suitable for assessment of target-site concentration of new chemical entities at easily accessible tissues (subcutaneous tissue).
Phase IIa	Confirms and explores the relationship between dose–concentration–effect in patients. Examines a variety of therapeutic endpoints to understand the most adequate for further. Not applicable due to low throughput of microdialysis modeling. Study of efficacy in the intended population.	Not applicable due to the low throughput of microdialysis.

TABLE 3 (*Continued*)

Stage of Drug Development	Benefits of PK–PD Modeling	Rationale of Microdialysis Sampling
Phase IIb	Determination of the dosing regimens for phase III studies. Prediction of the probability distribution of further clinical trial outcomes.	Not applicable due to the low throughput of microdialysis.
Phase III	Assessment of PK and PD changes or relationship in the patient population.	Not applicable due to the low throughput of a microdialysis.

for further modeling. Simulation can also be used to develop drug–disease models to understand the time course of disease progression and dose–response to interventions. In addition, by using a population PK–PD model, it is possible to assess the impact of covariates on drug response. Finally, PK–PD models determine dosing regimens for phase III studies [8,13].

PK–PD simulation during phase III studies is focused on the optimization of study design, reducing the risk of failed studies. Considering the large number of patients included in this phase of drug development, population PK–PD models are highly useful for the evaluation of the impact of covariates, including comorbidities, and concomitant medication on pharmacological response to new chemical entities [8,13].

5. ROLE OF MICRODIALYSIS IN DRUG DEVELOPMENT

The fact that assessment of target-site concentrations of new chemical entities is generally required to predict the clinical efficacy of lead compounds justifies the rationale of implementation of microdialysis during the drug development process. In addition, as regards the role of PK–PD modeling during all stages of drug development and the ability of microdialysis for continuous monitoring of tissue extracellular levels of drugs and their effect on biochemical markers, this technique allows an early proof of concept of the activity of new chemical entities in the first stages of drug development, especially in preclinical models of efficacy. The rationale for the use of microdialysis to improve drug development has been acknowledged by the American Association of Pharmaceutical Scientists (AAPS) and the U.S. Food and Drug Administration (FDA) through a Workshop White Paper [24]. Microdialysis could be used in various stages of early drug development, including estimation of plasma protein binding, in vivo pharmacodynamic models, in vivo pharmacokinetics, and in vivo PK–PD studies (Table 4).

Microdialysis sampling may be considered as a gold standard technique for the evaluation of in vivo pharmacokinetics of new chemical entities during the

TABLE 4 Applicability of Microdialysis During Drug Development

Phase of Drug Development	Applicability of Microdialysis
Drug discovery	In vitro protein binding
Preclinical	Assessment of extracellular concentrations of lead compounds at the target site
	Evaluation of intracellular concentrations of new chemical entities in combination with imaging techniques
	Estimation of in vivo pharmacodynamics by monitoring effects of new chemical entities on endogenous compounds levels
	Multiple PK–PD modeling
Phase I	Assessment of drug distribution at the target site of accessible tissues
	Estimation of in vivo pharmacodynamics by monitoring effects of new chemical entities on endogenous compounds levels at accessible tissues
	Assessment of PK–PD models
Phase II	Not applicable due to its low-throughput nature
Phase III	Not applicable due to its low-throughput nature

preclinical stage of drug development. To date, microdialysis is a unique technique that allows continuous measurement of extracellular target site concentrations of therapeutic agents, and therefore estimation of bioactive drug fraction. In addition, the possibility of chronic implantation of microdialysis probes permits monitoring of tissue drug levels for several days, allowing an accurate estimation of tissue pharmacokinetics [20]. Several works have also demonstrated the feasibility of multiprobe microdialysis sampling by implantation of several probes in different tissues [25–27]. This aspect of microdialysis technique is highly interesting for evaluation of the brain/plasma ratio in animal models of efficacy. Moreover, regional distribution in brain parenchyma of central-acting drugs could be assessed by means of implantation of several probes in different central nuclei.

It is important to mention that imaging techniques also permit assessment of the time profile of tissue pharmacokinetics of new chemical entities. Several imaging techniques, such as planar γ-scintigraphy, single photon-emission computed tomography (SPET), PET, and magnetic resonance spectroscopy (MRS), have been developed for the study of drug distribution in basic and clinical settings [28]. PET is a new nuclear imaging technique that employs molecules labeled with positron-emitting radioisotopes [29]. Although PET has some advantages with regard to microdialysis in drug development, including its noninvasive nature, high spatial resolution (1 to 5 mm), and time resolution (30 s), the utility of this imaging technique for tissue pharmacokinetic assessment of new chemical entities is restricted by several factors. In the first place, the physical half-life of the most used radioisotope, 11C (20.4 min), does not allow monitoring of tissue levels of radiolabeled drugs over several elimination half-lives as desired in pharmacokinetic studies [28]. However, the strongest limitation of PET for estimation of target-site distribution of new chemical entities is the fact that this methodology samples total tissue concentrations of drugs without discerning between extracellular biophase levels and intracellular drug concentrations. In addition, PET measures the tissue concentrations of new chemical entities and their metabolites but does not make it possible to differentiate between them [28].

Free extracellular target-site levels represent the best marker of the bioactive fraction of new chemical entities acting on receptors expressed at the cellular membrane (e.g., G-protein-coupled receptors, neurotransmitter transporters, and ion channels). However, some therapeutic agents, such as most antineoplastic drugs, and hormones and their antagonists, exert their pharmacological action by interaction with intracellular receptors. For these drugs, information regarding cellular drug accumulation is highly desirable. As noted above, while microdialysis sampling assesses extracellular tissue drug concentrations, PET imaging gives information regarding total tissue levels. Therefore, simultaneous microdialysis and PET studies allow a precise estimation of intracellular drug levels that may be highly relevant for drugs acting within the cellule. Langer et al. [30], using [18F]-labeled ciprofloxacin as a model drug, have found that in vivo intracellular ciprofloxacin pharmacokinetics was in accordance with previous in vitro data describing cellular ciprofloxacin uptake and retention. Therefore, a PET–microdialysis combination might be useful during the research and development of new drugs, for which knowledge of intracellular concentrations is of interest.

As microdialysis also samples extracellular levels of endogenous compounds, this technique could be a gold standard for the estimation of in vivo pharmacodynamic of new chemical entities. The effect of new chemical entities on metabolism can be monitored by means of the estimation of variation in glucose, lactate, and piruvate extracellular levels induced by the drug [23]. In addition, placing a microdialysis probe in the brain parenchyma allows an evaluation of the neurochemical effects of lead compounds. Although microdialysis sampling was traditionally restricted to the recovery of low-molecular-weight endogenous compounds, the availability of high-cutoff membranes (e.g., 100 kDa) also permits the assessment of drug effects on proteins, especially cytokines [31].

It is important to mention that microdialysis does not estimate in vivo binding of new chemical entities to receptors. PET imaging is a valuable tool for the estimation of parameters describing in vivo drug–receptor interactions [11]. Therefore, combining microdialysis with PET during drug development is also attractive because of the feasibility of the simultaneous evaluation of drug–receptor interactions and the effect of drugs on biochemical markers such as neurotransmitters. As shown by Schiffer et al. [32], while PET imaging allows study of the dopamine receptor-binding properties of [11C] raclopride, microdialysis assesses the effect of the drug on dopamine extracellular levels.

The fact that microdialysis sampling allows simultaneous monitoring of extra-cellular drug tissue levels and their effects on endogenous compounds with high temporal resolution makes this technique highly attractive for PK–PD modeling. Measurement of drug effects on other physiological parameters, such as blood pressure, electroencephalographic effects, and analgesia, inde-pendent of microdialysis sampling, is also feasible. Therefore, multiple PK–PD relationships may be estimated during a single experiment, increasing the knowledge of PK–PD properties in the preclinical phase of drug development (Figure 1).

Combining microdialysis with PET could be an attractive approach for PK–PD modeling during drug development. As a theoretical point of view, simultaneous assessment of target-site concentrations, in vivo drug–receptor

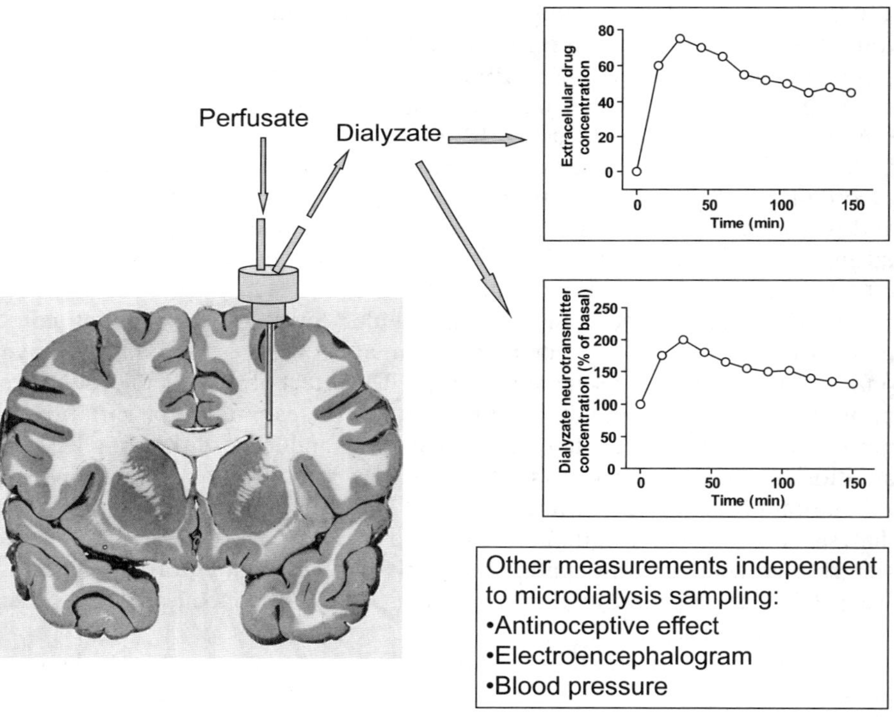

Figure 1 Applicability of microdialysis for drug development of centrally acting drugs. Placement of a concentric microdialysis probe in a specific brain nucleus allows the simultaneous assessment of free extracellular concentrations of new chemical entities and their effect on biophase levels of neurotransmitters, neuropeptides, and their metabolites. Additionally, monitoring of the effect of investigational drugs on physio-logical parameters is also possible during a microdialysis experiment.

binding, and the effects on endogenous compounds and therefore estimation of accurate PK–PD models of new chemical entities is possible by means of microdialysis–PET. Microdialysis is also well suited for the determination of drug protein binding during early drug development. The microdialysis technique allows the determination of in vivo protein binding using microdialysis sampling in blood and simultaneous blood sampling [33,34]. The in vivo determination of protein binding using the microdialysis method permits a more accurate determination of protein binding with regard to in vitro protocols, because it was found that in vitro determination systematically underestimated the unbound fraction [35]. In addition, microdialysis permits the determination of the temporal course of protein binding in the same animal to determine saturation of the plasma protein binding [33].

The utility of microdialysis sampling for estimation of the in vitro protein binding in time–kill curves of antimicrobials has recently been demonstrated [36]. Using a microdialysis technique, the authors have found that free antimicrobial concentrations differ substantially between plasma and protein supplements, correlating well with antibacterial efficacy. The authors concluded that free active levels of antimicrobials should be measured during in vitro time–kill curves for accurate estimation of the effective concentration yielding a half-maximal response (EC_{50}) of antimicrobials [36].

In conclusion, considering the fact that microdialysis allows continuous and simultaneous monitoring of both extracellular levels of new chemical entities at the target site and their effect on endogenous compounds, this technique could be considered to be the gold standard in the evaluation of in vivo pharmacokinetics, pharmacodynamics, and PK–PD modeling during early drug development. Use of microdialysis in the preclinical phase is also supported by the fact that microdialysis can be carried out in diverse laboratory animal species and in a great number of different tissues. In addition, an economical and ethical advantage is that 5 to 10 times fewer animal experiments have to be performed to determine the pharmacological profile of a drug [37].

Nevertheless, the applicability of microdialysis sampling during preclinical drug development has some restrictions. As a theoretical point of view, not all new chemical entities could be monitored with this technique. Large molecules are precluded to diffuse through the dialysis probe. Since proteins cannot pass through the membrane, only the free proportion of the drug is measured, and therefore, if the protein binding of the drug is high, only a very small amount of drug is available for analysis, requiring the existence of highly sensitive analytical methods [37]. In addition, highly lipophilic drugs suffer from sticking to tubing and probe components [38]. It is important to mention that recovery of these substances has been improved in recent years. For highly protein-bound drugs, the low recovery rate could be solved by use of new microdialysis membranes with a high-molecular-weight cutoff [31]. On the other hand, the addition of solubilizers to the perfusate could improve the recovery of lipophilic drugs [38–40]. Development of highly sensitive analytical methods may also provide significant progress in the use of microdialysis for drug development.

However, the most important restriction for the use of microdialysis in drug development is its low throughput, which restricts this technique for the evaluation of lead compounds. Therefore, microdialysis seems not to contribute to the initial screening of new chemical entities [24]. Conversely, as microdialysis assesses drug distribution in the biophase and multiple PK–PD relationships of lead compounds in animal models of efficacy, this technique allows an early determination of the proof of concept of new chemical entities during preclinical drug development and selection of the most adequate dosing interval for phase I clinical trials.

Although microdialysis could also contribute in early phases of clinical drug development, its applicability in human studies could be restricted by its invasive nature, by the need for technical expertise and additional laboratory, and for ethical reasons [24]. In recent years, microdialysis has been developed for monitoring of drug concentration in different human tissues, such as subcutaneous tissue, dermis, brain parenchyma, solid tumors, infection sites, and liver, among others (for a review, see [41–43]). Therefore, similar to preclinical applications, microdialysis could be useful for assessment of drug distribution at the target site and PK–PD modeling of drug effects. However, due to its invasive nature, the use of this technique for the assessment of drug distribution in brain parenchyma and solid malignancies is strongly limited. Conversely, the most attractive applications of microdialysis sampling during clinical drug development are estimation extracellular levels of antimicrobial agents at the site of infection [44] and the bioavailability of new chemical entities after topical application [45].

6. MICRODIALYSIS SAMPLING IN THE DRUG DEVELOPMENT OF SPECIFIC THERAPEUTIC GROUPS

As noted above, microdialysis sampling contributes greatly to increasing knowledge of pharmacological properties of new chemical entities in the early phases of drug development, especially in preclinical studies. However, the contribution of microdialysis to the reduction of attrition during drug development also depends on the therapeutic effect of the lead compound. Therefore, current applications and the outlook for microdialysis sampling in the drug development of specific therapeutic agents are discussed next.

6.1. Centrally Acting Drugs

Preclinical evaluation of centrally acting drugs has been the most attractive application of microdialysis in drug development (Table 5). As most cen-

TABLE 5 Role of Microdialysis in Drug Development of Centrally Acting Drugs

Therapeutic Group	Utility of Microdialysis Sampling
Antiepileptic drugs	Estimation of hippocampal bioavailability Assessment of compromise of efflux transporters in brain distribution Assessment of neurochemical effects Evaluation of PK–PD models by the study of the relationship between antiepileptic brain concentrations and their effect on neurotransmitter turnover and electroencephalogram
Antiparkinsonian drugs	Assessment of effects of new chemical entities on striatal dopamine levels
Antidepressive drugs	Estimation of the effects of lead compounds on serotonin turnover on prefrontal cortex
Anxiolytic drugs	Evaluation of effects of new chemical entities on GABAergic neurotransmission at the amygdala
Opioid analgesic drugs	Assessment of PK–PD modeling by the study of the relationship between brain concentrations of lead compounds and their antinoceptic effect
Neuroprotective agents	Estimation of the effects of lead compounds on glutamate brain extracellular concentrations

trally acting drugs exert their effect by affecting neurotransmitter turnover, continuous monitoring of extracellular levels of neurotransmitters and their metabolites during drug treatment represents an excellent biomarker of the pharmacological effects of centrally acting drugs (Figure 2). In addition, the blood–brain barrier expresses a high number of different drug efflux transporters, including P-glycoprotein, multidrug-resistance (MDR) proteins, nucleoside transporters, organic anion transporters, organic cation transporters, large amino acid transporters, and the scavenger receptors SB-AI and SB-BI [46,47]. Therefore, drug distribution in brain parenchyma could be greatly affected by the activity of these transporters, and estimation of the distribution of centrally acting new chemical entities is of great relevance (Figure 1). Moreover, recent studies have demonstrated that drug distribution in brain parenchyma is highly heterogeneous [48]. Considering that brain microdialysis has relatively high spatial resolution, microdialysis allows monitoring of drug levels in the specific central nuclei where drug effect is exerted.

In addition, the fact that microdialysis sampling allows simultaneous and continuous monitoring of target-site concentrations of centrally acting drugs and their effect on neurotransmitter turnover makes this technique attractive for PK–PD modeling during drug development of new chemical entities with a central mechanism of action. Monitoring of dopamine extracellular levels in the striatum through microdialysis sampling could be used for in

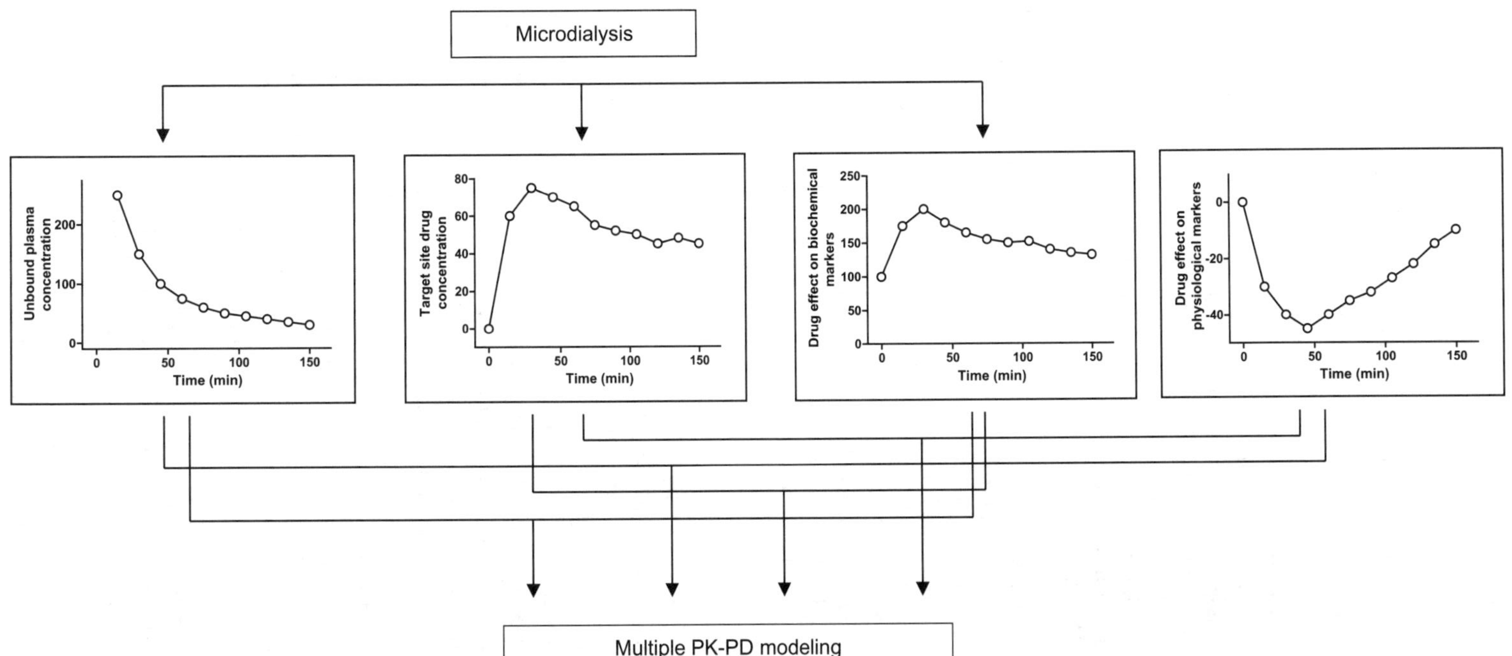

Figure 2 Applicability of microdialysis for PK–PD modeling. Microdialysis sampling allows estimation of multiple PK–PD models in a single experiment, because this technique is able to sample free extracellular target-site concentrations and their effect on endogenous compounds simultaneously. Additionally, placement of a second microdialysis probe in a vessel allows assessment of free plasma concentrations, so different pharmacological aspects of new chemical entities could be explored during a microdialysis PK–PD experiment. First, it is possible to establish the blood/brain ratio and the in vivo permeability of the blood–brain barrier. Second, by comparing the PK–PD parameters obtained from PK–PD relationship between plasma and central drug concentrations and their effects, the mechanism involved in the delay in pharmacological response could be clarified. Third, as microdialysis sampling allows the simultaneous assessment of drug effects on endogenous compounds and physiological parameters such as blood pressure, analgesia, and electroencephalogram data, this technique is feasible for exploring a possible link between the mechanism of action and the pharmacological response.

vivo evaluation of neurochemical actions of antiparkinsonian drugs. For example, in vivo microdialysis studies have shown that tolcapone effectively inhibits O-methylation of L-dopa, thereby improving its bioavailability and brain penetration and potentiating L-dopa antiparkinsonian effects [49]. In vivo microdialysis has also been used for the preclinical evaluation of rasagiline, showing that this compound increased extracellular dopamine levels following chronic treatment in the rat, at a dose that caused selective MAOB inhibition [50].

Drug development of antiepileptic drugs could also be improved by introducing microdialysis sampling during preclinical evaluation. Reduction in the central bioavailability of antiepileptic drugs by overexpression of efflux transporters has been established for several drugs in different models of experimental epilepsy as a mechanism of pharmacoresistance (for a review, see [51,52]). The need to study antiepileptic drug distribution at the target site is emphasized by the fact that overexpression of efflux transporters seems to affect drug distribution only in the biophase and not in other central nuclei [53]. Recently, Tong et al. [54] have found that brain distribution of vigabatrin is highly heterogeneous, considering that frontal cortex concentrations of this antiepileptic are twofold greater than those of the hippocampus.

Microdialysis has been used extensively for testing the pharmacokinetic hypothesis of antiepileptic drug resistance in different animal models of chronic epilepsy (for a review, see [51–53]). For example, we have demonstrated a critical role of P-glycoprotein overexpression in the development of pharmacoresistance to phenytoin in a model of epilepsy induced by 3-mercaptopropionic acid chronic administration, suggesting that administration of efflux transporters inhibitors could be an effective strategy to decrease pharmacoresistance to phenytoin antiepileptic treatment [55]. Neurochemical actions of antiepileptic drugs have also been assessed during preclinical testing. As mentioned in the package insert of zonisamide [56], the ability of this antiepileptic drug to enhance both dopaminergic and serotonergic neurotransmission has been proved by means of brain microdialysis.

As a theoretical point of view, microdialysis sampling is also attractive for PK–PD modeling of new chemical entities with anticonvulsant action. Although, to the best of our knowledge, microdialysis sampling has not been used to date for PK–PD modeling of anticonvulsive drugs, Chenel et al. [57] described the proconvulsive effect of norfloxacin by simultaneous monitoring of brain extracellular concentrations of norfloxacin by means of microdialysis and a quantitative electroencephalogram (EEG). Using a PK–PD model with an effect compartment, the authors demonstrated that the delayed EEG effect of norfloxacin is not due to BBB transport [57].

Microdialysis sampling seems to be a gold standard for preclinical evaluation of in vivo pharmacodynamics of therapeutic agents used for smoking cessation, considering that the efficacy of these agents is highly correlated with changes in dopamine turnover at the nucleus accumbens. During the nonclinical program of varenicline, a highly selective partial agonist of the nicotinic acetylcholine receptor $\alpha 4\beta 2$ subtype, in vivo microdialysis in freely moving rats showed that oral administration of varenicline caused moderate increases in dopamine release in the nucleus accumbens, inducing maximal response after 2 h of varenicline dosing [58]. In addition, it was found that maximal dopamine response to varenicline was around 63% of the full agonist nicotine [58].

As microdialysis allows monitoring of changes in the extracellular levels of monoamines, this technique is highly useful for antidepressant drug development. Noradrenaline and serotonin concentrations in brain dialysates are an indirect estimation of the activation of postsynaptic monoaminergic receptors and could be consider as a biomarker of a putative antidepressive effect of new chemical entities [59]. Neurochemical actions of tricyclic antidepressant and serotonin reuptake inhibitors have been studied extensively by means of brain microdialysis in laboratory animals (for a review, see [59]). For example, Hughes et al. [60] have recently found that WAY-20070, a selective agonist of estrogen receptor beta, could be beneficial in the treatment of depression and anxiety, considering that subcutaneous administration of the drug increases serotonin levels in striatal microdialysate. Current antidepressive drugs suffer from a delay in the onset of therapeutic effect, due to the time required for the 5-HT1A, and possibly 5-HT1B, autoreceptors to desensitize [59]. Thereby, an agent incorporating 5-HT reuptake inhibition coupled with 5-HT1A and/or 5-HT1B autoreceptor antagonism may provide a fast-acting antidepressant drug. By using microdialysis sampling, it was found that SB-649915 produced an acute increase in extracellular 5-HT in forebrain structures of the rat, providing a novel mechanism that could offer fast-acting antidepressant activity [61].

Anxiolytic-like activity of new chemical entities could be tested in preclinical models by evaluation of the effects on γ-aminobutyric acid (GABA)ergic neurotransmission. As an example of the utility of microdialysis in the development of anxiolytic drugs, Rajarao et al. [62] have found that intraperitoneal administration of galnon, a nonselective galanin receptor agonist, preferentially elevated levels of GABA in the rat amygdala, a brain area linked to fear and anxiety behaviors. In addition, galnon neurochemical action correlates with the efficacy of this compound on different preclinical models of anxiety.

In recent years, there has been a strong interest in the development and evaluation of neuroprotective agents. The protective effect of NGP1-01, a dual blocker of neuronal voltage- and ligand-operated calcium channels, was evaluated by monitoring choline release during N-methyl-D-aspartic acid (NMDA) infusion as a measure of excitotoxic membrane breakdown using in vivo microdialysis [63]. Intraperitoneal administration of NGP1-01 reduced NMDA-induced membrane breakdown, demonstrating that NGP1-01 blocks both major neuronal calcium channels simultaneously and is sufficiently brain-permeable. Therefore, NGP1-01 is a promising lead structure for a new class of dual-mechanism neuroprotective agents.

Finally, results obtained from PK–PD modeling of well-established centrally acting drugs suggest that brain microdialysis could be highly attractive for this approach during preclinical drug development [64]. For example, the effect of drug candidates on dopaminergic activity at different nuclei of the central nervous system (CNS) has been studied by means of PK–PD modeling coupled to microdialysis sampling [27,65]. The effect of benzatropine analogs on dopamine concentration in the nucleus accumbens after its intravenous administration was evaluated [65]. The authors fitted plasma concentration of the analogs and their effects on extracellular dopamine levels to two different PK–PD models, such as an effect compartment model and a model with indirect physiological response. The authors demonstrated that the indirect model is more suitable than the linked PK–PD model for PK–PD modeling of benzatropine analogs. These results are in accordance with the mechanism of action of the analogs because these drugs bind to the dopamine transporter, inhibiting the dopamine reuptake and consequently elevate dopamine extracellular levels. In an elegant study, Bouw et al. [66] simultaneously determined blood and brain concentrations of morphine-6-glucuronide and its antinociceptive effect by means of microdialysis sampling. By applying a PK–PD model with an effect compartment, the authors found a greater delay in the onset of the effect when antinociception was related to plasma morphine-6-glucuronide concentrations with regard to brain levels. Therefore, it was concluded that half of the effect delay could be explained by transport across the blood–brain barrier, suggesting that the remaining delay is a result of drug distribution in the brain parenchyma [66].

In conclusion, as expected, microdialysis sampling becomes a key methodology during preclinical drug development of centrally acting drugs, considering that this technique allows simultaneous and continuous monitoring of extracellular levels in specific brain nuclei and their effect on different neurotransmitters system. Thereby, microdialysis is also attractive because of the possibility of development of multiple PK–PD models of new chemical entities acting on the brain, increasing knowledge of the pharmacological properties of these compounds in animal models of efficacy.

6.2. Antimicrobial Drugs

Recent findings obtained from clinical microdialysis studies have demonstrated that tissue distribution of antimicrobials shows high intertissue and intersubject variability (for a review, see [67]). Traditionally, it was considered that total plasma concentrations and plasma protein binding can be used to predict free tissue levels of antibiotics, based on the assumption that unbound plasma concentrations and free tissue levels are equal at equilibrium, considering that tissue distribution is generally mediated only by passive diffusion [68]. However, many studies have shown lower tissue unbound levels than plasma concentrations [69–71]. On the other hand, time to reach equilibrium between plasma and tissue concentrations of antibiotics may range from minutes to days [68]. Therefore, pharmacokinetic assessment of antimicrobial agents was based largely on the measurement of total plasma concentrations as an inadequate surrogate marker of antimicrobial effect, and measurement of unbound drug concentrations in the interstitial fluid of the site of infection should be considered a gold standard for improvement of antimicrobial therapy and dose adjustment.

Microdialysis has been used to measure various antimicrobials agents in human and laboratory animal tissues, including aminoglucosides, penicillins, cephalosporines, fosfomycin, fluoroquinolones, and antiviral agents (for a review, see [72,73]). These studies have helped to evaluate drug distribution in several organs, including infective tissues, and to develop in vivo PK–in vitro PD models at the target site using the same parameters calculated in plasma: time (T) above the minimum inhibitory concentration (MIC) ($T > $ MIC), the ratio of the maximum concentration of drug in serum (C_{max}) to the MIC (C_{max}/ MIC), the area under the inhibitory curve, or the area under the curve (AUC)/ MIC ratio [44].

Considering the vast experience of microdialysis for evaluation of distribution in the site of infection of well-established antimicrobial agents in both laboratory animals and human beings, this technique becomes essential in preclinical and early clinical drug development of innovative anti-infective drugs. However, it is important to mention that although the feasibility of microdialysis for evaluation of interstitial fluid distribution of new antimicrobials is not restricted by the site of infection in preclinical development, the utility of microdialysis in early clinical development would be limited to infections at easily accessible soft tissues such as subcutaneous tissue.

Another attractive use of microdialysis sampling during the development of new antimicrobial agents is the design of in vivo PK–in vitro PD models (for a review, see [44,73]). A three-step approach has been used for the in vivo PK–in vitro PD modeling by means of microdialysis. First, interstitial fluid concentrations of the antibacterial drug at the site of infection are measured by means of microdialysis. Second, time versus drug concentration profile measured in vivo is simulated in an in vitro setting on bacterial cultures. Third, unbound antibiotic concentrations are linked to bacterial kill rates by means of a PK–PD model [74]. By using this approach, Delacher et al. [74] have demonstrated a significant correlation between the maximal bactericidal effect and several pharmacokinetic surrogate parameters, such as AUC/MIC, C_{max}/ MIC, and $T > $ MIC. The authors concluded that the therapeutic success or failure in antibacterial therapy depends on the target-site concentrations of the antimicrobial agent. Moreover, in vivo PK–in vitro PD modeling provides valuable guidance for drug antibacterial efficacy and dose selection during drug development [74].

It must be pointed out that most PK–PD studies of antimicrobial drugs by means of microdialysis have used a combined in vivo PK–in vitro PD simulation without applying mathematical PK–PD models in their analysis by relating pharmacokinetic parameter to MIC. However, MIC is a single static in vitro parameter that reduces information gained through PK–PD relationships. Conversely, kill curve approaches and subsequent pharmacokinetic–pharmacodynamic analysis may provide more meaningful information about the interaction between bacteria and antimicrobial agents, since these approaches describe this interaction by a dynamic integration of concentration and time, therefore using all the information available [75]. In this regard, Liu et al. [76] demonstrated that a PK–PD model based on unbound antibiotic concentrations at the site of infection, and a sigmoid E_{max} relationship, effectively described the antimicrobial efficacy of both cefpodoxime and cefixime. This approach offers more detailed information than the MIC does about the time course of antibacterial efficacy of antimicrobials under development [76]. Therefore, in vivo PK–in vitro PD modeling of anti-infective drugs allows the simulation of different dosing strategies without needing a large sample of experimental subjects, therefore reducing the cost of drug development.

In conclusion, microdialysis sampling allows an assessment of the distribution of novel antimicrobial agents at the interstitial fluid at the infection site and in vivo PK–in vitro PD simulation, providing early information of an anti-infective efficacy and dosing schedule of lead compounds with antimicrobial action. The use of microdialysis sampling is feasible for the study of target-site pharmacokinetics of new antimicrobial agents in both preclinical and early clinical development, although applicability in humans is restricted to infections in easily accessible tissues.

6.3. Antineoplastic Drugs

Measurement of target-site concentrations of antineoplastic drugs in malignancies and relating these levels to pharmacodynamic parameters is of great interest for the design of active new chemical entities with cytotoxic effects. Tumor drug exposure, a marker linked to clinical outcome, may be reduced dramatically, due to diffusion barriers in solid tumors [77]. Differences in tumor drug distribution do not make it possible to predict the antineoplastic response from plasma profiles [78]; thus, measurements of drug exposure in tumor interstitium may help to develop new antineoplastic drugs [79]. Microdialysis has been used to describe tissue pharmacokinetics of several antineoplastic drugs in both animal models and clinical settings (for a review, see [79,80]). Studies with 5-fluorouracil (5-FU) showed that plasma or subcutaneous levels of 5-FU failed to predict tumor response. Conversely, high interstitial tumor concentrations of 5-FU were associated with increased tumor response. In addition, this membrane-based technique allows assessment of the pharmacodynamics of chemotherapeutic agents [81]. For example, plasma concentrations of serotonin and 5-hydroxyindoleacetic acid during cisplatin treatment have been monitored by means of microdialysis in relation to the role of serotonin in the production of emesis associated with antineoplastic treatment [82]. Microdialysis has also been used to monitor the extracellular levels of growth factors, such as the vascular endothelial growth factor

(VEGF), during treatment with tamoxifen in a mouse model of human breast cancer [83]. Although to date, scarce PK–PD modeling studies using microdialysis have been made, integration of the pharmacological response with tumor PK profiles of the corresponding drug would help to define the PK–PD relationship, which is essential for the rational design of drug dosing during development. Thompson et al. [84] studied the clinical and biochemical responses to the time course of melphalan in the subcutaneous interstitial space and in tumor tissue from patients with various limb malignancies. The authors showed a significant correlation between the melphalan mean concentration in subcutaneous microdialysate and tumor response [84].

Another interesting approach to the development of antineoplastic agents is the evaluation of in vivo PK–in vitro PD models. For example, Müller et al. [85] have determined the unbound interstitial drug pharmacokinetics of 5-FU and methotrexate in solid tumor lesions of patients by means of in vivo microdialysis, then making a pharmacodynamic simulation of the time–drug concentration profile in an in vitro setting by exposing breast cancer cells to interstitial tumor concentration of the antineoplastic drugs. The authors concluded that in vivo PK–in vitro PD models might provide a rational approach to describing and predicting the pharmacodynamics of cytotoxic drugs at the target site [85].

Although microdialysis could also be considered a gold standard technique for preclinical evaluation of these therapeutic agents, several aspects limit its applicability. The rationale for microdialysis sampling in oncology drug development is restricted by the fact that most antineoplastic drugs act within cells, and the relationship between extracellular drug concentrations and intracellular drug levels remains unknown. Nevertheless, as noted above, this drawback could be overcome by employing an attractive approach based on simultaneous study of drug distribution by microdialysis and PET [30]. On the other hand, some antineoplastic drugs (e.g., 5-FU) require intracellular enzymatic conversion in order to exert their cytotoxic activity. In addition, other aspects, such as tumor location and accessibility for microdialysis probe implantation and the possibility of variation in interstitial concentrations of cytotoxic drugs in different metastases in a patient, restrict the utility of microdialysis for studies of novel antineoplastic drug distribution [41].

6.4. Other Therapeutic Groups

Microdialysis sampling is also attractive for PK–PD modeling of investigational antihypertensive drugs in animal models of hypertension to increase information gain during preliminary stages of drug development. Several animal models of hypertension have been developed in rat strains to mimic the various pathophysiological aspects of human hypertension [86]. However, PK–PD studies in preclinical drug development is limited by the fact that frequent plasma sampling could interfere with the pharmacokinetic and pharmacodynamic behavior of the antihypertensive drug under evaluation due to fluid loss, especially in small laboratory animals [87]. In this way, the use of intraarterial or intravenous microdialysis could be an interesting approach to overcoming this methodological limitation. Microdialysis allows continuous sampling of plasma drug concentrations without a need for frequent blood sample extraction [87]. In our laboratory we designed and validated a shunt intraarterial microdialysis probe with one vascular inlet and two vascular outlets [33]. The inlet and one outlet are inserted into the left carotid artery, and the remaining outlet is connected to a pressure transducer, allowing simultaneous monitoring of cardiovascular parameters. Therefore, PK–PD modeling of novel antihypertensive drugs in experimental hypertension allows the identification of biomarkers and animal models for efficacy and toxicity to establish the antihypertensive response in different pathophysiological states of hypertension [88].

In vivo selectivity of anti-inflammatory drugs could also be evaluated by means of microdialysis during drug development. Since prostaglandin E2 (PGE2) is a product of cyclooxygenase-1 (COX-1) and COX-2, whereas thromboxane B2 (TxB2) served as an indicator of COX-1 activity, continuous monitoring of tissue levels of these endogenous compounds allows assessment of the in vivo selectivity of nonsteroidal anti-inflammatory drugs. Khan et al. have demonstrated that oral administration of celecoxib suppressed PGE2 but not TxB2 dialysate concentrations, suggesting a relative selective in vivo COX-2 inhibition by celecoxib [89].

7. REGULATORY ASPECTS OF MICRODIALYSIS SAMPLING IN DRUG DEVELOPMENT

From a regulatory point of view, microdialysis is accepted increasingly by international regulatory agencies such as the U.S. Food and Drug Administration (FDA) and the European Medicines Agency (EMEA), considering that these agencies are receptive to including findings obtained from microdialysis studies as a part of a preclinical and clinical pharmacology package of drug development [24]. For example, microdialysis has been used for preclinical evaluation of the mechanism of action of zonisamide [56], rasagiline [90], and varenicline [58]. Microdialysis would also be used for preclinical evaluation of CPP-109, a new chemical entity for treatment of cocaine addiction [91]. In addition, regulatory authorities have encouraged the study of tissue distribution of antimicrobial agents in unaffected and infected target sites and the relationship of unbound drug concentrations at the site of action to the in vitro susceptibility of infecting microorganism [92]. Moreover, an FDA advisory committee has found that microdialysis is an attractive approach for clinical studies on the tissue distribution of antibiotics [93].

More recently, microdialysis has been included by the EMEA as an appropriate technique for blood sampling in the investigation of medicinal products in term and preterm neonates. Considering that preterm and term neonates have very limited blood volume and are often anemic, there is a need to limit extraction of blood samples by the use of alternative methods such as microdialysis, saliva sampling, or urinalysis [94]. Furthermore, the utility of microdialysis for preclinical evaluation of in vivo pharmacodynamics has been recognized in the EMEA guideline on the nonclinical investigation of the dependence potential of medicinal products [95]. As stated in the guideline: "Initial in vivo pharmacodynamic investigations could make use of neuropharmacological models, e.g., microdialysis (for example, dopamine release in nucleus accumbens), neurotransmitter turnover, head twitch, antinociception and locomotor activity. Such studies—like the in vitro studies mentioned above—are considered supportive and help to elucidate the profile and mechanism of action of the active substance" [95].

Microdialysis data obtained during preclinical evaluation of ceplene (histamine dihydrochloride) for the treatment of acute myeloid leukaemia has also been accepted by the EMEA [96]. Histamine concentrations were measured in the interstitial fluid sampled from normal and malignant tissues by microdialysis using a RIA assay following a single intravenous dose of 0.5 mg/kg. The highest radioactivity was found in plasma, liver, and liver tumor; the levels were lower in subcutis and subcutis tumor [97].

In conclusion, in recent years both the FDA and the EMEA have accepted the inclusion of findings obtained from microdialysis studies during preclinical evaluations of new chemical entities. Therefore, it is expected that applicability and acceptance of microdialysis will increase greatly in the next years.

8. CONCLUSIONS

Microdialysis sampling will became an attractive approach to early drug development of new chemical entities. Attrition during drug development is mostly a consequence of inadequate drug distribution at the target site. Microdialysis is a unique sampling technique that allows continuous monitoring of unbound extracellular concentrations at the site of action, and it could be therefore highly useful in selecting most adequate drug candidates during the preclinical stage of drug development. Continuous monitoring of changes on biochemical markers induced by lead compounds allows better understanding of in vivo pharmacodynamics during drug development, especially if biochemical markers are highly correlated with the clinical response. In this regard, microdialysis is highly attractive for assessment of neurochemical actions of centrally acting new chemical entities. In addition, as microdialysis simultaneously samples target-site concentrations of new chemical entities and their effect on biochemical markers with high temporal resolution, this technique makes possible the design of mechanism-based PK–PD models of lead compounds, thereby reducing costs in early drug development. Considering these aspects, acceptance of microdialysis data as a part of preclinical and clinical pharmacology packages of drug development by regulatory agencies is actually increasing.

REFERENCES

[1] DiMasi, J.A., Hansen, R.W., Grabowski, H.G. (2003). The price of innovation: new estimates of drug development costs. *Journal of Health Economy*, *22*, 151–185.

[2] DiMasi, J.A. (2002). The value of improving the productivity of the drug development process: faster times and better decisions. *Pharmacoeconomics*, *20*(Suppl. 3), 1–10.

[3] Prentis, R.A., Lis, Y., Walker, S.R. (1988). Pharmaceutical innovation by the seven UK-owned pharmaceutical companies (1964–1985). *British Journal of Clinical Pharmacology*, *25*, 387–396.

[4] Carlson, T.J., Fisher, M.B. (2008). Recent advances in high throughput screening for ADME properties. *Combinatorial Chemistry & High Throughput Screening*, *11*, 258–264.

[5] Höcht, C., Mayer, M., Opezzo, J.A.W., Bramuglia, G.F., Taira, C.A. (2009). New sampling techniques for pharmacokinetic–pharmacodynamic modeling. *Frontiers in Drug Design and Discovery*, *4*, 43–80.

[6] Danhof, M., de Lange, E.C., Della Pasqua, O.E., Ploeger, B.A., Voskuyl, R.A. (2008). Mechanism-based pharmacokinetic–pharmacodynamic (PK–PD) modeling in translational drug research. *Trends in Pharmacological Sciences*, *29*, 186–191.

[7] Mager, D.E., Jusko, W.J. (2008). Development of translational pharmacokinetic–pharmacodynamic models. *Clinical Pharmacology & Therapeutics*, *83*, 909–912.

[8] Chien, J.Y., Friedrich, S., Heathman, M.A., de Alwis, D.P., Sinha, V. (2005). Pharmacokinetics/pharmacodynamics and the stages of drug development: role of modeling and simulation. *AAPS Journal*, *7*, E544–E559.

[9] Eddershaw, P.J., Beresford, A.P., Bayliss, M.K. (2000). ADME/PK as part of a rational approach to drug discovery. *Drug Discovery Today*, *5*, 414.

[10] Wishart, D.S. (2007). Improving early drud discovery through ADME modelling. *Drugs in R & D*, *8*, 349–362.

[11] Fischman, A.J., Alpert, N.M., Rubin, R.H. (2002). Pharmacokinetic imaging a noninvasive method for determining drug distribution and action. *Clinical Pharmacokinetics*, *4*, 481–502.

[12] Danhof, M., de Jongh, J., de Lange, E.C., Della Pasqua, O., Ploeger, B.A., Voskuyl, R.A. (2007). Mechanism-based pharmacokinetic–pharmacodynamic modeling: biophase distribution, receptor theory, and dynamical systems analysis. *Annual Review of Pharmacology and Toxicology*, *47*, 357–400.

[13] Rajman, I. (2008). PK/PD modelling and simulations: utility in drug development. *Drug Discovery Today*, *13*, 341–346.

[14] Biomarkers Definitions Working Group (2001). Biomarkers and surrogate endpoints: preferred definitions and conceptual framework. *Clinical Pharmacology & Therapeutics*, *69*, 89–95.

[15] Marrer, E., Dieterle, F. (2007). Promises of biomarkers in drug development: a reality check. *Chemical Biology & Drug Design*, *69*, 381–394.

[16] Colburn, W.A. (1997). Selecting and validating biologic markers for drug development. *Journal of Clinical Pharmacology*, *37*, 355–362.

[17] Danhof, M., Alvan, G., Dahl, S.G., Kuhlmann, J., Paintaud, G. (2005). Mechanism-based pharmacokinetic–pharmacodynamic modeling: a new classification of biomarkers. *Pharmaceutical Research*, *22*, 1432–1437.

[18] Csajka, C., Verotta, D. (2006). Pharmacokinetic–pharmacodynamic modeling: history and perspectives. *Journal of Pharmacokinetics and Pharmacodynamics*, *33*, 227–279.

[19] Toutain, P.L. (2002). Pharmacokinetic/pharmacodynamic integration in drug development and dosage-regimen optimization for veterinary medicine. *AAPS Journal*, *4*, 1–29.

[20] Höcht, C., Opezzo, J.A.W., Bramuglia, G.F., Taira, C.A. (2006). Application of microdialysis for pharmacokinetic–pharmacodynamic (PK–PD) modeling. *Expert Opinion on Drug Discovery*, *1*, 289–301.

[21] Perez-Urizar, J., Granados-Soto, V., Flores-Murrieta, F.J., Castañeda-Hernandez, G. (2000). Pharmacokinetic–pharmacodynamic modeling: Why? *Archives of Medicine Research*, *31*, 539–545.

[22] Elmquist, W.F., Sawchuk, R.J. (1997). Application of microdialysis in pharmacokinetic studies. *Pharmaceutical Research*, *14*, 267–287.

[23] Li, Y., Peris, J., Zhong, L., Derendorf, H. (2006). Microdialysis as a tool in local pharmacodynamics. *AAPS Journal*, *8*, E222–E235.

[24] Chaurasia, C.S., Müller, M., Bashaw, E.D., Benfeldt, E., Bolinder, J., Bullock, R., Bungay, P.M., DeLange, E.C., Derendorf, H., Elmquist, W.F., et al. (2007). AAPS–FDA Workshop White Paper: Microdialysis Principles, Application and Regulatory Perspectives. *Pharmaceutical Research*, *24*, 1014–1025.

[25] Woo, K.L., Lunte C.E. (2008). The development of multiple probe microdialysis sampling in the stomach. *Journal of Pharmaceutical and Biomedical Analysis*, *48*, 20–26.

[26] Scott, D.O., Lunte, C.E. (2003). In vivo microdialysis sampling in the bile, blood, and liver of rats to study the disposition of phenol. *Pharmaceutical Research*, *10*, 335–342.

[27] Weikop, P., Egestad, B., Kehr, J. (2004). Application of triple-probe microdialysis for fast pharmacokinetic/pharmacodynamic evaluation of dopamimetic activity of drug candidates in rat brain. *Journal of Neuroscience Methods*, *140*, 59–65.

[28] Langer, O., Müller, M. (2004). Methods to assess tissue-specific distribution and metabolism of drugs. *Current Drug Metabolism*, *5*, 463–481.

[29] Nutt, R. (2002). 1999 ICP Distinguished Scientist Award: The history of positron emission tomography. *Molecular Imaging and Biology*, *4*, 11–26.

[30] Langer, O., Karch, R., Müller, U., Dobrozemsky, G., Abrahim, A., Zeitlinger, M., Lackner, E., Joukhadar, C., Dudczak, R., Kletter, K., Müller, M., Brunner, M. (2005). Combined PET and microdialysis for in vivo assessment of intracellular drug pharmacokinetics in humans. *Journal of Nuclear Medicine*, *46*, 1835–1841.

[31] Hutchinson, P.J., O'Connell, M.T., Nortje, J., Smith, P., Al-Rawi, P.G., Gupta, A.K., Menon, D.K., Pickard, J.D. (2005). Cerebral microdialysis methodology: evaluation of 20 kDa and 100 kDa catheters. *Physiological Measurement*, *26*, 423–428.

[32] Schiffer, W.K., Alexoff, D.L., Shea, C., Logan, J., Dewey, S.L. (2005). Development of a simultaneous PET/microdialysis method to identify the optimal dose of 11C-raclopride for small animal imaging. *Journal of Neuroscience Methods*, *144*, 25–34.

[33] Höcht, C., Opezzo, J.A.W., Taira, C.A. (2003). Validation of a new intraarterial microdialysis shunt probe for the estimation of pharmacokinetic parameters. *Journal of Pharmaceutical and Biomedical Analysis, 31*, 1105–1113.

[34] Tsai, T.H. (2003). Assaying protein unbound drugs using microdialysis techniques. *Journal of Chromatography B: Analytical Technologies in the Biomedical and Life Sciences, 797*, 161–173.

[35] Evrard, P.A., Cumps, J., Verbeeck, R.K. (1996). Concentration-dependent plasma protein binding of flurbiprofen in the rat: an in vivo microdialysis study. *Pharmaceutical Research, 13*, 18–22.

[36] Schmidt, S., Röck, K., Sahre, M., Burkhardt, O., Brunner, M., Lobmeyer, M.T., Derendorf, H. (2008). Effect of protein binding on the pharmacological activity of highly bound antibiotics. *Antimicrobial Agents and Chemotherapy, 52*, 3994–4000.

[37] Höcht, C., Opezzo, J.A.W., Taira, C.A. (2004). Microdialysis in drug discovery. *Current Drug Discovery Technologies, 1*, 269–285.

[38] Kurosaki, Y., Nakamura, S., Shiojiri, Y., Kawasaki, H. (1998). Lipo-microdialysis: a new microdialysis method for studying the pharmacokinetics of lipophilic substances. *Biological & Pharmaceutical Bulletin, 21*, 194–196.

[39] Schnetz, E., Fartasch, M. (2001). Microdialysis for the evaluation of penetration through the human skin barrier: a promising tool for future research? *European Journal of Pharmaceutical Sciences, 12*, 165–174.

[40] Sun, L., Stenken, J.A. (2003). Improving microdialysis extraction efficiency of lipophilic eicosanoids. *Journal of Pharmaceutical and Biomedical Analysis, 33*, 1059–1071.

[41] Höcht, C., Opezzo, J.A.W., Bramuglia, G.F., Taira, C.A. (2006). Application of microdialysis in clinical pharmacology. *Current Clinical Pharmacology, 1*, 163–183.

[42] Schmidt, S., Banks, R., Kumar, V., Rand, K.H., Derendorf, H. (2008). Clinical microdialysis in skin and soft tissues: an update. *Journal of Clinical Pharmacology, 48*, 351–364.

[43] Brunner, M., Langer, O. (2006). Microdialysis versus other techniques for the clinical assessment of in vivo tissue drug distribution. *AAPS Journal, 8*, E263–E271.

[44] Brunner, M., Derendorf, H., Müller, M. (2005). Microdialysis for in vivo pharmacokinetic/pharmacodynamic characterization of anti-infective drugs. *Current Opinion in Pharmacology, 5*, 495–499.

[45] Herkenne, C., Alberti, I., Naik, A., Kalia, Y.N., Mathy, F.X., Préat, V., Guy, R.H. (2008). In vivo methods for the assessment of topical drug bioavailability. *Pharmaceutical Research, 25*, 87–103.

[46] Cordon-Cardo, C., O'Brien, J.P., Casals, D., Rittman-Grauer, L., Biedler, J.L., Melamed, M.R., Bertino, J.R. (1989). Multidrug-resistance gene (P-glycoprotein) is expressed by endothelial cells at blood–brain barrier sites. *Proceedings of the National Academy of Science of the United States of America, 86*, 695–698.

[47] de Boer, A.G., van der Sandt, I.C., Gaillard, P.J. (2003). The role of drug transporters at the blood–brain barrier. *Annual Reviews of Pharmacology and Toxicology, 43*, 629–656.

[48] Venkatakrishnan, K., Tseng, E., Nelson, F.R., Rollema, H., French, J.L., Kaplan, I.V., Horner, W.E., Gibbs, M.A. (2007). Central nervous system pharmacokinetics of the Mdr1 P-glycoprotein substrate CP-615,003: intersite differences and implications for human receptor occupancy projections from cerebrospinal fluid exposures. *Drug Metabolism and Disposition, 35,* 1341–1349.

[49] European Medicines Agency (2004). http://www.emea.europa.eu/humandocs/ PDFs/EPAR/tasmar/034397en6.pdf. Accessed November 2008.

[50] Dashtipour, K., Chen, J.J., Lew, M.F. (2008). Rasagiline for the management of Parkinson's disease. *Therapy, 5,* 203–214.

[51] Remy, S., Beck, H. (2006). Molecular and cellular mechanisms of pharmacoresistance in epilepsy. *Brain, 129*(Pt. 1), 18–35.

[52] Lazarowski, A., Czornyj, L., Lubienieki, F., Girardi, E., Vazquez, S., D'Giano, C. (2007). ABC transporters during epilepsy and mechanisms underlying multidrug resistance in refractory epilepsy. *Epilepsia, 48*(Suppl. 5), 140–149.

[53] Löscher, W., Potschka, H. (2005). Drug resistance in brain diseases and the role of drug efflux transporters. *Nature Reviews: Neuroscience, 6,* 591–602.

[54] Tong, X., Ratnaraj, N., Patsalos, P.N. (2009). Vigabatrin extracellular pharmacokinetics and concurrent gamma-aminobutyric acid neurotransmitter effects in rat frontal cortex and hippocampus using microdialysis. *Epilepsia, 50,* 174–183.

[55] Höcht, C., Lazarowski, A., Gonzalez, N.N., Auzmendi, J., Opezzo, J.A., Bramuglia, G.F., Taira, C.A., Girardi, E. (2007). Nimodipine restores the altered hippocampal phenytoin pharmacokinetics in a refractory epileptic model. *Neuroscience Letters, 413,* 168–172.

[56] Physician's Desk Reference (2006). Zonegran clinical pharmacology FDA-approved label. http://www.thomsonhc.com/pdrel/librarian/PFDefaultActionId/ pdrcommon.IndexSearchTranslator. Accessed November 2008.

[57] Chenel, M., Marchand, S., Dupuis, A., Lamarche, I., Paquereau, J., Pariat, C., Couet, W. (2004). Simultaneous central nervous system distribution and pharmacokinetic–pharmacodynamic modelling of the electroencephalogram effect of norfloxacin administered at a convulsant dose in rats. *British Journal of Pharmacology, 142,* 323–330.

[58] European Medicines Agency (2006). http://www.emea.europa.eu/humandocs/ PDFs/EPAR/champix/H-699-en6.pdf. Accessed November 2008.

[59] Artigas, F., Adell, A. (2006). The use of brain microdialysis in antidepressant drug research. In: Wersterink, W.H.C., Cremers, T.I. (Eds.), *Handbook of Microdialysis: Methods, Applications and Perspectives,* Elsevier, Amsterdam, pp. 527–543.

[60] Hughes, Z.A., Liu, F., Platt, B.J., Dwyer, J.M., Pulicicchio, C.M., Zhang, G., Schechter, L.E., Rosenzweig-Lipson, S., Day, M. (2008). WAY-200070, a selective agonist of estrogen receptor beta as a potential novel anxiolytic/antidepressant agent. *Neuropharmacology, 54,* 1136–1142.

[61] Watson, J.M., Dawson, L.A. (2007). Characterization of the potent 5-HT(1A/B) receptor antagonist and serotonin reuptake inhibitor SB-649915: preclinical evidence for hastened onset of antidepressant/anxiolytic efficacy. *CNS Drug Reviews, 13,* 206–223.

[62] Rajarao, S.J., Platt, B., Sukoff, S.J., Lin, Q., Bender, C.N., Nieuwenhuijsen, B.W., Ring, R.H., Schechter, L.E., Rosenzweig-Lipson, S., Beyer, C.E. (2007). Anxiolytic-

like activity of the non-selective galanin receptor agonist, galnon. *Neuropeptides*, *41*, 307–320.

[63] Kiewert, C., Hartmann, J., Stoll, J., Thekkumkara, T.J., Van der Schyf, C.J., Klein, J. (2006). NGP1-01 is a brain-permeable dual blocker of neuronal voltage- and ligand-operated calcium channels. *Neurochemical Research*, *31*, 395–399.

[64] de Lange, E.C., Ravenstijn, P.G., Groenendaal, D., van Steeg, T.J. (2005). Toward the prediction of CNS drug-effect profiles in physiological and pathological conditions using microdialysis and mechanism-based pharmacokinetic–pharmacodynamic modeling. *AAPS Journal*, *7*, E532–E543.

[65] Raje, S., Cornish, J., Newman, A.H., Cao, J., Katz, J., Eddington, N.D. (2005). Pharmacodynamic assessment of the benztropine analogues AHN-2055 and AHN-2005 using intracerebral microdialysis to evaluate brain dopamine levels and pharmacokinetic/pharmacodynamic modeling. *Pharmaceutical Research*, *22*, 603–612.

[66] Bouw, M.R., Xie, R., Tunblad, K., Hammarlund-Udenaes, M. (2001). Blood–brain barrier transport and brain distribution of morphine-6-glucuronide in relation to the antinociceptive effect in rats: pharmacokinetic/pharmacodynamic modeling. *British Journal of Pharmacology*, *134*, 1796–1804.

[67] Müller, M., de la Peña, A., Derendorf, H. (2004). Issues in pharmacokinetics and pharmacodynamics of anti-infective agents: distribution in tissue. *Antimicrobial Agents and Chemotherapy*, *48*, 1441–1453.

[68] Liu, P., Derendorf, H. (2003). Antimicrobial tissue concentrations. *Infectious Disease Clinics North America*, *17*, 599–613.

[69] Nordbring, F. (1978). Tissue penetration of antibiotics. Introduction: focus on some problems involved in the treatment of infectious diseases. *Scandinavian Journal of Infectious Disease Supplementum*, *14*, 21–22.

[70] Fischman, A.J., Babich, J.W., Bonab, A.A., Alpert, N.M., Vincent, J., Callahan, R.J., Correia, J.A., Rubin, R.H. (1998). Pharmacokinetics of [18F]trovafloxacin in healthy human subjects studied with positron emission tomography. *Antimicrobial Agents and Chemotherapy*, *42*, 2048–2054.

[71] Heikkinen, T., Laine, K., Neuvonen, P.J., Ekblad, U. (2000). The transplacental transfer of the macrolide antibiotics erythromycin, roxithromycin and azithromycin. *International Journal of Obstetrics and Gynaecology*, *107*, 770–775.

[72] Joukhadar, C., Derendorf, H., Müller, M. (2001). Microdialysis: a novel tool for clinical studies of anti-infective agents. *European Journal of Clinical Pharmacology*, *57*, 211–219.

[73] Schuck, E.L., Derendorf, H. (2005). Pharmacokinetic/pharmacodynamic evaluation of anti-infective agents. *Expert Reviews on Anti-infective Therapy*, *3*, 361–373.

[74] Delacher, S., Derendorf, H., Hollenstein, U., Brunner, M., Joukhadar, C., Hofmann, S., Georgopoulos, A., Eichler, H.G., Müller, M. (2000). A combined in vivo pharmacokinetic–in vitro pharmacodynamic approach to simulate target site pharmacodynamics of antibiotics in humans. *Journal of Antimicrobial Chemotherapy*, *46*, 733–739.

[75] Müller, M., de la Peña, A., Derendorf, H. (2004). Issues in pharmacokinetics and pharmacodynamics of anti-infective agents: kill curves versus MIC. *Antimicrobial Agents and Chemotherapy*, *48*, 369–377.

[76] Liu, P., Rand, K.H., Obermann, B., Derendorf, H. (2005). Pharmacokinetic–pharmacodynamic modelling of antibacterial activity of cefpodoxime and cefixime in in vitro kinetic models. *International Journal of Antimicrobial Agents*, *25*, 120–129.

[77] Lin, J.H. (2006). Tissue distribution and pharmacodynamics: a complicated relationship. *Current Drug Metabolism*, *7*, 39–65.

[78] Presant, C.A., Wolff, W., Walush, V., Wiseman, C., Kennedy, P., Blayney, D., Brechner, R.R. (1994). Association of intratumoral pharmacokinetics of fluorouracil with clinical response. *Lancet*, *343*, 1184–1187.

[79] Brunner, M., Müller, M. (2002). Microdialysis: an in vivo approach for measuring drug delivery in oncology. *European Journal of Clinical Pharmacology*, *58*, 227–234.

[80] Kitzen, J.J., Verweij, J., Wiemer, E.A., Loos, W.J. (2006). The relevance of microdialysis for clinical oncology. *Current Clinical Pharmacology*, *1*, 255–263.

[81] Zhouh, Q., Gallo, J.M. (2005). In vivo microdialysis for PK and PD studies of anticancer drugs. *AAPS Journal*, *7*, E659–E667.

[82] Castejon, A.M., Paez, X., Hernandez, L., Luigi, X., Cubeddu, L.X. (1999). Use of intravenous microdialysis to monitor changes in serotonin release and metabolism induced by cisplatin in cancer patients: comparative effects of granisetron and ondansetron. *Journal of Pharmacological and Experimental Therapeutics*, *291*, 960–966.

[83] Garvin, S., Dabrosin, C. (2003). Tamoxifen inhibits secretion of vascular endothelial growth factor in breast cancer in vivo. *Cancer Research*, *63*, 8742–8748.

[84] Thompson, J.F., Siebert, G.A., Anissimov, Y.G., Smithers, B.M., Doubrovsky, A., Anderson, C.D., Roberts, M.S. (2001). Microdialysis and response during regional chemotherapy by isolated limb infusion of melphalan for limb malignancies. *British Journal of Cancer*, *85*, 157–165.

[85] Müller, M., Bockenheimer, J., Zellenger, U., Klein, N., Steger, G.G., Eichler, H.G., Mader, R.M. (2000). Relationship between in vivo drug exposure of the tumor interstitium and inhibition of tumor cell growth in vitro: a study in breast cancer patients. *Breast Cancer Research and Treatment*, *60*, 211–217.

[86] Lerman, L.O., Chade, A.R., Sica, V., Napoli, C. (2005). Animal models of hypertension: an overview. *Journal of Laboratory and Clinical Medicine*, *146*, 160–173.

[87] Höcht, C., DiVerniero, C., Opezzo, J.A., Taira, C.A. (2005). Applicability of microdialysis as a technique for pharmacokinetic–pharmacodynamic (PK/PD) modeling of antihypertensive beta-blockers. *Journal of Pharmacological and Toxicological Methods*, *52*, 244–250.

[88] Höcht, C., Mayer, M.A., Opezzo, J.A.W., Bertera, F.M., Taira, C.A. (2008). Pharmacokinetic–pharmacodynamic modeling of antihypertensive drugs: from basic research to clinical practice. *Current Hypertension Reviews*, *4*, 289–302.

[89] Khan, A.A., Brahim, J.S., Rowan, J.S., Dionne, R.A. (2002). In vivo selectivity of a selective cyclooxygenase 2 inhibitor in the oral surgery model. *Clinical and Pharmacological Therapeutics*, *72*, 44–49.

[90] European Medicines Agency (2005). http://www.emea.europa.eu/humandocs/PDFs/EPAR/Azilect/5289705en6.pdf. Accessed November 2008.

[91] Catalyst pharmaceuticals partners files investigational new drug application for CPP-109 to treat cocaine addiction (2005). http://www.bnl.gov/CTN/GVG/CPP.asp. Accessed November 2008.

[92] FDA Guidance for Industry: Developing antimicrobial drugs: a general consideration for clinical trials. http://www.fda.gov/downloads/Drugs/GuidanceCompliance RegulatoryInformation/Guidances/ucm070983.pdf. Accessed June 2010.

[93] Müller Reporting Company, Inc. Anti-infective drugs advisory committee meeting, 64th meeting. Department of Health and Human Services, Food and Drug Administration. Guidance documents on developing antimicrobial drugs: general considerations and individual indications. http://www.fda.gov/cder/present/anti-infective798/073198.pdf. Accessed November 2008.

[94] European Medicines Agency (2007). Guideline on the investigation of medicinal products in the term and preterm neonate. http://www.emea.europa.eu/pdfs/ human/paediatrics/26748407en.pdf. Accessed November 2008.

[95] European Medicines Agency (2006). Guideline on the non-clinical investigation of the dependence potential of medicinal products. http://www.emea.europa.eu/ pdfs/human/swp/9422704en.pdf. Accessed November 2008.

[96] European Medicines Agency (2008). Assessment Report for Ceplene. http:// www.emea.europa.eu/humandocs/PDFs/EPAR/Ceplene/H-796-en6.pdf. Accessed November 2008.

[97] Rizell, M., Naredi, P., Lindner, P., Hellstrand, K., Sarno, M., Jansson, P.A. (2002). Histamine pharmacokinetics in tumour and host tissues after bolus-dose administration in the rat. *Life Sciences*, *70*, 969–976.

3

ANALYTICAL CONSIDERATIONS FOR MICRODIALYSIS SAMPLING

PRADYOT NANDI, COURTNEY D. KUHNLINE,
AND SUSAN M. LUNTE

University of Kansas, Lawrence, Kansas

1. INTRODUCTION

Microdialysis (MD) is a powerful sampling technique that makes possible continuous monitoring of concentrations of biological molecules and other substances both in vivo and in vitro. This technique has been employed extensively for pharmacokinetic and pharmacodynamic studies as well as for drug delivery. Microdialysis has been used to monitor drugs and other biologically important compounds in virtually every tissue and organ in the body, including liver [1,2], heart [3,4], skin [5,6], blood [7,8], placenta [9], stomach [10,11], ear [12], and brain [13,14].

Microdialysis has many advantages for the continuous in vivo monitoring of drugs and neurotransmitters. Sampling can be performed on awake, freely moving animals. No fluid is removed from the tissue during the sampling process; therefore, it is possible to carry out long-term studies with minimal disruption of the physiological system. Because of the low-molecular-weight cutoff of the dialysis membranes, samples are protein-free and therefore can usually be injected directly into the analytical system. In addition, since enzymes are excluded by the dialysis membrane, analytes are protected from enzymatic degradation. The membrane also excludes protein-bound drugs;

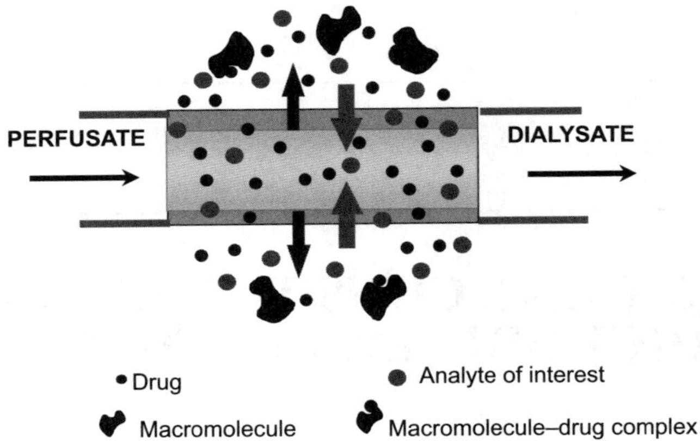

Figure 1 Microdialysis sampling process. Small molecules diffuse across the membrane based on their concentration gradient. Large molecules and protein-bound compounds are excluded by the microdialysis probe.

therefore, only the free (active) fraction of the drug is measured by microdialysis sampling.

Microdialysis is accomplished using a probe that consists of a short length of hollow-fiber dialysis membrane affixed to inlet and outlet tubing. Figure 1 shows the microdialysis sampling process. A solution that is similar in composition and ionic strength to the extracellular fluid (ECF) of the tissue of interest is pumped slowly through the probe. Small molecules in the extracellular space that are not present in the perfusate diffuse across the membrane based on their concentration gradient and are transported to the analysis system. Similarly, compounds in the perfusate that are not present in the ECF can be delivered directly to the physiological site of interest. Thus, it is also possible to deliver and recover compounds simultaneously from a single tissue site. This feature can be very useful for investigations involving site-specific release of neurotransmitters [15], regional metabolism of neuropeptides [14,16,17], or comparison of the metabolism of antineoplastic agents in tumor versus healthy tissue [18].

There are many different ways to analyze microdialysis samples. Samples can be analyzed directly online using sensors; however, most applications require a separation method prior to detection. The most commonly employed separation methods are liquid chromatography and electrophoresis (in either the capillary or microchip format). A variety of detection methods have also been used, including ultraviolet, fluorescence, electrochemical, and mass spectrometric detection. This chapter, which is focused on analytical aspects of microdialysis sampling, includes information on both sampling and analysis that has recently been reviewed by our group in two other publications [19,20].

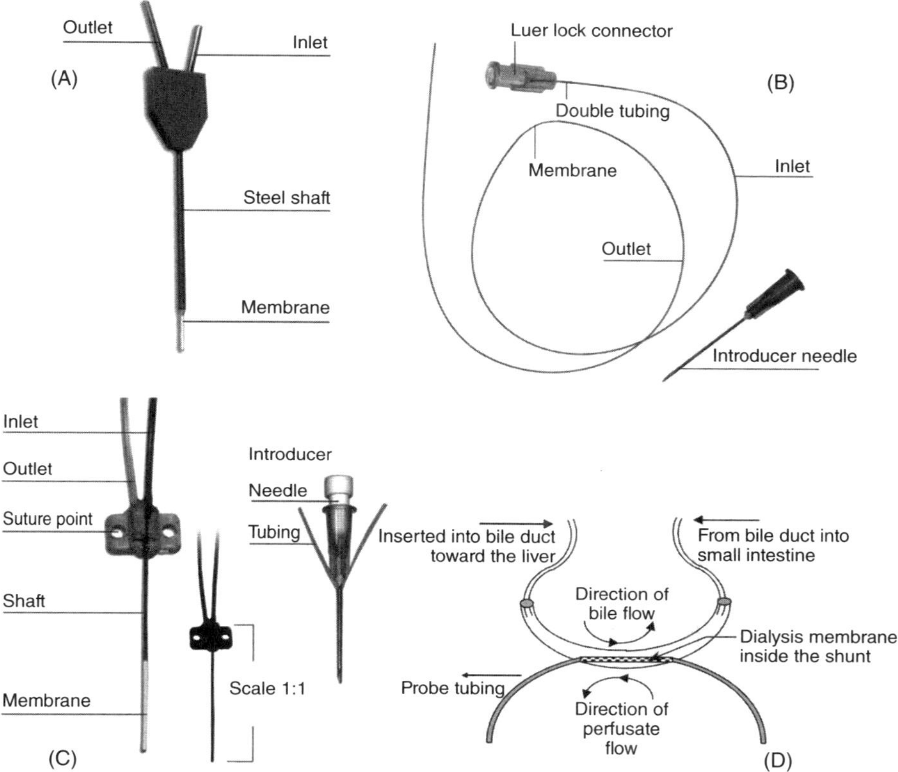

Figure 2 Probes used for microdialysis sampling: (A) concentric cannula design; (B) linear probe; (C) flexible probe (from CMA product catalog); (D) shunt probe. (From [27].)

1.1. Microdialysis Probes

Several different types of probes are available for microdialysis sampling (Figure 2). Probes have been designed specifically for brain, blood, tissue, and bile sampling. Special probes have also been developed for bioreactor and dissolution monitoring. Many different membrane materials are used to make microdialysis probes, including cellulose acetate, polyacrylanitrile (PAN) polycarbonate–polyether (PCE), polyether sulfone (PES), and cuprophan (CUP). Selection of the appropriate membrane for the application of interest depends on the target analyte and sample matrix.

Concentric Cannula The concentric cannula is the most commonly employed design for neurochemical studies (Figure 2A). These probes are composed of stainless steel and are implanted into the specific region of interest using a guide cannula. A typical probe used for rat brain studies is approximately

15 mm long, with an outer diameter between 200 and 500 μm. The dialysis membrane is located at the end of the concentric cannula and is usually 1 to 4 mm in length. Smaller probes are also available for sampling in mice. The mouse probes employ shorter shafts (7 to 12 mm) and have outer diameters between 220 and 380 μm. Cannula-style probes have also been designed specifically for bioreactor monitoring [21–23]. One example is a unique tunable probe developed by Laurell and Buttler that was used to optimize analyte recovery in bioreactors [24].

Linear Probe   The linear probe is used for soft tissue and dermal sampling and consists of a microdialysis membrane (4 to 10 mm) placed between two pieces of flexible tubing [25]. A diagram of the probe is shown in Figure 2B. This probe is normally threaded into the tissue using a guide needle. The linear probe can employ a larger membrane because soft tissues such as skin, liver, muscle, and heart are more homogeneous than brain tissue and therefore spatial resolution is not as important. Due to its large surface area and enhanced analyte recovery, this probe has also been employed for bioreactor and dissolution studies.

Flexible Probe   Figure 2C shows the flexible probe that was first described in 1992 by Telting-Diaz et al. [26]. This probe is ideal for blood sampling in pharmacokinetic studies. The configuration of the flexible probe is similar to that of the brain probe; it consists of two pieces of fused-silica tubing attached to the dialysis membrane. In contrast to the rigid brain cannula probe, the flexible probe can bend when the animal moves, minimizing damage to blood vessels.

Shunt Probe   The shunt probe was designed for sampling moving fluids both in vivo and in vitro [27]. A diagram of the probe is shown in Figure 2D. The most common application of this probe has been sampling bile in awake, freely moving animals. For these experiments, the bile duct is cannulated, and bile is sampled by a second dialysis membrane filled with fluid of similar ionic composition running in the opposite direction. This probe has also been found to be useful for desalting protein samples prior to introduction into a mass spectrometer [28].

1.2. Sampling Parameter Considerations

Recovery   Analyte recovery is a very important parameter in microdialysis sampling. At typical flow rates used in microdialysis sampling (1 to 5 μL/min), there is not enough time for complete equilibration between the perfusate and the surrounding environment to occur, and therefore the concentration of analyte in the perfusate reflects only a percentage of the total amount present in the extracellular space or sample. The recovery is defined as the ratio of the dialysate concentration to the actual tissue concentration and is controlled by the overall mass transport of the analyte across the probe membrane [29]. The percent recovery is calculated using the equation

$$\%R = \frac{C_d}{C_s} \times 100$$

where C_d is the concentration of the analyte in the dialysis perfusate and C_s is the concentration in the sample being interrogated.

Recovery is a function of the concentration gradient that is produced across the dialysis membrane. The magnitude of this concentration gradient is dependent on a number of parameters, including the chemical composition, length, thickness, and molecular-weight cutoff (MWCO) of the membrane that is employed, the probe design and dimensions, the composition of the perfusate as well as of the outer medium, the diffusion coefficient of the analyte of interest, the flow rate that is used for sampling, and whether the sampling is performed under hydrodynamic or static conditions. In general, analyte recoveries from blood are higher than that from the stagnant interstitial fluid, due to better mass transport [30].

More specifically, recovery can be described by

$$R = 1 - \exp\left[-\frac{1}{Q_d(R_d + R_m + R_e)}\right]$$

where Q_d is the volumetric flow rate of perfusate, and R_d, R_m, and R_e are the resistances of dialysate, membrane, and extracellular space, respectively, to the mass transfer of analyte(s). R_e includes factors such as the rate of diffusion of the analyte of interest through the sample matrix, as well as metabolism, reuptake, and other chemical or enzymatic reactions that the analyte may undergo during the sampling process The membrane resistance, R_m, takes into account probe-related factors, including probe dimensions and MWCO, as well as the diffusion coefficient of the analyte across the probe membrane. R_d takes into consideration the diffusion of the analyte in the microdialysate. This can be affected by the molecular weight and size of the compound of interest as well as the perfusate composition [28,31].

It is generally assumed that $R_m \gg R_e$, which implies that recovery information obtained for a probe in vitro can be applied to in vivo experiments. However, this is not always the case because diffusion of the analyte from distant extracellular space to the area around the probe can be slow and depends on the diffusion coefficient of the analyte(s) in the extracellular matrix. For large peptides or proteins, this slow diffusion can affect recovery. In particular, peptides can undergo metabolism and other kinetic processes in the brain before reaching the probe membrane. Figure 3 shows some of the processes that can affect recovery of neurotransmitters and other substances in the brain. Other considerations include loss of tissue integrity due to surgery,

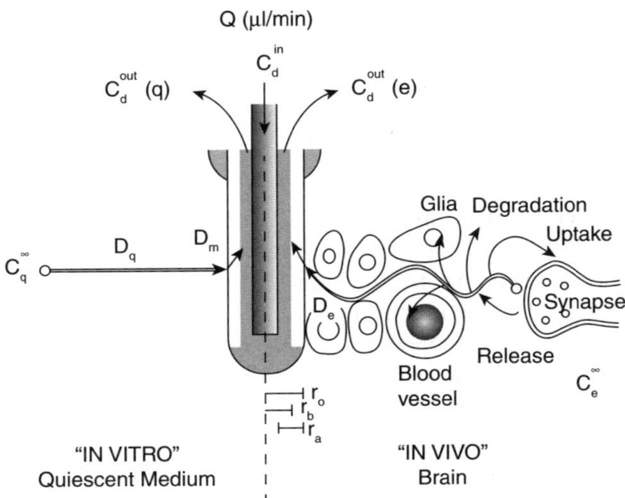

Figure 3 Considerations concerning recovery in microdialysis sampling. Comparison of diffusion paths for a hypothetical molecule in vivo (brain) and in vitro (quiescent medium). (From [45].)

changes in blood flow, and immunological reactions following insertion of the probe.

Physical characteristics of the membrane that affect recovery include charge, thickness, and length, as well as pore size and shape. In 1995, Zhao et al. investigated the effect of membrane type on in vitro recovery and delivery for 15 small-molecule analytes varying in isoelectric point and hydrophobicity [32]. They found that with a PAN membrane, anionic compounds exhibited much lower recoveries than those of neutral and basic compounds. As expected, recovery decreased with increasing membrane thickness and increased with increasing membrane length for all compounds studied.

The recoveries and apparent membrane diffusion coefficients for 10 different compounds spanning a molecular-weight range between 94 and 1355 Da and possessing different log P values were reported in a separate study [33]. For several compounds it was found that recovery was affected dramatically by the type of membrane used. In addition, the presence of 4% bovine serum albumin or 0.3% fibrinogen in the sample matrix did not significantly alter the membrane diffusion coefficients for most membrane–analyte combinations.

Most commercially available microdialysis probes exhibit molecular weight cutoffs between 20,000 and 60,000 Da. With probes of this type, recoveries decrease significantly for compounds greater than 10,000 Da in molecular weight. More recently, probes with a 100 kDa molecular-weight cutoff have become available with the goal of sampling larger biomolecules in vivo. Stenken's group investigated the mass transport properties of FITC-labeled dextrans ranging from 10 to 70 kDa using a 100 kDa commercially available probe. They found that as the molecules increased in size, the membrane became a significant barrier to mass transport, probably due to hindered diffusion [34].

Additives for Improving Recovery For compounds that are present at very low concentrations in the extracellular fluid, one way to lessen the dependence of the assay on the sensitivity of the analytical method is to increase recovery. This can be accomplished by decreasing the flow rate [31]. Another approach to improving recovery of the analyte of interest using microdialysis is to add a substance to the perfusate that has a strong affinity for the compound of interest [35]. This increases the overall flux of the analyte into the probe because the unbound analyte concentration in proximity to the probe membrane will be close to zero, driving mass transport into the probe.

Several compounds have been investigated as affinity agents for microdialysis, including cyclodextrins, antibodies, and complexing agents. Stenken's group has extensively evaluated the use of cyclodextrins as additives to increase the recovery of hydrophobic drugs [36]. In a model study using ibuprofen, they found that the recoveries improved substantially upon the addition of cyclodextrin to the perfusate. The phenyl ring on ibuprofen is known to interact strongly with the cyclodextrin cavity. Recoveries ranging from 10 to over 100% were obtained for this analyte and were dependent on the probe type, cyclodextrin concentration, and flow rate.

β-Cyclodextrin and hydroxypropyl β-cyclodextrin have also been investigated to improve the recovery of tricyclic antidepressants and structurally related analogs. Enhancements varied from 1.5- to almost 10-fold using a polycarbonate membrane and depended on the structure of the tricyclic compound investigated. The use of cyclodextrins as an additive to increase the recovery of enkephalins has also been investigated [37]. However, in this case, the recoveries were increased by less than a factor of 2.

Antibodies have been used to improve the recovery of peptides in microdialysis sampling. Use of a perfusate containing antibodies immobilized on microspheres was evaluated as an approach to improve recovery of cytokines [38], neuropeptides [37], and endocrine hormones in vitro using a 100-kDa MWCO polysulfone membrane. Although the antibodies were found to enhance the recovery of these peptides 3- to 20-fold in vitro, one potential drawback of this method in vivo is the possible saturation by the endogenous peptides of all the available binding sites on the microspheres, which would limit the dynamic range of the assay.

A significant disadvantage of using antibodies for microdialysis studies is that they are difficult to produce and can be expensive. Therefore, less expensive alternative affinity ligands have been explored. Heparin is an inexpensive and highly soluble reagent that has been shown to bind many of the human cytokines in vivo [39]. The relative recoveries for IL-6, IL-7, MCP-1, and TNF-α were all improved with the addition of heparin. Heparin also did not interfere with the ELISA assay of the peptides following collection.

Quantitation in Microdialysis   Quantitation in microdialysis sampling can be a controversial subject, and several excellent reviews have been published in recent years concerning the advantages and disadvantages of different approaches [29,39–46]. Before discussing specific calibration methods, it should be kept in mind that for many in vivo and in vitro applications, enough information can be obtained by monitoring the percent change of analyte with respect to the basal concentration by using the untreated animal (or sample) as its own control. An example of this type of experiment is monitoring a rapid change in neurotransmitters in the brain due to the administration of a drug known to affect a neurochemical pathway [30]. During such experiments it is generally assumed that the probe is performing consistently throughout the duration of the experiment.

In experiments where it is necessary to determine the in vivo concentrations precisely (e.g., in pharmacokinetic studies), quantitation becomes more important and is also more complicated. Frequently, the recoveries using in vitro calibration do not match those obtained in vivo. This is because, as was shown in Figure 3, there are many in vivo processes, such as metabolism, elimination in the blood, and uptake, that can influence the concentration gradient near the dialysis probe [47,48]. Therefore, in addition to optimizing probe membrane type and size, location of sampling probe, and flow rate for the study of interest, the probe must also be calibrated in vivo [31].

One technique that has been employed as the gold standard for determining in vivo concentrations of compounds using microdialysis is the *no net flux* (NNF) *method* [49,50]. This procedure can be used to determine steady-state concentrations of a drug or endogenous compound in tissue. The protocol involves adding a known concentration of analyte to the perfusate. This concentration is then varied over a range that is both higher and lower than the extracellular concentration expected for the exogenous or endogenous compound. The concentration of the analyte in the dialysate is measured and considered to be at the NNF condition when there is no exchange of analyte between extracellular space and perfusate. The concentration at this point is considered to be the extracellular fluid concentration. Figure 4 shows the no-net (zero-flux) experiment used to determine the concentration of cocaine in the rat striatum during a 0.3 mg/kg per minute intravenous (i.v.) infusion of cocaine [31].

The NNF method becomes more complicated if dynamic changes in concentration of an endogenous or exogenous compound are to be monitored [47]. Administration of a drug or other substance to an animal can produce changes in the extracellular environment around the probe, along with the pharmacological action of interest. This change in extracellular environment can change the recovery of the analyte through the probe and hence the accuracy of the estimation of the extracellular concentration.

In 1993, Olson and Justice described a procedure for obtaining extracellular concentrations of endogenous neurotransmitters such as dopamine under transient conditions [47]. This method, termed *dynamic no net flux* (DNNF),

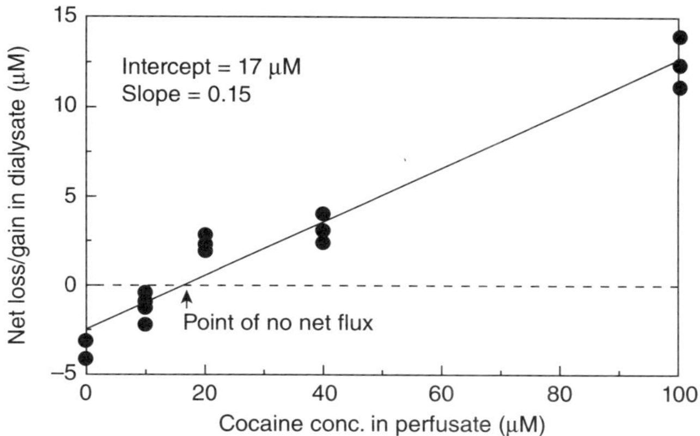

Figure 4 Non-net-flux plot for cocaine in brain microdialysates during a continuous i.v. infusion (0.3 mg·kg·min). The difference between the analyte concentration in the dialysate and the perfusate is plotted as a function of perfusate concentration. In this rat, the point of no net flux (concentration in the brain) was estimated to be 17 µM. The slope of the regression line is the recovery, which is estimated to be 15% for this subject. (From [31].)

uses groups of subjects instead of a single animal to measure the analyte concentration and recovery as a function of time. In their experiment, three groups of four rats were used to measure the effect of cocaine on dopamine release. The perfusate for each of the three groups contained 0, 10, and 40 nM dopamine, respectively. The change in dopamine concentration in the three groups of rats following cocaine administration was plotted versus time. The no net flux plots (dopamine in/dopamine out) for each time point could then be employed to calculate the extracellular concentration of dopamine. These investigators found that the extracellular dopamine increased but recovery decreased after cocaine administration. This is consistent with the microdialysis recovery model, which says that any process generating a sink for the compound of interest will affect probe recovery. Processes include synthesis or metabolism and release and uptake, as well as transport of the substance under investigation in and out of the blood vessels.

A less tedious method of probe calibration is called *retrodialysis* or *reverse dialysis*; it is particularly applicable for in vivo experiments involving exogenous compounds such as drugs [51,52]. The underlying assumption of this method is that diffusion across the probe membrane is quantitatively equal in both directions. This method uses an internal standard whose physical, biological, and pharmacokinetic properties closely resemble those of the compound under scrutiny. Such a compound is added at a known concentration to the perfusate, and its rate of disappearance is calculated from the equation

$$\text{recovery } (\%) = 100 - \left(100 \times \frac{C_{\text{dialysate}}}{C_{\text{perfusate}}} \right).$$

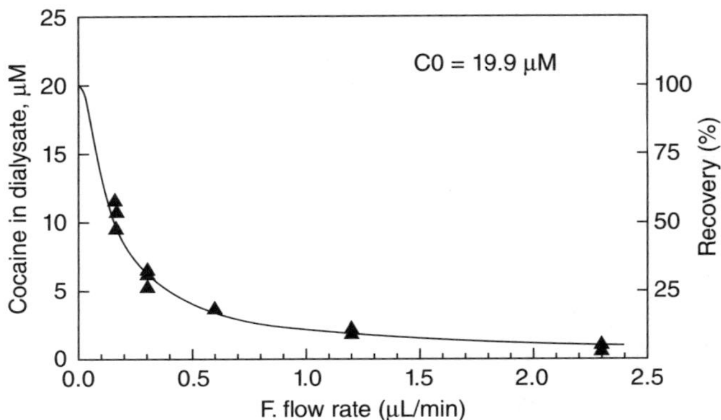

Figure 5 Extrapolation to zero-flow-rate method in a single rat for a continuous i.v. infusion (0.3 mg/kg·min). In this rat, the estimate of the concentration of the extracellular fluid in the brain was 19.9 µM. (From [31].)

Calibration should ideally be performed before the drug is administered so that there is no analyte present in the tissue. This will preserve the membrane concentration gradient during the pharmacokinetic (or other drug delivery) experiment. Also, the probe should be perfused thoroughly with the physiological solution to remove the drug delivered to the tissue during retrodialysis before administration of the exogenous compound of interest [40,53].

Song and Lunte compared retrodialysis to no net flux for exogenous compounds both in vivo and in vitro in several different tissues [54–55]. Using acetaminophen and caffeine as model compounds, they found no difference in the numbers obtained using recovery, delivery, or no-net-flux experiments in vitro. The delivery (retrodialysis) method of calibration was then compared to that of no net flux in vivo. No difference was found between the two approaches in muscle tissue for either acetaminophen or caffeine. However, in brain tissue the extraction efficiency determined by delivery was higher for caffeine and lower for acetaminophen than the value obtained using no net flux. For caffeine this was determined to be due to saturable active transport across the blood–brain barrier. This active transport resulted in the extraction efficiency being dependent on the concentration of caffeine in the brain [55].

Menacherry et al. reported that at flow rates of less than 100 nL/min, recovery of the analyte across the probe is essentially 100%. This makes it possible to quantitate analytes in the extracellular fluid without an internal or external reference. Figure 5 shows the low-flow-rate method for the determination of cocaine in the brain at the same concentration as in Figure 4. When the flow rate is extrapolated to zero flow, the concentration of the perfusate matches that of the extracellular fluid [31]. Sampling at these low flow rates can generate significant analytical challenges regarding the manipulation and analysis of submicroliter samples. It also generally prohibits the application of microdialysis to awake, freely moving animals because the swivels and tubing used in these studies produce large dead volumes. However, recent advances in using capillary electrophoresis both online and off-line have made it possible to analyze these small volumes in near real time [19,56].

Recently, Westerink's group developed a microdialysis probe for general use that samples at a flow rate of less than 200 nL/min and employs a makeup flow generated within the probe to produce sample at a rate of 1 to 2 µL/min. This approach takes advantage of the quantitative recovery obtained at very low flow rates combined with the need for larger flow rates (volumes) for analysis and awake animal studies [57]. After correcting for dilution in the probe, an accurate determination of the in vivo concentration of drugs and neurotransmitters can be obtained.

Another semiquantitative approach has been described by the Pawliszyn group. This method employs two probes in close proximity with one probe perfused at a flow rate that is one-half that of the other [58]. The use of two probes makes it possible to calculate a concentration correction factor based on experimental data that can be used to approximate the initial analyte concentration in the sample matrix. This approach was first evaluated using an agarose gel spiked with pesticide as a model system and was later employed to determine the approximate concentrations of pesticides in the leaves of Jade plants.

2. ANALYTICAL METHODOLOGIES

2.1. Non-Separation-Based Methods of Analysis

Biosensors Biosensors are analytical devices with a biological recognition element that produce an electrical signal in response to a biological change. An ideal biosensor should be able to perform continuous and reliable monitoring of analyte from complex body fluids over a significant period of time. There are many types of biosensors. Most of these are enzyme-based and employ either electrochemical or optical detection. Online microdialysis–biosensor systems should be able to monitor concentrations in small sample volumes (microliter) if high temporal resolution is required. Also, the sensor should ideally exhibit high sensitivity and specificity for the analyte(s) of interest in the presence of other endogenous interferences [59].

Online microdialysis sampling coupled to biosensors has been reported for analytes such as ascorbate [60], glucose [61–69], lactate [61–64,66,67,69], and glutamate [61,70–72]. The simultaneous monitoring of glucose and lactate in rats under hypoxic conditions was recently reported by Jones et al. [63]. Yao et al. also reported an online system employing a triple-enzyme electrode for the detection of glucose, L-lactate, and pyruvate [69]. Similar flow injection–based online systems were described by the same group for L-glutamate, acetylcholine, and dopamine [72] and D-/L-lactic acid [73].

Gramsbergen et al. developed an online system for monitoring glucose and lactate during ischemia in freely moving rats. The analytes were monitored by

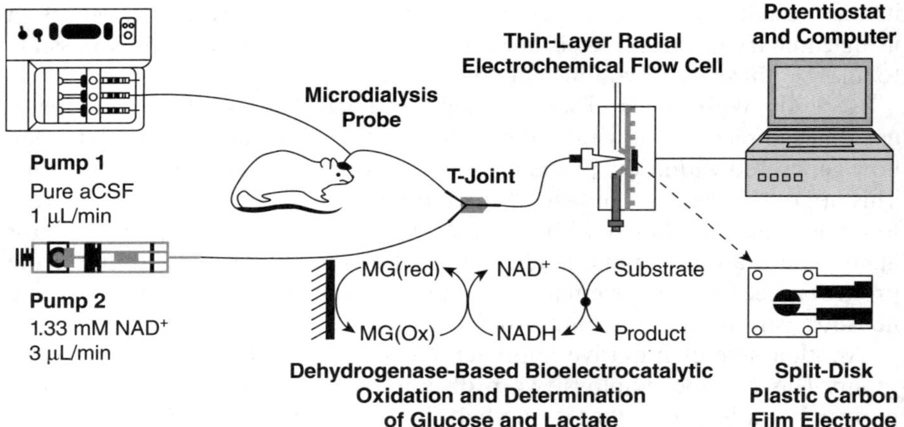

Figure 6 Online electroanalytical system for continuous and simultaneous monitoring of striatum glucose and lactate. (From [66].)

flow injection analysis with enzyme-based amperometric detection [62]. The development of a flow-through sensor with chemiluminescence detection for glucose monitoring in an awake rabbit has also been described [65]. Recently, online monitoring of glucose and lactate from rat brain dialysate was performed following ischemia and reperfusion. The sensor employed methylene green adsorbed on single-walled carbon nanotubes for detection. The experimental setup and the redox reaction with methylene green are shown in Figure 6. The application of online microdialysis with a biosensor has recently been commercialized by Glucoday for continuous monitoring of glucose in diabetics [68].

Biosensors can yield good temporal resolution since no separation step is required. However, most biosensors are developed for a single analyte, so it is not possible to analyze several compounds of a similar class (e.g., catecholamines and amino acids) simultaneously as it is with separation-based methods. Microdialysis does have an advantage over placing the sensor directly into the tissue of interest in that the probe membrane can reduce fouling and an immune response that could be generated by sensor implantation.

Immunoassay Immunoassays have been employed following microdialysis sampling for measuring peptides and some drug substances in microdialysis samples. Neuropeptides are present at nanomolar-to-picomolar concentrations in the extracellular fluid of the brain. In addition, the recovery of peptides across the dialysis membrane is generally much lower than that of smaller-molecular-weight neurotransmitters, generating extremely dilute samples of analyte. To maintain adequate temporal resolution, highly sensitive methods requiring small sample volumes are needed, and immunoassays meet these requirements.

The earliest studies of neuropeptide release using microdialysis sampling employed radioimmunoassay (RIA), and this approach continues to be popular. RIA has been used for the determination of a number of different peptides, including neuropeptide Y [74] and substance P [75], neurotensin [76], and the opioid peptides [77]. Although immunoassays are highly sensitive, the assay itself is time consuming, and the antibodies often exhibit undesirable cross-reactivity, leading to erroneous results in some matrices. In addition, RIAs require disposal of radioactive material.

Mass Spectrometry Mass spectrometric detection has the advantages of conclusive analyte identification based on molecular weight as well as the sensitivity to detect low concentrations of analyte(s) often present in microdialysis samples. This is particularly useful for the determination of neuropeptides due to their low abundance endogenously. As mentioned previously, the traditional means for quantitating neuropeptides is radioimmunoassay, and cross-reactivity can be a significant problem, especially in distinguishing between the peptide of interest, its precursor, and metabolites [78–80]. Matrix-assisted laser desorption ionization (MALDI) has been shown to be an attractive choice for off-line analysis of peptides in microdialysis samples. The Kreek group has utilized microdialysis and MALDI to investigate the metabolism of several significant opioid peptides including β-endorphin [14] and dynorphin A [81] in the rat striatum.

Microdialysis sampling with online mass spectrometric (MS) detection has also been utilized for pharmaceutical applications for the detection of exogenous compounds. A novel central analgesic R-84760 was sampled via microdinlysis from rat blood after i.v. administration and analyzed by MS/MS in a quantitative fashion [82]. Direct coupling of the microdialysis flow to the ionization source is reported and compared to a previously optimized liquid chromatographic (LC)–MS/MS method. To ensure compatibility with the electrospray ionization method, the perfusate consisted of 50:50 ethanol/H_2O. The extent of protein binding of this drug was also determined by LC–MS and compared to results obtained by ultracentrifugation. This same approach was applied to the in vitro analysis of CS-866, a prodrug angiotensin II receptor antagonist, with human and rat plasma as well as liver and intestine microsomes [83]. More frequently, mass spectrometry is coupled to a separation method. The use of LC-MS for the analysis of microdialysis samples is covered in the following section.

2.2. Separation-Based Methods of Analysis

Liquid Chromatography Liquid chromatography is the most commonly used analytical method for the separation of analytes present in microdialysis samples. The detection method employed for LC analysis is dependent on the analyte of interest [15,19]. Ultraviolet detection has been a popular detection scheme for online monitoring of drugs for pharmacokinetic studies.

Electrochemical detection has been employed for the detection of catecholamines and other redox-active compounds of biological interest such as thiols and aromatic amines [49,84]. Fluorescence detection (FL) has also been popular, but usually requires derivatization of analytes prior to analysis. Finally, mass spectrometry is becoming increasingly popular for off-line analysis of microdialysis samples following liquid chromatographic separations.

The type of stationary phase employed for the separation of analytes in a microdialysis sample is dependent on the physiochemical properties of the analytes of interest. Reversed-phase columns are the most popular, due to the aqueous nature of the microdialysis samples, but other types of stationary phases, such as ion exchange, have also been used. The other characteristics of the column (length, particle size, and internal diameter) are determined by the sampling interval desired and the required sensitivity.

Most liquid chromatographic assays require 5 to 10 µL of sample. If a flow rate of 1 µL/min is employed in the microdialysis study, the temporal resolution is 5 to 10 min. If lower flow rates are employed in an effort to increase analyte recovery, the temporal resolution is compromised unless modifications to the chromatographic system are performed. To improve both sensitivity and temporal resolution, microbore and capillary columns have become very popular for the analysis of microdialysis samples [56]. These smaller-diameter columns provide separations that are equivalent to those obtained with larger-inside-diameter (i.d.) conventional columns. More important, since peak dispersion is proportional to the square of the column diameter, it is possible to achieve much higher sensitivity with microbore and capillary columns. However, this is often at the expense of analysis time.

Short microbore columns have become particularly popular for the analysis of microdialysis samples because they provide the optimum combination of high sensitivity and rapid analysis. One challenge of microbore and capillary LC separations is the need for low flow rate pulse-free pumps. Also, dead volumes in the system must be kept very small to minimize analyte dispersion and the resulting band broadening. Dead volume in microbore column systems can be reduced by minimizing the length and the internal diameter of all tubing. The detector flow cell for analysis also needs to have a low volume to avoid mixing and dispersion, all of which can increase band broadening.

Capillary columns have very small internal diameters and offer the advantage of high mass sensitivity, low-flow-rate requirements, and low consumption of sample and reagents. These columns are particularly well suited for online microdialysate analysis [85,86]. However, a drawback of this approach for online systems is that the separation times can be relatively long. This can be detrimental if a system with very high near real-time temporal resolution is desired. In addition, the robustness of capillary LC columns is less than that of conventional LC columns, and specialized equipment is required to load samples onto such columns.

Off-Line Analysis Using Liquid Chromatography   Off-line analysis of microdialysis samples using liquid chromatography is normally accomplished using conventional columns because samples are collected in fraction collectors prior to analysis. Due to surface tension and evaporation considerations, it is impractical to collect samples smaller than 5 µL.

Some of the most common applications of off-line analysis of microdialysis samples using liquid chromatography are determinations of catecholamines [87], amino acid neurotransmitters [15], and neuropeptides [88]. Using microbore and capillary columns in conjunction with electrochemical or fluorescence detection makes it possible to detect femtomoles of catecholamines and amino acid neurotransmitters in brain dialysates [89–91]. Off-line analysis of peptides using LC coupled to radioimmunoassays has also been employed to circumvent the cross-reactivity of the immunoassays [76]. Pharmacokinetic studies will also frequently employ off-line analysis, especially if multiple probes are used in a single animal [43,92].

Over the past 10 years, mass spectrometry has become an extremely popular method of detection for the off-line analysis of microdialysis samples following liquid chromatographic separations. Electrospray is the most commonly employed method for ionization of the chromatographic effluent. However, the high ionic strength of the microdialysis perfusate can present a challenge when electrospray ionization (ESI) sources are used, making direct online coupling difficult. With off-line analysis, several methods are available to desalt the sample, including solid-phase extraction (SPE) prior to analysis and diversion of column flow in the first few minutes of the LC separations to divert the salt to waste before the flow is directed through the ionization source. Another option is to use capillary and nano-LC for the separation. The low flow rates associated with these columns pair well with those necessary for improved extraction efficiencies from microdialysis probes. In addition, the capillary and nano-LC minimize the dilution of the small sample volumes collected via microdialysis.

Andren's group reported off-line analysis of the neuropeptides LVV-hemorphin-7 and dynorphin A 1-17 by LC with a C_{18} packed capillary column and subsequent mass spectrometric analysis by ESI–time-of-flight (TOF) MS [93,94]. To improve sensitivity and aid in the identification of larger peptides, tryptic digests can also be performed prior to analysis [88]. Most work with neuropeptides to date concerns microdialysis sampling in rats. The Li group reported the first use of microdialysis sampling from crustaceans [95]. Specifically, 10 peptide families were identified from the hemolymph of the crab, cancer boralis. Analysis was achieved by nano-LC–MS and MALDI–TOF/TOF, and three previously unidentified neuropeptides were reported. The microdialysis setup and a representative mass spectrum obtained from the dialysate are shown in Figure 7.

Small-molecule neurotransmitters have also been measured by microdialysis with mass spectrometry. Most reports have focused on the analysis of acetylcholine using electrospray ionization with either a triple quadrupole or

(A)

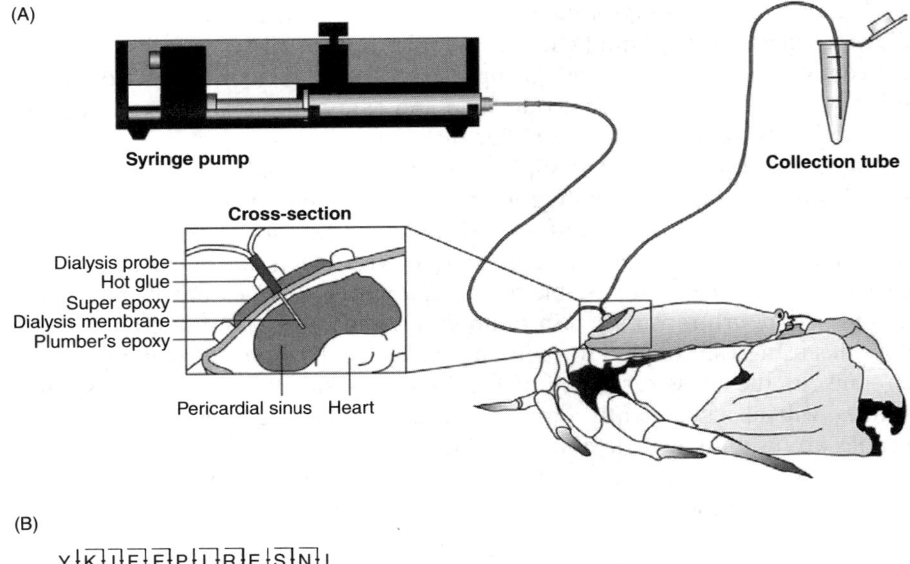

(B)

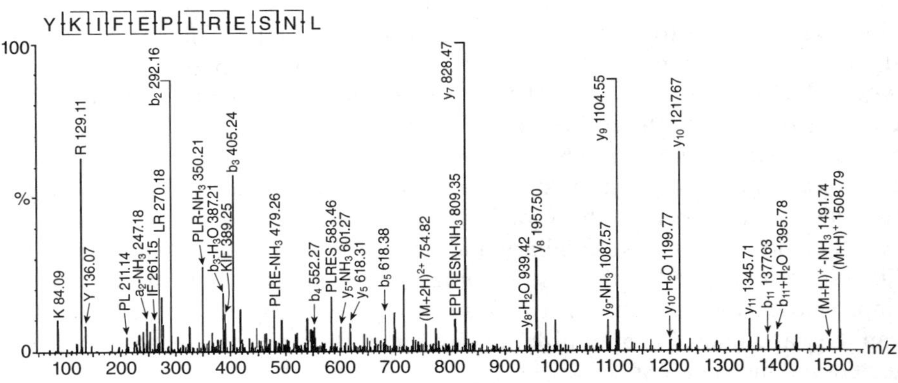

Figure 7   (A) Microdialysis of the pericardial sinus of the *C. borealis* hemolymph. (B) MS–MS sequencing of the *m/z* 1508.8 from the dialysate. (From [95].)

ion-trap mass spectrometer [96–100]. Acetylcholine and choline were also determined in mouse brains via microdialysis utilizing MALDI–TOF MS [101]. Hows et al. were able to measure dopamine, noradrenaline, serotonin, and cocaine simultaneously using two RP columns coupled to an ESI source with a triple-quadrupole MS [102].

The metabolism of oxymatrine and its metabolite matrine was investigated using LC–MS/MS [103]. Another report uses microdialysis sampling and LC–MS to probe the extracellular environment of tumors in mice bearing human melanoma xenografts. Pharmacokinetic parameters were determined in both the plasma and the tumor following treatment with a STEALTH liposomal formulation of a camptothecin analog [104].

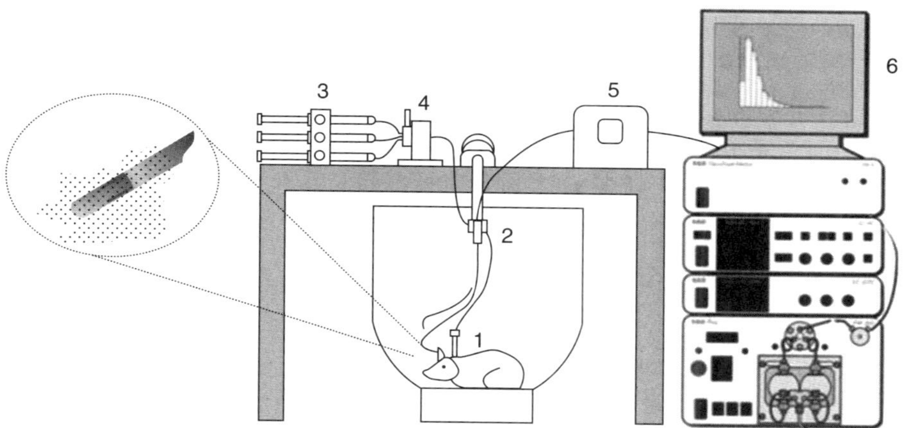

Figure 8   Online microdialysis–LC system. (From BAS product information.)

Online Analysis Using Liquid Chromatography   The direct coupling of liquid chromatography to the microdialysis sampling system has several advantages, and this has recently been reviewed by several groups [15,19,56,105,106]. The use of online injectors allows the manipulation of submicroliter sample volumes, thereby improving temporal resolution in comparison to off-line systems. For online microdialysis–LC, the temporal resolution will often be dependent on the separation time, not the injection volume. A typical online microdialysis–LC system consists of LC pumps, an online injector with injection loop, LC column, detector, and data acquisition system (Figure 8). Online sampling is usually accomplished using automated valves that reduce sample loss and make it possible to perform continuous monitoring.

An early report by Steele and Lunte employed multiple sample loops to inject microdialysis samples continuously without having to divert any sample to waste [107]. In this setup, no temporal information was lost by discarding sample, and each sample represented an integrated time period, depending on the time of sample collection. This system was well suited for pharmacokinetic studies [108].

The use of online microdialysis with microbore liquid chromatography for the continuous analysis of tirazepam and its reduced metabolites in blood and muscle was reported by McLaughlin et al. [18]. The online system allowed the investigators to improve the temporal resolution of the experiment by a factor of 2. Online in vivo analysis of fluconazole in blood and dermis was performed by Mathy et al. These investigators also reported a considerable improvement in temporal resolution using the online system with a microbore column [109].

Capillary liquid chromatography has also been employed with online systems employing MS detection. In addition to low sample volume requirements, these columns produced better ionization efficiencies, resulting in improved limits of detection for MS [96]. Shackman et al. used MD–LC–MS

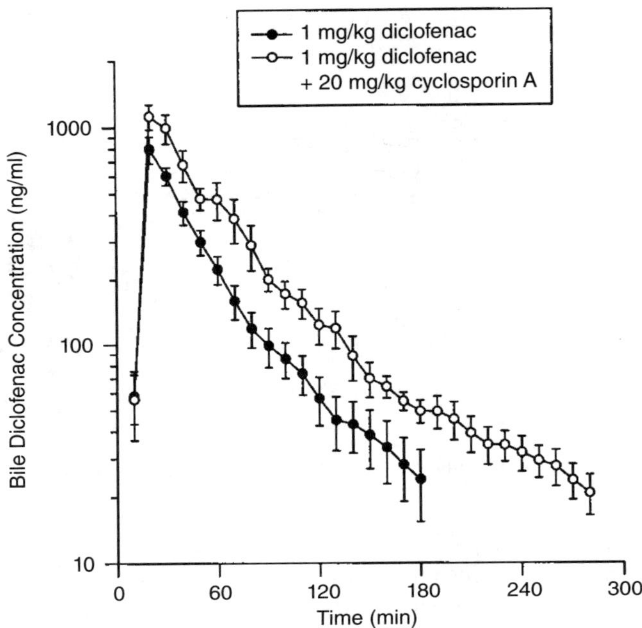

Figure 9   Online microdialysis–LC determination of the effect of cyclosporin A on the clearance of dichlofenac in rat bile using a shunt probe. (From [119].)

to monitor acetylcholine in vivo with limits of detection of 8 amol [96]. In another study, the pharmacokinetics of melatonin (administered i.v. in rats) was monitored over 15 h using online microdialysis with LC–MS/MS [110].

A recent example of online MD–LC–FL was reported by Yoshitake et al. in which serotonin sampled from brain was derivatized postcolumn with benzylamine in the presence of potassium hexaferrocyanate [111]. There have been several applications of online MD–LC systems reported over the past decade by the Tsai group and others for the in vivo monitoring of drugs [112–116], neurotransmitters, neuropeptides, and other important biomolecules, including reactive oxygen species [117,118]. Applications included primarily pharmacokinetic, metabolism, and neurochemical studies. Figure 9 shows an example of the online analysis of bile microdialysis samples using a shunt probe following i.v. administration of diclofenac with or without coadministration of cyclosporin A [119]. A list of some recent applications of online systems is provided in recent reviews [19,105].

Capillary Electrophoresis   Capillary electrophoresis (CE) is particularly attractive for the analysis of microdialysis samples because it has very low sample volume requirements (nanoliters to picoliters) and can perform extremely fast separations. In CE, both the analysis speed and separation efficiency improve with field strength (in the absence of Joule heating).

Therefore, by using very short capillaries and high field strengths, very fast, highly efficient separations can be accomplished. A variety of detectors can be used, including ultraviolet, laser–induced fluorescence (LIF), electrochemical (amperometric and conductivity), and mass spectrometry.

Off-Line Analysis As mentioned in Section 2.21, it is difficult to collect and analyze samples smaller than 1 µL, due to issues with evaporation, surface tension, and difficulties with actually manipulating submicroliter samples. Therefore, volumes of 1 to 5 µL are typically required for off-line analysis. Smaller volumes can be analyzed if the analytes are derivatized and the sample is collected in the derivatization buffer. This strategy has been employed for the determination of amino acids in brain microdialysates using LIF. Dilution of the sample can be a concern for analytes that are present at extremely low concentrations, such as peptides.

Off-line analysis using CE–LIF has been performed for the detection of amino acids [120], neurotransmitters [121], and neuropeptides [16,17]. The Philips lab has employed microdialysis with off-line CE for the analysis of cytokines and chemokines from cultured astrocytes. The astrocytes were stimulated with vasoactive intestinal peptide (VIP), and sampling was performed via a microdialysis probe with fractions collected every 5 min (20 µL) [122]. Samples were derivatized off-line and then injected onto an immunoaffinity capillary that was previously prepared in-house. Figure 10 shows the microdialysis sampling setup and a representative electropherogram that was obtained using this procedure.

Electrochemical detection has been employed for the analysis of amino acids in brain microdialysis [123–125] as well as for the analysis of small monoamine neurotransmitters. Analytes such as dopamine and norepinephine readily undergo oxidation–reduction at an electrode surface, making this technique advantageous and selective for electroactive components in microdialysis samples. Figure 11 shows the detection of a biomarker of oxidative stress, 8-oxyguanine, in brain microdialysates using capillary electrophoresis with electrochemical detection.

Online Analysis When microdialysis sampling is interfaced directly to CE, it is possible to make reproducible nanoliter injections of sample into the CE capillary. The electrophoretic separations range in duration from a few minutes to several seconds. This makes it possible to monitor biological processes continuously with excellent temporal resolution. Laser-induced fluorescence detection is the most popular method for online MD–CE, primarily because of its high sensitivity and the availability of inexpensive lasers and laser-based detectors. There are also a large number of reagents commercially available for fluorescence derivatization of amines, thiols, carbohydrates, carboxylic acids, and other functional groups. NDA (naphthalene-2,3-dicarboxaldehyde) and OPA (*o*-phthalaldehyde) are the most common reagents employed for the detection of amino acid neurotransmitters and other primary amines in

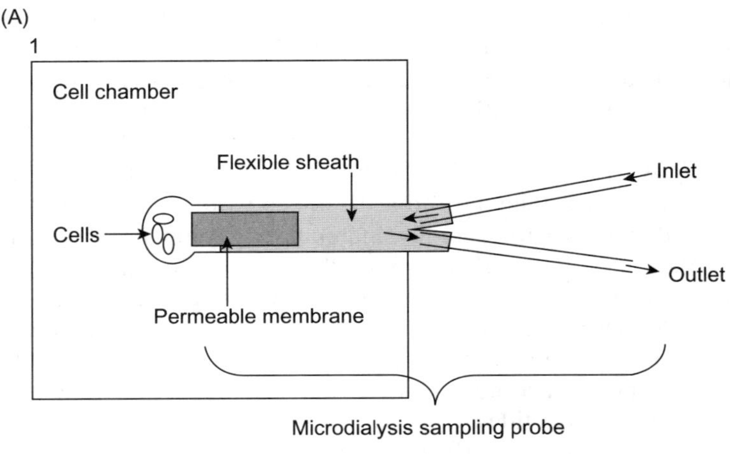

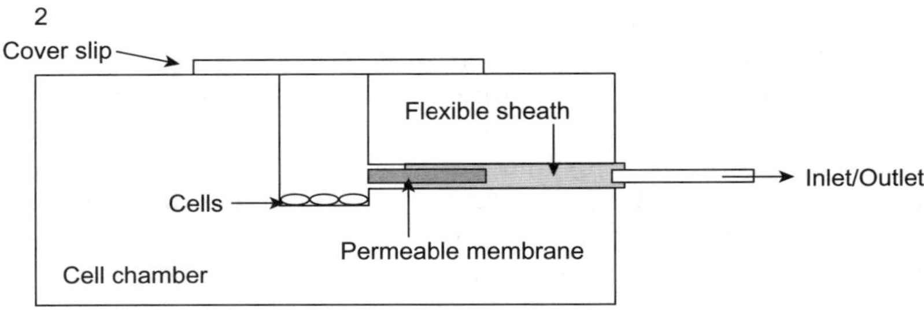

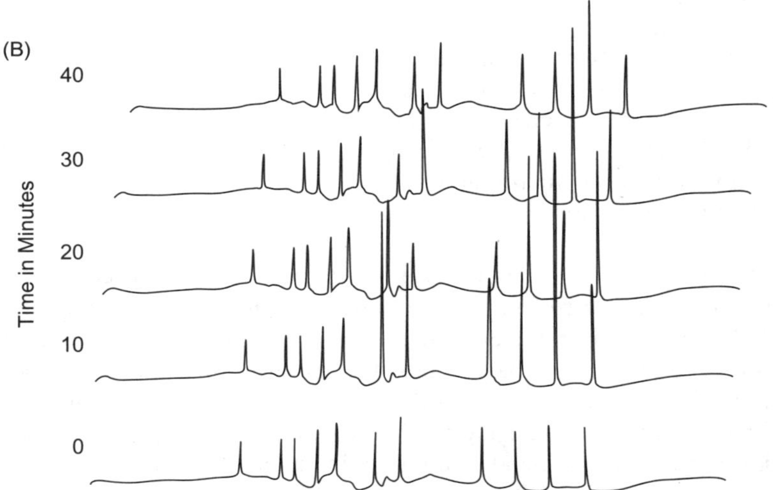

Figure 10 (A) Diagram of (1) the top view and (2) the side view of the single-cell microincubation chamber with the microdialysis probe cemented in place; (B) electropherogram of cytokine–chemokine secretion from VIP-simulated astrocytes. (From [122].)

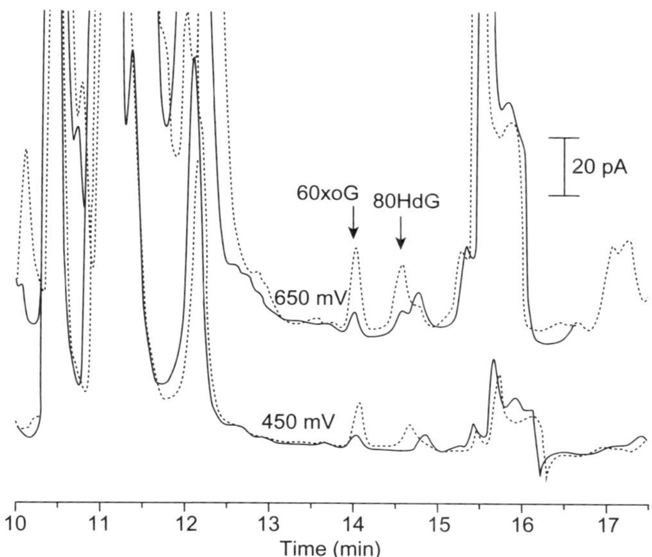

Figure 11 Identification of 8-oxyguanine (8oxoG) and 8-hydroxy-2'-deoxyguanosine (8OHdG) in rat brain dialysate by capillary electrophoresis with electrochemical detection. (From [117].)

vivo. Both of these reagents are fluorogenic and exhibit fast reaction kinetics, making them well suited for precolumn derivatization of amino acids prior to analysis by CE–LIF. Online reaction times for OPA/β-ME and NDA–CN with primary amines have been reported to be 10 to 30 s and 120 to 240 s, respectively [126].

A key component of online microdialysis–CE systems is the development of an interface that is capable of injecting discrete nanoliter-size sample plugs from continuous hydrodynamic flow from the microdialysis probe (μL). The first report of an online MD–CE system, by Hogan et al., used a nanoliter injection valve as the interface between the MD system and the CE separation capillary with continuously running CE separation [127]. A diagram of the experimental setup is shown in Figure 12. Perfusate from the dialysis experiment was collected in the nanoliter injection loop of the valve, and a special transfer line delivered the sample from the valve to the separation capillary inlet The system was used to monitor antineoplastic agents in blood with a temporal resolution of 90 s [127]. A similar system incorporating online derivatization with NDA–CN was used for monitoring aspartate and glutamate release in the brain [128]. Later, the transdermal delivery of nicotine was monitored using an online microdialysis–CE system with electrochemical detection [129]. The system incorporated a carbon fiber working electrode and a cellulose acetate decoupler before the separation capillary to shield the animal from high voltage. The cutaneous nicotine concentration was

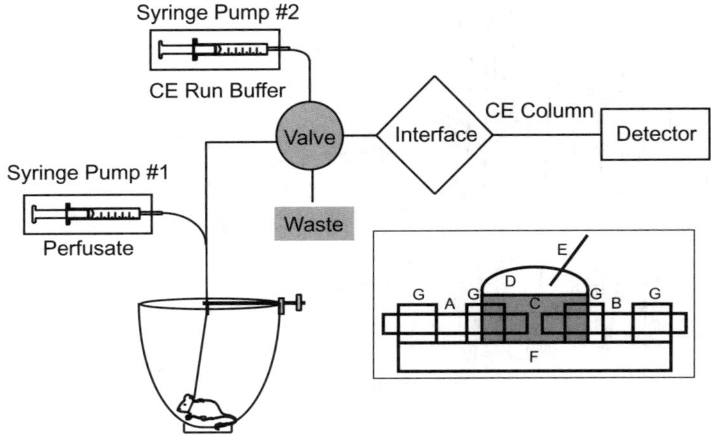

Figure 12   Online microdialysis–capillary electrophoresis system. (From [127].)

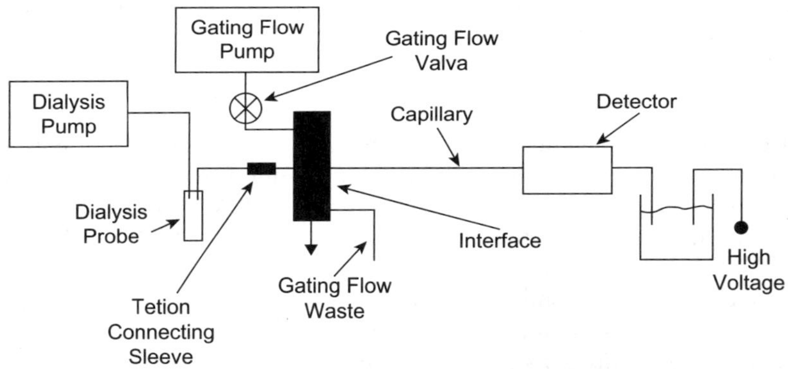

Figure 13   Flow-gated interface for microdialysis coupled to capillary electrophoresis. (From [130].)

monitored for 24 h following application of a Nicotrol patch. The temporal resolution was 10 min.

Lada et al. used a flow-gated interface instead of a valve to inject microdialysis samples into the separation capillary (Figure 13). This interface made it possible to perfuse the microdialysis probe at submicroliter per minute flow rates and still obtain temporal resolutions of 65 to 85 s [130]. Subsequently, the same group demonstrated extremely fast separations of glutamate and aspartate using a short capillary (6.5 cm) with a temporal resolution of 12 s [131]. A later version of this system was utilized to separate and perform quantitative analysis of multiple neurotransmitters in the brain (45 s temporal resolution)

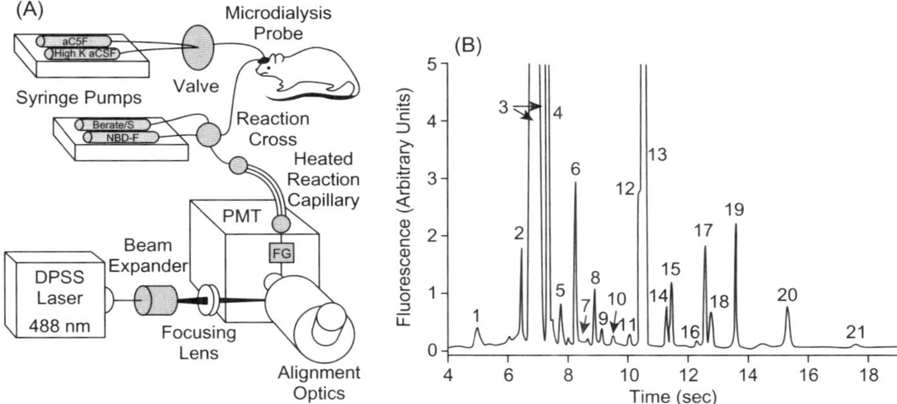

Figure 14 (A) Online microdialysis–capillary electrophoresis system employing NBD-F as a derivatizing reagent; (B) electropherogram of samples obtained from the rat striatum. (From [138].)

using micellar electrokinetic chromatography (MEKC) [132]. A system designed with a sheath-flow cuvette for improved sensitivity (15-fold) was reported by Bowser and Kennedy [133]. This system was used to monitor 17 amino acids in the brain of an awake, freely moving animal [126].

The most popular application of online MD–CE systems in vivo has been measuring amino acid neurotransmitters in the brain. In general, the change in concentration from basal is measured following some sort of external stimulus. In most cases, a drug or other substance is added to the perfusate to evoke the response. A common experiment is that of high-potassium stimulation, which causes the release of excitatory amino acids into the extracellular fluid [128,132–135]. The glutamate reuptake inhibitor L-*trans*-pyrrolidine-2,4-dicarboxylic acid (PDC) has also been administered via the microdialysis probe for monitoring glutamate and aspartate [131,136]. Other experiments include monitoring changes in ascorbic acid following subcutaneous injection of amphetamine [130], monitoring brain amino acids following intraperitoneal injections of saline (0.9% v/v) and ethanol (20% v/v) [126], and monitoring dopamine in the brain following injection of cocaine [137]. Recently, Bowser's group employed online microdialysis–CE with online derivatization using 4-fluoro-7-nitrobenzofurazan (NBD-F) for the continuous monitoring of primary and secondary amines in brain dialysates [138]. A diagram of the experimental setup and the resulting electropherograms is shown in Figure 14.

A major drawback of online systems using fused-silica capillaries is the need to connect tubing between the microdialysis system and the capillary electrophoresis system that can lead to band broadening and loss of temporal resolution. This can be a problem particularly for monitoring neurotransmitter release in awake, freely moving animals with commercial awake animal microdialysis systems that use liquid swivels. The online systems using conventional capillary electrophoresis are also quite large and do not lend themselves to the development of on-animal or portable sensing systems. In contrast, the microchip systems (next section) have the potential to be completely miniaturized.

Microchip Electrophoresis   Over the past decade, microchip electrophoresis has evolved to become an attractive analytical platform for the online analysis of microdialysis samples [19]. Microchips have several advantages for online analysis, including the ability to manipulate samples on-chip using a combination of electroosmotic and hydrodynamic flow, integrated sample preparation or derivatization, nanoliter-to-picoliter sample volume requirements, fast analysis times, and the ability to integrate the detector directly into the chip. Most of the research in this area has exploited one or both of the two major advantages of microchip-based devices. The first is the ability to perform very fast separations on-chip for monitoring fast biochemical processes. The second is miniaturization of the system for on-animal sensing.

Most of the work over the past several years has been focused on the development of interfaces between the microdialysis sampling system and the microchip electrophoresis system. The major challenge has been to develop an interface that can inject nanoliter-to-picoliter volume samples continuously and reproducibly into the electrophoresis channel while still achieving adequate temporal resolution (frequency of injecting and analyzing samples) for the experiment of interest. Other challenges include interfacing the external tubing used for microdialysis sampling with the analysis chip. In addition, fast and efficient derivatization strategies must be incorporated online for the detection of amino acid neurotransmitters and peptides.

Microdialysis sampling coupled to microchip electrophoresis for online analysis was first reported by Huyhn et al. [139]. The microchip device was fabricated from soda-lime glass and consisted of a twin T-channel. A continuous stream of perfusate was delivered to the microchip using a syringe pump that continuously perfused the cylindrical 4-mm CMA/12 microdialysis probe with 20 mM boric acid buffer. A commercially available microtight union (Upchurch Scientific) was used to connect the PEEK tubing from the microdialysis sampling system to the electrophoresis chip. Injection was accomplished using gated voltage that cut off the hydrodynamic flow at the injection cross, allowing introduction of a plug of sample into the separation channel for analysis. Such an injection scheme made it possible to inject discrete plugs of microdialysis sample continuously into the separation channel for fast electrophoretic separation of analyte. The system was used to monitor the activity of the enzyme β-galactosidase. This enzyme catalyzes the hydrolysis of fluorescein mono-β-galactoside (FMG) to produce fluorescein. Both the substrate and the product were monitored using an online system with LIF detection. The lag time (time needed for the device to respond to a concentration change) on this device was 5 to 7 min.

A similar setup was employed for online sampling, derivatization and detection of an in vitro mixture of amino acids and peptides [140]. On-chip labeling of analytes was accomplished using NDA and 2-mercaptoethanol (2ME). The NDA and 2ME were dissolved in 20 mM borate buffer and added to the run buffer reservoir. Following a gated injection scheme, the plug of sample reacted with the derivatization reagent at the beginning of the electrophoretic separation. The analytes were separated and detected with a temporal resolution of 30 to 40 s.

Kennedy's group employed a gated injection scheme to monitor in vivo amino acid neurotransmitters continuously using a microdialysis microchip setup with an effective temporal resolution of 2 to 4 min [141]. The chip device was fabricated from borosilicate glass. Reaction channels were incorporated for the derivatization of the sample collected via a side-by-side microdialysis probe (fabricated in-house) implanted in the brain of an anesthetized rat. Precolumn reaction was achieved by mixing OPA and 2ME with the stream of microdialysate. Following reaction, a gated injection scheme introduced plugs of sample into the separation channel for analysis by LIF. The interface in this device consisted of capillary tubing from the probe outlet connected to the chip via an Upchurch fitting. More recently, online microdialysis with on-chip precolumn derivatization using NDA–CN has been reported. This all-PDMS device uses a simple injection scheme to inject the derivatized samples into the microchip electrophoresis system. The system was employed to monitor amino acids and blood–brain barrier permeability in the rat striatum simultaneously [142] (Figure 15).

Cellar et al. employed low-flow, push–pull-based perfusion with online derivatization and conventional CE separation using a very short fused-silica capillary and LIF detection [143]. The pump used for sampling was chip based and consisted of pressure-actuated valves fabricated using a multilayer soft lithography technique. Use of a solenoid-controlled pneumatic valving method for sample injection in an online microdialysis microchip electrophoresis system has been described by the Martin group [144,145]. The general design consists of one or two nitrogen-actuated polydimethylsiloxane (PDMS) micro-valves that are used to inject or manipulate the flow of fluid within the chip. The CE separation channel is situated at a right angle to the flow channels. Sample injection is achieved by actuation of the valve at the flow channel/separation channel junction and is followed by electrophoresis and LIF or electrochemical detection

Li et al. used the online microdialysis–microchip electrophoresis system described above for continuous injection and separation of fluorescein and dichlorofluorescein with temporal resolution of 20 s [144]. All fluids were delivered from the syringe pump or from the probe using a capillary. The lag and rise times (time needed for the response to change from 10% to 90% of the total change) were reported to be approximately 6 and 2 min, respectively. More recently, amperometric detection has been incorporated into the system by Mecker and Martin. In this study, the stimulated release of dopamine from

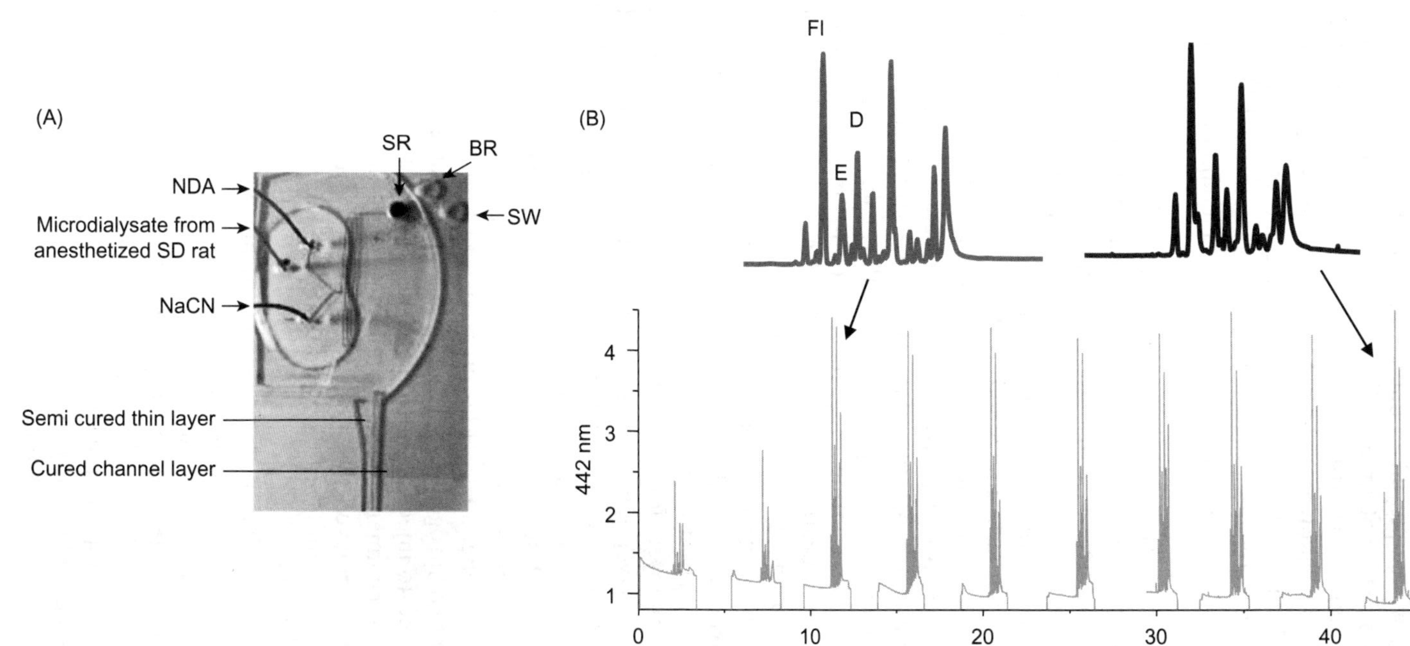

Figure 15   (A) Online microdialysis with on-chip derivatization and electrophoresis separation; (B) detection of amino acids in the rat striatum. (From [142].)

PC12 cells cultured in a petri dish was monitored using a 5-mm cylindrical microdialysis probe placed over the cells. In this case, the lag time was approximately 5 min and the rise time was 2 min [145]. Figure 16 shows the online microdialysis–microchip system used for these studies and the detection of dopamine using amperometric detection.

The use of segmented flow to create reaction chambers for derivatization has also been investigated for online microdialysis–microchip electrophoresis. Wang et al. developed a system in which aqueous sample plugs were created in a stream of oil (perfluorodecalin) [146,147]. The frequency and size of the droplets were dependent on the relative perfusion rates of oil and water from the respective syringe pumps. The plug of aqueous sample was interrogated at the end of a 7-cm-long capillary connected to the chip. With this system, glucose was measured online using glucose oxidase and Amplex Red. The same reaction scheme was employed for monitoring extracellular glucose in rat brain by microdialysis sampling; a change in glucose concentration in the brain was brought about by perfusion of 100 mM potassium through the probe.

This interface has also been employed to perform electrophoretic separations from a continuous segmented flow, and has been used for the electrophoretic separation of microdialysis samples [148]. For injection of discrete plugs for CE separation, two modes of injection were described. For the *discrete injector*, a series of sample plugs was injected; the injection volume was dependent on the electrophoresis field strength and temporal width of the plug. For the *desegmenting injector*, the sample plugs coalesce at the K junction, from which electrokinetic injections can be made repeatedly for analysis. Primary amines were derivatized on-chip with NDA–CN and then analyzed using the injection schemes described above. A concentration change experiment was also performed with the amino acids in which the change in concentration from 1 µM to a blank solution was registered on-chip in 20 s. These methods offer a more reliable way to analyze sample plugs by electrophoresis while preserving temporal information. Wang et al. used this device to perform online derivatization of the dialysate samples in the droplets followed by injection into the electrophoresis channel [147]. Figure 17 shows the microfluidic chip employing segmented flow as well as a typical electropherogram of amino acids in the rat brain microdialysate.

These online microchip-based systems are ideal for studies requiring high temporal resolution. However, there can be several challenges associated with building such devices. First, the separation must be optimized, and careful consideration needs to be given to the separation and identification of the analyte(s) of interest in the matrix being sampled. Interfacing the electrophoretic separation with the hydrodynamic flow can be a tedious process in which extensive investigation is required to determine the optimal microchannel design and dimensions for the specific application.

Second, fabricating glass-based chips can be challenging in many lab facilities, as it involves a high-temperature thermal bonding step that can yield

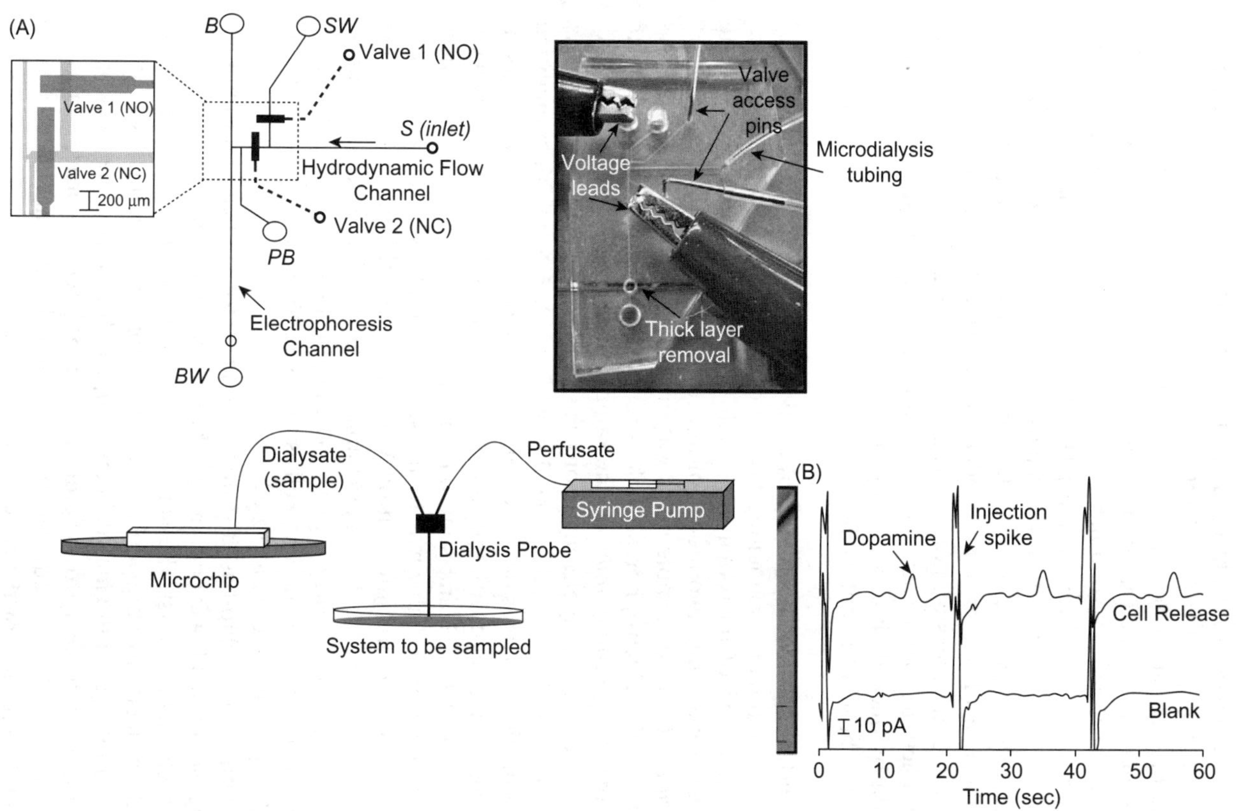

Figure 16   (A) Online microdialysis with microchip electrophoresis with electrochemical detection using a pneumatic valve; (B) detection of dopamine release from PC-12 cells using the device. (From [145].)

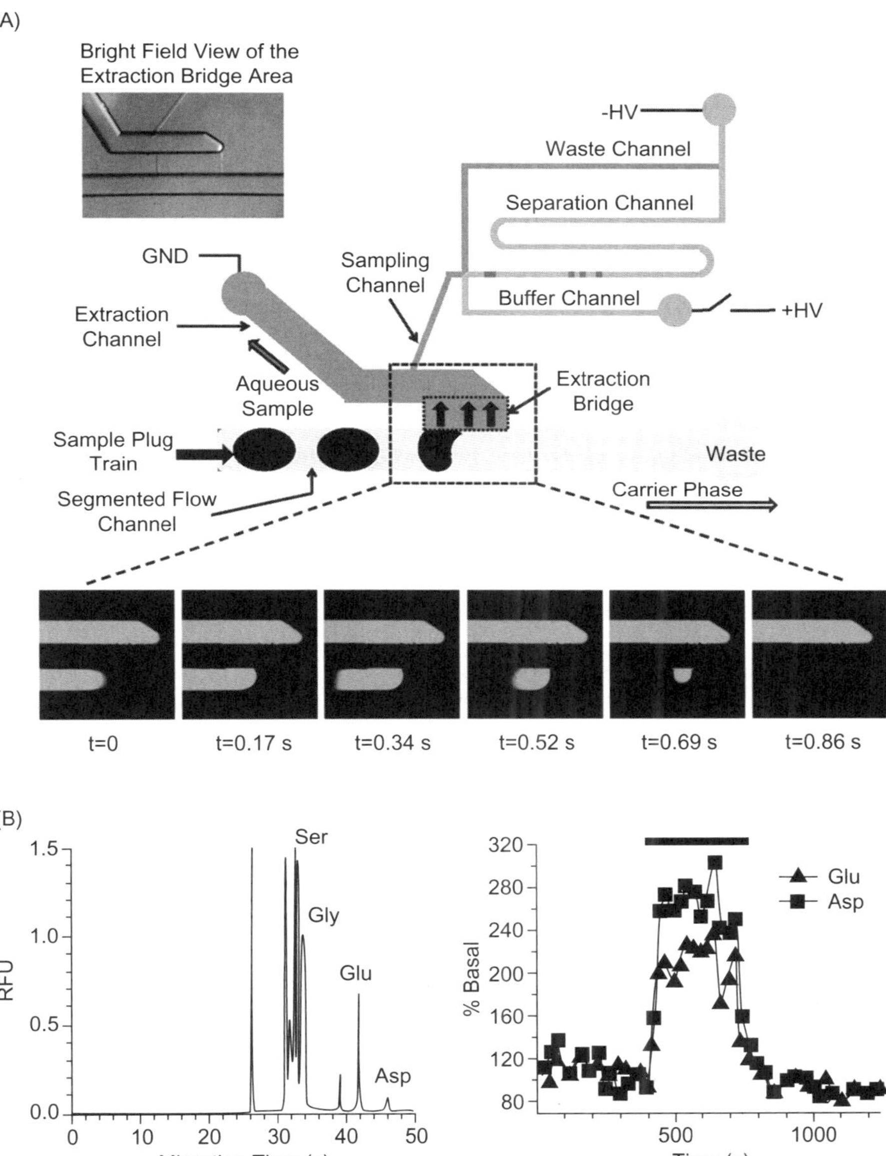

Figure 17 (A) Microfluidic chip for high-efficiency electrophoretic analysis of segmented flow from a microdialysis probe and in vivo chemical monitoring; (B) typical electropherogram and release profile obtained in vivo from the rat striatum with the online derivatization setup. (From [146].)

poor success rates. Glass microchips are available commercially for such applications, but for many labs, these are cost prohibitive. Although PDMS microchip devices are easy to fabricate, they suffer from analyte adsorption and changing electroosmotic flow (EOF). Several complex modification approaches have been suggested to reduce adsorption and maintain the EOF for a longer period of time on such devices [149].

Third, the performance of the chip is dependent on a number of factors, such as removal of waste (reduce hydrostatic pressure) and frequent replenishment of buffer; integrating such steps on-chip may require a fair degree of automation. Finally, much of the equipment used for such systems must be custom built or is lab specific (e.g., software, power supplies). Therefore, a wide variety of technical resources and expertise needs to be available for developing these systems.

Considerations Regarding Online Versus Off-Line Analysis   Online sample analysis offers several potential advantages over off-line analysis. In an online system, the sample collection, manipulation, injection, and analysis steps are all integrated in a continuous, streamlined fashion. Therefore, problems related to handling submicroliter volumes of sample (sample loss, mislabeling, evaporation, and surface tension) as well as sample degradation that can occur with sample exposure to air (e.g., ascorbic acid and catecholamines) can be avoided [150,151]. Also, such systems are usually capable of manipulating and analyzing submicroliter sample volumes, which allow high temporal resolution analysis to be performed. Such studies can yield continuous near real-time data, which provide immediate feedback on the biological process under investigation.

One important consideration in deciding whether to perform an analysis online or off-line is temporal resolution. This can be defined as the smallest increment of time over which the change in a dynamic process can be observed. Especially in the case of neurochemical experiments, microdialysis is usually coupled online with the objective of improving temporal resolution compared to off-line analysis. Many neurochemical events, such as neurotransmitter release, occur at a time scale of seconds or less. This results in very fast and transient changes in the concentration of substances in the extracellular fluid. Therefore, the primary objective in such experiments is to analyze samples as frequently as possible in order to detect these fast changes; otherwise, the concentration change could be missed due to dilution and averaging of the signal in the sample.

The first parameter that needs to be considered is whether the method is sensitive enough to actually detect the quantity of analyte(s) present in the small volume of sample generated by very fast sampling. For example, if a probe is perfused at 1 µL/min and the perfusate contains an analyte at micromolar concentrations, the analytical system must have sufficient mass sensitivity to detect 1 pmol of analyte if 1 min of temporal resolution is desired. If the analytical method is capable of detecting only 10 pmol or higher, this means that a much larger sample (10 µL) must be collected to measure the analyte of interest. This limits the temporal resolution to 10 min at a flow rate of 1 μL/min.

In many cases with online systems, the temporal resolution of the technique is defined by the analysis time. If the analysis time is longer than the duration of the event being measured, the change will appear digital and show up in the next injection. On the other hand, in those cases where the analysis step is much faster than the event being measured, it is possible to detect the change in concentration as a function of time. In this case, the *rise time* is defined as the time required for the signal to increase from 10% to 90% of maximum intensity [145]. For very fast analyses, the rise time becomes dependent on the rate of diffusion of analyte across the probe membrane [131]. However, in most cases, it is the dead volume in the system, injection method, and the flow rate of the dialysate that determine how fast a concentration change can be measured with an online system.

For most online separation-based microdialysis systems, analysis is performed on analytes such as amino acid neurotransmitters whose in vivo concentrations are relatively high and well within the detection limits of laser-induced fluorescence or other methods of detection. In such cases, the factor that dictates temporal resolution is the time that is required to separate the compounds so that serial analyses can be performed without overlapping the analysis peaks from two different runs.

Online microdialysis systems can also be useful for studies where high temporal resolution is not essential. A classic example is pharmacokinetic (PK) experiments, in which analysis is required every few minutes over a period of hours. Most drugs exhibit pharmacokinetic profiles (absorption–distribution–metabolism–excretion) that last for a few hours to several days. Therefore, measuring the average concentration over a period of time (10 to 20 min) is sufficient for PK modeling studies. Another application that does not require high temporal resolution but where continuous online monitoring is useful is the in vitro monitoring of products of bioreactors. Here, the time course of the experiment is usually several hours or days. In this case, sample collection and analysis every 15 min can provide adequate information regarding the progress of the reaction system [152].

Representative Applications of MD to Pharmaceutical Analysis

Neurochemical Studies   The most popular application of microdialysis is sampling of the extracellular fluid in the brain. This is evident from the numerous books and review papers on this topic, including a recent book by Westerink and Cremers [84] and the often-cited monograph by Robinson and Justice [153]. A thorough review of all the applications of microdialysis to brain sampling is beyond the scope of this chapter. In this section we discuss only sampling issues directly related to brain tissue.

Brain microdialysis is almost always accomplished using a cannula probe, whose size depends on the size of the brain of the animal being sampled. Probes for microdialysis studies of rats and mice are commercially available from a number of vendors. Probes for use in other animals can be custommade either by a vendor or in-house. A significant advantage of microdialysis for neurochemical studies is that it can be performed on awake, freely moving animals. This makes it possible to correlate the concentrations of drugs and/or neurotransmitters in the extracellular fluid of the brain with behavior. Because of the small size and relatively noninvasive nature of the microdialysis probes, it is also possible to have multiple probes in a single animal. It is therefore possible to measure blood, brain, and tissue concentrations of drugs or endogenous substances simultaneously. If the animal is awake, these measurements can be correlated temporally with behavior.

In a very nice example of the use of multiple probes in a single animal, a rat was given an intravenous injection of methylphenidate (Ritalin) [154]. Brain and blood sampling were accomplished using a concentric cannula probe and a flexible probe, respectively. Dialysates were collected off-line, and the concentrations of methylphenidate and dopamine in both the brain and blood were determined using liquid chromatography with electrochemical detection. In this manner it was possible to measure the transport of Ritalin across the blood brain barrier (BBB) as well as its effect on catecholamine release. Finally, by using a RatTurn, the extracellular concentration of these substances could be correlated directly with the overall activity level of the rat.

A major concern with microdialysis sampling in the brain is changes in the environment around the probe during the sampling process. If the microdialysis studies will be performed over a fairly long period of time, tissue damage associated with probe implantation and the potential for an immune response must be taken into consideration. Fibrosis or gliosis has been reported following several days of probe implantation [53]. Grabb et al. compared the effects of acute (2 to 4 h) and chronic (24 h) implantation of microdialysis probes in brain tissue. Inflammation, hemorrhage, and edema in the area around the microdialysis probe were observed 24 h after implantation. The development of fibrinlike polymer (gliosis) was also observed around the probe. Both of these factors can adversely affect recovery of the probe. The extracellular edema increases the diffusional distance between the probe and the extracellular fluid, and the fibrinlike polymer generated via gliosis can create a physical barrier between the dialysis probe and the extracellular fluid surrounding the cells [155]. In a separate study, based on local cerebral blood flow (LCBF) and local cerebral glucose metabolism (LCGM), Benveniste et al. recommended a 24-h period of recovery after probe implantation [156].

The integrity of the BBB following probe implantation has also been an important and controversial issue in brain microdialysis [157]. Studies conducted using autoradiography with [^{14}C]AIB (which does not cross the BBB under normal conditions) as well as transport characteristics of hydrophilic and moderately lipophilic drugs (following i.v. injection) postsurgery indicate that the BBB integrity is maintained overall [157,158]. However, other studies have shown a significant effect of probe implantation on BBB permeability using [^{51}Cr]EDTA transport [159].

The combination of microdialysis and in vivo voltammetry has been employed to investigate the effects of the microdialysis probe on measured dopamine concentrations in brain tissue as well as the integrity of the BBB. The microelectrode used for the voltammetry studies is only 7 μm in diameter and is therefore substantially smaller than the brain probe cannula (280 μm). These studies show that the probe does cause injury to the tissue and generates a diffusional barrier for the transport of dopamine to the probe [160–164]. An estimate of the extent of "injury" based on the combined studies is approximately 1 mm from the probe membrane. This increase in diffusion distance explains the difference in response times for release of dopamine by microdialysis in comparison to in vivo voltammetry studies.

As mentioned earlier, applications of microdialysis in the brain are too numerous to be included in this chapter. Examples include monitoring oxidative metabolism in the brain [165], understanding addiction [166], central nervous system (CNS) disorders [167], stress [168], behavior [169], proteomics [170], brain trauma [171], and stroke [53], as well as many others. Many different analytical methods have also been employed for the determination of neurotransmitters and neuropeptides in the brain [15], including liquid chromatography with electrochemical detection, capillary and microchip electrophoresis [19,51], and mass spectrometry [170,172–174]. Readers are referred to the book by Westerink and Cremers [84] or the recent reviews cited above for further information on applications and analytical methods related to brain sampling [15,19].

Pharmacokinetic and ADME Studies   One of the most popular applications of microdialysis for tissue sampling is pharmacokinetic studies. In this case, microdialysis is used to monitor the concentration of the drug (and/or metabolite) in the tissue or organ of interest over time. Compared to traditional pharmacokinetics, where the data are collected at each time point, microdialysis data reflect an average concentration over the time period in which sampling is done, and this needs to be taken into consideration for AUC calculations [175]. The time period of sampling can be reduced (fraction of a second; see Section 2.1) such that the concentration data collected during this time period will be very close to the data collected by manual sampling at a given time point. However, achieving such small collection times depends on the sensitivity of the detector as well as the capability of the analysis system to perform fast injections.

The first successful pharmacokinetic study using microdialysis was performed by Craig Lunte's group in 1990. Continuous blood sampling was achieved using a flexible probe following the administration of acetaminophen [176–178]. The pharmacokinetics of aspirin were also investigated by microdialysis compared to manual blood sampling [179]. In assessing pharmacokinetic parameters using microdialysis, the fact that protein-bound drugs cannot cross the membrane must be taken into account. Therefore, the concentration obtained with microdialysis experiments is representative of only the free

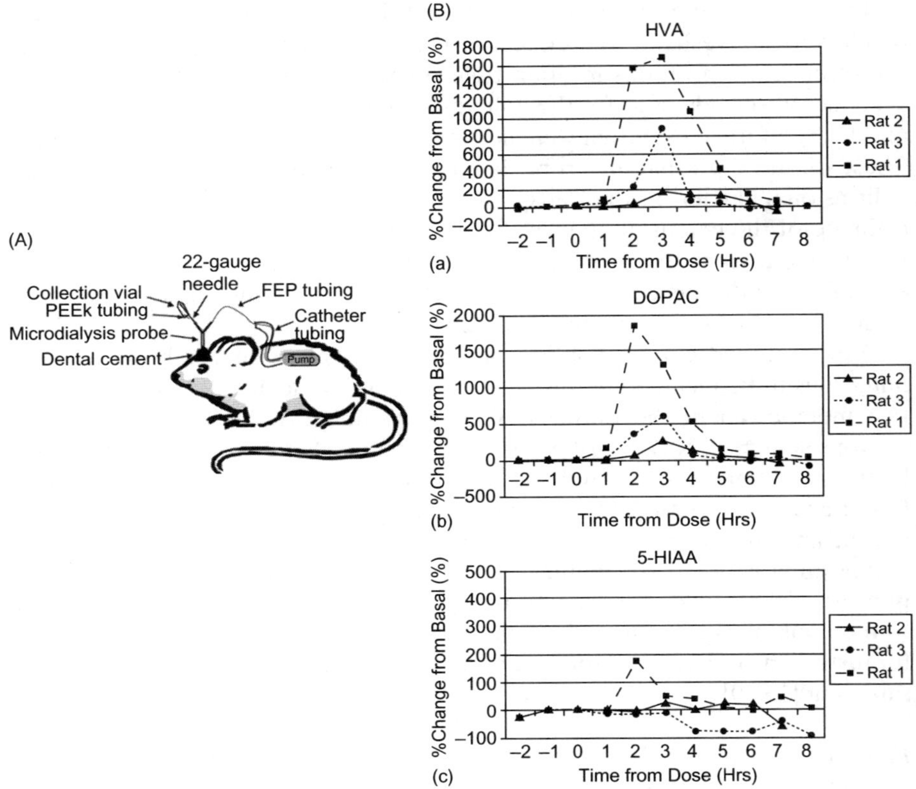

Figure 18 (A) Schematic of the on-rat collection system incorporating an osmotic pump; (B) in vivo release of the neurotransmitters HVA (a), dopac (b), and 5-HIAA (c) following administration of benserazide and L-dopa. (From [186].)

fraction of the drug [180]. Later, the same group demonstrated the use of a linear probe for transdermal delivery studies [25,181]. This same probe design was then widely utilized for studies of various tissues, including liver [182], tumor [183], stomach [10,11], and muscle [184].

Microdialysis has been used extensively to monitor blood–brain barrier transport. A recent example is the use of microdialysis for the investigation of the differential transport of oxycodone and morphine into the brain [185]. The transport and metabolism of L-dopa in the brain was monitored in an awake, freely moving rat using microdialysis sampling with an osmotic pump and on-rat collection. The appearance of 3,4-dihydroxyphenylacetic acid (dopac), homovanillic acid (HVA), and 5-hydroxyindole-3-acetic acid (5-HIAA) in the brain was monitored by LCEC following peripheral administration of benserazide and L-3,4-dihydroxyphenylalanine (L-dopa) [186]. The experimental setup and plots of the L-dopa metabolites in the brain as a function of time are shown in Figure 18.

Microdialysis has been used to investigate drug delivery to the eye. In a recent study, Hosoya et al. investigated the in vivo transport of anionic drugs to characterize the organic anion transporters (OATs) in the rat retina; in this study, a linear microdialysis probe was placed in the vitreous chamber of the eye [187]. Several studies have also been performed by the Mitra group concerned with corneal drug delivery and pharmacokinetics as well as characterization of transporters present in the eye. In all of these studies, a linear probe was implanted in the anterior chamber of the eye for sampling [188]. Several additional reports in the literature demonstrate the popularity of ocular microdialysis sampling for pharmacokinetic studies [189–191].

Muscle is another popular site for microdialysis sampling in both humans and rats. It has been used to monitor change in biochemistry due to exercise [192–194]. For example, the concentrations of bradykinin and kallidin in muscle were compared in female individuals at rest, during a 20-min repetitive low-force exercise, and following recovery [195]. In another study, Hamrin and Henriksson found that glucose concentration in the interstitial fluid of insulin-resistant skeletal muscle is markedly decreased for several hours following a single exercise session [196]. Lönnroth et al. have investigated the effect of probes on muscle tissue and reported no major damage to tissues by the microdialysis probe following histological examination [197,198].

A number of muscle microdialysis studies have also been performed in rats. The concentration of acetaminophen in muscle was determined following topical administration and compared with that of intramuscular administration [199]. Awake rat studies were performed by Marchand et al. in which they investigated the distribution of imipenem in muscle extracellular fluid [200].

Microdialysis is a particularly powerful technique for the investigation of transdermal delivery. It is possible to monitor endogenous and exogenous compounds in the dermal tissue continuously with minimal trauma [41]. Dermal microdialysis has been employed for a number of studies aimed at investigating the distribution of drugs in subcutaneous tissue following topical application [199,201,202]. An additional utilization delivers compounds to the dermal tissue via the microdialysis probe while measuring changes in the concentration of endogenous analytes [203].

Derendorf and others have extensively used microdialysis to monitor the concentration of antibiotics in peripheral tissue [193,204,205]. Microdialysis has also been applied to lung [206], kidney [207], liver [208], pancreas [209], blood [210], and inner ear [211,212]. By 2007, microdialysis had evolved enough as a tool for pharmacokinetic and drug metabolism studies that the U.S. Food and Drug Administration sponsored a workshop on the role of microdialysis in the drug evaluation process [41]. Recently, Woo and Lunte used multiple probes to monitor the absorption of caffeine by a healthy versus ulcerated stomach in a rat model [10,11]. Figure 19 shows the average caffeine concentration in the blood and in ulcerated and healthy stomach tissue following an oral dose of caffeine.

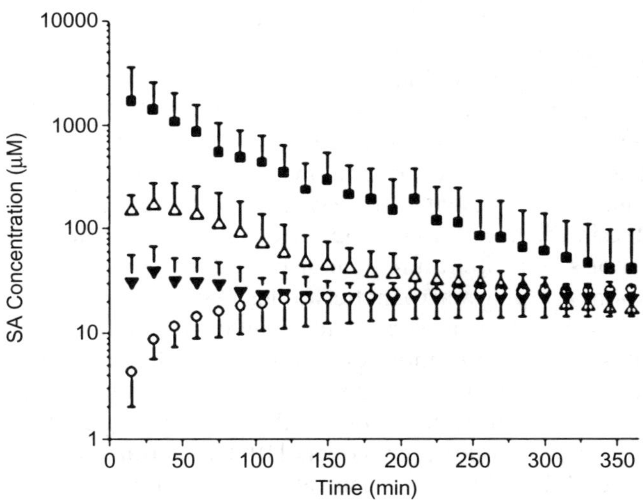

Figure 19 Use of multiple probes to investigate healthy and ulcerated stomach tissue. Plot of the average caffeine concentration in the lumen (filled squares), in the submucosa of ulcerated (empty triangles), and in blood (empty circles) determined by microdialysis sampling as a function of time following a 5 mM oral dose to the ulcerated stomach. (From [10].)

Dissolution Testing   Dissolution is an important test used in the pharmaceutical industry to determine the potential bioavailability of drugs. Normally, the drug is placed in a dissolution apparatus and aliquots are removed at distinct time points to monitor the amount of drug that is dissolved. Dissolution time profiles can be particularly important for sustained-release pharmaceuticals. Microdialysis has several advantages as a monitoring system for dissolution studies [213]. First, it is possible to monitor the concentration of drug in solution continuously without having to remove aliquots from the apparatus, thereby disturbing the equilibrium [214,215]. Second, by coupling microdialysis online to the analytical system, it is possible to obtain the dissolution profile in near real-time. Finally, using automated valving systems, it is possible to monitor multiple dissolution apparatus simultaneously.

Shah et al. described an online microdialysis LC–UV system using a loop probe suspended in the dissolution medium to monitor the dissolution of acetaminophen and Sulfatrim tablets, and compared these to the profiles obtained with manual sampling [214]. The dissolution was performed in 0.1 M HCl, but the use of microdialysis sampling allowed the use of a buffered perfusate to make the sample compatible with the online LC-based analysis system. The profiles obtained using microdialysis were similar to those obtained via manual sampling.

Microdialysis has also been used to investigate the dissolution of sustained-release pharmaceuticals such as Accutrim [215]. Dash et al. used microdialysis sampling to investigate the dissolution of implantable drug delivery systems using ciprofloxacin microcapsules in poly(lactic acid) (PLA) and poly(lactic-glycolic acid) (PLGA) as a model system [216]. Later, Fang's group developed a stopped-flow microdialysis sampling system for multivessel dissolution testing with high temporal resolution [216,217].

Bioreactor and Cell Culture Monitoring Microdialysis has many advantages for monitoring bioprocesses [23,218,219]. These include the ability to monitor small molecules (such as glucose) in the presence of cells and extracellular proteins without additional sample preparation steps. The probes can be sterilized and inserted directly into the bioreactor for continuous monitoring without fluid loss. Since microdialysis is a generic sampling technique, samples can be monitored either on- or off-line, depending on the requirements of the overall system [122].

Some of the important issues that have to be addressed with bioreactors are probe membrane type and geometry [24,220]. The type of membrane used can dramatically affect the recovery of analytes from the complex bioreactor sample [22,218]. Laurell and Butler developed a probe with an adjustable cannula that can be used to maximize recovery of analytes for bioreactor studies [24]. Membranes must also be able to withstand the higher temperatures that are employed in some bioreactors. The composition of the perfusate can also dramatically affect analyte recovery.

Several groups have used microdialysis sampling to monitor enzyme reactions. Torto et al. investigated extensively the use of microdialysis for monitoring the enzymatic hydrolysis of starch [221]. The solution can be stirred and the membrane excludes the enzyme, making it possible to measure the reactant and the products using the analytical method of choice. Modi and LaCourse used microdialysis sampling coupled to HPLC–UV to monitor carbohydrate enzymatic reactions. One application was the identification of complex carbohydrates present in willow bark tea by adding β-galactosidase to the sample [222]. Huyhn used online microdialysis coupled to microchip electrophoresis to monitor the reaction of FMG with β-galacotsidase in vitro with a temporal resolution of approximately 15 s [139].

Other Applications Microdialysis has been used to desalt protein solutions prior to mass spectrometric analysis [28,223–226]. Affinity dialysis in a microfluidic format has been used as a sample preparation method prior to ESI–MS for the trace analysis of food residues as well as for high-throughput drug assays [227]. A recent review from our group outlines various applications of microdialysis as a sample preparation method [20]. Since microdialysis measures only the non-protein-bound fraction, it has also been used to determine the degree of protein binding of drugs [228,229].

3. CONCLUSIONS

This chapter addresses important analytical considerations concerning microdialysis sampling; specifically, issues regarding recovery and quantitation are highlighted. Both separation- and non-separation-based analysis techniques are discussed with a more detailed discussion concerning online versus off-line analysis. The final pages highlight some key applications of microdialysis with an emphasis on the most prevalent probes, analysis systems, and detectors employed.

Acknowledgments

The authors wish to thank Nancy Harmony for her assistance in the preparation of this manuscript. Funding provided by National Institutes of Health grant NINDS R56-NS042929 is gratefully acknowledged. P.N. was a recipient of an American Heart Association predoctoral fellowship, and C.D.K. gratefully acknowledges the support of a predoctoral fellowship from Pfizer.

REFERENCES

[1] Richards, D.A., Silva, M.A., Murphy, N., Wigmore, S.J., Mirza, D.F. (2007). Extracellular amino acid levels in the human liver during transplantation: a microdialysis study from donor to recipient. *Amino Acids*, *33*, 429–437.

[2] Davies, M.I., Lunte, C.E. (1996). Simultaneous microdialysis sampling from multiple sites in the liver for the study of phenol metabolism. *Life Sciences*, *59*, 1001–1013.

[3] Gilinsky, M.A., Faibushevish, A.A., Lunte, C.E. (2001). Determination of myocardial norepinephrine in freely moving rats using in vivo microdialysis sampling and liquid chromatography with dual-electrode amperometric detection. *Journal of Pharmaceutical and Biomedical Analysis*, *24*, 929–935.

[4] Price, K.E., Vandaveer, S.S., Lunte, C.E., Larive, C.K. (2005). Tissue-targeted metabonomics: metabolic profiling by microdialysis sampling and microcoil NMR. *Journal of Pharmaceutical and Biomedical Analysis*, *38*, 904–909.

[5] Ault, J.M., Riley, C.M., Meltzer, N.M., Lunte, C.E. (1994). Dermal microdialysis sampling in vivo. *Pharmaceutical Research*, *11*, 1631–1639.

[6] Bielecka-Grzela, S., Klimowicz, A. (2003). Application of cutaneous microdialysis to evaluate metronidazole and its main metabolite concentrations in the skin after a single oral dose. *Journal of Clinical Pharmacy and Therapeutics*, *28*, 465–469.

[7] Huang, H., Zhang, Y., Yang, R., Tang, X. (2008). Determination of baicalin in rat cerebrospinal fluid and blood using microdialysis coupled with ultra-performance liquid chromatography–tandem mass spectrometry. *Journal of Chromatography B: Analytical Technologies in the Biomedical and Life Sciences*, *874*, 77–83.

[8] Lin, L.-C., Hung, L.-C., Tsai, T.-H. (2004). Determination of (−)-epigallocatechin gallate in rat blood by microdialysis coupled with liquid chromatography. *Journal of Chromatography A*, *1032*, 125–128.

[9] Ward, K.W., Pollack, G.M. (1996). Use of intrauterine microdialysis to investigate methanol-induced alterations in uteroplacental blood flow. *Toxicology and Applied Pharmacology*, *140*, 203–210.

[10] Woo, K.L., Lunte, C.E. (2008). The direct comparison of health and ulcerated stomach tissue: a multiple probe microdialysis sampling approach. *Journal of Pharmaceutical and Biomedical Analysis*, *48*, 85–91.

[11] Woo, K.L., Lunte, C.E. (2008). The development of multiple probe microdialysis sampling in the stomach. *Journal of Pharmaceutical and Biomedical Analysis*, *48*, 20–26.

[12] Zhu, T., Cheung, B.W.Y., Cartier, L.L., Giebink, G.S., Sawchuk, R.J. (2003). Simultaneous intravenous and intramiddle-ear dosing to determine cefditoren influx and efflux clearances in middle ear fluid in freely moving chinchillas. *Journal of Pharmaceutical Sciences*, *92*, 1947–1956.

[13] Freed, A.L., Audus, K.L., Lunte, S.M. (2001). Investigation of the metabolism of substance P at the blood–brain barrier using capillary electrophoresis with laser-induced fluorescence detection. *Electrophoresis*, *22*, 3778–3784.

[14] Reed, B., Bidlack, J.M., Chait, B.T., Kreek, M.J. (2008). Extracellular biotransformation of β-endorphin in rat striatum and cerebrospinal fluid. *Journal of Neuroendocrinology*, *20*, 606–616.

[15] Perry, M., Li, Q., Kennedy, R.T. (2009). Review of recent advances in analytical techniques for the determination of neurotransmitters. *Analytica Chimica Acta*, *653*, 1–22.

[16] Freed, A.L., Cooper, J.D., Davies, M.I., Lunte, S.M. (2001). Investigation of the metabolism of substance P in rat striatum by microdialysis sampling and capillary electrophoresis with laser-induced fluorescence detection. *Journal of Neuroscience Methods*, *109*, 23–29.

[17] Kostel, K.L., Lunte, S.M. (1997). Evaluation of capillary electrophoresis with post-column derivatization and laser-induced fluorescence detection for the determination of substance P and its metabolites. *Journal of Chromatography B: Biomedical Sciences and Applications*, *695*, 27–38.

[18] McLaughlin, K.J., Faibushevich, A.A., Lunte, C.E. (2000). Microdialysis sampling with on-line microbore HPLC for the determination of tirapazamine and its reduced metabolites in rats. *Analyst*, *125*, 105–110.

[19] Nandi, P., Lunte, S.M. (2009). Recent trends in microdialysis sampling integrated with conventional and microanalytical systems for monitoring biological events: A review. *Analytica Chimica Acta*, *651*, 1–14.

[20] Nandi, P., Lunte, S.M. (2010). *Microdialysis Sampling as a Sample Preparation Method*, Wiley, Hoboken, NJ.

[21] Torto, N., Marko-Varga, G., Gorton, L., Staalbrand, H., Tjerneld, F. (1996). Online quantitation of enzymic mannan hydrolyzates in small-volume bioreactors by microdialysis sampling and column liquid chromatography-integrated pulsed electrochemical detection. *Journal of Chromatography A*, *725*, 165–175.

[22] Torto, N., Bang, J., Richardson, S., Nilsson, G.S., Gorton, L., Laurell, T., Marko-Varga, G. (1998). Optimal membrane choice for microdialysis sampling of oligosaccharides. *Journal of Chromatography A*, *806*, 265–278.

[23] Torto, N., Gorton, L., Laurell, T., Marko-Varga, G. (1999). Technical issues of in vitro microdialysis sampling in bioprocess monitoring. *Trends in Analytical Chemistry*, *18*, 252–260.

[24] Laurell, T., Buttler, T. (1995). A microdialysis probe offering arbitrary membrane length and an in situ tunable relative recovery. *Analytical Methods and Instrumentation*, *2*, 197–201.

[25] Ault, J.M., Lunte, C.E., Meltzer, N.M., Riley, C.M. (1992). Microdialysis sampling for the investigation of dermal drug transport. *Pharmaceutical Research, 9,* 1256–1261.

[26] Telting-Diaz, M., Scott, D.O., Lunte, C.E. (1992). Intravenous microdialysis sampling in awake, freely-moving rats. *Analytical Chemistry, 64,* 806–810.

[27] Huff, J.K., Heppert, K.E., Davies, M.I. (1999). The microdialysis shunt probe: profile of analytes in rats with erratic bile flow or rapid changes in analyte concentration in the bile. *Current Separations, 18,* 85–90.

[28] Wu, Q., Liu, C., Smith, R.D. (1996). Online microdialysis desalting for electrospray ionization–mass spectrometry of proteins and peptides. *Rapid Communications in Mass Spectrometry, 10,* 835–838.

[29] Stenken, J.A., Lunte, C.E., Southard, M.Z., Staahle, L. (1997). Factors that influence microdialysis recovery: comparison of experimental and theoretical microdialysis recoveries in rat liver. *Journal of Pharmaceutical Sciences, 86,* 958–966.

[30] Stenken, J.A. (1999). Methods and issues in microdialysis calibration. *Analytica Chimica Acta, 379,* 337–357.

[31] Menacherry, S., Hubert, W., Justice, J.B., Jr. (1992). In vivo calibration of microdialysis probes for exogenous compounds. *Analytical Chemistry, 64,* 577–583.

[32] Zhao, Y., Liang, X., Lunte, C.E. (1995). Comparison of recovery and delivery in vitro for calibration of microdialysis probes. *Analytica Chimica Acta, 316,* 403–410.

[33] Snyder, K.L., Nathan, C.E., Yee, A., Stenken, J.A. (2001). Diffusion and calibration properties of microdialysis sampling membranes in biological media. *Analyst (Cambridge, U.K.), 126,* 1261–1268.

[34] Wang, X., Stenken, J.A. (2006). Microdialysis sampling membrane performance during in vitro macromolecule collection. *Analytical Chemistry, 78,* 6026–6034.

[35] Duo, J., Fletcher, H., Stenken, J.A. (2006). Natural and synthetic affinity agents as microdialysis sampling mass transport enhancers: current progress and future perspectives. *Biosensors & Bioelectronics, 22,* 449–457.

[36] Khramov, A.N., Stenken, J.A. (1999). Enhanced microdialysis extraction efficiency of ibuprofen in vitro by facilitated transport with beta-cyclodextrin. *Analytical Chemistry, 71,* 1257–1264.

[37] Fletcher, H.J., Stenken, J.A. (2008). An in vitro comparison of microdialysis relative recovery of Met- and Leu-enkephalin using cyclodextrins and antibodies as affinity agents. *Analytica Chimica Acta, 620,* 170–175.

[38] Ao, X., Sellati, T.J., Stenken, J.A. (2004). Enhanced microdialysis relative recovery of inflammatory cytokines using antibody-coated microspheres analyzed by flow cytometry. *Analytical Chemistry, 76,* 3777–3784.

[39] Wang, Y., Stenken, J.A. (2009). Affinity-based microdialysis sampling using heparin for in vitro collection of human cytokines. *Analytica Chimica Acta, 651,* 105–111.

[40] Cano-Cebrian, M.J., Zornoza, T., Polache, A., Granero, L. (2005). Quantitative in vivo microdialysis in pharmacokinetic studies: some reminders. *Current Drug Metabolism, 6,* 83–90.

[41] Chaurasia, C.S., Müller, M., Bashaw, E.D., Benfeldt, E., Bolinder, J., Bullock, R., Bungay, P.M., DeLange, E.C.M., Derendorf, H., Elmquist, W.F., et al. (2007).

AAPS–FDA Workshop White Paper: Microdialysis principles, application and regulatory perspectives. *Pharmaceutical Research, 24,* 1014–1025.

[42] Dai, H., Elmquist, W.F. (2003). Drug transport studies using quantitative microdialysis. *Methods in Molecular Biology, 89,* 249–264.

[43] Davies, M.I., Cooper, J.D., Desmond, S.S., Lunte, C.E., Lunte, S.M. (2000). Analytical considerations for microdialysis sampling. *Advanced Drug Delivery Reviews, 45,* 169–188.

[44] Gonzales, R.A., Tang, A., Robinson, D.L. (2002). Quantitative microdialysis for in vivo studies of pharmacodynamics. In *Methods in Alcohol-Related Neuroscience Research* (Yuan li and David Lovinger, eds.), CRC Press Online, pp. 287–317.

[45] Kehr, J. (1993). A survey on quantitative microdialysis: theoretical models and practical implications. *Journal of Neuroscience Methods, 48,* 251–261.

[46] Peters, J.L., Yang, H., Michael, A.C. (2000). Quantitative aspects of brain microdialysis. *Analytica Chimica Acta, 412,* 1–12.

[47] Olson, R.J., Justice, J.B., Jr. (1993). Quantitative microdialysis under transient conditions. *Analytical Chemistry, 65,* 1017–1022.

[48] Cosford, R.J.O., Vinson, A.P., Kukoyi, S., Justice, J.B., Jr. (1996). Quantitative microdialysis of serotonin and norepinephrine: pharmacological influences on in vivo extraction fraction. *Journal of Neuroscience Methods, 68,* 39–47.

[49] Justice, J.B., Jr. (1993). Quantitative microdialysis of neurotransmitters. *Journal of Neuroscience Methods, 48,* 263–276.

[50] Lönnroth, P., Jansson, P.A., Smith, U. (1987). A microdialysis method allowing characterization of intercellular water space in humans. *American Journal of Physiology, 253,* E228–E231.

[51] Scheller, D., Kolb, J. (1991). The internal reference technique in microdialysis: a practical approach to monitoring dialysis efficiency and to calculating tissue concentration from dialyzate samples. *Journal of Neuroscience Methods, 40,* 31–38.

[52] Yokel, R.A., Allen, D.D., Burgio, D.E., McNamara, P.J. (1992). Antipyrine as a dialyzable reference to correct differences in efficiency among and within sampling devices during in vivo microdialysis. *Journal of Pharmacological and Toxicological Methods, 27,* 135–142.

[53] Brunner, M., Derendorf, H. (2006). Clinical microdialysis: current applications and potential use in drug development. *Trends in Analytical Chemistry, 25,* 674–680.

[54] Song, Y., Lunte, C.E. (1999). Calibration methods for microdialysis sampling in vivo: muscle and adipose tissue. *Analytica Chimica Acta, 400,* 143–152.

[55] Song, Y., Lunte, C.E. (1999). Comparison of calibration by delivery versus no net flux for quantitative in vivo microdialysis sampling. *Analytica Chimica Acta, 379,* 251–262.

[56] Schultz, K.N., Kennedy, R.T. (2008). Time-resolved microdialysis for in vivo neurochemical measurements and other applications. *Annual Review of Analytical Chemistry, 1,* 627–661.

[57] Cremers, T.I.F.H., de Vries, M.G., Huinink, K.D., van Loon, J.P., van der Hart, M., Ebert, B., Westerink, B.H.C., De Lange, E.C.M. (2009). Quantitative microdialysis using modified ultraslow microdialysis: direct rapid and reliable determination

of free brain concentrations with the MetaQuant technique. *Journal of Neuroscience Methods, 178,* 249–254.

[58] Zhou, S.N., Oakes, K.D., Servos, M.R., Pawliszyn, J. (2009). Use of simultaneous dual-probe microdialysis for the determination of pesticide residues in a jade plant (Crassula ovata). *Analyst, 134,* 748–754.

[59] Zhang, M., Mao, L. (2005). Enzyme-based amperometric biosensors for continuous and on-line monitoring of cerebral extracellular microdialysate. *Frontiers in Bioscience, 10,* 345–352.

[60] Miele, M., Fillenz, M. (1996). In vivo determination of extracellular brain ascorbate. *Journal of Neuroscience Methods, 70,* 15–19.

[61] Boutelle, M.G., Fellows, L.K., Cook, C. (1992). Enzyme packed bed system for the on-line measurement of glucose, glutamate, and lactate in brain microdialysate. *Analytical Chemistry, 64,* 1790–1794.

[62] Gramsbergen, J.B., Skjoth-Rasmussen, J., Rasmussen, C., Lambertsen, K.L. (2004). On-line monitoring of striatum glucose and lactate in the endothelin-1 rat model of transient focal cerebral ischemia using microdialysis and flow-injection analysis with biosensors. *Journal of Neuroscience Methods, 140,* 93–101.

[63] Jones, D.A., Ros, J., Landolt, H., Fillenz, M., Boutelle, M.G. (2000). Dynamic changes in glucose and lactate in the cortex of the freely moving rat monitored using microdialysis. *Journal of Neurochemistry, 75,* 1703–1708.

[64] Kaptein, W.A., Zwaagstra, J.J., Venema, K., Korf, J. (1998). Continuous ultraslow microdialysis and ultrafiltration for subcutaneous sampling as demonstrated by glucose and lactate measurements in rats. *Analytical Chemistry, 70,* 4696–4700.

[65] Li, B., Zhang, Z., Jin, Y. (2001). Chemiluminescence flow sensor for in vivo on-line monitoring of glucose in awake rabbit by microdialysis sampling. *Analytica Chimica Acta, 432,* 95–100.

[66] Lin, Y., Zhu, N., Yu, P., Su, L., Mao, L. (2009). Physiologically relevant online electrochemical method for continuous and simultaneous monitoring of striatum glucose and lactate following global cerebral ischemia/reperfusion. *Analytical Chemistry, 81,* 2067–2074.

[67] Rhemrev-Boom, M.M., Jonker, M.A., Venema, K., Tiessen, R., Korf, J., Jobst, G. (2001). On-line continuous monitoring of glucose or lactate by ultraslow microdialysis combined with a flow-through nanoliter biosensor based on poly(*m*-phenylenediamine) ultra-thin polymer membrane as enzyme electrode. *Analyst, 126,* 1073–1079.

[68] Ricci, F., Moscone, D., Palleschi, G. (2008). Ex vivo continuous glucose monitoring with microdialysis technique: the example of GlucoDay. *IEEE Sensors Journal, 8,* 63–70.

[69] Yao, T., Yano, T., Nishino, H. (2004). Simultaneous in vivo monitoring of glucose, L-lactate, and pyruvate concentrations in rat brain by a flow-injection biosensor system with an on-line microdialysis sampling. *Analytica Chimica Acta, 510,* 53–59.

[70] Berners, M.O.M., Boutelle, M.G., Fillenz, M. (1994). Online measurement of brain glutamate with an enzyme/polymer-coated tubular electrode. *Analytical Chemistry, 66,* 2017–2021.

[71] Miele, M., Berners, M., Boutelle, M.G., Kusakabe, H., Fillenz, M. (1996). The determination of the extracellular concentration of brain glutamate using quantitative microdialysis. *Brain Research, 707,* 131–133.

[72] Yao, T., Okano, G. (2008). Simultaneous determination of L-glutamate, acetylcholine and dopamine in rat brain by a flow-injection biosensor system with microdialysis sampling. *Analytical Sciences, 24,* 1469–1473.

[73] Nanjo, Y., Yano, T., Hayashi, R., Yao, T. (2006). Optically specific detection of D- and L-lactic acids by a flow-injection dual biosensor system with on-line microdialysis sampling. *Analytical Sciences, 22,* 1135–1138.

[74] Lambert, P.D., Wilding, J.P., Turton, M.D., Ghatei, M.A., Bloom, S.R. (1994). Effect of food deprivation and streptozotocin-induced diabetes on hypothalmic neuropeptide Y release as measured by a radioimmunoassay-linked microdialysis procedure. *Brain Research, 656,* 135–140.

[75] Brodin, E., Linderoth, B., Gazelius, B., Ungerstedt, U. (1987). In vivo release of substance P in cat dorsal horn studied with microdialysis. *Neuroscience Letters, 76,* 357–362.

[76] Maidment, N.T., Siddall, B.J., Rudolph, V.R., Erdelyi, E., Evans, C.J. (1991). Dual determination of extracellular cholecystokinin and neurotensin fragments in rat forebrain: microdialysis combined with a sequential multiple antigen radioimmunoassay. *Neuroscience (Oxford), 45,* 81–93.

[77] Maidment, N.T., Brumbaugh, D.R., Rudolph, V.D.E., Evans, C.J. (1989). Microdialysis of extracellular endogenous opioid peptides from rat brain in vivo. *Neuroscience, 33,* 549–557.

[78] Blakeman, K.H., Wiesenfeld-Hallin, Z., Alster, P. (2001). Microdialysis of galanin in rat spinal cord: in vitro and in vivo studies. *Experimental Brain Research, 139,* 354–358.

[79] Consolo, S., Baldi, G., Russi, G., Civenni, G., Bartfai, T., Vezzani, A. (1994). Impulse flow dependency of galanin release in vivo in the rat ventral hippocampus. *Proceedings of the National Academy of Sciences of the United States of America, 91,* 8047–8051.

[80] Maidment, N.T., Brumbaugh, D.R., Rudolph, V.D., Erdelyi, E., Evans, C.J. (1989). Microdialysis of extracellular endogenous opioid peptides from rat brain in vivo. *Neuroscience, 33,* 549–557.

[81] Reed, B., Zhang, Y., Chait, B.T., Kreek, M.J. (2003). Dynorphin A (1–17) biotransformation in striatum of freely moving rats using microdialysis and matric-assisted laser desorption/ionization mass spectrometry. *Journal of Neurochemistry, 86,* 815–823.

[82] Kobayashi, N., Kazui, M., Ikeda, T. (2000). Rapid, real-time sampling of R-84760 in blood by in vivo microdialysis with tandem mass spectrometry. *Journal of Pharmaceutical and Biomedical Analysis, 21,* 1233–1242.

[83] Kobayashi, N., Fujimori, I., Watanabe, M., Ikeda, T. (2000). Real-time monitoring of metabolic reactions by microdialysis in combination with tandem mass spectrometry: hydrolysis of CS-866 in vitro in human and rat plasma, livers, and small intestines. *Analytical Biochemistry, 287,* 272–278.

[84] Westerink, B., Cremers, T.I.F.H. (2007). *Handbook of Microdialysis: Methods, Applications and Perspectives,* Elsevier, Amsterdam, pp. 1–697.

[85] Baseski, H.M., Watson, C.J., Cellar, N.A., Shackman, J.G., Kennedy, R.T. (2005). Capillary liquid chromatography with MS3 for the determination of enkephalins in microdialysis samples from the striatum of anesthetized and freely-moving rats. *Journal of Mass Spectrometry*, *40*, 146–153.

[86] Shackman, H.M., Shou, M., Cellar, N.A., Watson, C.J., Kennedy, R.T. (2006). Microdialysis coupled on-line to capillary liquid chromatography with tandem mass spectrometry for monitoring acetylcholine in vivo. *Journal of Neuroscience Methods*, *159*, 86–92.

[87] Sarre, S., Michotte, Y., Herregodts, P., Deleu, D., De Klippel, N., Ebinger, G. (1992). High-performance liquid chromatography with electrochemical detection for the determination of levodopa, catecholamines and their metabolites in rat brain dialysates. *Journal of Chromatography B: Biomedical Sciences and Applications*, *575*, 207–212.

[88] Li, Q., Zubieta, J.-K., Kennedy, R.T. (2009). Practical aspects of in vivo detection of neuropeptides by microdialysis coupled off-line to capillary LC with multi-stage MS. *Analytical Chemistry*, *81*, 2242–2250.

[89] Boyd, B.W., Kennedy, R.T. (2002). Ultrahigh sensitivity analysis of amino acids and peptides by capillary liquid chromatography with electrochemical detection. In *Electroanalytical Methods of Biological Materials* (A. Brajter-toth and J.G. Chambers, eds.), CRC Press Online, pp. 491–521.

[90] Huang, T., Kissinger, P.T. (1998). Determination of neurotransmitters in rat brain microdialyzates by microbore liquid chromatography/electrochemistry. *Fenxi Huaxue*, *26*, 748–751.

[91] Ye, W., Yin, P., Mei, Z. (1997). Microbore liquid chromatography with dual electrode electrochemical detection of monoamine transmitters in brain microdialyzates. *Sepu*, *15*, 185–188.

[92] Davies, M.I. (1999). A review of microdialysis sampling for pharmacokinetic applications. *Analytica Chimica Acta*, *379*, 227–249.

[93] Nydahl, K.S., Pierson, J., Nyberg, F., Caprioli, R.M., Andren, P.E. (2003). *In vivo* processing of LVV-hemorphin-7 in rat brain and blood utilizing microdialysis combined with electrospray mass spectrometry. *Rapid Communications in Mass Spectrometry*, *17*, 838–844.

[94] Klintenberg, R., Andren, P.E. (2005). Altered extracellular striatal in vivo bio-transformation of the opioid neuropeptide dynorphin A (1–17) in the unilateral 6-OHDA rat model of Parkinson's disease. *Journal of Mass Spectrometry*, *40*, 261–270.

[95] Behrens, H.L., Chen, R., Li, L. (2008). Combining microdialysis, nanoLC-MS, and MALDI–TOF/TOF to detect neuropeptides secreted in the crab, cancer borealis. *Analytical Chemistry*, *80*, 6949–6958.

[96] Shackman, H.M., Shou, M., Cellar, N.A., Watson, C.J., Kennedy, R.T. (2007). Microdialysis coupled on-line to capillary liquid chromatography with tandem mass spectrometry for monitoring acetylcholine in vivo. *Journal of Neuroscience Methods*, *159*, 86–92.

[97] Hows, M.E.P., Organ, A.J., Murray, S., Dawson, L.A., Foxton, R., Heidbreder, C., Hughes, Z.A., Lacroix, L., Shah, A.J. (2002). High-performance liquid chromatography/tandem mass spectrometry assay for the rapid high sensitivity measurement of basal acetylcholine from microdialysates. *Journal of Neuroscience Methods*, *121*, 33–39.

[98] Lacrois, L.O., Ceolin, L., Zocchi, A., Barnier, G., Garzottie, M., Curcuruto, O. (2006). Selective dopamine D3 receptor antagonists enhance cortical acetylcholine levels measured with high-performance liquid chromatography/tandem mass spectrometry without anti-cholinesterases. *Journal of Neuroscience Methods, 157*, 25–31.

[99] Zhang, M.-Y., Hughes, Z.A., Edward, H., Lin, Q., Beyer, C.E. (2007). Development of a liquid chromatography/tandem mass spectrometry method for the quantitation of acetylcholine and related neurotransmitters in brian microdialysis samples. *Journal of Pharmaceutical and Biomedical Analysis, 44*, 586–593.

[100] Uutela, P., Reinila, R., Piepponen, P., Ketola, R.A., Kostiainen, R. (2005). Analysis of acetylcholine and choline in microdialysis samples by liquid chromatography/tandem mass spectrometry. *Rapid Communications in Mass Spectrometry, 19*, 2950–2956.

[101] Persike, M., Zimmermann, M., Klein, J., Karas, M. (2010). Quantitative determination of acetylcholine and choline in microdialysis samples by MALDI–TOF MS. *Analytical Chemistry (Washington, D.C.), 82*, 922–929.

[102] Hows, M.E. P., Lacroix, L., Heidbreder, C., Organ, A.J., Shah, A.J. (2004). High-performance liquid chromatography/tandem mass spectrometric assay for the simultaneous measurement of dopamine, norepinephrine, 5-hydroxytryptamine and cocaine in biological samples. *Journal of Neuroscience Methods, 138*, 123–132.

[103] Zheng, H., Chen, G., Shi, L., Lou, Z., Chen, F., Hu, J. (2009). Determination of oxymatrine and its metabolite matrine in rat blood and dermal microdialysates by high throughput liquid chromatography/tandem mass spectrometry. *Journal of Pharmaceutical and Biomedical Analysis, 49*, 427–433.

[104] Zamboni, W.C., Strychor, S., Joseph, E., Walsh, D.R., Zamboni, B.A., Parise, R.A., Tonda, M.E., Yu, N.Y., Engbers, C., Eiseman, J.L. (2007). Plasma, tumor, and tissue disposition of STEALTH liposomal CKD-602 (S-CKD602) and nonliposomal CKD-602 in mice bearing A375 human melanoma xenografts. *Clinical Cancer Research, 13*, 7217–7223.

[105] Cheng, G.-W., Hsu, K.-C., Lee, C.-F., Wu, H.-L., Huang, Y.-L. (2009). On-line microdialysis coupled with liquid chromatography for biomedical analysis. *Journal of Chromatographic Science, 47*, 624–630.

[106] Guihen, E., O'Connor, W.T. (2009). Current separation and detection methods in microdialysis the drive towards sensitivity and speed. *Electrophoresis, 30*, 2062–2075.

[107] Steele, K.M., Lunte, C.E. (1995). Microdialysis sampling coupled to online microbore liquid chromatography for pharmacokinetic studies. *Journal of Pharmaceutical and Biomedical Analysis, 13*, 149–154.

[108] Chen, A., Lunte, C.E. (1995). Microdialysis sampling coupled online to fast microbore liquid chromatography. *Journal of Chromatography A, 691*, 29–35.

[109] Mathy, F.-X., Vroman, B., Ntivunwa, D., De Winne, A.J., Verbeeck, R.K., Preat, V. (2003). On-line determination of fluconazole in blood and dermal rat microdialysates by microbore high-performance liquid chromatography. *Journal of Chromatography B: Analytical Technologies in the Biomedical and Life Sciences, 787*, 323–331.

[110] Wong, P.S.H., Yoshioka, K., Xie, F., Kissinger, P.T. (1999). *In vivo* microdialysis/ liquid chromatography/tandem mass spectrometry for the online monitoring of melatonin in rat. *Rapid Communications in Mass Spectrometry, 13*, 407–411.

[111] Yoshitake, T., Iizuka, R., Kehr, J., Nohta, H., Ishida, J., Yamaguchi, M. (2001). Determination of serotonin in microdialysis samples from rat brain by microbore column liquid chromatography with post-column derivatization and fluorescence detection. *Journal of Neuroscience Methods, 109*, 91–96.

[112] Chang, Y.L., Chou, M.H., Lin, M.F., Chen, C.-F., Tsai, T.-H. (2001). Determination and pharmacokinetic study of unbound cefepime in rat bile by liquid chromatography with on-line microdialysis. *Journal of Chromatography A, 914*, 77–82.

[113] Tsai, T.-H., Cheng, F.-C., Hung, L.-C., Chen, C.-F. (1999). Determination of unbound ceftriaxone in rat blood by on-line microdialysis and microbore liquid chromatography. *International Journal of Pharmaceutics, 193*, 21–26.

[114] Tsai, T.-H., Tsai, T.-R., Chen, Y.-F., Chou, C.-J., Chen, C.-F. (1999). Determination of unbound 20(*S*)-camptothecin in rat bile by on-line microdialysis coupled to microbore liquid chromatography with fluorescence detection. *Journal of Chromatography B: Biomedical Sciences and Applications, 732*, 221–225.

[115] Tsai, T.-H., Cheng, F.-C., Hung, L.-C., Chen, C.-F. (1998). Online microdialysis coupled with microbore liquid chromatography for the determination of unbound chloramphenicol and its glucuronide in rat blood. *Journal of Chromatography B: Biomedical Sciences and Applications, 720*, 165–169.

[116] Tsai, T.-R., Cheng, F.-C., Hung, L.-C., Chen, C.-F., Tsai, T.-H. (1999). Determination of unbound cefmetazole in rat blood by on-line microdialysis and microbore liquid chromatography: a pharmacokinetic study. *Journal of Chromatography B: Biomedical Sciences and Applications, 736*, 129–134.

[117] Arnett, S.D., Osbourn, D.M., Moore, K.D., Vandaveer, S.S., Lunte, C.E. (2005). Determination of 8-oxoguanine and 8-hydroxy-2'-deoxyguanosine in the rat cerebral cortex using microdialysis sampling and capillary electrophoresis with electrochemical detection. *Journal of Chromatography B: Analytical Technologies in the Biomedical and Life Sciences, 827*, 16–25.

[118] Tsai, T.-H., Cheng, F.-C., Hung, L.-C., Chen, C.-F. (1999). Measurement of hydroxyl radical in rat blood vessel by microbore liquid chromatography and electrochemical detection: an on-line microdialysis study. *Journal of Chromatography B: Biomedical Sciences and Applications, 734*, 277–283.

[119] Liu, S.-C., Tsai, T.-H. (2002). Determination of diclofenac in rat bile and its interaction with cyclosporin A using on-line microdialysis coupled to liquid chromatography. *Journal of Chromatography B: Analytical Technologies in the Biomedical and Life Sciences, 769*, 351–356.

[120] Sauvinet, V., Parrot, S., Benturquia, N., Bravo-Moraton, E., Renaud, B., Denoroy, L. (2003). In vivo simultaneous monitoring of gamma-aminobutyric acid, glutamate, and ʟ-aspartate using brain microdialysis and capillary electrophoresis iwth laser-induced fluorescence detection: analytical developments and in vitro/in vivo validations. *Electrophoresis, 24*, 3187–3196.

[121] Benturquia, N., Couderc, F., Sauvinet, V., Orset, C., Parrot, S., Bayle, C., Renaud, B., Denoroy, L. (2005). Analysis of serotonin in brian microdialysates using capillary electrophoresis and native laser-induced fluorescence detection. *Electrophoresis, 26*, 1071–1079.

[122] Kalish, H., Phillips, T.M. (2009). Application of immunaffinity capillary electrophoresis to the measurements of secreted cytokines by cultured astrocytes. *Journal of Separation Science, 32*, 1605–1612.

[123] Malone, M., Zuo, H., Lunte, S.M., Smyth, M.R. (1995). Determination of tryptophan and kynurenine in brain microdialysis samples by capillary electrophoresis with electrochemical detection. *Journal of Chromatography A, 700*, 73–80.

[124] O'Shea, T.J., Weber, P.L., Bammel, B.P., Lunte, C.E., Lunte, S.M., Smyth, M.R. (1992). Monitoring excitatory amino acid release in vivo by microdialysis with capillary electrophoresis-electrochemistry. *Journal of Chromatography, 608*, 189–195.

[125] Zhou, J., Lunte, S.M. (1995). Direct amperometric detection of amino acids by capillary electrophoresis–electrochemistry using a copper microelectrode and switterionic buffers. *Electrophoresis, 16*, 498–503.

[126] Shou, M., Smith, A.D., Shackman, J.G., Peris, J., Kennedy, R.T. (2004). In vivo monitoring of amino acids by microdialysis sampling with on-line derivatization by naphthalene-2,3-dicarboxyaldehyde and rapid micellar electrokinetic capillary chromatography. *Journal of Neuroscience Methods, 138*, 189–197.

[127] Hogan, B.L., Lunte, S.M., Stobaugh, J.F., Lunte, C.E. (1994). Online coupling of in vivo microdialysis sampling with capillary electrophoresis. *Analytical Chemistry, 66*, 596–602.

[128] Zhou, S.Y., Zuo, H., Stobaugh, J.F., Lunte, C.E., Lunte, S.M. (1995). Continuous in vivo monitoring of amino acid neurotransmitters by microdialysis sampling with online derivatization and capillary electrophoresis separation. *Analytical Chemistry, 67*, 594–599.

[129] Zhou, J., Heckert, D.M., Zuo, H., Lunte, C.E., Lunte, S.M. (1999). Online coupling of in vivo microdialysis with capillary electrophoresis/electrochemistry. *Analytica Chimica Acta, 379*, 307–317.

[130] Lada, M.W., Schaller, G., Carriger, M.H., Vickroy, T.W., Kennedy, R.T. (1995). On-line interface between microdialysis and capillary zone electrophoresis. *Analytica Chimica Acta, 307*, 217–225.

[131] Lada, M.W., Vickroy, T.W., Kennedy, R.T. (1997). High temporal resolution monitoring of glutamate and aspartate in vivo using microdialysis online with capillary electrophoresis with laser-induced fluorescence detection. *Analytical Chemistry, 69*, 4560–4565.

[132] Lada, M.W., Kennedy, R.T. (1996). Quantitative in vivo monitoring of primary amines in rat caudate nucleus using microdialysis coupled by a flow-gated interface to capillary electrophoresis with laser-induced fluorescence detection. *Analytical Chemistry, 68*, 2790–2797.

[133] Bowser, M.T., Kennedy, R.T. (2001). In vivo monitoring of amine neurotransmitters using microdialysis with on-line capillary electrophoresis. *Electrophoresis, 22*, 3668–3676.

[134] Lada, M.W., Kennedy, R.T. (1995). Quantitative in vivo measurements using microdialysis on-line with capillary zone electrophoresis. *Journal of Neuroscience Methods, 63*, 147–152.

[135] Lada, M.W., Kennedy, R.T. (1997). In vivo monitoring of glutathione and cysteine in rat caudate nucleus using microdialysis online with capillary zone

electrophoresis-laser induced fluorescence detection. *Journal of Neuroscience Methods*, *72*, 153–159.

[136] Robert, F., Bert, L., Parrot, S., Denoroy, L., Stoppini, L., Renaud, B. (1998). Coupling on-line brain microdialysis, precolumn derivatization and capillary electrophoresis for routine minute sampling of *O*-phosphoethanolamine and excitatory amino acids. *Journal of Chromatography A*, *817*, 195–203.

[137] Shou, M., Ferrario, C.R., Schultz, K.N., Robinson, T.E., Kennedy, R.T. (2006). Monitoring dopamine in vivo by microdialysis sampling and on-line CE–laser-induced fluorescence. *Analytical Chemistry*, *78*, 6717–6725.

[138] Klinker, C.C., Bowser, M.T. (2007). 4-Fluoro-7-nitro-2,1,3-benzooxadiazole as a fluorogenic labeling reagent for the in vivo analysis of amino acid neurotransmitters using online microdialysis-capillary electrophoresis. *Analytical Chemistry*, *79*, 8747–8754.

[139] Huynh, B.H., Fogarty, B.A., Martin, R.S., Lunte, S.M. (2004). On-line coupling of microdialysis sampling with microchip-based capillary electrophoresis. *Analytical Chemistry*, *76*, 6440–6447.

[140] Huynh, B.H., Fogarty, B.A., Nandi, P., Lunte, S.M. (2006). A microchip electrophoresis device with on-line microdialysis sampling and on-chip sample derivatization by naphthalene 2,3-dicarboxaldehyde/2-mercaptoethanol for amino acid and peptide analysis. *Journal of Pharmaceutical and Biomedical Analysis*, *42*, 529–534.

[141] Sandlin, Z.D., Shou, M., Shackman, J.G., Kennedy, R.T. (2005). Microfluidic electrophoresis chip coupled to microdialysis for in vivo monitoring of amino acid neurotransmitters. *Analytical Chemistry*, *77*, 7702–7708.

[142] Nandi, P., Desai, D.P., Lunte, S.M. (2010). Development of a PDMS-based microchip electrophoresis device for continuous online in vivo monitoring of microdialysis samples. *Electrophoresis*, *31*, 1414–1422.

[143] Cellar, N.A., Burns, S.T., Meiners, J.-C., Chen, H., Kennedy, R.T. (2005). Microfluidic chip for low-flow push-pull perfusion sampling in vivo with on-line analysis of amino acids. *Analytical Chemistry*, *77*, 7067–7073.

[144] Li, M.W., Huynh, B.H., Hulvey, M.K., Lunte, S.M., Martin, R.S. (2006). Design and characterization of poly(dimethylsiloxane)-based valves for interfacing continuous-flow sampling to microchip electrophoresis. *Analytical Chemistry*, *78*, 1042–1051.

[145] Mecker, L.C., Martin, R.S. (2008). Integration of microdialysis sampling and microchip electrophoresis with electrochemical detection. *Analytical Chemistry*, *80*, 9257–9264.

[146] Wang, M., Roman, G.T., Perry, M.L., Kennedy, R.T. (2009). Microfluidic chip for high efficiency electrophoretic analysis of segmented flow from a microdialysis probe and in vivo chemical monitoring. *Analytical Chemistry*, *81*, 9072–9078.

[147] Wang, M., Roman, G.T., Schultz, K., Jennings, C., Kennedy, R.T. (2008). Improved temporal resolution for in vivo microdialysis by using segmented flow. *Analytical Chemistry*, *80*, 5607–5615.

[148] Roman, G.T., Wang, M., Shultz, K.N., Jennings, C., Kennedy, R.T. (2008). Sampling and electrophoretic analysis of segmented flow streams using virtual walls in a microfluidic device. *Analytical Chemistry*, *80*, 8231–8238.

[149] Miyaki, K., Zeng, H.L., Nakagama, T., Uchiyama, K. (2007). Steady surface modification of polydimethylsiloxane microchannel and its application in simultaneous analysis of homocysteine and glutathione in human serum. *Journal of Chromatography A, 1166,* 201–206.

[150] Jin, G., Cheng, Q., Feng, J., Li, F. (2008). On-line microdialysis coupled to analytical systems. *Journal of Chromatographic Science, 46,* 276–287.

[151] Tsai, P.-J., Wu, J.-P., Lin, N.-N., Kuo, J.-S., Yang, C.-S. (1996). In vivo, continuous and automatic monitoring of extracellular ascorbic acid by microdialysis and online liquid chromatography. *Journal of Chromatography B: Biomedical Applications, 686,* 151–156.

[152] Torto, N., Laurell, T., Gorton, L., Marko-Varga, G. (1999). Recent trends in the application of microdialysis in bioprocesses. *Analytica Chimica Acta, 379,* 281–305.

[153] Robinson, T.E., Justice, J.B., Jr. (Eds.) (1991). *Techniques in the Behavioral and Neural Sciences, Vol. 7, Microdialysis in the Neurosciences,* Elsevier, Amsterdam.

[154] Huff, J.K., Davies, M.I. (2002). Microdialysis monitoring of methylphenidate in blood and brain correlated with changes in dopamine and rat activity. *Journal of Pharmaceutical and Biomedical Analysis, 29,* 767–777.

[155] Grabb, M.C., Sciotti, V.M., Gidday, J.M., Cohen, S.A., van Wylen, D.G. (1998). Neurochemical and morphological responses to acutely and chronically implanted brain microdialysis probes. *Journal of Neuroscience Methods, 82,* 25–34.

[156] Benveniste, H., Drejer, J., Schousboe, A., Diemer, N.H. (1987). Regional cerebral glucose phosphorylation and blood flow after insertion of a microdialysis fiber through the dorsal hippocampus in the rat. *Journal of Neurochemistry, 49,* 729–734.

[157] de Lange, E.C., Danhof, M., de Boer, A.G., Breimer, D.D. (1994). Critical factors of intracerebral microdialysis as a technique to determine the pharmacokinetics of drugs in rat brain. *Brain Research, 666,* 1–8.

[158] Benveniste, H., Drejer, J., Schousboe, A., Diemer, N.H. (1984). Elevation of the extracellular concentrations of glutamate and aspartate in rat hippocampus during transient cerebral ischemia monitored by intracerebral microdialysis. *Journal of Neurochemistry, 43,* 1369–1374.

[159] Major, O., Shdanova, T., Duffek, L., Nagy, Z. (1990). Continuous monitoring of blood–brain barrier opening to Cr^{51}-EDTA by microdialysis following probe injury. *Acta Neurochirurgica Supplement (Wien), 51,* 46–48.

[160] Borland, L.M., Shi, G., Yang, H., Michael, A.C. (2005). Voltammetric study of extracellular dopamine near microdialysis probes acutely implanted in the striatum of the anesthetized rat. *Journal of Neuroscience Methods, 146,* 149–158.

[161] Lu, Y., Peters, J.L., Michael, A.C. (1998). Direct comparison of the response of voltammetry and microdialysis to electrically evoked release of striatal dopamine. *Journal of Neurochemistry, 70,* 584–593.

[162] Mitala, C.M., Wang, Y., Borland, L.M., Jung, M., Shand, S., Watkins, S., Weber, S.G., Michael, A.C. (2008). Impact of microdialysis probes on vasculature and dopamine in the rat striatum: a combined fluorescence and voltammetric study. *Journal of Neuroscience Methods, 174,* 177–185.

[163] Peters, J.L., Michael, A.C. (1998). Modeling voltammetry and microdialysis of striatal extracellular dopamine: the impact of dopamine uptake on extraction and recovery ratios. *Journal of Neurochemistry, 70,* 594–603.

[164] Yang, H., Peters, J.L., Allen, C., Chern, S.-S., Coalson, R.D., Michael, A.C. (2000). A theoretical description of microdialysis with mass transport coupled to chemical events. *Analytical Chemistry, 72,* 2042–2049.

[165] Zielke, H.R., Zielke, C.L., Baab, P.J. (2009). Direct measurement of oxidative metabolism in the living brain by microdialysis: a review. *Journal of Neurochemistry, 109,* 24–29.

[166] Torregrossa, M.M., Kalivas, P.W. (2008). Microdialysis and the neurochemistry of addiction. *Pharmacology, Biochemistry and Behavior, 90,* 261–272.

[167] McAdoo, D.J., Wu, P. (2008). Microdialysis in central nervous system disorders and their treatment. *Pharmacology, Biochemistry and Behavior, 90,* 282–296.

[168] Linthorst, A.C.E., Reul, J.M. (2008). Stress and the brain: solving the puzzle using microdialysis. *Pharmacology, Biochemistry and Behavior, 90,* 163–173.

[169] Fillenz, M. (2005). In vivo neurochemical monitoring and the study of behaviour. *Neuroscience & Biobehavioral Reviews, 29,* 949–962.

[170] Maurer, M.H., Haux, D., Unterberg, A.W., Sakowitz, O.W. (2008). Proteomics of human cerebral microdialysate: from detection of biomarkers to clinical application. *Proteomics: Clinical Applications, 2,* 437–443.

[171] Ungerstedt, U., Rostami, E. (2004). Microdialysis in neurointensive care. *Current Pharmaceutical Design, 10,* 2145–2152.

[172] Buck, K., Voehringer, P., Ferger, B. (2009). Rapid analysis of GABA and glutamate in microdialysis samples using high performance liquid chromatography and tandem mass spectrometry. *Journal of Neuroscience Methods, 182,* 78–84.

[173] Lanckmans, K., Sarre, S., Smolders, I., Michotte, Y. (2008). Quantitative liquid chromatography/mass spectrometry for the analysis of microdialysates. *Talanta, 74,* 458–469.

[174] Zhang, M.-Y., Beyer, C.E. (2006). Measurement of neurotransmitters from extracellular fluid in brain by in vivo microdialysis and chromatography–mass spectrometry. *Journal of Pharmaceutical and Biomedical Analysis, 40,* 492–499.

[175] Elmquist, W.F., Sawchuk, R.J. (1997). Application of microdialysis in pharmacokinetic studies. *Pharmaceutical Research, 14,* 267–288.

[176] Scott, D.O., Lunte, C.E. (1993). In vivo microdialysis sampling in the bile, blood, and liver of rats to study the disposition of phenol. *Pharmaceutical Research, 10,* 335–342.

[177] Scott, D.O., Sorensen, L.R., Lunte, C.E. (1990). In vivo microdialysis sampling coupled to liquid chromatography for the study of acetaminophen metabolism. *Journal of Chromatography A, 506,* 461–469.

[178] Scott, D.O., Sorenson, L.R., Steele, K.L., Puckett, D.L., Lunte, C.E. (1991). In vivo microdialysis sampling for pharmacokinetic investigations. *Pharmaceutical Research, 8,* 389–392.

[179] Steele, K.L., Scott, D.O., Lunte, C.E. (1991). Pharmacokinetic studies of aspirin in rats using in vivo microdialysis sampling. *Analytica Chimica Acta, 246,* 181–167.

[180] Lunte, C.E., Scott, D.O., Kissinger, P.T. (1991). Sampling living systems using microdialysis probes. *Analytical Chemistry*, *63*, 773A–774A, 776A–778A, 780A.

[181] McDonald, S., Lunte, C. (2003). Determination of the dermal penetration of esterom components using microdialysis sampling. *Pharmaceutical Research*, *20*, 1827–1834.

[182] Davies, M.I., Lunte, C.E. (1995). Microdialysis sampling for hepatic metabolism studies: impact of microdialysis probe design and implantation technique on liver tissue. *Drug Metabolism and Disposition*, *23*, 1072–1079.

[183] Palsmeier, R.K., Lunte, C.E. (1994). Microdialysis sampling in tumor and muscle: study of the disposition of 3-amino-1,2,4-benzotriazine-1,4-di-*N*-oxide (SR 4233). *Life Sciences*, *55*, 815–825.

[184] Palsmeier, R.K., Lunte, C.E. (1994). In the MD literature: microdialysis; microdialysis sampling in tumor and muscle—study of the disposition of 3-amino-1,2,4-benzotriazine-1,4-di-*N*-oxide (SR 4233). *Current Separations*, *13*, 93.

[185] Bostrom, E., Hammarlund-Udenaes, M., Simonsson, U.S.H. (2008). Blood–brain barrier transport helps to explain discrepancies in in vivo potency between oxycodone and morphine. *Anesthesiology*, *108*, 495–505.

[186] Copper, J.D., Heppert, K.E., Davies, M.I., Lunte, S.M. (2007). Evaluation of an osmotic pump for microdialysis sampling in an awake and untethered rat. *Journal of Neuroscience Methods*, *160*, 269–275.

[187] Hosoya, K., Makihara, A., Tsujikawa, Y., Yoneyama, D., Mori, S., Terasaki, T., Akanuma, S., Tomi, M., Tachikawa, M. (2009). Roles of inner blood–retinal barrier organic anion transporter 3 in the vitreous/retina-to-blood efflux transport of *p*-aminohippuric acid, benzylpenicillin, and 6-mercaptopurine. *Journal of Pharmacology and Experimental Therapeutics*, *329*, 87–93.

[188] Dey, S., Gunda, S., Mitra, A.K. (2004). Pharmacokinetics of erythromycin in rabbit corneas after single-dose infusion: role of P-glycoprotein as a barrier to in vivo ocular drug absorption. *Journal of Pharmacology and Experimental Therapeutics*, *311*, 246–255.

[189] Anand, B.S., Katragadda, S., Gunda, S., Mitra, A.K. (2006). In vivo ocular pharmacokinetics of acyclovir dipeptide ester prodrugs by microdialysis in rabbits. *Molecular Pharmacology*, *3*, 431–440.

[190] Macha, S., Mitra, A.K. (2001). Ocular pharmacokinetics in rabbits using a novel dual probe microdialysis technique. *Experimental Eye Research*, *72*, 289–299.

[191] Waga, J., Nilsson-Ehle, I., Ljungberg, B., Skarin, A., Stahle, L., Ehinger, B. (1999). Microdialysis for pharmacokinetic studies of ceftazidime in rabbit vitreous. *Journal of Ocular Pharmacology and Therapeutics*, *15*, 455–463.

[192] Barbour, A., Schmidt, S., Rout, W.R., Ben-David, K., Burkhardt, O., Derendorf, H. (2009). Soft tissue penetration of cefuroxime determined by clinical microdialysis in morbidly obese patients undergoing abdominal surgery. *International Journal of Antimicrobial Agents*, *34*, 231–235.

[193] de la Peña, A., Liu, P., Derendorf, H. (2000). Microdialysis in peripheral tissues. *Advanced Drug Delivery Reviews*, *45*, 189–216.

[194] Flodgren, G.M., Crenshaw, A.G., Gref, M., Fahlstrom, M. (2009). Changes in interstitial noradrenaline, trapezius muscle activity and oxygen saturation during low-load work and recovery. *European Journal of Applied Physiology*, *107*, 31–42.

[195] Gerdle, B., Hilgenfeldt, U., Larsson, B., Kristiansen, J., Sogaard, K., Rosendal, L. (2008). Bradykinin and kallidin levels in the trapezius muscle in patients with work-related trapezius myalgia, in patients with whiplash associated pain, and in healthy controls: a microdialysis study of women. *Pain, 139*, 578–587.

[196] Hamrin, K., Henriksson, J. (2008). Interstitial glucose concentration in insulin-resistant human skeletal muscle: influence of one bout of exercise and of local perfusion with insulin or vanadate. *European Journal of Applied Physiology, 103*, 595–603.

[197] Lönnroth, P., Carlsten, J., Johnson, L., Smith, U. (1991). Measurements by micro-dialysis of free tissue concentrations of propranolol. *Journal of Chromatography: Biomedical Applications, 568*, 419–425.

[198] Lönnroth, P., Smith, U. (1990). Microdialysis: a novel technique for clinical investigations. *Journal of Internal Medicine, 227*, 295–300.

[199] Kurosaki, Y., Tagawa, M., Omoto, A., Suito, H., Komori, Y., Kawasaki, H., Aiba, T. (2007). Evaluation of intramuscular lateral distribution profile of topically administered acetaminophen in rats. *International Journal of Pharmaceutics, 343*, 190–195.

[200] Marchand, S., Dahyot, C., Lamarche, I., Plan, E., Mimoz, O., Couet, W. (2005). Lack of effect of experimental hypovolemia on imipenem muscle distribution in rats assessed by microdialysis. *Antimicrobial Agents and Chemotherapy, 49*, 4974–4979.

[201] Klimowicz, A., Farfal, S., Bielecka-Grzela, S. (2007). Evaluation of skin penetration of topically applied drugs in humans by cutaneous microdialysis: acyclovir vs. salicylic acid. *Journal of Clinical Pharmacy and Therapeutics, 32*, 143–148.

[202] Mathy, F.X., Ntivunwa, D., Verbeeck, R.K., Préat, V. (2005). Fluconazole distribution in rat dermis following intravenous and topical application: a microdialysis study. *Journal of Pharmaceutical Sciences, 94*, 770–780.

[203] Obreja, O., Rukwied, R., Steinhoff, M., Schmelz, M. (2006). Neurogenic components of trypsin- and thrombin-induced inflammation in rat skin, in vivo. *Experimental Dermatology, 15*, 58–65.

[204] Delacher, S., Derendorf, H., Hollenstein, U., Brunner, M., Joukhadar, C., Hofmann, S., Georgopoulos, A., Eichler, H.G., Müller, M. (2000). A combined in vivo pharmacokinetic–in vitro pharmacodynamic approach to simulate target site pharmacodynamics of antibiotics in humans. *Journal of Antimicrobial Chemotherapy, 46*, 733–739.

[205] Joukhadar, C., Derendorf, H., Müller, M. (2001). Microdialysis: a novel tool for clinical studies of anti-infective agents. *European Journal of Clinical Pharmacology, 57*, 211–219.

[206] Joukhadar, C., Thallinger, C., Poppl, W., Kovar, F., Konz, K.H., Joukhadar, S.M., Traunmuller, F. (2009). Concentrations of voriconazole in healthy and inflamed lung in rats. *Antimicrobial Agents and Chemotherapy, 53*, 2684–2686.

[207] Wesson, D.E., Simoni, J. (2009). Increased tissue acid mediates a progressive decline in the glomerular filtration rate of animals with reduced nephron mass. *Kidney International, 75*, 929–935.

[208] Price, K.E., Larive, C.K., Lunte, C.E. (2009). Tissue-targeted metabonomics: biological considerations and application to doxorubicin-induced hepatic oxidative stress. *Metabolomics, 5*, 219–228.

[209] Lefeuvre, S., Marchand, S., Pariat, C., Lamarche, I., Couet, W. (2008). Microdialysis study of imipenem distribution in the peritoneal fluid of rats with experimental acute pancreatitis. *Antimicrobial Agents and Chemotherapy*, *52*, 1516–1518.

[210] Wu, Y.-T., Lin, L.-C., Tsai, T.-H. (2009). Measurement of free hydroxytyrosol in microdialysates from blood and brain of anesthetized rats by liquid chromatography with fluorescence detection. *Journal of Chromatography A*, *1216*, 3501–3507.

[211] Huang, Y., Yang, Z., Cartier, L., Cheung, B., Sawchuk, R.J. (2007). Estimating amoxicillin influx/efflux in chinchilla middle ear fluid and simultaneous measurement of antibacterial effect. *Antimicrobial Agents and Chemotherapy*, *51*, 4336–4341.

[212] Sawchuk, R.J., Cheung, B. W.Y., Ji, P., Cartier L.L. (2005). Microdialysis studies of the distribution of antibiotics into chinchilla middle ear fluid. *Pharmacotherapy*, *25*, 140S–145S.

[213] Fang, Z.-L., Fang, Q., Liu, X.-Z., Chen, H.-W., Liu, C.-L. (1999). Continuous monitoring in drug dissolution testing using flow injection systems. *Trends in Analytical Chemistry*, *18*, 261–271.

[214] Shah, K.P., Chang, M., Riley, C.M. (1994). Automated analytical systems for drug development studies: II. A system for dissolution testing. *Journal of Pharmaceutical and Biomedical Analysis*, *12*, 1519–1527.

[215] Shah, K.P., Chang, M., Riley, C.M. (1995). Automated analytical systems for drug development studies: 3. Multivessel dissolution testing system based on microdialysis sampling. *Journal of Pharmaceutical and Biomedical Analysis*, *13*, 1235–1241.

[216] Dash, A.K., Haney, P.W., Garavalia, M.J. (1999). Development of an in vitro dissolution method using microdialysis sampling technique for implantable drug delivery systems. *Journal of Pharmaceutical Sciences*, *88*, 1036–1040.

[217] Fang, Q., Liu, S.-S., Wu, J.-F., Sun, Y.-Q., Fang, Z.-L. (1999). A stopped-flow microdialysis sampling-flow injection system for automated multivessel high resolution drug dissolution testing. *Talanta*, *49*, 403–414.

[218] Buttler, T., Nilsson, C., Gorton, L., Marko-Varga, G., Laurell, T. (1996). Membrane characterization and performance of microdialysis probes intended for use as bioprocess sampling units. *Journal of Chromatography A*, *725*, 41–56.

[219] Torto, N., Laurell, T., Gorton, L., Marko-Varga, G. (1998). Recent trends in the application of microdialysis in bioprocesses. *Analytica Chimica Acta*, *374*, 111–135.

[220] Wisniewski, N., Torto, N. (2002). Optimization of microdialysis sampling recovery by varying inner cannula geometry. *Analyst (Cambridge, U.K.)*, *127*, 1129–1134.

[221] Torto, N., Laurell, T., Gorton, L., Varga, G.M. (1997). A study of a polysulfone membrane for use in an in-situ tunable microdialysis probe during monitoring of starch enzymic hydrolyzates. *Journal of Membrane Science*, *130*, 239–248.

[222] Modi, S.J., LaCourse, W.R. (2006). Monitoring carbohydrate enzymatic reactions by quantitative in vitro microdialysis. *Journal of Chromatography A*, *1118*, 125–133.

[223] Liu, C., Wu, Q., Harms, A.C., Smith, R.D. (1996). On-line microdialysis sample cleanup for electrospray ionization mass spectrometry of nucleic acid samples. *Analytical Chemistry*, *68*, 3295–3299.

[224] Xiang, F., Lin, Y., Wen, J., Matson, D.W., Smith, R.D. (1999). An integrated micro-fabricated device for dual microdialysis and online ESI–ion trap mass spectrometry for analysis of complex biological samples. *Analytical Chemistry*, *71*, 1485–1490.

[225] Xu, N., Lin, Y., Hofstadler, S.A., Matson, D., Call, C.J., Smith, R.D. (1998). A microfabricated dialysis device for sample cleanup in electrospray ionization mass spectrometry. *Analytical Chemistry*, *70*, 3553–3556.

[226] Yang, L., Lee, C.S., Hofstadler, S.A., Smith, R.D. (1998). Characterization of microdialysis acidification for capillary isoelectric focusing–microelectrospray ionization mass spectrometry. *Analytical Chemistry*, *70*, 4945–4950.

[227] Jiang, Y., Wang, P.-C., Locascio, L.E., Lee, C.S. (2001). Integrated plastic micro-fluidic devices with ESI–MS for drug screening and residue analysis. *Analytical Chemistry*, *73*, 2048–2053.

[228] Herrera, A.M., Scott, D.O., Lunte, C.E. (1990). Microdialysis sampling for determination of plasma protein binding of drugs. *Pharmaceutical Research*, *7*, 1077–1081.

[229] Lunte, C.E., Scott, D.O., Herrera, A.M. (1991). Determination of drug binding to proteins by microdialysis perfusion sampling. *Current Separations*, *10*, 41–44.

4

MONITORING DOPAMINE IN THE MESOCORTICOLIMBIC AND NIGROSTRIATAL SYSTEMS BY MICRODIALYSIS: RELEVANCE FOR MOOD DISORDERS AND PARKINSON'S DISEASE

GIUSEPPE DI GIOVANNI

Department of Physiology and Biochemistry, University of Malta, Msida, Malta; School of Biosciences, Cardiff University, Cardiff, UK

MASSIMO PIERUCCI

Departmenr of Physiology and Biochemistry, University of Malta, Msida, Malta

VINCENZO DI MATTEO

Istituto di Ricerche Farmacologiche Consorzio Mario Negri Sud , Santa Maria Imbaro, Italy

1. INTRODUCTION

Dopamine (DA) is a versatile neurotransmitter that has a fundamental role in almost all behavioral aspects, from motor control to mood regulation, cognition, drug addiction, and reward. For this reason the pathophysiology of DAergic systems is one of the most investigated topics in neuroscience [1]. Dopamine pathways in the brain are generally divided into the well-characterized mesostriatal (nigrostriatal) system, which originates in the substantia nigra pars compacta (SNc) and projects to the dorsal striatum, and the mesocorticolimbic system, which starts in the ventral tegmental area (VTA)

Applications of Microdialysis in Pharmaceutical Science, First Edition. Edited by Tung-Hu Tsai.
© 2011 John Wiley & Sons, Inc. Published 2011 by John Wiley & Sons, Inc.

and projects to the frontal cortex and limbic areas, including the amygdala and the nucleus accumbens [2–4] (Figure 1). Degeneration of the nigrostriatal neurons results in the motor deficits of Parkinson's disease (PD), whereas dysfunction of the mesocorticolimbic system leads to various mood disorders, including depression, schizophrenia, and drug abuse [1,5,6]. These neuropsychaitric diseases are in reality multifactorial disorders, and other neurotransmitter dysfunction is likely to occur.

Serotonin (5-HT)-containing neurons originating from the medial and dorsal raphe nuclei innervate both the substantia nigra and the VTA [7]. In addition, terminal areas of the SNc and VTA receive input from 5-HT-ergic neurons originating in the raphe nuclei [8]. Thus, at the neuroanatomical levels there is a close relationship between 5-HT and DA-containing neurons, and this suggests that 5-HT could regulate the function of DA neurons via actions on midbrain DA cell bodies and on DA terminals. Microdialysis, coupled to high-performance liquid chromatography (HPLC), is an established technique for studying physiological, pharmacological, and pathological changes of a wide range of low-molecular-weight substances in the brain extracellular fluid. It is based on the evidence that a probe made of a hollow fiber permeable to solutes of low molecular weight inserted into the brain tissue mimics blood capillaries in exchanging material from and to the extracellular fluid [9,10] (Figure 2). Microdialysis has been employed over the last 25 years by several authors, primarily to study brain function and changes in levels of endogenous compounds such as neurotransmitters or metabolites [11]. Nevertheless, in central nervous system studies, reverse microdialysis has been used extensively for the study of the effects of diverse pharmacological and toxicological agents, such as antidepressants, antipsychotics, antiparkinsonians, hallucinogens, drugs of abuse, and experimental drugs, on local effects on neurotransmission at different central nuclei. Thus, the microdialysis approach has contributed largely not only to clarification of the physiological role of the serotonergic and dopaminergic neuronal systems but also to the development of therapeutic strategies for the treatment of a number of neuropsychiatric disorders. In this chapter we therefore focus on the microdialysis studies that have extensively explored the role of DA central systems in the pathophysiology of the main neuropsychiatric disorders.

2. PATHOPHYSIOLOGY OF SEROTONIN–DOPAMINE INTERACTION: IMPLICATION FOR MOOD DISORDERS

5-HT by itself is involved both directly and indirectly via actions on complex neuronal circuitry in the regulation of DA release through multiple 5-HT receptors, and plays a critical role in the development of normal and abnormal behaviors. These receptors are presently divided into seven classes (5-HT$_1$ to 5-HT$_7$), which are then subdivided into subclasses with a total of at least 14 different receptors, based on their pharmacological profiles, cDNA-deduced

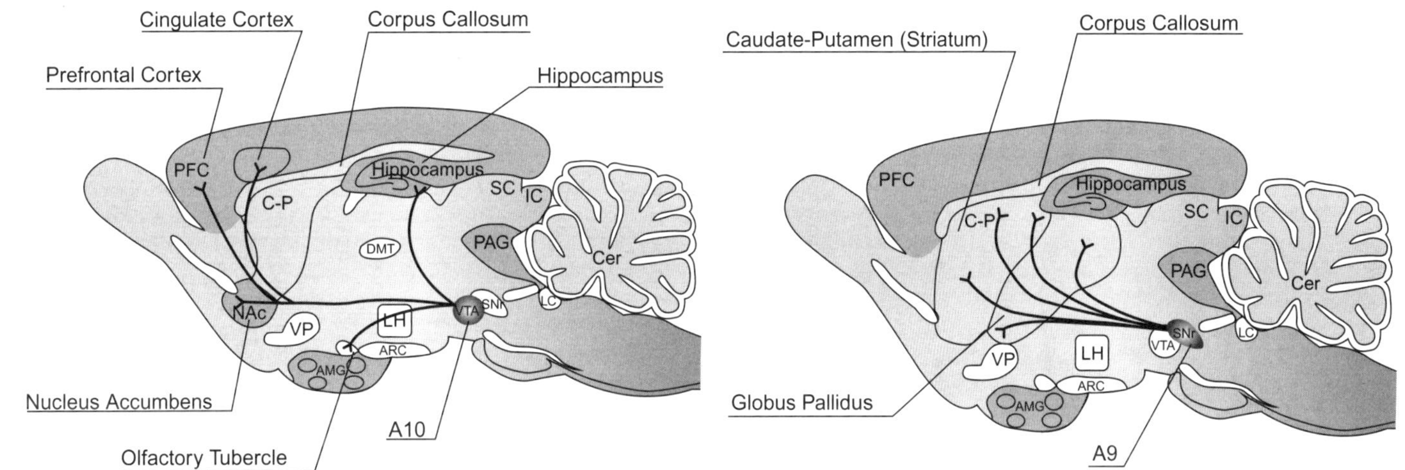

Figure 1 Midsagittal view of the rat brain DAergic systems. The mesolimbic pathway originates in the ventral tegmental area (VTA) A10 and terminates in the nucleus accumbens (NAc). The mesocortical pathway also originates in the VTA but projects to the prefrontal cortex and other cortical structures, where it is thought to regulate cognitive processes (left side). The nigrostriatal pathway originates from cell bodies, which reside in the substantia nigra pars compacta (SNc) A9 and project to the dorsal striatum (caudate-putamen) (right side). Degeneration of these neurons results in the motor deficits of Parkinson's disease.

Figure 2   "In vivo freely moving" microdialysis setup.

primary sequences, and signal transduction mechanisms [12–14]. Several 5-HT receptor subtypes, including the 5-HT_{1A}, 5-HT_{1B}, 5-HT_{2A}, 5-HT_3, and 5-HT_4 receptors, act to facilitate DA release, while the 5-HT_{2C} receptor mediates an inhibitory effect of 5-HT on DA release.

The 5-HT_{1A} receptor agonists have been shown to have complex effects on DA neurotransmission, in a region-specific manner. The selective 5-HT_{1A} receptor agonist 8-hydroxy-2-(di-n-propylamino)tetralin (8-OH-DPAT) reduced 5-HT levels and increased those of DA in the medial prefrontal cortex (mPFC) and hippocampus of rats, while the selective 5-HT_{1A} antagonist, WAY 100635 [N-[2-[4-(2-methoxyphenyl)-1-piperazinyl]ethyl]-N-(2-pyridinyl) cyclohexanecarboxamide], which had little or no effect on monoamine levels alone, suggesting that 5-HT_{1A} receptors do not have a role in the modulation of tonic DA release in mPFC, abolished the influence of 8-OH-DPAT upon 5-HT and DA levels in the same area [15–21]. A different situation emerged when stimulation of 5-HT_{1A} receptors had no effect, or decreased accumbal and striatal DA release [15–17,19,22,23].

In line with these findings, 5-{3-[((2S)-1,4-benzodioxan-2-ylmethyl)amino] propoxy}-1,3-benzodioxole (MKC-242), another potent and selective 5-HT_{1A} receptor agonist with anxiolytic and antidepressant-like effects, increased DA release in the prefrontal cortex and hippocampus but not in the striatum or nucleus accumbens. Furthermore, the 5-HT_{1A} receptor agonist–induced increase in DA release was greater in the hippocampus than in the prefrontal cortex. This region-specific effect was explained by the idea that DA outflow is modulated by postsynaptic 5-HT_{1A} receptors [24], as these are dense in the hippocampus and mPFC but are sparse in the striatum and nucleus accumbens [25]. In addition, treatment with 5,7-DHT, which destroys presynaptic serotoninergic nerve fibers, did not alter the effect of MKC-242 in increasing cortical DA release, suggesting that DA efflux is modulated in part by postsynaptic 5-HT_{1A} receptors in the prefrontal cortex [24]. The involvement of the postsynaptic 5-HT_{1A} receptors in MKC-242-induced cortical dopamine release is supported further by employing local application of 5-HT_{1A} receptor agonists and antagonists. The effect of MKC-242 in increasing cortical DA release was blocked by local application of WAY100635, and local application of 8-OH-DPAT increased DA release in the cortex. A long duration of WAY100635 or 8-OH-DPAT perfusion was required, suggesting that the site of action of these drugs was away from the position of the probe; thus, postsynaptic 5-HT_{1A} receptors modulating cortical DA release appeared to be localized on sites other than dopaminergic nerve terminals [24,26].

That the activity of DAergic neurons in the VTA and the mesocortical DA release are modulated primarily by postsynaptic 5-HT_{1A} receptors, whereas 5-HT release is both pre- and postsynaptically controlled, was confirmed in a recent study of Díaz-Mataix and co-workers [27]. Systemic administration of the highly selective 5-HT_{1A} agonist $R\text{-}(-)\text{-}2\text{-}(4\text{-}[(\text{chroman-2-ylmethyl})\text{amino}]$ butyl)-1,1-dioxobenzo[d] isothiazolone hydrochloride (BAY) increased the firing rate and bursting activity of DAergic neurons in the VTA and enhanced DA release in both VTA and mPFC in rats and mice, effects reversed by WAY 10063. Interestingly, BAY did not alter DA cell activity or DA release in the VTA of cortically transected rats, further suggesting the involvement of mPFC 5-HT_{1A} receptors. On the other hand, local application of BAY produced a biphasic effect on cortical DA release: A low concentration increased DA release and was blocked by bicuculline, a $GABA_A$ (γ-aminobutyric acid) receptor antagonist, while a higher concentration decreased DA efflux; both effects appeared to be due to the activation of 5-HT_{1A} receptors. In the first case, low BAY concentrations preferentially activated 5-HT_{1A} receptors located on GABAergic interneurons, resulting in a disinhibition of pyramidal neurons projecting to the VTA. A higher BAY concentration might overcome this effect, activating pyramidal 5-HT_{1A} receptors directly and reducing the prefrontal excitatory output to DA neurons. Further, these effects were not observed in 5-HT_{1A} knockout mice [27].

Interestingly, the superior clinical efficacy of clozapine, a prototype atypical antipsychotic drug, may be related to its ability to increase DA release selectively in the prefrontal cortex and hippocampus rather than in the nucleus accumbens and striatum. This in due in part to stimulation of the 5-HT_{1A} receptors, as demonstrated by measuring the decrease in extracellular 5-HT in the rat hippocampus, a region that has been widely used to assess activation of somatodendritic 5-HT_{1A} autoreceptors in the raphe nuclei, and by selective antagonism at 5-HT_{1A} receptors on 5-HT and DA efflux in both areas [19–21,27–34]. Like clozapine, several atypical antipsychotic drugs, such as amperozide, olanzapine, risperidone, loxapine, ziprasidone, BIMG 80 [19,20,27,29,31,35], quetiapine, iloperone, melperone [36], and aripiprazole [37–39], produced a greater increase in extracellular DA in the mPFC or hippocampus than in the nucleus accumbens and striatum, attenuated by WAY 10063. Since 8-OH-DPAT potentiated sulpiride-induced increase of DA release in mPFC and nucleus accumbens but not in the striatum [22], and ritanserin that of raclopride [40], as Ichikawa and colleagues [20] demonstrated, the combination of 5-HT_{2A} and D_2 receptor blockade increases DA release in the mPFC, via activation of 5-HT_{1A} receptors, a common feature of most atypical antipsycotics to improve negative symptoms and cognitive dysfunctions in schizophrenia.

Interestingly, 8-OH-DPAT (or amperozide) and M 100907, two $5HT_{2A}$ receptor antagonists, inhibited the ability of amphetamine to increase DA release in rat nucleus accumbens and striatum [41–43]. Thus, attenuation of stimulated DA release in the nucleus accumbens and striatum by $5\text{-}HT_{1A}$ receptor agonism and/or $5\text{-}HT_{2A}$ antagonism may contribute to reverse neuroleptic-induced catalepsy in rats. In this respect, combining antagonist/ partial agonist activity at dopamine D_2 and agonist activity at serotonin $5\text{-}HT_{1A}$ receptors is one of the approaches that has recently been chosen to develop the new generation of antipsychotics, including bifeprunox, SSR181507, and SLV313, in that $5\text{-}HT_{1A}$ receptor activation greatly reduces or prevents the cataleptogenic potential of these novel antipsychotics [21,33,44]. Furthermore, serotonergic regulation of the mesocorticolimbic DAergic pathway plays an important role in the effects of antidepressants [17]. Valproic acid, carbamazepine, and zonisamide, three anticonvulsant mood stabilizers, have been reported to increase DA release preferentially in the mPFC of rats, sharing a common mechanism of action mediated by $5\text{-}HT_{1A}$ receptor activation [45,46]. Thus, anticonvulsant mood stabilizers, as well as atypical antipsychotic drugs, would be expected to ameliorate or prevent depression, at least in part, via reversal of decreased prefrontal cortical activity by facilitating $5\text{-}HT_{1A}$ activation and its resulting increase in mPFC DA release. Moreover, a number of antidepressant agents, such as flibanserin [47], ipsapirone [48], mirtazapine [49], fluoxetine, and buspirone [24,50], all drugs showing agonistic properties at $5\text{-}HT_{1A}$ receptors, raised extracellular DA in mPFC markedly, attenuated by pretreatment with WAY 100635. Also, the combination of atypical antipsychotic drugs in addition to serotonin reuptake inhibitors has recently proven to be beneficial in a number of neuropsychiatric disorders, such as resistant depression, schizophrenia, and obsessive–compulsive disorder, as this route markedly potentiates mPFC DA release compared to that elicited by the administration of a single drug alone. Thus, activation of $5\text{-}HT_{1A}$ receptors secondary to the combined blockade of $5\text{-}HT_{2A}$ and $D_{2/3}$ receptors seems to be relevant for this action [39,50–57].

Neurochemical studies suggest the important role of the $5\text{-}HT_{1B}$ receptor in modulating the activity of mesoaccumbens and mesostriatal DAergic neurons. Evidence that activation of these receptors facilitates DA neurotransmission has also been obtained. Administration of the $5\text{-}HT_{1B}$ receptor agonist 3-(1,2,5,6-tetrahydro-4-pyridyl)-5-propoxy-pyrrolo[3,2-b]pyridine (CP 93129) into the VTA has been shown to increase DA levels in the nucleus accumbens [58–60] and concurrently decrease GABA in the VTA [59,60], both antagonized by co-infusion of N-[3-[3-(dimethylamino)ethoxy]-4-methoxyphenyl]-2′-methyl-4′-(5-methyl-1,2,4-oxadiazol-3-yl)-[1,1′-biphenyl]-4-carboxamide hydrochloride (SB 216641), a $5\text{-}HT_{1B}$ selective antagonist, but not by WAY 100635 a $5\text{-}HT_{1A}$ antagonist, or 4-[3-chlorophenyl]-α-[diphenylmethyl]-1-piperazineethanol hydrochloride (BRL 15572), a $5\text{-}HT_{1D/1A}$ antagonist [59].

Systemic 5-HT$_{1B}$ agonism also decreased VTA GABA [61] and increased DA release in the nucleus accumbens [62]. These studies suggest that 5-HT$_{1B}$ receptors within the VTA regulate mesolimbic DA activity by inhibiting GABA release. Administration of CP 93129 into the nucleus accumbens also resulted in a local increase in DA release [58]. While stimulation of accumbal 5-HT$_{1B}$ receptors phasically increased DA release, administration of the 5-HT$_{1B}$ antagonist [4-(5-methoxy-3-(4-methyl-piperazin-1-yl)phenyl]amide (GR 127935) alone into the nucleus accumbens had no effect on basal DA levels [63], indicating that accumbal 5-HT$_{1B}$ receptors do not tonically modulate mesolimbic DA release. Increases in DA levels in the dorsal striatum are also observed in response to 5-HT$_{1B}$ receptor stimulation [64–66]. These effects have been attributed to an inhibition of GABA release and a disinhibition of DA neuronal activity [67].

Less is known about 5-HT$_{1B}$ receptor regulation of the mesocortical pathway, but some studies have suggested a facilitative role. Local application of 5-HT or a 5-HT$_{1B}$ receptor agonist (CP 93129 or CP 94253) in the mPFC increased cortical DA release, which was blocked by the 5-HT$_{1B}$ receptor antagonist GR 127935 [68]. In addition, local pretreatment with GR 127935 has been shown to attenuate the increase in mPFC DA release seen in response to intracortical administration of fluoxetine. This result suggests that fluoxetine-induced increases in synaptic 5-HT levels activate 5-HT$_{1B}$ receptors and thereby act to facilitate DA release in the mPFC [69]. Other pharmacological studies have shown that the acute administration of 5-HT$_{1B}$ receptor agonists augments cocaine-evoked DA overflow within the nucleus accumbens [61]. The ability of extracellular 5-HT to facilitate mesolimbic DA release through 5-HT$_{1B}$ receptors has implications for psychostimulant abuse.

Similarly, studies have shown that systemic or intra-VTA 5-HT$_{1B}$ receptor agonism (CP 93129 and RU 24969, respectively) potentiated cocaine-induced increase in DA efflux in the nucleus accumbens and decreased GABA release in the VTA [60,61]. Dopaminergic activity from the mesolimbic pathway may then be disinhibited by the stimulation of 5-HT$_{1B}$ receptors on GABAergic projection neurons from the nucleus accumbens to the VTA, resulting in a potentiated response to cocaine. There is also evidence that VTA 5-HT$_{1B}$ receptors may be involved, in part, in mediating the activating effects of ethanol on mesolimbic DA neurons, in that activation and blockade of VTA 5-HT$_{1B}$ receptors potentiated and attenuated, respectively, the ethanol-induced increases in extracellular DA concentrations in both the VTA and the ipsilateral nucleus accumbens [70]. Taken together, these data suggest that 5-HT$_{1B}$ mesoaccumbal receptors are implicated in the effects of these abused drugs, and antagonism at 5-HT$_{1B}$ receptors may have a role in the consequent therapy.

It has been reported that stimulation of 5-HT$_{2A}$ receptors by (±)-1-(2,5-dimethoxy-4-iodophenyl)-2-aminopropane hydrochloride (DOI), a mixed 5-HT$_{2A/2C}$ receptor agonist, increased DA release in the mPFC [71–74], and in the posterior nucleus accumbens [75], an effect abolished completely by the selective 5-HT$_{2A}$ receptor antagonist R-(+)-α-(2,3-dimethoxyphenyl)-

1-[2-(4-fluorophenylethyl)]-4-piperidine-methanol (M 100907), confirming that DOI increases DA release in the mPFC primarily via 5-HT$_{2A}$ receptor stimulation. Although blockade of 5-HT$_{2A}$ receptors by itself did not significantly affect basal DA release in the mPFC [71–74,76,77], local infusion of M 100907 into the mPFC attenuated K$^+$ and DOI-induced DA release, indicating that under activated conditions, cortical 5-HT$_{2A}$ receptors potentiate the phasic DA release [73,74]. Systemic or intracortical administration of DOI also increased DAergic activity and glutamate efflux in the VTA [74,78], and infusions of M 100907 directly into the mPFC blocked this increase, suggesting that regulation of mesocortical DA by cortical 5-HT$_{2A}$ receptors may involve a polysynaptic neural circuit from mPFC to the VTA. Thus, 5-HT$_{2A}$ receptor–mediated stimulation of corticotegmental projections may result in an enhanced glutamate efflux in the VTA, which subsequently stimulates glutamate receptors on VTA mesocortical neurons, increasing DA neuronal activity and, consequently, DA release in the mPFC.

In contrast with the aforementioned data, no significant effect of DOI on basal DA release in the cortex, nucleus accumbens, and striatum was reported, although it potentiated amphetamine or 3,4-metylenedioxymethamphetamine (MDMA)-induced DA efflux in these three major DAergic terminal areas [79–81]. Thus, compelling evidence from biochemical and electrophysiological studies indicates that 5-HT$_{2A}$ receptors may modulate either independent impulse flow or dependent release of DA through mechanisms involving regulation of either DA synthesis or the DA neuron firing rate [79,82,83]. On the other hand, M 100907, which by itself did not affect DA outflow in all three areas [43,77,82,84], inhibited the ability of amphetamine and MDMA to increase DA release in the nucleus accumbens and striatum [79,82,84]. M 100907 also attenuated the *N*-methyl-D-aspartate (NMDA) receptor channel blocker dizocilpine-induced DA release in the nucleus accumbens but not in the mPFC [83]. Furthermore, infusion of the 5-HT$_{2A/2C}$ receptor antagonist ritanserin into either the striatum or the ipsilateral substantia nigra attenuated MDMA-induced DA release in the striatum. MDMA treatment also decreased GABA release in the striatum, and this was blocked by local ritanserin infusions in either brain site, indicating a role for striatal and nigral 5-HT$_{2A}$ receptors in MDMA-evoked nigrostriatal DA release, mediated by GABAergic input to the substantia nigra [85]. Systemic administration of the preferential 5-HT$_{2A}$ receptor antagonist {*trans*-4-[(3*Z*) 3-[(2-dimethylaminoethyl)oxyimino]-3-(2-fluorophenyl)propen-1-yl] phenol hemifumarate]} (SR 46349B) and ritanserin attenuated the ability of electrical stimulation of the dorsal raphe nucleus to increase DA release in the nucleus accumbens [86]. SR 46349B, devoid of any effect on basal DA release, also blocked amphetamine and haloperidol-induced DA release in both the nucleus accumbens and striatum [77,87–90].

Taken together, these results suggest that 5-HT$_{2A}$ receptor antagonism may inhibit the stimulated DA release. Conversely, 5-HT$_{2A}$ receptor antagonism had no influence on the enhancement of DA release induced by morphine in either nucleus accumbens or striatum [89]. It was shown that 5-HT$_{2A}$ and 5-HT$_{2C}$ receptors specifically regulate the activation of midbrain DA neurons induced by amphetamine or morphine, respectively [89]. This differential contribution may be conditioned by the specific mechanism of the action of the drug considered and/or by the neuronal circuitry involved in its effect on DA neurons. Therefore, the fact that drugs of abuse stimulate DA release through different cellular mechanisms leads to the possibility that their effect on DA function could be modulated differentially by each of the 5-HT$_2$ receptor subtypes [89]. Thus, amphetamine-induced DA release, which occurs independent of the DA neuron firing rate and involves an increase in DA synthesis [91,92], could be sensitive to 5-HT$_{2A}$ but not 5-HT$_{2C}$ receptor regulation. Conversely, morphine-stimulated DA release, thought to be a consequence of its excitatory effect on the DA neuron firing rate [93], could be controlled by 5-HT$_{2C}$ receptors. Moreover, it has been reported that the increased locomotor activity and accumbal DA release elicited by phencyclidine are enhanced further by the blockade of 5-HT$_{2C}$ receptors [94], while antagonism at 5-HT$_{2A}$ receptors had the opposite effect [95]. A similar picture emerges when considering the influence of these receptors on 3,4-methylenedioxymethamphetamine (MDMA, "ecstasy")–induced effects on DA neuron activity. Thus, the selective 5-HT$_{2A}$ antagonist MDL 100907 significantly reduced hyperlocomotion and stimulated DA release produced by MDMA, while the selective 5-HT$_{2C}$ antagonists SB 242084 and SB 206553 potentiated it [82,96–98].

Numerous studies have shown that compared to the typical antipsychotics haloperidol and or (−)sulpiride, the so-called "atypical antipsychotics," such as clozapine, amperozide, olanzapine, and risperidone, stimulate the release of DA more potently in the mPFC and mesocorticolimbic innervated areas than in the striatum [20,28,29,31,37,99–101]. This selective action is associated with a lower incidence of extrapyramidal side effects and with a greater ability to improve negative symptoms and cognitive functions in schizophrenia [102–105]. Since a common property of these drugs, which distinguishes them from the typical antipsychotics, is their high affinity for the 5-HT$_{2A}$ receptor, it has been suggested that potent 5-HT$_{2A}$ antagonism, in relation to a weaker DA D$_2$ receptor antagonism, contributes to their beneficial effects. Thus, pretreatment with the selective 5-HT$_{2A}$ antagonist M100907 before administration of the D$_2$ antagonists haloperidol, sulpiride, or raclopride produced an increase in mPFC DA release which was not observed when these compounds were administered alone [20,43,77,101]. Interestingly, M 100907 potentiated low- but not high-dose haloperidol-induced DA release in the mPFC and inhibited it in the nucleus accumbens [43,77]. Thus, weak D$_2$ and potent 5-HT$_{2A}$ receptor blockade may exert an important influence on the preferential increase of mPFC DA release by the atypical antipsychotics and on their clinical effectiveness.

Further, evidence has been provided that this effect may be mediated by actions of released 5-HT interacting with 5-HT$_{1A}$ receptors. In fact, reversal by WAY 100635 of the potentiation of DA release in mPFC induced by selective antagonism at 5-HT$_{2A}$ and D$_2$ receptors suggested that facilitation of 5-HT$_{1A}$ receptor stimulation is essential to the simultaneous blockade of 5-HT$_{2A}$ and D$_2$ receptors to increase cortical DA release [20,77]. Agents acting at multireceptor sites appear to be more promising as antipsychotic drugs, and recent data show that blockade of DA receptors and combined antagonism at 5-HT$_{2A}$ as well as 5-HT$_{2C}$ receptors may be involved in the therapeutic effects of novel antipsychotics [77,103,104,106]. Earlier studies demonstrated that administration of ritanserin, a mixed 5-HT$_{2A/2C}$ receptor antagonist, increased nigrostriatal and mesocorticolimbic DA efflux [107–109]. Interestingly, ritanserin, has been reported to potentiate the D$_{2/3}$ receptor antagonist raclopride-induced DA release in the mPFC and nucleus accumbens, but not in the striatum [40]. Another putative atypical antipsychotic drug, SR 46349B, which shares both 5-HT$_{2A}$ and 5-HT$_{2C}$ receptor antagonism, increased cortical DA release and potentiated haloperidol-induced DA release in both mPFC and nucleus accumbens, suggesting that 5-HT$_{2C}$ receptor antagonism may also contribute to the potentiation of DA release produced by haloperidol [77]. A novel putative atypical antipsychotic ACP-103, an inverse agonist at both 5-HT$_{2A}$ and 5-HT$_{2C}$ receptors, increased DA relese in the mPFC but not in the nucleus accumbens, and potentiated low-dose haloperidol-induced DA release in the mPFC while inhibiting release in the nucleus accumbens [110]. Taken together, these data suggest that combined 5-HT$_{2A/2C}$ receptor antagonism may be more advantageous than selective 5-HT$_{2A}$ antagonism alone as an adjunct to D$_2$ antagonism to improve both cogniton and negative symptoms in schizophrenia.

Several studies have focused on the role of 5-HT$_{2C}$ receptors in the regulation of forebrain DA function and highlighted their potential as a target for improved treatment of neuropsychiatric disorders related to central DA neuron dysfunction [111–116]. The involvement of 5-HT$_{2C}$ receptor subtypes in the control of mesocorticolimbic and nigrostriatal DA neuron activity is now well established, and evidence has been provided that they exert both tonic and phasic modulation of central dopaminergic function [71,86–89,94,117–131].

In our laboratory it was found initially that the firing rate of DA neurons in the VTA was reduced by mCPP and trifluoromethylphenylpiperazine (TFMPP), two mixed 5-HT$_{1B/2A/2B/2C}$ receptor agonists [12], whereas these neurons were stimulated by mesulergine [117]. Based on those findings, it was suggested that 5-HT could exert an inhibitory action on DA neurons in the VTA by acting through 5-HT$_2$ receptors [117]. However, these data did not allow us to distinguish the relative contribution of each 5-HT$_2$ receptor subtype in the control of central DA function. Subsequently, our and other studies clearly indicated a selective involvement of 5-HT$_{2C}$ receptors for the suppressive influence of 5-HT on the activity of mesocorticolimbic and nigrostriatal DAergic pathways. In fact, a series of in vivo electrophysiological and neurochemical studies showed that 5-methyl-1-(3-pyridylcarbamoyl)-1,2,3,

5-tetrahydropyrrolo[2,3-*f*]indole) (SB 206553), a selective 5-HT$_{2C/2B}$ receptor inverse agonist [128,132], and 6-chloro-5-methyl-l-[2-(2-methylpyridiyl-3-oxy)-pyrid-5-yl carbomoyl] indoline (SB 242084), the most potent and selective 5-HT$_{2C}$ receptor antagonist available [133], increased the basal firing rate and the bursting activity of VTA DA neurons, and enhanced DA release in both rat nucleus accumbens and prefrontal cortex [71,86,119,122,124].

Conversely, systemic administration of (*S*)-2-(chloro-5-fluoro-indo-l-yl)-l-methylethylamine 1:1 C$_4$H$_4$O$_4$ (RO 60-0175), a selective 5-HT$_{2C}$ receptor agonist [134], had the opposite effects [119–122,124]. SB 206553 and SB 242084 were also found to potentiate pharmacological-induced accumbal DA release [89,94,129] and stress-stimulated DA outflow in the rat prefrontal cortex [126], while stimulation of 5-HT$_{2C}$ receptors by RO 60-0175 in the VTA suppressed it [126], suggesting a role for these receptors in evoked accumbal and mPFC DA release as well. On the other hand, 5-HT$_{2C}$ receptor agonists such as mCPP, MK 212 [6-chloro-2-(1-piperazinyl)piperazine], and RO 60-0175 did not significantly affect the activity of SNc DA neurons and in vivo DA release in the striatum [119,123]. Moreover, the mixed 5HT$_{2B/2C}$ antagonist SB 206553 caused only a slight increase in the basal activity of DA neurons in the nigrostriatal pathway [122], suggesting that the serotonergic system controls both basal and stimulated impulse flow–dependent release of DA preferentially in the mesocorticolimbic system by acting through 5-HT$_{2C}$ receptors.

A study carried out in our laboratory has shown consistently that mCPP excites non-DA (presumably GABA-containing) neurons in both the SNr and the VTA by activating 5-HT$_{2C}$ receptors [135]. One interesting finding of that study was the differential effect exerted by mCPP on subpopulations of SNr neurons. Thus, mCPP caused a marked excitation of presumed GABAergic SNr projection neurons, whereas it did not affect SNr GABA-containing interneurons that exert a direct inhibitory influence on DA neurons in the substantia nigra [135]. On the other hand, all non-DA neurons in the VTA were equally excited by mCPP. It is tempting to speculate that this differential response to mCPP might be the basis of the preferential inhibitory effect of 5-HT$_{2C}$ agonists on the mesocorticolimbic versus the nigrostriatal DA function. Other in vivo electrophysiological and neurochemical studies have confirmed and extended the aforementioned data, indicating that 5-HT exerts a direct excitatory effect on GABAergic neurons in the substantia nigra pars reticulata and VTA by acting on 5-HT$_{2C}$ receptors [136,137]. In fact, about 50% of SNr neurons are excited by the selective 5-HT$_{2C}$ receptors agonist RO 60-0175, and this effect is counteracted by the new and selective 5-HT$_{2C}$ inverse agonist SB 243213 (5-methyl-1-[[-2-[(2-methyl-3-pyridyl)oxy]-5-pyridyl] carbamoyl]-6-trifluoromethylindoline hydrochloride) [138,139].

In addition, microiontophoretic application of RO 60-0175 clearly showed a direct effect of the 5-HT$_{2C}$ receptors on the SNr neurons, antagonized by SB 243213. Infusion of RO 60-0175 and mCPP by reverse dialysis significantly increased extracellular levels of GABA in the SNr [137]. Nevertheless, intra-VTA infusion of SB 206553 has been shown to attenuate MDMA-induced increase GABA levels in the VTA and to potentiate the concurrent increase in accumbal DA release [136]. Although recent studies showed that systemic administration of 5-HT$_{2C}$ receptor agonists, including RO 60-0175, does not significantly decrease the activity of nigrostriatal DAergic neurons [119,123], such treatment decreases DA efflux in the striatum [124,127,129], while systemic administration of SB 206553 and SB 242084 enhance it [86,89,122,129]. A recent study has shown that the 5-HT$_{2C}$ receptor inverse agonist–induced increase in accumbal and striatal DA release is insensitive to the depletion of extracellular 5-HT, suggesting that constitutive activity of the 5-HT$_{2C}$ receptors participates in the tonic inhibitory control that they exert upon DA release in both the nucleus accumbens and striatum [128]. Furthermore, biochemical evidence indicates that both VTA and accumbal 5-HT$_{2C}$ receptors participate in the phasic inhibitory control exerted by central 5-HT$_{2C}$ receptors on meso-accumbens DA neurons [130,131], and that the nucleus accumbens shell region constitutes the major site for the expression of the tonic inhibitory control involving the constitutive activity of 5-HT$_{2C}$ receptors [130]. There is also evidence that 5-HT$_{2C}$ receptors can modulate the phasic activity of the DAergic nigrostriatal system. Indeed, SB 206553 has been shown to potentiate cocaine-, morphine-, and haloperidol-induced increases of DA outflow in the rat striatum [89,129,140] and systemic administration of RO 60-0175 was found to attenuate haloperidol-induced DA release in the same area [129], as well as nicotine-induced increase in DA activity in the nigrostriatal system [141,142].

Thus, the disinhibitory effect of SB 206553 and SB 242084 on the mesocorticolimbic DA system might open new possibilities for the employment of 5-HT$_{2C}$ receptor antagonists as antidepressants [71,86,112,118,119,122,124,143,144]. This hypothesis is consistent with the suggestion that 5-HT$_{2C}$ receptor blockers might exert antidepressant activity [111,112,114,143–145] and that the disinhibition of the mesocorticolimbic DA system underlies the mechanism of action of several antidepressant drugs [143,144,146–148]. In this respect it is interesting to note that several antidepressant drugs have been shown to bind with submicromolar affinity to 5-HT$_{2C}$ receptors in the pig brain and to antagonize mCPP-induced penile erections in rats, an effect mediated through the stimulation of central 5-HT$_{2C}$ receptors [111,149,150]. Based on those findings, Di Matteo et al. [144] have carried out experiments showing that acute administration of amitriptyline and mianserin, two antidepressants with high affinity for 5-HT$_{2C}$ receptors, enhances DA release in the rat nucleus accumbens by blocking these receptor subtypes, in addition to their other pharmacological properties.

Interestingly, amitriptyline and mianserin have been tested in the chronic, mild, stress-induced anhedonia model of depression and were found to be effective in reversing the stress effects [151,152]. The ability of antidepressants such as tricyclics, SSRIs, and mianserin to affect DA systems via indirect mechanisms was also reported by studies of Tanda et al. [17,153] suggesting that potentiation of DA release in the rat cortex may play a role in the therapeutic action of antidepressants. Two different antidepressant treatments [72-h

REM (rapid eye movement) sleep deprivation and 10-day administration of moclobemide, a reversible inhibitor of monoamine oxidase type A] resulted in a reduction of this 5-HT_{2C} receptor–mediated function [154]. This was interpreted as an indication that the 5-HT_{2C} receptor may be altered, and presumably may exist in a dysregulated (hypersensitive) state in depressive illness. Thus, adaptive processes resulting from chronic antidepressant treatment (i.e., desensitization and/or downregulation of 5-HT_{2C} receptors) may play an important role in reversing the 5-HT_{2C} receptor system supersensitivity resulting from a depressive state [111,155].

In contrast to most other receptors, 5-HT_{2C} is not classically regulated. Indeed, 5-HT_{2C} receptors appear not only to decrease their responsiveness upon chronic agonist stimulation, but also, paradoxically, after chronic treatment with antagonists [156,157]. This mechanism appears to be related to an internalization process that removes activated cell surface receptors from the plasma membrane involving a phosphorylation step and possible degradation in lysosomes [156]. As a large number of psychotropic drugs, including atypical antipsychotics, antidepressants, and anxiolitics, can all induce downregulation of 5-HT_{2C} receptors, it has been suggested that this receptor adaptation plays a role in the therapeutic action of these drugs [156,157]. Therefore, it is interesting to note that chronic treatment with 5-HT_2 agonists or antagonists resulted in a paradoxical downregulation at the 5-HT_{2A} and 5-HT_{2C} receptors [156–160], and it seems that the downregulation state occurring after chronic exposure to mianserin in isolated systems as well as in cell cultures is a direct receptor-mediated mechanism of this drug at these receptors [160]. Thus, the downregulating capacity of 5-HT_{2C} agonists and antagonists may play a particularly important role in treating the supersensitivity of 5-HT_{2C} receptors resulting from a depressive state [111,155,157].

The possible involvement of 5-HT_{2C} receptors in the pathogenesis of depressive disorders and in the mode of action of antidepressants is further substantiated by several other observations. For example, by employing complementary electrophysiological and neurochemical approaches, and both acute and chronic administration routes, it was found that mirtazapine, nefazodone, and agomelatine, three effective and innovative antidepressants, elicited a robust and pronounced enhancement in the activity of mesocorticolimbic DA pathways. These actions were ascribed to their antagonistic properties at inhibitory, tonically active 5-HT_{2C} receptors, which desensitize after repeated drug administration [161–163]. Similarly, the novel benzourea derivative S32006 showed potent 5-HT_{2C} receptor antagonistic properties and enhanced the activity of mesocortical dopaminergic and adrenergic projections, displaying a broad-based profile of antidepressant properties upon acute and/or repeated administration, and exhibited both rapid and sustained anxiolytic actions [164].

Intracortical administration of the 5-HT_3 agonist 1-phenylbiguanide or *n*-methylquipazine (NMQ) has been shown to increase mPFC DA release [165,166], suggesting a facilitative role of 5-HT_3 cortical receptors on DA release. Local 1-phenylbiguanide or *m*-chlorophenylbiguanide (mCPBG) and systemic administration of the 5-HT$_3$ agonist 2-methylserotonin (2-Me-5HT) increased DA release in the nucleus accumbens also, and this effect, dependent on the impulse flow of DAergic cells [167], was blocked by locally applied BRL-43694 (granisetron), zacopride, or GR 38032F, all selective 5-HT$_3$ antagonists [167–169]. Further, this effect was also observed in 5-HT-depleted rats, suggesting that 5-HT$_3$ receptors are located presynaptically on DA terminals in the nucleus accumbens [168]. On the other hand, a role for VTA 5-HT$_3$ receptors in facilitating somatodendritic DA release in the same area has also been described [170]. Interestingly, 5-HT$_3$ antagonism by systemic ondansetron or (S)-zacopride, without affecting basal DA efflux [171–173], significantly attenuated dorsal raphe nucleus–stimulated DA release in the nucleus accumbens [173], suggesting that endogenous 5-HT, via 5-HT$_3$ receptors, may exert a facilitatory role on accumbal DA outflow.

In line with this finding, the elevation of DA release induced by the selective 5-HT uptake inhibitors fluoxetine or mazindol was attenuated by selective 5-HT$_3$ antagonism in the mPFC [174] and in the nucleus accumbens [175], respectively. Within the nigrostriatal system, intrastriatal 5-HT$_3$ antagonism by 3-tropanyl-indole-3-carboxylate (ICS 205930), MDL 72222, or ondansetron, without effect on basal DA efflux, attenuated 5-HT, or morphine-induced striatal DA release, suggesting that serotonin acts at 5-HT$_3$ receptors to facilitate DA release also in the striatum [176,177]. Furthermore, systemic treatment with a variety of 5-HT$_3$ antagonists did not affect DA efflux in the striatum, suggesting that 5-HT$_3$ receptors do not regulate tonic DA levels [177,178], but appeared to regulate evoked nigrostriatal DA release. Indeed, systemic administration of 5-HT$_3$ receptor antagonists ICS 205930, ondansetron, or MDL 72222, attenuated striatal DA release induced by morphine [177] and ethanol [179], but not that induced by haloperidol, amphetamine, or cocaine [177]. On the contrary, when haloperidol was coadministered with citalopram, known to elevate 5-HT tone, the resulting increase in DA release was attenuated by 5-HT$_3$ receptor blockade [177]. On the basis of these results, and knowing that enhanced DA release induced by cocaine and amphetamine is due to blocking or inverting the function of the DA transporter, respectively, while increased DA release by haloperidol or morphine is thought to be a consequence of an increase in DA neuron firing rate, it was suggested that 5-HT$_3$ receptors modulate nigrostriatal [177] or mesolimbic [171,175,180,181] DA function only when the stimulated DA release is depolarization-dependent and both DA and 5-HT tones are elevated concomitantly.

Significantly, 5-HT$_3$ antagonism has been shown to counteract the increase in accumbal DA release induced by various drugs of abuse, such as ethanol, nicotine, morphine, or cocaine [171,172,175,181–183]. Systemically, ICS 205930, ondansetron, and MDL 72222 attenuated morphine-induced DA release in the nucleus accumbens [171,172,181,184]; nevertheless, intra-VTA, but not intra-accumbal infusion of ICS 205930 was able to counteract the action of morphine [184], suggesting that selective antagonism on VTA 5-HT$_3$ receptors is able to modulate morphine's action in the mesolimbic system. Furthermore, intra-accumbal infusion of ondansetron strongly reduced the enhancement of DA release elicited by a high but not a low dose of morphine in the same area [181]. Therefore, it was proposed that in addition to increased DA tone, increased 5-HT release is required to trigger the excitatory action of accumbal 5-HT$_3$ receptors on DA release. Systemic pretreatment with ICS 205930 also attenuated ethanol-induced increase of DA efflux in the nucleus accumbens [171,179]. Furthermore, local infusion of ICS 205930 into the VTA [170] or the nucleus accumbens [169] prevented ethanol's action in both areas. In addition, the 5-HT$_3$ agonist mCPBG had an additive effect on DA release when infused in the nucleus accumbens concomitantly to the systemic injection of ethanol in rats [169].

There are, however, contrasting results in the case of amphetamine or cocaine. Systemic administration of the 5-HT$_3$ antagonists MDL 72222 and zacopride has been shown to attenuate both cocaine- or amphetamine-induced DA release in the nucleus accumbens [175,182,183], while other studies have shown that systemic 5-HT$_3$ receptor antagonism had no effect on DA efflux enhanced by these drugs in the same area [171,180,181]. Therefore, the fact that drugs of abuse stimulate DA release through different cellular mechanisms leads to the possibility that their effect on DA function could be modulated by the 5-HT$_3$ receptor only under specific conditions, in that it requires a concomitant increase in both endogenous DA and 5-HT tones and operates selectively on the depolarization-dependent exocytosis of DA.

The majority of studies on 5-HT$_4$ receptor control of dopamine focus on the nigrostriatal DA system. Striatal DA release has been shown to be unaltered after systemic administration of preferential 5-HT$_4$ agonists [185], but increased after local perfusion of 5-HT or 5-HT$_4$ agonists by reverse dialysis [176,186–188]. Conversely, selective 5-HT$_4$ antagonists, administered systemically [189,190] and locally, into the striatum [186,188,189,191] or in the substantia nigra [192], had no effect under basal conditions, indicating that 5-HT$_4$ receptors do not tonically modulate nigrostriatal DA, but reduce DA release under conditions of increased 5-HT and DA output. Indeed, co-perfusion in the substantia nigra of 5-HT and the 5-HT$_4$ antagonist RS 39604, blocked 5-HT-induced nigral DA efflux [192], and injection of GR 113808, another 5-HT$_4$ blocker, into the nigra attenuated the enhancement of striatal DA release induced by morphine [191]. Such a state-dependent control exerted by 5-HT$_4$ receptors seemed to be restricted to the nigrostriatal DA pathway, in that morphine-stimulated accumbal DA output was insensitive to 5-HT$_4$ antagonism [191]. Further, the increase in striatal but not accumbal DA release induced by morphine was potentiated by the 5-HT$_4$ receptor agonist prucalopride and reduced by either GR 125487 or SB 204070, two selective 5-HT$_4$ receptor antagonists [190], and GR 125487 attenuated the increase in both striatal DA release and nigral DA neuron firing induced by haloperidol, a compound known to elicit an impulse flow–dependent release of DA [189].

Interestingly, 5-HT$_4$ agents did not modulate the increase in striatal DA release induced by cocaine or amphetamine [190]. Therefore, these findings indicate that enhanced DA neuron activity, although necessary, is not sufficient per se to trigger 5-HT$_4$ receptor–mediated control. Further, these data suggest that the ability of 5-HT$_4$ receptors to control DA neuron activity might be dependent on the specific mechanism of action of a given drug to activate DA neurons, and ultimately on the means of DA release (i.e., exocytotic versus nonexocytotic). Indeed, whereas morphine and cocaine elicit an exocytotic release of DA consequent to their action on the DA neuron firing rate and DA reuptake sites, respectively, amphetamine elicits a nonexocytotic release of DA which occurs independent of DA neuron impulse flow and involves an increase in DA synthesis [190]. Thus, 5-HT$_4$ receptors selectively modulate striatal but not accumbal DA exocytosis associated with an increased DA neuron firing rate.

Interest in the 5-HT$_6$ receptor has been based, in part, on the finding that some atypical antipsychotic drugs, as well as some antidepressant agents, are relatively potent 5-HT$_6$ receptor antagonists [193,194]. However, there is considerable confusion over the extent to which 5-HT$_6$ inhibition affects dopamine transmission, especially cortical DA release. Recently published studies demonstrated that systemic administration of selective 5-HT$_6$ receptor antagonists increased dialysate levels of DA, but not 5-HT, in the rat prelimbic/infralimic subregion of the prefrontal cortex [195] and hippocampus [196]. On the contrary, other studies failed to find a direct effect of 5-HT$_6$ antagonism on cortical [196–199] and hippocampal DA efflux [198]. Also, in other DAergic regions, such as the nucleus accumbens and striatum, the selective blockade of 5-HT$_6$ receptors had no effect on basal DA release [197,198,200].

Rather, it appears that 5-HT$_6$ blockade serves to potentiate dopamine transmission from stimulatory drugs such as amphetamine [201]. For example, DA-induced outflow by peripherally administered amphetamine was enhanced by 5-HT$_6$ receptor antagonism, in the cortex, nucleus accumbens [201], or striatum [200], although more robustly in the frontal cortex. The 5-HT$_6$ receptor antagonist SB 399885 alone increased DA efflux in the hippocampus but not in the medial prefrontal cortex [196]. It potentiated cortical or hippocampal DA efflux produced by haloperidol and risperidone, suggesting that combined blockade of 5-HT$_6$, in addition to that of 5-HT$_{2A}$ and D$_2$ receptors, stimulates DA efflux in the medial prefrontal cortex, whereas combined blockade of 5-HT$_6$ and D$_2$ receptors is sufficient to enhance DA release in the hippocampus [196], supporting a possible therapeutic role for 5-HT$_6$ receptor antagonists in the treatment of cognitive dysfunctions in schizophrenia.

It is difficult to identify any physiological role for 5-HT$_7$ receptors in the modulation of central DA function because of a lack of selective ligands. On the other hand, the fact that several antipsychotic drugs have high affinity for these receptors [194] has led to the hypothesis that occupancy of 5-HT$_7$ receptors could contribute to certain therapeutic actions of several atypical antipsychotics. Thus, a combined neurochemical and behavioral study [202] found that decreases in exploration were correlated with decreases in DA and 5-HT turnover in the amygdala after systemic 5-HT$_7$ receptor inhibition in mice, suggesting a role for these receptors in the modulation of certain aspects of emotionality.

3. DOPAMINE DEPLETION IN THE NIGROSTRIATAL SYSTEM: PARKINSON'S DISEASE

The correct execution of voluntary movements is controlled by the *motor circuit* of the brain. Voluntary movements appear to be initiated at the cortical level, where there is activation of excitatory glutamatergic and inhibitory GABAergic projections to the basal ganglia [203], which process these signals, producing an output signal that returns to the cortex, through the thalamus, to modulate movement execution [203,204]. The basal ganglia are a group of related subcortical nuclei, including the neostriatum (caudate nucleus and putamen), the ventral striatum, the external and internal segments of the globus pallidus (GPe and GPi, respectively), the subthalamic nucleus (STN), and the substantia nigra pars reticulata and compacta (SNr and SNc, respectively). These structures are components of circuits involving portions of the cerebral cortex, thalamus, and brainstem [203,204].

Parkinson's disease (PD) is a progressive neurodegenerative disorder in which the ability to control voluntary movements is lost as a consequence of profound changes in the functional organization of the basal ganglia nuclei [205]. The neuropathological hallmark of PD is the selective degeneration of dopaminergic neurons in the nigrostriatal system (Figure 1) [206,207]. The striatum (or caudate-putamen) is the main input nucleus of the basal ganglia, which receives topographical excitatory projections from almost the entire cerebral cortex, especially from the sensorimotor and frontal cortex [208]. The striatum and the downstream structures in the basal ganglia are organized in topographically and functionally segregated pathways (Figure 3). The cortical inputs to the striatum are convergent, such that, for example, sensory and motor cortex areas converge into single striatal zones [209]. Close to the striatum is located the GPi and the SNr, the main output nuclei of the basal ganglia [210]. They project, via various thalamic nuclei, to most cortical areas of the frontal lobe [211]. This architecture means that the basal ganglia are part of extensive loops, basal ganglia–thalamocortical circuits, which link almost the entire cerebral cortex to the frontal lobe. The striatum projects to the output structures (GPi and SNr) by two pathways, the direct and indirect pathways. The *indirect pathway* also includes the STN. All the projections from the striatum, the GPe, GPi, and SNr release GABA and are inhibitory, whereas projections from the cortex, STN, and thalamus are excitatory, and use glutamate as their neurotransmitter [203,204]. The GABA-containing neurons in the Gpi and the SNr are tonically active; they project to the ventral tier of thalamus (ventrolateral, ventromedial, ventral anterior nuclei) and form inhibitory

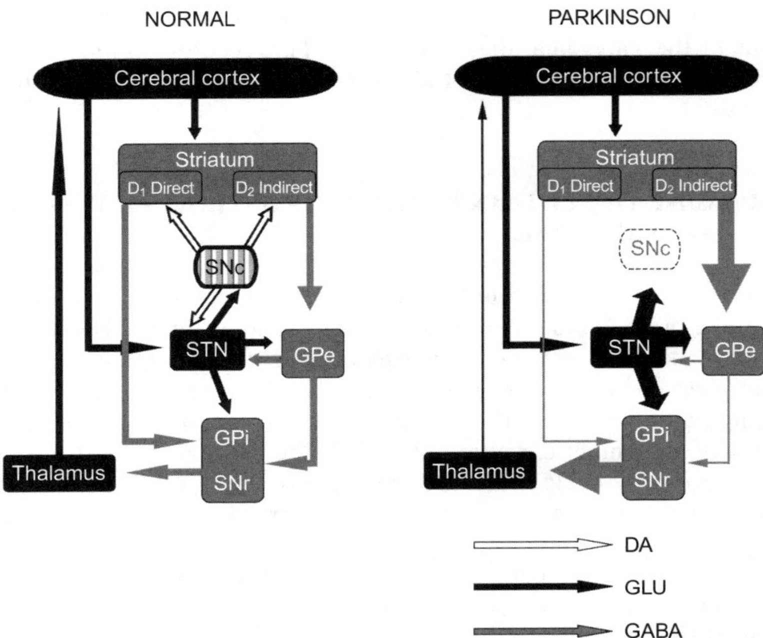

Figure 3 Simplified diagram illustrating the changes occurring in the basal ganglia functional organization in Parkinson's disease with respect to normal condition. Relative thickness of arrows indicate the degrees of activation of the transmitter pathways. The basal ganglia participate in larger circuits that also include cortex and thalamus. The striatum is the principal input structure of the basal ganglia, and the internal segment of the globus pallidus (GPi) and the substantia nigra pars reticulata (SNr) are the major output structures, projecting toward the thalamus and brainstem. According to conventional anatomical models, basal ganglia input and output structures are linked via a monosynaptic "direct" pathway and a polysynaptic "indirect" pathway that involves the external pallidal segment (GPe) and the subthalamic nucleus (STN). Dopamine released from terminals of the nigrostriatal (SNc) projection is thought to modulate basal ganglia activity by inhibiting activity along the indirect pathway through stimulation of dopamine D_2 receptors and enhancing activity along the direct pathway by the stimulation of the dopamine D_1 receptor. The same model has been applied to explain aspects of the pathophysiology of parkinsonism. Loss of striatal dopamine is believed to result in increased striatal inhibition of GPe, leading to disinhibition of STN neurons and to increased basal ganglia output from GPi and SNr. Increased and altered basal ganglia output to the thalamus is thought to disturb cortical processing, which is ultimately responsible for the development of many of the parkinsonian motor signs. Dopamine (DA): unfilled arrows; glutamate (GLU): black arrows; γ-aminobutyric acid (GABA): gray arrows. (Modified from [204].)

synaptic contacts with thalamocortical neurons that project to the motor and premotor cortex (Figure 3). Activation of the direct pathway inhibits GPi/SNr neurons, which in turn disinhibit thalamic neurons, finally resulting in excitation of the cortical neurons. Activation of the *indirect pathway* has the opposite effect, activating the GPi/SNr and thereby inhibiting the cortex [203,204,210]. In this way, the two pathways balance each other, modulating cortical activity.

Among the basal ganglia nuclei, the striatum seems to play a prominent role in determining when a given motor program should be selected and called into action. For the hypothesized function of the striatum in the selection of motor programs, a certain level of tonic DA activity is required. Dopamine projections from the SNc to the striatum modulate the activity of striatal neurons in a complex way. According to a simplified model, the striatal neurons forming the direct pathway mainly express excitatory D_1 receptors, whereas the striatal neurons in the indirect pathway have mainly inhibitory D_2 receptors. This means that DA would facilitate motor behaviors through activation of the direct pathway and, conversely, through inhibition of the indirect pathway [203,204,211,212]. Indeed, reduced DA innervation of the striatum results in hypokinesia and difficulty in initiating different motor patterns, including facial expression [213]. Dopamine depletion in PD causes a series of changes in the basal ganglia that may not begin with alterations in the striatopallidal circuits but rather with changes at the level of the subthalamic motor nuclei (i.e., glutamatergic) (Figure 3).

The capacity of tonic DA release to modulate basal ganglia circuitry permits compensation for the effects of nigral dopamine cell degeneration in the early stages of parkinsonism [214]. In the early stages, excess stimulation of GPi/SNr resulting from overactivity in the STN may be compensated for by increased inhibitory input from the GPe [214]. The functional hallmark of the parkinsonian state is an increased neuronal activity in the output nuclei of the basal ganglia, leading to excess inhibition of the thalamocortical and brainstem motor systems [210,215] (Figure 3). According to the classic pathophysiologic model of the basal ganglia, dopamine deficiency leads to the loss of the D_1-receptor-mediated stimulation of the direct pathway that becomes underactive, while the indirect pathway becomes overactive, due to the loss of a D_2-receptor-mediated inhibition, with consequently reduced inhibition of GABAergic striatal neurons in the indirect pathway and decreased facilitation of GABAergic neurons in the direct pathway [210,215]. Reduced inhibition of neurons in the indirect pathway would lead sequentially to over-inhibition of the GPe, disinhibition of the STN, and excessive glutamatergic drive of the GPi and SNr. Similarly, decreased activation of neurons in the direct pathway reduces its inhibitory influence on GPi/SNr and contributes to the excessive basal ganglia output activity (Figure 3). With continued neuro-degeneration, the striatal dopamine deficit increases, leading to excess inhibition of the GPe. In turn, hypoactivity of the GPe reduces its inhibitory output onto the GPi and STN, which now become more overtly hyperactive with consequent aggravation of the parkinsonian motor states [203,204,214]. These changes are thought to represent the neural substrate for parkinsonian motor symptoms, and microdialysis studies have largely contributed to this knowledge. Most of the progress in this field has been gained thanks to the toxin models of PD.

6-Hydroxydopamine (6-OHDA) was the first chemical agent discovered that had specific neurotoxic effects on the catecholaminergic pathways [216]. 6-OHDA uses the same catecholamine transport system as those of dopamine and norepinephrine, and produces degeneration of catecholaminergic neurons when injected stereotactically into the nigrostriatal system because it does not cross the blood–brain barrier [217,218].

Several studies have shown that in the striatum of 6-OHDA-lesioned rats, dialysate DA and its metabolite, dihydroxyphenylacetic acid (dopac), levels decrease only moderately, despite an 80 to 95% loss of dopaminergic neurons [219–224]. These authors suggested that presynaptic compensatory changes in the remaining dopaminergic neurons exist to normalize extracellular DA concentrations, and this contributes to the recovery of function. When the loss of DAergic cells exceeds the limit of 95%, the compensatory mechanisms are no longer sufficient and DA levels decrease dramatically. Although Hoffman et al. [225] observed reduced dialysate DA and dopac levels in the substantia nigra of 6-OHDA-lesioned rats, other authors reported that basal dialysate DA and dopac levels in the SN were similar to those observed in normal rats [224,226,227], as seen in the striata of lesioned rats. Compensatory mechanisms also exist in the SN to maintain extracellular DA and dopac after dopaminergic denervation. Indeed, the remaining DAergic neurons in the SN increase their rate of discharge, and electrophysiological data suggest, in contrast to the striatum, that diffusion, rather than uptake, is the most important determinant of the dopamine time course in the SN [228].

Two different pools contribute to basal DA levels in the SN: a fast sodium channel–dependent portion and a TTX-insensitive portion originating from diffusion of dopamine [227], supporting the idea that volume transmission might contribute to the nigral DA levels. In addition, reduced dopamine reuptake and catabolism with a simultaneous increase of DA synthesis in the SN [227], as in the striatum [223], contributes to the maintainance of extracellular DA and dopac after dopaminergic denervation. Interestingly, amphetamine-induced DA release in both SN [225,227] and striatum [219,220] is attenuated by the lesion, compared to the intact side. As long as the lesion size did not exceed 95%, the amphetamine-induced DA release in dopamine-depleted striatum was comparable to that in intact striatum, when the dopamine depletion was >95%; dialysate DA levels no longer increased after systemic administration of amphetamine [219,220], proving that at least 5% of nigral cells are required to maintain a relatively normal extracellular concentration of DA, necessary for any significant recovery of function. The loss of dopaminergic neurons in the SNc after 6-OHDA lesions (and in PD) is associated with an imbalance in the activity of the other structures of the basal ganglia circuity, which is reflected in changes in the release of the other neurotransmitters, particularly glutamate and GABA.

Striatal glutamate release appears to be under inhibitory control of DA [229,230]; therefore, variable and time-dependent effects on basal glutamate levels following 6-OHDA in the striatum have been found. Increased basal levels of striatal glutamate in the lesioned rats were observed by several authors [231–233]. Moreover, effects of dopamine denervation on extracellular striatal glutamate levels were reported to undergo biphasic changes: namely, an increase after three weeks and a decrease 11 weeks post-lesioning [234,237]. The contrary was reported in the SNr: a decrease after a month and an increase three months later [236]. These time-dependent changes of glutamate overflow were explained as a consequence of an initial compensatory response following the first month after the lesion that induced activation of the thalamocorticostriatal pathway, leading to a decrease in output of this pathway after three months [236]. However, some groups reported unchanged extracellular glutamate levels after 6-OHDA in the striatum and SNr [237–240]. Others have shown glutamate levels to increase two weeks post-lesioning in the SNr [241] and entopeduncular nucleus, functionally similar to the SNr, in rats [242]. Striatal GABA concentrations were either unaffected [243] or enhanced [231,232,237,244,245] after DAergic denervation. No changes occurred in extracellular levels of both glutamate and GABA in the STN [246]. SNc lesions induced a dramatic increase in extracellular glutamate level in both the SNr and the GP [244,247,248]. Such observations thus provide neurochemical confirmation of the hyperactivity of subthalamopallidal and subthalamonigral pathways induced by DA depletion. An increase in GABA levels was also measured in the GP of hemiparkinsonian rats [243,245,248]. On the other hand, no change of basal GABA concentrations in the SNr was shown: this result may be attributable to the complex GABAergic systems in the SNr arising from its anatomical features [243,247,248].

Oxidative stress may contribute to nigral cell death in Parkinson's disease. Several investigations show increased iron levels, decreased levels of reduced glutathione (GSH), and impaired mitochondrial function [249]. This leads to oxidative damage because lipid peroxidation is increased in substantia nigra, and there is a widespread increase in protein and DNA oxidation in the brain in Parkinson's disease [250]. When perfused into the striatum by reverse microdialysis, 6-OHDA is accumulated by DAergic terminals and then retrogradely transported in the cell bodies of DAergic neurons, where it is oxidized [218,251], inducing a massive release of hydroxyl free radicals ($^{\bullet}$OH) [252–254] and peroxynitrite [255]. The $^{\bullet}$OH production observed is associated with a massive parallel increase in extracellular striatal DA [252,256], likely to effect an insult to synaptic function [252]. On the other hand, this massive overflow of DA is likely to be the earliest response of DA neurons to 6-OHDA, and endogenous DA may exacerbate 6-OHDA toxicity either by aggravating oxidative stress or through direct interactions with cellular regulatory processes [252]. Furthermore, oxidation of DA by enzymatic and nonenzymatic mechanisms produces neuromelanin, which potentiates ˙OH formation when combined with iron [257]. In this early stage, 6-OHDA produces enhanced levels of oxidative DNA base modifications, leading to damage of the DNA [256], and also acts as a potent inhibitor of mitochondrial respiratory complexes I and IV [251], which then starts a sequence of toxic processes, resulting in cell death.

MPTP is another potent neurotoxin that produces selective destruction of nigrostriatal dopaminergic neurons in humans and animals and therefore elicits neurochemical and neuropathological changes that are similar to idiopathic Parkinson's disease. It seems, however, that the mouse MPTP model provides the most useful animal model of PD to study neuropathological and neurochemical changes [258]. In contrast, rats are relatively insensitive to MPTP [259], since when injected with doses of MPTP or MPP^+ comparable to those in mice, rats exhibited at least an order of magnitude less DAergic neurodegeneration [259,260]: this is due in part to a more pronounced vesicular sequestration of MPP^+ in rats [261]. MPTP itself is highly lipophilic and readily crosses the blood–brain barrier. Its toxicity is induced through conversion to MPP^+ in astrocytes by monoamine oxidase B [262], which is then actively accumulated into dopaminergic neurons [263–265]. Its specific selectivity seems to be closely related to the DA uptake system (DAT); indeed, intrastriatal infusion of MPP^+ caused an immediate and massive 300-fold release of DA in DAT+/+ mice, whereas the same concentration produced only a sixfold increase of DA output in the striatum of DAT–/– mice, demonstrating that the accessibility of the neurotoxin to dopaminergic cells via the DAT represents an absolute requirement for MPTP toxicity [265]. Once inside the neuron, MPP^+ accumulates within mitochondria, where it acts by inhibiting the electron transport system of the mitochondrial complex I (NADPH-ubiquinone oxidoreductase I), leading to an impairment in ATP production [266], an elevation of the intracellular calcium concentration, the generation of free radicals [267,268], resulting in cellular energy failure [269], and the formation of superoxide anions (O_2^-) [270]. These, coupled with the generation of both neuronal- and microglial-derived nitric oxide (NO), form peroxynitrite ($ONOO^-$), which oxidatively injures DA neurons, perhaps through DNA damage, and activation of the highly energy-consuming DNA repair enzyme poly(ADP-ribose) polymerase (PARP) promotes further ATP depletion, thus contributing to DA neuron death [271,272].

The major part of MPP^+ toxicity may be attributed to the increase in oxidative stress, as studied by microdialysis [273–275], in which free iron is involved. The damage is produced when these ˙OH react with transition metals, such as iron, with the creation of even more reactive oxygen species (ROS) [273,274,276–279]. Furthermore, acute MPP^+-induced destruction of DAergic nerve terminals or cell bodies is also associated with a massive increase in the extracellular content of DA [263,277,280–284], which itself produces toxic effects through its metabolism and autooxidation, leading to the production of hydrogen peroxide, which if not reduced by cellular mechanisms can react with transition metal such as iron to form further hydroxyl radicals, thus priming a vicious circle with the consequent cell damage [275,276,278,279,284–288]. In this respect it is interesting to note that systemic MPTP caused a significant depletion of glutathione (GSH) in both SNc and striatum of mice [289]: on the other hand, during the perfusion of MPP$^+$ into the rat striatum or SNc, GSH maintained normal basal concentrations in quantities sufficient to scavenge most $^•$OH [290]. The fact that MPTP/MPP$^+$ causes the loss of nigrostriatal GSH without corresponding increases of glutathione disulfide (GSSG) [291] suggests that factors other than excessive ROS production are also able to induce MPTP/MPP$^+$ parkinsonism. Intrastriatal infusion of MPP$^+$, however, also causes a massive release of glutamate and other excitatory amino acids [292–295]. Consequently, glutamate may exert its neurotoxic properties in a dual way: direct excitotoxicity caused by its excessive release with consequent activation of glutamate receptors [in particular of the *N*-methyl-D-aspartate (NMDA) type] that leads to an intracellular accumulation of Ca^{2+}, which in turn, initiates a cascade of alterations, resulting in the formation of ROS [293,295] as well as to a reduced intracellular glutathione synthesis [258] and indirect excitotoxicity based on impaired mitochondrial function and a cascade of events that enable normally "nontoxic levels" or lower levels of glutamate to become cytotoxic [258,296]. As a consequence, the biochemical changes due to MPTP/MPP$^+$ are reflected in marked depletion of DA and its free acid metabolites, homovanillic acid (HVA) and dopac in the nigrostriatal system [230,282,283,297], which leads to an increased release of GABA in the GPe [298,299], which in turn reduces the release of GABA in the STN [300], resulting in the disinhibition of STN and increased basal ganglia output, according to the classic pathophysiological model of the basal ganglia (Fig. 3) described above.

As already reported, dopamine autoxidation and sustained DA turnover can lead to free-radical formation, which in turn causes oxidant damage in the iron-enriched nigral neurons during senescence and in Parkinson's disease. Moreover, several compounds have demonstrated their antioxidant properties in preclinical and clinical studies. Monoamine oxidase is one of the most important enzymes in neurotransmitter metabolism. Although dopamine is metabolized by both MAO-A and MAO-B, most therapeutic attention has focused on MAO-B, as it is the predominant form in the basal ganglia, and a number of studies have suggested neuroprotective properties of MAO-B inhibitors. MAO-B regulates both the free intraneuronal concentration of DA and its releasable stores by converting it to its corresponding carboxylic acid (3,4-dihydroxyphenylacetic acid, dopac); therefore, inhibition of this metabolic pathway prolongs the activity of both endogenously and exogenously derived DA. Thus, these drugs are considered specific for the treatment of Parkinson's disease [301]. In vivo microdialysis studies on animal models of PD also suggested neuroprotective effects on DAergic neurons by inhibition of MAO activity [302,303].

On the other hand, several authors suggested that not only the inhibition of MAO-B by L-deprenil (selegiline), but also other mechanisms may contribute to its neuroprotective actions [221,304–306]. Indeed, Chiueh et al. [304] found that selegiline, in combination with the MAO-A inhibitor clorgyline, inhibited the enhanced formation of $^{\bullet}$OH induced by the 2′-methyl analog of MPTP. Similarly, Wu et al. [305] found that pretreatment with selegiline, administered into the striatum, decreased the formation of $^{\bullet}$OH elicited by intrastriatal injection of MPP^{+}. The enzyme xanthine oxidase (XO) is also thought to be a source of the superoxide anion radical that is involved in the generation of $^{\bullet}$OH species. Thus, allopurinol, an inhibitor of XO, suppressed $^{\bullet}$OH generation induced by MPP^{+} [307,308]. There is considerable evidence that intracellular iron mediates the toxicity of excess of ROS, such as superoxide anion, hydrogen peroxide, and hydroxyl free radicals to the cells, and it was reported that desferrioxamine, a strong iron(III) chelator, protects the nigrostriatal system against the neurotoxic effect of MPP^{+} [276,277]. Moreover, Obata [279] showed that the elevation of DA release by intrastriatal infusion of MPP^{+} in the presence of oxygen and iron may cause $^{\bullet}$OH generation through its extracellular autooxidation, and this may be counteracted by local application of iron(III) and not by desferrioxamine.

Interestingly, reserpine-induced DA depletion reduced MPP^{+}-stimulated $^{\bullet}$OH formation in the rat striatum [287], and several microdialysis studies demonstrated potential neuroprotective properties of DA agonists by scavenging free radicals [252,256,309]. Nevertheless, R-apomorphine has been suggested to be neuroprotective because of its effects on presynaptic DA receptors, thereby normalizing DA turnover and therefore the free-radical formation associated with its metabolism [310]. Pergolide in association with L-dopa was shown to restore striatal DA levels in hemiparkinsonian rats [311], and the coadministration of a dopamine D_2 receptor agonist such as quinpirole could reduce the excessively high dopamine levels produced after L-dopa administration, thereby possibly avoiding further neuronal degeneration [222], probably by keeping extracellular levels of DA at more "physiological" levels. In this respect, it was shown that systemic L-dopa underwent autoxidation in the striatal extracellular compartment of freely moving rats [312], promoting nonenzymatic oxidation of released DA. Also, endogenous melatonin might play an active role in maintaining the oxidative homeostasis in the striatum [313,314]; thus, coadministration of melatonin at pharmacological doses might be beneficial as an adjuvant of L-dopa therapy [314].

It is well known that angiotensin II–converting enzyme (ACE) stimulates striatal DA release and ACE inhibitors (SH-group-containing captopril and non-SH-group-containing enalaprilat or imidaprilat) may suppress DA and $^{\bullet}$OH efflux induced by *para*-nonylphenol and MPP^{+}, suggesting that the antioxidant effects of these drugs probably does not depend only on the SH group in their structure [315–317]. It is known that the angiotensin II–induced DA release is Ca^{2+}-dependent; therefore, the antioxidant effect of ACE inhibitors may also be due to the suppression of Ca^{++}-dependent release of DA [275]. Indeed, diltiazem, an L-type calcium channel antagonist, suppressed $^{\bullet}$OH generation induced by bisphenol A and MPP$^+$ [268]. As bisphenol A, *para*-nonylphenol is an environmental compound that disrupts various tissues via steroid receptor [307,318], and tamoxifen, a synthetic non-steroidal antiestrogen, was shown to have antioxidant properties in the rat striatum [288,319]. The blockage of DA oxidation by histidine, a free amino acid with singlet oxygen ($^1O_2$) scavenger activity, further prevented *para*-nonylphenol-induced $^{\bullet}$OH generation in MPP$^+$-treated rats [318]. Antioxidant activity was also found by nicotine [320] or inhibiting low-density-lipoprotein (LDL) oxidation by fluvastatin [321]. Pretreatment with sodium salicylate significantly and almost completely protected MPTP-induced striatal DA depletion, locomotor activity, and loss of nigral dopaminergic neurons [289]. Similar protective effects of aspirin were observed in the parkinsonian model induced by both MPP$^+$ and 6-OHDA [322]. Sodium salicylate and aspirin had no effects on the uptake of DA and monoamine oxidase activity in the striatum [289,322]: therefore, their protective effects against DAergic toxin-induced neurotoxicity may be based on the free-radical scavenging activity or other pharmacological properties, but not by their COX-inhibiting action [289,322].

A direct role of NO in the neurotoxic effects of MPTP has been evidenced, since pharmacological inhibition of NOS in the brain rendered neuroprotection against MPTP [272,323,324]. Pretreatment with 7-nitroindazole (7-NI) or L-NAME, neuronal inhibitors of NOS, was also reported to be protective against DAergic damage induced by in vivo infusion of MPP$^+$ in freely moving rats [325–327] by suppressing toxin-induced $^{\bullet}$OH production [325,327]. The MPP$^+$ effect was also blocked completely by the intrastriatal perfusion of dizocilpine (MK801) a NMDA ion-channel antagonist [324]. Indeed, MPP$^+$ may activate NOS by increasing glutamate release, which through NMDA channels, increases calcium influx [292] and activates constitutive neuronal NOS [324]. A growing body of evidence indicates that an increase of NMDA transmission in the striatum is the major functional consequence of the nigrostriatal DAergic degeneration during PD; thus, there are numerous reports regarding the protective effects of the NMDA receptor antagonist MK-801, or NR2B-selective NMDA antagonists, that show reduced undesired side effects with respect to MK-801 against the loss of nigrostriatal DA [292,324,328–331]. Activation of group II mGlu receptors also produced protective effects against the neurotoxic action of MPP$^+$ as a consequence of the release of brain-derived neurotrophic factor (BDNF), a protein with neurotrophic-like properties able to activate glial cells that, in turn, show scavenger activity and tissue repair [332]. Furthermore, antagonism at mGlu 5 receptors provided specific protection against dyskinesia, due to an excessive responsivenes of direct pathway neurons to L-dopa, through normalization of GABA overflow in the SNr of a rat model of PD [333].

There is also evidence that coadministration of NMDA receptor antagonists in conjunction with L-dopa has seen some success [224,235,331,334,335] in normalizing excessive striatal DA and glutamate release due to L-dopa, a cause of dyskinesias and "on–off" phenomena [224,335]. Since L-dopa is not able per se to induce dyskinesia or motor fluctuations, these side effects are due to an unregulated DA efflux into the striatal extracellular space, consequent on the lack of high-affinity reuptake by the DA plasma membrane transporter and by missing nigrostriatal autoreceptors for regulation of synthesis and DA release; these predispose degenerated neurons to these side effects, as a physiological balance of DA transmission cannot be achieved, leading to the high fluctuations of extracellular DA levels induced after repeated exposure to L-dopa [221,223,335–339].

Inhibition of the enzyme aromatic amino acid decarboxylase, which converts L-dopa to DA, by benserazide was also able to prevent or reduce these undesired side effects when coadministered with L-dopa, by inhibiting the rapid changes of extracellular DA concentration due to an administration of L-dopa alone [339–341]. In a series of studies it was demonstrated that exogenous L-dopa-derived DA is mainly stored in, and released from, serotonergic neurons when nigrostriatal neurons are denervated [342–344], and that fluoxetine attenuated L-dopa-derived extracellular DA levels in the striatum of 6-OHDA lesionded rats, due to its indirect 5-HT$_{1A}$ agonistic property [345], as also demonstrated by another study, which showed that stimulation of 5-HT$_{1A}$ receptors reduced an increase in extracellular DA derived from exogenous L-dopa [344]. Interestingly, the 5-HT$_{1A}$ agonist R-(+)-8-OHDPAT also decreased extracellular glutamate and aspartate in DA-denervated striatum of rats, showing antiparkinsonian efficacy [346].

Further, other serotonergic agents, such as 5-HT$_{2A/2C}$ receptor antagonists, may also provide an attractive nondopaminergic target for improving PD therapies [347]. Reduction of glutamate release in the SNr may represent the mechanism by which nociceptin/orfanin FQ (NOP) receptor antagonists reverse parkinsonism, because this class of compounds normalize haloperidol-evoked glutamate levels in the SNr, an effect that correlates with attenuation of akinesia [348,349], whereas NOP receptor antagonists elevate SNr GABA release [350] and nigrostriatal DA transmission [351]. Combined administration of a NOP receptor antagonist (J-113397) and L-dopa produced additive attenuation of parkinsonian-like symptoms through an increase in SNr GABA release and consequent over-inhibition of SNr GABAergic neurons projecting to the ventromedial thalamus, leading to the disinhibition of thalamocortical glutamatergic projections and locomotion [350]. This also appears relevant in humans, in which motor improvement induced by deep brain stimulation in the subthalamic nucleus was associated with lowered GABA release in the motor thalamus [352–354]. As already discussed, a balance of DA, GABA, and glutamate is critical for maintaining the normal function of basal ganglia (Figure 3).

It is important to remember that the degeneration of DAergic neurons in the SNc results in decreased DAergic activity in the striatum while increasing glutamatergic activity in the STN. This overactivity can lead to increased glutamate-mediated excitation in basal ganglia output regions, such as the SNr and GPi, which may in turn lead to reduced thalamocortical feedback (Figure 3) and the subsequent appearance of akinesia [204]. Therefore, enhancing the activity of DA in the direct pathway and reducing the activity of glutamate in the indirect pathway have been proposed as two therapeutic strategies for PD [204,243]. In this respect, iptacalim, a novel ATP-sensitive potassium channel opener, inhibited both an increase of extracellular glutamate and a decrease of extracellular DA in the lesioned-side striatum of rats [294,233,355]. Furthermore, systemic KW-6002, an A_{2A} receptor selective antagonist, increased both GABA and glutamate in the SNr [247]. Since A_{2A} receptor antagonism decreased pallidal (GPe) GABA concentrations in 6-OHDA lesioned rats [245], this decrease might induce disinhibition of pallidal GABAergic neurons, in turn increasing nigral GABA levels, via GABAergic neurons from the GPe to SNr, resulting in stimulation of thalamic glutamatergic projections to the cortex and therefore to the striatum [245,247,356]. On the other hand, an increase in nigral glutamate release by KW-6002 would be apparently inconsistent with the current standard model of basal ganglia pathophysiology [210,215]. However, it is consistent with recent analyses of deep brain stimulation effects that are accompanied by increased release of glutamate in the SNr [357,358] and recent results that show an increase of nigral glutamate release by L-dopa [359]. An interesting possibility linked to the role of A_{2A} receptors in control of movement in PD is the potential modification of the A_{2A} receptor. As suggested by Antonelli et al. [360], in 6-OHDA-lesioned rats the presence of an abnormal increase in A_{2A} signaling, which then prevails over D_2 signaling in the control of GABA release in GP, might diminish the therapeutic efficacy of L-dopa at the D_2 receptor site; therefore, blockade of the A_{2A} receptor prevailing tone could be one of the factors underlying the positive effects produced by A_{2A} antagonists in PD [360,361].

High-frequency stimulation of the subthalamic nucleus (STN-HFS) has emerged as a powerful therapeutic approach to the treatment of PD patients, and it has been suggested that STN-HFS acts by increasing striatal DA release, TH activity, and DA metabolism [362–365]. Interestingly, in DA-depleted rats, STN-HFS stabilized the L-dopa-induced increase of striatal DA levels [366], probably by modulating DA turnover, uptake, and synthesis, thus suggesting that adaptive mechanisms, such as the restoration of autoregulation for pre-synaptic DA release in the striatum, are involved in the stabilization of striatal DA concentrations potentially involved in alleviating L-dopa-related motor fluctuations [366,367]. STN-HFS also increases striatal glutamate levels in both normal and hemiparkinsonian rats, suggesting an increase in the activity of striatal excitatory glutamatergic afferents [368]. Indeed, the inhibition of basal ganglial output activity by STN-HFS [248,358] may overcome inhibition of the thalamocortical pathway activating bilateral corticostriatal projections via cortical collaterals and, probably, also the thalamostriatal pathway, as shown by the increase in extracellular GABA levels in the SNr after STN-HFS in lesioned rats [248,358]. Also microdialysis and behavioral studies showed that

GABAergic tone in SNr plays a role in the regulation of motor functions of the basal ganglia. Indeed, stimulation of locomotor activity and reduction of tremulous jaw movements, that have many of the characteristics of PD tremor, were linked to an increase in SNr GABA release [369–371].

Although current therapies for Parkinson's disease focus on treating the symptoms of the disorder, the potent dopaminergic trophic factor glial cell line–derived neurotrophic factor (GDNF) offers a promising approach potentially to repair and restore function to damaged DAergic neurons. Thus, both the induction of endogenous GDNF with long-term 1,25-dihydroxyvitamin D_3 (calcitriol) treatment [372] and the injection of vector coding for GDNF protein [373,374], or exogenous GDNF into the nigra of 6-OHDA lesioned rats [225,375] and aged or lesioned monkeys [376,377] lead to similar changes in nigrostriatal DA neuron functioning, resulting in an increase in extracellular DA and improved motor function.

Cellular therapy with stem cells also appears to be an opportunity for restoration of the DAergic function in Parkinson's disease. Indeed, the perfusate from the grafted striatum with mesencephalic DAergic neurons showed levels of DA, dopac, and HVA that were not statistically different from those of the intact contralateral striatum [378–381]. In addition, the response of DA released from these grafted DA neurons to apomorphine and nomifensine suggests that the DA-receptor-mediated autoregulatory control and the normal DA reuptake mechanism was also functioning in these cells [379], and the rise of DA release by amphetamine or dihydrokainic acid further indicates that grafted DA neurons become functionally integrated into the host circuitry [378,380,381]. More recent studies also demonstrated that embryonic mouse–derived DAergic stem cells or rat adult bone-marrow mesenchymal stem cells grafted into the DA-depleted striatum may survive and ameliorate amphetamine-induced ipsilateral rotation for a long period of time. Moreover, this functional recovery was correlated to increased extracellular levels of DA and its metabolites in the grafted striatum, as determined by microdialysis [382,383], as well as increased DA release in response to pharmacological challenges in vivo, including K^+-induced depolarization, nomifensine-induced inhibition of DA reuptake, and amphetamine-induced DA release [382] and to an increased density of DAergic markers such as TH, DAT, and vesicular monoamine transporter ($VMAT_2$) [383], suggesting that these cells show DA release and reuptake, and stimulate appropriate postsynaptic responses for long periods after implantation.

4. CONCLUSIONS

The complex network of neurotransmitter systems, which influence each other, forms the neurological substrate for the activity of the entire brain, and its dysfunction contributes to various neurological disorders. Over the last three decades, a great deal of information has been discovered about chemical neurotransmission, since the development of the ability to measure extracellular basal levels of neurotransmitters in the brain of awake animals through the microdialysis technique has made possible determination of the effects of various systemic challenges (pharmacological or physiological) to the central nervous system. Furthermore, precise introduction of drugs through the microdialysis probe has allowed for refined work on site specificity in a compound's mechanism of action. Since its introduction, microdialysis has become a popular method for the measurement of brain chemistry and is used widely in the fields of neuropharmacology, neuroanatomy, and neurophysiology. Moreover, microdialysis studies have proven to be valuable in the investigation of neurodegenerative and psychiatric disease pathology, as well as in identifying novel drugs to treat such disorders. Since dopamine has a fundamental role in almost all aspects of behavior, from motor control to mood regulation, cognition, addiction, and reward, we have provided an overview of numerous microdialysis studies on the neurochemistry of this neurotransmitter and its interactions with other systems, particularly the regulation of dopaminergic activity in the mesocorticolimbic DA pathways by serotonin, as well as the complex role of DA transmission in the basal ganglia.

Acknowledgments

The authors wish to thank Clare Austen for the English revision and Barbara Mariani for her help in preparing the manuscript.

REFERENCES

[1] Björklund, A., Dunnett, S.B. (2007). Fifty years of dopamine research. *Trends in Neurosciences*, *30*(5), 185–187.

[2] Björklund, A., Dunnett, S.B. (2007). Dopamine neuron systems in the brain: an update. *Trends in Neurosciences*, *30*(5), 194–202.

[3] Grace, A., Bunney, B. (1985). Dopamine. In: Rogawski, M.A., Barker J.L. (Eds.), *Neurotransmitter Action in the Vertebrate Nervous System*, Plenum Press, New York, pp. 285–319.

[4] Le Moal, M., Simon, H. (1991). Mesocorticolimbic dopaminergic network: functional and regulatory roles. *Physiological Reviews*, *71*, 155–234.

[5] Iversen, S.D., Iversen, L.L. (2007). Dopamine: 50 years in perspective. *Trends in Neurosciences*, *30*, 188–193.

[6] Sulzer, D. (2007). Multiple hit hypotheses for dopamine neuron loss in Parkinson's disease. *Trends in Neurosciences*, *30*, 244–250.

[7] Hervé, D., Pickel, V.M., Joh, T.H., Beaudet, A. (1987). Serotonin axon terminals in the ventral tegmental area of the rat: fine structure and synaptic input to dopaminergic neurons. *Brain Research*, *435*, 71–83.

[8] Azmitia, E.C., Segal, M. (1978). An autoradiographic analysis of the differential ascending projections of the dorsal and median raphé nuclei in the rat. *Journal of Comparative Neurology*, *179*, 641–668.

[9] Benveniste, H. (1989). Brain microdialysis. *Journal of Neurochemistry*, *52*, 1667–1679.

[10] Ungerstedt, U. (1991). Microdialysis: principles and applications for studies in animals and man. *Journal of Internal Medicine*, *230*, 365–373.

[11] Höcht, C., Opezzo, J.A., Taira, C.A. (2007). Applicability of reverse microdialysis in pharmacological and toxicological studies. *Journal of Pharmacological and Toxicological Methods*, *55*, 3–15.

[12] Hoyer, D., Clarke, D.E., Fozard, J.R., Harting, P.R., Martin, G.R., Mylecharane, E.J., Saxena, P.R., Humphrey, P.P.A. (1994). VII. International union of pharmacology classification of receptors for 5-hydroxytryptamine (serotonin). *Pharmacological Reviews*, *46*, 157–203.

[13] Hoyer, D., Hannon, J.P., Martin, G.R. (2002). Molecular, pharmacological and functional diversity of 5-HT receptors. *Pharmacology Biochemistry and Behavior*, *71*, 533–554.

[14] Barnes, N.M., Sharp, T. (1999). A review of central 5-HT receptors and their function. *Neuropharmacology*, *38*, 1083–1152.

[15] Arborelius, L., Nokimos, G.G., Hacksell, U., Svensson, T.H. (1993). (*R*)-8-OH-DPAT preferentially increases dopamine release in rat medial prefrontal cortex. *Acta Physiologica Scandinavica*, *148*, 465–466.

[16] Rasmusson, A.M., Goldstein, L.E., Deutch, A.Y., Bunney, B.S., Roth, R.H. (1994). 5-HT$_{1A}$ agonist $\pm$ 8-OH-DPATmodulates basal and stress-induced changes in medial prefrontal cortical dopamine. *Synapse*, *18*, 218–224.

[17] Tanda, G., Carboni, E., Frau, R., Di Chiara, G. (1994). Increase of extracellular dopamine in the prefrontal cortex: a trait of drugs with antidepressant potential? *Psychopharmacology*, *155*, 285–288.

[18] Gobert, A., Rivet, J.M., Audinot, C., Newman-Tancredi, A.N., Cistarelli, L., Millan, M.J. (1998). Simultaneous quantification of serotonin, dopamine and noradrenaline levels in single frontal cortex dialysates of freely-moving rats reveals a complex pattern of reciprocal auto- and heteroreceptor-mediated control of release. *Neuroscience*, *84*, 413–429.

[19] Rollema, H., Lu, Y., Schmidt, A.W., Sprouse, J., Zorn, S.H. (2000). 5-HT$_{1A}$ receptor activation contributes to ziprasidone-induced dopamine release in rat prefrontal cortex. *Biological Psychiatry*, *48*, 229–237.

[20] Ichikawa, J., Ishii, H., Bonaccorso, S., Fowler, W.L., O'Laughlin, I.A., Meltzer, H.Y. (2001). 5-HT$_{2A}$ and D$_2$ receptor blockade increases cortical DA release via 5-HT$_{1A}$ receptor activation: a possible mechanism of atypical antipsychotic-induced cortical dopamine release. *Journal of Neurochemistry*, *76*, 1521–1531.

[21] Assié, M.-B., Ravailhe, V., Faucillon, V., Newman-Tancredi, A. (2005). Contrasting contribution of 5-hydroxytryptamine 1A receptor activation to neurochemical profile of novel antipsychotics: frontocortical dopamine and hippocampal serotonin release in rat brain. *Journal of Pharmacology and Experimental Therapeutics*, *315*, 265–272.

[22] Ichikawa, J., Meltzer, H.Y. (1999). *R*(+)-8-OH-DPAT, a serotonin$_{1A}$ receptor agonist, potentiated *S*(−)-sulpiride-induced dopamine release in rat medial prefrontal cortex and nucleus accumbens but not striatum. *Journal of Pharmacology and Experimental Therapeutics*, *291*, 1227–1232.

[23] Ichikawa, J., Meltzer, H.Y. (2000). The effect of serotonin$_{1A}$ receptors on antipsychotic drug-induced dopamine release in rat striatum and nucleus accumbens. *Brain Research*, *858*, 252–263.

[24] Sakaue, M., Somboonthum, P., Nishihara, B., Koyama, Y., Hashimoto, H., Baba, A., Matsuda, T. (2000). Postsynaptic 5-hydroxytryptamine$_{1A}$ receptor activation increases in vivo dopamine release in rat prefrontal cortex. *British Journal of Pharmacology*, *129*, 1028–1034.

[25] Pompeiano, M., Palacios, J.M., Mengod, G. (1992). Distribution and cellular localization of mRNA coding for 5-HT$_{1A}$ receptor in the rat brain: correlation with receptor binding. *Journal of Neuroscience*, *12*, 440–453.

[26] Ago, Y., Koyama, Y., Baba, A., Matsuda, T. (2003). Regulation by 5-HT$_{1A}$ receptors of the in vivo release of 5-HT and DA in mouse frontal cortex. *Neuropharmacology*, *45*, 1050–1056.

[27] Díaz-Mataix, L., Scorza, M.C., Bortolozzi, A., Toth, M., Celada, P., Artigas, F. (2005). Involvement of 5-HT$_{1A}$ receptors in prefrontal cortex in the modulation of dopaminergic activity: role in atypical antipsychotic action. *Journal of Neuroscience*, *25*, 10831–10843.

[28] Moghaddam, B., Bunney, B.S. (1990). Acute effects of typical and atypical antipsychotic drugs on the release of dopamine from prefrontal cortex, nucleus accumbens, and striatum of the rat: an in vivo microdialysis study. *Journal of Neurochemistry*, *54*, 1755–1760.

[29] Volonté, M., Monferini, E., Cerutti, M., Fodritto, F., Borsini, F. (1997). BIMG 80, a novel potential antipsychotic drug: evidence for multireceptor actions and preferential release of dopamine in prefrontal cortex. *Journal of Neurochemistry*, *69*, 182–190.

[30] Millan, M.J., Gobert, A., Newman-Tancredi, A., Audinot, V., Lejeune, F., Rivet, J.M., Cussac, D., Nicolas, J.P., Müller, O., Lavielle, G. (1998). S16924 ((R)-2-{1-[2-(2,3-dihydrobenzo[1,4]dioxin-5-yloxy)-ethyl]-pyrrolidin-3yl}-1-(4-fluoro-phenyl)-ethanone), a novel, potential antipsychotic with marked serotonin (5-HT)$_{1A}$ agonist properties: I. Receptorial and neurochemical profile in comparison with clozapine and haloperidol. *Journal of Pharmacology and Experimental Therapeutics*, *286*, 1341–1355.

[31] Kuroki, T., Meltzer, H.Y., Ichikawa, J. (1999). Effects of antipsychotic drugs on extracellular dopamine levels in rat medial prefrontal cortex and nucleus accumbens. *Journal of Pharmacology and Experimental Therapeutics*, *288*, 774–781.

[32] Hagino, Y., Watanabe, M. (2002). Effects of clozapine on the efflux of serotonin and dopamine in the rat brain: the role of 5-HT$_{1A}$ receptors. *Canadian Journal of Physiology and Pharmacology*, *80*, 1158–1166.

[33] Claustre, Y., De Peretti, D., Brun, P., Gueudet, C., Allouard, N., Alonso, R., Lourdelet, J., Oblin, A., Damoiseau, G., Françon, D., et al. (2003). SSR181507, a dopamine D$_2$ receptor antagonist and 5-HT$_{1A}$ receptor agonist: I. Neurochemical and electrophysiological profile. *Neuropsychopharmacology*, *28*, 2064–2076.

[34] Chung, Y.-C., Li, Z., Dai, J., Meltzer, H.Y., Ichikawa, J. (2004). Clozapine increases both acetylcholine and dopamine release in rat ventral hippocampus: role of 5-HT$_{1A}$ receptor agonism. *Brain Research*, *1023*, 54–63.

[35] Li, Z., Ichikawa, J., Meltzer, H.Y. (2003). A comparison of the effects of loxapine with ziprasidone and thioridazine on the release of dopamine and acetylcholine

in the prefrontal cortex and nucleus accumbens. *Psychoparmacology, 167,* 315–323.

[36] Ichikawa, J., Li, Z., Dai, J., Meltzer, H.Y. (2002). Atypical antipsychotic drugs, quetiapine, iloperidone, and melperone, preferentially increase dopamine and acetylcholine release in rat medial prefrontal cortex: role of 5-HT1A receptor agonism. *Brain Research,* 956, 349–357.

[37] Li, Z., Ichikawa, J., Dai, J., Meltzer, H.Y. (2004). Aripiprazole, a novel antipsychotic drug, preferentially increases dopamine release in the prefrontal cortex and hippocampus in rat brain. *European Journal of Pharmacology, 493,* 75–83.

[38] Zocchi, A., Fabbri, D., Heidbreder, C.A. (2005). Aripiprazole increases dopamine but not noradrenaline and serotonin levels in the mouse prefrontal cortex. *Neuroscience Letters, 387,* 157–161.

[39] Bortolozzi, A., Díaz-Mataix, L., Toth, M., Celada, P., Artigas, F. (2007). In vivo actions of aripiprazole on serotonergic and dopaminergic systems in rodent brain. *Psychopharmacology, 191,* 745–758.

[40] Andersson, J.L., Nomikos, G.G., Marcus, M., Hertel, P., Mathé, J.M., Svensson, T.H. (1995). Ritanserin potentiates the stimulatory effects of raclopride on neuronal activity and dopamine release selectively in the mesolimbic dopaminergic system. *Naunyn-Schmiedeberg's Archives of Pharmacology, 352,* 374–385.

[41] Ichikawa, J., Meltzer, H.Y. (1992). Amperozide, a novel antipsychotic drug, inhibits the ability of D-amphetamine to increase dopamine release in vivo in rat striatum and nucleus accumbens. *Journal of Neurochemistry, 58,* 2285–2291.

[42] Ichikawa, J., Kuroki, T., Kitchen, M.T., Meltzer, H.Y. (1995). *R*(+)-8-OH-DPAT, a 5-HT1A receptor agonist, inhibits amphetamine-induced dopamine release in rat striatum and nucleus accumbens. *European Journal of Pharmacology, 287,* 179–184.

[43] Liégeois, J.F., Ichikawa, J., Meltzer, H.Y. (2002). 5-HT$_{2A}$ receptor antagonism potentiates haloperidol-induced dopamine release in rat medial prefrontal cortex and inhibits that in the nucleus accumbens in a dose-dependent manner. *Brain Research, 947,* 157–165.

[44] McCreary, A.C., Glennon, J.C., Ashby, C.R., Meltzer, H.Y., Li, Z., Reinders, J.H., Hesselink, M.B., Long, S.K., Herremans, A.H., van Stuivenberg, H., Feenstra, R.W., Kruse, C.G. (2007). SLV313 (1-(2,3-dihydro-benzo[1,4]dioxin-5-yl)-4-[5-(4-fluoro-phenyl)pyridin-3-ylmethyl]-piperazine monohydrochloride): a novel dopamine D$_2$ receptor antagonist and 5-HT$_{1A}$ receptor agonist potential antipsychotic drug. *Neuropsychopharmacology, 32,* 78–94.

[45] Ichikawa, J., Meltzer, H.Y. (1999). Valproate and carbamazepine increase prefrontal dopamine release by 5-HT$_{1A}$ receptor activation. *European Journal of Pharmacology, 380,* R1–R3.

[46] Ichikawa, J., Dai, J., Meltzer, H.Y. (2005). Lithium differs from anticonvulsant mood stabilizers in prefrontal cortical and accumbal dopamine release: role of 5-HT$_{1A}$ receptor agonism. *Brain Research, 1049,* 182–190.

[47] Invernizzi, R.W., Sacchetti, G., Parini, S., Acconcia, S., Samanin, R. (2003). Flibanserin, a potential antidepressant drug, lowers 5-HT and raises dopamine and noradrenaline in the rat prefrontal cortex dialysate: role of 5-HT$_{1A}$ receptors. *British Journal of Pharmacology, 139,* 1281–1288.

[48] Wędzony, K., Maćkowiak, M., Fijal, K., Golembiowska, K. (1996). Ipsapirone enhances the dopamine outflow via 5-HT$_{1A}$ receptors in the rat prefrontal cortex. *European Journal of Pharmacology, 305*, 73–78.

[49] Nakayama, K., Sakurai, T., Katsu, H. (2004). Mirtazapine increases dopamine release in prefrontal cortex by 5-HT$_{1A}$ receptor activation. *Brain Research Bulletin, 63*, 237–241.

[50] Gobert, A., Rivet, J.M., Cistarelli, L., Melon, C., Millan, M.J. (1999). Buspirone modulates basal and fluoxetine-stimulated dialysate levels of dopamine, noradrenaline and serotonin in the frontal cortex of freely moving rats: activation of serotonin$_{1A}$ receptors and blockade of alpha$_2$-adrenergic receptors underlie its actions. *Neuroscience, 93*, 1251–1262.

[51] Gobert, A., Rivet, J.M., Cistarelli, L., Millan, M.J. (1997). Buspirone enhances duloxatine- and fluoxetine-induced increases in dialysate levels of dopamine and noradrenaline, but not serotonin, in the frontal cortex of freely moving rats. *Journal of Neurochemistry, 68*, 1326–1329.

[52] Gobert, A., Millan, M.J. (1999). Modulation of dialysate levels of dopamine, noradrenaline, and serotonin (5 HT) in the frontal cortex of freely-moving rats by (–)-pindolol alone and in association with 5-HT reuptake inhibitors: comparative roles of beta-adrenergic, 5-HT$_{1A}$, and 5-HT$_{1B}$ receptors. *Neuropsychopharmacology, 21*, 268–284.

[53] Yoshino, T., Nisijima, K., Katoh, S., Yui, K., Nakamura, M. (2002). Tandospirone potentiates the fluoxetine-induced increases in extracellular dopamine via 5-HT$_{1A}$ receptors in the rat frontal cortex. *Neurochemistry International, 40*, 355–360.

[54] Yoshino, T., Nisijima, K., Shioda, K., Yui, K., Katoh, S. (2004). Perospirone, a novel atypical antipsychotic drug, potentiates fluoxetine-induced increases in dopamine levels via multireceptor actions in the rat medial prefrontal cortex. *Neuroscience Letters, 364*, 16–21.

[55] Denys, D., Klompmakers, A.A., Westenberg, H.G. (2004). Synergistic dopamine increase in the rat prefrontal cortex with the combination of quetiapine and fluvoxamine. *Psychopharmacology, 176*, 195–203.

[56] Ago, Y., Nakamura, S., Baba, A., Matsuda, T. (2005). Sulpiride in combination with fluvoxamine increases in vivo dopamine release selectively in rat prefrontal cortex. *Neuropsychopharmacology, 1*, 43–51.

[57] Huang, M., Ichiwaka, J., Li, Z., Dai, J., Meltzer, H.Y. (2006). Augmentation by citalopram of risperidone-induced monoamine release in rat prefrontal cortex. *Psychopharmacology, 185*, 274–281.

[58] Yan, Q.S., Yan, S.E. (2001). Activation of 5-HT$_{1B/1D}$ receptors in the mesolimbic dopamine system increases dopamine release from the nucleus accumbens: a microdialysis study, *European Journal of Pharmacology, 418*, 55–64.

[59] Yan, Q.S., Zheng, S.Z., Yan, S.E. (2004). Involvement of 5-HT$_{1B}$ receptors within the ventral tegmental area in regulation of mesolimbic dopaminergic neuronal activity via GABA mechanisms: a study with dual-probe microdialysis. *Brain Research, 1021*, 82–91.

[60] O'Dell, L., Parsons, L. (2004). Serotonin$_{1B}$ receptors in the ventral tegmental area modulate cocaine-induced increases in nucleus accumbens dopamine levels. *Journal of Pharmacology and Experimental Therapeutics, 311*, 711–719.

[61] Parsons, L., Koob, G.F., Weiss, F. (1999). RU24969, a 5-HT$_{1B/1A}$ receptor agonist, potentiates cocaine-induced increases in nucleus accumbens dopamine. *Synapse*, *32*, 132–135.

[62] Boulenguez, P., Rawlins, J.N.P., Chauveau, J., Joseph, M.H., Mitchell, S.N., Gray, J.A. (1996). Modulation of dopamine release in the nucleus accumbens by 5-HT$_{1B}$ agonists: involvement of the hippocampo–accumbens pathway. *Neuropharmacology*, *35*, 1521–1529.

[63] Hållbus, M., Magnusson, T., Magnusson, O. (1997). Influence of 5-HT$_{1B/1D}$ receptors on dopamine release in the guinea pig nucleus accumbens: a microdialysis study. *Neuroscience Letters*, *225*, 57–60.

[64] Benloucif, S., Keegan, M.J., Galloway, M.P. (1993). Serotonin-facilitated dopamine release in vivo: pharmacological characterization. *Journal of Pharmacology and Experimental Therapeutics*, *265*, 373–377.

[65] Galloway, M.P., Suchowski, C.S., Keegan, M.J., Hjorth, S. (1993). Local infusion of the selective 5HT-1b agonist, CP-93,128 facilitates striatal dopamine release in vivo. *Synapse*, *15*, 90–92.

[66] Bentué-Ferrer, D., Reymann, J.-M., Rousselle, J.-C., Massot, O., Bourin, M., Allain, H., Fillion, G. (1998). 5-HT-moduline, a 5-HT$_{1B/1D}$ receptor endogenous modulator, interacts with dopamine release measured in vivo by microdialysis. *European Journal of Pharmacology*, *358*, 129–137.

[67] Johnson, S.W., Mercuri, N.B., North, R.A. (1992). 5-Hydroxytryptamine$_{1B}$ receptors block the GABA$_B$ synaptic potential in rat dopamine neurons. *Journal of Neuroscience*, *12*, 2000–2006.

[68] Iyer, R.N., Bradberry, C.W. (1996). Serotonin-mediated increase in prefrontal cortex dopamine release: pharmacological characterization. *Journal of Pharmacology and Experimental Therapeutics*, *277*, 40–47.

[69] Matsumoto, M., Togashi, H., Mori, K., Ueno, K.I., Miyamoto, A., Yoshioka, M. (1999). Characterization of endogenous serotonin-mediated regulation of dopamine release in the rat prefrontal cortex. *European Journal of Pharmacology*, *383*, 39–48.

[70] Yan, Q.S., Zheng, S.Z., Feng, M.J., Yan, S.E. (2005). Involvement of 5-HT$_{1B}$ receptors within the ventral tegmental area in ethanol-induced increases in mesolimbic dopaminergic transmission. *Brain Research*, *1060*, 126–137.

[71] Gobert, A., Millan, M.J. (1999). Serotonin (5-HT)$_{2A}$ receptor activation enhances dialysate levels of dopamine and noradrenaline, but not 5-HT, in the frontal cortex of freely-moving rats. *Neuropharmacology*, *38*, 315–317.

[72] Ichikawa, J., Dai, J., Meltzer, H.Y. (2001). DOI, a 5-HT$_{2A/2C}$ receptor agonist, attenuates clozapine-induced cortical dopamine release. *Brain Research*, *907*, 151–155.

[73] Pehek, E.A., McFarlane, H.G., Maguschak, K., Price, B., Pluto, C.P. (2001). M100,907, a selective 5-HT$_{2A}$ antagonist, attenuates dopamine release in the rat medial prefrontal cortex. *Brain Research*, *888*, 51–59.

[74] Pehek, E.A., Nocjar, C., Roth, B.L., Byrd, T.A., Mabrouk, O.S. (2006). Evidence for the preferential involvement of 5-HT2A serotonin receptors in stress- and drug-induced dopamine release in the rat medial prefrontal cortex. *Neuropsychopharmacology*, *31*, 265–277.

[75] Bowers, B.J., Henry, M.B., Thielen, R.J., McBride, W.J. (2000). Serotonin 5-HT$_2$ receptor stimulation of dopamine release in the posterior but not anterior nucleus accumbens of the rat. *Journal of Neurochemistry, 75,* 1625–1633.

[76] Kuroki, T., Kawahara, T., Yonezawa, Y., Tashiro, N. (1999). Effects of the serotonin$_{2A/2C}$ receptor agonist and antagonist on phencyclidine-induced dopamine release in rat medial prefrontal cortex. *Progress in Neuro-Psychopharmacology & Biological Psychiatry, 23,* 1259–1275.

[77] Bonaccorso, S., Meltzer, H.Y., Li, Z., Dai, J., Alboszta, A.R., Ichikawa, J. (2002). SR46349-B, a 5-HT$_{2A/2C}$ receptor antagonist, potentiates haloperidol-induced dopamine release in rat medial prefrontal cortex and nucleus accumbens. *Neuropsychopharmacology, 27,* 430–441.

[78] Bortolozzi, A., Díaz-Mataix, L., Scorza, M.C., Celada, P., Antigas, F. (2005). The activation of 5-HT$_{2A}$ receptors in prefrontal cortex enhances dopaminergic activity. *Journal of Neurochemistry, 95,* 1597–1607.

[79] Gudelsky, G.A., Yamamoto, B.K., Nash, J.F. (1994). Potentiation of 3, 4 methylenedioxymethamphetamine-induced dopamine release and serotonin neurotoxicity by 5-HT$_2$ receptor agonists. *European Journal of Pharmacology, 264,* 325–330.

[80] Ichikawa, J., Meltzer, H.Y. (1995). DOI, a 5-HT$_{2A/2C}$ receptor agonist, potentiates amphetamine-induced dopamine release in rat striatum. *Brain Research, 698,* 204–208.

[81] Kuroki, T., Meltzer, H.Y., Ichikawa, J. (2003). 5-HT$_{2A}$ receptor stimulation by DOI, a 5-HT$_{2A/2C}$ receptor agonist, potentiates amphetamine-induced dopamine release in rat medial prefrontal cortex and nucleus accumbens. *Brain Research, 972,* 216–221.

[82] Schmidt, C.J., Fadayel, G.M., Sullivan, C.K., Taylor, V.L. (1992). 5-HT$_2$ receptors exert a state-dependent regulation of dopaminergic function: studies with MDL 100,907 and the amphetamine analogue, 3,4-methylenedioxymethamphetamine. *European Journal of Pharmacology, 223,* 65–74.

[83] Schmidt, C.J., Fadayel, G.M. (1996). Regional effects of MK-801 on dopamine release: effects of competitive NMDA or 5-HT$_{2A}$ receptor blockade. *Journal of Pharmacology and Experimental Therapeutics, 277,* 1541–1549.

[84] Schmidt, C.J., Sullivan, C.K., Fadayel, G.M. (1994). Blockade of striatal 5-hydroxytryptamine$_2$ receptors reduces the increase in extracellular concentrations of dopamine produced by the amphetamine analogue 3,4-methylenedioxymethamphetamine. *Journal of Neurochemistry, 62,* 1382–1389.

[85] Yamamoto, B.K., Nash, J.F., Gudelsky, G.A. (1995). Modulation of methylenedioxymethamphetamine-induced striatal dopamine release by the interaction between serotonin and gamma-aminobutyric acid in the substantia nigra. *Journal of Pharmacology and Experimental Therapeutics, 273,* 1063–1070.

[86] De Deurwaerdère, P., Spampinato, U. (1999). Role of serotonin$_{2A}$ and serotonin$_{2B/2C}$ receptor subtypes in the control of accumbal and striatal dopamine release elicited in vivo by dorsal raphe nucleus electrical stimulation. *Journal of Neurochemistry, 73,* 1033–1042.

[87] Lucas, G., De Deurwaerdère, P., Caccia, S., Spampinato, U. (2000). The effect of serotonergic agents on haloperidol-induced striatal dopamine release in vivo:

opposite role of 5-HT$_{2A}$ and 5-HT$_{2C}$ receptor subtypes and significance of the haloperidol dose used. *Neuropharmacology, 39*, 1053–1063.

[88] Lucas, G., Spampinato, U. (2000). Role of striatal serotonin$_{2A}$ and serotonin$_{2C}$ receptor subtypes in the control of in vivo dopamine outflow in the rat striatum. *Journal of Neurochemistry, 74*, 693–701.

[89] Porras, G., Di Matteo, V., Fracasso, C., Lucas, G., De Deurwaerdère, P., Caccia, S., Esposito, E., Spampinato, U. (2002). 5-HT$_{2A}$ and 5-HT$_{2C/2B}$ receptor subtypes modulate dopamine release induced in vivo by amphetamine and morphine in both the rat nucleus accumbens and striatum. *Neuropsychopharmacology, 26*, 311–324.

[90] Auclair, A., Blanc, G., Glowinski, J., Tassin, J.-P. (2004). Role of serotonin$_{2A}$ receptors in the D-amphetamine-induced release of dopamine: comparison with previous data on (1b-adrenergic receptors. *Journal of Neurochemistry, 91*, 318–326.

[91] Seiden, L.S., Sabol, K.E., Ricaurte, G.A. (1993). Amphetamine: effects on catecholamine systems and behavior. *Annual Review of Pharmacology and Toxicology, 32*, 639–677.

[92] Cadoni, C., Pinna, A., Russi, G., Consolo, S., Di Chiara, G. (1995). Role of vesicular dopamine in the in vivo stimulation of striatal dopamine transmission by amphetamine: evidence from microdialysis and Fos immunohistochemistry. *Neuroscience, 65*, 1027–1039.

[93] Di Chiara, G., North, R.A. (1992). Neurobiology of opiate abuse. *Trends in Pharmacological Sciences, 13*, 185–192.

[94] Hutson, P.H., Barton, C.L., Jay, M., Blurton, P., Burkamp, F., Clarkson, R., Bristow, L.J. (2000). Activation of mesolimbic dopamine function by phencyclidine is enhanced by 5-HT2C/2B receptor antagonists: neurochemical and behavioural studies. *Neuropharmacology, 39*, 2318–2328.

[95] Maurel-Remy, S., Bervoets, K., Millan, M.J. (1995). Blockade of phencyclidine-induced hyperlocomotion by clozapine and MDL 100,907 in rats reflects antagonism of 5-HT$_{2A}$ receptors. *European Journal of Pharmacology, 280*, R9–R11.

[96] Kehne, J.H., Ketteler, H.J., McCloskey, T.C., Sullivan, C.K., Dudley, M.W., Schmidt, C.J. (1996). Effects of the selective 5-HT$_{2A}$ receptor antagonist MDL 100,907 on MDMA-induced locomotor stimulation in rats. *Neuropsychopharmacology, 15*, 116–124.

[97] Bankson, G.M., Cunningham, K.A. (2002). Pharmacological studies of the acute effects of (+)-3,4-Methylenedioxymethamphetamine on locomotor activity: role of 5-HT$_{1B/1D}$ and 5-HT$_2$ receptors. *Neuropsychopharmacology, 26*, 40–52.

[98] Fletcher, P.J., Korth, K.M., Robinson, S.R., Baker, G.B. (2002). Multiple 5-HT receptors are involved in the effects of acute MDMA treatment: studies on locomotor activity and responding for conditioned reinforcement. *Psychopharmacology, 162*, 282–291.

[99] Nomikos, G.G., Iurlo, M., Andersson, J.L., Kimura, K., Svensson, T.H. (1994). Systemic administration of amperozide, a new atypical antipsychotic drug, preferentially increases dopamine release in the rat medial prefrontal cortex. *Psychopharmacology, 115*, 147–156.

[100] Hertel, P., Nomikos, G.G., Iurlo, M., Swensson, T.H. (1996). Risperidone: regional effects in vivo on release and metabolism of dopamine and serotonin in the rat brain. *Psychopharmacology, 124*, 74–86.

[101] Westerink, B.H.C., Kawahara, Y., De Boer, P., Geels, C., De Vries, J.B., Wikström, H.V., Van Kalkeren, A., Van Vliet, B., Kruse, C.G., Long, S.K. (2001). Antipsychotic drugs classified by their effects on the release of dopamine and noradrenaline in the prefrontal cortex and striatum. *European Journal of Pharmacology, 412*, 127–138.

[102] Meltzer, H.Y., Nash, J.F. (1991). VII. Effects of antipsychotic drugs on serotonin receptors. *Pharmacological Reviews, 43*, 587–604.

[103] Meltzer, H.Y. (1999). The role of serotonin in antipsychotic drug action. *Neuropsychopharmacology, 21*, 106S–115S.

[104] Meltzer, H.Y., Li, Z., Kaneda, Y., Ichikawa J. (2003). Serotonin receptors: their key role in drugs to treat schizophrenia. *Progress in Neuro-Psychopharmacology & Biological Psychiatry, 27*, 1159–1172.

[105] Roth, B.L., Roland, D., Ciaranello, D., Meltzer, H.Y. (1992). Binding of typical and atypical antipsychotic agents to transiently expressed 5-HT$_{1C}$ receptors. *Journal of Pharmacology and Experimental Therapeutics, 260*, 1361–1365.

[106] Jones, B.J., Blackburn, T.P. (2002). The medical benefit of 5-HT research. *Pharmacology Biochemistry and Behavior, 71*, 555–568.

[107] Devaud, L.L., Hollingsworth, E.B., Cooper, B.R. (1992). Alterations in extracellular and tissue levels of biogenic amines in rat brain induced by the serotonin$_2$ receptor antagonist, ritanserin. *Journal of Neurochemistry, 59*, 1459–1466.

[108] Pehek, E.A. (1996). Local infusion of the serotonin antagonists ritanserin or ICS 205,930 increases in vivo dopamine release in the rat medial prefrontal cortex. *Synapse, 24*, 12–18.

[109] Pehek, E.A., Bi, Y. (1997). Ritanserin administration potentiates amphetamine-stimulated dopamine release in the rat prefrontal cortex. *Progress in Neuro-Psychopharmacology & Biological Psychiatry, 21*, 671–682.

[110] Li, Z., Ichikawa, J., Huang, M., Prus, A.J., Dai, J., Meltzer, H.Y. (2005). ACP-103, a 5-HT$_{2A/2C}$ inverse agonist, potentiates haloperidol-induced dopamine release in rat medial prefrontal cortex and nucleus accumbens. *Psychopharmacology, 183*, 144–153.

[111] Jenck, F., Bös, J., Wichmann, J., Stadler, H., Martin, J.R., Moreau, J.L. (1998). The role of 5-HT$_{2C}$ receptors in affective disorders. *Expert Opinion on Investigational Drugs, 7*, 1587–1599.

[112] Di Matteo, V., De Blasi, A., Di Giulio, C., Esposito, E. (2001). Role of 5-HT$_{2C}$ receptors in the control of central dopamine function. *Trends in Pharmacological Sciences, 22*, 229–232.

[113] Higgins, G.A., Fletcher, P.J. (2003). Serotonin and drug reward: focus on 5-HT$_{2C}$ receptors. *European Journal of Pharmacology, 480*, 151–162.

[114] Giorgetti, M., Tecott, L. (2004). Contribution of 5-HT$_{2C}$ receptors to multiple action of central serotonin systems. *European Journal of Pharmacology, 488*, 1–9.

[115] Di Giovanni, G., Di Matteo, V., Pierucci, M., Benigno, A., Esposito, E. (2006). Central serotonin$_{2C}$ receptor: from physiology to pathology. *Current Topics in Medicinal Chemistry, 6*, 1909–1925.

[116] Alex, K.D., Pehek, E.A. (2007). Pharmacologic mechanisms of serotonergic regulation of dopamine neurotransmission. *Pharmacology & Therapeutics, 113*, 296–320.

[117] Prisco, S., Pagannone, S., Esposito, E. (1994). Serotonin-dopamine interaction in the rat ventral tegmental area: an electrophysiological study in vivo. *Journal of Pharmacology and Experimental Therapeutics*, *271*, 83–90.

[118] Di Matteo, V., Di Giovanni, G., Di Mascio, M., Esposito, E. (1998). Selective blockade of serotonin$_{2C/2B}$ receptors enhances dopamine release in the rat nucleus accumbens. *Neuropharmacology*, *37*, 265–272.

[119] Di Matteo, V., Di Giovanni, G., Di Mascio, M., Esposito, E. (1999). SB 242084, a selective serotonin$_{2C}$ receptor antagonist, increases dopaminergic transmission in the mesolimbic system. *Neuropharmacology*, *38*, 1195–1205.

[120] Di Matteo, V., Di Giovanni, G., Di Mascio, M., Esposito, E. (2000). Biochemical and electrophysiological evidence that RO 60–0175 inhibits mesolimbic dopaminergic function through serotonin$_{2C}$ receptors. *Brain Research*, *865*, 85–90.

[121] Millan, M.J., Dekene, A., Gobert, A. (1998). Serotonin (5-HT)$_{2C}$ receptors tonically inhibit dopamine (DA) and noradrenaline (NA), but not 5-HT release in the frontal cortex in vivo. *Neuropharmacology*, *37*, 953–955.

[122] Di Giovanni, G., De Deurwaerdère, P., Di Mascio, M., Di Matteo, V., Esposito, E., Spampinato, U. (1999). Selective blockade of serotonin$_{2C/2B}$ receptors enhances mesolimbic and mesostriatal dopaminergic function: a combined in vivo electrophysiological and microdialysis study. *Neuroscience*, *91*, 587–597.

[123] Di Giovanni, G., Di Matteo, V., Di Mascio, M., Esposito, E. (2000). Preferential modulation of mesolimbic versus nigrostriatal dopaminergic function by serotonin$_{2C/2B}$ receptor agonists: a combined in vivo electrophysiological and microdialysis study. *Synapse*, *35*, 53–61.

[124] Gobert, A., Rivet, J.M., Lejeune, F., Newman-Tancredi, A., Adhumeau-Auclair, A., Nicolas, J.-P., Cistarelli, L., Melon, C., Millan, M.J. (2000). Serotonin2C receptors tonically suppress the activity of mesocortical dopaminergic and adrenergic, but not serotonergic, pathways: a combined dialysis and electrophysiological analysis in the rat. *Synapse*, *36*, 205–221.

[125] Blackburn, T.P., Minabe, Y., Middlemiss, D.N., Shirayama, Y., Hashimoto, K., Ashby, C.R. (2002). Effect of acute and chronic administration of the selective 5-HT$_{2C}$ receptor antagonist SB-243213 on midbrain dopamine neurons in the rat: an in vivo extracellular single cell study. *Synapse*, *46*, 129–139.

[126] Pozzi, L., Acconcia, S., Ceglia, I., Invernizzi, R.W., Samanin, R. (2002). Stimulation of 5-hydroxytryptamine (5-HT2C) receptors in the ventrotegmental area inhibits stress-induced but not basal dopamine release in the rat prefrontal cortex. *Journal of Neurochemistry*, *82*, 93–100.

[127] Alex, K.D., Yavanian, G.J., McFarlane, H.G., Pluto, C.P., Pehek, E.A. (2005). Modulation of dopamine release by striatal 5-HT2C receptors. *Synapse*, *55*, 242–251.

[128] De Deurwaerdère, P., Navailles, S., Berg, K.A., Clarke, W.P., Spampinato, U. (2004). Constitutive activity of the serotonin2C receptor inhibits in vivo dopamine release in the rat striatum and nucleus accumbens. *Journal of Neuroscience*, *24*, 3235–3241.

[129] Navailles, S., De Deurwaerdère, P.D., Porras, G., Spampinato, U. (2004). In vivo evidence that 5-HT$_{2C}$ receptor antagonist but not agonist modulates cocaine-induced dopamine outflow in the rat nucleus accumbens and striatum. *Neuropsychopharmacology*, *29*, 319–326.

[130] Navailles, S., Moison, D., Ryczko, D., Spampinato, U. (2006). Region-dependent regulation of mesoaccumbens dopamine neurons in vivo by the constitutive activity of central serotonin2C receptors. *Journal of Neurochemistry*, *99*, 1311–1319.

[131] Navailles, S., Moison, D., Cunningham, K.A., Spampinato, U. (2008). Differential regulation of the mesoaccumbens dopamine circuit by serotonin2C receptors in the ventral tegmental area and the nucleus accumbens: an in vivo microdialysis study with cocaine. *Neuropsychopharmacology*, *33*, 237–246.

[132] Kennett, G.A., Wood, M.D., Bright, F., Cilia, J., Piper, D.C., Gager, T., Thomas, D.R., Baxter, G.S., Forbes, I.T., Ham, P., Blackburn, T.P. (1996). In vitro and in vivo profile of SB 206553, a potent $5\text{-}HT_{2C}/5HT_{2B}$ receptor antagonist with anxiolytic-like properties. *British Journal of Pharmacology*, *117*, 427–434.

[133] Kennett, G.A., Wood, M.D., Bright, F., Trail, B., Riley, G., Holland, V., Avenel, K.J., Stean, T., Upton, N., Bromidge, S., et al. (1997). SB 242084, a selective and brain penetrant $5\text{-}HT_{2C}$ receptor antagonist. *Neuropharmacology*, *36*, 609–620.

[134] Martin, J.R., Bös, M., Jenck, F., Moreau, J.L., Mutel, V., Sleight, A.J., Wichmann, J., Andrews, J.S., Berendsen, H.H.G., Broekkamp, C.L.E., et al. (1998). $5\text{-}HT_{2C}$ agonists: pharmacological characterisytics and therapeutical potential. *Journal of Pharmacology and Experimental Therapeutics*, *286*, 913–924.

[135] Di Giovanni, G., Di Matteo, V., La Grutta, V., Esposito, E. (2001). *m*-Chlorophenylpiperazine excites non-dopaminergic neurons in the rat substantia nigra and ventral tegmental area by activating serotonin-2C receptors. *Neuroscience*, *103*, 111–116.

[136] Bankson, M.G., Yamamoto, B.K. (2004). Serotonin–GABA interactions modulate MDMA-induced mesolimbic dopamine release. *Journal of Neurochemistry*, *91*, 852–859.

[137] Invernizzi, R.W., Pierucci, M., Calcagno, E., Di Giovanni, G., Di Matteo, V., Benigno, A., Esposito, E. (2007). Selective activation of $5\text{-}HT_{2C}$ receptors stimulates GABA-ergic function in the rat substantia nigra pars reticulata: a combined in vivo electrophysiological and neurochemical study. *Neuroscience*, *144*, 1523–1535.

[138] Wood, M.D., Reavill, C., Trail, B., Wilson, A., Stean, T., Kennett, G.A., Lightowler, S., Blackburn, T.P., Thomas, D., Gager, T.L., et al. (2001). SB-243213; a selective $5\text{-}HT_{2C}$ receptor inverse agonist with improved anxiolytic profile: lack of tolerance and withdrawal anxiety. *Neuropharmacology*, *41*, 186–199.

[139] Berg, K.A., Navailles, S., Sanchez, T.A., Silva, Y.M., Wood, M.D., Spampinato, U., Clarke, W.P. (2006). Differential effects of 5-methyl-1-[[2-[(2-methyl-3-pyridyl) oxyl]-5-pyridyl] carbamoyl]-6-trifluoromethylindone (SB 243213) on 5-hydroxytryptamine$_{2C}$ receptor–mediated responses. *Journal of Pharmacology and Experimental Therapeutics*, *319*, 260–268.

[140] Navailles, S., De Deurwaerdère, P.D., Spampinato, U. (2006). Clozapine and haloperidol differentially alter the constitutive activity of central serotonin$_{2C}$ receptors in vivo. *Biological Psychiatry*, *59*, 568–575.

[141] Di Matteo, V., Pierucci, M., Esposito, E. (2004). Selective stimulation of serotonin$_{2C}$ receptors blocks the enhancement of striatal and accumbal dopamine release induced by nicotine administration. *Journal of Neurochemistry*, *89*, 418–429.

[142] Pierucci, M., Di Matteo, V., Esposito, E. (2004). Stimulation of serotonin$_{2C}$ receptors blocks the hyperactivation of midbrain dopamine neurons induced by nicotine administration. *Journal of Pharmacology and Experimental Therapeutics*, *309*, 109–118.

[143] Di Matteo, V., Di Giovanni, G., Esposito, E. (2000). SB 242084: a selective 5-HT$_{2C}$ receptor antagonist. *CNS Drug Reviews*, *6*, 195–205.

[144] Di Matteo, V., Di Mascio, M., Di Giovanni, G., Esposito, E. (2000). Acute administration of amitriptyline and mianserin increases dopamine release in the rat nucleus accumbens: possible involvement of serotonin$_{2C}$ receptors. *Psychopharmacology*, *150*, 45–51.

[145] Baxter, G.S., Kennett, G.A., Blaney, F., Blackburn, T. (1995). 5-HT2 receptor subtypes: a family reunited? *Trends in Pharmacological Sciences*, *16*, 105–110.

[146] Cervo, L., Samanin, R. (1987). Evidence that dopamine mechanisms in the nucleus accumbens are selectively involved in the effect of desipramine in the forced swimming test. *Neuropharmacology*, *26*, 1469–1472.

[147] Cervo, L., Samanin, R. (1988). Repeated treatment with imipramine and amitriptyline reduced the immobility of rats in the swimming test by enhancing dopamine mechanisms in the nucleus accumbens. *Journal of Pharmacy and Pharmacology*, *40*, 155–156.

[148] Cervo, L., Grignaschi, G., Samanin, R. (1990). The role of the mesolimbic dopaminergic system in the desipramine effect in the forced swimming test. *European Journal of Pharmacology*, *178*, 129–133.

[149] Jenck, F., Moreau, J.L., Mutel, V., Martin, J.R., Haefely, W.E. (1993). Evidence for a role of 5-HT$_{1C}$ receptors in the antiserotonergic properties of some antidepressant drugs. *European Journal of Pharmacology*, *231*, 223–229.

[150] Jenck, F., Moreau, J.L., Mutel, V., Martin, J.R. (1994). Brain 5-HT1C receptors and antidepressants. *Progress in Neuro-Psychopharmacology & Biological Psychiatry*, *18*, 563–574.

[151] Sampson, D., Muscat, R., Willner, P. (1991). Reversal of antidepressant action by dopamine antagonists in an animal model of depression. *Psychopharmacology*, *104*, 491–495.

[152] Moreau, J.L., Bourson, A., Jenck, F., Martin, J.R., Mortas, P. (1994). Curative effects of the atypical antidepressant mianserin in the chronic mild stress-induced anhedonia model of depression. *Journal of Psychiatry & Neuroscience*, *19*, 51–56.

[153] Tanda, G., Bassareo, V., Di Chiara, G. (1996). Mianserin markedly and selectively increases extracellular dopamine in the prefrontal cortex as compared to the nucleus accumbens of the rat. *Psychopharmacology*, *123*, 127–130.

[154] Moreau, J.L., Jenck, F., Martin, J.R., Perrin, S., Haefely, W.E. (1993). Effect of repeated mild stress and two antidepressant treatments on the behavioral response to 5-HT$_{1C}$ receptor activation in rats. *Psychopharmacology*, *110*, 140–144.

[155] Moreau, J.L., Bös, M., Jenck, F., Martin, J.R., Mortas, P., Wichmann, J. (1996). 5-HT$_{2C}$ receptor agonists exhibit antidepressant—like properties in the anhedonia model of depression in rats. *European Neuropsychopharmacology*, *6*, 169–175.

[156] Van Oekelen, D., Luyten, W.H., Leysen, J.E. (2003). 5-HT$_{2A}$ and 5-HT$_{2C}$ receptors and their atypical regulation properties. *Life Sciences*, *72*, 2429–2449.

[157] Serretti, A., Artioli, P., De Ronchi, D. (2004). The 5-HT$_{2C}$ receptor as a target for mood disorders. *Expert Opinion on Therapeutic Targets*, *8*, 1–9.

[158] Barker, E.L., Sanders-Bush, E. (1993). 5-Hydroxytryptamine$_{1C}$ receptor density and mRNA levels in choroid plexus epithelial cells after treatment with mianserin and (−)-1-(4-bromo-2,5-dimethoxyphenyl)-2-aminopropane. *Molecular Pharmacology*, *44*, 725–730.

[159] Pranzatelli, M.R., Murthy, J.N., Tailor, P.T. (1993). Novel regulation of 5-HT$_{1C}$ receptors: down-regulation induced both by 5-HT$_{1C/2}$ receptor agonists and antagonists. *European Journal of Pharmacology*, *244*, 1–5.

[160] Newton, R.A., Elliott, J.M. (1997). Mianserin-induced down-regulation of human 5-hydroxytryptamine$_{2A}$ and 5-hydroxytryptamine$_{2C}$ receptors stably expressed in the human neuroblastoma cell line SH-SY5Y. *Journal of Neurochemistry*, *69*, 1031–1038.

[161] Millan, M.J., Gobert, A., Rivet, J.M., Adhumeau-Auclair, A., Cussac, D., Newman-Tancredi, A., Dekeyne, A., Nicolas, J.P., Lejeune, F. (2000). Mirtazapine enhances frontocortical dopaminergic and corticolimbic adrenergic, but not serotonergic, transmission by blockade of α_2-adrenergic and serotonin$_{2C}$ receptors: a comparison with citalopram. *European Journal of Neuroscience*, *12*, 1079–1095.

[162] Millan, M.J., Gobert, A., Lejeune, F., Dekeyne, A., Newman-Tancredi, A., Pasteau, V., Rivet, J.-M., Cussac, D. (2003). The novel melatonin agonist agomelatine (S20098) is an antagonist at 5-hydroxytryptamine$_{2C}$ receptors, blockade of which enhances the activity of frontocortical dopaminergic and adrenergic pathways. *Journal of Pharmacology and Experimental Therapeutics*, *306*, 954–964.

[163] Dremencov, E., Newman, M.E., Kinor, N., Blatman-Jan, G., Schindler, C.J., Overstreet, D.H., Yadid, G. (2005). Hyperfunctionality of serotonin-2C receptor-mediated inhibition of accumbal dopamine release in an animal model of depression is reversed by antidepressant treatment. *Neuropharmacology*, *48*, 34–42.

[164] Dekeyne, A., Mannoury la Cour, C., Gobert, A., Brocco, M., Lejeune, F., Serres, F., Sharp, T., Daszuta, A., Soumier, A., Papp, M., et al. (2008). S32006, a novel 5-HT$_{2C}$ receptor antagonist displaying broad-based antidepressant and anxiolytic properties in rodent models. *Psychopharmacology*, *19*, 549–568.

[165] Chen, J., Paredes, W., Van Praag, H.M., Lowinson, J.H., Gardner, E.L. (1992). Presynaptic dopamine release is enhanced by 5-HT$_3$ receptor activation in medial prefrontal cortex of freely moving rats. *Synapse*, *10*, 264–266.

[166] Kurata, K., Ashby, C.R., Oberlender, R., Tanii, Y., Kurakchi, M., Rini, N.J., Strecker, R.E. (1996). The characterization of the effect of locally applied *n*-methylquipazine, a 5-HT$_3$ receptor agonist, on extracellular dopamine levels in the anterior medial prefrontal cortex in the rat: an in vivo microdialysis study. *Synapse*, *24*, 313–321.

[167] Jiang, L.H., Ashby, C.R.J., Kasser R.J., Wang, R.Y. (1990). The effect of intraventricular administration of the 5-HT3 receptor agonist 2-methylserotonin on the release of dopamine in the nucleus accumbens: an in vivo chronocoulometric study. *Brain Research*, *513*, 156–160.

[168] Chen, J., van Praag, H.M., Gardner, E.L. (1991). Activation of 5-HT$_3$ receptor by 1-phenylbiguanide increases dopamine release in the rat nucleus accumbens. *Brain Research*, *543*, 354–357.

[169] Campbell, W., McBride, W.J. (1995). Serotonin-3 receptor and ethanol-stimulated dopamine release in the nucleus accumbens. *Pharmacology Biochemistry and Behavior*, *51*, 835–842.

[170] Campbell, A., Kohl, R., McBride, W. (1996). Serotonin-3 receptor and ethanol-stimulated somatodendritic dopamine release. *Alcohol*, *13*, 569–574.

[171] Carboni, E., Acquas, E., Frau, R., Di Chiara, G. (1989). Differential inhibitory effects of a 5-HT$_3$ antagonist on drug-induced stimulation of dopamine release. *European Journal of Pharmacology*, *164*, 515–519.

[172] Pei, Q., Zetterstrom, T., Leslie, R., Grahame-Smith, D. (1993). 5-HT$_3$ receptor antagonists inhibit morphine-induced stimulation of mesolimbic dopamine release and function in the rat. *European Journal of Pharmacology*, *230*, 63–68.

[173] De Deurwaerdère, P., Stinus, L., Spampinato, U. (1998). Opposite change of in vivo dopamine release in the rat nucleus accumbens and striatum that follows electrical stimulation of dorsal raphe nucleus: role of 5-HT$_3$ receptors. *Journal of Neuroscience*, *18*, 6528–6538.

[174] Tanda, G., Frau, R., Di Chiara, G. (1995). Local 5-HT$_3$ receptors mediate fluoxetine but not desipramine-induced increase of extracellular dopamine in the prefrontal cortex. *Psychopharmacology*, *119*, 15–19.

[175] Kankaanpää, A., Meririnne, E., Seppälä, T. (2002). 5-HT$_3$ receptor antagonist MDL 72222 attenuates cocaine- and mazindol-, but not methylphenidate-induced neurochemical and behavioral effects in the rat. *Psychopharmacology*, *159*, 341–350.

[176] Benloucif, S., Keegan, M.J., Galloway, M.P. (1993). Serotonin-facilitated dopamine release in vivo: pharmacological characterization. *Journal of Pharmacology and Experimental Therapeutics*, *265*, 373–377.

[177] Porras, G., De Deurwaerdère, P., Moison, D., Spampinato, U. (2003). Conditional involvement of striatal serotonin$_3$ receptors in the control of in vivo dopamine outflow in the rat striatum. *European Journal of Neuroscience*, *17*, 771–781.

[178] Invernizzi, R., Pozzi, L., Samanin, R. (1995). Selective reduction of extracellular dopamine in the rat nucleus accumbens following chronic treatment with DAU6215, a 5-HT$_3$ receptor antagonist. *Neuropharmacology*, *34*, 211–215.

[179] Wozniak, K.M., Pert, A., Linnoila, M. (1990). Antagonism of 5-HT$_3$ receptors attenuates the effects of ethanol on extracellular dopamine. *European Journal of Pharmacology*, *187*, 287–289.

[180] Cervo, L., Pozzi, L., Samanin, R. (1996). 5-HT$_3$ receptor antagonists do not modify cocaine place conditioning or the rise in extracellular dopamine in the nucleus accumbens of rats. *Pharmacology Biochemistry and Behavior*, *55*, 33–37.

[181] De Deurwaerdère, P., Moison, D., Navailles, S., Porras, G., Spampinato, U. (2005). Regionally and functionally distinct serotonin3 receptors control in vivo dopamine outflow in the rat nucleus accumbens. *Journal of Neurochemistry*, *94*, 140–149.

[182] McNeish, C.S., Svingos, A.L., Hitzemann, R., Strecker, R.E. (1993). The 5-HT$_3$ antagonist zacopride attenuates cocaine-induced increases in extracellular dopamine in rat nucleus accumbens. *Pharmacology Biochemistry and Behavior*, *45*, 759–763.

[183] Kankaanpää, A., Lillsunde, P., Ruotsalainen, M., Ahtee, L., Seppälä, T. (1996). 5-HT$_3$ receptor antagonist MDL 72222 dose-dependently attenuates

cocaine- and amphetamine-induced elevations of extracellular dopamine in the nucleus accumbens and the dorsal striatum. *Pharmacology & Toxicology, 78,* 317–321.

[184] Imperato, A., Angelucci, L. (1989). 5-HT$_3$ receptors control dopamine release in the nucleus accumbens of freely moving rats. *Neuroscience Letters, 101,* 214–217.

[185] Taylor, S.G., Routledge, C. (1996). Lack of effect of systemically administered 5-HT$_4$ agonists on dopamine levels measured from the nucleus accumbens and striatum: an in vivo microdialysis study in freely-moving rats. *British Journal of Pharmacology, 118,* 326P. (suppl.).

[186] Bonhomme, N., De Deurwaerdère, P., Le Moal, M., Spampinato, U. (1995). Evidence for 5-HT$_4$ receptor subtype involvement in the enhancement of striatal dopamine release induced by serotonin, a microdialysis study in the halothane-anesthetized rat. *Neuropharmacology, 34,* 269–279.

[187] Steward, L.J., Ge, J., Stowe, R.L., Brown, D.C., Bufton, R.K., Stokes, P.R.A., Barnes, N.M. (1996). Ability of 5-HT$_4$ receptor ligands to modulate rat striatal dopamine release in vitro and in vivo. *British Journal of Pharmacology, 117,* 55–62.

[188] De Deurwaerdère, P., L'hirondel, M., Bonhomme, N., Lucas, G., Cheramy, A., Spampinato, U. (1997). Serotonin stimulation of 5-HT$_4$ receptors indirectly enhances in vivo dopamine release in the rat striatum. *Journal of Neurochemistry, 68,* 195–203.

[189] Lucas, G., Di Matteo, V., De Deurwaerdère, P., Porras, G., Martin-Ruiz, R., Artigas, F., Esposito, E., Spampinato, U. (2001). Neurochemical and electrophysiological evidence that 5-HT$_4$ receptors exert a state-dependent facilitatory control in vivo on nigrostriatal, but not mesoaccumbal, dopaminergic function. *European Journal of Neuroscience, 13,* 889–898.

[190] Porras, G., Di Matteo, V., De Deurwaerdère, P., Esposito, E., Spampinato, U. (2002). Central serotonin$_4$ receptors selectively regulate the impulse-dependent exocytosis of dopamine in the rat striatum: in vivo studies with morphine, amphetamine and cocaine. *Neuropharmacology, 43,* 1099–1109.

[191] Pozzi, L., Trabace, L., Invernizzi, R., Samanin, R. (1995). Intranigral GR 113808, a selective 5-HT$_4$ receptor antagonist, attenuates morphine-stimulated dopamine release in the rat striatum. *Brain Research, 692,* 265–268.

[192] Thorré, K., Ebinger, G., Michotte, Y. (1998). 5-HT$_4$ receptor involvement in the serotonin-enhanced dopamine efflux from the substantia nigra of the freely moving rat: a microdialysis study. *Brain Research, 796,* 117–124.

[193] Monsma, F.J., Jr., Shen, Y., Ward, R.P., Hamblin, M.W., Sibley, D.R. (1993). Cloning and expression of a novel serotonin receptor with high affinity for tricyclic psychotropic drugs. *Molecular Pharmacology, 43,* 320–327.

[194] Roth, B.L., Craigo, S.C., Choudhary, M.S., Uluer, A., Monsma, F.J., Jr., Shen, Y., Meltzer, H.Y., Sibley, D.R. (1994). Binding of typical and atypical antipsychotic agents to 5-hydroxytryptamine-6 and 5-hydroxytryptamine-7 receptors. *Journal of Pharmacology and Experimental Therapeutics, 268,* 1403–1410.

[195] Lacroix, L.P., Dawson, L.A., Hagan, J.J., Heidbreder, C.A. (2004). 5-HT$_6$ receptor antagonist SB-271046 enhances extracellular levels of monoamines in the rat medial prefrontal cortex. *Synapse, 51,* 158–164.

[196] Li, Z., Huang, M., Prus, A.J., Dai, J., Meltzer, H.Y. (2007). 5-HT$_6$ receptor antagonist SB-399885 potentiates haloperidol and risperidone-induced dopamine efflux in the medial prefrontal cortex or hippocampus. *Brain Research, 1134,* 70–78.

[197] Dawson, L.A., Nguyen, H.Q., Li, P. (2000). *In vivo* effects of the 5-HT$_6$ antagonist SB-271046 on striatal and frontal cortex extracellular concentrations of noradrenaline, dopamine, 5-HT, glutamate and aspartate. *British Journal of Pharmacology, 130,* 23–26.

[198] Dawson, L.A., Nguyen, H.Q., Li, P. (2001). The 5-HT$_6$ receptor antagonist SB-271046 selectively enhances excitatory neurotransmission in the rat frontal cortex and hippocampus. *Neuropsychopharmacology, 25,* 662–668.

[199] Dawson, L.A., Li, P. (2003). Effects of 5-HT$_6$ receptor blockade on the neurochemical outcome of antidepressant treatment in the frontal cortex of the rat. *Journal of Neural Transmission, 110,* 577–590.

[200] Dawson, L.A., Nguyen, H.Q., Li, P. (2003). Potentiation of amphetamine-induced changes in dopamine and 5-HT by a 5-HT$_6$ receptor antagonist. *Brain Research Bulletin, 59,* 513–521.

[201] Frantz, K.J., Hansson, K.J., Stouffer, D.G., Parsons, L.H. (2002). 5-HT$_6$ receptor antagonism potentiates the behavioral and neurochemical effects of amphetamine but not cocaine. *Neuropharmacology, 42,* 170–180.

[202] Takeda, H., Tsuji, M., Ikoshi, H., Yamada, T., Masuya, J., Iimori, M., Matsumiya, T. (2005). Effects of a 5-HT$_7$ receptor antagonist DR4004 on the exploratory behavior in a novel environment and on brain monoamine dynamics in mice. *European Journal of Pharmacology, 518,* 30–39.

[203] Wichmann, T., DeLong, M.R. (2006). Neurotransmitters and disorders of the basal ganglia. In: Georg, J., Sieghel, R., Albers, W., Brady, S.T., Donald, L. (Eds.), *Basic Neurochemistry: Molecular, Cellular and Medical Aspects,* 7th ed., Academic Press, San Diego, CA.

[204] Blandini, F., Nappi, G., Tassorelli, C., Martignoni, E. (2000). Functional changes of the basal ganglia circuitry in Parkinson's disease. *Progress in Neurobiology, 62,* 63–88.

[205] Ehringer, H., Hornykiewicz, O. (1960). Distribution of noradrenaline and dopamine (3-hydroxytyramine) in the human brain and their behavior in diseases of the extrapyramidal system. *Klinische Wochenschrift, 15,* 1236–1239.

[206] Jellinger, K. (1989). Pathology of Parkinson's disease. In: Calne, D.B. (Ed.), *Handbook of Experimental Pharmacology,* Springer-Verlag, Berlin.

[207] Scherman, D., Desnos, C., Darchen, F., Pollak, P., Javoy-Agid, F., Agid, Y. (1989). Striatal dopamine deficiency in Parkinson's disease: role of aging. *Annals of Neurology, 26,* 551–557.

[208] Parent, A. (1996). *Carpenter's Human Neuroanatomy,* 9th ed., Williams & Wilkins, Baltimore.

[209] Flaherty, A.W., Graybiel, A.M. (1991). Corticostriatal transformations in the primate somatosensory system: projections from physiologically mapped body-part representations. *Journal of Neurophysiology, 66,* 1249–1263.

[210] DeLong, M.R. (1990). Primate models of movement disorders of basal ganglia origin. *Trends in Neurosciences, 13,* 281–285.

[211] Alexander, G.E., Crutcher, M.D., DeLong, M.R. (1990). Basal ganglia-thalamocortical circuits: parallel substrates for motor, oculomotor, "prefrontal" and "limbic" functions. *Progress in Brain Research*, *85*, 119–146.

[212] O'Connor, W.T. (1998). Functional neuroanatomy of the basal ganglia as studied by dual-probe microdialysis. *Nuclear Medicine and Biology*, *25*, 743–746.

[213] Blair, R.J. (2003). Facial expressions, their communicatory functions and neuro-cognitive substrates. *Philosophical Transactions of the Royal Society B: Biological Sciences*, *358*, 561–572.

[214] Obeso, J.A., Rodriguez-Oroz, M., Marin, C., Alonso, F., Zamarbide, I., Lanciego, J.L., Rodriguez-Diaz, M. (2004). The origin of motor fluctuations in Parkinson's disease: importance of dopaminergic innervation and basal ganglia circuits. *Neurology*, *62*(Suppl. 1), S17–S30.

[215] Albin, R.L., Young, A.B., Penney, J.B. (1989). The functional anatomy of basal ganglia disorders. *Trends in Neurosciences*, *12*, 366–375.

[216] Ungerstedt, U. (1968). 6-Hydroxy-dopamine induced degeneration of central dopamine neurons. *European Journal of Pharmacology*, *5*, 107–110.

[217] Beal, M.F. (2001). Experimental models of Parkinson's disease. *Nature Reviews Neuroscience*, *2*, 325–334.

[218] Blum, D., Torch, S., Lambeng, N., Nissou, M.-F., Benabid, A.-L., Sadoul, R., Verna, J.-M. (2001). Molecular pathways involved in the neurotoxicity of 6-OHDA, dopamine and MPTP: contribution to the apoptotic theory in Parkinson's disease. *Progress in Neurobiology*, *65*, 135–172.

[219] Robinson, T.E., Whishaw, I.Q. (1988). Normalization of extracellular dopamine in striatum following recovery from a partial unilateral 6-OHDA lesion of the substantia nigra: a microdialysis study in freely moving rats. *Brain Research*, *450*, 209–224.

[220] Castañeda, E., Whishaw, I.Q., Robinson, T.E. (1990). Changes in striatal dopamine neurotransmission assessed with microdialysis following recovery from a bilateral 6-OHDA lesion: variation as a function of lesion size. *Journal of Neuroscience*, *10*, 1847–1854.

[221] Wachtel, S.R., Abercrombie, E.D. (1994). L-3,4-Dihydroxyphenylalanine-induced dopamine release in the striatum of intact and 6-hydroxydopamine-treated rats: differential effects of monoamineoxidase A and B inhibitors. *Journal of Neurochemistry*, *63*, 108–117.

[222] Sarre, S., Ebinger, G., Michotte, Y. (1996). Levodopa biotransformation in hemi-Parkinson rats: effect of dopamine receptor agonists and antagonists. *European Journal of Pharmacology*, *296*, 247–260.

[223] Miller, D.W., Abercrombie, E.D. (1999). Role of high-affinity dopamine uptake and impulse activity in the appearance of extracellular dopamine in striatum after administration of exogenous L-dopa: studies in intact and 6-hydroxydopamine-treated rats. *Journal of Neurochemistry*, *72*, 1516–1522.

[224] Jonkers, N., Sarre, S., Ebinger, G., Michotte, Y. (2000). MK801 influences L-dopa–induced dopamine release in intact and hemi-Parkinson rats. *European Journal of Pharmacology*, *407*, 281–291.

[225] Hoffman, A.F., van Horne, C.G., Eken, S., Hoffer, B.J., Gerhardt, G.A. (1997). In vivo microdialysis studies of somatodendritic dopamine release in the rat

substantia nigra: effects of unilateral 6-OHDA lesions and GDNF. *Experimental Neurology, 147*, 130–141.

[226] Bergquist, F., Shahabi, H.N., Nissbrandt, H. (2003). Somatodendritic dopamine release in rat substantia nigra influences motor performance in the accelerating rod. *Brain Research, 973*, 81–91.

[227] Sarre, S., Yuan, H., Jonkers, N., Van Hemelrijck, A., Ebinger, G., Michotte, Y. (2004). In vivo characterization of somatodendritic dopamine release in the substantia nigra of 6-hydroxydopamine-lesioned rats. *Journal of Neurochemistry, 90*, 29–39.

[228] Cragg, S.J., Nicholson, C., Kume-Kick, J., Tao, L., Rice, M.E. (2001). Dopamine-mediated volume transmission in midbrain is regulated by distinct extracellular geometry and uptake. *Journal of Neurophysiology, 85*, 1761–1771.

[229] Yamamoto, B.K., Davy, S. (1992). Dopaminergic modulation of glutamate release in striatum as measured by microdialysis. *Journal of Neurochemistry, 58*, 1736–1742.

[230] Robinson, S., Freeman, P., Moore, C., Touchon, J.C., Krentz, L., Meshul, C.K. (2003). Acute and subchronic MPTP administration differentially affects striatal glutamate synaptic function. *Experimental Neurology, 180*, 74–87.

[231] Tossman, U., Segovia, J., Ungerstedt, U. (1986). Extracellular levels of amino acids in striatum and globus pallidus of 6-hydroxydopamine-lesioned rats measured with microdialysis. *Acta Physiologica Scandinavica, 127*, 547–551.

[232] Lindefors, N., Ungerstedt, U. (1990). Bilateral regulation of glutamate tissue and extracellular levels in caudate-putamen by midbrain dopamine neurons. *Neuroscience Letters, 115*, 248–252.

[233] Yang, J., Hu, L.F., Liu, X., Zhou, F., Ding, J.H., Hu, G. (2006). Effects of iptakalim on extracellular glutamate and dopamine levels in the striatum of unilateral 6-hydroxydopamine-lesioned rats: a microdialysis study. *Life Sciences, 78*, 1940–1944.

[234] Meshul, C.K., Emre, N., Nakamura, C.M., Allen, C., Donohue, M.K., Buckman, J.F. (1999). Time-dependent changes in striatal glutamate synapses following a 6-hydroxydopamine lesion. *Neuroscience, 88*, 1–16.

[235] Jonkers, N., Sarre, S., Ebinger, G., Michotte, Y. (2002). MK801 suppresses the L-dopa–induced increase of glutamate in striatum of hemi-parkinson rats. *Brain Research, 926*, 149–155.

[236] Touchon, J.C., Holmer, H.K., Moore, C., McKee, B.L., Frederickson, J., Meshul, C.K. (2005). Apomorphine-induced alterations in striatal and substantia nigra pars reticulata glutamate following unilateral loss of striatal dopamine. *Experimental Neurology, 193*, 131–140.

[237] Abarca, J., Bustos, G. (1999). Differential regulation of glutamate, aspartate and gamma-amino-butyrate release by *N*-methyl-D-aspartate receptors in rat striatum after partial and extensive lesions to the nigro-striatal dopamine pathway. *Neurochemistry International, 35*, 19–33.

[238] Marti, M., Mela, F., Bianchi, C., Beani, L., Morari, M. (2002). Striatal dopamine-NMDA receptor interactions in the modulation of glutamate release in the substantia nigra pars reticulata in vivo: opposite role for D_1 and D_2 receptors. *Journal of Neurochemistry, 83*, 635–644.

[239] Galeffi, F., Bianchi, L., Bolam, J.P., Della Corte, L. (2003). The effect of 6-hydroxydopamine lesions on the release of amino acids in the direct and indirect pathways of the basal ganglia: a dual microdialysis probe analysis. *European Journal of Neuroscience, 18*, 856–868.

[240] Robelet, S., Melon, C., Guillet, B., Salin, P., Kerkerian-Le Goff, L. (2004). Chronic L-dopa treatment increases extracellular glutamate levels and GLT1 expression in the basal ganglia in a rat model of Parkinson's disease. *European Journal of Neuroscience, 20*, 1255–1266.

[241] You, Z.-B., Herrera-Marschitz, M., Petterson, E., Nylander, I., Goiny, M., Shou, H.-Z., Kehr, J., Godukhin, O., Hokfelt, T., Terenius, L., Ungerstedt, U. (1996). Modulation of neurotransmitter release by cholecystokinin in the neostriatum and substantia nigra of the rat: regional and receptor specificity. *Neuroscience, 74*, 793–804.

[242] Biggs, C.S., Fowler, L.J., Whitton, P.S., Starr, M.S. (1997). Extracellular levels of glutamate and aspartate in the entopeduncular nucleus of the rat determined by microdialysis: regulation by striatal D_2 receptors via the indirect striatal output pathway. *Brain Research, 753*, 163–175.

[243] Bianchi, L., Galeffi, F., Bolam, J.P., Della Corte, L. (2003). The effect of 6-hydroxydopamine lesions on the release of amino acids in the direct and indirect pathways of the basal ganglia: a dual microdialysis probe analysis. *European Journal of Neuroscience, 18*, 856–868.

[244] Lindefors, N., Brodin, E., Tossman, U., Segovia, H., Ungerstedt, U. (1989). Tissue levels and in vivo release of tachykinins and GABA in striatum and substantia nigra of rat brain after unilateral striatal dopamine denervation. *Experimental Brain Research, 74*, 527–534.

[245] Ochi, M., Koga, K., Kurokawa, M., Kase, H., Nakamura, J., Kuwana, Y. (2000). Systemic administration of adenosine A_{2A} receptor antagonist reverses increased GABA release in the globus pallidus of unilateral 6-hydroxydopamine-lesioned rats: a microdialysis study. *Neuroscience, 100*, 53–62.

[246] Ampe, B., Massie, A., D'Haens, J., Ebinger, G., Michotte, Y., Sarre, S. (2007). NMDA-mediated release of glutamate and GABA in the subthalamic nucleus is mediated by dopamine: an in vivo microdialysis study in rats. *Journal of Neurochemistry, 103*, 1063–1074.

[247] Ochi, M., Shiozaki, S., Kase, H. (2004). Adenosine A_{2A} receptor–mediated modulation of GABA and glutamate release in the output regions of the basal ganglia in a rodent model of Parkinson's disease. *Neuroscience, 127*, 223–231.

[248] Windels, F., Carcenac, C., Poupard, A., Savasta, M. (2005). Pallidal origin of GABA release within the substantia nigra pars reticulate during high-frequency stimulation of the subthalamic nucleus. *Journal of Neuroscience, 25*, 5079–5086.

[249] Sayre, M.L., Perry, G., Smith, M.A. (2008). Oxidative stress and neurotoxicity. *Chemical Research in Toxicology, 21*, 172–188.

[250] Jenner, P. (2003). Oxidative stress in Parkinson's disease. *Annals of Neurology, 53*(Suppl. 3), S26–S38.

[251] Glinka, Y., Gassen, M., Youdim, M.B.H. (1997). Mechanism of 6-Hydroxydopamine neurotoxicity. *Journal of Neural Transmission, 50*, 55–66.

[252] Opacka-Juffry, J., Wilson, A.W., Blunt, S.B. (1998). Effects of pergolide treatment on in vivo hydroxyl free radical formation during infusion of 6-hydroxydopamine in rat striatum. *Brain Research, 810*, 27–33.

[253] Ferger, B., Rose, S., Jenner, A., Halliwell, B., Jenner, P. (2001). 6-Hydroxydopamine increases hydroxyl free radical production and DNA damage in rat striatum. *Neuroreport, 12*, 1155–1159.

[254] Themann, C., Teismann, P., Kuschinsky, K., Ferger, B. (2001). Comparison of two independent aromatic hydroxylation assays in combination with intracerebral microdialysis to determine hydroxyl free radicals. *Journal of Neuroscience Methods, 108*, 57–64.

[255] Ferger, B., Themann, C., Rose, S., Halliwell, B., Jenner, P. (2001). 6-Hydroxydopamine increases the hydroxylation and nitration of phenylalanine in vivo: implication of peroxynitrite formation. *Journal of Neurochemistry, 78*, 509–514.

[256] Ferger, B., Teismann, P., Mierau, J. (2000). The dopamine agonist pramipexole scavenges hydroxyl free radicals induced by striatal application of 6-hydroxydopamine in rats: an in vivo microdialysis study. *Brain Research, 883*, 216–223.

[257] Jenner, P., Dexter, D.T., Sian, J., Schapira, A.H.V., Marsden, C.D. (1992). Oxidative stress as a cause of nigral cell death in Parkinson's disease and incidental Lewy body disease. The Royal Kings and Queens Parkinson's Disease Research Group. *Annals of Neurology, 32*(Suppl.), S82–S87.

[258] Schmidt, N., Ferger, B. (2001). Neurochemical findings in the MPTP model of Parkinson's disease. *Journal of Neural Transmission, 108*, 1263–1282.

[259] Giovanni, A., Sieber, B.A., Heikkila, R.E., Sonsalla, P.K. (1994). Studies on species sensitivity to the dopaminergic neurotoxin 1-methyl-4-phenyl-1,2,3,6-tetrahydropyridine: 1. Systemic administration. *Journal of Pharmacology and Experimental Therapeutics, 270*, 1000–1007.

[260] Giovanni, A., Sonsalla, P.K., Heikkila, R.E. (1994). Studies on species sensitivity to the dopaminergic neurotoxin 1-methyl-4-phenyl-1,2,3,6-tetrahydropyridine: 2. Central administration of 1-methyl-4-phenylpyridinium. *Journal of Pharmacology and Experimental Therapeutics, 270*, 1008–1014.

[261] Staal, R.G., Sonsalla, P.K. (2000). Inhibition of brain vesicular monoamine transporter (VMAT2) enhances 1-methyl-4-phenylpyridinium neurotoxicity in vivo in rat striata. *Journal of Pharmacology and Experimental Therapeutics, 293*, 336–342.

[262] Chiba, K., Trevor, A., Castagnoli, N., Jr. (1984). Metabolism of the neurotoxic tertiary amine, MPTP, by brain monoamine oxidase. *Biochemical and Biophysical Research Communications, 120*, 574–578.

[263] Rollema, H., De Vries, J.B., Damsma, G., Westerink, B.H.C., Kranenborg, G.L., Kuhr, W.G., Horn, A.S. (1988). The use of in vivo dialysis of dopamine, acetylcholine, aminoacids and lactic acid in studies on the neurotoxin 1-methyl-4-phenyl-1,2,3,6-tetrahydropyridine (MPTP). *Toxicology, 49*, 503–511.

[264] Santiago, M., Granero, L., Machado, A., Cano, J. (1995). Complex I inhibitor effect on the nigral and striatal release of dopamine in the presence and absence of nomifensine. *European Journal of Pharmacology, 280*, 251–256.

[265] Gainetdinov, R.R., Fumagalli, F., Jones, S.R., Caron, M.G. (1997). Dopamine transporter is required for in vivo MPTP neurotoxicity: evidence from mice lacking the transporter. *Journal of Neurochemistry, 69*, 1322–1325.

[266] Fabre, E., Monserrat, J., Herrero, A., Barja, G., Leret, M.L. (1999). Effect of MPTP on brain mitochondrial H_2O_2 and ATP production and on dopamine and dopac in the striatum. *Journal of Physiology and Biochemistry, 55*, 325–331.

[267] Sun, C.J., Johannessen, J.N., Gessner, W., Namura, I., Singhaniyom, W., Brossi, A., Chiueh, C.C. (1988). Neurotoxic damage to the nigrostriatal system in rats following intranigral administration of MPDP$^+$ and MPP$^+$. *Journal of Neural Transmission*, *74*, 75–86.

[268] Obata, T., Kinemuchi, H., Aomine, M. (2002). Protective effect of diltiazem, a L-type calcium channel antagonist, on bisphenol A–enhanced hydroxyl radical generation by 1-methyl-4-phenylpyridinium ion in rat striatum. *Neuroscience Letters*, *334*, 211–213.

[269] Nicklas, W.J., Youngster, S.K., Kindt, M.V., Heikkila, R.E. (1987). MPTP, MPP$^+$ and mitochondrial function. *Life Sciences*, *40*, 721–729.

[270] Dawson, T.M. (2000). New animal models for Parkinson's disease. *Cell*, *101*, 115–118.

[271] Grünewald, T., Beal, M.F. (1999). NOS knockouts and neuroprotection. *Nature Medicine*, *5*, 1354–1355.

[272] Obata, T. (2006). Nitric oxide and MPP$^+$-induced hydroxyl radical generation. *Journal of Neural Transmission*, *113*, 1131–1144.

[273] Chiueh, C.C., Krishna, G., Tulsi, P., Obata, T., Lang, K., Huang, S.J., Murphy, D.L. (1992). Intracranial microdialysis of salicylic acid to detect hydroxyl radical generation through dopamine autooxidation in the caudate nucleus: effects of MPP$^+$. *Free Radical Biology & Medicine*, *13*, 581–583.

[274] Chiueh, C.C., Wu, R.M., Mohanakumar, K.P., Sternberger, L.M., Krishna, G., Obata, T., Murphy, D.L. (1994). In vivo generation of hydroxyl radicals and MPTP-induced dopaminergic toxicity in the basal ganglia. *Annals of the New York Academy of Sciences*, *738*, 25–36.

[275] Obata, T. (2002). Role of hydroxyl radical formation in neurotoxicity as revealed by in vivo free radical trapping. *Toxicology Letters*, *132*, 83–93.

[276] Matarredona, E.R., Santiago, M., Cano, J., Machado, A. (1997). Involvement of iron in MPP$^+$ toxicity in substantia nigra: protection by desferrioxamine. *Brain Research*, *773*, 76–81.

[277] Santiago, M., Matarredona, E.R., Granero, L., Cano, J., Machado, A. (1997). Neuroprotective effect of the iron chelator desferrioxamine against MPP$^+$ toxicity on striatal dopaminergic terminals. *Journal of Neurochemistry*, *68*, 732–738.

[278] Santiago, M., Matarredona, E.R., Granero, L., Cano, J., Machado, A. (2000). Neurotoxic relationship between dopamine and iron in the striatal dopaminergic nerve terminals. *Brain Research*, *858*, 26–32.

[279] Obata, T. (2006). Effect of desferrioxamine, a strong iron(III) chelator, on 1-methyl-4-phenylpyridinium ion (MPP$^+$)-induced hydroxyl radical generation in the rat striatum. *European Journal of Pharmacology*, *539*, 34–38.

[280] Rollema, H., Damsma, G., Horn, A.S., De Vries, J.B., Westerink, B.H.C. (1986). Brain dialysis in conscious rats reveals an instantaneous massive release of striatal dopamine in response to MPP$^+$. *European Journal of Pharmacology*, *126*, 345–346.

[281] Rollema, H., Kuhr, W.G., Kranenborg, G., De Vries, J., Van den Berg, C. (1988). MPP$^+$-induced efflux of dopamine and lactate from rat striatum have similar time courses as shown by in vivo brain dialysis. *Journal of Pharmacology and Experimental Therapeutics*, *245*, 858–866.

[282] Santiago, M., Rollema, H., de Vries, J.B., Westerink, B.H.C. (1991). Acute effects of intranigral application of MPP$^+$ on nigral and bilateral striatal release of dopamine simultaneously recorded by microdialysis. *Brain Research*, *538*, 226–230.

[283] Santiago, M., Westerink, B.H.C., Rollema, H. (1991). Responsiveness of striatal dopamine release in awake animals after chronic 1-methyl-4-phenylpyridinium ion-induced lesions of the substantia nigra. *Journal of Neurochemistry*, *56*, 1336–1342.

[284] Obata, T., Yamanaka, Y., Kinemuchi, H., Oreland, L. (2001). Release of dopamine by perfusion with 1-methyl-4-phenylpyridinium ion (MPP$^+$) into the striatum is associated with hydroxyl free radical generation. *Brain Research*, *906*, 170–175.

[285] Obata, T., Chiueh, C.C. (1992). In vivo trapping of hydroxyl free radicals in the striatum utilizing intracranial microdialysis perfusion of salicylate: effects of MPTP, MPDP$^+$, and MPP$^+$. *Journal of Neural Transmission (General Section)*, *89*, 139–145.

[286] Chiueh, C.C., Miyake, H., Peng, M.T. (1993). Role of dopamine autoxidation, hydroxyl radical generation, and calcium overload in underlying mechanisms involved in MPTP-induced parkinsonism. *Advances in Neurology*, *60*, 251–258.

[287] Obata, T. (1999). Reserpine prevents hydrohyl radical formation by MPP$^+$ in rat striatum. *Brain Research*, *828*, 68–73.

[288] Obata, T., Kubota, S. (2001). Protective effect of tamoxifen on 1-methyl-4-phenylpyridine–induced hydroxyl radical generation in the rat striatum. *Neuroscience Letters*, *308*, 87–90.

[289] Mohanakumar, K.P., Muralikrishnan, D., Thomas, B. (2000). Neuroprotection by sodium salicylate against 1-methyl-4-phenyl-1,2,3,6-tetrahydropyridine–induced neurotoxicity. *Brain Research*, *864*, 281–290.

[290] Han, J., Cheng, F.C., Yang, Z., Dryhurst, G. (1999). Inhibitors of mitochondrial respiration, iron(II), and hydroxyl radical evoke release and extracellular hydrolysis of glutathione in rat striatum and substantia nigra: potential implications to Parkinson's disease. *Journal of Neurochemistry*, *73*, 1683–1695.

[291] Foster, S.B., Wrona, M.Z., Han, J., Dryhurst, G. (2003). The parkinsonian neurotoxin 1-methyl-4-phenylpyridinium (MPP$^+$) mediates release of l-3,4-dihydroxyphenylalanine (*l*-dopa) and inhibition of *l*-dopa decarboxylase in the rat striatum: a microdialysis study. *Chemical Research in Toxicology*, *16*, 1372–1384.

[292] Carboni, S., Melis, F., Pani, L., Hadjiconstantinou, M., Rossetti, Z.L. (1990). The noncompetitive NMDA-receptor antagonist MK-801 prevents the massive release of glutamate and aspartate from rat striatum induced by 1-methyl-4-phenylpyridinium (MPP$^+$). *Neuroscience Letters*, *117*, 129–133.

[293] Yang, C.S., Tsai, P.J., Lin, N.N., Liu, L., Kuo, J.S. (1995). Elevated extracellular glutamate levels increased the formation of hydroxyl radical in the striatum of anesthetized rat. *Free Radical Biology & Medicine*, *19*, 453–459.

[294] Yang, Y.J., Wang, Q.M., Hu, L.F., Sun, X.L., Ding, J.H., Hu, G. (2006). Iptakalim alleviated the increase of extracellular dopamine and glutamate induced by 1-methyl-4-phenylpyridinium ion in rat striatum. *Neuroscience Letters*, *404*, 187–190.

[295] Ferger, B., van Amsterdam, C., Seyfried, C., Kuschinsky, K. (1998). Effects of (-phenyl-*tert*-butylnitrone and selegiline on hydroxyl free radicals in rat striatum produced by local application of glutamate. *Journal of Neurochemistry, 70,* 276–280.

[296] Beal, M.F., Brouillet, E., Jenkins, B.G., Ferrante, R.J., Kowall, N.W., Miller, J.M., Storey, E., Srivastava, R., Rosen, B.R., Hyman, B.T. (1993). Neurochemical and histologic characterization of striatal excitotoxic lesions produced by the mitochondrial toxin 3-nitropropionic acid. *Journal of Neuroscience, 13,* 4181–4192.

[297] Skirboll, S., Wang, J., Mefford, I., Hsiao, J., Bankiewicz, K.S. (1990). In vivo changes of catecholamines in hemiparkinsonian monkeys measured by microdialysis. *Experimental Neurology, 110,* 187–193.

[298] Robertson, R.G., Graham, W.C., Sambrook, M.A., Crossman, A.R. (1991). Further investigations into the pathophysiology of MPTP-induced parkinsonism in the primate: an intracerebral microdialysis study of gamma-aminobutyric acid in the lateral segment of the globus pallidus. *Brain Research, 563,* 278–280.

[299] Schroeder, J.A., Schneider, J.S. (2002). GABA-opioid interactions in the globus pallidus: [D-Ala^2]-Met-enkephalinamide attenuates potassium-evoked GABA release after nigrostriatal lesion. *Journal of Neurochemistry, 82,* 666–673.

[300] Soares, J., Kliem, M.A., Betarbet, R., Greenamyre, J.T., Yamamoto, B., Wichmann, T. (2004). Role of external pallidal segment in primate parkinsonism: comparison of the effects of 1-methyl-4-phenyl-1,2,3,6-tetrahydropyridine-induced parkinsonism and lesions of the external pallidal segment. *Journal of Neuroscience, 24,* 6417–6426.

[301] Riederer, P., Gerlach, M., Müller, T., Reichmann, H. (2007). Relating mode of action to clinical practice: dopaminergic agents in Parkinson's disease. *Parkinsonism & Related Disorders, 13,* 466–479.

[302] Nomoto, M., Fukuda, T. (1993). A selective MAO$_B$ inhibitor Ro19–6327 potentiates the effects of levodopa on parkinsonism induced by MPTP in the common marmoset. *Neuropharmacology, 32,* 473–477.

[303] Wu, W.-R., Zhu, Z.-T., Zhu, X.-Z. (2000). Differential effects of L-deprenyl on MPP$^+$- and MPTP-induced dopamine overflow in microdialysates of striatum and nucleus accumbens. *Life Sciences, 67,* 241–250.

[304] Chiueh, C.C., Huang, S.-J., Murphy, D.L. (1992). Enhanced hydroxyl radical generation by 2'-methyl analog of MPTP: suppression by clorgyline and deprenyl. *Synapse, 11,* 346–348.

[305] Wu, R.-M., Chiueh, C.C., Pert, A., Murphy, D.L. (1993). Apparent antioxidant effect of l-deprenyl on hydroxyl radical formation and nigral injury elicited by MPP$^+$ in vivo. *European Journal of Pharmacology, 243,* 241–247.

[306] Matsubara, K., Senda, T., Uezono, T., Awaya, T., Ogawa, S., Chiba, K., Shimizu, K., Hayase, N., Kimura, K. (2001). L-Deprenyl prevents the cell hypoxia induced by dopaminergic neurotoxins, MPP$^+$ and (-carbolinium: a microdialysis study in rats. *Neuroscience Letters, 302,* 65–68.

[307] Obata, T., Kubota, S., Yamanaka, Y. (2001). Allopurinol suppresses para-nonylphenol and 1-methyl-4-phenylpyridinium ion (MPP$^+$)-induced hydroxyl radical generation in rat striatum. *Neuroscience Letters, 306,* 9–12.

[308] Obata, T. (2006). Allopurinol suppresses 2-bromoethylamine and 1-methyl-4-phenylpyridinium ion (MPP$^+$)-induced hydroxyl radical generation in rat striatum. *Toxicology, 218,* 75–79.

[309] Cassarino, D.S., Fall, C.P., Smith, T.S., Bennett, J.P., Jr. (1998). Pramipexole reduces reactive oxygen species production in vivo and in vitro and inhibits the mitochondrial permeability transition produced by the parkinsonian neurotoxin methylpyridinium ion. *Journal of Neurochemistry, 71,* 295–301.

[310] Yuan, H., Sarre, S., Ebinger, G., Michotte, Y. (2004). Neuroprotective and neurotrophic effect of apomorphine in the striatal 6-OHDA-lesion rat model of Parkinson's disease. *Brain Research, 1026,* 95–107.

[311] Dethy, S., Laute, M.A., Damhaut, P., Goldman, S. (1999). Pergolide potentiates L-dopa-induced dopamine release in rat striatum after lesioning with 6-hydroxydopamine. *Journal of Neural Transmission, 106,* 145–158.

[312] Serra, P.A., Esposito, G., Enrico, P., Mura, M.A., Migheli, R., Delogu, M.R., Miele, M., Desole, M.S., Grella, G., Miele, E. (2000). Manganese increases L-dopa autoxidation in the striatum of the freely moving rat: potential implications to L-dopa long-term therapy of Parkinson's disease. *British Journal of Pharmacology, 130,* 937–945.

[313] Rocchitta, G., Migheli, R., Mura, M.P., Esposito, G., Marchetti, B., Desole, M.S., Miele, E., Serra, P.A. (2005). Role of endogenous melatonin in the oxidative homeostasis of the extracellular striatal compartment: a microdialysis study in PC12 cells in vitro and in the striatum of freely moving rats. *Journal of Pineal Research, 39,* 409–418.

[314] Rocchitta, G., Migheli, R., Esposito, G., Marchetti, B., Desole, M.S., Miele, E., Serra, P.A. (2006). Endogenous melatonin protects L-dopa from autoxidation in the striatal extracellular compartment of the freely moving rat: potential implication for long-term L-dopa therapy in Parkinson's disease. *Journal of Pineal Research, 40,* 204–213.

[315] Obata, T. (1999). Protective effect of imidaprilat, a new angiotensin-converting enzyme inhibitor against 1-methyl-4-phenylpyridinium ion-induced •OH generation in rat striatum. *European Journal of Pharmacology, 378,* 39–45.

[316] Obata, T. (2006). Imidaprilat suppresses nonylphenol and 1-methyl-4-phenylpyridinium ion (MPP$^+$)-induced hydroxyl radical generation in rat striatum. *Neuroscience Research, 54,* 192–196.

[317] Obata, T., Takahashi, S., Kashiwagi, Y., Kubota, S. (2008). Protective effect of captopril and enalaprilat, angiotensin-converting enzyme inhibitors, on paranonylphenol-induced •OH generation and dopamine efflux in rat striatum. *Toxicology, 250,* 96–99.

[318] Obata, T., Kubota, S., Yamanaka, Y. (2001). Protective effect of histidine on *para*-nonylphenol-enhanced hydroxyl free radical generation induced by 1-methyl-4-phenylpyridinium ion (MPP$^+$) in rat striatum. *Biochimica et Biophysica Acta, 1568,* 171–175.

[319] Obata, T. (2006). Tamoxifen protect against hydroxyl radical generation induced by phenelzine in rat striatum. *Toxicology, 222,* 46–52.

[320] Obata, T., Aomine, M., Inada, T., Kinemuchi, H. (2002). Nicotine suppresses 1-methyl-4-phenylpyridinium ion-induced hydroxyl radical generation in rat striatum. *Neuroscience Letters, 330,* 122–124.

[321] Obata, T., Yamanaka, Y. (2000). Protective effect of fluvastatin, a new inhibitor of 3-hydroxy-3-methylglutaryl coenzyme A reductase, on MPP$^+$-induced hydroxyl radical in the rat striatum. *Brain Research, 860,* 166–169.

[322] Di Matteo, V., Pierucci, M., Di Giovanni, G., Di Santo, A., Poggi, A., Benigno, A., Esposito, E. (2006). Aspirin protects striatal dopaminergic neurons from neurotoxin-induced degeneration: an in vivo microdialysis study. *Brain Research, 1095,* 167–177.

[323] Smith, T.S., Swerdlow, R.H., Parker, W.D., Jr., Bennett, J.P., Jr. (1994). Reduction of MPP$^+$-induced hydroxyl radical formation and nigrostriatal MPTP toxicity by inhibiting nitric oxide synthase. *Neuroreport, 5,* 2598–2600.

[324] Smith, T.S., Bennett, J.P., Jr. (1997). Mitochondrial toxins in models of neurodegenerative diseases: I. In vivo brain hydroxyl radical production during systemic MPTP treatment or following microdialysis infusion of methylpyridinium or azide ions. *Brain Research, 765,* 183–188.

[325] Rose, S., Hindmarshand, J.G., Jenner, P. (1999). Neuronal nitric oxide synthase inhibition reduces MPP$^+$-evoked hydroxyl radical formation but not dopamine efflux in rat striatum. *Journal of Neural Transmission, 106,* 477–486.

[326] Obata, T., Yamanaka, Y. (2001). Nitric oxide enhances MPP$^+$-induced hydroxyl radical generation via depolarization activated nitric oxide synthase in rat striatum. *Brain Research, 902,* 223–228.

[327] Di Matteo, V., Benigno, A., Pierucci, M., Giuliano, D.A., Crescimanno, G., Esposito, E., Di Giovanni, G. (2006). 7-nitroindazole protects striatal dopaminergic neurons against MPP$^+$-induced degeneration: an in vivo microdialysis study. *Annals of the New York Academy of Sciences, 1089,* 462–471.

[328] Santiago, M., Venero, J.L., Machado, A., Cano, J. (1992). In vivo protection of striatum from MPP$^+$ neurotoxicity by N-methyl-D-aspartate antagonists. *Brain Research, 586,* 203–207.

[329] Richard, M.G., Bennett, J.P., Jr. (1995). NMDA receptor blockade increases in vivo striatal dopamine synthesis and release in rats and mice with incomplete, dopamine-depleting, nigrostriatal lesions. *Journal of Neurochemistry, 64,* 2080–2086.

[330] Marti, M., Sbrenna, S., Fuxe, K., Bianchi, C., Beani, L., Morari, M. (2000). Increased responsivity of glutamate release from the substantia nigra pars reticulata to striatal NMDA receptor blockade in a model of Parkinson's disease: a dual probe microdialysis study in hemiparkinsonian rats. *European Journal of Neuroscience, 12,* 1848–1850.

[331] Sarre, S., Lanza, M., Makovec, F., Artusi, R., Caselli, G., Michotte, Y. (2008). In vivo neurochemical effects of the NR2B selective NMDA receptor antagonist CR 3394 in 6-hydroxydopamine lesioned rats. *European Journal of Pharmacology, 584,* 297–305.

[332] Matarredona, E.R., Santiago, M., Venero, J.L., Cano, J., Machado, A. (2001). Group II metabotropic glutamate receptor activation protects striatal dopaminergic nerve terminals against MPP$^+$-induced neurotoxicity along with brain-derived neurotrophic factor induction. *Journal of Neurochemistry, 76,* 351–360.

[333] Mela, F., Marti, M., Dekundy, A., Danysz, W., Morari, M., Cenci, M.A. (2007). Antagonism of metabotropic glutamate receptor type 5 attenuates *l*-dopa-

induced dyskinesia and its molecular and neurochemical correlates in a rat model of Parkinson's disease. *Journal of Neurochemistry*, *101*, 483–497.

[334] Arai, A., Kannari, K., Shen, H., Maeda, T., Suda, T., Matsunaga, M. (2003). Amantadine increases L-dopa-derived extracellular dopamine in the striatum of 6-hydroxydopamine-lesioned rats. *Brain Research*, *972*, 229–234.

[335] Abercrombie, E.D., Bonatz, A.E., Zigmond, M.J. (1990). Effects of L-dopa on extracellular dopamine in striatum of normal and 6-hydroxydopamine-treated rats. *Brain Research*, *525*, 36–44.

[336] Maeda, T., Kannari, K., Suda, T., Matsunaga, M. (1999). Loss of regulation by presynaptic dopamine D_2 receptors of exogenous L-dopa-derived dopamine release in the dopaminergic denervated striatum. *Brain Research*, *817*, 185–191.

[337] Meissner, W., Ravenscroft, P., Reese, R., Harnack, D., Morgenstern, R., Kupsch, A., Klitgaard, H., Bioulac, B., Gross, C.E., Bezard, E., Boraud, T. (2006). Increased slow oscillatory activity in substantia nigra pars reticulata triggers abnormal involuntary movements in the 6-OHDA lesioned rat in the presence of excessive extracellular striatal dopamine. *Neurobiology of Disease*, *22*, 586–598.

[338] Rodriguez, M., Morales, I., Gonzalez-Mora, J.L., Gomez, I., Sabate, M., Dopico, J.G., Rodriguez-Oroz, M.C., Obeso, J.A. (2007). Different levodopa actions on the extracellular dopamine pools in the rat striatum. *Synapse*, *61*, 61–71.

[339] Buck, K., Ferger, B. (2008). Intrastriatal inhibition of aromatic amino acid decarboxylase prevents *l*-dopa-induced dyskinesia: a bilateral reverse in vivo microdialysis study in 6-hydroxydopamine lesioned rats. *Neurobiology of Disease*, *29*, 210–220.

[340] Jonkers, N., Sarre, S., Ebinger, G., Michotte, Y. (2001). Benserazide decreases central AADC activity, extracellular dopamine levels and levodopa decarboxylation in striatum of the rat. *Journal of Neural Transmission*, *108*, 559–570.

[341] Shen, H., Kannari, K., Yamato, H., Arai, A., Matsunaga, M. (2003). Effects of benserazide on L-dopa-derived extracellular dopamine levels and aromatic L-amino acid decarboxylase activity in the striatum of 6-hydroxydopamine-lesioned rats. *Tohoku Journal of Experimental Medicine*, *199*, 149–159.

[342] Tanaka, H., Kannari, K., Maeda, T., Tomiyama, M., Suda, T., Matsunaga, M. (1999). Role of serotonergic neurons in L-dopa-derived extracellular dopamine in the striatum of 6-OHDA-lesioned rats. *Neuroreport*, *10*, 631–634.

[343] Kannari, K., Tanaka, H., Maeda, T., Tomiyama, M., Suda, T., Matsunaga, M. (2000). Reserpine pretreatment prevents increases in extracellular striatal dopamine following L-dopa administration in rats with nigrostriatal denervation. *Journal of Neurochemistry*, *74*, 263–269.

[344] Kannari, K., Yamato, H., Shen, H., Tomiyama, M., Suda, T., Matsunaga, M. (2001). Activation of $5\text{-}HT_{1A}$ but not $5\text{-}HT_{1B}$ receptors attenuates an increase in extracellular dopamine derived from exogenously administered L-dopa in the striatum with nigrostriatal denervation. *Journal of Neurochemistry*, *76*, 1346–1353.

[345] Yamato, H., Kannari, K., Shen, H., Suda, T., Matsunaga, M. (2001). Fluoxetine reduces L-dopa-derived extracellular DA in the 6-OHDA-lesioned rat striatum. *Neuroreport*, *12*, 1123–1126.

[346] Mignon, L.J., Wolf, W.A. (2005). 8-Hydroxy-2-(di-*n*-propylamino)tetralin reduces striatal glutamate in an animal model of Parkinson's disease. *Neuroreport*, *16*, 699–703.

[347] Nowak, P., Szczerbak, G., Biedka, I., Drosik, M., Kostrzewa, R.M., Brus, R. (2006). Effect of ketanserin and amphetamine on nigrostriatal neurotransmission and reactive oxygen species in parkinsonian rats: in vivo microdialysis study. *Journal of Physiology and Pharmacology, 57,* 583–597.

[348] Marti, M., Mela, F., Guerrini, R., Calo, G., Bianchi, C., Morari, M. (2004). Blockade of nociceptin/orphanin FQ transmission in rat substantia nigra reverses haloperidol-induced akinesia and normalizes nigral glutamate release. *Journal of Neurochemistry, 91,* 1501–1504.

[349] Marti, M., Mela, F., Fantin, M., Zucchini, S., Brown, J.M., Witta, J., Di Benedetto, M., Buzas, B., Reinscheid, R.K., Salvadori, S., et al. (2005). Blockade of nociceptin/orphanin FQ transmission attenuates symptoms and neurodegeneration associated with Parkinson's disease. *Journal of Neuroscience, 25,* 9591–9601.

[350] Marti, M., Trapella, C., Viaro, R., Morari, M. (2007). The nociceptin/orphanin FQ receptor antagonist J-113397 and L-dopa additively attenuate experimental parkinsonism through overinhibition of the nigrothalamic pathway. *Journal of Neuroscience, 27,* 1297–1307.

[351] Marti, M., Mela, F., Veronesi, C., Guerrini, R., Salvadori, S., Federici, M., Mercuri, N.B., Rizzi, A., Franchi, G., Beani, L., Bianchi, C., Morari, M. (2004). Blockade of nociceptin/orphanin FQ receptor signaling in rat substantia nigra pars reticulata stimulates nigrostriatal dopaminergic transmission and motor behavior. *Journal of Neuroscience, 24,* 6659–6666.

[352] Stefani, A., Fedele, E., Galati, S., Pepicelli, O., Frasca, S., Pierantozzi, M., Peppe, A., Brusa, L., Orlacchio, A., Hainsworth, A.H., et al. (2005). Subthalamic stimulation activates internal pallidus: evidence from cGMP microdialysis in PD patients. *Annals of Neurology, 57,* 448–452.

[353] Stefani, A., Fedele, E., Galati, S., Raiteri, M., Pepicelli, O., Brusa, L., Pierantozzi, M., Peppe, A., Pisani, A., Gattoni, G., et al. (2006). Deep brain stimulation in Parkinson's disease patients: biochemical evidence. *Journal of Neural Transmission Supplement, 70,* 401–408.

[354] Galati, S., Mazzone, P., Fedele, E., Pisani, A., Peppe, A., Pierantozzi, M., Brusa, L., Troppi, D., Moschella, V., Raiteri, M., et al. (2006). Biochemical and electrophysiological changes of substantia nigra pars reticulata driven by subthalamic stimulation in patients with Parkinson's disease. *European Journal of Neuroscience, 11,* 2923–2928.

[355] Wang, S., Hu, L.F., Zhang, Y., Sun, T., Sun, Y.H., Liu, S.Y., Ding, J.H., Wu, J., Hu, G. (2006). Effects of systemic administration of iptakalim on extracellular neurotransmitter levels in the striatum of unilateral 6-hydroxydopamine-lesioned rats. *Neuropsychopharmacology, 31,* 933–940.

[356] Corsi, C., Pinna, A., Gianfriddo, M., Melani, A., Morelli, M., Pedata, F. (2003). Adenosine A_{2A} receptor antagonism increases striatal glutamate outflow in dopamine-denervated rats. *European Journal of Pharmacology, 464,* 33–38.

[357] Windels, F., Bruet, N., Poupard, A., Urbain, N., Chouvet, G., Feuerstein, C., Savasta, M. (2000). Effects of high frequency stimulation of subthalamic nucleus on extracellular glutamate and GABA in substantia nigra and globus pallidus in the normal rat. *European Journal of Neuroscience, 12,* 4141–4146.

[358] Boulet, S., Lacombe, E., Carcenac, C., Feuerstein, C., Sgambato-Faure, V., Poupard, A., Savasta, M. (2006). Subthalamic stimulation-induced forelimb dyskinesias are

linked to an increase in glutamate levels in the substantia nigra pars reticulata. *Journal of Neuroscience, 26*, 10768–10776.

[359] Ochi, M., Shiozaki, S., Kase, H. (2004). L-dopa-induced modulation of GABA and glutamate release in substantia nigra pars reticulata in a rodent model of Parkinson's disease. *Synapse, 52*, 163–165.

[360] Antonelli, T., Fuxe, K., Agnati, L., Mazzoni, E., Tanganelli, S., Tomasini, M.C., Ferraro, L. (2006). Experimental studies and theoretical aspects on A2A/D2 receptor interactions in a model of Parkinson's disease: relevance for L-dopa induced dyskinesias. *Journal of the Neurological Sciences, 248*, 16–22.

[361] Tanganelli, S., Sandager Nielsen, K., Ferraro, L., Antonelli, T., Kehr, J., Franco, R., Ferré, S., Agnati, L.F., Fuxe, K., Scheel-Krüger, J. (2004). Striatal plasticity at the network level: focus on adenosine A_{2A} and D_2 interactions in models of Parkinson's disease. *Parkinsonism & Related Disorders, 10*, 273–280.

[362] Bruet, N., Windels, F., Bertrand, A., Feuerstein, C., Poupard, A., Savasta, M. (2001). High frequency stimulation of the subthalamic nucleus increases the extracellular contents of striatal dopamine in normal and partially dopaminergic denervated rats. *Journal of Neuropathology & Experimental Neurology, 60*, 15–24.

[363] Meissner, W., Reum, T., Paul, G., Harnack, D., Sohr, R., Morgenstern, R., Kupsch, A. (2001). Striatal dopaminergic metabolism is increased by deep brain stimulation of the subthalamic nucleus in 6-hydroxydopamine lesioned rats. *Neuroscience Letters, 303*, 165–168.

[364] Meissner, W., Harnack, D., Paul, G., Reum, T., Sohr, R., Morgenstern, R., Kupsch, A. (2002). Deep brain stimulation of subthalamic neurons increases striatal dopamine metabolism and induces contralateral circling in freely moving 6-hydroxydopamine-lesioned rats. *Neuroscience Letters, 328*, 105–108.

[365] Meissner, W., Harnack, D., Reese, R., Paul, G., Reum, T., Ansorge, M., Kusserow, H., Winter, C., Morgenstern, R., Kupsch, A. (2003). High frequency stimulation of the subthalamic nucleus enhances striatal dopamine release and metabolism in rats. *Journal of Neurochemistry, 85*, 601–609.

[366] Lacombe, E., Carcenac, C., Boulet, S., Feuerstein, C., Bertrand, A., Poupard, A., Savasta, M. (2007). High-frequency stimulation of the subthalamic nucleus prolongs the increase in striatal dopamine induced by acute L-3,4-dihydroxyphenylalanine in dopaminergic denervated rats. *European Journal of Neuroscience, 26*, 1670–1680.

[367] Nimura, T., Yamaguchi, K., Ando, T., Shibuya, S., Oikawa, T., Nakagawa, A., Shirane, R., Itoh, M., Tominaga, T. (2005). Attenuation of fluctuating striatal synaptic dopamine levels in patients with Parkinson disease in response to subthalamic nucleus stimulation: a positron emission tomography study. *Journal of Neurosurgery, 103*, 968–973.

[368] Bruet, N., Windels, F., Carcenac, C., Feuerstein, C., Bertrand, A., Poupard, A., Savasta, M. (2003). Neurochemical mechanisms induced by high frequency stimulation of the subthalamic nucleus: increase of extracellular striatal glutamate and GABA in normal and hemiparkinsonian rats. *Journal of Neuropathology & Experimental Neurology, 62*, 1228–1240.

[369] Morari, M., O'Connor, W.T., Ungerstedt, U., Bianchi, C., Fuxe, K. (1996). Functional neuroanatomy of the nigrostriatal and striatonigral pathways as

studied with dual probe microdialysis in the awake rat: II. Evidence for striatal N-methyl-D-aspartate receptor regulation of striatonigral GABAergic transmission and motor function. *Neuroscience, 72*, 89–97.

[370] Trevitt, T., Carlson, B., Correa, M., Keene, A., Morales, M., Salamone, J.D. (2002). Interactions between dopamine D1 receptors and gamma-aminobutyric acid mechanisms in substantia nigra pars reticulata of the rat: neurochemical and behavioral studies. *Psychopharmacology, 159*, 229–237.

[371] Ishiwari, K., Mingote, S., Correa, M., Trevitt, J.T., Carlson, B.B., Salamone, J.D. (2004). The GABA uptake inhibitor beta-alanine reduces pilocarpine-induced tremor and increases extracellular GABA in substantia nigra pars reticulata as measured by microdialysis. *Journal of Neuroscience Methods, 140*, 39–46.

[372] Smith, M.P., Fletcher-Turner, A., Yurek, D.M., Cass, W.A. (2006). Calcitriol protection against dopamine loss induced by intracerebroventricular administration of 6-hydroxydopamine. *Neurochemical Research, 31*, 533–539.

[373] Gerin, C. (2002). Behavioral improvement and dopamine release in a parkinsonian rat model. *Neuroscience Letters, 330*, 5–8.

[374] Smith, A.D., Kozlowski, D.A., Bohn, M.C., Zigmond, M.J. (2005). Effect of AdGDNF on dopaminergic neurotransmission in the striatum of 6-OHDA-treated rats. *Experimental Neurology, 193*, 420–426.

[375] Opacka-Juffry, J., Ashworth, S., Hume, S.P., Martin, D., Brooks, D.J., Blunt, S.B. (1995). GDNF protects against 6-OHDA nigrostriatal lesion: in vivo study with microdialysis and PET. *Neuroreport, 7*, 348–352.

[376] Grondin, R., Cass, W.A., Zhang, Z., Stanford, J.A., Gash, D.M., Gerhardt, G.A. (2003). Glial cell line-derived neurotrophic factor increases stimulus-evoked dopamine release and motor speed in aged rhesus monkeys. *Journal of Neuroscience, 23*, 1974–1980.

[377] Gerhardt, G.A., Cass, W.A., Huettl, P., Brock, S., Zhang, Z., Gash, D.M. (1999). GDNF improves dopamine function in the substantia nigra but not the putamen of unilateral MPTP-lesioned rhesus monkeys. *Brain Research, 817*, 163–171.

[378] Zetterström, T., Brundin, P., Gage, F.H., Sharp, T., Isacson, O., Dunnett, S.B., Ungerstedt, U., Björklund, A. (1986). In vivo measurement of spontaneous release and metabolism of dopamine from intrastriatal nigral grafts using intracerebral dialysis. *Brain Research, 362*, 344–349.

[379] Strecker, R.E., Sharp, T., Brundin, P., Zetterström, T., Ungerstedt, U., Björklund, A. (1987). Autoregulation of dopamine release and metabolism by intrastriatal nigral grafts as revealed by intracerebral dialysis. *Neuroscience, 22*, 169–178.

[380] Rioux, L., Gaudin, D.P., Bui, L.K., Grégoire, L., DiPaolo, T., Bédard, P.J. (1991). Correlation of functional recovery after a 6-hydroxydopamine lesion with survival of grafted fetal neurons and release of dopamine in the striatum of the rat. *Neuroscience, 40*, 123–131.

[381] Kondoh, T., Low, W.C. (1994). Glutamate uptake blockade induces striatal dopamine release in 6-hydroxydopamine rats with intrastriatal grafts: evidence for host modulation of transplanted dopamine neurons. *Experimental Neurology, 127*, 191–198.

[382] Rodríguez-Gómez, J.A., Lu, J.Q., Velasco, I., Rivera, S., Zoghbi, S.S., Liow, J.S., Musachio, J.L., Chin, F.T., Toyama, H., Seidel, J., et al. (2007). Persistent dopamine

functions of neurons derived from embryonic stem cells in a rodent model of Parkinson disease. *Stem Cells*, 25, 918–928.

[383] Bouchez, G., Sensebé, L., Vourc'h, P., Garreau, L., Bodard, S., Rico, A., Guilloteau, D., Charbord, P., Besnard, J.C., Chalon, S. (2008). Partial recovery of dopaminergic pathway after graft of adult mesenchymal stem cells in a rat model of Parkinson's disease. *Neurochemistry International*, 52, 1332–1342.

5

MONITORING NEUROTRANSMITTER AMINO ACIDS BY MICRODIALYSIS: PHARMACODYNAMIC APPLICATIONS

SANDRINE PARROT AND BERNARD RENAUD
Université de Lyon and Lyon Neuroscience Research Center, NeuroChem, Lyon, France; Université Lyon 1, Villeurbanne, France

LUC ZIMMER AND LUC DENOROY
Université de Lyon and Lyon Neuroscience Research Center, BioRaN Team, Lyon, France; Université Lyon 1, Villeurbanne, France

1. INTRODUCTION

Glutamic acid (Glu) and aspartic acid (Asp) are the predominant excitatory amino acid neurotransmitters in the central nervous system, and γ-aminobutyric acid (GABA) and glycine (Gly) are the major inhibitory amino acid neurotransmitters [1–4]. These neurotransmitters are present in all neuron networks throughout the neuraxis and therefore exhibit a very widespread anatomical distribution throughout the brain, being present in pathways as well as within short interneurons. They play a crucial role in all the neurophysiological processes, being involved in breathing pattern generation, control of the autonomic system, motor function, pain control, cognitive processes, learning and memory, and many other brain functions. Experiments carried out on animal models of human diseases as well as clinical or pathological studies have revealed alterations in excitatory and/or inhibitory neurotransmission in various pathologies, such as epilepsy, ischemia, neurodegenerative syndromes such as Parkinson's and Huntington's diseases, and psychiatric illness such as schizophrenia [5–8]. In that context, pharmacological agents targeting glutamatergic or GABAergic neurotransmission have been developed to treat epilepsy and to elicit neuroprotection from an excitotoxic insult as in ischemia, as shown later in this chapter and in other chapters of the book. The involvement of these amino acid neurotransmitters in various physiological processes and pathophysiological or pharmacodynamic mechanisms can be studied by monitoring changes in their respective extracellular concentrations using in vivo brain microdialysis. This chapter is devoted to brain microdialysis studies of amino acid neurotransmitters, mostly Glu/Asp and GABA, and describes the interest being displayed in such techniques. After reviewing the latest technical developments in microdialysis and in techniques for analyzing amino acids in microdialysis samples, four areas of pharmacological research in which monitoring amino acids by microdialysis is widely used are reviewed: basic studies on control of amino acidergic neurons by receptors, psychostimulants and addictive drugs, analgesia, and ischemia/anoxia.

2. MONITORING NEUROTRANSMITTER AMINO ACIDS BY MICRODIALYSIS

2.1. General Considerations

Microdialysis has become a conventional method for the continuous sampling of amino acid compounds present in the extracellular fluid of a selected brain area of animals or humans [9]. It has been used to monitor glutamate (Glu)/ aspartate (Asp) and GABA/glycine (Gly), responsible for excitatory and inhibitory neurotransmissions, respectively, but also nonneurotransmitter amino acids such as arginine, tyrosine, and histidine. The brain microdialysis relies on a membrane that allows free diffusion of solutes between the brain extracellular space and an artificial fluid. A microdialysis probe consists of two concentric tubes with the distal part (1 to 5 mm) covered by a dialysis (i.e., semipermeable) hollow fiber. Such a probe is inserted into living tissue and is perfused by an isotonic physiological fluid. Molecules diffuse down their concentration gradient across the dialysis membrane in a bidirectional way (*dialysis* for collecting endogenous compounds or *reverse dialysis* for applying exogenous molecules). In the case of collection, the relative recovery across the probe membrane, defined as the ratio between the extracellular concentration and the concentration of a compound at the outlet of the probe, depends on several factors: It increases with the surface of the membrane, decreases with higher flow rate of the perfusion fluid, and varies with the chemical and physical characteristics of the membrane [10]. Thus, two membranes may not exhibit the same recovery toward a given compound even if they have a similar cutoff.

TABLE 1 In Vitro Recoveries (%) Obtained at Ambient Temperature with Three 3-mm-Length Membranes for Three Neurotransmitter Amino Acids[a]

	Cuprophan Membrane CMA/11		Polycarbonate Membrane CMA/12		Regenerated Cellulose Membrane-Spectra/Por	
	Total Recovery	%/mm^2	Total Recovery	%/mm^2	Total Recovery	%/mm^2
Glutamate	27.6	12.4	27.5	5.9	28.0	13.2
GABA	2.3	1.3	20.9	4.4	22.9	15.5
Aspartate	12.6	5.7	18.5	4.0	20.9	9.9

[a]The probes were infused at 1 µL/min and plunged in a solution containing 10^{-6} mol/L GABA and 10^{-5} mol/L Glu and Asp.

For example, an in vitro comparison made in our laboratory between probes having a 3-mm-long membrane either in polycarbonate (CMA/12) or in Cuprophan, an unmodified cellulose (CMA/11), showed that their respective total recovery for Glu was similar (around 27%, Table 1). However, both the geometries of these two probes and their molecular cutoffs are different; the polycarbonate membrane diameter is larger (0.5 mm vs. 0.24 mm) and its cutoff is higher (20 kDa instead of 6 kDa for cellulose). Thus, despite the higher porosity of polycarbonate membrane and its larger area of diffusion, the recovery of Glu/mm^2 of membrane surface was twice as low as that for cellulose probes. In contrast, the recovery of GABA/mm^2 was more greatly enhanced with polycarbonate than with cellulose (Table 1). However, the best recoveries for both GABA and Glu were obtained using homemade probes with a Spectra/Por regenerated cellulose (Table 1). The performance of dialysis can be explained both by a larger cutoff (13 kDa instead of 6 kDa) and by the special physicochemical properties of the cellulose.

Therefore, the choice of the dialysis membrane can be crucial when trace compounds are sampled (or when the limit of detection of the analytical method is relatively high). This may also occur when the length of the membrane is too short to sample very small brain areas in rodents, such as the periaqueductal gray matter, the locus coeruleus, the amygdala, and hypothalamic nuclei. The characteristics of the probe membrane can also be primordial when applying drugs by reverse dialysis, since their diffusion toward the tissue will depend on their relative hydrophobicity and lack of interaction with the semipermeable membrane.

Another point to take into account is the choice of the geometry of the probe tubings when considering the sampling rate. Indeed, the higher the temporal resolution, the better the monitoring of rapid extracellular variations. Figure 1 illustrates the value of the high sampling rate, which makes it possible to improve the description of changes already known or to reveal changes never described before. For example, the glutamatergic receptor agonist

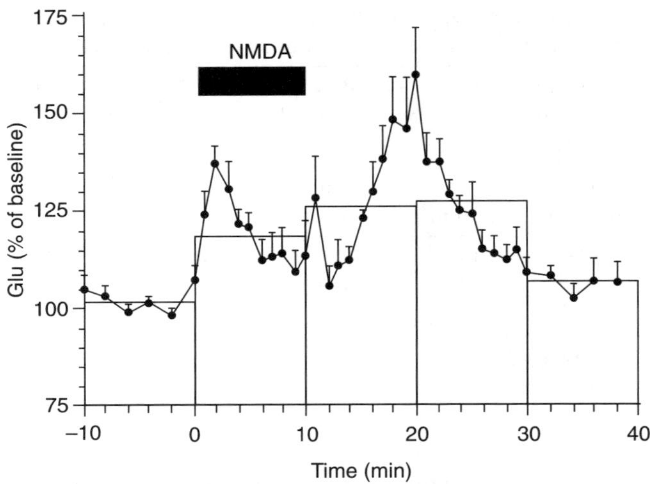

Figure 1   Effect of reverse dialysis of 1 mmol/L NMDA on dorsolateral striatal Glu concentrations. NMDA was administered for 10 min (black bar). Full circles represent microdialysis experiments carried out with a 1-min sampling rate. Histograms represent a 10-min simulated sampling rate of NMDA experiments. Glu concentration is expressed as percent (mean ± S.E.M.) of baseline levels: 4.04 ± 1.08 μmol/L ($n = 5$). Note that the biphasic increase of Glu concentrations is revealed only with a high sampling rate. (From [17], with permission from Springer Science and Business Media.)

N-methyl-D-aspartate (NMDA; 1 mmol/L for 10 min) is known to induce an increase in Glu levels of rat striatum, from earlier studies carried out with a 10-min (or longer) resolution microdialysis [11,12]. Using a 1-min sampling rate, we recently revealed that reverse dialysis of NMDA induced the first increase at the beginning of NMDA application followed by a return to baseline before the end of NMDA administration. Surprisingly, a second increase occurred after the end of the application (see Figure 4). We also demonstrated that the second increase was linked to NMDA removal [13]. This biphasic change in Glu could not have been observed without the use of a high sampling rate.

However, a high sampling rate requires appropriate probe tubings since solutes could undergo more longitudinal diffusion in the outlet probe tubing if the interval of time between dialysis and collection were superior to the sampling time. As a consequence, to avoid mixing of analytes between successive samples, the dead volume of these tubings (i.e., between the active dialysis membrane and the outlet of the probe) has to be minimized. This is particularly relevant when microdialysis experiments are carried out on awake animals because the setting requires long inlet and outlet probetubings in order to let the animals move freely. In this case, the dead volume can be greatly minimized by using capillary tubings, as demonstrated previously [14]. We also showed that an experimental determination of the dead volume is necessary to adapt each setup to the sampling rate required (Figure 2).

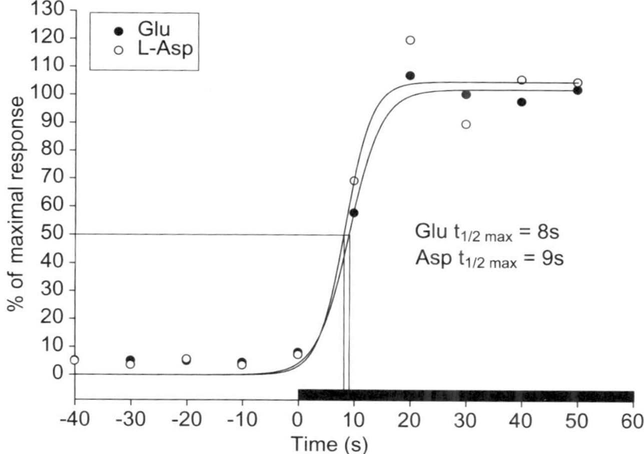

Figure 2 Determination of the time of response. Representation of the fitted curves corresponding to the increases in Glu and Asp concentration after the probe was plunged in a standard solution containing 5×10^{-6} mol/l Glu and Asp (black bar). Data are expressed as a percent of the maximal response. $t_{1/2max}$ corresponds to the half-time between the beginning of the dialysis in Glu/Asp-concentrated solution and the reach of the maximal response. Samples were collected every 10 s. In this case the length of the outlet tubing needed for the experimental animal chamber was 38 cm; the dead volume calculated for the 38 cm long $\times$ 40 μm i.d. outlet tubing was 0.47 μL and corresponded to a theoretical delay of 14 s with a 2-μL/min flow rate. The experimental dead time varied from 18 to 23 s (20.4 s $\pm$ 1.0, mean $\pm$ S.E.M., $n = 5$) between experiments and was always higher than the theoretical dead time. (From [14], with permission from Elsevier.)

Using optimized microdialysis setups, our group demonstrated previously that appropriate sampling conditions allow more accurate monitoring of pharmacologically induced increases [15] or behaviorally induced variations [14] in extracellular levels of amino acid neurotransmitters. Eventually, high temporal resolution can be useful in no-net-flux experiments, which can be shortened from several hours to less than 2 h [16–18]. This aspect can be significant when one determines two or three amino acid neurotransmitter concentrations on the same rat [16].

Finally, as the probe is perfused continuously at a constant flow rate, continuous sampling of extracellular amino acids is possible with no fluid loss. Amino acids present in the microdialysate can be analyzed directly without cleanup procedures, as high-molecular-weight proteins cannot cross the dialysis membrane. However, manipulation of microdialysis samples requires the use of sterile tubes, filtered artificial cerebrospinal fluid (aCSF), and gloves, to avoid contamination due to the ubiquitous presence of free amino acids on labware and skin. Samples may be analyzed by radioenzymatic methods or, more usually, by a separative method such as high-performance liquid chromatography (HPLC) or capillary electrophoresis (CE) [17,19]. Microdialysates can be analyzed online (i.e., at the outlet of the probe, through an analytical interface) or off-line (i.e., after sample collection in microtubes, in combination with a separative method) [17,20,21]. Thus, the determination of concentrations in each sample reflects the average concentration over the sampling duration defined.

2.2. Analytical Considerations

Microdialysis sample analysis has conventionally used classical HPLC with electrochemical [22,23] or fluorometric detection [24,25], and also enzymatic methods [26]. However, these analytical techniques exhibit poor mass sensitivity and require large-volume samples to determine neurotransmitter contents, leading to lengthy dialysis sampling times (15 to 20 min), although this requirement has been reduced by the breakthrough of microbore HPLC [24,25,27]. This temporal resolution, poor compared with the duration of neurobiological events, is even worse when samples have to be split for the simultaneous determination of different classes of neurotransmitters. As a consequence, most of the earlier microdialysis experiments were severely limited by the temporal resolution of microdialysis (5 to 30 min) compared to rapid changes occurring in the extracellular concentrations of neurotransmitters. In contrast, microdialysis coupled to CE, a more recent technique, makes it possible to monitor rapid changes in the extracellular concentration of neurotransmitters. Indeed, CE is able to analyze nanovolume samples with low limits of detection and appears to be suitable for microdialysis with a high sampling rate. At present, capillary electrophoresis (CE), in conjunction with laser-induced fluorescence (LIF) detection, has become one of the most powerful analytical tools for the determination of amino acid neurotransmitters in brain microdialysates because it offers the advantage of rapidity, high resolution, and sensitivity while requiring very small sample sizes [17,28,29]. In brain microdialysates, excitatory amino acids (EAAs) (i.e., Glu and Asp) and γ-aminobutyric acid (GABA) [14,16,30–36] were often analyzed. However, as amino acid neurotransmitters are not fluorescent at the wavelengths of most commercially available lasers, derivatization is needed prior to separation. Fluorescent reagents such as naphthalene-2,3-dicarboxaldehyde (NDA), orthophtaldehyde (OPA), or fluorescein isothiocyanate, reacting with the primary amine function of neurotransmitters, allow their detection following laser excitation at 442, 325, or 488 nm, respectively [35–37].

Several groups, including our own, have developed methodologies for the CE–LIF analysis of brain microdialysates after precapillary derivatization of samples with fluorogenic agents in these methods, the derivatization reaction is generally carried out manually and batch-processed before the derivatized sample is introduced into the separation capillary. The main advantage of these manual derivatizations is a relatively free choice of the reaction conditions. Their main disadvantage is the "open vial" procedure, which can cause analyte losses through solvent evaporation or by adsorption to the walls of the vials and/or which can cause contamination of the sample. Such drawbacks are reduced when precapillary derivatization is performed through an online mode, by means of a microreactor connected to the microdialysis probe outlet [14,36,38], but such an approach requires construction of special equipment, making its use difficult in ordinary practice. The in-capillary derivatization has been introduced as an approach to eliminate the limitations of these precapillary derivatizations. Indeed, as CE allows free solution analysis in an open tubular capillary, a part of the inner space of a capillary can be regarded as a place where reaction derivatization could occur prior to electrophoretic analysis.

The major advantage of such a procedure is that only small volumes of reagent are needed and sample dilution is reduced to a minimum, since the front end of the capillary is used as a reaction chamber. As a result, these approaches should be especially suitable for the analysis of extremely small sample volumes, such as brain microdialysates [39]. Our group recently reported the first in-capillary NDA derivatization for CE–LIF analysis of brain microdialysis amino acids using a commercially available CE–LIF instrument and allowing a short run time [40].

When considering these derivatization procedures, one may note that there is still a large difference between the total volume of sample and the volume actually injected into the CE instrument. In both cases the volume effectively introduced into the capillary (several nanoliters) is one thousandth to one hundredth of the solution. The rest remains unused or discarded. To hamper the loss of samples and to increase the temporal resolution of microdialysis, new strategies based on miniaturization are being developed. Recently, analytical research has been focused on micro total analysis systems (μTASs), also called labs-on-chips. The lab-on-chip appears to be a powerful and promising approach which should overcome the limitations of the CE–LIF instrument, mainly in term of analysis throughput, ease of automation, and smaller size, which allows its use at the "point of care." In that context, a lab-on-chip should automate the processing steps, including separation and LIF detection for the identification and quantification of AAs in biological samples with higher accuracy, repeatability, and reproducibility.

For example, recently, a glass microfluidic electrophoresis chip was coupled to microdialysis for in vivo monitoring of amino acid (AA) levels [41]. This microsystem exhibited a high degree of integration and incorporated microdialysis sampling, on-chip derivatization, electrophoretic separation, and LIF detection using a confocal microscope. A detection limit of $0.2\,\mu$M (for phenylalanine) was reached using a separation length of $9.4\,$cm instead of the 50 to $60\,$cm used in classical CE–LIFD. OPA reacts rapidly with AAs (the half-time of reaction for alanine is $4\,$s [42]); however, the fluorescent products degrade rapidly (lifetime $10\,$min [43]). To solve this problem, other fluorogenic molecules, such as DTAF (dichlorotriazine fluorescein) and NDA, were also tested for on-chip derivatization. The authors demonstrated on-chip labeling of biogenic amines using DTAF in glass microchips using a separation length of $3.5\,$cm [44].

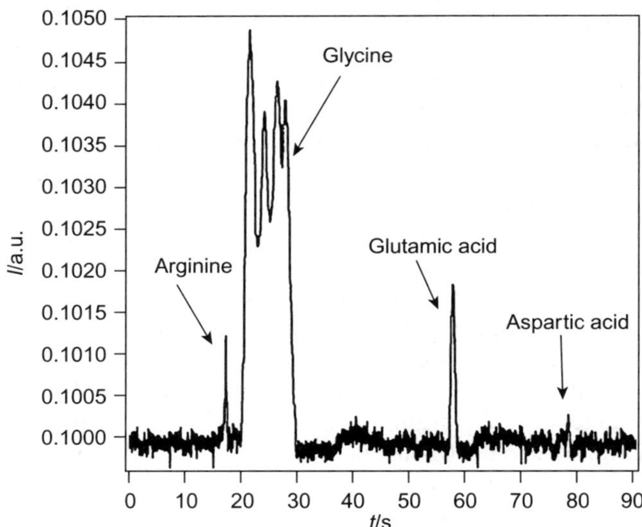

Figure 3 Electropherogram obtained for a sample of a rat hippocampus microdialysate using the microchip reported. (From [45], with permission from Elsevier.)

Quite recently, our group and collaborators developed a poly(dimethylsiloxane) (PDMS) electrophoresis device able to carry out on-chip derivatization and quantification of amino acids using NDA as a fluorogenic agent. A chemical modification of the PDMS surface was found compulsory for achieving the derivatization of standard AAs with NDA and a limit of detection (LOD) of 40 nM was reached for glycine [45]. An example of the applicability of this microdevice for the analysis of real biological samples such as a rat hippocampus microdialysate is shown in Figure 3. The features of such a TAS are its ease in use, ability to handle high numbers of sample, high mass sensitivity, and ready disposability, while maintaining accuracy, repeatability, and reproducibility and these standardization of the assay. Although developed for a given application, µTASs can be versatile enough to be used for any determination of a given compound.

Finally, compared to other approaches, the mass spectrometry (MS) detector is a highly selective tool since it allows the identification of compounds through a structure-based analysis. In the typical liquid chromatography/tandem mass spectrometry (LC–MS/MS) assay, an analyte can be identified by its retention time, molecular weight, and characteristic fragmentation ions. It is an attractive choice for determining trace-level neurotransmitters from the aCSF and is expected to reduce interference from endogenous compounds significantly compared to LIF or electrochemical detectors. HPLC–MS, as an alternative to HPLC–EC detection, has been used to detect amino acids from microdialysates after derivatization. For example, GABA was separated by means of precolumn derivatization with 7-fluoro-4-nitrobenzoxadiazole and then sensitively detected by MS [46]. The tool can be particularly relevant when quantifying dialysate Glu with ^{13}C isotope enrichment [47].

Finally, one of the most interesting and useful features of microdialysis is the ability to apply drugs locally in order to study the effect of pharmacological agents acting on various receptors (see Section 3). The literature related to this field is abundant, but most works reported the dynamics of amino acid variations without studying the pharmacokinetics of the drug that induced these variations. Simultaneous measurement of both neurotransmitters and drugs within the same microdialysis samples can be carried out if a suitable analytical methodology is available. For example, if the drug is structurally similar to the target amino acid, a protocol of separation can be developed more easily. In that respect, we have investigated the effect of the antiepileptic drug vigabatrin, an analog of GABA, on Glu, Asp, and GABA variations in rats [48]. For that purpose, we built a double-probe system whose aim was to apply vigabatrin through the first probe and to collect vigabatrin and the three neurotransmitter amino acids via the second probe. Even if such information can be useful in pharmacokinetic–pharmacodynamic studies, the generalization to other drugs will need some analytical development for each new drug tested.

2.3. Signification of Microdialysates Glu and GABA

In vivo microdialysis is a useful method for monitoring the neurotransmitters present in brain extracellular fluid [49]. However, many studies have questioned to what extent dialysate Glu concentration is related to the amount of Glu released by the presynaptic neuron. Indeed, due to the ubiquitous localization of its metabolism enzymes in all brain cells and its role in protein synthesis and general metabolism in addition to neurotransmission, numerous studies have attempted to determine the origin of extracellular Glu. If the proportion of extracellular Glu taken up by glia is about 80 to 90% of the Glu pool available as a substrate for glutamine synthesis as confirmed by in vivo nuclear magnetic resonance and microdialysis [50], the part of the neuronal Glu released for neurotransmission purpose is still questioned. Moreover, because Glu may come from many sources of effluxes, such as neuronal "classical" release [1], exchange via cysteine–glutamate transporters [51–54], inversion of transporters [55], and glia release via exocytosis or nonexocytosis [56–58], former methodological strategies used for monoamine neurotransmitters provided limited answers. In the 1990s and early 2000s, the majority of studies used tools such as calcium-low or calcium-free aCSF to impair the vesicular neuronal release and tetrodotoxine (TTX) used to block Na^+-dependent channels. Several studies reported decreases in the basal level of dialysate Glu after Ca^{2+} removal or TTX application, suggesting that part of the basal Glu may originate from neuronal release. In contrast, other studies described no change or even increases under such experimental conditions [59]. The discrepancies of the results observed in all the studies fail to characterize the neuronal versus glial origin of extracellular Glu [59].

Consequently, it was suggested that most neurotransmitter Glu released into a synaptic cleft under basal conditions may be taken up into surrounding glia and diffuses poorly to the dialysis probe. Increasing the microdialysis sampling rate has been proposed in order to observe the rapid variations in Glu extracellular level which are expected to occur in neurophysiological events [15,60]. Thus, recent studies using high-sampling-rate microdialysis showed that NMDA application [13] or electrical stimulation of the prefrontal cortex [60] increases dialysate Glu concentrations in brain regions receiving projections from this area, and that the increase is suppressed or partly altered by TTX [13,61]. These studies strongly suggest that the transient increase in dialysate Glu detected under these conditions represents evoked neurotransmitter Glu release.

However, when considering basal conditions in the absence of any stimulus, one may still question to what extend microdialysate Glu represents neurotransmitter released into synaptic cleft. The same question can be asked for extracellular GABA and Asp, because they have both neurotransmitter properties and various sources or because their metabolism is tightly linked to Glu. Moreover, Asp shares one receptor and one transporter with Glu, but its presence in presynaptic vesicles as a classical neurotransmitter is still questioned, even if evidence for such a presence has been seen in the hippocampus [62]. As for Glu, unsuccessful conclusions were also reported regarding the origin of dialysate GABA and Asp in basal conditions, so that the significance of dialysate Glu, Asp, and GABA is still under debate [17,59].

Thus, the extracellular (or dialysate) concentrations of Glu and GABA may not provide a reliable index of their synaptic exocytotic release. Moreover, a strict area between the intra- and extrasynaptic compartments was evidenced, with poor spillover from the synaptic to the extrasynaptic compartment [63,64], due mainly to a strong glial uptake. Thus, the changes in extracellular concentrations of Glu and GABA under specific pharmacological and behavioral stimuli could be interpreted not only as a consequence of the activation of specific neurochemical circuits, but as an expression of the activity of the neurone–astrocyte unit in specific circuits of the brain. Several authors proposed that dialysate changes in Glu and GABA could be used as an index of volume transmission–mediated actions of these two neurotransmitters. This hypothesis is based, first, on the assumption that the activity of neurones is functionally linked to the activity of astrocytes, which can release Glu and GABA to the extracellular space [56,65]; second, on the existence of extrasynaptic Glu and GABA receptors with functional properties different from those of receptors located in the synapse [66,67]; and third, on the experimental evidence reporting specific electrophysiological and neurochemical effects of Glu and GABA when their levels in the extracellular space are increased [68]. Thus, Glu and GABA, once released into the extracellular compartment, could diffuse and have long-lasting effects modulating glutamatergic and/or GABAergic neurone–astrocytic networks and interacting with neurons containing other neurotransmitters and located in the same areas of the brain (Figure 4).

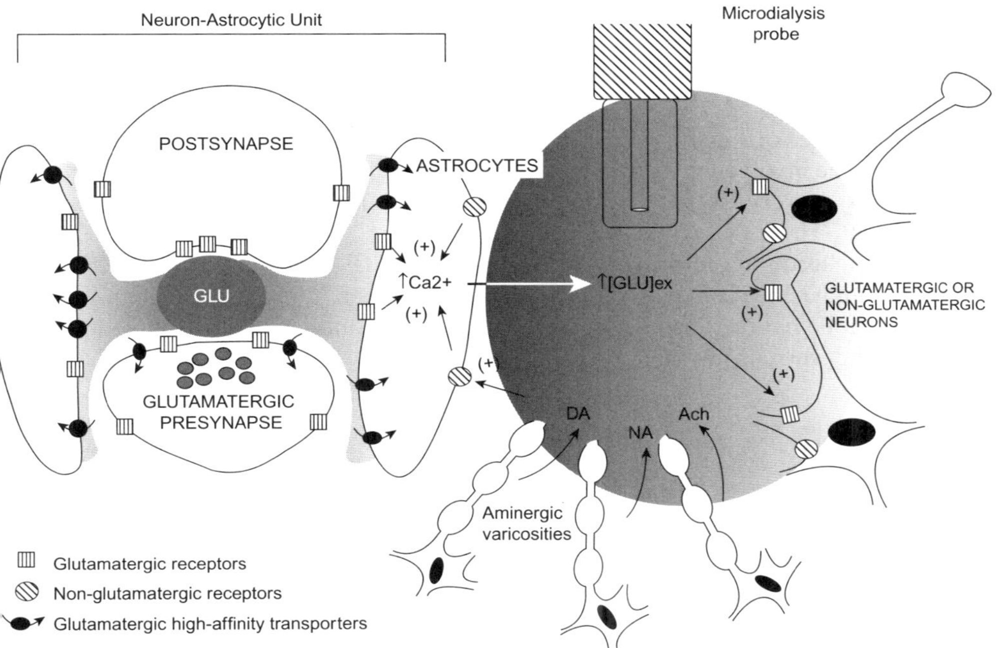

Figure 4 Scheme depicting the hypothesis proposed by Del Arco and colleagues [68] in which the changes of extracellular concentrations of neurotransmitter amino acids [only the case of glutamate (Glu) is presented here] monitored by microdialysis are an index of the volume transmission signal in the brain. Astrocytes could be specifically involved in modulating extracellular concentrations of these amino acids under specific pharmacological treatments or behavioral stimuli. On the one hand, synaptic glutamate could act on extrasynaptic glutamatergic receptors located on neurons and astrocytes. In the case of astrocytes, it is shown that the increase of calcium (Ca^{2+}) levels induces astrocytic glutamate release. On the other hand, other neurotransmitters, such as dopamine (DA), noradrenaline (NA), or acetylcholine (Ach), could activate nonglutamatergic receptors located on astrocytes and in turn induce (i.e., via intracellular pathways) the release of astrocytic glutamate to the extracellular space, where calcium can again play a central role. Glutamate once increased into the extracellular space ([Glu]ex) would diffuse through the extracellular space to reach the microdialysis probe; [Glu]ex would also activate glutamatergic extrasynaptic receptors to modulate the activity of glial-neuronal assemblies around the microdialysis probe. Astrocytic glutamate and GABA, as well as neuronal amines released from nonsynaptic sites, would create an extracellular microenvironment that would modulate the activity of neighboring neuron–astrocyte assemblies via volume transmission. The effect produced by one particular neurotransmitter would depend on other neurotransmitters existing in the extracellular microenvironment through integration of these signals in astrocytes and neurons. (From [68], with permission from Wiley-Blackwell.)

Similar assumptions could be made for Asp [69], as experiments carried out with synaptosomes have suggested that Asp is released primarily outside the presynaptic active zones and may therefore serve as the predominant agonist for extrasynaptic NMDA receptors. Other recent studies suggest that glycine, the amino acid co-agonist of NMDA receptor, and taurine, an amino acid known to be involved in osmoregulation and a partial agonist of Gly receptors, could also be very relevant in modulating neurotramission as a volume transmitter [70,71].

In conclusion, monitoring of Glu, Asp, and GABA concentrations in the extracellular space using microdialysis may provide an indirect index of amino acid synaptic neurotransmission while giving direct indications of amino acid volume neurotransmission.

3. BASIC RESEARCH ON RECEPTORS

Microdialysis provides in vivo evidence regarding modulation of the extracellular levels of amino acid neurotransmitters by both homo- and heteroreceptors. Due to an abundant literature on this subject, we chose to review only recent microdialysis data dealing with the effect of drugs or pharmacologically active compounds on amino acid neurotransmitters. Our aim here is not to give an exhaustive review of the numerous papers but to put forward typical examples of recent pharmacological research based on microdialysis studies.

3.1. Glutamate Receptors

Ionotropic Receptors   Ionotropic receptors mediate the excitatory neurotransmission in the nervous system: Activation of AMPA/KA receptors induces depolarization by an inward flux of Na^+ generating the nerve influx, followed by activation of the NMDA, triggering entry of Ca^{2+} and subsequent modulations of intracellular metabolism. Many studies have reported on the crucial role of ionotropic receptors in brain functioning, including microdialysis studies focused on the neurotoxicity of Glu. Whereas the earlier microdialysis studies dealt with intact animals, the most recent studies are more devoted to animal models of human diseases, such as Parkinson's disease. In that respect, recent studies emphasized perspectives on new potential methods of treatment as well as a better understanding of the neuronal circuitry involved in physiopathological mechanisms. First, multiple probe locations and the use of NMDA antagonists that target either NMDA 2A or NMDA 2B subtypes, NVP-AAM077 and Ro 25-6981, respectively, allowed the investigation of striatal output pathways in models of Parkinson's disease. Striatal NMDA receptors containing NR2A and NR2B subunits preferentially regulate the striatopallidal and striatonigral projection neurons, respectively [72]. Under experimental parkinsonian conditions, NVP-AAM077 was able to attenuate motor deficits in rats associated with a reduction of GABA levels in globus pallidus but not in substantia nigra pars reticulata, while reverse dialysis of Ro 25-6981 failed to change GABA release in these areas. NR2A subunits could facilitate striatopallidal neurons, while NR2B subunits do not participate in modulation of striatal output pathways. Systemic administration of low doses of NVP-AAM077 or Ro 25-6981 produced mild improvement of motor performance and reduced pallidal GABA release. The authors hypothesize that motor improvement induced by NVP-AAM077 is accomplished by a blockade of NR2A subunits on striatopallidal neurons. Modulation of pallidal activity may also underlie motor improvement induced by Ro 25-6981, although in this case, blockade of extrastriatal NR2B subunits may be involved [73]. However, the complexity of basal ganglia has to take into account the multiple interactions between Glu, GABA, and dopamine and their respective receptor subtypes [74,75].

Nowadays, the pharmacological works dealing with brain ionotropic receptors investigate interactions with other neuromodulators, due to the ubiquitous localization of those receptors throughout the brain neurons and glial cells. For example, one can cite here a possible *ménage-à-trois* among cannabinoid CB1 receptors, NMDA receptors, and oxytocin neurons in the paraventricular nucleus of the hypothalamus mediating penile erection [76]. However, most microdialysis studies reported on the pharmacodynamics effects of antipsychotic drugs (e.g., [77,78]), hallucinogens [79], and drugs of abuse (see Section 4) on amino acid neurotransmitters.

Metabotropic Receptors   The metabotropic receptors (mGluRs), which can be expressed on both glutamatergic and nonglutamatergic neurons, can regulate excitatory and inhibitory neurotransmissions. They are divided into three groups: group I, with mGluR1 and mGluR5 coupled to Gq/11 protein located mainly extrasynaptically on postsynaptic neurons; group II, with mGluR2 and mGluR3 coupled to Gi/0 protein located extrasynaptically on presynaptic neurons; and group III, coupled to Gi/0 with mGluR4, mGluR6, mGluR7, and mGluR8 located in the synaptic cleft on terminals [80]. Thanks to recent compounds used as selective agonists or antagonists, their implications in several pathologies have been studied, and therefore pieces of evidence could be available for putative treatments.

Focus can be placed on the microdialysis studies aiming to evaluate the interactions between ionotropic and metabotropic receptors, especially in the context of schizophrenia. Recent preclinical discoveries have brought into focus the potential use of mGluR2/3 agonists in the treatment of schizophrenia, where overactivity of Glu transmission has been suggested. For example, Lorrain's group has investigated the effects of LY379268, a selective mGluR2/3 agonist, in a model of schizophrenia-like symptoms induced by the NMDA antagonist ketamine [81]. These authors showed that the stimulation of mGluR2/3 located in the mPFC can prevent the ketamine-evoked Glu release. The same results have been found previously with PCP, another NMDA antagonist that has hallucinogen properties [82].

Metabotropic receptors from groups I, II, and III are also involved in the modulation of hyperexcitability occurring in epilepsy. An interesting study put forward the role of mGluR4 in the genesis of absence epilepsy. The effluxes of GABA and Glu [83] were studied in mutant mGluR$_4$–/– mice resistant to chemically induced absence seizures. The authors demonstrated that the basal levels and K$^+$-evoked releases of GABA and Glu were altered only in several areas of the thalamocortical circuitry, a loop where spike and wave discharges are generated. The effects of several metabotropic agonists on hippocampal Glu and GABA levels have been investigated in the model of pilocarpine-induced epilepsy [84]. Intrahippocampal mGluR group I agonist [(R,S)-DHPG] perfusion via a microdialysis probe induced seizures and concomitant increases in amino acid dialysate levels. The mGlu1a receptor antagonist LY367385, which exhibits anticonvulsant properties, decreased baseline Glu but not GABA concentrations, suggesting a tonic activation of these receptors by endogenous Glu. The mGlu5 receptor antagonist 2-methyl-6-(phenylethynyl) pyridine (MPEP) also clearly abolished pilocarpine-induced seizures. The data found with group I antagonists have not been observed with LY379268, a mGluR2/3 agonist, that decreased basal hippocampal GABA but not Glu levels, concomitant to poor anticonvulsant protection.

The role of mGluR5 receptors was also emphasized in models of neurodegenerative diseases such as Parkinson's disease or in chemically induced neurotoxicity. For example, a selective and potent mGluR5 antagonist, 3-[(2-methyl-1,3-thiazol-4-yl)ethynyl]pyridine (MTEP), was tested for its ability to modulate behavioral and neurochemical correlates of dyskinesia in 6-hydroxydopamine–lesioned rats treated with L-dopa. The compound significantly attenuated both the induction of abnormal involuntary movements and the increased GABA outflow within the substantia nigra pars reticulata [85]. Another study observed that MPEP, antagonist of mGluR5, applied through a dialysis probe, significantly reduced the increase in Glu striatal levels induced by quinolinic acid [86].

Finally, interesting data obtained in periaqueductal gray (PAG), a brain area involved in the ascending pathways of pain integration, showed that the activation of mGluR7 and mGluR8 receptors belonging to the same group can lead to differential effluxes of Glu and GABA. Application of AMN082, a mGluR7 agonist, induces a decrease in Glu levels without affecting GABA concentrations, whereas (S)-3,4-DCPG, a mGluR8 agonist, evokes an increase in Glu concentrations and a reduction in GABA levels [87,88].

3.2. Nonglutamatergic Receptors

Dopamine The interactions between dopamine (DA) and Glu or GABA have been studied extensively for several years and are still the subject of research, especially in the field of addiction (see Section 4), Parkinson's disease, and schizophrenia. One may emphasize the microdialysis work carried out by several groups, where both limbic areas and basal ganglia components were targeted, such as the prefrontal cortex [89], the striatum/ventral tegmental area [90,91], the substantia nigra [91], the nucleus accumbens [92], and the globus pallidus [93].

Serotonine The latest studies dealing with the modulation of extracellular amino acids by serotonergic receptors are devoted to drugs that deteriorate cognition as hallucinogens or recent molecules developed for the improvement of cognitive functions.

As the 5-hydroxytryptamine (5-HT)1A receptor is involved in cognitive processes by a facilitating glutamatergic transmission, a novel 5-HT1A receptor antagonist, 4-cyano-*N*-[2*R*-[4-(2,3-dihydrobenzo[1,4]-dioxin-5-yl)-piperazin-1-yl]-propyl]-*N*-pyridin-2-yl-benzamide (lecozotan), was tested in several species. Using in vivo microdialysis, lecozotan [0.3 mg/kg, subcutaneously (s.c.)] significantly potentiated the K^+-stimulated release of Glu and acetylcholine in the dentate gyrus of the hippocampus, while its chronic administration improved learning deficits induced by the glutamatergic antagonist MK-801 [94].

As mentioned previously in this chapter, intracortical injection of the selective and competitive NMDA receptor antagonist 3-(*R*)-2-carboxypiperazin-4-propyl-1-phosphonic acid (CPP) impaired attention performance in rats; however, a blockade of 5-HT2A receptors or stimulation of 5-HT1A receptors antagonized this effect, as demonstrated by Ceglia et al. [95] and Calcagno et al. [96]. This group used microdialysis in awake rats to study the effect of CPP on extracellular Glu in the medial prefrontal cortex and the dose-dependent regulation of this effect by an antagonist of 5-HT2A receptors, M100907 or the 5-HT1A agonist 8-hydroxy-2-(di-*n*-propylamino)tetralin, 8-OH-DPAT. M100907 or 8-OH-DPAT blocks the rise in extracellular Glu induced by CPP. Similar results were obtained with another hallucinogen, LSD [97].

Finally, several studies showed the possible in vivo modulation of basal or evoked GABA/Glu levels by 5-HT6 receptor agonists in various brain areas receiving inhibitory and excitatory projections [98,99].

Adenosine The functional role of adenosine in the brain and the physiological implication of adenosine A2 receptors, particularly in basal ganglia, have been reviewed [100]. The interaction between neurotransmitter amino acids and adenosine has also been studied by microdialysis. For example, an increase in striatal adenosine is reported to be an index of mitochondrial dysfunction observed in the postmortem brain of patients suffering from Huntington's disease. In this respect, works carried out on transgenic R6/2 mice, a model of Huntington's disease, revealed that striatal adenosine extracellular levels, assessed by microdialysis, are increased significantly in comparison to wild-type mice, while no changes in Glu and GABA levels were revealed [101]. However, the reverse dialysis of the selective A2A receptor antagonist SCH 58261 antagonized that increase in adenosine and affected levels of Glu alone.

On the other hand, some A2A receptor antagonists may be putative agents for the treatment of Parkinson's disease (PD). A study examined the modulation of GABA and Glu effluxes in the substantia nigra pars reticulata (SNr) using in vivo microdialysis in 6-hydroxydopamine (6-OHDA)-lesioned rats, an experimental model of Parkinson's disease (PD) [102]. In 6-OHDA-lesioned rats compared with normal rats, basal extracellular GABA levels in the SNr show no change, whereas basal Glu levels are increased significantly. To clarify the adenosine/amino acids interaction, oral administration of the A2A receptor-selective antagonist, (*E*-1,3-diethyl-8-(3,4-dimethoxystyryl)-7-methyl-3,7-dihydro-1*H*-purine-2,6-dion (KW-6002), to 6-OHDA-lesioned rats at 1 mg/kg caused a marked and sustained increase of GABA and Glu levels in the SNr. The increase of nigral Glu by KW-6002 was abolished by a kainic acid–induced lesion of the globus pallidus (GP) or subthalamic nucleus (STN) in 6-OHDA-lesioned rats, whereas the increase in nigral GABA was blocked completely by the GP lesion but only partially by the STN lesion. These results indicate that changes in neurotransmitter release in the SNr brought about by KW-6002 are largely attributable to blockade of A2A receptor–mediated modulation of striatopallidal neurons.

Cannabinoid   In a few microdialysis studies, the modulation of excitatory amino acids by the cannabinoid (CB) receptors has also been examined in rats. Application of the CB1 receptor agonist WIN 55,212-2 induces an increase in Glu level in the striatum. Besides, in the quinolinic acid–induced excitoxicity model of Huntington's disease, reverse dialysis of WIN 55,212-2 prevented any increase in extracellular striatal Glu induced by QA. Parallel studies using electrophysiological recordings in corticostriatal slices showed that application of WIN 55,212-2 prevented QA-induced reduction of the field potential amplitude. In in vivo experiments, intrastriatal WIN 55,212-2 significantly attenuated the striatal damage induced by QA, although no significant effects were observed in the behavioral tests [103].

3.3. Peptidergic Receptors

The influence of peptides on amino acid neurotransmitters have been far less documented than classical neurotransmitters such as dopamine, acetylcholine, or serotonin. However, peptide receptors can be expressed or recruited specifically in natural behaviors such as maternal care as well as in stressor situations, epilepsy, or pain. Several groups attempted to understand the links between peptide receptors and their effects on both excitatory and inhibitory neuronal pathways.

Oxytocin and Vasopressin   Kendrick's team was one of the first groups working on oxytocin and vasopressin using microdialysis in sheep. They demonstrated that oxytocin (OXT), but not vasopressin (AVP), induces GABA efflux in the olfactory bulb, without affecting Glu, in both nulliparous and multiparous ewes via both OXT and V1 receptors [104]. Singewald's group studied the influence of OXT in amygdala, where a substantial number of oxytocinergic fibers and OXT receptors are localized. They explored the role they play in the response to swim stress in male rats [105]. A release of Glu and Asp, induced by the local application of the OXT receptor antagonist des-Gly-NH$_2$$d$(CH$_2$)$_5$(Tyr(Me)$_2Thr_4$)OVT, was found to be elevated within the central amygdala, whereas local GABA release remained unchanged. However, this effect was present only under stimulation (i.e., during exposure to stress) when local release of OXT is enhanced, but could not be found under basal conditions. Thus, the interaction between OXT and the excitatory neurotransmitter might functionally modulate the emotional behavior. If amygdala GABA had no noticeable influence in swim stress, it has been strongly involved in the social defeat. Other data from the same group demonstrate that GABA is released within the suprachiasmatic nucleus of the hypothalamus (SON) during emotional stress to act as a selective inhibitor of both central and peripheral OXT secretion, as revealed by the use of bicuculline, a GABAB receptor antagonist [106]. Finally, the contribution of AVP on the stress-induced release of amino acids within the septum was also investigated. The data showed that a 10-min forced swimming session stimulated the release of Glu, Asp, arginine, GABA, and taurine, but only taurine efflux was affected by local administration of the AVP V1 receptor antagonist d(CH$_2$)$_5$Tyr(Me) AVP by reverse microdialysis.

Bradykinin Microdialysis data suggest that the bradykinin system contributes to the neuronal hyperexcitability in epilepsy through B1 receptors. Indeed, B1 receptors, which are physiologically absent, are expressed at high levels in the rat brain after completion of kindling, a model of temporal lobe epilepsy. Microdialysis experiments showed that Glu efflux was increased in kindled rats after B1 receptor agonist Lys-des-Arg$_9$-BK administration, whereas selective B1 receptor antagonist R-715 prevented that effect [107].

Angiotensins Two types of angiotensins have recently been studied by in vivo microdialysis in two different paradigms. First, the involvement of angiotensin in the production of NO and prostaglandins, which is well known in peripheral tissues, has been hypothetized in brain, especially for angiotensin(1–7) which does not have affinity for AT1 and AT2 receptors nor for insulin-regulated aminopeptidase (IRAP), another target for angiotensins. A work shows that angiotensin(1–7) induces an enhancement of GABA efflux in striatum via NO production as the nitric oxide synthase inhibitor L-NAME blocks the Ang(1–7)-induced GABA release [108]. Second, the co-localization of angiotensins in both brain catecholaminergic and noncatecholaminergic neurons led to an investigation of the effects of an injection of Ang II in one area involved in the control of sympathetic nerve activity and arterial blood pressure, the rostral ventrolateral medulla (RVLM) [109]. The authors showed that the stimulation of RVLM by Ang II led to an increase in Glu and GABA effluxes in its target projection, the intermediolateral column of the thoracic spinal cord.

Somatostatin   Somatostatin (SST) is widely distributed both within the brain and in peripheral tissues. It is involved in many functions, such as neurotransmission, secretion, and proliferation. However, some microdialysis studies have been devoted to understanding its role in the modulation of basal ganglia. Two main studies can be emphasized here showing that SST may control the level of extracellular amino acids in the striatum. First, Hathway et al. [110] gave evidence that SST can dose-dependently increase in vivo extracellular levels of Asp, Glu, dopamine, and GABA via both ionotropic and GABAB receptors. More recently, the same authors showed that SST-evoked dopamine and Glu release was attenuated in the striatum of Sstr2lacZ/lacZ mice; however, levels of Glu were significantly higher in transgenic than in wild-type mice. The authors concluded that SSTR2 is the only SST receptor that mediates the effects of SST on striatal afferent neurotransmitters [111]. Finally, the anticonvulsant properties of SST were examined in the model of pilocarpine-induced epilepsy in rats by inhibiting the insulin-regulated aminopeptidase (IRAP), the SST-degradating enzyme. Ang IV and SST [10 nmol/h, intracerebroventicularly (i.c.v.)] were able to completely abolish the pilocarpine-induced increase of Glu in hippocampus, suggesting direct or indirect interaction between SST and Glu release [112].

Cholecystokinin   Cholecystokinin (CCK) has been demonstrated to increase both Glu and Asp concentrations in frontoparietal cortex and striatum via the specific activation of CCK2 receptors [113], but the contribution of CCK1 receptors cannot be totally excluded for Glu because another study found that CCK-8S-induced Asp efflux was prevented by systemic pretreatment with the CCK2 antagonist L-365,260, but not with the CCK1 antagonist L-364,718, while Glu efflux was antagonized by both antagonists. [114]. Besides, CCK also modulates GABA release as demonstrated [115], as intra-accumbens perfusion with CCK-8S exerts CCK2 receptor-mediated inhibition of ventral pallidal GABA levels. Furthermore, the TTX and bicuculline sensitivity of this effect suggests that it may be mediated via CCK2 receptors probably located on local GABA interneurons.

Neurotensin   Neurotensin is a peptide encountered ubiquitously in striatum, prefrontal cortex, substantia nigra, or hippocampus, for example. When administered locally by reverse dialysis, this peptide can exert differential effects on extracellular levels of Glu, Asp, or GABA, depending on the dose tested and the brain area [116–119]. The NT effects on AA levels are of neural origin, at least in the prefrontal cortex, as they are TTX-sensitive [119].

4. PSYCHOSTIMULANTS AND ADDICTIVE DRUGS

Brain microdialysis has been shown to be a powerful tool for determining the mechanisms of action of psychostimulants and the neurobiological mechanisms underlying drug addiction. Indeed, microdialysis makes possible monitoring of neurochemical changes, often referred to as *drug-induced neuroplasticity*, that occur after initial exposure as well as after repeated exposures to a drug of abuse [120]. The most prominent data on the neurochemical changes induced by amphetamine, cocaine, heroin, or alcohol and obtained using brain microdialysis are reviewed below.

4.1. Amphetamine

Acute Effects The effects of acute systemic d-amphetamine on extracellular Glu levels have been studied by in vivo microdialysis in several brain regions of awake, freely moving rats. Glu levels were increased by d-amphetamine [2 to 5 mg/kg, intraperitoneally (i.p.)] in the striatum, nucleus accumbens (NAcc, corresponding to ventral striatum), prefrontal cortex, and ventral tegmental area (VTA) [121,122]. In the NAcc, the increase in Glu was greater in rats raised under enrichment conditions (ECs) during development than in rats raised under impoverished conditions (ICs) [123]. This suggests that alterations in Glu in the NAcc may be involved in the environment-dependent effects of amphetamine, since EC rats are more sensitive to the locomotor stimulant effects of acute amphetamine.

Lowering the Ca^{2+} concentration in the aCSF perfusing the probe significantly reduced the increase in Glu level in the striatum and ventral striatum evoked by i.p. amphetamine (2.5 mg/kg), distinguishing a Ca^{2+}-dependent and a Ca^{2+}-independent component of release [124,125]. Moreover, the systemically administered amphetamine-evoked striatal Glu efflux appeared to be modulated by opioid receptors. Indeed, the delta opioid antagonist naltrindole (perfused by reverse dialysis) decreased amphetamine-evoked Glu levels, whereas the delta agonist [D-Pen(2,5)]-enkephalin reversed the effect of naltrindole. When the aCSF Ca^{2+} concentration was lowered, naltrindole had no effect on the Ca^{2+}-independent component of amphetamine-evoked Glu levels [125]. On the other hand, the κ-opioid agonist U-69593 (subcutaneously) significantly decreased the amphetamine-evoked increase in Glu levels in the ventral striatum, and reverse dialysis of the selective kappa-opioid receptor antagonist nor-binaltorphimine antagonized these effects; however, U-69593 did not alter the Ca^{2+}-independent component of amphetamine-evoked Glu levels [124]. Taken together, these microdialysis studies show that intrastriatal δ-opioid receptors facilitate, while κ receptors inhibit, the Ca^{2+}-dependent component of amphetamine-induced Glu efflux.

Like systemic administration, the reversed microdialysis infusion of amphetamine increases the extracellular concentrations of Glu and Asp in the striatum. An increase in GABA was also found. It was both Ca^{2+}-dependent and high-affinity GABA transporter-mediated, whereas the increase in Glu was independent of Ca^{2+} in the perfusion medium but significantly attenuated by specific blockers of the high-affinity Glu transporters [126,127]. In the more ventral NAcc, in contrast, the local retrodialysis of amphetamine may lead to a dose-dependent reduction in basal Glu, which is prevented by the systemic administration of the dopamine D_1 receptor antagonist SCH 23390 or the D_2 antagonist sulpiride [128].

The mechanism underlying systemic amphetamine-induced Glu efflux in the VTA has been studied [129,130]. The NMDA receptor antagonist MK-801 or the D_1 dopamine receptor antagonist SCH 23390, given i.p. 30 min before acute amphetamine (5 mg/kg, i.p.), prevented the amphetamine-induced Glu efflux in the VTA. This Glu efflux was also abolished by the Glu uptake inhibitor dihydrokainate, but was unaffected by perfusion with a low Ca^{2+}/high Mg^{2+} solution, implicating Glu transporters. Moreover, as it was also prevented by trapping agents perfused locally (D-phenylalanine) or injected systemically (α-phenyl-*N-tert*-butyl nitrone), it may represent a response to oxidative stress. Like peripheral administration, amphetamine administered through the dialysis probe into the VTA also produced a gradual and sustained increase in Glu efflux which began after return to normal perfusion solution. It may be preceded during amphetamine perfusion by an apparent decrease in Glu efflux [131].

Like the systemic injection, the intracerebral infusion of different doses of amphetamine produced a dose-related increase in extracellular concentrations of Glu in the medial prefrontal cortex of the rat [132], and this effect was found to be independent of Ca^{2+} in the perfusion medium.

Sensitization   A critical event in the development of behavioral sensitization to amphetamine is a transient increase in the excitatory drive to dopamine neurons of the VTA. To investigate whether neurons of the laterodorsal tegmentum (LDT) which provide an important source of glutamatergic drive to the VTA are involved in such sensitization, AMPA (0.4 nmol) was microinjected into the LDT and Glu monitored by microdialysis in the ipsilateral VTA of rats, 2 days after discontinuing repeated saline or amphetamine injections. Glu efflux was transiently increased to a higher extent in amphetamine rats, suggesting that neuronal plasticity in the LDT may contribute to behavioral sensitization [133]. The enhanced glutamatergic drive to the VTA may also involve a transient increase in AMPA receptor responsiveness in this area. Indeed, in rats treated for 5 days with amphetamine, intra-VTA AMPA (but not NMDA) increased VTA Glu levels in animals tested from 3 to 14 days after the last injection of amphetamine, but not in saline controls [134]. However, the glutamatergic drive to the VTA may also be modulated by GABA receptors in the same experimental conditions. Administration of the $GABA_B$ receptor antagonist CGP 55845A into the VTA produced a robust increase in Glu levels in rats tested 3 days (but not 10 to 14 days) after discontinuing repeated amphetamine injections, whereas it did not induce any change in saline-treated rats. These results suggest that repeated amphetamine administration enhances $GABA_B$ receptor transmission in the VTA during the early withdrawal period, increasing the inhibitory tone on Glu levels [135].

The in vivo Glu output is increased upon amphetamine challenge in the ventral pallidum as well as in the NAcc of behaviorally sensitized rats, and Glu in these areas is thought to play a significant role in the expression of behavioral sensitization to amphetamine. Indeed, the activation of group II mGluRs reduced both the enhanced overflow of Glu in the NAcc and the expression of motor sensitization by amphetamine [136]. On the other hand, microinjections of the NMDA receptor antagonist MK-801 in the ventral pallidum can significantly block the locomotion hyperactivity induced by an amphetamine challenge in sensitized rats [137].

4.2. Cocaine

Systemic Administration Intracerebral microdialysis was used to assess the effects of cocaine (7.5 to 30 mg, i.p.) on extracellular concentrations of Asp and Glu in the NAcc of awake, freely moving rats. Glu levels were stimulated by cocaine, whereas Asp was found enhanced [138] or unchanged [122]. These acute changes were Ca^{2+}-dependent and were decreased significantly by local retrodialysis of tetrodotoxin. They were also blocked by i.p. pretreatment with the dopamine antagonists haloperidol, SCH 23390, and raclopride, as well as local 6-OHDA lesions of the NAcc.

The chronic administration of cocaine could also elicit long-term changes in group I mGluRs functions. Three weeks after having discontinuated one week of daily cocaine injections, the capacity of the group I mGluR agonist (*R*,*S*)-3,5-dihydroxyphenylglycine (DHPG) to induce Glu release after retrodialysis infusion into the NAcc was markedly reduced. Similarly, the increase in locomotor activity induced by microinjection of DHPG into the NAcc was blunted significantly. These data show that repeated cocaine administration produces an enduring inhibition of the neurochemical and behavioral consequences of stimulating mGluR1 [139]. Three weeks after discontinuing a 7-day cocaine treatment, no-net-flux microdialysis revealed a significant increase in the basal levels of extracellular GABA in the NAcc of cocaine-treated rats. The elevated extracellular GABA was normalized by blocking voltage-dependent Na^+ channels and provided increased tone on $GABA_B$ presynaptic autoreceptors and heteroreceptors, because blocking $GABA_B$ receptors produced a greater elevation in extracellular GABA and Glu in cocaine-treated compared with control subjects [140].

In the striatum, Glu and Asp levels were not affected by an acute i.p. injection of cocaine [122]. In contrast, Glu level has been found increased in the dorsolateral striatum one day after a single cocaine injection and decreased afterward from 3 to 14 days following the acute cocaine injection, these changes being partially Ca^{2+}-dependent [141].

In the VTA, systemic administration of cocaine induces an increase in extracellular Glu associated with a reduction in GABA concentration; the change in Glu, but not in GABA, could be prevented by retrodialysis of the D_1 antagonist SCH 23390 into the VTA [142].

In the prefrontal cortex, Glu levels were also stimulated by cocaine (15 to 30 mg/kg, i.p.), and such an acute increase was found Ca^{2+}-dependent and blocked by i.p. pretreatment with SCH 23390 but not with haloperidol or raclopride [122]. After 4 days of repeated cocaine treatment, rats were withdrawn for various durations and challenged with cocaine, whereas in vivo microdialysis of the median prefrontal cortex (mPFC) was conducted with concurrent analysis of locomotor activity. Animals that were withdrawn for 1 day and 7 days displayed an augmentation in cocaine-induced mPFC Glu levels compared to saline and acute control subjects. At the 7-day time point, a subset of animals that received repeated cocaine did not express behavioral sensitization, nor did these animals exhibit enhancement in mPFC Glu in response to cocaine challenge. In contrast to these early effects, 30 days of withdrawal resulted in no significant changes in cocaine-induced mPFC Glu levels, regardless of the pretreatment or behavioral response. These data suggest that repeated cocaine administration transiently increases cocaine-induced Glu levels in the mPFC during the first week of withdrawal, which may play an important role in the development of behavioral sensitization to cocaine [143]. To investigate these cocaine-induced neurochemical changes in mPFC further, in vivo microdialysis studies were conducted to monitor Glu and GABA levels in the mPFC and Glu levels in the ipsilateral NAcc, and VTA during the infusion of baclofen into mPFC. Baclofen minimally affected Glu levels in the mPFC, NAcc, or VTA of control animals, but dose-dependently increased Glu levels in each of these regions in animals sensitized to cocaine. These data suggest that alterations in $GABA_B$ receptor modulation of mPFC excitatory output may play an important role in the development of sensitization to cocaine [144]. Sensitized rats exhibited a significant increase in extracellular GABA levels in the mPFC after 1 and 7 days but not 28 days following repeated cocaine exposure [145]. The transient increase in GABA was blocked by mPFC infusion of the AMPA/KA antagonists 6,7-dinitroquinoxaline-2,3-dione (DNQX), but not the NMDA antagonist 3-[(R)-2 carboxypiperazin-4-yl]-propyl-1-phosphonic acid (CPP), indicating that the enhanced mPFC GABA transmission seen in cocaine-sensitized animals involves glutamatergic stimulation of AMPA receptors [146].

Self-Administration and Drug-Induced Reinstatement   In order to define the time course of cocaine self-administration and extinction effects on Glu level in the NAcc, rats were trained to self-administer cocaine for 20 days, and the levels of extracellular Glu were measured by in vivo microdialysis both during cocaine self-administration and after a priming cocaine injection at different time points after extinction. After 1 or 5 days, but not 10 days, of extinction, the Glu levels were reduced and the ability of i.v. cocaine priming injections to increase Glu levels followed a similar time course. These data suggest that glutamatergic transmission could be involved in the maintenance of cocaine self-administration and in the early phases of abstinence [147]. Such elevation in extracellular levels of NAcc Glu during reinstatement of extinguished cocaine self-administration by a cocaine priming injection appeared as a specific response for a drug-induced reinforcement of drug-seeking behavior and the source of Glu was shown to be glutamatergic afferents from the prefrontal cortex [148].

Cocaine-primed reinstatement of drug seeking is also associated with a decrease in extracellular GABA in the ventral pallidum (VP). Microdialysis was performed to measure extracellular GABA in the VP while simultaneously either a combination of the GABA agonists baclofen and muscimol was microinjected into the prefrontal cortex (PFC) or the AMPA/kainate Glu receptor antagonist CNQX was microinjected into the NAcc core. Inhibition of the PFC with GABA agonists and blockade of AMPA receptors in the NAcc core were both sufficient to prevent the cocaine-induced decrease in VP GABA, further implicating increased activity of the corticostriatopallidal circuit in relapse to drug seeking [149]. On the other hand, in vivo microdialysis revealed that the neurotensin agonist neurotensin peptide fragment 8–13 [NT(8–13)] increased GABA in the VP in a tetrodotoxin-dependent manner and blocked the cocaine-induced decrease in GABA. Moreover, NT(8–13) microinjected into the VP prevented cue-induced reinstatement without affecting cocaine self-administration. Thus, neurotensin which is co-localized with GABA antagonizes cocaine-induced decreases in GABA in the VP and prevents reinstatement of cocaine seeking [150].

The projection from the NAcc to the VP regulates the reinstatement of cocaine seeking in rats extinguished from cocaine self-administration. This projection co-expresses GABA and enkephalin, posing a role for μ-opioid receptors in the VP in mediating the reinstatement of cocaine seeking. In rats extinguished from cocaine self-administration, the reinstatement of cocaine seeking was associated with reduced extracellular GABA in the VP and the reinstatement and the reduction in GABA were prevented by blocking μ receptors with the antagonist Cys–Tyr–D–Trp–Arg–Thr–Pen–Thr–NH$_2$ (CTAP). The capacity of morphine to reduce VP levels of extracellular GABA was augmented in rats extinguished from cocaine self-administration. These data are consistent with the reinstatement of cocaine seeking, being modulated in part by co-released enkephalin and GABA from the NAcc–VP projection, a modulation that may involve the inhibition of GABA release by presynaptic receptors [151].

Cocaine-conditioned locomotion in rats has been used as an animal model for cocaine-conditioned responses that contribute to drug craving and relapse in human addicts. The GABA$_B$ agonist baclofen attenuated conditioned locomotion previously associated with administration of cocaine. Considering the importance of Glu transmission in the NAcc during associative responses to reward-related stimuli, the effect of baclofen on extracellular levels of Glu in the NAcc was tested with microdialysis. Baclofen did not alter basal Glu levels, but prevented the predatory odor, 2,5-dihydro-2,4,5-trimethylthiazoline, from increasing Glu levels, suggesting that it may have general utility for suppressing stimulus-evoked increases in NAc Glu levels. This could explain its ability to prevent cocaine-conditioned responses [152].

Exposure to a stressful situation is another cue leading to relapse. Footshock stress led to elevated Glu in the VTA of cocaine-experienced but not cocaine-naive rats [153]. This increase in local Glu release is due to the stress-induced release of corticotropin-releasing factor (CRF) and may play an important role in stress-induced relapse to drug use. Indeed, footshock-induced reinstatement of cocaine seeking and release of VTA Glu were blocked by selective blockade of VTA CRF(2) receptors [CRF(2)Rs] but not CRF(1)Rs. VTA perfusion of CRF or CRF(2)R agonists that have strong affinity for CRF-binding protein mimicked the effects induced by footshock, while CRFR agonists that do not bind CRF-BP were ineffective. CRF(6-33), which competes for the CRF binding site on CRF-BP, attenuated the effects of CRF on VTA Glu release and on reinstatement of cocaine seeking. This study revealed a role of VTA CRF-BP and suggest an involvement of CRF(2)R in the effectiveness of stress in triggering Glu release and cocaine seeking in drug-experienced animals [154].

More generally, Glu could be at least one of the sources of VTA signals from reward-associated environmental stimuli [155]. Indeed, i.p. injections of cocaine methiodide (a cocaine analog that does not cross the blood–brain barrier) were ineffective in cocaine-naive animals but stimulated VTA Glu release in rats previously trained to lever-press for cocaine [156]. This peripherally triggered Glu input was sufficient to reinstate cocaine seeking in previously trained animals that had undergone extinction of the habit.

4.3. Heroin

Concentrations of GABA in VP dialysates were significantly reduced by heroin self-administration. Dialysate Glu levels were unaltered during the first hour of heroin intake, but increased significantly to a stable level of approximately 120% pre-session values during the subsequent 2 h of self-administration. These observations are consistent with the hypothesis that heroin self-administration results in a disinhibition and/or excitation of the VP [157].

Microdialysis was conducted in the NAcc core during reinstatement of heroin seeking in animals extinguished from heroin self-administration. Reinstatement by either heroin or cue increased extracellular Glu in the NAcc core. The increase in Glu during heroin-induced drug seeking was abolished by inhibiting synaptic transmission in the NAcc core with tetrodotoxin or by inhibiting glutamatergic afferents to the NAcc core from the prelimbic cortex. Supporting critical involvement of Glu release, the heroin seeking induced by cue or heroin was blocked by inhibiting AMPA/kainate receptors in the NAcc core [158].

4.4. Ethanol

Extracellular Glu in the NAcc of freely moving rats was monitored by microdialysis after i.p. injection of ethanol (2 g/kg). Ethanol significantly decreased basal extracellular Glu and attenuated K^+-stimulated Glu release, suggesting that ethanol may suppress glutamatergic transmission in the NAcc by lowering presynaptic Glu release [159].

In contrast, amino acid adaptations produced in the NAcc by repeated bouts of high alcohol consumption may be characterized by enhanced Glu. Mice consuming a 5% alcohol solution for the first or for the sixth time consumed comparable amounts of alcohol; however, animals consuming for the sixth time exhibit reduced alcohol-mediated increases in GABA but enhanced alcohol-mediated increases in Glu in the NAcc. A second set of microdialysis experiments assessed the neurotransmitter response to an alcohol challenge injection and showed that mice having consumed alcohol for the sixth time exhibited sensitized Glu release after alcohol injection. Thus, repeated bouts of high alcohol consumption induce an imbalance between inhibitory and excitatory amino acids within the NAcc that may drive excessive drinking behavior [160].

The central nucleus of the amygdala (CeA) is important in regulating alcohol consumption and plays a major role in the anxiogenic response to ethanol withdrawal. It has been shown that infusion of ethanol (0.1 to 1 M) via reverse microdialysis significantly increased Glu release into the CeA, but only after chronic ethanol treatment, suggesting that chronic ethanol treatment lead to neuroadaptations of glutamatergic transmission in CeA [161]. However, ethanol produced a dose-dependent increase in GABA release in the CeA dialysate in both chronic ethanol treated and naive rats. Moreover, chronic ethanol-treated rats presented a fourfold increase in baseline GABA dialysate content compared with naive rats. This suggests that acute and chronic ethanol increases GABA release in CeA, supporting the belief that the behavioral actions of ethanol are mediated, in part, by increased GABAergic transmission in the CeA [162].

Furthermore, microdialysis was used to assay the changes in neurotransmitter amino acids after withdrawal from chronic ethanol. Extracellular levels of Glu were assayed in the NAcc 24 h after repeated ethanol exposure (1 g/kg, i.p. daily for 7 days) with use of in vivo no-net-flux microdialysis and found to be higher. Parallel in vitro experiments revealed that Na$^+$-dependent [^{3}H]Glu uptake was reduced significantly. The in vivo and in vitro ethanol-induced changes in Glu levels and uptake returned to control levels 14 days after discontinuing 7 days of repeated ethanol exposure [163]. In another study, rats were made physically dependent on ethanol by vapor inhalation for 4 weeks, and the basal concentration of GABA was significantly decreased in the NAcc, although there were no significant changes in Glu. During the first 12 h after withdrawal from ethanol, only Glu increased significantly at 6 h and for the duration of the study period of 12 h. However, i.p. injection of taurine significantly blocked the increased Glu release during ethanol withdrawal, suggesting that taurine may interact with Glu during ethanol withdrawal [164]. Baum et al. [165] have used a binge-drinking model to ethanol-intoxicate rats: recurrent cycle of 4 days of intoxication, followed by a 3-day recovery, a 2-day intoxication period, and subsequent abstinence. After the last oral intake, microdialysis was performed in the NAcc for a 16-h withdrawal period. The basal values of extracellular Glu were found to be decreased in the ethanol-intoxicated rats before withdrawal. Cessation of ethanol significantly increased Glu levels, with a peak between 4 and 10 h after the last oral intake and 16 h after ethanol withdrawal, the same level as in the control group was achieved.

Extracellular Glu was measured by microdialysis in the striatum of ethanol-dependent, freely behaving rats following withdrawal from chronic ethanol treatment. Within 12 h of withdrawal, extracellular Glu rose, remained elevated for the subsequent 12 h, and returned to control levels within 36 h. The changes in Glu were time-locked to the overt physical signs of withdrawal. In 12-h ethanol-withdrawn rats, an ethanol challenge suppressed the withdrawal signs and reduced the extracellular Glu. The NMDA receptor antagonist, dizocilpine, reduced both the physical signs of withdrawal and Glu output. In ethanol-withdrawn animals, the local application of NMDA into the striatum increased Glu levels to 598% of baseline values, compared to 268% in chronic sucrose-treated control rats. These findings suggest that the increased extra-neuronal Glu and upregulated NMDA receptors reflect overactivity of excitatory neurotransmission during withdrawal [166,167].

Changes in amino acid levels in the hippocampus during repeated ethanol withdrawal were studied in rats made ethanol-dependent by four-week vapor inhalation. After this first cycle of chronic ethanol treatment, rats underwent repeated and alternate cycles of 24 h of withdrawal followed by one week of chronic ethanol treatment. During the first cycle of ethanol withdrawal, increases in Glu levels were observed and were further enhanced during the third withdrawal period, showing that excitatory amino acid levels increased with subsequent withdrawal periods [168].

As repeated alcohol administration alters NAcc basal Glu content and sensitizes the capacity of alcohol to increase NAcc extracellular Glu levels, some microdialysis studies aimed to examine the relationship between genetic variance in alcohol consumption and alcohol-induced neuroadaptations within the NAcc. Repeated alcohol treatment elevated basal Glu content in the alcohol-preferring C57BL/6J mice but not in the alcohol-avoiding DBA2/J mice. Moreover, C57BL/6J mice exhibited a sensitized Glu and GABA response to alcohol followed repeated treatment, which was not observed in DBA2/J mice. Finally, intra-NAcc infusion of the Glu reuptake inhibitor *d,l*-threo-β-benzyloxyaspartic acid (TBOA) elevated alcohol intake selectively in C57BL/6J mice. These data indicate an active role for NAcc Glu in regulating alcohol consumption in mice and support the hypothesis that predisposition to high alcohol intake involves genetic factors that facilitate alcohol-induced adaptations in Glu release within the NAcc [169].

In selectively bred high-alcohol-sensitive (HAS) and low-alcohol-sensitive (LAS) rats, the basal level of extracellular Glu, as determined by microdialysis, was statistically similar, but differed in the rats' response to ethanol administration: Glu was significantly decreased in HAS rats, whereas Glu was significantly increased in LAS rats by the end of the experiment. These data suggest that HAS and LAS rats differ in their behavioral as well as in their neuro-chemical responses to ethanol [170]. Effects of ethanol on the NAcc extracel-lular concentrations of neurotransmitter amino acids were also studied in other strains of alcohol-tolerant (AT) and alcohol-nontolerant (ANT) rats selected for low and high sensitivity to ethanol-induced motor impairment. The smaller dose of ethanol decreased the output of GABA only in the AT rats, whereas the larger dose of ethanol decreased the output of GABA in rats of both lines. Ethanol at a dose of 2 g/kg decreased the output of Glu in rats of both lines slightly, but statistically significantly, but a larger dose of ethanol decreased the output of Glu only in the AT rats [171].

5. ANALGESIA

Pain is defined as an unpleasant sensory and emotional experience associated with actual or potential tissue damage and can be associated with nociception, neuralgia/neuropathy (due to nerve damage/dysfunction), analgesia, hyperal-gesia, allodynia (due to innocuous stimulus), and paresthesia (abnormal sensa-tion). Many studies involving neurotransmitter amino acids have been reported and dealt with pain-induced aversion, headaches, postoperative pain, arthritis, inflammation, plantar incision, spinal cord injury, chronic pain, and so on. Neurotransmitter amino acids (i.e., Gly, GABA, and Glu) are thought to play a key role in pain control [172,173]. Therefore, various areas which are parts of both ascending and descending pathways in the nervous system have been explored by microdialysis, such as spinal cord dorsal horn, brain stem nuclei, thalamic nucleus, cortex, hypothalamus, and periaqueductal gray (PAG), but also nucleus accumbens and amygdala, in reinforcement contexts or aversions. A common finding is that noxious stimuli in acute or chronic pain increase amino acid levels in the spinal cord, but also in ascending pathways such as thalamic nuclei and in brain areas of the descending modulation of pain such as PAG (for a review, see [174]). Here, we focus mainly on works reporting the use of microdialysis to study the involvement of amino acid neurotransmit-ters in the mechanisms of action of analgesics or molecules with analgesic effect, whose aim is to reduce the response to pain, or at least to increase the threshold of responce to a noxious stimulus.

5.1. Antinociceptive Activity of Morphine and Opioid Tolerance

Morphine, a μ-opioid receptor agonist, is a commonly prescribed treatment for pain. The impact of morphine on amino acidergic transmission has been studied in animals and revealed tight interaction between opioid and Glu systems in models of pain. In a recent study, the release of Glu and Asp was examined in the subcutaneous perfusate of the rat hind instep by in vivo microdialysis. Antidromic stimulation of the sciatic nerve and noxious stimuli in the form of heat stimulation and local application of capsaicin cream (1%)

to the instep caused an increase in the Glu and Asp microdialysate level. Both systemic (10 mg/kg, i.v.) and reverse dialysis of morphine inhibited the increase in excitatory amino acid release evoked by local application of capsaicin cream to the instep. This inhibitory effect of morphine was antagonized by naloxone. These results suggest that Glu and Asp are released from small-diameter afferent fibers by heat stimulation in the periphery or local application of capsaicin cream, and that activation of opioid receptors, present on the peripheral endings of small-diameter afferent fibers, can regulate noxious stimulus-induced excitatory amino acid release [175].

Intracerebroventricular morphine produces antinociception by evoking GABA release through activation of 5-hydroxytryptamine 3 receptors in the spinal cord [176]. Using the spinal microdialysis method, those authors showed that evoked spinal GABA release induced by i.c.v. morphine administration was reversed by intracerebroventricular naloxone (40 nmol) or spinal perfusion of 3-tropanyl-indole-3-carboxylate methiodide, a 5-HT3 receptor antagonist (1 mmol/L). The neurochemical data were correlated with a reverse of morphine-induced antinociceptive effect revealed by a tail-flick test.

The effect of continuous spinal infusion of a μ-opioid agonist with a shorter lasting effect than morphine, remifentanil, on behavior and spinal Glu release evoked by subcutaneous formalin in the rat was investigated using the injection of formalin into the hind paw of the rat. In this animal model of postoperative pain in humans, the initial response is caused by activation of peripheral nociceptors and is followed by a second phase attributed to ongoing activity in primary afferents and increased sensitivity of dorsal horn neurones. Remifentanil was delivered intrathecally during the phase of peripheral nociceptors and totally inhibited the early behavioral response without affecting the subsequent behavior associated with the increased sensitivity of DH neurons. This inhibition was associated with a decrease in spinal Glu level [177].

Although highly efficient, morphine has many unwanted side effects. Neurochemical studies showed that neurotransmitter amino acids are involved in disruption of sleep [178] or in the development of opioid tolerance. For the latter aspect, Wen et al. [179] showed the involvement of NMDA receptors in the loss of morphine's antinociceptive effect in morphine-tolerant rats. Rats were continuously infused i.t. for 5 days with morphine (15 μg/h). The analgesic dose was 63-fold higher in morphine-tolerant rats than in control rats, while it was 8.4-fold higher in rats that had received the NMDA antagonist, MK-801, plus morphine. No significant differences were observed in CSF amino acid levels between the groups from day 1 to day 5, but morphine started to lose its analgesic effect on day 2, and this effect was overcome by MK-801. After 5 days of treatment, and after a morphine challenge, morphine-tolerant rats showed a significant increase in the release of Glu and Asp and a tail-flick latency test revealed no antinociceptive effect, while MK-801/morphine co-infused rats showed no increase in morphine-induced EAA release and a partial antinociceptive effect.

Intrathecal gabapentin (GBP), an antiepileptic drug, is also shown to attenuate morphine tolerance. At 10 μg intrathecally (i.t.), GBP did not enhance morphine's antinociception in naive rats. In morphine-tolerant rats, however, acute injection potentiated morphine's antinociception. When GBP was co-infused with morphine, GBP attenuated the development of morphine tolerance. Acute injection of GBP, morphine, or GBP followed by morphine had no significant effect on CSF spinal EAA concentration in naive rats; however, in tolerant rats, a morphine challenge increased Asp and Glu levels to 221 and 296%, respectively, and those increases were inhibited by GBP co-infusion [180].

Another class of drug was tested against morphine tolerance in rats. The tricyclic antidepressant amitriptyline was shown to suppress neuroinflammation by upregulating Glu transporters (GLAST, GLT-1, and EAAC1) in the spinal cord dorsal horn of morphine-tolerant rats linked to an increase in levels of the excitatory amino acids Glu and Asp in aCSF [181].

Another side effect of morphine is pain induced by a high dose. It is commonly known that injection of a high dose of morphine into the spinal lumbar intrathecal space of rats induces pain vocalization and agitation. Intrathecal substance P(1–7) prevents morphine-evoked pain behavior [182]. Indeed, the behavioral responses of morphine-induced pain that were induced were attenuated dose-dependently by i.t. pretreatment with the N-terminal fragment substance P(1–7). The inhibitory effect of substance P(1–7) was reversed significantly by pretreatment with [D-Pro2, *d*-Phe7] substance P(1–7), a D-isomer and antagonist of substance P(1–7). Neurochemical data showed a significant increase in extracellular Glu and NO metabolites in the spinal cord after i.t. injection of high-dose morphine (500 nmol). Pretreatment with substance P(1–7) produced a significant reduction in the elevated concentrations of Glu and NO metabolites evoked by i.t. morphine. The reduced levels of Glu and NO metabolites were reversed significantly by the substance P(1–7) antagonist. The authors suggested that i.t. substance P(1–7) might attenuate the pain behavior of high-dose i.t. morphine by inhibiting the presynaptic release of Glu and reducing NO production in the dorsal spinal cord.

5.2. Antinociceptive Activity by Enhancing Inhibitory Transmission

The neurochemical basis of the antinociceptive activity of GABA uptake inhibitors and/or antiepileptic drugs has been explored with the aim of reducing neuronal excitability. For example, tiagabine, an uptake GABA inhibitor currently used as an antiepileptic drug, has antinociceptive effects in mechanical (paw pressure), chemical (abdominal constriction), and thermal (hot plate) noxious stimuli [183]. Indeed, systemic administration of tiagabine, 30 mg/kg i.p., increased nearly twofold the dialysate GABA levels in medial thalamus of freely moving rats and increased significantly the rat paw pressure nociceptive threshold in a time-correlated manner. Dose-related significant tiagabine-induced antinociception was also observed at doses of 1 and 3 mg/kg i.p. in the mouse hot plate and abdominal constriction tests. The tiagabine antinociception was completely antagonized by pretreatment with the selective GABA(B) receptor antagonist, CGP 35348 (3-aminopropyldiethoxymethylphosphonic acid), but not by naloxone. Tiagabine was thought to cause antinociception due to raised endogenous GABA levels, which in turn activate GABA(B) receptors.

To clarify the mechanisms leading to antinociceptive activity of GABA reuptake inhibitors, microdialysis in the anaesthetized rat was used to examine the effect of selective GABA transporter type 1 (GAT-1) inhibition on basal and evoked amino acid release in the dorsal spinal cord. Reverse dialysis of the selective GAT-1 inhibitor NO-711 induced a concentration-related increase in extracellular GABA without affecting the other neurotransmitter amino acids. Effluxes of Asp and Glu, but not GABA or Gly, which were evoked by brief high K^+-induced depolarization were found to be reduced significantly by reverse dialysis of NO-711. Coadministration of selective antagonists for GABA(A) or GABA(B) receptors (+)-bicuculline or SCH 50911 prevented the GAT-1 inhibition-induced reduction of evoked Asp. In contrast, whereas (+)-bicuculline also antagonized the reduction of evoked Glu, SCH 50911 was without effect. These data suggest that an enhancement of GABAergic neurotransmission by reuptake inhibition activates Asp and Glu releases differentially in the dorsal spinal cord [184].

If antinociceptive activity of GABA uptake inhibitors seems to involve several GABAergic mechanisms, the impact of drugs acting on voltage-gated ion channels on Glu release was also investigated. The effects of treatment with the anticonvulsant agents lamotrigine and riluzole were compared with those for gabapentin, an antiepileptic drug with antihyperalgesic properties used clinically to treat certain neuropathic pain states, in chronic constriction injury (CCI) of the sciatic nerve, a rat experimental model of neuropathic pain, using formalin-induced nociceptive behavioral scores [185]. Formalin-induced increases in Glu levels in spinal cord dorsal horn were lowered significantly compared with controls, with all drugs.

Moreover, gabapentin (100 mg/kg, i.p.) markedly reduced acetic acid–induced visceral nociception. The antinociceptive effect of gabapentin correlates with the suppression of noxious-evoked release of Glu, Asp, and serine, in the spinal cord, whereas increases in glutamine and Gly were unaffected by gabapentin treatment [186].

5.3. Glutamate Blockade

As mentioned previously, ionotropic and metabotropic receptors can increase the release of amino acids in the nociceptive pathways. For example, group I metabotropic glutamate receptor antagonists block secondary thermal hyperalgesia in rats with knee joint inflammation [187]. Rats pretreated with spinal microdialysis infusion of group I mGlu receptor antagonists LY393053, LY367385, or AIDA before knee joint injection exhibited paw withdrawal latencies significantly longer in the presence of group I mGlu receptor antagonists than those of the arthritic control group. Post-treatment with the group I mGlu receptor antagonists LY367385 and AIDA allowed significant recovery of the paw withdrawal latencies after the onset of the knee joint inflammation. Those results are in accordance with the neurochemical data obtained using nociceptive behavioral tests showing that the stimulation of group I metabotropic glutamate receptor (mGluR) by the agonist (R,S)-3,5-dihydroxyphenylglycine [(R,S)-3,5-DHPG] induces spinal Glu release in a dose-dependent manner [188], confirming the nociceptive action of Glu.

5.4. Antidepressants

Increasing the serotoninergic neurotransmission at the level of the dorsal spinal cord is proposed to relieve pain. In this context, the analgesic activity of selective serotonin (5-HT) reuptake inhibitors (SSRIs) has been evaluated. For example, reverse dialysis of paroxetine (1 to 30 μM) in the dorsal horn of spinal cord of anaesthetized rats reduced the K^+-induced increase in Glu in a concentration-dependent manner. Administration of citalopram (300 μM) also reduced depolarization-induced Glu release, whereas Asp levels were poorly affected by any SSRI tested. Co-perfusion of paroxetine (30 μM) with the selective 5-HT(1A) receptor antagonist WAY 100635 did not prevent a reduction in depolarization-induced Glu efflux. These results demonstrate that local administration of SSRIs has an inhibitory influence on evoked release of Glu in the dorsal horn, independent of the autoreceptors 5-HT1A [189].

5.5. Prostanoids

Prostaglandins facilitate nociceptive transmission. The use of specific antagonists gave new evidence of the tonic role of the four prostanoid EP1, EP2, EP3, and EP4 receptors in the PAG for the appearance of formalin-induced hyperalgesia [190]. The authors observed that hyperalgesia induced either by prostaglandin receptor stimulation or by formalin-induced inflammatory pain generated an increase in Glu and a biphasic change in GABA extracellular concentrations.

The effect of the inositol trisphosphate analog α-trinositol was examined in unanesthetized rats using intrathecal microdialysis probes. In that study, subcutaneous injection of 50 μL of a 5% formalin solution produced painlike behavior and significant elevation of Glu, Asp, Gly, taurine, and serine at the early stage of pain. PGE2 concentrations were also increased but for longer. Intraperitoneal delivery of 300 mg/kg α-trinositol significantly suppressed formalin-induced behavior and the concomitant increase in both amino acids and PGE2 in the spinal cord [191]. Previously, the same group showed that the inhibition of cyclooxygenase by ibuprofen (administration of 10 mg/kg i.p., but not 1 mg/kg) also suppressed paw formalin-induced behavior, PGE2 immunoreactivity, as well as Glu and Asp releases monitored in the spinal cord of unanesthetized rats [192].

5.6. Potential Strategies

A recent study reports the first data dealing with prokineticin (Bv8), a small protein secreted by the skin of the *Bombina variegata* frog, and acting on PK1 and PK2 receptors. In combination with the plantar test, dialysis data showed that intra-PAG perfusion with Bv8 increased GABA but not Glu extracellular levels, and decreased thermoceptive thresholds [193]. These findings show that stimulation of PAG PK receptors might worsen pain perception, so that the use of potential antagonists might help to reduce pain.

Use of the thermoreceptor transient receptor potential vanilloid 1 (TRPV1) antagonists to relieve pain is based on the fact that TRPV1 agonists as capsaicin provide pain in certain conditions [194]. A study conducted in PAG showed surprisingly that injection of capsaicin increased the threshold of thermal pain sensitivity using a plantar test, whereas the selective thermoreceptor antagonist 5′-iodo-resiniferatoxin (I-RTX) facilitated nociceptive responses and blocked the capsaicin analgesic effect at a dose inactive per se. Intra-VL PAG capsaicin evoked a robust release of Glu in RVM microdialysates. I-RTX, at a dose inactive per se, blocked the effect of capsaicin and inhibited Glu release at a higher dose [195].

6. ISCHEMIA–ANOXIA

6.1. Changes in Glu and GABA Microdialysates in Experimental Models of Ischemia–Anoxia

Since the first paper by Benveniste and colleagues [196] using microdialysis to demonstrate ischemia-evoked releases of Glu, Asp, and taurine in the rat hippocampus, there have been a lot of in vivo microdialysis studies. They were focused primarily on ischemia-induced changes in extracellular amino acids levels in the cerebral cortex, hippocampus, and striatum of rats subjected to four vessel occlusion-elicited forebrain ischemia or middle cerebral artery occlusion-evoked focal ischemia, with variations in the duration of ischemia (for a review, see [197]). For example, in conscious rats, a 30-min global ischemia obtained through the four-vessel occlusion model leads to enhancement in striatal extracellular Glu and Asp, which is inhibited by perfusion of the NMDA receptor blockers dizocilpine (MK-801) and Mg^{2+} [198]. Microdialysis can also be used to apply a model of local ischemia in any brain region of a conscious animal. The reverse dialysis of an aCSF containing NaCN and 2-deoxyglucose creates a local histotoxic anoxia/aglycemia, which in rat striatum leads to a dramatic efflux of Glu and Asp [199].

In anesthetized rats subjected to hypoxia/ischemia for 30 min (ventilation with 10% O_2 and unilateral carotid artery occlusion), the concentrations of Glu monitored by microdialysis increased during hypoxia–ischemia, followed by continuous recovery during reperfusion. Moreover, the time course of the increase in Glu correlates with the time course of the reduction in extracellular space volume fraction. Thus, the decrease in the extracellular space volume fraction can contribute to enhanced accumulation of Glu, which may aggravate functional deficits and lead to damage [200].

A marked increase in extracellular Glu in the rat striatum was observed by microdialysis immediately after microsphere embolism (ME) induction, and Glu remained elevated at least 12 h after ischemia. Concomitantly, impairment of high KCl-induced Glu release was observed in the striatum 12 h after ME [200].

Although the role of excitotoxic amino acids, particularly that of Glu, has been described in ischemic stroke and head trauma, no information exists regarding their possible contribution to the pathogenesis of neuronal injury in intracerebral hemorrhage. Intracerebral hemorrhage was introduced in anesthetized New Zealand rabbits by injecting autologous blood under arterial pressure into the deep gray matter of the cerebrum. Extracellular fluid samples were collected from the perihematoma region and contralateral hemisphere by in vivo microdialysis. Glu concentrations were significantly higher in the hemisphere ipsilateral to the hematoma than in the contralateral hemisphere 30 min after hematoma creation, and persisted, but nonsignificantly, between 1 and 5 h after hematoma creation. In the hemisphere ipsilateral to the hematoma, a threefold increase in the concentration of glycine was observed 30 min after hematoma; a similar transient increase was observed in taurine and asparagine concentrations. Thus, Glu and other amino acids accumulate transiently in extracellular fluids in the perihematoma region during the early period of intracerebral hemorrhage [201].

To assess the effects of amino acids in the nucleus tractus solitarii (NTS) on the ventilatory response to hypoxia at one and two weeks of age, amino acids were sampled by microdialysis in unanesthetized and chronically instrumented piglets. A biphasic ventilatory response to hypoxia was observed in weeks 1 and 2, but the decrease in minute ventilation was more marked in week 1. This was associated with an increase in NTS GABA and glycine levels during hypoxia in week 1. In contrast, Glu and Asp levels were elevated during hypoxia in weeks 1 and 2. These data suggest that the larger depression in the ventilatory response to hypoxia observed in younger piglets is mediated by the predominance of the inhibitory amino acids in the NTS [202].

Amino acids were measured using microdialysis in gray matter (GM) and white matter (WM) of cats during 10 min of global ischemia and 120 min of reperfusion. Glu, Asp, and GABA increased in GM but not in WM during ischemia and reperfusion. Contrasting with GM, deleterious processes such as Glu accumulation do not occur in WM during short-term ischemia and reperfusion [203].

Non-neurotransmitter amino acids may also be altered by ischemia. The concentrations of Glu, Asp, taurine, GABA, glycine, alanine, serine, threonine, and asparagine measured in the rat striatum by microdialysis increased, whereas glutamine concentrations decreased, during global forebrain ischemia (two-vessel occlusion with hypotension for 20 min) [204].

In contrast to mammals, some poïkilothermic animals could, more or less, resist hypoxia. In the moderately anoxia-resistant leopard frog *Rana pipiens*, extracellular levels of forebrain Asp, taurine, and GABA increased significantly, whereas Glu levels decreased during anoxia. The maintenance of low extracellular Glu was the most significant difference between the frog and the anoxic–ischemic mammalian brain, although aspartate did increase 215% over a 4-h period of anoxia [205].On the other hand, the freshwater turtle *Trachemys scripta* is exceptionally resistant to anoxia, and microdialysis has revealed a large increase in extracellular GABA in the striatum during 240 min of anoxia, reaching 90 times the normoxic level, whereas no substantial release of Glu occurred; moreover, Gly and taurine also displayed increased levels [206]. Increased intracellular levels of inhibitory amino acids may be one of the hitherto elusive mechanisms that underlie the decreased activity and energy consumption characterizing the anoxic turtle brain. The crucian carp is another one of the few vertebrates that has the ability to survive long periods of anoxia. Using microdialysis, Hylland and Nilsson [207] have shown that this species avoids a release of Glu in the telencephalon during anoxia (6 h), while the extracellular level of GABA was doubled. Perfusing the microdialysis probe with a high-K^+ Ringer's solution, however, showed that the telencephalon had the ability to release both Glu and GABA. Moreover, if energy deficiency was produced further during anoxia by inhibiting glycolysis with iodoacetate added to the perfusion fluid, the resulting release of GABA was more rapid and profound than that of Glu, possibly reflecting a second line of anoxia defense aimed at minimizing the effect of a temporary energy failure.

In contrast to ischemia and anoxia, extracellular Glu is not enhanced during moderate hypoxia, which does not lead to a dramatic fall in ATP and neuron loss. For example, breating 7% O_2 did not induced any change in extracellular Glu and Asp in the anaesthetized piglet [208], whereas a 1-h exposure to 10% O_2 failed to induce any change in extracellular Glu and Asp in the striatum of conscious rats [199]. However, 60 min of cerebral oligaemic hypoxia, induced by bilateral clamping of the carotid arteries in anaesthetized rats, induced in the hippocampus a moderate increase in microdialysate Glu and Asp during the occlusion, followed by a more important increase during the reperfusion period. GABA is increased markedly only during the reperfusion period [209]. In the striatum, Glu decreased during the occlusion, but was markedly increased together with Asp during the reperfusion period. GABA concentration increased during occlusion and early reperfusion.

6.2. Receptor Modulation of Ischemia-Induced Changes in Glu and GABA Microdialysates

The combination of microdialysis, use of neuropharmacological agents, and/or genetically modified rodents has made it possible to decipher the mechanisms at the origin of the anoxia–ischemia-induced efflux in amino acid neurotransmitters. Ca^{2+}-evoked exocytotic release account for the initial (1 to 2 min) efflux of Glu following the onset of ischemia and has been revealed by coupling microdialysis to a Glu-specific enzymatic biosensor [210]. Afterward, extracellular Ca^{2+}-independent nonvesicular release is responsible for much of the subsequent efflux of amino acid neurotransmitters and is mediated by Na^+-dependent amino acid transporters in the plasma membrane operating in the reverse mode and by the opening of a volume-sensitive organic anion channel (VSOAC) which allows the passage of amino acids down their concentration gradients. It is also possible that a disruption of plasma membrane integrity by phospholipases is involved in ischemia-induced Ca^{2+}-independent amino acid release [197].

The efflux in striatal Glu and Asp induced by 30 min of global ischemia is still present when Ca^{2+} is omitted from the aCSF perfusing the probe, showing that most of this efflux is Ca^{2+}-independent [198]. The function of the glial Glu transporter (GLT-1) during brain ischemia was investigated by using an in vivo brain microdialysis technique in the hippocampal CA1 of GLT-1 mutant and wild-type mice. Glu levels in mice lacking GLT-1 were significantly higher than the corresponding Glu levels in wild-type mice during 5 min of ischemia. Delayed neuronal death was induced in the CA1 of the mice lacking GLT-1 but not in the CA1 of the wild-type mice. When ischemia was elongated, the Glu levels in wild-type mice were significantly higher than in mice lacking GLT-1 during the last 12.5 min of a 20-min ischemia. Acute neuronal death was also observed in the CA1 of wild-type mice. These results suggest that GLT-1 takes up extracellular Glu to protect neurons in the early stage of ischemia and then releases Glu, triggering acute neuronal death when ischemic conditions are elongated [211].

The sources and mechanisms of excitatory amino acid release during incomplete ischemia (perfusion reduced by 50 to 80%) were investigated using microdialysis in rat brain regions by applying inhibitors of VSOAC and the GLT-1 Glu transporter. Concentrations of Glu, Asp, and taurine were measured before, for 2 h during, and for 2 h after reversible middle cerebral artery occlusion (rMCAO). Dihydrokainate, a GLT-1 inhibitor, or tamoxifen, a VSOAC inhibitor, were administered by retrodialysis starting 1 h prior to ischemia. During ischemia, dialysate Glu levels were significantly lower in the tamoxifen group than in the control and dihydrokainate groups. As perfusion returned toward baseline levels, Glu levels declined in the vehicle and tamoxifen-treated animals but remained elevated in the dihydrokainate-treated animals. In contrast to severely ischemic regions, dihydrokainate did not reduce Glu release in less severely ischemic brain, suggesting a diminished role for Glu transporter reversal in these areas. These findings also support the hypothesis that in regions of incomplete ischemia, release of Glu via VSOACs may play a larger role than reversal of the GLT-1 transporter [212]. Indeed, the involvement of reversal of transporters in amino acid efflux may vary between models of ischemia and along the time course of changes. In this context, a downregulation of Glu transporters after ME was associated with elevation of basal Glu concentrations and KCl-induced Glu release in the brain [213].

Models of hypoglycemia in which energy deficiency is brought without disruption in oxygen availability or blood flow in contrast to ischemia have the importance of energy availability in the mechanisms of Glu overflow. The intrahippocampal administration of the glycolysis inhibitor, iodoacetate (IOA), induces the accumulation of Glu and neuronal death. The role of exocytosis, Glu transporters, and VSOAC on IOA-induced EAA release was investigated using microdialysis. Microdialysis made it possible to show that the early component of Glu release is inhibited by riluzole, a voltage-dependent Na^+ channel blocker, and by the VSOAC inhibitor tamoxifen, while the early and late components are blocked by the Glu transport inhibitors L-*trans*-pyrrolidine 2,4-dicarboxylate (PDC) and D,L-threo-β-benzyloxyaspartate (D,L-TBOA) and by the VSOAC blocker 4,4′-dinitrostilbene-2,2′-disulfonic acid (DNDS) [214].

6.3. Glutamate and GABA Brain Microdialysis in Preclinical Studies of Neuroprotective Drugs

As elevation of extracellular levels of amino acids has been implicated in the pathogenesis of stroke, intracerebral microdialysis has been used to monitor the effects of acute 17β-estradiol treatment on the release of Glu and Asp in rats subjected to 90 min of rMCAO followed by a 24-h reperfusion. Acute systemic treatment with 17 β-estradiol at the time of occlusion reduced the ischemia-induced increases in levels of Glu and Asp rapidly and significantly and reduced the ischemic infarct by about 50% [215].

Chlormethiazole administered i.p. 5 min after the onset of focal ischemia in the cerebral cortex of rats reduced the ischemia-induced neurodegeneration by approximately 30% measured histologically 24 h later and reduced the rise in the concentration of extracellular Glu, Asp, and GABA by 30 to 60% during the first 4 h following the onset of ischemia [216]. However, analysis of the time course of the changes in amino acid suggests that chlormethiazole is not neuroprotective because of the inhibition of excitatory amino acid release but, rather, that the attenuated rise in the concentration of all the amino acids is reflective of neuroprotection involving another mechanism.

Neuroprotection and Glu release inhibition by the potent VSOAC inhibitor 4-(2-butyl-6,7-dichloro-2-cyclopentyl-indan-1-on-5-yl) oxobutyric acid (DCPIB) have been studied in an rMCAO model in adult rats. Rats given DCPIB intracisternally had significantly better neurobehavioral scores after 24 h and showed significantly reduced infarct volumes. Microdialysis demonstrated significant reduced brain extracellular Glu when DCPIB was present in the fluid perfusing the probe, supporting the hypothesis that Glu released via VRAC contributes to later ischemic-induced damage. However, a high dose of DCPIB given intravenously did not reduce infarct volume or improve behavior, indicating that this drug does not cross the blood–brain barrier [217].

The effects of acetylsalicylic acid (ASA) on ischemia-induced Glu release were assessed by microdialysis in the striatum of rats. Permanent middle cerebral artery occlusion (pMCAO) led to large infarct volumes which were unaffected by treatment with ASA (administered as a bolus 30 min after stroke onset). In contrast, ASA therapy in rMCAO reduced Glu significantly 90, 120, and 150 min after occlusion and reduced infarct size. These results suggest that the neuroprotective effect of ASA is reflected by Glu attenuation even if given after stroke onset, but only if reperfusion is achieved [218].

The effects of the α2 adrenergic agonist clonidine on nearly complete cerebral ischemia-evoked release of Glu and Asp from normo- and hyperglycemic rat brains were investigated using microdialysis. Glu and Asp levels were greatly increased over control values in both normo- and hyperglycemic groups during the 40-min ischemia, and clonidine pretreatment suppressed this release of excitatory amino acids. Significant neuroprotection of cells in the cortex, striatum, and hippocampus was also observed in drug-treated animals 48 h postischemia. The neuroprotective effect of clonidine during ischemia may be ascribed to a reduced release of Glu, thereby reducing NMDA receptor activation and neuronal damage [219].

The agent ONO-2506 [(R)-(−)-2-propyloctanoic acid] was shown to mitigate post-ischemia-delayed infarct expansion through inhibition of the enhanced production of S-100β. To elucidate whether ONO-2506 also involved modulation of Glu release, the extracellular Glu levels in the cerebral cortex of rats subjected to rMCAO and receiving ONO-2506 or saline were monitored by intracerebral microdialysis. In the untreated rats, the Glu level increased during rMCAO, returned to normal on reperfusion, increased again around 5 h, and increased further from 15 h on, to reach about 280% of the normal level at 24 h. This secondary increase in the Glu level in the late phase of reperfusion was prevented by ONO-2506. The intracerebral infusion of Glu transporter inhibitor, l-*trans*-pyrrolidine-2,4-dicarboxylic acid, 24 h after rMCAO induced an increase in the Glu level, which was more marked in the ONO-2506-treated group than in the saline-treated group. Since ONO-2506 has been shown to enhance mRNA expression of glial Glu transporters GLT1 and GLAST, one may suggest that functional modulation of activated astrocytes by ONO-2506 may inhibit the secondary rise of Glu level in the late phase of reperfusion, leading to amelioration of delayed infarct expansion and neurological deficits [220].

To assess the separate and combined effects of moderate hypothermia and of brain-derived neurotrophic factor (BDNF) on striatal Glu release in the acute phase of stroke, microdialysis probes were inserted into the striatum of rats 2 h before pMCAO. Four treatment strategies were begun 30 min after pMCAO: (1) hypothermia at 33°C, (2) intravenous BDNF infusion, (3) combination of hypothermia and BDNF, and (4) control group. Total infarct volume was reduced significantly in the hypothermia and BDNF group compared to the control group and reduced further in the combination group. Postischemic Glu concentrations in the control group constantly remained significantly higher than in all other treatment groups. At 255 and 270 min after pMCAO, striatal Glu in the combination group decreased significantly more than in animals treated with hypothermia or BDNF alone [221].

The ability of granulocyte-colony stimulating factor (G-CSF) to inhibit Glu release was investigated by microdialysis in a rat stroke model. Animals were treated with either intravenous saline or G-CSF 30 min after rMCAO. G-CSF attenuated the release of Glu in the infarcted striatum significantly from 30 to 180 min after rMCAO compared with controls while the infarct volume was significantly reduced in the G-CSF-treated group 24 h after rMCAO, suggesting that G-CSF possesses the ability to inhibit excitotoxicity after ischemic stroke [213].

The neuroprotective effect of intrathecal ketorolac pretreatment against ischemic spinal cord injury was investigated in rats using microdialysis. One hour before spinal cord ischemic injury induced by balloon inflation of a catheter in the thoracic aorta, animals received saline or ketorolac. Intrathecal ketorolac attenuated spinal cord ischemic injury assessed by hind limb motor function. Dialysate concentrations of Glu and Asp were increased after spinal cord ischemia, and this effect was inhibited by intrathecal administration of ketorolac [222].

The buckwheat polyphenol (BWP; continuous 21-day p.o.) significantly ameliorated not only the impairment of spatial memory but also necrosis and TUNEL-positive cells in the CA1 area of rats subjected to repeated cerebral ischemia. To investigate the mechanism of BWP protective action, the release of Glu induced by repeated cerebral ischemia in the rat dorsal hippocampus was measured using microdialysis. A 14-day BWP treatment significantly inhibited the excess release of Glu after the second occlusion. However, the 14-day treatment did not affect hippocampal blood flow in either intact rats or rats subjected to repeated ischemia measured by a laser Doppler flowmeter. These results suggested that BWP might ameliorate spatial memory impairment by inhibiting glutamate release in rats subjected to repeated cerebral ischemia [223].

7. CONCLUSIONS AND PERSPECTIVES

The recent development of high-sampling-rate microdialysis coupled with high-throughput separative microtechniques able to handle submicroliter sample allows monitoring rapid changes in extracellular levels of amino acid neurotransmitters. Such a platform allows in vivo investigating synaptic as well as volumic amino acid neurotransmission in anaesthetized or freely moving animals. Moreover, separation protocols that allow simultaneous measurement of amino acids and monoamines give way to study interactions between neurotransmitters involved in drug action. Furthermore, neurochemical measurements through microdialysis can be coupled to the monitoring of other neurobiological indexes of physiopathological or pharmacological mechanisms, such as behavioral measurement and EEG recordings or measurement of the drug concentration at the site of action. Thus, high-sampling-rate microdialysis of amino acids maker it possible to reveal rapid change in extracellular levels of glutamate or GABA linked to specific neurobiological events, such as epileptic seizures [14], or associated with the application of determined-concentration vigabatrin [48]. Brain microdialysis is particularly suitable for an accurate investigation of the mechanisms of action of a drug thanks to its capability for local application of drugs through reverse dialysis. It has brought invaluable information to the involvement of amino acid neurotransmitters in the mechanisms of action of neuropharmacological agents from various classes.

REFERENCES

[1] Fonnum, F. (1984). Glutamate: a neurotransmitter in mammalian brain. *Journal of Neurochemistry*, *42*, 1–11.

[2] Farber, N.B., Newcomer, J.W., Olney, J.W. (1998). The glutamate synapse in neuropsychiatric disorders: focus on schizophrenia and Alzheimer's disease. *Progress in Brain Research*, *116*, 421–437.

[3] Legendre, P. (2001). The glycinergic inhibitory synapse. *Cellular and Molecular Life Sciences*, *58*, 760–793.

[4] Owens, D.F., Kriegstein, A.R. (2002). Is there more to GABA than synaptic inhibition? *Nature Reviews Neuroscience*, *3*, 715–727.

[5] Avoli, M., Louvel, J., Pumain, R., Kohling, R. (2005). Cellular and molecular mechanisms of epilepsy in the human brain. *Progress in Neurobiology*, *77*, 166–200.

[6] Javitt, D.C. (2004). Glutamate as a therapeutic target in psychiatric disorders. *Molecular Psychiatry*, *9*, 979, 984–997.

[7] Gubellini, P., Pisani, A., Centonze, D., Bernardi, G., Calabresi, P. (2004). Metabotropic glutamate receptors and striatal synaptic plasticity: implications for neurological diseases. *Progress in Neurobiology*, *74*, 271–300.

[8] Bevan, M.D., Atherton, J.F., Baufreton, J. (2006). Cellular principles underlying normal and pathological activity in the subthalamic nucleus. *Current Opinion in Neurobiology*, *16*, 621–628.

[9] Ungerstedt, U. (1991). Microdialysis: principles and applications for studies in animals and man. *Journal of Internal Medicine*, *230*, 365–373.

[10] Benveniste, H., Huttemeier, P.C. (1990). Microdialysis: theory and application. *Progress in Neurobiology*, *35*, 195–215.

[11] Morari, M., O'Connor, W.T., Ungerstedt, U., Fuxe, K. (1993). *N*-Methyl-D-aspartic acid differentially regulates extracellular dopamine, GABA, and glutamate levels in the dorsolateral neostriatum of the halothane-anesthetized rat: an in vivo microdialysis study. *Journal of Neurochemistry*, *60*, 1884–1893.

[12] Young, A.M., Bradford, H.F. (1991). *N*-Methyl-D-aspartate releases excitatory amino acids in rat corpus striatum in vivo. *Journal of Neurochemistry*, *56*, 1677–1683.

[13] Parrot, S., Bert, L., Renaud, B., Denoroy, L. (2003). Glutamate and aspartate do not exhibit the same changes in their extracellular concentrations in the rat

striatum after *N*-methyl-D-aspartate local administration. *Journal of Neuroscience Research*, *71*, 445–454.

[14] Parrot, S., Sauvinet, V., Riban, V., Depaulis, A., Renaud, B., Denoroy, L. (2004). High temporal resolution for in vivo monitoring of neurotransmitters in awake epileptic rats using brain microdialysis and capillary electrophoresis with laser-induced fluorescence detection. *Journal of Neuroscience Methods*, *140*, 29–38.

[15] Bert, L., Parrot, S., Robert, F., Desvignes, C., Denoroy, L., Suaud-Chagny, M.F., Renaud, B. (2002). In vivo temporal sequence of rat striatal glutamate, aspartate and dopamine efflux during apomorphine, nomifensine, NMDA and PDC in situ administration. *Neuropharmacology*, *43*, 825–835.

[16] Touret, M., Parrot, S., Denoroy, L., Belin, M.F., Didier-Bazes, M. (2007). Glutamatergic alterations in the cortex of genetic absence epilepsy rats. *BMC Neuroscience*, *8*, 69.

[17] Parrot, S., Bert, L., Mouly-Badina, L., Sauvinet, V., Colussi-Mas, J., Lambas-Senas, L., Robert, F., Bouilloux, J.P., Suaud-Chagny, M.F., Denoroy, L., Renaud, B. (2003). Microdialysis monitoring of catecholamines and excitatory amino acids in the rat and mouse brain: recent developments based on capillary electrophoresis with laser-induced fluorescence detection: a mini-review. *Cellular and Molecular Neurobiology*, *23*, 793–804.

[18] Brun, P., Begou, M., Andrieux, A., Mouly-Badina, L., Clerget, M., Schweitzer, A., Scarna, H., Renaud, B., Job, D., Suaud-Chagny, M.F. (2005). Dopaminergic transmission in STOP null mice. *Journal of Neurochemistry*, *94*, 63–73.

[19] Kehr, J. (1999). Monitoring chemistry of brain microenvironment: biosensors, microdialysis and related techniques. In: Windhorst, U., Johansson, H. (Eds.), *Modern Techniques in Neuroscience Research*, Springer-Verlag, Berlin, pp. 1149–1198.

[20] Westerink, B.H. (1995). Brain microdialysis and its application for the study of animal behaviour. *Behavioural Brain Research*, *70*, 103–124.

[21] Watson, C.J., Venton, B.J., Kennedy, R.T. (2006). In vivo measurements of neurotransmitters by microdialysis sampling. *Analytical Chemistry*, *78*, 1391–1399.

[22] Abercrombie, E.D., Finlay, J.M. (1991). Monitoring extracellular norepinephrine in brain using in vivo microdialysis and HPLC–EC. In: Robinson, T.E., Justice, J.B.J. (Eds.), *Microdialysis in Neurosciences*, Elsevier, Amsterdam, pp. 253–274.

[23] Smolders, I., Sarre, S., Michotte, Y., Ebinger, G. (1995). The analysis of excitatory, inhibitory and other amino acids in rat brain microdialysates using microbore liquid chromatography. *Journal of Neuroscience Methods*, *57*, 47–53.

[24] Kehr, J. (1998). Determination of gamma-aminobutyric acid in microdialysis samples by microbore column liquid chromatography and fluorescence detection. *Journal of Chromatography B: Biomedical Sciences and Applications*, *708*, 49–54.

[25] Kehr, J. (1998). Determination of glutamate and aspartate in microdialysis samples by reversed-phase column liquid chromatography with fluorescence and electrochemical detection. *Journal of Chromatography B: Biomedical Sciences and Applications*, *708*, 27–38.

[26] Obrenovitch, T.P., Zilkha, E. (2001). Microdialysis coupled to online enzymatic assays. *Methods*, *23*, 63–71.

[27] McKenzie, J.A., Watson, C.J., Rostand, R.D., German, I., Witowski, S.R., Kennedy, R.T. (2002). Automated capillary liquid chromatography for simultaneous determination of neuroactive amines and amino acids. *Journal of Chromatography A*, *962*, 105–115.

[28] Lunte, S.M., Lunte, C.E. (1996). Microdialysis sampling for pharmacological studies: HPLC and CE analysis. *Advances in Chromatography*, *36*, 383–432.

[29] Kennedy, R.T., Watson, C.J., Haskins, W.E., Powell, D.H., Strecker, R.E. (2002). In vivo neurochemical monitoring by microdialysis and capillary separations. *Current Opinion in Chemical Biology*, *6*, 659–665.

[30] Bergquist, J., Vona, M.J., Stiller, C.O., O'Connor, W.T., Falkenberg, T., Ekman, R. (1996). Capillary electrophoresis with laser-induced fluorescence detection: a sensitive method for monitoring extracellular concentrations of amino acids in the periaqueductal grey matter. *Journal of Neuroscience Methods*, *65*, 33–42.

[31] Dawson, L.A., Stow, J.M., Palmer, A.M. (1997). Improved method for the measurement of glutamate and aspartate using capillary electrophoresis with laser induced fluorescence detection and its application to brain microdialysis. *Journal of Chromatography B: Biomedical Sciences and Applications*, *694*, 455–460.

[32] Lada, M.W., Vickroy, T.W., Kennedy, R.T. (1997). High temporal resolution monitoring of glutamate and aspartate in vivo using microdialysis on-line with capillary electrophoresis with laser-induced fluorescence detection. *Analytical Chemistry*, *69*, 4560–4565.

[33] Li, Y.M., Qu, Y., Vandenbussche, E., Arckens, L., Vandesande, F. (2001). Analysis of extracellular gamma-aminobutyric acid, glutamate and aspartate in cat visual cortex by in vivo microdialysis and capillary electrophoresis-laser induced fluorescence detection. *Journal of Neuroscience Methods*, *105*, 211–215.

[34] Robert, F., Bert, L., Parrot, S., Denoroy, L., Stoppini, L., Renaud, B. (1998). Coupling on-line brain microdialysis, precolumn derivatization and capillary electrophoresis for routine minute sampling of *O*-phosphoethanolamine and excitatory amino acids. *Journal of Chromatography A*, *817*, 195–203.

[35] Tucci, S., Rada, P., Sepulveda, M.J., Hernandez, L. (1997). Glutamate measured by 6-s resolution brain microdialysis: capillary electrophoretic and laser-induced fluorescence detection application. *Journal of Chromatography B: Biomedical Sciences and Applications*, *694*, 343–349.

[36] Zhou, S.Y., Zuo, H., Stobaugh, J.F., Lunte, C.E., Lunte, S.M. (1995). Continuous in vivo monitoring of amino acid neurotransmitters by microdialysis sampling with on-line derivatization and capillary electrophoresis separation. *Analytical Chemistry*, *67*, 594–599.

[37] Hernandez, L., Joshi, N., Murzi, E., Verdeguer, P., Mifsud, J.C., Guzman, N. (1993). Collinear laser-induced fluorescence detector for capillary electrophoresis: analysis of glutamic acid in brain dialysates. *Journal of Chromatography A*, *652*, 399–405.

[38] Shou, M., Smith, A.D., Shackman, J.G., Peris, J., Kennedy, R.T. (2004). In vivo monitoring of amino acids by microdialysis sampling with on-line derivatization

by naphthalene-2,3-dicarboxyaldehyde and rapid micellar electrokinetic capillary chromatography. *Journal of Neuroscience Methods, 138*, 189–197.

[39] Wang, C., Zhao, S., Yuan, H., Xiao, D. (2006). Determination of excitatory amino acids in biological fluids by capillary electrophoresis with optical fiber light-emitting diode induced fluorescence detection. *Journal of Chromatography B: Analytical Technologies in the Biomedical and Life Sciences, 833*, 129–134.

[40] Denoroy, L., Parrot, S., Renaud, L., Renaud, B., Zimmer, L. (2008). In-capillary derivatization and capillary electrophoresis separation of amino acid neurotransmitters from brain microdialysis samples. *Journal of Chromatography A, 1205*, 144–149.

[41] Sandlin, Z.D., Shou, M., Shackman, J.G., Kennedy, R.T. (2005). Microfluidic electrophoresis chip coupled to microdialysis for in vivo monitoring of amino acid neurotransmitters. *Analytical Chemistry, 77*, 7702–7708.

[42] Jacobson, S.C., Koutny, L.B., Hergenroeder, R., Moore, A.W., Ramsey, J.M. (1994). Microchip capillary electrophoresis with an integrated postcolumn reactor. *Analytical Chemistry, 66*, 3472–3476.

[43] Lindroth, P., Mopper, K. (1979). High performance liquid chromatographic determination of subpicomole amounts of amino acids by precolumn fluorescence derivatization with *o*-phthaldialdehyde. *Analytical Chemistry, 51*, 1667–1674.

[44] Beard, N.P., Edel, J.B., deMello, A.J. (2004). Integrated on-chip derivatization and electrophoresis for the rapid analysis of biogenic amines. *Electrophoresis, 25*, 2363–2373.

[45] Yassine, O., Morin, P., Dispagne, O., Renaud, L., Denoroy, L., Kleimann, P., Faure, K., Rocca, J.L., Ouaini, N., Ferrigno, R. (2008). Electrophoresis PDMS/glass chips with continuous on-chip derivatization and analysis of amino acids using naphthalene-2,3-dicarboxaldehyde as fluorogenic agent. *Analytica Chimica Acta, 609*, 215–222.

[46] Song, Y., Shenwu, M., Dhossche, D.M., Liu, Y.M. (2005). A capillary liquid chromatographic/tandem mass spectrometric method for the quantification of gamma-aminobutyric acid in human plasma and cerebrospinal fluid. *Journal of Chromatography B: Analytical Technologies in the Biomedical and Life Sciences, 814*, 295–302.

[47] Kondrat, R.W., Kanamori, K., Ross, B.D. (2002). In vivo microdialysis and gas-chromatography/mass-spectrometry for ^{13}C-enrichment measurement of extracellular glutamate in rat brain. *Journal of Neuroscience Methods, 120*, 179–192.

[48] Benturquia, N., Parrot, S., Sauvinet, V., Renaud, B., Denoroy, L. (2004). Simultaneous determination of vigabatrin and amino acid neurotransmitters in brain microdialysates by capillary electrophoresis with laser-induced fluorescence detection. *Journal of Chromatography B: Analytical Technologies in the Biomedical and Life Sciences, 806*, 237–244.

[49] Benveniste, H. (1989). Brain microdialysis. *Journal of Neurochemistry, 52*, 1667–1679.

[50] Kanamori, K., Ross, B.D., Kondrat, R.W. (2002). Glial uptake of neurotransmitter glutamate from the extracellular fluid studied in vivo by microdialysis and (13) C NMR. *Journal of Neurochemistry, 83*, 682–695.

[51] Melendez, R.I., Vuthiganon, J., Kalivas, P.W. (2005). Regulation of extracellular glutamate in the prefrontal cortex: focus on the cystine glutamate exchanger and group I metabotropic glutamate receptors. *Journal of Pharmacology and Experimental Therapeutics, 314*, 139–147.

[52] Xi, Z.X., Shen, H., Baker, D.A., Kalivas, P.W. (2003). Inhibition of non-vesicular glutamate release by group III metabotropic glutamate receptors in the nucleus accumbens. *Journal of Neurochemistry, 87*, 1204–1212.

[53] Baker, D.A., Xi, Z.X., Shen, H., Swanson, C.J., Kalivas, P.W. (2002). The origin and neuronal function of in vivo nonsynaptic glutamate. *Journal of Neuroscience, 22*, 9134–9141.

[54] Xi, Z.X., Ramamoorthy, S., Baker, D.A., Shen, H., Samuvel, D.J., Kalivas, P.W. (2002). Modulation of group II metabotropic glutamate receptor signaling by chronic cocaine. *Journal of Pharmacology and Experimental Therapeutics, 303*, 608–615.

[55] Nicholls, D., Attwell, D. (1990). The release and uptake of excitatory amino acids. *Trends in Pharmacological Sciences, 11*, 462–468.

[56] Araque, A., Li, N., Doyle, R.T., Haydon, P.G. (2000). SNARE protein-dependent glutamate release from astrocytes. *Journal of Neuroscience, 20*, 666–673.

[57] Bezzi, P., Gundersen, V., Galbete, J.L., Seifert, G., Steinhauser, C., Pilati, E., Volterra, A. (2004). Astrocytes contain a vesicular compartment that is competent for regulated exocytosis of glutamate. *Nature Neuroscience, 7*, 613–620.

[58] Malarkey, E.B., Parpura, V. (2008). Mechanisms of glutamate release from astrocytes. *Neurochemistry International, 52*, 142–154.

[59] Timmerman, W., Westerink, B.H. (1997). Brain microdialysis of GABA and glutamate: What does it signify? *Synapse, 27*, 242–261.

[60] Drew, K.L., Pehek, E.A., Rasley, B.T., Ma, Y.L., Green, T.K. (2004). Sampling glutamate and GABA with microdialysis: suggestions on how to get the dialysis membrane closer to the synapse. *Journal of Neuroscience Methods, 140*, 127–131.

[61] Lada, M.W., Vickroy, T.W., Kennedy, R.T. (1998). Evidence for neuronal origin and metabotropic receptor-mediated regulation of extracellular glutamate and aspartate in rat striatum in vivo following electrical stimulation of the prefrontal cortex. *Journal of Neurochemistry, 70*, 617–625.

[62] Gundersen, V., Chaudhry, F.A., Bjaalie, J.G., Fonnum, F., Ottersen, O.P., Storm-Mathisen, J. (1998). Synaptic vesicular localization and exocytosis of L-aspartate in excitatory nerve terminals: a quantitative immunogold analysis in rat hippocampus. *Journal of Neuroscience, 18*, 6059–6070.

[63] Obrenovitch, T.P., Urenjak, J., Zilkha, E., Jay, T.M. (2000). Excitotoxicity in neurological disorders: the glutamate paradox. *International Journal of Developmental Neuroscience, 18*, 281–287.

[64] Rusakov, D.A., Kullmann, D.M. (1998). Extrasynaptic glutamate diffusion in the hippocampus: ultrastructural constraints, uptake, and receptor activation. *Journal of Neuroscience, 18*, 3158–3170.

[65] Schousboe, A., Waagepetersen, H.S. (2006). Glial modulation of GABAergic and glutamatergic neurotransmission. *Current Topics in Medicinal Chemistry, 6*, 929–934.

[66] Sierra-Paredes, G., Sierra-Marcuno, G. (2007). Extrasynaptic GABA and glutamate receptors in epilepsy. *CNS & Neurological Disorders: Drug Targets, 6,* 288–300.

[67] Galvan, A., Kuwajima, M., Smith, Y. (2006). Glutamate and GABA receptors and transporters in the basal ganglia: What does their subsynaptic localization reveal about their function? *Neuroscience, 143,* 351–375.

[68] Del Arco, A., Segovia, G., Fuxe, K., Mora, F. (2003). Changes in dialysate concentrations of glutamate and GABA in the brain: an index of volume transmission mediated actions? *Journal of Neurochemistry, 85,* 23–33.

[69] Bradford, S.E., Nadler, J.V. (2004). Aspartate release from rat hippocampal synaptosomes. *Neuroscience, 128,* 751–765.

[70] Dopico, J.G., Gonzalez-Hernandez, T., Perez, I.M., Garcia, I.G., Abril, A.M., Inchausti, J.O., Rodriguez Diaz, M. (2006). Glycine release in the substantia nigra: interaction with glutamate and GABA. *Neuropharmacology, 50,* 548–557.

[71] Garcia Dopico, J., Perdomo Diaz, J., Alonso, T.J., Gonzalez Hernandez, T., Castro Fuentes, R., Rodriguez Diaz, M. (2004). Extracellular taurine in the substantia nigra: taurine–glutamate interaction. *Journal of Neuroscience Research, 76,* 528–538.

[72] Fantin, M., Marti, M., Auberson, Y.P., Morari, M. (2007). NR2A and NR2B subunit containing NMDA receptors differentially regulate striatal output pathways. *Journal of Neurochemistry, 103,* 2200–2211.

[73] Fantin, M., Auberson, Y.P., Morari, M. (2008). Differential effect of NR2A and NR2B subunit selective NMDA receptor antagonists on striato-pallidal neurons: relationship to motor response in the 6-hydroxydopamine model of parkinsonism. *Journal of Neurochemistry, 106,* 957–968.

[74] Ampe, B., Massie, A., D'Haens, J., Ebinger, G., Michotte, Y., Sarre, S. (2007). NMDA-mediated release of glutamate and GABA in the subthalamic nucleus is mediated by dopamine: an in vivo microdialysis study in rats. *Journal of Neurochemistry, 103,* 1063–1074.

[75] Marti, M., Manzalini, M., Fantin, M., Bianchi, C., Della Corte, L., Morari, M. (2005). Striatal glutamate release evoked in vivo by NMDA is dependent upon ongoing neuronal activity in the substantia nigra, endogenous striatal substance P and dopamine. *Journal of Neurochemistry, 93,* 195–205.

[76] Succu, S., Mascia, M.S., Sanna, F., Melis, T., Argiolas, A., Melis, M.R. (2006). The cannabinoid CB1 receptor antagonist SR 141716A induces penile erection by increasing extra-cellular glutamic acid in the paraventricular nucleus of male rats. *Behavioural Brain Research, 169,* 274–281.

[77] Lopez-Gil, X., Babot, Z., Amargos-Bosch, M., Sunol, C., Artigas, F., Adell, A. (2007). Clozapine and haloperidol differently suppress the MK-801–increased glutamatergic and serotonergic transmission in the medial prefrontal cortex of the rat. *Neuropsychopharmacology, 32,* 2087–2097.

[78] Abekawa, T., Ito, K., Koyama, T. (2006). Role of the simultaneous enhancement of NMDA and dopamine D1 receptor-mediated neurotransmission in the effects of clozapine on phencyclidine-induced acute increases in glutamate levels in the rat medial prefrontal cortex. *Naunyn-Schmiedeberg's Archives of Pharmacology, 374,* 177–193.

[79] Lambe, E.K., Aghajanian, G.K. (2006). Hallucinogen-induced UP states in the brain slice of rat prefrontal cortex: role of glutamate spillover and NR2B–NMDA receptors. *Neuropsychopharmacology, 31*, 1682–1689.

[80] Takumi, Y., Bergersen, L., Landsend, A.S., Rinvik, E., Ottersen, O.P. (1998). Synaptic arrangement of glutamate receptors. *Progress in Brain Research, 116*, 105–121.

[81] Lorrain, D.S., Baccei, C.S., Bristow, L.J., Anderson, J.J., Varney, M.A. (2003). Effects of ketamine and *N*-methyl-D-aspartate on glutamate and dopamine release in the rat prefrontal cortex: modulation by a group II selective metabotropic glutamate receptor agonist LY379268. *Neuroscience, 117*, 697–706.

[82] Moghaddam, B., Adams, B.W. (1998). Reversal of phencyclidine effects by a group II metabotropic glutamate receptor agonist in rats. *Science, 281*, 1349–1352.

[83] Wang, X., Ai, J., Hampson, D.R., Snead, O.C., 3rd (2005). Altered glutamate and GABA release within thalamocortical circuitry in metabotropic glutamate receptor 4 knockout mice. *Neuroscience, 134*, 1195–1203.

[84] Smolders, I., Lindekens, H., Clinckers, R., Meurs, A., O'Neill, M.J., Lodge, D., Ebinger, G., Michotte, Y. (2004). In vivo modulation of extracellular hippocampal glutamate and GABA levels and limbic seizures by group I and II metabotropic glutamate receptor ligands. *Journal of Neurochemistry, 88*, 1068–1077.

[85] Mela, F., Marti, M., Dekundy, A., Danysz, W., Morari, M., Cenci, M.A. (2007). Antagonism of metabotropic glutamate receptor type 5 attenuates *l*-dopa–induced dyskinesia and its molecular and neurochemical correlates in a rat model of Parkinson's disease. *Journal of Neurochemistry, 101*, 483–497.

[86] Popoli, P., Pintor, A., Tebano, M.T., Frank, C., Pepponi, R., Nazzicone, V., Grieco, R., Pezzola, A., Reggio, R., Minghetti, L., et al. (2004). Neuroprotective effects of the mGlu5R antagonist MPEP towards quinolinic acid-induced striatal toxicity: involvement of pre- and post-synaptic mechanisms and lack of direct NMDA blocking activity. *Journal of Neurochemistry, 89*, 1479–1489.

[87] Marabese, I., de Novellis, V., Palazzo, E., Mariani, L., Siniscalco, D., Rodella, L., Rossi, F., Maione, S. (2005). Differential roles of mGlu8 receptors in the regulation of glutamate and gamma-aminobutyric acid release at periaqueductal grey level. *Neuropharmacology, 49* (Suppl. 1), 157–166.

[88] Marabese, I., Rossi, F., Palazzo, E., de Novellis, V., Starowicz, K., Cristino, L., Vita, D., Gatta, L., Guida, F., Di Marzo, V., Maione, S. (2007). Periaqueductal gray metabotropic glutamate receptor subtype 7 and 8 mediate opposite effects on amino acid release, rostral ventromedial medulla cell activities, and thermal nociception. *Journal of Neurophysiology, 98*, 43–53.

[89] Del Arco, A., Mora, F. (2008). Prefrontal cortex-nucleus accumbens interaction: in vivo modulation by dopamine and glutamate in the prefrontal cortex. *Pharmacology Biochemistry and Behavior, 90*, 226–235.

[90] McGinty, J.F. (1999). Regulation of neurotransmitter interactions in the ventral striatum. *Annals of the New York Academy of Sciences, 877*, 129–139.

[91] Morari, M., Marti, M., Sbrenna, S., Fuxe, K., Bianchi, C., Beani, L. (1998). Reciprocal dopamine–glutamate modulation of release in the basal ganglia. *Neurochemistry International, 33*, 383–397.

[92] Del Arco, A., Mora, F. (2005). Glutamate–dopamine in vivo interaction in the prefrontal cortex modulates the release of dopamine and acetylcholine in the nucleus accumbens of the awake rat. *Journal of Neural Transmission, 112,* 97–109.

[93] Lindefors, N. (1993). Dopaminergic regulation of glutamic acid decarboxylase mRNA expression and GABA release in the striatum: a review. *Progress in Neuro-psychopharmacology and Biological Psychiatry, 17,* 887–903.

[94] Schechter, L.E., Smith, D.L., Rosenzweig-Lipson, S., Sukoff, S.J., Dawson, L.A., Marquis, K., Jones, D., Piesla, M., Andree, T., Nawoschik, S., et al. (2005). Lecozotan (SRA-333): a selective serotonin 1A receptor antagonist that enhances the stimulated release of glutamate and acetylcholine in the hippocampus and possesses cognitive-enhancing properties. *Journal of Pharmacology and Experimental Therapeutics, 314,* 1274–1289.

[95] Ceglia, I., Carli, M., Baviera, M., Renoldi, G., Calcagno, E., Invernizzi, R.W. (2004). The 5-HT receptor antagonist M100,907 prevents extracellular glutamate rising in response to NMDA receptor blockade in the mPFC. *Journal of Neurochemistry, 91,* 189–199.

[96] Calcagno, E., Carli, M., Invernizzi, R.W. (2006). The 5-HT(1A) receptor agonist 8-OH-DPAT prevents prefrontocortical glutamate and serotonin release in response to blockade of cortical NMDA receptors. *Journal of Neurochemistry, 96,* 853–860.

[97] Muschamp, J.W., Regina, M.J., Hull, E.M., Winter, J.C., Rabin, R.A. (2004). Lysergic acid diethylamide and [–]-2,5-dimethoxy-4-methylamphetamine increase extracellular glutamate in rat prefrontal cortex. *Brain Research, 1023,* 134–140.

[98] Schechter, L.E., Lin, Q., Smith, D.L., Zhang, G., Shan, Q., Platt, B., Brandt, M.R., Dawson, L.A., Cole, D., Bernotas, R., et al. (2008). Neuropharmacological profile of novel and selective 5-HT6 receptor agonists: WAY-181187 and WAY-208466. *Neuropsychopharmacology, 33,* 1323–1335.

[99] Dawson, L.A., Nguyen, H.Q., Li, P. (2001). The 5-HT(6) receptor antagonist SB-271046 selectively enhances excitatory neurotransmission in the rat frontal cortex and hippocampus. *Neuropsychopharmacology, 25,* 662–668.

[100] Kase, H. (2001). New aspects of physiological and pathophysiological functions of adenosine A2A receptor in basal ganglia. *Bioscience, Biotechnology, and Biochemistry, 65,* 1447–1457.

[101] Gianfriddo, M., Melani, A., Turchi, D., Giovannini, M.G., Pedata, F. (2004). Adenosine and glutamate extracellular concentrations and mitogen-activated protein kinases in the striatum of Huntington transgenic mice: selective antagonism of adenosine A2A receptors reduces transmitter outflow. *Neurobiology of Disease, 17,* 77–88.

[102] Ochi, M., Shiozaki, S., Kase, H. (2004). Adenosine A(2A) receptor-mediated modulation of GABA and glutamate release in the output regions of the basal ganglia in a rodent model of Parkinson's disease. *Neuroscience, 127,* 223–231.

[103] Pintor, A., Tebano, M.T., Martire, A., Grieco, R., Galluzzo, M., Scattoni, M.L., Pezzola, A., Coccurello, R., Felici, F., Cuomo, V., et al. (2006). The cannabinoid

receptor agonist WIN 55,212-2 attenuates the effects induced by quinolinic acid in the rat striatum. *Neuropharmacology, 51,* 1004–1012.

[104] Levy, F., Kendrick, K.M., Goode, J.A., Guevara-Guzman, R., Keverne, E.B. (1995). Oxytocin and vasopressin release in the olfactory bulb of parturient ewes: changes with maternal experience and effects on acetylcholine, gamma-aminobutyric acid, glutamate and noradrenaline release. *Brain Research, 669,* 197–206.

[105] Ebner, K., Bosch, O.J., Kromer, S.A., Singewald, N., Neumann, I.D. (2005). Release of oxytocin in the rat central amygdala modulates stress-coping behavior and the release of excitatory amino acids. *Neuropsychopharmacology, 30,* 223–230.

[106] Engelmann, M., Bull, P.M., Brown, C.H., Landgraf, R., Horn, T.F., Singewald, N., Ludwig, M., Wotjak, C.T. (2004). GABA selectively controls the secretory activity of oxytocin neurons in the rat supraoptic nucleus. *European Journal of Neuroscience, 19,* 601–608.

[107] Mazzuferi, M., Binaschi, A., Rodi, D., Mantovani, S., Simonato, M. (2005). Induction of B1 bradykinin receptors in the kindled hippocampus increases extracellular glutamate levels: a microdialysis study. *Neuroscience, 135,* 979–986.

[108] Stragier, B., Hristova, I., Sarre, S., Ebinger, G., Michotte, Y. (2005). In vivo characterization of the angiotensin-(1–7)-induced dopamine and gamma-aminobutyric acid release in the striatum of the rat. *European Journal of Neuroscience, 22,* 658–664.

[109] Hu, L., Zhu, D.N., Yu, Z., Wang, J.Q., Sun, Z.J., Yao, T. (2002). Expression of angiotensin II type 1 (AT(1)) receptor in the rostral ventrolateral medulla in rats. *Journal of Applied Physiology, 92,* 2153–2161.

[110] Hathway, G.J., Emson, P.C., Humphrey, P.P., Kendrick, K.M. (1998). Somatostatin potently stimulates in vivo striatal dopamine and gamma-aminobutyric acid release by a glutamate-dependent action. *Journal of Neurochemistry, 70,* 1740–1749.

[111] Allen, J.P., Hathway, G.J., Clarke, N.J., Jowett, M.I., Topps, S., Kendrick, K.M., Humphrey, P.P., Wilkinson, L.S., Emson, P.C. (2003). Somatostatin receptor 2 knockout/lacZ knockin mice show impaired motor coordination and reveal sites of somatostatin action within the striatum. *European Journal of Neuroscience, 17,* 1881–1895.

[112] Stragier, B., Clinckers, R., Meurs, A., De Bundel, D., Sarre, S., Ebinger, G., Michotte, Y., Smolders, I. (2006). Involvement of the somatostatin-2 receptor in the anti-convulsant effect of angiotensin IV against pilocarpine-induced limbic seizures in rats. *Journal of Neurochemistry, 98,* 1100–1113.

[113] Ge, J., Long, S.K., Kilpatrick, I.C. (1998). Preferential blockade of cholecystokinin-8S-induced increases in aspartate and glutamate levels by the CCK(B) receptor antagonist, L-365,260, in rat brain. *European Journal of Pharmacology, 345,* 163–170.

[114] You, Z.B., Godukhin, O., Goiny, M., Nylander, I., Ungerstedt, U., Terenius, L., Hokfelt, T., Herrera-Marschitz, M. (1997). Cholecystokinin-8S increases dynorphin B, aspartate and glutamate release in the fronto-parietal cortex of the rat via different receptor subtypes. *Naunyn-Schmiedeberg's Archives of Pharmacology, 355,* 576–581.

[115] Ferraro, L., O'Connor, W.T., Glennon, J., Tomasini, M.C., Bebe, B.W., Tanganelli, S., Antonelli, T. (2000). Evidence for a nucleus accumbens CCK2 receptor regulation of rat ventral pallidal GABA levels: a dual probe microdialysis study. *Life Sciences*, *68*, 483–496.

[116] Ferraro, L., Antonelli, T., O'Connor, W.T., Fuxe, K., Soubrie, P., Tanganelli, S. (1998). The striatal neurotensin receptor modulates striatal and pallidal glutamate and GABA release: functional evidence for a pallidal glutamate-GABA interaction via the pallidal–subthalamic nucleus loop. *Journal of Neuroscience*, *18*, 6977–6989.

[117] Ferraro, L., Tomasini, M.C., Fernandez, M., Bebe, B.W., O'Connor, W.T., Fuxe, K., Glennon, J.C., Tanganelli, S., Antonelli, T. (2001). Nigral neurotensin receptor regulation of nigral glutamate and nigroventral thalamic GABA transmission: a dual-probe microdialysis study in intact conscious rat brain. *Neuroscience*, *102*, 113–120.

[118] Rakovska, A., Giovannini, M.G., Della Corte, L., Kalfin, R., Bianchi, L., Pepeu, G. (1998). Neurotensin modulation of acetylcholine and GABA release from the rat hippocampus: an in vivo microdialysis study. *Neurochemistry International*, *33*, 335–340.

[119] Petkova-Kirova, P., Rakovska, A., Della Corte, L., Zaekova, G., Radomirov, R., Mayer, A. (2008). Neurotensin modulation of acetylcholine, GABA, and aspartate release from rat prefrontal cortex studied in vivo with microdialysis. *Brain Research Bulletin*, *77*, 129–135.

[120] Torregrossa, M.M., Kalivas, P.W. (2008). Microdialysis and the neurochemistry of addiction. *Pharmacology Biochemistry and Behavior*, *90*, 261–272.

[121] Xue, C.J., Ng, J.P., Li, Y., Wolf, M.E. (1996). Acute and repeated systemic amphetamine administration: effects on extracellular glutamate, aspartate, and serine levels in rat ventral tegmental area and nucleus accumbens. *Journal of Neurochemistry*, *67*, 352–363.

[122] Reid, M.S., Hsu, K., Jr., Berger, S.P. (1997). Cocaine and amphetamine preferentially stimulate glutamate release in the limbic system: studies on the involvement of dopamine. *Synapse*, *27*, 95–105.

[123] Rahman, S., Bardo, M.T. (2008). Environmental enrichment increases amphetamine-induced glutamate neurotransmission in the nucleus accumbens: a neurochemical study. *Brain Research*, *1197*, 40–46.

[124] Gray, A.M., Rawls, S.M., Shippenberg, T.S., McGinty, J.F. (1999). The kappa-opioid agonist, U-69593, decreases acute amphetamine-evoked behaviors and calcium-dependent dialysate levels of dopamine and glutamate in the ventral striatum. *Journal of Neurochemistry*, *73*, 1066–1074.

[125] Rawls, S.M., McGinty, J.F. (2000). Delta opioid receptors regulate calcium-dependent, amphetamine-evoked glutamate levels in the rat striatum: an in vivo microdialysis study. *Brain Research*, *861*, 296–304.

[126] Del Arco, A., Castaneda, T.R., Mora, F. (1998). Amphetamine releases GABA in striatum of the freely moving rat: involvement of calcium and high affinity transporter mechanisms. *Neuropharmacology*, *37*, 199–205.

[127] Del Arco, A., Gonzalez-Mora, J.L., Armas, V.R., Mora, F. (1999). Amphetamine increases the extracellular concentration of glutamate in striatum of the awake

rat: involvement of high affinity transporter mechanisms. *Neuropharmacology*, *38*, 943–954.

[128] Kalivas, P.W., Duffy, P. (1997). Dopamine regulation of extracellular glutamate in the nucleus accumbens. *Brain Research*, *761*, 173–177.

[129] Wolf, M.E., Xue, C.J. (1999). Amphetamine-induced glutamate efflux in the rat ventral tegmental area is prevented by MK-801, SCH 23390, and ibotenic acid lesions of the prefrontal cortex. *Journal of Neurochemistry*, *73*, 1529–1538.

[130] Wolf, M.E., Xue, C.J., Li, Y., Wavak, D. (2000). Amphetamine increases glutamate efflux in the rat ventral tegmental area by a mechanism involving glutamate transporters and reactive oxygen species. *Journal of Neurochemistry*, *75*, 1634–1644.

[131] Wolf, M.E., Xue, C.J. (1998). Amphetamine and D1 dopamine receptor agonists produce biphasic effects on glutamate efflux in rat ventral tegmental area: modification by repeated amphetamine administration. *Journal of Neurochemistry*, *70*, 198–209.

[132] Del Arco, A., Martinez, R., Mora, F. (1998). Amphetamine increases extracellular concentrations of glutamate in the prefrontal cortex of the awake rat: a microdialysis study. *Neurochemical Research*, *23*, 1153–1158.

[133] Nelson, C.L., Wetter, J.B., Milovanovic, M., Wolf, M.E. (2007). The laterodorsal tegmentum contributes to behavioral sensitization to amphetamine. *Neuroscience*, *146*, 41–49.

[134] Giorgetti, M., Hotsenpiller, G., Ward, P., Teppen, T., Wolf, M.E. (2001). Amphetamine-induced plasticity of AMPA receptors in the ventral tegmental area: effects on extracellular levels of dopamine and glutamate in freely moving rats. *Journal of Neuroscience*, *21*, 6362–6369.

[135] Giorgetti, M., Hotsenpiller, G., Froestl, W., Wolf, M.E. (2002). In vivo modulation of ventral tegmental area dopamine and glutamate efflux by local GABA(B) receptors is altered after repeated amphetamine treatment. *Neuroscience*, *109*, 585–595.

[136] Kim, J.H., Austin, J.D., Tanabe, L., Creekmore, E., Vezina, P. (2005). Activation of group II mGlu receptors blocks the enhanced drug taking induced by previous exposure to amphetamine. *European Journal of Neuroscience*, *21*, 295–300.

[137] Chen, J.C., Liang, K.W., Huang, Y.K., Liang, C.S., Chiang, Y.C. (2001). Significance of glutamate and dopamine neurons in the ventral pallidum in the expression of behavioral sensitization to amphetamine. *Life Sciences*, *68*, 973–983.

[138] Smith, J.A., Mo, Q., Guo, H., Kunko, P.M., Robinson, S.E. (1995). Cocaine increases extraneuronal levels of aspartate and glutamate in the nucleus accumbens. *Brain Research*, *683*, 264–269.

[139] Swanson, C.J., Baker, D.A., Carson, D., Worley, P.F., Kalivas, P.W. (2001). Repeated cocaine administration attenuates group I metabotropic glutamate receptor-mediated glutamate release and behavioral activation: a potential role for Homer. *Journal of Neuroscience*, *21*, 9043–9052.

[140] Xi, Z.X., Ramamoorthy, S., Shen, H., Lake, R., Samuvel, D.J., Kalivas, P.W. (2003). GABA transmission in the nucleus accumbens is altered after withdrawal from repeated cocaine. *Journal of Neuroscience*, *23*, 3498–3505.

[141] McKee, B.L., Meshul, C.K. (2005). Time-dependent changes in extracellular glutamate in the rat dorsolateral striatum following a single cocaine injection. *Neuroscience, 133,* 605–613.

[142] Kalivas, P.W., Duffy, P. (1995). D1 receptors modulate glutamate transmission in the ventral tegmental area. *Journal of Neuroscience, 15,* 5379–5388.

[143] Williams, J.M., Steketee, J.D. (2004). Cocaine increases medial prefrontal cortical glutamate overflow in cocaine-sensitized rats: a time course study. *European Journal of Neuroscience, 20,* 1639–1646.

[144] Jayaram, P., Steketee, J.D. (2004). Effects of repeated cocaine on medial prefrontal cortical GABAB receptor modulation of neurotransmission in the mesocorticolimbic dopamine system. *Journal of Neurochemistry, 90,* 839–847.

[145] Jayaram, P., Steketee, J.D. (2005). Effects of cocaine-induced behavioural sensitization on GABA transmission within rat medial prefrontal cortex. *European Journal of Neuroscience, 21,* 2035–2039.

[146] Jayaram, P., Steketee, J.D. (2006). Cocaine-induced increases in medial prefrontal cortical GABA transmission involves glutamatergic receptors. *European Journal of Pharmacology, 531,* 74–79.

[147] Miguens, M., Del Olmo, N., Higuera-Matas, A., Torres, I., Garcia-Lecumberri, C., Ambrosio, E. (2008). Glutamate and aspartate levels in the nucleus accumbens during cocaine self-administration and extinction: a time course microdialysis study. *Psychopharmacology, 196,* 303–313.

[148] McFarland, K., Lapish, C.C., Kalivas, P.W. (2003). Prefrontal glutamate release into the core of the nucleus accumbens mediates cocaine-induced reinstatement of drug-seeking behavior. *Journal of Neuroscience, 23,* 3531–3537.

[149] Torregrossa, M.M., Tang, X.C., Kalivas, P.W. (2008). The glutamatergic projection from the prefrontal cortex to the nucleus accumbens core is required for cocaine-induced decreases in ventral pallidal GABA. *Neuroscience Letters, 438,* 142–145.

[150] Torregrossa, M.M., Kalivas, P.W. (2008). Neurotensin in the ventral pallidum increases extracellular gamma-aminobutyric acid and differentially affects cue- and cocaine-primed reinstatement. *Journal of Pharmacology and Experimental Therapeutics, 325,* 556–566.

[151] Tang, X.C., McFarland, K., Cagle, S., Kalivas, P.W. (2005). Cocaine-induced reinstatement requires endogenous stimulation of mu-opioid receptors in the ventral pallidum. *Journal of Neuroscience, 25,* 4512–4520.

[152] Hotsenpiller, G., Wolf, M.E. (2003). Baclofen attenuates conditioned locomotion to cues associated with cocaine administration and stabilizes extracellular glutamate levels in rat nucleus accumbens. *Neuroscience, 118,* 123–134.

[153] Wang, B., Shaham, Y., Zitzman, D., Azari, S., Wise, R.A., You, Z.B. (2005). Cocaine experience establishes control of midbrain glutamate and dopamine by corticotropin-releasing factor: a role in stress-induced relapse to drug seeking. *Journal of Neuroscience, 25,* 5389–5396.

[154] Wang, B., You, Z.B., Rice, K.C., Wise, R.A. (2007). Stress-induced relapse to cocaine seeking: roles for the CRF(2) receptor and CRF-binding protein in the ventral tegmental area of the rat. *Psychopharmacology, 193,* 283–294.

[155] You, Z.B., Wang, B., Zitzman, D., Azari, S., Wise, R.A. (2007). A role for conditioned ventral tegmental glutamate release in cocaine seeking. *Journal of Neuroscience, 27*, 10546–10555.

[156] Wise, R.A., Wang, B., You, Z.B. (2008). Cocaine serves as a peripheral interoceptive conditioned stimulus for central glutamate and dopamine release. *PLoS One, 3*, e2846.

[157] Caille, S., Parsons, L.H. (2004). Intravenous heroin self-administration decreases GABA efflux in the ventral pallidum: an in vivo microdialysis study in rats. *European Journal of Neuroscience, 20*, 593–596.

[158] LaLumiere, R.T., Kalivas, P.W. (2008). Glutamate release in the nucleus accumbens core is necessary for heroin seeking. *Journal of Neuroscience, 28*, 3170–3177.

[159] Yan, Q.S., Reith, M.E., Yan, S.G., Jobe, P.C. (1998). Effect of systemic ethanol on basal and stimulated glutamate releases in the nucleus accumbens of freely moving Sprague–Dawley rats: a microdialysis study. *Neuroscience Letters, 258*, 29–32.

[160] Szumlinski, K.K., Diab, M.E., Friedman, R., Henze, L.M., Lominac, K.D., Bowers, M.S. (2007). Accumbens neurochemical adaptations produced by binge-like alcohol consumption. *Psychopharmacology, 190*, 415–431.

[161] Roberto, M., Schweitzer, P., Madamba, S.G., Stouffer, D.G., Parsons, L.H., Siggins, G.R. (2004). Acute and chronic ethanol alter glutamatergic transmission in rat central amygdala: an in vitro and in vivo analysis. *Journal of Neuroscience, 24*, 1594–1603.

[162] Roberto, M., Madamba, S.G., Stouffer, D.G., Parsons, L.H., Siggins, G.R. (2004). Increased GABA release in the central amygdala of ethanol-dependent rats. *Journal of Neuroscience, 24*, 10159–10166.

[163] Melendez, R.I., Hicks, M.P., Cagle, S.S., Kalivas, P.W. (2005). Ethanol exposure decreases glutamate uptake in the nucleus accumbens. *Alcoholism: Clinical and Experimental Research, 29*, 326–333.

[164] Dahchour, A., De Witte, P. (2000). Taurine blocks the glutamate increase in the nucleus accumbens microdialysate of ethanol-dependent rats. *Pharmacology Biochemistry and Behavior, 65*, 345–350.

[165] Baum, S., Huebner, A., Krimphove, M., Morgenstern, R., Badawy, A.A., Spies, C.D. (2006). Nicotine stimulation on extracellular glutamate levels in the nucleus accumbens of ethanol-withdrawn rats in vivo. *Alcoholism: Clinical and Experimental Research, 30*, 1414–1421.

[166] Rossetti, Z.L., Carboni, S. (1995). Ethanol withdrawal is associated with increased extracellular glutamate in the rat striatum. *European Journal of Pharmacology, 283*, 177–183.

[167] Rossetti, Z.L., Carboni, S., Fadda, F. (1999). Glutamate-induced increase of extracellular glutamate through N-methyl-D-aspartate receptors in ethanol withdrawal. *Neuroscience, 93*, 1135–1140.

[168] Dahchour, A., De Witte, P. (2003). Excitatory and inhibitory amino acid changes during repeated episodes of ethanol withdrawal: an in vivo microdialysis study. *European Journal of Pharmacology, 459*, 171–178.

[169] Kapasova, Z., Szumlinski, K.K. (2008). Strain differences in alcohol-induced neurochemical plasticity: a role for accumbens glutamate in alcohol intake. *Alcoholism: Clinical and Experimental Research*, *32*, 617–631.

[170] Dahchour, A., Hoffman, A., Deitrich, R., de Witte, P. (2000). Effects of ethanol on extracellular amino acid levels in high-and low-alcohol sensitive rats: a microdialysis study. *Alcohol and Alcoholism*, *35*, 548–553.

[171] Piepponen, T.P., Kiianmaa, K., Ahtee, L. (2002). Effects of ethanol on the accumbal output of dopamine, GABA and glutamate in alcohol-tolerant and alcohol-nontolerant rats. *Pharmacology Biochemistry and Behavior*, *74*, 21–30.

[172] Millan, M.J. (1999). The induction of pain: an integrative review. *Progress in Neurobiology*, *57*, 1–164.

[173] Millan, M.J. (2002). Descending control of pain. *Progress in Neurobiology*, *66*, 355–474.

[174] Stiller, C.O., Brodin, E., Taylor, B.K. (2007). Microdialysis in pain research. In: Westerink, B.H., Cremers, T. (Eds.), *Handbook of Microdialysis: Methods, Applications and Perspectives*, Elsevier, Amsterdam, pp. 473–481.

[175] Jin, Y.H., Nishioka, H., Wakabayashi, K., Fujita, T., Yonehara, N. (2006). Effect of morphine on the release of excitatory amino acids in the rat hind instep: Pain is modulated by the interaction between the peripheral opioid and glutamate systems. *Neuroscience*, *138*, 1329–1339.

[176] Kawamata, T., Omote, K., Toriyabe, M., Kawamata, M., Namiki, A. (2002). Intracerebroventricular morphine produces antinociception by evoking gamma-aminobutyric acid release through activation of 5-hydroxytryptamine 3 receptors in the spinal cord. *Anesthesiology*, *96*, 1175–1182.

[177] Buerkle, H., Marsala, M., Yaksh, T.L. (1998). Effect of continuous spinal remifentanil infusion on behaviour and spinal glutamate release evoked by subcutaneous formalin in the rat. *British Journal of Anaesthesia*, *80*, 348–353.

[178] Watson, C.J., Lydic, R., Baghdoyan, H.A. (2007). Sleep and GABA levels in the oral part of rat pontine reticular formation are decreased by local and systemic administration of morphine. *Neuroscience*, *144*, 375–386.

[179] Wen, Z.H., Chang, Y.C., Cherng, C.H., Wang, J.J., Tao, P.L., Wong, C.S. (2004). Increasing of intrathecal CSF excitatory amino acids concentration following morphine challenge in morphine-tolerant rats. *Brain Research*, *995*, 253–259.

[180] Lin, J.A., Lee, M.S., Wu, C.T., Yeh, C.C., Lin, S.L., Wen, Z.H., Wong, C.S. (2005). Attenuation of morphine tolerance by intrathecal gabapentin is associated with suppression of morphine-evoked excitatory amino acid release in the rat spinal cord. *Brain Research*, *1054*, 167–173.

[181] Tai, Y.H., Wang, Y.H., Wang, J.J., Tao, P.L., Tung, C.S., Wong, C.S. (2006). Amitriptyline suppresses neuroinflammation and up-regulates glutamate transporters in morphine-tolerant rats. *Pain*, *124*, 77–86.

[182] Sakurada, T., Komatsu, T., Kuwahata, H., Watanabe, C., Orito, T., Sakurada, C., Tsuzuki, M., Sakurada, S. (2007). Intrathecal substance P (1–7) prevents morphine-evoked spontaneous pain behavior via spinal NMDA-NO cascade. *Biochemical Pharmacology*, *74*, 758–767.

[183] Ipponi, A., Lamberti, C., Medica, A., Bartolini, A., Malmberg-Aiello, P. (1999). Tiagabine antinociception in rodents depends on GABA(B) receptor activation: parallel antinociception testing and medial thalamus GABA microdialysis. *European Journal of Pharmacology, 368*, 205–211.

[184] Smith, C.G., Bowery, N.G., Whitehead, K.J. (2007). GABA transporter type 1 (GAT-1) uptake inhibition reduces stimulated aspartate and glutamate release in the dorsal spinal cord in vivo via different GABAergic mechanisms. *Neuropharmacology, 53*, 975–981.

[185] Coderre, T.J., Kumar, N., Lefebvre, C.D., Yu, J.S. (2007). A comparison of the glutamate release inhibition and anti-allodynic effects of gabapentin, lamotrigine, and riluzole in a model of neuropathic pain. *Journal of Neurochemistry, 100*, 1289–1299.

[186] Feng, Y., Cui, M., Willis, W.D. (2003). Gabapentin markedly reduces acetic acid-induced visceral nociception. *Anesthesiology, 98*, 729–733.

[187] Zhang, L., Lu, Y., Chen, Y., Westlund, K.N. (2002). Group I metabotropic glutamate receptor antagonists block secondary thermal hyperalgesia in rats with knee joint inflammation. *Journal of Pharmacology and Experimental Therapeutics, 300*, 149–156.

[188] Lorrain, D.S., Correa, L., Anderson, J., Varney, M. (2002). Activation of spinal group I metabotropic glutamate receptors in rats evokes local glutamate release and spontaneous nociceptive behaviors: effects of 2-methyl-6-(phenylethynyl)-pyridine pretreatment. *Neuroscience Letters, 327*, 198–202.

[189] Langman, N.J., Smith, C.G., Whitehead, K.J. (2006). Selective serotonin re-uptake inhibition attenuates evoked glutamate release in the dorsal horn of the anaesthetised rat in vivo. *Pharmacological Research, 53*, 149–155.

[190] Oliva, P., Berrino, L., de Novellis, V., Palazzo, E., Marabese, I., Siniscalco, D., Scafuro, M., Mariani, L., Rossi, F., Maione, S. (2006). Role of periaqueductal grey prostaglandin receptors in formalin-induced hyperalgesia. *European Journal of Pharmacology, 530*, 40–47.

[191] Malmberg, A.B., Hedner, T., Fallgren, B., Calcutt, N.A. (1997). The effect of alpha-trinositol (D-myo-inositol 1,2,6-trisphosphate) on formalin-evoked spinal amino acid and prostaglandin E2 levels. *Brain Research, 747*, 160–164.

[192] Malmberg, A.B., Yaksh, T.L. (1995). Cyclooxygenase inhibition and the spinal release of prostaglandin E2 and amino acids evoked by paw formalin injection: a microdialysis study in unanesthetized rats. *Journal of Neuroscience, 15*, 2768–2776.

[193] de Novellis, V., Negri, L., Lattanzi, R., Rossi, F., Palazzo, E., Marabese, I., Giannini, E., Vita, D., Melchiorri, P., Maione, S. (2007). The prokineticin receptor agonist Bv8 increases GABA release in the periaqueductal grey and modifies RVM cell activities and thermoceptive reflexes in the rat. *European Journal of Neuroscience, 26*, 3068–3078.

[194] Szallasi, A., Cortright, D.N., Blum, C.A., Eid, S.R. (2007). The vanilloid receptor TRPV1: 10 years from channel cloning to antagonist proof-of-concept. *Nature Reviews Drug Discovery, 6*, 357–372.

[195] Starowicz, K., Maione, S., Cristino, L., Palazzo, E., Marabese, I., Rossi, F., de Novellis, V., Di Marzo, V. (2007). Tonic endovanilloid facilitation of glutamate

release in brainstem descending antinociceptive pathways. *Journal of Neuroscience*, *27*, 13739–13749.

[196] Benveniste, H., Drejer, J., Schousboe, A., Diemer, N.H. (1984). Elevation of the extracellular concentrations of glutamate and aspartate in rat hippocampus during transient cerebral ischemia monitored by intracerebral microdialysis. *Journal of Neurochemistry*, *43*, 1369–1374.

[197] Phillis, J.W., O'Regan, M.H. (2003). Characterization of modes of release of amino acids in the ischemic/reperfused rat cerebral cortex. *Neurochemistry International*, *43*, 461–467.

[198] Ghribi, O., Callebert, J., Verrecchia, C., Plotkine, M., Boulu, R.G. (1995). Blockers of NMDA-operated channels decrease glutamate and aspartate extracellular accumulation in striatum during forebrain ischaemia in rats. *Fundamental & Clinical Pharmacology*, *9*, 141–146.

[199] Parrot, S., Cottet-Emard, J.M., Sauvinet, V., Pequignot, J.M., Denoroy, L. (2003). Effects of acute hypoxic conditions on extracellular excitatory amino acids and dopamine in the striatum of freely-moving rats. *Advances in Experimental Medicine and Biology*, *536*, 433–444.

[200] Homola, A., Zoremba, N., Slais, K., Kuhlen, R., Sykova, E. (2006). Changes in diffusion parameters, energy-related metabolites and glutamate in the rat cortex after transient hypoxia/ischemia. *Neuroscience Letters*, *404*, 137–142.

[201] Qureshi, A.I., Ali, Z., Suri, M.F., Shuaib, A., Baker, G., Todd, K., Guterman, L.R., Hopkins, L.N. (2003). Extracellular glutamate and other amino acids in experimental intracerebral hemorrhage: an in vivo microdialysis study. *Critical Care Medicine*, *31*, 1482–1489.

[202] Hehre, D.A., Devia, C.J., Bancalari, E., Suguihara, C. (2008). Brainstem amino acid neurotransmitters and ventilatory response to hypoxia in piglets. *Pediatric Research*, *63*, 46–50.

[203] Dohmen, C., Kumura, E., Rosner, G., Heiss, W.D., Graf, R. (2005). Extracellular correlates of glutamate toxicity in short-term cerebral ischemia and reperfusion: a direct in vivo comparison between white and gray matter. *Brain Research*, *1037*, 43–51.

[204] Molchanova, S., Koobi, P., Oja, S.S., Saransaari, P. (2004). Interstitial concentrations of amino acids in the rat striatum during global forebrain ischemia and potassium-evoked spreading depression. *Neurochemical Research*, *29*, 1519–1527.

[205] Milton, S.L., Manuel, L., Lutz, P.L. (2003). Slow death in the leopard frog *Rana pipiens*: neurotransmitters and anoxia tolerance. *Journal of Experimental Biology*, *206*, 4021–4028.

[206] Nilsson, G.E., Lutz, P.L. (1991). Release of inhibitory neurotransmitters in response to anoxia in turtle brain. *American Journal of Physiology*, *261*, R32–37.

[207] Hylland, P., Nilsson, G.E. (1999). Extracellular levels of amino acid neurotransmitters during anoxia and forced energy deficiency in crucian carp brain. *Brain Research*, *823*, 49–58.

[208] Pastuszko, A. (1994). Metabolic responses of the dopaminergic system during hypoxia in newborn brain. *Biochemical Medicine and Metabolic Biology*, *51*, 1–15.

[209] Heim, C., Zhang, J., Lan, J., Sieklucka, M., Kurz, T., Riederer, P., Gerlach, M., Sontag, K.H. (2000). Cerebral oligaemia episode triggers free radical formation and late cognitive deficiencies. *European Journal of Neuroscience, 12,* 715–725.

[210] Satoh, M., Asai, S., Katayama, Y., Kohno, T., Ishikawa, K. (1999). Real-time monitoring of glutamate transmitter release with anoxic depolarization during anoxic insult in rat striatum. *Brain Research, 822,* 142–148.

[211] Mitani, A., Tanaka, K. (2003). Functional changes of glial glutamate transporter GLT-1 during ischemia: an in vivo study in the hippocampal CA1 of normal mice and mutant mice lacking GLT-1. *Journal of Neuroscience, 23,* 7176–7182.

[212] Feustel, P.J., Jin, Y., Kimelberg, H.K. (2004). Volume-regulated anion channels are the predominant contributors to release of excitatory amino acids in the ischemic cortical penumbra. *Stroke, 35,* 1164–1168.

[213] Han, J.L., Kollmar, R., Tobyas, B., Schwab, S. (2008). Inhibited glutamate release by granulocyte-colony stimulating factor after experimental stroke. *Neuroscience Letters, 432,* 167–169.

[214] Camacho, A., Montiel, T., Massieu, L. (2006). The anion channel blocker, 4,4′-dinitrostilbene-2,2′-disulfonic acid prevents neuronal death and excitatory amino acid release during glycolysis inhibition in the hippocampus in vivo. *Neuroscience, 142,* 1005–1017.

[215] Ritz, M.F., Schmidt, P., Mendelowitsch, A. (2004). Acute effects of 17beta-estradiol on the extracellular concentration of excitatory amino acids and energy metabolites during transient cerebral ischemia in male rats. *Brain Research, 1022,* 157–163.

[216] Baldwin, H.A., Williams, J.L., Snares, M., Ferreira, T., Cross, A.J., Green, A.R. (1994). Attenuation by chlormethiazole administration of the rise in extracellular amino acids following focal ischaemia in the cerebral cortex of the rat. *British Journal of Pharmacology, 112,* 188–194.

[217] Zhang, Y., Zhang, H., Feustel, P.J., Kimelberg, H.K. (2008). DCPIB, a specific inhibitor of volume regulated anion channels (VRACs), reduces infarct size in MCAo and the release of glutamate in the ischemic cortical penumbra. *Experimental Neurology, 210,* 514–520.

[218] Berger, C., Stauder, A., Xia, F., Sommer, C., Schwab, S. (2008). Neuroprotection and glutamate attenuation by acetylsalicylic acid in temporary but not in permanent cerebral ischemia. *Experimental Neurology, 210,* 543–548.

[219] Jellish, W.S., Murdoch, J., Kindel, G., Zhang, X., White, F.A. (2005). The effect of clonidine on cell survival, glutamate, and aspartate release in normo- and hyperglycemic rats after near complete forebrain ischemia. *Experimental Brain Research, 167,* 526–534.

[220] Mori, T., Tateishi, N., Kagamiishi, Y., Shimoda, T., Satoh, S., Ono, S., Katsube, N., Asano, T. (2004). Attenuation of a delayed increase in the extracellular glutamate level in the peri-infarct area following focal cerebral ischemia by a novel agent ONO-2506. *Neurochemistry International, 45,* 381–387.

[221] Berger, C., Schabitz, W.R., Wolf, M., Mueller, H., Sommer, C., Schwab, S. (2004). Hypothermia and brain-derived neurotrophic factor reduce glutamate synergistically in acute stroke. *Experimental Neurology, 185,* 305–312.

[222] Hsieh, Y.C., Cheng, H., Chan, K.H., Chang, W.K., Liu, T.M., Wong, C.S. (2007). Protective effect of intrathecal ketorolac in spinal cord ischemia in rats: a microdialysis study. *Acta Anaesthesiologica Scandinavica*, *51*, 410–414.

[223] Pu, F., Mishima, K., Egashira, N., Iwasaki, K., Kaneko, T., Uchida, T., Irie, K., Ishibashi, D., Fujii, H., Kosuna, K., Fujiwara, M. (2004). Protective effect of buckwheat polyphenols against long-lasting impairment of spatial memory associated with hippocampal neuronal damage in rats subjected to repeated cerebral ischemia. *Journal of Pharmacological Sciences*, *94*, 393–402.

6

MICRODIALYSIS AS A TOOL TO UNRAVEL NEUROBIOLOGICAL MECHANISMS OF SEIZURES AND ANTIEPILEPTIC DRUG ACTION

ILSE SMOLDERS, RALPH CLINCKERS, AND YVETTE MICHOTTE
Vrije Universiteit Brussels, Brussels, Belgium

1. INTRODUCTION

Epilepsies are a family of chronic neurological disorders characterized by the spontaneous recurrence of seizures that disrupt periods of more-or-less normal electroencephalographic (EEG) activity and behavior. An epileptic seizure is expressed clinically as an intermittent, stereotyped disturbance of consciousness, motor function, behavior, emotion, and/or sensation. It is the result of simultaneous hypersynchronous high-frequency activation of a large population of neurons, which leads to a transient, self-sustained interruption of normal brain function. Most epileptic seizures are due to discharges generated in cortical and hippocampal structures, although subcortical structures are also involved in some seizure types. The clinical expression of a seizure depends on its site of origin, time course, and discharge propagation.

Most of what we know about seizures and epilepsy is derived directly or indirectly from animal models, which all mimic certain features of the human disorder. The large number of models of seizures and of epilepsy [1] derives from two reasons. First, there are numerous types of epileptic seizures and syndromes to be modeled. Second, important findings need confirmation in several seizure models since none of them fully imitates clinical epilepsy. The chronic epilepsy models with an initial precipitating injury followed by a latent phase of epileptogenesis and spontaneous recurrent seizures give the best representation of the human situation but are also the most complex models to work with. Animal models have been and are still used extensively to elucidate the neurobiological and pathological changes associated with seizures and epilepsy. The use of animal models also remains a major drug discovery approach to elucidate innovative antiepileptic drug targets.

Besides plenty of electrophysiological neuromethods, which remain a gold standard to understand mechanisms of seizures and epilepsy, in vivo microdialysis also became widely used in animal models. To ascertain the presence or absence of epileptic seizures within the different animal models, monitoring the electrophysiological activity of the brain's cortex or one of the regions involved in seizure generation and/or semiquantifying the characteristic seizure-related behavior provides a decisive answer. These two approaches are routinely combined within the microdialysis laboratories. Microdialysis in seizure animal models can also be combined with other complementary neurotechniques, such as brain imaging and immunological detections methods, to answer scientific questions. Since microdialysis is preferably performed in conscious animals, having ad libitum access to food and water, and minimal restrictions in their movement, their seizure-related behavioral alterations and responses can be quantified and correlated with the simultaneously monitored biochemical changes. Microdialysis causes minimal pain and distress to the experimental animal, reduces the number of animals used in biomedical research, and even allows chronic assessment of neurobiological parameters in chronic epilepsy models. Microdialysis made it possible to collect high-quality, reproducible, seizure-related data in a wide range of laboratory animal species, such as mice, rats, guinea pigs [2], rabbits [3,4], cats [5], dogs [6], sheep [7], pigs [8], and even humans [9]. The discovery of transgenic animals and the elegance of the transgenes to clarify functions of the knocked-in or knocked-out gene made the mice recently more popular than ever before.

In this chapter we further illustrate how intracerebral microdialysis has become an indispensable tool for the neuroscientist interested in seizure- or epilepsy-related neurobiological mechanisms for unraveling mechanisms of antiepileptic drug action. The dialysates collected from the extracellular space of the brain region of interest can be analyzed for a wide range of endogenous substances, such as amino acids, monoamines, histamine, neuropeptides, hormones, ions, cyclic nucleotides, oxidative stress components, and metabolic markers. Antiepileptic drugs, drugs in development, various receptor ligands, or other exogenous substances can be administered systemically or can simultaneously be perfused through the probe in order to study their pharmacodynamic effects. An enormous advantage of the microdialysis technique, which is both a sampling and a drug delivery tool, is thus its ease to discriminate between direct local and indirect mechanisms of action. Intracerebral microdialysis is also an elegant tool to elucidate mechanisms of electrical brain stimulation and to gain insights into neuronal circuits involved in the generation, spread, and control of epileptic seizures.

Microdialysis is a first-choice method for pharmacokinetic studies of antiepileptic drugs. Indeed, microdialysis gives rise to relatively clean dialysates that require no further cleanup, it samples the pharmacodynamically active free drug fraction, and it allows the study of the biophase disposition of antiepileptic drugs and blood–brain barrier transport. Not only the brain but various peripheral tissues can be sampled. Simultaneous sampling from various tissue sites in a single animal avoids problems associated with intra-animal variability. Given its importance for solving both pharmacodynamic and pharmacokinetic problems, microdialysis is consequently an important tool in pharmacodynamic–pharmacokinetic modeling.

Finally, microdialysis is part of the neuroscientist's toolbox to study relationships between seizures or epilepsy and its comorbidities, such as depression and anxiety. Our aim in the present chapter is to convince the epileptologist that many interesting processes and mechanisms in seizure and epilepsy animal models can be unraveled with the microdialysis technique.

2. MICRODIALYSIS TO CHARACTERIZE SEIZURE-RELATED NEUROBIOLOGICAL AND METABOLIC CHANGES IN ANIMAL MODELS AND IN HUMANS

2.1. Monitoring Extracellular Amino Acid Levels Before and During Seizures

The first studies using microdialysis to investigate seizure-related mechanisms were published in the late 1980s. A pioneer study by Wade et al. [10] studied extracellular amino acid levels in the rat piriform cortex. Seizures were induced by systemic administration of either a potent inhibitor of acetylcholinesterase, soman, or the excitotoxin kainic acid. Extracellular cortical glutamate levels increased significantly in the soman-treated rats and tended to augment in the kainic acid–treated animals. The seizures were also clearly accompanied by elevated extracellular taurine levels. Butcher et al. [11] used microdialysis to deliver the chemoconvulsants kainic acid (10 to 100 µM) and dihydrokainic acid (1 to 10 mM) to the dentate gyrus of the rat and to assess simultaneously the effect of these compounds on the extracellular hippocampal amino acid levels. An increase in extracellular hippocampal glutamate and aspartate levels accompanied the dihydrokainate-induced epileptiform activity. Both chemoconvulsants also evoked increases in taurine, alanine, and phosphoethanolamine dialysate levels.

From the beginning of the 1990s to date, in vivo intracerebral microdialysis has been widely used to characterize changes in the major excitatory and inhibitory amino acids, glutamate, and γ-aminobutyric acid (GABA), evoked by seizures generated in different animal models and by the spontaneous seizures observed in the brains of epilepsy patients subjected to epilepsy surgery. A majority of these studies focused on the hippocampus, because of its involvement in the pathophysiology of temporal lobe epilepsy, the most common type of refractory epilepsy in adults. Increases in hippocampal glutamate levels have been observed consistently just preceding and/or during spontaneous epileptic seizures in the human brain [9,12–15]. In vitro microdialysis was also used recently to sample glutamate and GABA from en bloc hippocampal resection slices removed from patients with medically intractable seizures who underwent surgery for mesial temporal lobe epilepsy with hippocampal sclerosis [16]. This method may serve as an experimental model for human brain to study reciprocal neurotransmitter interactions.

We, among others, have demonstrated seizure-related increases in hippocampal glutamate dialysate levels within different epilepsy models. Indeed, significantly elevated extracellular glutamate levels were reported in rat hippocampus following maximal electroshock [17], during the recurrent spontaneous seizure phase of the chronic kainate model [13], following systemic administration of the chemoconvulsants pilocarpine and kainic acid [18], during focally evoked pilocarpine-induced seizures [19–21], during focally evoked aminopyridine-induced seizures [22,23], during kainate-evoked neonatal seizures precipitating after hypoxia [24], and during the seizures evoked by high doses of norfloxacin [25]. Enhanced glutamate levels sampled with microdialysis have also been shown in the striatum of rats displaying 4-aminopyridine-induced seizures [26], in the amydala of seizure-prone rats during kindling [27], in the parietotemporal cortex of rats with post-traumatic epilepsy induced by intracortical iron injection [28], and in the frontal cortex of rats exhibiting pentylenetetrazole-induced seizures [29]. Other studies, however, showed no alterations in extracellular glutamate levels during kainic acid–evoked seizures [30,31], picrotoxin-evoked seizures [32], okadaic acid–evoked seizures [33], and pilocarpine-induced convulsions [34]. In the latter study, however, in the presence of the glutamate uptake inhibitor 1-*trans*-pyrrolidine-2,4-dicarboxylic acid, short-lasting increases in hippocampal glutamate levels preceded pilocarpine-induced seizure onset. Some authors reported decreases in extracellular glutamate levels during seizures elicited by systemic picrotoxin [35] or local picrotoxin administration [36].

Although the studies performed on the human brain clearly demonstrated a link between spontaneous epileptic seizures and enhanced glutamate dialysate levels, conflicting data derived from various animal models has led to interesting debates about high extracellular glutamate levels being causal or just an epiphenomenon of the seizure activity or excitotoxicity [32]. We have been using the increased hippocampal glutamate dialysate levels routinely as biomarkers of increased network activity during seizures, and demonstrated in several studies that these increases were abolished by both anticonvulsant drugs and ligands under development [20,37–40]. Despite that microdialysis is widely used for in vivo monitoring of extracellular neurotransmitter concentrations, its poor temporal resolution may explain that short-lasting seizure-related changes in extracellular glutamate concentrations can be overlooked. Capillary electrophoresis coupled to laser-induced fluorescence detection significantly improves the temporal resolution of microdialysis sampling and is a powerful analytical method used to study the dynamics of in vivo seizure-related neurotransmitter release [41]. An online enzyme fluorometric detection assay of glutamate in the dialysates was described by Ueda and co-workers. The dialysate sampled was mixed with a reactant containing glutamate dehydrogenase. This method allowed detection of short-lasting changes in extracellular hippocampal glutamate concentrations by measuring at 1-min intervals the fluorometric intensities of β-nicotinamide adenine dinucleotide (NADH) resulting from the reaction of glutamate and NAD^+ catalyzed by glutamate dehydrogenase [42]. Indeed, transient kindling-like phenomena evoked in the rat ventral hippocampus were consistently accompanied by short-term calcium-dependent increases in extracellular hippocampal glutamate levels [42]. Another approach was to measure the kainic acid and pilocarpine seizure-related elevations in glutamate concentrations in real time by applying a glutamate biosensor consisting of a glutamate oxidase–based electrochemical detector contained within a standard microdialysis probe [18]. Efficient clearing of released glutamate by the various glutamate transporters may also explain that in some animal models no alterations in extracellular glutamate levels have been reported. Nevertheless, since the extracellular glutamate levels did not change significantly during local bicuculline- or systemic picrotoxin-induced seizure activity, in either the absence or presence of glutamate uptake inhibitors, Millan et al. [43] concluded that seizure activity is not necessarily accompanied by an overall increase in the extracellular glutamate concentration.

In the central newous system (CNS), GABA maintains the inhibitory tone that counterbalances neuronal excitation. When this balance is perturbed, seizures may ensue [44]. In a key publication in 1993, During and Spencer reported that the GABA concentrations in patients with complex partial epilepsy investigated before surgery were lower in the epileptogenic hippocampus than in the non-epileptogenic hippocampus [12]. While the glutamate increases preceded seizure onset, the GABA concentrations remained unaltered just before the seizures but increased during them, with a greater rise in the non-epileptogenic hippocampus [12]. Increases in extracellular GABA levels in the human hippocampus during seizures have also been observed in other studies [13], although these rises in GABA were generally reported to be delayed in comparison with the changes in excitatory amino acids [45]. Several animal studies supported that the extracellular levels of GABA remained unaltered or even decreased before seizures, thereby facilitating seizure generation, whereas during and following seizure activity, GABA levels rose, probably as a mechanism to curtail ongoing seizure activity. Following a maximal electroshock in the rat, there was a rapid and sustained decline in the GABA levels of the ventral hippocampus that preceded the glutamate increases [17]. Progressively decreasing GABA concentrations were monitored in the amygdala of seizure-prone rats during massed amygdala kindling [46]. Kainic acid- or bicuculline-evoked status epilepticus did not significantly affect hippocampal glutamate and GABA dialysate levels, probably due to efficient reuptake mechanisms [47]. Hippocampal GABA concentrations were elevated following focally evoked 4-aminopyridine- and pilocarpine-induced convulsions [19,21,22,39] as well as during the recurrent spontaneous seizure phase of the chronic kainate model [13]. The 4-aminopyridine-elicited high-frequency seizure discharges were also accompanied by increases in the extracellular striatal GABA concentrations [26].

Coupling of microdialysis to capillary electrophoresis with laser-induced fluorescence detection could reveal previously unobserved short-lasting increases in thalamic GABA concentrations concomitant with the seizures in rats with genetic absence epilepsy [48]. Biosensors for real-time monitoring of neuronally released tetrodotoxin-dependent glutamate [49] or GABA levels can be used complementary to microdialysis that samples extracellular amino acids originating from both synaptic and nonsynaptic neurotransmission. Specific changes in glutamate and GABA dialysate concentrations independent of neuronal activity indicate that these amino acids are also essential signaling molecules of extrasynaptic brain transmission [50], which can help to better understand the astroglial–neuron communication. These alterations should be considered as important as the changes in neuronal amino acid release, because various drugs are able to interfere with extrasynaptic signals in vivo [50].

2.2. Seizure-Related Changes in Brain Monoamine Concentrations

Another controversial subject is the role of brain monoamines in seizure mechanisms. It is generally accepted that excessive levels of brain monoamines aggravate epileptic seizures. As a result, every pharmacology textbook states that one of the adverse effects of antidepressant therapy is lowering of the seizure threshold in patients prone to epilepsy. Similarly, cocaine-induced convulsions are also well known, although the mechanisms underlying these seizures are much broader than solely monoamine enhancements in the brain [51]. One of the reasons that high levels of hippocampal monoamines can worsen seizures is the fact that these elevations can be accompanied by significantly enhanced glutamate concentrations [52]. A closer look at the preclinical literature currently available, however, often points to the anticonvulsant effects of the brain's monoaminergic systems. We showed further that extracellular glutamate increases per se do not necessarily induce seizures but that they can definitely modulate the anticonvulsant effects exerted by hippocampal monoamines [53].

Various authors unraveled possible involvements of the monoaminergic systems in a variety of seizure and epilepsy models. Pentylenetetrazole-induced kindling enhanced both basal and stimulated extracellular levels of dopamine in the prefrontal cortex, nucleus accumbens, and striatum in freely moving rats [54]. Pioneer work by Kokaia and co-workers published in 1989 [55] showed that both focal and generalized hippocampal seizures evoked by electrical kindling increased noradrenaline markedly, but not serotonin release in the hippocampus. Levels of noradrenaline and dopamine, but again not of serotonin, were also increased in seizure-prone and seizure-resistant rats throughout amygdala kindling [27]. These results indicate that kindling mechanisms enhance the activity and sensitivity of monoaminergic neurons differentially, suggesting that the latter contribute to the central alterations seen within the experimental epilepsy models.

Several microdialysis studies also monitored monoamine concentrations during acutely evoked seizures. The seizures induced by electroconvulsive shock and by the convulsant agent flurothyl both enhanced the extracellular serotonin concentrations markedly in the rat hippocampus [56]. The same group showed that electroconvulsive shock also increased dopamine release in the striatum. They suggested that the latter increase in dopamine was not related to the seizure activity since the flurothyl-induced seizures did not influence striatal dopamine dialysate levels [57]. Enhancements in extracellular hippocampal dopamine and serotonin levels were measured during pilocarpine-induced convulsions [19–21,52]. Also, 4-aminopyridine-evoked seizures significantly increased striatal serotonin, noradrenaline, and dopamine dialysate concentrations [26]. Most of these authors believe that increased levels of dopamine, noradrenaline, or serotonin mediate an inhibitory response which probably contributes to the cessation or suppression of the ongoing seizure activity.

In support of this overall inhibitory effect of central monoamines are studies showing that a lack of monoaminergic inhibition leads to increased brain excitability. Indeed, lower baseline concentrations of noradrenaline, dopamine, and serotonin in amygdala and the locus coeruleus of young cats correlated with subsequent increases in duration of focal and generalized after-discharges as well as the number of behavioral seizures [58]. Jobe and co-workers [59] also used microdialysis as one of their tools to show that noradrenergic and/or serotoninergic deficits may contribute to predisposition to epilepsies and depression.

Noradrenaline and dopamine as well as serotonin can affect the brain's excitability via opposing actions at their various receptor subtypes. Within the abundant literature there are a few studies using microdialysis to elucidate the role of certain monoaminergic receptors in seizure mechanisms. Intrahippocampally applied dopamine and serotonin concentrations via the microdialysis probe were able to protect rats from limbic pilocarpine-induced seizures as long as extracellular dopamine or serotonin concentrations ranged, respectively, between 70–400% and 80–350% increases compared with the baseline levels [52]. These anticonvulsant effects were clearly abolished by co-perfusion with the selective D_2 receptor antagonist remoxipride or the selective 5-HT$_{1A}$ receptor antagonist WAY 100635, indicating that hippocampal dopamine and serotonin can mediate anticonvulsant actions independently via D$_2$ and 5-HT$_{1A}$ receptors. Soman-evoked seizures in the freely moving guinea pig could be blocked by systemic administration of the D$_1$ receptor antagonist SCH 23390, while blockade of the D$_2$ receptor with sulpiride augmented the seizure activity evoked [2]. These findings are in accordance with the fact that activation of the various dopamine receptor subtypes regulates seizure thresholds and epileptic activity oppositely [60,61]; that is, D$_1$ receptor activation can facilitate seizures, whereas D$_2$ receptor activation is anticonvulsant.

2.3. Dialysate Sampling of the Excitatory Neurotransmitter Acetylcholine and the Inhibitory Neuromodulator Adenosine

Besides the measurement of amino acids and monoamines in the dialysates obtained from animals exhibiting seizures, other transmitters and neuromodulators have also been determined. Acetylcholine is known to have potent excitatory actions, and cholinergic agonists are used as powerful convulsants. Soman-induced seizures in rats were shown to evoke a biphasic increase in the extracellular acetylcholine levels in the septum and hippocampus and a progressive acetylcholine increase in the amygdala [62]. The authors concluded that in septohippocampal areas, the glutamatergic system is recruited after an early accumulation of extracellular acetylcholine following systemic administration of soman. Electrical amygdaloid kindling was associated with increases in amygdala acetylcholine levels both during kindling development and in fully kindled rats [63]. Short-term treatment with pentylenetetrazole or isoniazid also increased hippocampal acetylcholine release. Nevertheless, in kindled rats treated chronically with pentylenetetrazole, the basal hippocampal acetylcholine dialysate concentrations were reduced compared to the levels in vehicle-treated rats [64].

In the mid-1980s, Dragunow and co-workers proposed that adenosine is the brain's natural anticonvulsant [65]. Microdialysis in the hippocampus of kindled rats demonstrated that kindling increases the release of adenosine and its major purinergic metabolites [66]. Dialysate hippocampal adenosine levels and metabolites increased during bicuculline-, kainic acid-, and pentylenetetrazole-induced seizure activity [67], supporting the hypothesis that adenosine levels increase in an attempt to curtail ongoing hippocampal seizures and to alter the subsequent seizures pattern. Also, uridine can be released under sustained depolarization and can inhibit hippocampal neuronal activity [23]. Indeed, Slézia and co-workers showed that hippocampal uridine release correlated well with the 3-aminopyridine-induced seizure activity and suggested that uridine may contribute to epilepsy-related neuronal activity changes.

2.4. Activation of Peptidergic Systems Following Epileptic Seizures

Microdialysis experiments combined with determination of dialysate neuropeptide content by radioimmunoassays revealed that several peptidergic systems are altered markedly before, during, or following seizures. Dialysis experiments coupled with liquid chromatography linked to radioimmunoassay showed that enkephalin release was enhanced in the amygdala during the early amygdala kindling stages, which might mediate a suppressive effect trying to avoid seizure spread [68]. The extracellular enkephalin levels subsequently decreased after onset of stage V kindled seizures, which could reflect a general impairment of inhibitory mechanisms in the amygdala [68]. Amygdala kindling increased the extracellular opioid peptide content in hippocampus immediately following kindling stimulation, while extracellular hippocampal opioid levels were lower in fully kindled rats than in sham-kindled rats [69]. The extracellular opioid peptide levels in the amygdala were elevated following a subconvulsant dose of pentylenetetrazole [70]. The same researchers also revealed an important activation of the opioid peptide systems by kainic acid–induced status epilepticus, but showed reduced hippocampal extracellular opioid peptide levels 28 days after kainic acid administration [71]. There is much evidence for an excitatory effect of μ- and δ-opioid receptor activation (e.g., by inhibiting GABA release from inhibitory hippocampal interneurons), while inhibitory and anticonvulsant effects can be mediated by κ receptors. Kainic acid–evoked status epilepticus caused a biphasic increase in nociceptin–orphanin FQ release in both the rat hippocampus and thalamus [72]. The exact physiological role with regard to seizure modulation of the nociceptin–orphanin FQ neuropeptide is not yet clear. On the other hand, both neuropeptide Y and somatostatin are well known to exhibit clear anticonvulsant effects in various seizure models. Systemic administration of the chemoconvulsant kainic acid or a single electroconvulsive stimulation delivered to the rat via ear clips significantly enhanced the neuropeptide Y-like immunoreactivity in dialysates obtained from the rat dorsal hippocampus [73,74]. Basal and evoked hippocampal somatostatin-like immunoreactivity release was enhanced markedly in the amygdala kindling model [75]. Initially, intracortically applied γ-hydroxybutyrate increased the release of somatostatin-like immunoreactivity in dialysates sampled from the frontal cortex [76]. However, further exposure to γ-hydroxybutyrate led to a manifestation of epileptic spikes and seizures, during which the somatostatin release was attenuated significantly [76].

Monitoring dialysate peptide immunoreactivity instead of total tissue peptide immunoreactivity was already a first big step toward a better understanding of the temporal changes in neuropeptide levels before, during, and following epileptic seizures. Improvements in this domain can still be made to eliminate antibody-related cross-reactivity for structurally related peptide fragments as well as to monitor more closely the various peptide fragments' release and metabolization. Nano-liquid chromatography coupled to tandem mass spectrometry fulfills these criteria and is an interesting future avenue to broaden our knowledge in this field.

2.5. Seizure Activity and Increased Oxidative Stress

Oxidative stress is presumed to play a role in epileptogenesis, seizure generation, and postseizure neuronal death. Microdialysis can thus be used to monitor reactive oxygen species in seizure animal models. Systemically evoked kainic acid seizures were demonstrated to be accompanied by an increase in reactive oxidant species in dialysates obtained from rat piriform cortex that were assayed for isoluminol-dependent chemiluminescence [77]. On the other hand, there were no significant differences in hydroxyl radical production in striatal dialysates analyzed by salicylate trapping methods during hyperbaric oxygen convulsions in rats [78]. Dialysate levels of the nitric oxide end products nitrite and nitrate were used as indices of nitric oxide synthesis in the contralateral hippocampus following intrahippocampal kainate injection [79]. The nitric oxide end product levels increased immediately after kainate injection, and this elevation preceded the seizure discharges, suggesting that remote seizure activity caused by the transneuronal spread of kainate-induced discharges may be related to nitric oxide production. These authors further clarified that nitric oxide originated from neuronal nitric oxide synthase activity [79]. The nitric oxide donor sodium nitroprusside, administered locally in the rat hippocampus via a microdialysis probe, was also able to induce epileptic seizures per se in the absence of other chemoconvulsants [80]. In the latter study a recording electrode was inserted into the stratum pyramidale of the CA1 hippocampal region 0.5 mm lateral to the microdialysis probe to record the population spikes and the background EEG. Direct evidence for progressive lipid peroxidation of the neuronal membranes during kainic acid–induced seizure activity was obtained by brain microdialysis coupled to direct electron spin resonance analysis of the dialysate samples [81].

2.6. Seizures and Enhanced Energy Metabolism

Preclinical animal work has suggested that generation of extracellular lactate, as measured by microdialysis, is an index of local glucose utilization and is dependent on neuronal activity. Maximal electroshock convulsive seizures were associated with marked changes in extracellular glucose, lactate, and pyruvate levels in the striatum and in peripheral subcutaneous tissue during the post-ictal state [82]. Limbic seizures induced by systemic injection of pilocarpine or focally elicited by microinfusions of bicuculline or cyclothiazide within the anterior piriform cortex also enhanced the extracellular lactate concentrations [83]. In dialysates collected from temporal and frontal lobes of epilepsy patients, large increases in lactate/glucose ratios were observed close to the epileptogenic region [84]. Similarly, During et al. [85] reported that spontaneous complex partial seizures were accompanied by increases in extracellular lactate levels that remained restricted to the site of seizure activity. As acidification of the extracellular compartment has an inhibitory effect on neuronal excitability, the rise in extracellular lactate levels may be a mechanism of seizure arrest and postictal refractoriness [85].

3. MICRODIALYSIS AS A CHEMOCONVULSANT DELIVERY TOOL IN ANIMAL SEIZURE MODELS

One of the major advantages of in vivo microdialysis is that in addition to being a well-established sampling tool, it is a very elegant drug delivery tool. Researchers working in the field of epilepsy have long utilized microdialysis to deliver chemoconvulsants locally to the brain region of interest and in this way to mimic an epileptic focus. The most frequently used locally perfused chemoconvulsants are glutamate receptor agonists such as kainic acid [11], GABA receptor antagonists such as picrotoxin [36,86], muscarinic receptor agonists such as pilocarpine [19,34,39,52], or voltage-dependent potassium channel blockers such as 4-aminopyridine [87]. An advantage of such a local delivery approach is that seizure-related neurobiological mechanisms can be studied at the site of seizure onset. A disadvantage is that chemoconvulsant drug-induced pharmacodynamic effects can interfere with the seizure-related effects. To better characterize the relationship between seizure activity and changes in hippocampal glutamate, GABA, dopamine, and serotonin, we compared three limbic seizure models that differed only in the chemoconvulsant applied and thus in the pharmacodynamic drug–receptor mechanism used to evoke seizures [21]. Seizures induced by intrahippocampal administration of pilocarpine, picrotoxin, or (R,S)-3,5-dihydroxyphenylglycine were all of comparable severity. Seizure activity was always accompanied by increased extracellular concentrations of hippocampal glutamate, GABA, and dopamine but not serotonin [21].

Microdialysis has also been used to deliver chemoconvulsant drugs with an entirely new mechanism of action (e.g., okadaic acid and latrunculin A). Okadaic acid is a complex polyether compound extracted from a marine sponge which is a potent inhibitor of the serine–threonine protein phosphatases types 1 and 2A and thus a powerful tool for studying the regulatory mechanisms associated with protein phosphorylation [33]. Intrahippocampal administration of okadaic acid produced intense behavioral and persistent EEG seizure activity in conscious rats, showing that protein phosphatase inhibition in vivo can result in epileptic discharges [33]. The authors suggested further that pathological alterations of the phosphorylation–dephosphorylation balance of hippocampal NMDA receptors may be a factor in the etiology of epilepsy. The latrunculins are a family of natural products and toxins produced by certain sponges that can bind actin monomers, prevent them from polymerizing, and finally, result in disruption of the actin filaments of the cytoskeleton. Latrunculin A microperfusion in the hippocampus of awake rats induced acute epileptic seizures and long-term changes in neuronal excitability, leading to the onset of spontaneous seizures approximately one month following latrunculin A treatment [88,89]. These studies showed that actin disruption can be a possible cause of epileptic seizures and that intrahippocampal latrunculin A–induced seizures are a novel experimental rat model to study whether biochemical changes might lead to chronic epilepsy.

4. MICRODIALYSIS USED TO ELUCIDATE MECHANISMS OF ELECTRICAL BRAIN STIMULATION AND NEURONAL CIRCUITS INVOLVED IN SEIZURES

Deep brain stimulation is a promising therapy for intractable epilepsy, yet the optimal target regions as well as the underlying mechanisms of action remain controversial. Stimulation of the histaminergic tuberomammillary nucleus provided anticonvulsant effects against pentylenetetrazole-induced seizures and gave rise to a strong increase in the dialysate levels of histamine sampled from rat frontal cortex [90]. These anticonvulsant effects were reversed in a dose-dependent manner by systemic administration of a histamine H_1 receptor antagonist pyrilamine [90]. Bilateral anterior thalamic nucleus deep brain stimulation also clearly delayed the onset of pentylenetetrazole-induced seizures [91]. Intravenous pentylenetetrazole infusion per se, or together with electrical stimulation, gave rise to a steady increase in noradrenaline dialysate levels of the anterior thalamic nucleus and the posterior thalamus. The extracellular levels of the serotonin metabolite 5-hydroxyindoleacetic acid were, however, selectively enhanced in the anterior thalamic nucleus and solely after stimulation [91], suggesting that modulation of the serotoninergic activity in the anterior thalamic nucleus may be critical in altering pentylenetetrazole seizure threshold and may underlie the efficacy of the anticonvulsant thalamic deep brain stimulation. Electrical stimulation of the subthalamic nucleus also induced increases in the extracellular GABA levels of the substantia nigra pars reticulata in both epileptic and control rats, although the rise in extracellular GABA levels was more pronounced in the kainic acid–treated epileptic rats [92]. This study demonstrated differential neurochemical modifications in subthalamic nucleus target regions during electrical subthalamic nucleus stimulation, since the GABA levels of the globus pallidus remained unaltered [92].

Microdialysis is also an elegant tool for elucidating mechanisms and neuronal cicuits involved in the generation and control of epileptic seizures. In the latter respect, Nail-Boucherie and co-workers [93] performed a microdialysis study to investigate whether the nucleus parafascicularis of the thalamus could be a relay of the nigral control of epileptic seizures by the superior colliculus. For this purpose, they worked with rats with genetic absence seizures (GAERS) and measured extracellular GABA and glutamate levels within the parafascicular nucleus. They showed that disinhibition of the caudal part of the superior colliculus results in suppression of absence seizures and is associated with a significant increase of glutamate dialysate levels of the nucleus parafascicularis, whereas no changes in GABA levels were observed [93]. These data suggest that glutamatergic projections to the parafascicular nucleus of the thalamus could be involved in the nigral control of seizures by the superior colliculus. Another microdialysis study in GAERS demonstrated increased basal GABA levels in the primary motor cortex and the ventrolateral thalamus of GAERS versus Wistar rats and confirmed previous studies pointing to an increased GABAergic tonus in primary motor cortex and thalamus of GAERS [94].

5. MICRODIALYSIS USED TO UNRAVEL THE MECHANISMS OF ACTION OF ESTABLISHED ANTIEPILEPTIC DRUGS AND NEW THERAPEUTIC STRATEGIES

The search for innovative antiepileptic drugs remains relevant since a large number of epileptic patients remain refractory to the medication currently available, and since several adverse effects are known for most of the antiepileptic drugs on the market. In addition to the search for new drug targets, microdialysis is often used to investigate yet unidentified mechanisms of action of the antiepileptic drugs on the market.

5.1. Mechanisms of Action of Established Antiepileptic Drugs

Microdialysis has been widely applied to study mechanisms of action of both old and new-generation antiepileptic drugs as well as of their active metabolites. One of the current strategies is indeed to develop improved derivatives of existing antiepileptic drugs. In that context, we compared, for example, the effectiveness of valproic acid with two of its amide derivatives, valpromide and valnoctamide, against intrahippocampally evoked pilocarpine-induced convulsions in the rat. The derivatives were more potent than valproic acid within this model, and we suggested that enhancement of hippocampal GABA-mediated inhibition may not be part of their anticonvulsant mechanism of action [40]. In line with these findings, Khongsombat and co-workers reported that both valproic acid and another analog, N-(2-propylpentanoyl) urea, were able to protect rats against systemically evoked pilocarpine-induced seizures, the analog being six times more potent than its parent compound. They showed further that the anticonvulsant activity of N-(2-propylpentanoyl) urea against pilocarpine-induced seizures was due at least partly to a marked reduction of the enhanced extracellular levels of the excitatory amino acids aspartate and glutamate. These effects were generally more pronounced than were the effects on the extracellular hippocampal GABA concentrations [95]. On the other hand, Rowley et al. showed that sodium valproate was effective in the maximal electroshock model, thereby enhancing the immediate seizure-related GABA release and preventing secondary seizure-related post-ictal sustained reduction in GABA levels of the rat hippocampus [96]. In the same study, effective anticonvulsive treatment with phenytoin attenuated the immediate seizure-related increases in hippocampal glutamate dialysate levels.

Eslicarbazepine acetate shares with carbamazepine and oxcarbazepine the anticonvulsant potency as well as the basic chemical structure of a dibenzazepine nucleus with the 5-carboxamide substituent but is structurally different at the 10,11-position. Eslicarbazepine acetate is a prodrug of eslicarbazepine, also known as the S-enantiomer of oxcarbazepine's active metabolite 10,11-dihydro-10-hydroxycarbamazepine. Oral gavage of eslicarbazepine acetate in rats showed an excellent anticonvulsant effect against intrahippocampally evoked latrunculin A–induced seizures and prevented the increases in extracellular hippocampal glutamate and aspartate levels induced by latrunculin A [97]. Systemic or intrahippocampal carbamazepine administration, in doses known to be effective against maximal electroshock seizures, caused significant and dose-related increases in hippocampal serotonin dialysate levels [98]. Since these elevations could not be suppressed by tetrodotoxin or zero calcium in the microdialysis perfusion fluid, it was concluded that the serotonin-releasing effect of carbamazepine was not mediated via exocytosis [98]. In contrast, carbamazepine as well as zonisamide were proposed to exert their monoamine enhancing effects at least partly by modulating two functional pathways for exocytosis: the N-type voltage-sensitive calcium channel/syntaxin pathway during baseline conditions and the P-type voltage-sensitive calcium channel/synaptobrevin pathway during neuronal depolarizing stages [99]. A more recent microdialysis paper by the latter group clarified that carbamazepine can also alter hippocampal glutamate and GABA levels differentially during resting and neuronal depolarizing conditions via interaction with the ryanodine receptor–sensitive calcium-induced calcium-releasing system [100]. Microdialysis coupled to electron paramagnetic resonance spectroscopy revealed another mechanism of action of zonisamide and demonstrated that systemic administration of zonisamide was able to reduce the oxidative stress generated during kainic acid–induced status epilepticus [101]. Similar to carbamazepine, oxcarbazepine and 10,11-dihydro-10-hydroxycarbamazepine also exerted important monoamine-enhancing effects in rat hippocampus [102]. Their anticonvulsant activity against pilocarpine-induced seizures was always accompanied by significant increases in extracellular hippocampal dopamine and serotonin levels and was dependent on selective dopamine D_2 or serotonin 5-HT_{1A} receptor activation. The hippocampal monoamine elevations were therefore proposed as pharmacodynamic markers for the anticonvulsant activity of oxcarbazepine and the racemate of eslicarbazepine [102]. Marked increases in extracellular hippocampal dopamine concentrations were also demonstrated during intrahippocampal lamotrigine perfusion via the microdialysis probe and were suggested to be part of the anticonvulsant mechanism of action of lamotrigine [19].

Pilocarpine-induced convulsions were completely abolished following both systemic and intrahippocampal pretreatment with phenobarbital, although paradoxically, intrahippocampal phenobarbital perfusion per se induced an enhancement of the extracellular hippocampal glutamate levels, which were attenuated significantly during and following pilocarpine administration [103]. We showed that intrahippocampal perfusion of high concentrations of topiramate increased hippocampal glutamate and GABA dialysate levels. It seems unlikely that these locally elicited changes contributed to topiramate's anticonvulsant effect, since the intrahippocampal topiramate perfusion did not attenuate the pilocarpine-induced limbic seizures [104]. Prompted by the latter observation, we identified the substantia nigra pars reticulata as a site of action of topiramate and showed in the same microdialysis study that the nigral GABAergic system is central to topiramate's anticonvulsant effect. Interestingly, intranigral topiramate administration again increased hippocampal glutamate but not GABA levels. These increases in extracellular hippocampal glutamate levels following intranigral topiramate administration suggested the existence of a nigrohippocampal circuit, which may be involved in the control of limbic seizures [104]. Our finding that intraperitoneal injection of topiramate was able to attenuate the seizure-related pilocarpine-induced increases in hippocampal glutamate levels was in agreement with a study by Kanda et al., who demonstrated that topiramate reduced abnormally high extracellular levels of glutamate and aspartate in the hippocampus of spontaneously epileptic rats [105].

Some other studies investigated the effects of several GABA-modulating drugs on the extracellular GABA levels in various brain areas. The anticonvulsant effects of the selective inhibitor of the GABA transporter 1, tiagabine, and two other nonmarketed nonselective GABA-uptake inhibitors were assessed within the maximal electroshock model in rats and mice, against audiogenic seizures in DBA/2 mice and against pentylentetrazole-induced tonic convulsions in NMRI mice [106]. The authors showed different effects of the selective versus nonselective GABA-uptake inhibitors in these animal models and on the GABA dialysate levels obtained from halothane-anaesthetized rat hippocampus and thalamus [106]. Systemic vigabatrin administration (1000 mg/kg) was shown to increase the extracellular levels of GABA in the substantia nigra pars reticulata but only partially protected the rats from pentylenetetrazole-induced seizures [107]. Much lower systemic vigabatrin doses (30 mg/kg), however, clearly protected rats against pilocarpine-induced seizures, while intrahippocampal or intranigral vigabatrin perfusion showed at best only a partial anticonvulsant effect in the pilocarpine model [108], suggesting that these two brain regions are not the major sites of vigabatrin's anticonvulsant action. Nevertheless, clear increases in GABA dialysate levels and decreases in the extracellular glutamate (but also dopamine) concentrations were observed following local vigabatrin perfusion [108]. A single dose of intraperitoneal diazepam (5 mg/kg) was administered to rats exhibiting full-blown pilocarpine-induced convulsions and produced immediate inhibition of seizure activity that was associated with a prompt attenuation of the elevated extracellular hippocampal glutamate overflow without concurrent alteration of pilocarpine-induced increases in GABA levels [20]. An additional study with the benzodiazepine receptor antagonist flumazenil demonstrated that this diazepam-elicited attenuation of the extracellular glutamate levels was indeed reversed and thus mediated at the level of the benzodiazepine–GABA$_A$ receptor complex [39].

The well-established anti-absence drug ethosuximide was able to suppress the enhanced GABA levels in the primary motor cortex of GAERS but did not modulate the higher GABA levels in their ventrolateral thalamus, implying that attenuation of the GABA-mediated transmission in the frontal cortex of GAERS is implicated in the anticonvulsant effects of ethosuximide [94]. Another study in GAERS, which used microdialysis mainly as a drug delivery tool, confirmed that targeting the thalamus alone is insufficient for an immediate and full anti-absence action of ethosuximide as measured by the incidence and durantion of the spike and wave discharges on an EEG [109].

Finally, microdialysis was used to demonstrate that combining lamotrigine and valproate gave rise to various and complex neurochemical profiles of the basal and stimulated extracellular amino acid and monoamine concentrations in rat hippocampus not seen with either drug alone. The authors concluded that a lamotrigine and valproate combination can lead to quite profound pharmacodynamic interactions, whereas pharmacokinetic interactions of similar doses of these two antiepileptic drugs in plasma, whole brain, and hippocampal dialysates seemed to be absent [110].

5.2. Microdialysis as an Elegant Tool in the Search for New Therapeutic Strategies

In view of the strong glutamatergic effects on the brain's excitability, several studies have focused on future therapeutic strategies targeting both ionotropic (iGluR) and metabotropic glutamate receptors (mGluR). mGlu$_4$ knockout mice were fully resistant to absence seizures induced by low doses of GABA$_A$ receptor antagonists [111]. Microdialysis data obtained in these mice indicated that deletion of the mGlu$_4$ gene resulted in a selective perturbation of glutamate and GABA release within the thalamocortical circuitry involved in the pathogenesis of absence seizures [111]. This study offered perspectives for possible future therapeutic effects of mGlu$_4$ antagonists against absences. Modulation of mGluRs also represents an interesting new approach for the future treatment of limbic seizures. We observed full anticonvulsant effects of mGlu$_{1a}$ and mGlu$_5$ antagonists and partial anticonvulsant effects of mGlu$_{2/3}$ agonists against pilocarpine-induced limbic seizures. The mGlu$_{1a}$ receptor antagonist LY367385 decreased baseline glutamate but not GABA dialysate concentrations, while agonist-mediated actions at mGlu$_{2/3}$ receptors by LY379268 decreased basal hippocampal GABA but not glutamate levels, which may provide a neurochemical basis for complete and partial anticonvulsant protection, respectively, within our epilepsy model [38]. However, systemic administration of selective and brain-penetrable $mGlu_1$ and $mGlu_5$ antagonists at doses appropriate for $mGlu_1$ or $mGlu_5$ receptor-mediated effects in rodent models of partial seizures lacked anticonvulsant efficacy in two models of difficult-to-treat partial epilepsy: the 6-Hz electroshock model of partial seizures in mice and the amygdala-kindling model in rats [112]. Competitive blockade of the N-methyl-D-aspartate (NMDA) receptor has been shown to exert strong anticonvulsive but adverse cognitive and behavioral effects; therefore, alternative approaches of antagonizing iGluRs have been considered. Both kynurenic acid and its synthetic derivative, 7-chlorokynurenic acid, are antagonists of the glycine co-agonist site of the NMDA receptor. In vivo microdialysis revealed that peripheral administration of the transportable precursors kynurenine and 4-chlorokynurenine elevated extracellular kynurenic acid levels and enhanced *de novo* formation of 7-chlorokynurenic acid, respectively, in the entorhinal cortex and hippocampus of rats with spontaneously recurring seizures following pilocarpine-induced status epilepticus [113]. The same group further demonstrated that administration of 4-chlorokynurenine to spontaneous epileptic rats, following electrical stimulation status epilepticus, ameliorated epileptiform evoked potentials in vivo, suggesting that the use of glial cells for the neosynthesis and local delivery of neuroactive compounds may be a feasible strategy for the treatment of limbic epilepsy [114]. Agmatine, an endogenous neuromodulator and antagonist of NMDA receptors, has anticonvulsive effects in the pentylenetetrazole-induced seizure rat model and reduced the seizure-related enhanced extracellular glutamate levels in the frontal cortex [29]. Selective antagonists at the GLU_{K5} subtype of the kainate class of iGluRs prevented and interrupted limbic seizures induced by intrahippocampal pilocarpine perfusion in rats and attenuated the accompanying rises in extracellular glutamate and GABA concentrations. This anticonvulsant activity occurred without overt side effects [37]. Dose-dependent anticonvulsive activity of the subtype-selective GLU_{K5} antagonist LY377770 was also shown in the 6-Hz corneal stimulation model in mice [37], sustaining the fact that even blockade of solely one subtype of glutamate receptors can have profound anticonvulsive effects while reducing the adverse effects to a minimum.

Another interesting avenue for the future pharmacotherapeutic drug treatment of epilepsy is the search for blood–brain-permeable and stable ligands for various neuropeptide receptors. Microdialysis has frequently been applied to reveal the mechanisms of anticonvulsant action of several of these neuropeptides, their receptor agonists or antagonists. We unraveled a novel mechanism of anticonvulsant action of neuropeptide Y that is separate from, and may be complementary to, the well-established Y_2 receptor-mediated inhibition of hippocampal excitability [115]. Indeed, our results indicated that neuropeptide Y-induced increases in hippocampal dopamine are mediated via sigma 1 receptors and contribute to the anticonvulsant effect of neuropeptide Y via increased activation of hippocampal D_2 receptors. A thyrotropin-releasing hormone-related analog CNK-602A suppressed absence-like seizures and tonic convulsions in spontaneously epileptic rats and increased the extracellular dopamine levels in the nucleus caudatus, making the authors conclude that this analog probably inhibits epileptic seizures in a manner similar to thyrotropin-releasing hormone, by increasing the release of dopamine [116].

Intracerebroventricular administration of angiotensin IV or somatostatin-14 protected rats from intrahippocampally evoked pilocarpine-induced seizures and induced concomitant increases in the hippocampal extracellular dopamine and serotonin levels, which also revealed a possible role for both monoamines in the anticonvulsant effect of these peptides [117]. Gastrin-releasing peptide, a selective agonist for the bombesin BB_2 receptor subtype, was reported to increase the extracellular levels of GABA significantly in the hippocampus of the freely moving rat and to exhibit anticonvulsant properties in the audio-genic seizure-prone DBA/2 mouse [118]. These gastrin-releasing peptide effects were both blocked by a selective BB_2 receptor antagonist, providing pharmacological evidence that BB_2 receptor agonism can functionally modulate hippocampal GABAergic transmission and seizure thresholds. Activation of hippocampal B1 bradykinin receptors, almost absent in control rats but expressed at high levels in kindled rats, contributed to the glutamate-mediated hyperexcitability of the kindled hippocampus, as shown by enhanced glutamate dialysate levels [119], and may represent a novel therapeutic drug target.

As described above and as known from the mechanism of anticonvulsant action of topiramate, phosphorylation and dephosphorylation of receptors may be involved in the development of epileptic seizures. This hypothesis is supported further by the in vivo anticonvulsant effect of the intrahippocampally applied calcineurin/protein phosphatase 2B inhibitor ascomycin against seizures induced by intrahippocampal perfusion of picrotoxin [120].

6. MICRODIALYSIS STUDIES IN THE SEARCH FOR MECHANISMS OF ADVERSE EFFECTS OF CLINICALLY USED DRUGS, DRUGS OF ABUSE, AND TOXINS

Microdialysis has occasionally been used to unravel possible mechanisms of convulsant or proconvulsant actions of clinically used drugs. Fluoroquinolones are frequently prescribed antibiotics that display undesired central excitatory effects, such as epileptic seizures in certain patients. These adverse effects have generally been attributed to inhibition of GABA binding to $GABA_A$ receptors. We showed in a microdialysis study that not all rats receiving norfloxacin displayed seizures. Nevertheless, when norfloxacin-induced convulsions did occur, they were severe and did not affect the extracellular hippocampal GABA levels but were accompanied by enhanced glutamate dialysate concentrations [25]. Cyclosporin A is a widely used standard immunosuppressant with frequently reported neurological complications. It was suggested that cyclosporin A induced epileptic convulsions as a result of an interaction between nitric oxide and the GABAergic system in the hippocampus [121]. Indeed, systemic cyclosporin A administration facilitated the intensity of bicuculline-induced seizures in mice, a proconvulsant effect that was blocked by nitric oxide synthase inhibition. Cyclosporin A also dose-dependently increased NO_2 levels in microdialysates obtained from the rat dorsal hippocampus [121].

Other researchers have used microdialysis to get insights into possible mechanisms of the withdrawal seizures exhibited by alcohol, cocaine, or amphetamines and their potential treatment options. Microdialysis was used, for example, to measure basal and potassium-stimulated septohippocampal acetylcholine release in a genetic mouse model of ethanol-withdrawal severity following the administration of several cholinergic chemoconvulsants [122]. These authors showed that alcohol-naive withdrawal seizure-prone mice had a lower sensitivity to the cholinergic chemoconvulsants than did withdrawal seizure-resistant mice. They demonstrated further that the baseline hippocampal acetylcholine dialysate levels of the two groups of mice were not different, but the withdrawal seizure-prone mice displayed a significantly higher acetylcholine release in response to a first potassium challenge [122]. Another study confirmed the implication of cholinergic mechanisms in alcohol-withdrawal seizure mechanisms and showed, among other findings, significantly elevated baseline levels of hippocampal extracellular choline and acetylcholine in ethanol-withdrawn rats versus ethanol-naive control rats [123]. Coadministration of phentermine with the serotonin precursor 5-hydroxy-L-tryptophan reduced alcohol intake, suppressed the alcohol withdrawal seizures in rats, and was accompanied by elevations in dopamine and serotonin microdialysate concentrations sampled from the rat nucleus accumbens, a brain region crucially involved in rewarding [124]. The GABA analog gabapentin failed to inhibit cocaine self-administration under a fixed-ratio reinforcement schedule or cocaine-triggered reinstatement of cocaine-seeking behavior in rats [125]. In vivo microdialysis in the nucleus accumbens showed that gabapentin produced only a modest increase in extracellular GABA levels but failed to alter either basal or cocaine-enhanced dopamine concentrations in this brain region, suggesting a limited potential of gabapentin in treating cocaine dependence in humans [125]. As described earlier in this chapter, cocaine- or amphetamine-induced seizures have often been related to high monoamine levels in the brain. However, Tor-Agbidye and co-workers showed that exposure to *d*-amphetamine correlated dose-dependently with hyperactivity, behavioral stereotypies, hyperthermia, and extracellular monoamine increases in the rat amygdala. Nevertheless, the convulsive activity and epileptic seizures were noticed only at high *d*-amphetamine exposure and when excessive dopamine and serotonin peak dialysate concentrations were reached [126].

Finally, other studies have tried to shed light on the convulsant mechanisms of certain food supplements or the seizures induced by food poisoning. 5-Hydroxyindole is a putative tryptophan metabolite requiring tryptophanase, an enzyme present in gut bacteria, for its synthesis. Tryptophan is, besides a common dietary protein, also used frequently as a dietary supplement. 5-Hydroxyindole caused seizures when injected systemically into rodents, while its local perfusion into the CA1 region of the rat hippocampus via a microdialysis probe increased glutamate concentrations significantly in the dialysates [127]. Another microdialysis study revealed that *Clostridium perfringens* ε toxin, best known as a food poison, caused seizures as well as excessive glutamate release in the hippocampus [128].

7. COMBINING MICRODIALYSIS WITH OTHER COMPLEMENTARY NEUROTECHNIQUES TO UNRAVEL MECHANISMS OF SEIZURES AND EPILEPSY

Since the major hallmark of all epilepsies is the disruption of normal EEG activity, microdialysis as a drug delivery tool and/or sampling method in animals and humans has very frequently been combined with the recording of the electrical activity by depth electrodes (e.g., hippocampal, striatal registrations) or by surface electrodes placed on the cortex (also called *electrocorticography*). Many of the aforementioned studies as well as other investigations have applied this methodology to acertain the presence or absence of seizures in their animal models or in the human brain [20,23,26,85,90,93,129–131]. Monitoring the paroxysmal discharges indeed remains the best correlate of seizure activity and other possible neurobiological alterations in the brain.

By combining microdialysis with immunoblotting techniques, several human and animal studies were able to demonstrate involvement of the plasma membrane transporters in mechanisms of seizures and epilepsy. This approach was applied, for example, to unravel hippocampal GABA transporter function in human temporal lobe epilepsy [132]. An increase in the neuronal EAAC-1 transporter expression of the rat hippocampal CA1 region was observed in kindled rats, which probably compensated for the decrease in total extracellular glutamate and aspartate transporter activity, as indicated by microdialysis studies with the high-affinity transporter blocker L-*trans*-pyrrolidine-2,4-dicarboxylate [133]. Decreased expression of the glial glutamate transporter GLT-1 accompanied higher basal dialysate levels of glutamate, lower levels of glutamine, and low arginine/citrulline ratios in the iron-induced animal model of posttraumatic epilepsy [134]. The same study showed that the radical scavenger α-phenyl-*tert*-*N*-butyl nitrone was partially able to restore this process of impaired astrocytic uptake of glutamate. Chronic seizures induced by amygdalar injection of FeCl$_3$ were also accompanied by a downregulation of glial glutamate transporters GLT-1 and GLAST as well as an increased expression of the neuronal glutamate transporter EAAC-1. The GABA transporter GAT-3 expression was also enhanced in this limbic epilepsy model, with recurrent seizures and further impairment of glutamate transport again appearing to be due to oxidative stress [135].

Other studies combining microdialysis and immunobiotechnological methods showed a specific decrease in glutamate uptake in the cortex of GAERS and an enhanced cortical expression of the vesicular glutamate transporter VGLUT2, thereby contributing to the hyperexcitability in this model for absence epilepsy [136]. In the model of 4-aminopyridine-induced seizures, an increase in striatal c-fos-immunostained cell nuclei correlated best with the increases in extracellular striatal glutamate levels, causing the authors to conclude that glutamate plays the key role in the extracellular signaling during 4-aminopyridine convulsions, leading to further intracellular cascades and c-fos gene expression [26]. In intrahippocampally 4-aminopyridine-treated rats, the inducible heat shock protein 70, detected immunocytochemically, was expressed primarily in neurons of the contralateral hippocampus. NMDA receptor antagonists prevented the seizures as well as the expression of heat shock protein 70 and neurodegeneration in the contralateral hippocampus [137]. These results sustain that the 4-aminopyridine-induced NMDA receptor-mediated hyperexcitability propagated to the contralateral hippocampus to upregulate this particular stress protein generally linked to neuroprotective mechanisms.

Adén et al. [66] combined the in vivo determination of dialysate changes in purine concentrations in hippocampus-kindled rats with a receptor binding assay, demonstrating that kindling induces adaptive changes in adenosine receptors. Microdialysis in the kainate model of status epilepticus combined with Timm's staining of the brain for the detection of zinc demonstrated that kainate-induced seizures were associated with an impaired zinc homeostasis in the brain [138].

Last but not least, the team of Dennis D. Spencer has provided a large number of exciting manuscipts in which microdialysis in the human brain of epileptic patients undergoing simultaneous intracranial EEG monitoring for seizure localization or epilepsy surgery was also combined with quantitative magnetic resonance spectroscopic imaging and 13C intraoperative glucose turnover studies (for a review, see [139]). By focusing on the energy metabolism of the amino acid neurotransmitters, GABA, glutamate, and their metabolites, they were able to show that numerous aspects of mitochondrial and energetic states link significantly with electrophysiologic and microdialysis measures of brain excitability in human epilepsy. For example, this combination of neurotechniques unraveled the fact that the epileptogenic gliotic human hippocampus appears to be characterized metabolically by slow rates of glutamate–glutamine cycling [140], increased extracellular glutamate levels, increased lactate production, and poor glucose utilization [141]. They demonstrated a significant association between the increased hippocampal glutamate dialysate levels and a decreased ipsilateral epileptogenic hippocampal volume in refractory temporal lobe epilepsy patients [142]. Within the hippocampal seizure focus of the human brain, extracellular GABA levels, as measured by microdialysis, increased with declining mitochondrial function, as determined by the spectroscopic measures of neuronal mitochondrial function (i.e., a decrease in the ratio of N-acetyl aspartate to creatine) [143].

8. THE ADVANTAGE OF MICRODIALYSIS USED TO SAMPLE ANTIEPILEPTIC DRUG LEVELS AND TO MONITOR NEUROTRANSMITTERS AS MARKERS FOR ANTICONVULSANT ACTIVITY

8.1. Investigating Plasma Pharmacokinetics of Antiepileptic Drugs Using Microdialysis

In general, serum concentration–time relationships of drugs are used as the principal basis in determining the pharmacokinetic properties of a compound or to compare the bioequivalence between different routes of administration. In clinical practice, serum concentration monitoring of antiepileptic drugs is used widely in the management of epileptic patients. Its primary purpose is as an index of brain/action site concentrations. As protein binding can have a major impact on the pharmacokinetics and pharmacodynamics of a drug, it is important to be able to quantify plasma protein binding or to selectively measure free drug concentrations only. The microdialysis technique can be used to sample drugs in serum. Due to the molecular cutoff of a probe, the unbound drug is separated from the fraction bound to proteins and/or other macromolecules and collected in the dialysate [144,145].

Several investigators have used microdialysis to sample serum for antiepileptic compounds to investigate the pharmacokinetics of phenytoin [146], and gastrodin, an anticonvulsant isolated from *Gastrodelia elata* Blume [147]. We and others have used microdialysis successfully to obtain the free carbamazepine plasma levels as a function of time following systemic exposure in rats [148,149]. We used the internal standard technique as an in vivo calibration method to correct each dialysis sample for probe recovery, yielding the exact free carbamazepine plasma concentrations. Patsalos and colleagues monitored the same compound together with its principal metabolite, but also phenytoin, primidone, and phenobarbitone, in epilepsy patients using plasma microdialysis [150]. They demonstrated that plasma dialysis is a reliable alternative for regular blood sampling, since both approaches resulted in comparable pharmacokinetic profiles. Moreover, patients preferred microdialysis sampling over regular blood sampling. The group of Stahle and co-workers investigated the feasibility of using subcutaneous microdialysis in patients as an alternative to plasma sampling [151,152]. Valproic acid was monitored subcutaneously in epilepsy patients and healthy volunteers. An acceptable correlation was demonstrated between total concentration in plasma and free concentration in plasma and dialysate. It was therefore concluded that microdialysis can be used to sample subcutaneous extracellular valproic acid in a clinical setting giving reliable estimates of the unbound plasma concentrations. Since the probes were connected to a portable pump, this approach makes it possible to monitor drug concentrations under nonhospital conditions in order to study the influence of daily activities on drug concentration. In another series of studies by this group the same protocol was repeated with the addition of an extra microdialysis probe in the subdural cerebrospinal fluid to study the time course of the distribution of different antiepileptic drugs to different body compartments [153–155]. Additionally, they demonstrated the applicability of cerebrospinal fluid microdialysis to obtain reliable estimations of protein-unbound drug concentrations. Hack et al. [156] demonstrated the feasibility and reliability of using subcutaneous microdialysis for close and extensive monitoring of metabolism with a ketogenic diet.

8.2. Investigating Biophase Pharmacokinetics in Physiological Conditions Using Microdialysis

Drug effects, side effects, and toxic effects are generally assumed to be correlated with free-drug concentrations in plasma [157]. It is believed that only the free drug in plasma is available for transport across the blood–brain barrier, drives the biophase (brain) kinetics, and eventually determines the in vivo drug effects (i.e., the free-drug hypothesis) [158]. However, apart from plasma pharmacokinetics, it is anticipated that blood–brain barrier transport and the processes of distribution and elimination within the brain strongly influence biophase pharmacokinetics of CNS compounds [159,160]. Blood–brain barrier transport and brain distribution often do not occur instantaneously and to a full extent, resulting in significantly different plasma versus brain kinetics [160–162]. Under these circumstances, serum pharmacokinetics will be a very poor index of brain pharmacokinetics [163]. Therefore, an understanding of the neuropharmacokinetics is imperative to accurately determine the temporal relationship between drug concentration and the intensity of its pharmacodynamic and pharmacologic effects. This information is crucial for therapy optimization and to ascertain drug modes and mechanisms of action. Additionally, the brain cannot simply be viewed as a homogeneous tissue, because it is composed of many anatomical structures with different characteristics [162,164]. The main compartments are the extracellular fluid, the brain intracellular space, and the cerebrospinal fluid. After blood–brain barrier passage, a drug will enter the extracellular fluid and may thereafter distribute into the brain intracellular space and the cerebrospinal fluid [159]. The drug action site will determine which compartment's kinetics is of most relevance and should be considered as the biophase. Many CNS-active drugs, including antiepileptic drugs, have their target at extracellular recognition sites, and therefore their extracellular brain concentrations are most closely related to the biophase concentrations [163]. Intracerebral microdialysis can provide important data to determine the blood–brain barrier transport characteristics and extracellular and biophase concentrations of a compound. A prerequisite for the use of intracerebral microdialysis for the determination of blood–brain barrier transport characteristics of a compound is that the intracerebral implantation of a microdialysis does not significantly affect the barrier properties. Based on a series of studies in which the effect of probe implantation on active and passive blood–brain barrier transport were evaluated, it can be concluded that this prerequisite appears to hold, provided that well-controlled surgical and experimental conditions are respected [165].

Upon introduction of the intracerebral microdialysis technique, numerous investigators have implemented this technique to sample the extracellular fluid for antiepileptic compounds in discrete brain areas. Gabapentin was monitored in the prefrontal cortex of nonepileptic rats by Rada and colleagues [166]. Wang and Welty [161] reported that cortical gabapentin levels corresponded to striatal levels and are approximately eight times lower than plasma levels. They used a microdialysis–pharmacokinetic modeling approach to estimate simultaneously the influx and efflux blood–brain barrier permeability of gabapentin. In the same study, they demonstrated that the total brain tissue concentration is significantly higher than the extracellular fluid concentration at steady state, due to intracellular accumulation and tissue binding. In a separate microdialysis study the same group demonstrated that the maximal anticonvulsant effect of gabapentin lagged behind both the plasma and brain extracellular fluid concentrations, due to time-dependent events other than distribution from blood to brain [167]. Luer et al. [168] combined regular plasma sampling and intracerebral microdialysis to show that gabapentin crosses the blood–brain barrier via a carrier-mediated transport mechanism that is saturable, especially at higher plasma concentrations.

Although the brain extracellular fluid and cerebrospinal fluid are produced independently, they are in direct communication. Therefore, cerebrospinal fluid sampling is often, but sometimes wrongly, used as an indirect index of extracellular fluid brain concentrations [169]. Walker et al. [170] investigated the rate of lamotrigine penetration into and the intrarelationship among the serum, cerebrospinal fluid, and hippocampal and frontal cortex extracellular fluid compartments following systemic administration in nonepileptic rats. The cerebrospinal fluid concentration did not turn out to be a reliable index for extracellular fluid concentration, as it clearly overestimated the extracellular fluid concentration. In the same study, the researchers also countered the aforementioned free-drug hypothesis by showing that the cerebrospinal fluid/serum concentration ratio was greater than the unbound/total serum concentration ratio. This result suggests that serum protein bound lamotrigine is possibly available for transport into the brain, a process that had already been described for other compounds [171–173]. Finally, the lamotrigine concentration in extracellular hippocampus and frontal cortex fluid were identical. This contrasts to other reports from the same group in which intracerebral microdialysis in the same brain areas was used, demonstrating ubiquitous brain extracellular distribution of phenytoin [174] or vigabatrin [175]. A few years later the same group performed an identical study to investigate the pharmacokinetic interrelationship of tiagabine in blood, cerebrospinal fluid, and brain extracellular fluid [176]. Again a considerable difference between serum and brain kinetics was observed, and cerebrospinal fluid concentrations did not reflect free drug concentrations. These studies recognize and emphasize the importance of determining both the peripheral and central kinetics of antiepileptic drugs in order to employ appropriate dosing strategies to study the pharmacodynamics of a drug and to aid appropriate interpretation of experimental data.

Two studies have even taken this one step further. We studied the biophase, plasma, and liver kinetics of carbamazepine simultaneously and its metabolic interaction with the antidepressant clomipramine in rats using quantitative microdialysis [149]. By combining the dialysis approach in different tissues in the same animal we were able to acquire information concerning the transport of carbamazepine and its metabolite across the blood–brain barrier, liver, and blood distribution kinetics and metabolic drug interaction. In a separate treatment group, carbamazepine was perfused in the liver via a microdialysis probe to obtain the liver metabolic ratio expressed as the ratio of the epoxy metabolite formed to the carbamazepine concentration administered. Lin et al. [147] demonstrated the feasibility of combining brain, plasma, and bile microdialysis for pharmacokinetic evaluation of the antiepileptic compound gastrodin.

8.3. Investigating Biophase Pharmacokinetics in Epileptic Conditions Using Microdialysis

Studying biophase kinetics of antiepileptic drugs is thus important, but it should be emphasized that it is even more relevant to do so in epileptic conditions. As mentioned above, the blood–brain barrier is a crucial determinant in the distribution of CNS compounds to their biophase. It is well documented that the delicate homeostasis of the blood–brain barrier is under continuous physiological control by surrounding astrocytes, pericytes, neurons, and plasma components [177]. This dynamic regulation of barrier functionality can, however, be disturbed during different pathological conditions. Epileptic seizures have a devastating impact on the ultrastructure and permeability of the blood–brain barrier [178–180]. They acutely induce a temporary leaky barrier, due to a lessening of the endothelial tightness, allowing increased passive drug entry into the brain [181]. The improved passive entry may be tempered by compensatory mechanisms which are triggered by blood–brain barrier disruption, such as vasogenic edema in brain tissue, which curtails passive diffusion of lipophilic compounds, and multidrug transporter overexpression, which results in increased active extrusion of its substrates out of the brain [181–183]. The extent of these compensatory processes can be highly variable, and its impact on brain entry of compounds cannot be predicted. The implications of these epilepsy-induced changes in blood–brain barrier functionality on the biophase disposition of antiepileptic drugs should be investigated thoroughly to be able to optimally adjust antiepileptic pharmacotherapy in anticipation of pharmacoresistance or intoxication due to the attainment, respectively, of sub- or supratherapeutic biophase drug levels.

Blood–brain barrier multidrug transporter overexpression has emerged as a potential source of impaired drug penetration into the epileptic brain, leading to pharmacoresistance [184]. Co-therapy of antiepileptic drugs with efflux transporter inhibitors has often been suggested as an approach to overcome this potential source of treatment refractoriness. Obviously, this theory only holds if pharmacoresistance is developed against antiepileptic drugs that are transporter substrates. The Löscher group has published a series of articles where they investigate the impact of different multidrug efflux transporter systems (P-glycoprotein and multidrug resistance-associated protein) on the brain distribution of a variety of antiepileptic drugs in rats. After systemic administration of different antiepileptic drugs, extracellular drug levels were repeatedly determined via microdialysis probes implanted into the right and left motor cortex. Multiple multidrug transporter inhibitors were administered locally via one of the probes, thus allowing a comparison of extracellular drug levels in the absence and presence of the inhibitors in the same animal.

Efflux transporter inhibition will only increase the extracellular concentrations of its substrates. Although systemic transporter inhibition seems clinically most relevant, this approach does not allow drawing clear-cut conclusions concerning the impact of blood–brain barrier efflux mechanisms solely on brain partition of a substrate, since brain pharmacokinetics can be substantially contaminated by inhibitor-induced alterations in plasma pharmacokinetics. Intracerebral microdialysis enables in situ efflux transporter inhibitor administration. Via this approach they were able to show that the biophase kinetics of phenytoin [185,186], carbamazepine [185], phenobarbital, lamotrigine, and felbamate [187] were influenced by P-glycoprotein and/or multidrug resistance–associated protein-mediated efflux. Via a comparable protocol we were able to demonstrate that hippocampal extracellular oxcarbazepine levels increased significantly during efflux transporter blockade [188]. Löscher's group was not able to reproduce the results for carbamazepine, lamotrigine, and felbamate in mice lacking the multidrug resistance-associated protein transporter [189]. Efflux mechanisms were shown not to influence the blood–brain barrier penetration of valproic acid [190] and levetiracetam [191]. Scism et al. [3] confirmed that the extracellular valproic acid concentrations cannot be affected by blood–brain barrier efflux transporter inhibition in rabbits. They did, however, demonstrate important inhibitable efflux transport of valproic acid in brain parenchymal cells, resulting in elevated intracellular levels. Another microdialysis study demonstrating the therapeutic potential of using multidrug transport inhibitor co-therapy to overcome pharmacoresistance in epilepsy was published by Hesselink et al. [192]. Inhibition of the probenecid-sensitive organic acid transporter was demonstrated to increase the striatal extracellular levels of a group of NMDA receptor glycine$_B$-site antagonists and to prolong their anticonvulsant activity against maximal electroshock-induced convulsions in mice.

The studies summarized above investigated the impact of active efflux mechanisms on biophase disposition only by reducing the amount of available transporters. However, in the clinical situation an overexpression of the transporters is anticipated in combination with a temporal loss of the selective permeability of the blood–brain barrier. The net result of these epilepsy-related changes on biophase pharmacokinetics is difficult to predict. To address this issue, we performed a microdialysis study in which we determined the free plasma and hippocampal concentrations of 10,11-dihydro-10-hydroxycarbamazepine, the active metabolite of oxcarbazepine, during control conditions, acute limbic seizures, and efflux transport inhibition [193]. To quantify the impact of these experimental conditions on drug distribution unambiguously, we performed simultaneous population pharmacokinetic modeling of plasma and brain concentrations of 10,11-dihydro-10-hydroxycarbamazepine [193]. The experimental conditions did not affect plasma pharmacokinetics but had a major impact on the biophase concentrations. Both seizures and efflux transport inhibition resulted in significantly increased extracellular drug concentrations. A distributional process was shown to underlie the increased brain drug levels following acute seizures and efflux transport inhibition. It was concluded that characterization of biophase pharmacokinetics is critical to assessments of the impact of efflux transporters and acute seizures on brain disposition of antiepileptic drugs and, consequently, on their effects. Löscher and colleagues [112] reported the lack of anticonvulsant efficacy of mGlu$_1$ and mGlu$_5$ receptor antagonists in rodent models of difficult-to-treat partial epilepsy. They suggested that increased active efflux leading to subtherapeutic drug levels may be responsible. By determining the biophase concentrations, however, they substantiated that effective (i.e., producing complete selective receptor occupancy in vivo) brain concentrations were reached. Therefore, they were able to conclude unambiguously that these compounds lacked anticonvulsant efficacy. In an elegant study, Rizzi et al. [194] investigated the temporal expression of the mdr1 gene encoding for P-glycoprotein following chemoconvulsant-induced limbic seizures and determined hippocampal extracellular phenytoin levels as well as tissue levels. They showed that limbic seizures caused a transient P-glycoprotein overexpression which was functionally relevant to significantly influence local phenytoin concentrations. Höcht and co-workers [146] used plasma and hippocampal dialysis to demonstrate decreased brain penetration of phenytoin in epileptic rats in comparison to nonepileptic control rats. Potschka et al. [195] used microdialysis to test the hypothesis of lowered penetration of antiepileptic drugs into epileptogenic brain tissue. Seizures in amygdala-kindled rats are less responsive to treatment with phenytoin than are seizures in nonkindled rats, whereas the central adverse effects are at least comparable. They therefore compared phenytoin levels in amygdala and hippocampus of kindled and nonkindled rats [196,197] and found lower but nonsignificant average dialysate levels. These levels were comparable to those described by Scheyer et al. [198], who determined dialysate levels of phenytoin in the hippocampus of refractory epilepsy patients. The steady-state extracellular phenytoin concentrations corresponded closely to the unbound serum concentrations. A major disadvantage of this and other human studies is that no comparison is available between dialysate levels in epileptogenic and nonepileptogenic tissue. Tisdall et al. [199] also found a significant correlation between brain microdialysate and free plasma phenytoin concentration in two patients with acute brain injury. In an extensive study, Rambeck and colleagues [200] compared brain extracellular fluid, brain tissue, cerebrospinal fluid, and serum concentrations of different antiepileptic drugs measured intraoperatively in 22 pharmacoresistant epilepsy patients with complex partial seizures or secondary generalized seizures. Intraoperative microdialysis was only performed in the brain parts that had to be resected for therapeutic reasons. The data showed considerable intra- and interindividual variation of the antiepileptic drug concentrations in the extracellular space of cortical regions. Extracellular drug concentrations were significantly lower than the brain tissue and cerebrospinal fluid concentrations. Because no data are available for responsive patients, it is not possible to judge whether the extracellular drug levels are reduced considerably in the epileptogenic brain tissue of nonresponders.

Different approaches to minimize the impact of the blood–brain barrier on biophase drug levels in neurological diseases are currently under investigation [201,202]. Structural modifications of compounds and the construction of prodrugs to facilitate barrier transport have been evaluated by Battaglia et al. [203] using intracerebral microdialysis. Alternative delivery is another putative strategy to enhance efficacy and the tolerability of chronic antiepileptic pharmacotherapy [204]. Nasal drug administration is considered to provide a putative means for targeted CNS delivery. Intranasal administration of benzodiazepines has already been used successfully for the treatment of status epilepticus [205]. To validate the bioequivalence and therapeutic value of alternative methods for antiepileptic drug delivery, biophase monitoring of antiepileptic drugs can provide decisive information. Czapp and co-workers [206] have investigated the brain penetration of intranasal phenobarbital in rat frontal cortex using microdialysis rather than intravenous administration. The same dose resulted in significantly higher phenobarbital dialysate levels following intranasal administration. Therapeutically relevant biophase levels could be attained, as decreased seizure severity and increased seizure threshold was observed in fully amygdala kindled rats.

8.4. Investigating the Pharmacokinetic–Pharmacodynamic Correlation of Antiepileptic Drugs Using Intracerebral Microdialysis

Together with the biophase drug concentrations, the microdialysis technique enables the monitoring of (potential) extracellular biomarkers of drug effects or disease. Based on this unique information, the pharmacokinetic–

pharmacodynamic correlation of a compound can be determined. Graumlich et al. [148] designed a study to ascribe dose–response differences among three rat strains to pharmacokinetic or pharmacodynamic factors. Following equal weight-adjusted systemic carbamazepine administration, plasma and hippocampal dialysate analysis yielded marked differences in unbound carbamazepine levels in both compartments. In separate experiments, carbamazepine-induced serotonin release from the hippocampus was used as a pharmacodynamic marker for carbamazepine's pharmacological activity. Intrahippocampal carbamazepine infusions produced similar concentration–response relations for the different rat strains. These results supported the hypothesis that the dose–response differences among the rat strains are pharmacokinetic in nature. The group of Whitton described the plasma and brain kinetics of lamotrigine alone or in combination with valproate in correlation with the basal and stimulated extracellular amino acids and monoamines in the hippocampus of freely moving rats [110,207]. In the former studies the pharmacokinetic and pharmacodynamic data were generated in separate experiments. By collecting the pharmacokinetic and pharmacodynamic information from the same animal, one can improve the accuracy and quality of the data by minimizing the interanimal variation. We correlated free plasma and extracellular hippocampal oxcarbazepine levels with compound-innate-increased pharmacodynamic response on hippocampal dopamine and serotonine dialysate levels in the same animal [188]. These hippocampal monoamine effects were defined as biomarkers for the anticonvulsant efficacy of oxcarbazepine and its active metabolite 10,11-dihydro-10-hydroxycarbamazepine [102]. Oxcarbazepine administration alone did not affect hippocampal monoaminergic neurotransmission and did not prevent limbic seizures in rats. In contrast, multidrug transporter inhibition resulted in increased biophase extracellular drug levels and drug responses on hippocampal monoamines. The improved pharmacological response mediated potent anticonvulsant effects. Chenel and co-workers [208] performed pharmacokinetic–pharmacodynamic modeling of the EEG effect of norfloxacin in combination with intracerebral glutamate determinations using hippocampal microdialysis. Norfloxacin is a fluoroquinolone antibiotic that can induce convulsions that are associated with hippocampal glutamate increases [25]. They used the increased total power of the EEG signal as a surrogate pharmacodynamic endpoint for the investigation of the convulsant activity of antibiotics in rats. The fact that the increased power effect was observed consistently without glutamate-level modifications proved that norfloxacin-induced seizures are not triggered by intracerebral glutamate enhancement. Crick et al. [131] also correlated biophase 3-mercaptopropionic acid levels with electroencephalogram responses by simultaneous monitoring, in real time, of the brain concentration of this proconvulsant and the corresponding electroencephalogram activity. This information led to the development of a steady-state chemical seizure model. Feng and colleagues [209] used brain microdialysis to investigate the pharmacokinetic–pharmacodynamic correlation of pregabalin in rats. Brain pregabalin levels were determined to calculate the blood–brain barrier influx and efflux permeability. A pharmacokinetic–pharmacodynamic link model was used to describe the counterclockwise hysteresis relationship between pregabain brain extracellular concentration and the anticonvulsant effect in rats.

9. MICRODIALYSIS USED TO STUDY RELATIONSHIPS BETWEEN EPILEPSY AND ITS COMORBIDITIES

Refractory epilepsy has an enormous impact on the daily, social, and psychosocial life of the patients. Mood disorders, such as major depression or anxiety, are the most frequent psychiatric comorbidities in temporal lobe epilepsy [210]. Microdialysis is one of the interest-gaining neuromethods applicable to unravel some of the neurobiological relationships between epilepsy and the affective disorders in animal models. For example, we studied whether several selective monoamine reuptake inhibitors and glutamate receptor ligands were anticonvulsant against pilocarpine-induced seizures, exhibited antidepressant-like actions within the rat forced-swim and mouse tail suspension tests, and were able to alter the hippocampal dopamine or serotonin levels determined by in vivo microdialysis in rats [211]. In this way we showed that the compounds with combined antidepressant–anticonvulsant properties all directly enhanced the extracellular hippocampal dopamine or serotonin concentrations and proposed this direct elevation of hippocampal monoamines as pharmacodynamic measures of combined antidepressant–anticonvulsant action [211]. Another recent manuscript described that vagus nerve stimulation therapy, effective for treatment-resistant epilepsy as well as treatment-resistant depression, was able to increase the noradrenaline microdialysate concentrations obtained from the rat prefrontal cortex and to induce alterations in the hippocampal and cortical mRNA levels of various growth factors [212]. Selectively bred amydala kindling seizure-prone and seizure-resistant rats also exhibit differences in anxiety tests. Indeed, Merali and colleagues showed a differential impact of predator or immobilization stressors on the endocrinological and neurochemical response patterns in these two lines of rats. Predator exposure elicited a larger increase in the plasma adrenocorticotropic hormone concentrations (measured with radioimmunoassay) in the seizure-resistant rats, while immobilization stress provoked a greater rise in the plasma adrenocorticotropic hormone levels in the seizure-prone rats [213]. In vivo microdialysis experiments within the same study further revealed that predator exposure also induced a greater rise in amygdala dialysate levels of corticotropin-releasing hormone in the seizure-resistant rats, while immobilization elicited a more pronounced release of corticotropin-releasing hormone in the seizure-prone rats. Maciejak et al. analyzed the effects of hippocampal mGlu$_1$ receptors on the consolidation of a fear-conditioned response and on the hippocampal glutamate and GABA dialysate concentrations obtained from rats subjected to pentylenetetrazole-induced kindling of seizures. They showed that kindling led to the potentiation of excitatory processes in the hippocampus, thereby altering the mGlu$_1$ receptor–mediated freezing response to the aversively conditioned context [214]. The results of Ando and co-workers demonstrated attenuated basal striatal extracellular dopamine concentrations and a hypersensitivity of the striatal dopamine system in the chronic phase of the kainate-induced temporal lobe epilepsy model, which may explain the increased incidence of schizophrenia-like psychosis in temporal lobe epilepsy patients [215].

Finally, the hypothalamus plays a vital role in the homeostatic processes of blood pressure and heart rate regulation via GABAergic and glutamatergic neurotransmission. It has been hypothesized that dysfunctions of these autonomic neuronal mechanisms could be a cause of sudden unexpected death in epilepsy patients. Some microdialysis studies have been undertaken to investigate possible neurotransmitter alterations in the hypothalamus of epileptic rats. There were no alterations in the basal hypothalamic glutamate or GABA dialysate levels in amygdala kindled epileptic rats versus control animals; the blood pressure and heart rate responses to chemoconvulsants were also not affected by amygdale kindling [216]. On the other hand, similar experiments in genetic absence rats showed that the group of GAERS animals displayed an altered mode of central cardiovascular regulation compared with nonepileptic Wistar rats [217].

REFERENCES

[1] Pitkänen, A., Schwarzkroin, P.A., Moshé, S.L. (2005). *Models of Seizures and Epilepsy*, Academic Press, San Diego, CA.

[2] Bourne, J.A., Fosbraey, P., Halliday, J. (2001). SCH 23390 affords protection against soman-evoked seizures in the freely moving guinea-pig: a concomitant neurochemical, electrophysiological and behavioural study. *Neuropharmacology*, *40*, 279–288.

[3] Scism, J.L., Powers, K.M., Artru, A.A., Lewis, L., Shen, D.D. (2000). Probenecid-inhibitable efflux transport of valproic acid in the brain parenchymal cells of rabbits: a microdialysis study. *Brain Research*, *884*, 77–86.

[4] Young, R.S., During, M.J., Aquila, W.J., Tendler, D., Ley, E. (1992). Hypoxia increases extracellular concentrations of excitatory and inhibitory neurotransmitters in subsequently induced seizure: in vivo microdialysis study in the rabbit. *Experimental Neurology*, *117*, 204–209.

[5] Shouse, M.N., Staba, R.J., Saquib, S.F., Farber, P.R. (2001). Long-lasting effects of feline amygdala kindling on monoamines, seizures and sleep. *Brain Research*, *892*, 147–165.

[6] Morita, T., Takahashi, M., Takeuchi, T., Hikasa, Y., Ikeda, S., Sawada, M., Sato, K., Shibahara, T., Shimada, A. (2005). Changes in extracellular neurotransmitters in the cerebrum of familial idiopathic epileptic shetland sheepdogs using an intracerebral microdialysis technique and immunohistochemical study for glutamate metabolism. *Journal of Veterinary Medical Science*, *67*, 1119–1126.

[7] Tan, W.K., Williams, C.E., During, M.J., Mallard, C.E., Gunning, M.I., Gunn, A.J., Gluckman, P.D. (1996). Accumulation of cytotoxins during the development of seizures and edema after hypoxic–ischemic injury in late gestation fetal sheep. *Pediatric Research*, *39*, 791–797.

[8] Thoresen, M., Hallström, A., Whitelaw, A., Puka-Sundvall, M., Loberg, E.M., Satas, S., Ungerstedt, U., Steen, P.A., Hagberg, H. (1998). Lactate and pyruvate changes in the cerebral gray and white matter during posthypoxic seizures in newborn pigs. *Pediatric Research*, *44*, 746–754.

[9] Carlson, H., Ronne-Engstrom, E., Ungerstedt, U., Hillered, L. (1992). Seizure-related elevations of extracellular amino acids in human focal epilepsy. *Neuroscience Letters*, *140*, 30–32.

[10] Wade, J.V., Samson, F.E., Nelson, S.R., Pazdernik, T.L. (1987). Changes in extracellular amino acids during soman- and kainic acid–induced seizures. *Journal of Neurochemistry*, *49*, 645–650.

[11] Butcher, S.P., Jacobson, I., Hamberger, A. (1988). On the epileptogenic effects of kainic acid and dihydrokainic acid in the dentate gyrus of the rat. *Neuropharmacology*, *27*, 375–381.

[12] During, M.J., Spencer, D.D. (1993). Extracellular hippocampal glutamate and spontaneous seizure in the conscious human brain. *Lancet*, *341*, 1607–1610.

[13] Wilson, C.L., Maidment, N.T., Shomer, M.H., Behnke, E.J., Ackerson, L., Fried, I., Engel, J., Jr. (1996). Comparison of seizure related amino acid release in human epileptic hippocampus versus a chronic, kainate rat model of hippocampal epilepsy. *Epilepsy Research*, *26*, 245–254.

[14] Ronne-Engstrom, E., Hillered, L., Flink, R., Spannare, B., Ungerstedt, U., Carlson, H. (1992). Intracerebral microdialysis of extracellular amino acids in the human epileptic focus. *Journal of Cerebral Blood Flow & Metabolism*, *12*, 873–876.

[15] Thomas, P.M., Phillips, J.P., Delanty, N., O'Connor, W.T. (2003). Elevated extracellular levels of glutamate, aspartate and gamma-aminobutyric acid within the intraoperative, spontaneously epileptiform human hippocampus. *Epilepsy Research*, *54*, 73–79.

[16] Goren, M.Z., Onat, F., Ozkara, C., Ozyurt, E., Eskazan, E., Aker, R. (2001). GABA and L-glutamic acid release in en bloc resection slices of human hippocampus: an in vitro microdialysis study. *Neurological Sciences*, *22*, 297–302.

[17] Rowley, H.L., Martin, K.F., Marsden, C.A. (1995). Decreased GABA release following tonic–clonic seizures is associated with an increase in extracellular glutamate in rat hippocampus in vivo. *Neuroscience*, *68*, 415–422.

[18] Liu, Z., Stafstrom, C.E., Sarkisian, M.R., Yang, Y., Hori, A., Tandon, P., Holmes, G.L. (1997). Seizure-induced glutamate release in mature and immature animals: an in vivo microdialysis study. *Neuroreport*, *8*, 2019–2023.

[19] Smolders, I., Khan, G.M., Manil, J., Ebinger, G., Michotte, Y. (1997). NMDA receptor-mediated pilocarpine-induced seizures: characterization in freely moving rats by microdialysis. *British Journal of Pharmacology*, *121*, 1171–1179.

[20] Khan, G.M., Smolders, I., Lindekens, H., Manil, J., Ebinger, G., Michotte, Y. (1999). Effects of diazepam on extracellular brain neurotransmitters in pilocarpine-induced seizures in rats. *European Journal of Pharmacology*, *373*, 153–161.

[21] Meurs, A., Clinckers, R., Ebinger, G., Michotte, Y., Smolders, I. (2008). Seizure activity and changes in hippocampal extracellular glutamate, GABA, dopamine and serotonin. *Epilepsy Research*, *78*, 50–59.

[22] Pena, F., Tapia, R. (1999). Relationships among seizures, extracellular amino acid changes, and neurodegeneration induced by 4-aminopyridine in rat hippocampus: a microdialysis and electroencephalographic study. *Journal of Neurochemistry*, *72*, 2006–2014.

[23] Slézia, A., Kekesi, A.K., Szikra, T., Papp, A.M., Nagy, K., Szente, M., Magloczky, Z., Freund, T.F., Juhasz, G. (2004). Uridine release during aminopyridine-induced epilepsy. *Neurobiology of Disease*, *16*, 490–499.

[24] Yager, J.Y., Armstrong, E.A., Miyashita, H., Wirrell, E.C. (2002). Prolonged neonatal seizures exacerbate hypoxic–ischemic brain damage: correlation with cerebral energy metabolism and excitatory amino acid release. *Developmental Neuroscience*, *24*, 367–381.

[25] Smolders, I., Gousseau, C., Marchand, S., Couet, W., Ebinger, G., Michotte, Y. (2002). Convulsant and subconvulsant doses of norfloxacin in the presence and absence of biphenylacetic acid alter extracellular hippocampal glutamate but not gamma-aminobutyric acid levels in conscious rats. *Antimicrobial Agents and Chemotherapy*, *46*, 471–477.

[26] Kovacs, A., Mihaly, A., Komaromi, A., Gyengesi, E., Szente, M., Weiczner, R., Krisztin-Peva, B., Szabo, G., Telegdy, G. (2003). Seizure, neurotransmitter release, and gene expression are closely related in the striatum of 4-aminopyridine-treated rats. *Epilepsy Research*, *55*, 117–129.

[27] Shin, R.S., Anisman, H., Merali, Z., McIntyre, D.C. (2004). Amygdala amino acid and monoamine levels in genetically fast and slow kindling rat strains during massed amygdala kindling: a microdialysis study. *European Journal of Neuroscience*, *20*, 185–194.

[28] Engstrom, E.R., Hillered, L., Flink, R., Kihlstrom, L., Lindquist, C., Nie, J.X., Olsson, Y., Silander, H.C. (2001). Extracellular amino acid levels measured with intracerebral microdialysis in the model of posttraumatic epilepsy induced by intracortical iron injection. *Epilepsy Research*, *43*, 135–144.

[29] Feng, Y., LeBlanc, M.H., Regunathan, S. (2005). Agmatine reduces extracellular glutamate during pentylenetetrazole-induced seizures in rat brain: a potential mechanism for the anticonvulsive effects. *Neuroscience Letters*, *390*, 129–133.

[30] Tanaka, K., Graham, S.H., Simon, R.P. (1996). The role of excitatory neurotransmitters in seizure-induced neuronal injury in rats. *Brain Research*, *737*, 59–63.

[31] Bruhn, T., Cobo, M., Berg, M., Diemer, N.H. (1992). Limbic seizure-induced changes in extracellular amino acid levels in the hippocampal formation: a microdialysis study of freely moving rats. *Acta Neurologica Scandinavica*, *86*, 455–461.

[32] Obrenovitch, T.P., Urenjak, J., Zilkha, E. (1996). Evidence disputing the link between seizure activity and high extracellular glutamate. *Journal of Neurochemistry*, *66*, 2446–2454.

[33] Ramirez-Munguia, N., Vera, G., Tapia, R. (2003). Epilepsy, neurodegeneration, and extracellular glutamate in the hippocampus of awake and anesthetized rats treated with okadaic acid. *Neurochemical Research*, *28*, 1517–1524.

[34] Millan, M.H., Chapman, A.G., Meldrum, B.S. (1993). Extracellular amino acid levels in hippocampus during pilocarpine-induced seizures. *Epilepsy Research*, *14*, 139–148.

[35] Nyitrai, G., Kekesi, K.A., Szilagyi, N., Papp, A., Juhasz, G., Kardos, J. (2002). Neurotoxicity of lindane and picrotoxin: neurochemical and electrophysiological correlates in the rat hippocampus in vivo. *Neurochemical Research*, *27*, 139–145.

[36] Sierra-Paredes, G., Galan-Valiente, J., Vazquez-Illanes, M.D., Aguilar-Veiga, E., Sierra-Marcuno, G. (2000). Effect of ionotropic glutamate receptors antagonists on the modifications in extracellular glutamate and aspartate levels during picrotoxin seizures: a microdialysis study in freely moving rats. *Neurochemistry International*, *37*, 377–386.

[37] Smolders, I., Bortolotto, Z.A., Clarke, V.R., Warre, R., Khan, G.M., O'Neill, M.J., Ornstein, P.L., Bleakman, D., Ogden, A., Weiss, B., et al. (2002). Antagonists of GLU(K5)-containing kainate receptors prevent pilocarpine-induced limbic seizures. *Nature Neuroscience*, *5*, 796–804.

[38] Smolders, I., Lindekens, H., Clinckers, R., Meurs, A., O'Neill, M.J., Lodge, D., Ebinger, G., Michotte, Y. (2004). In vivo modulation of extracellular hippocampal glutamate and GABA levels and limbic seizures by group I and II metabotropic glutamate receptor ligands. *Journal of Neurochemistry*, *88*, 1068–1077.

[39] Khan, G.M., Smolders, I., Ebinger, G., Michotte, Y. (2000). Anticonvulsant effect and neurotransmitter modulation of focal and systemic 2-chloroadenosine against the development of pilocarpine-induced seizures. *Neuropharmacology*, *39*, 2418–2432.

[40] Lindekens, H., Smolders, I., Khan, G.M., Bialer, M., Ebinger, G., Michotte, Y. (2000). In vivo study of the effect of valpromide and valnoctamide in the pilocarpine rat model of focal epilepsy. *Pharmaceutical Research*, *17*, 1408–1413.

[41] Parrot, S., Bert, L., Mouly-Badina, L., Sauvinet, V., Colussi-Mas, J., Lambas-Senas, L., Robert, F., Bouilloux, J.P., Suaud-Chagny, M.F., Denoroy, L., Renaud, B. (2003). Microdialysis monitoring of catecholamines and excitatory amino acids in the rat and mouse brain: recent developments based on capillary electrophoresis with laser-induced fluorescence detection: a mini-review. *Cellular and Molecular Neurobiology*, *23*, 793–804.

[42] Ueda, Y., Doi, T., Tokumaru, J., Mitsuyama, Y., Willmore, L.J. (2000). Kindling phenomena induced by the repeated short-term high potassium stimuli in the ventral hippocampus of rats: on-line monitoring of extracellular glutamate overflow. *Experimental Brain Research*, *135*, 199–203.

[43] Millan, M.H., Obrenovitch, T.P., Sarna, G.S., Lok, S.Y., Symon, L., Meldrum, B.S. (1991). Changes in rat brain extracellular glutamate concentration during seizures induced by systemic picrotoxin or focal bicuculline injection: an in vivo dialysis study with on-line enzymatic detection. *Epilepsy Research*, *9*, 86–91.

[44] Treiman, D.M. (2001). GABAergic mechanisms in epilepsy. *Epilepsia*, *42* (Suppl. 3), 8–12.

[45] Sherwin, A.L. (1999). Neuroactive amino acids in focally epileptic human brain: a review. *Neurochemical Research*, *24*, 1387–1395.

[46] Shin, R.S., Anisman, H., Merali, Z., McIntyre, D.C. (2002). Changes in extracellular levels of amygdala amino acids in genetically fast and slow kindling rat strains. *Brain Research*, *946*, 31–42.

[47] Lehmann, A., Hagberg, H., Jacobson, I., Hamberger, A. (1985). Effects of status epilepticus on extracellular amino acids in the hippocampus. *Brain Research, 359,* 147–151.

[48] Parrot, S., Sauvinet, V., Riban, V., Depaulis, A., Renaud, B., Denoroy, L. (2004). High temporal resolution for in vivo monitoring of neurotransmitters in awake epileptic rats using brain microdialysis and capillary electrophoresis with laser-induced fluorescence detection. *Journal of Neuroscience Methods, 140,* 29–38.

[49] Oldenziel, W.H., Dijkstra, G., Cremers, T.I., Westerink, B.H. (2006). In vivo monitoring of extracellular glutamate in the brain with a microsensor. *Brain Research, 1118,* 34–42.

[50] Nyitrai, G., Kekesi, K.A., Juhasz, G. (2006). Extracellular level of GABA and Glu: in vivo microdialysis–HPLC measurements. *Current Topics in Medicinal Chemistry, 6,* 935–940.

[51] Lason, W. (2001). Neurochemical and pharmacological aspects of cocaine-induced seizures. *Polish Journal of Pharmacology, 53,* 57–60.

[52] Clinckers, R., Smolders, I., Meurs, A., Ebinger, G., Michotte, Y. (2004). Anticonvulsant action of hippocampal dopamine and serotonin is independently mediated by D and 5-HT receptors. *Journal of Neurochemistry, 89,* 834–843.

[53] Clinckers, R., Gheuens, S., Smolders, I., Meurs, A., Ebinger, G., Michotte, Y. (2005). In vivo modulatory action of extracellular glutamate on the anticonvulsant effects of hippocampal dopamine and serotonin. *Epilepsia, 46,* 828–836.

[54] Dazzi, L., Serra, M., Porceddu, M.L., Sanna, A., Chessa, M.F., Biggio, G. (1997). Enhancement of basal and pentylenetetrazol (PTZ)-stimulated dopamine release in the brain of freely moving rats by PTZ-induced kindling. *Synapse, 26,* 351–358.

[55] Kokaia, M., Kalen, P., Bengzon, J., Lindvall, O. (1989). Noradrenaline and 5-hydroxytryptamine release in the hippocampus during seizures induced by hippocampal kindling stimulation: an in vivo microdialysis study. *Neuroscience, 32,* 647–656.

[56] Zis, A.P., Nomikos, G.G., Brown, E.E., Damsma, G., Fibiger, H.C. (1992). Neurochemical effects of electrically and chemically induced seizures: an in vivo microdialysis study in the rat hippocampus. *Neuropsychopharmacology, 7,* 189–195.

[57] Zis, A.P., Nomikos, G.G., Damsma, G., Fibiger, H.C. (1991). In vivo neurochemical effects of electroconvulsive shock studied by microdialysis in the rat striatum. *Psychopharmacology, 103,* 343–350.

[58] Shouse, M.N., Staba, R.J., Ko, P.Y., Saquib, S.F., Farber, P.R. (2001). Monoamines and seizures: microdialysis findings in locus ceruleus and amygdala before and during amygdala kindling. *Brain Research, 892,* 176–192.

[59] Jobe, P.C., Dailey, J.W., Wernicke, J.F. (1999). A noradrenergic and serotonergic hypothesis of the linkage between epilepsy and affective disorders. *Critical Reviews in Neurobiology, 13,* 317–356.

[60] Barone, P., Palma, V., de Bartolomeis, A., Cicarelli, G., Campanella, G. (1992). Dopaminergic regulation of epileptic activity. *Neurochemistry International, 20*(Suppl.), 245S–249S.

[61] Barone, P., Palma, V., de Bartolomeis, A., Tedeschi, E., Muscettola, G., Campanella, G. (1991). Dopamine D1 and D2 receptors mediate opposite functions in seizures

induced by lithium–pilocarpine. *European Journal of Pharmacology, 195,* 157–162.

[62] Lallement, G., Carpentier, P., Collet, A., Baubichon, D., Pernot-Marino, I., Blanchet, G. (1992). Extracellular acetylcholine changes in rat limbic structures during soman-induced seizures. *Neurotoxicology, 13,* 557–567.

[63] Baptista, T., Weiss, S.R., Zocchi, A., Sitcoske, M., Post, R. (1994). Electrical kindling is associated with increases in amygdala acetylcholine levels: an in vivo microdialysis study. *Neuroscience Letters, 167,* 133–136.

[64] Serra, M., Dazzi, L., Cagetti, E., Chessa, M.F., Pisu, M.G., Sanna, A., Biggio, G. (1997). Effect of pentylenetetrazole-induced kindling on acetylcholine release in the hippocampus of freely moving rats. *Journal of Neurochemistry, 68,* 313–318.

[65] Dragunow, M., Goddard, G.V., Laverty, R. (1985). Is adenosine an endogenous anticonvulsant? *Epilepsia, 26,* 480–487.

[66] Adén, U., O'Connor, W.T., Berman, R.F. (2004). Changes in purine levels and adenosine receptors in kindled seizures in the rat. *Neuroreport, 15,* 1585–1589.

[67] Berman, R.F., Fredholm, B.B., Adén, U., O'Connor, W.T. (2000). Evidence for increased dorsal hippocampal adenosine release and metabolism during pharmacologically induced seizures in rats. *Brain Research, 872,* 44–53.

[68] Rocha, L., Maidment, N.T., Evans, C.J., Ackermann, R.F., Engel, J., Jr. (1994). Microdialysis reveals changes in extracellular opioid peptide levels in the amygdala induced by amygdaloid kindling stimulation. *Experimental Neurology, 126,* 277–283.

[69] Rocha, L.L., Evans, C.J., Maidment, N.T. (1997). Amygdala kindling modifies extracellular opioid peptide content in rat hippocampus measured by microdialysis. *Journal of Neurochemistry, 68,* 616–624.

[70] Rocha, L., Cano, A., Cruz, C., Omana-Zapata, I., Villalobos, R., Maidment, N.T. (1999). Opioid peptide systems following a subconvulsant dose of pentylenetetrazol in rats. *Epilepsy Research, 37,* 141–150.

[71] Rocha, L., Maidment, N.T. (2003). Opioid peptide release in the rat hippocampus after kainic acid–induced status epilepticus. *Hippocampus, 13,* 472–480.

[72] Aparicio, L.C., Candeletti, S., Binaschi, A., Mazzuferi, M., Mantovani, S., Di Benedetto, M., Landuzzi, D., Lopetuso, G., Romualdi, P., Simonato, M. (2004). Kainate seizures increase nociceptin/orphanin FQ release in the rat hippocampus and thalamus: a microdialysis study. *Journal of Neurochemistry, 91,* 30–37.

[73] Husum, H., Gruber, S.H., Bolwig, T.G., Mathé, A.A. (2002). Extracellular levels of NPY in the dorsal hippocampus of freely moving rats are markedly elevated following a single electroconvulsive stimulation, irrespective of anticonvulsive Y1 receptor blockade. *Neuropeptides, 36,* 363–369.

[74] Husum, H., Mikkelsen, J.D., Mork, A. (1998). Extracellular levels of neuropeptide Y are markedly increased in the dorsal hippocampus of freely moving rats during kainic acid–induced seizures. *Brain Research, 781,* 351–354.

[75] Marti, M., Bregola, G., Morari, M., Gemignani, A., Simonato, M. (2000). Somatostatin release in the hippocampus in the kindling model of epilepsy: a microdialysis study. *Journal of Neurochemistry, 74,* 2497–2503.

[76] Lahtinen, H., Brankack, J., Koivisto, E., Riekkinen, P.J. (1992). Somatostatin release in rat neocortex during gamma-hydroxybutyrate–provoked seizures:

microdialysis combined with EEG recording. *Brain Research Bulletin, 29,* 837–841.

[77] Layton, M.E., Pazdernik, T.L. (1999). Reactive oxidant species in piriform cortex extracellular fluid during seizures induced by systemic kainic acid in rats. *Journal of Molecular Neuroscience, 13,* 63–68.

[78] Amiridze, N., Dang, Y., Brown, O.R. (1999). Hydroxyl radicals detected via brain microdialysis in rats breathing air and during hyperbaric oxygen convulsions. *Redox Report, 4,* 165–170.

[79] Yasuda, H., Fujii, M., Fujisawa, H., Ito, H., Suzuki, M. (2001). Changes in nitric oxide synthesis and epileptic activity in the contralateral hippocampus of rats following intrahippocampal kainate injection. *Epilepsia, 42,* 13–20.

[80] Kaku, T., Jiang, M.H., Hada, J., Morimoto, K., Hayashi, Y. (2001). Sodium nitroprusside–induced seizures and adenosine release in rat hippocampus. *European Journal of Pharmacology, 413,* 199–205.

[81] Ueda, Y., Yokoyama, H., Niwa, R., Konaka, R., Ohya-Nishiguchi, H., Kamada, H. (1997). Generation of lipid radicals in the hippocampal extracellular space during kainic acid–induced seizures in rats. *Epilepsy Research, 26,* 329–333.

[82] Darbin, O., Risso, J.J., Carre, E., Lonjon, M., Naritoku, D.K. (2005). Metabolic changes in rat striatum following convulsive seizures. *Brain Research, 1050,* 124–129.

[83] Fornai, F., Bassi, L., Gesi, M., Giorgi, F.S., Guerrini, R., Bonaccorsi, I., Alessandri, M.G. (2000). Similar increases in extracellular lactic acid in the limbic system during epileptic and/or olfactory stimulation. *Neuroscience, 97,* 447–458.

[84] Cornford, E.M., Shamsa, K., Zeitzer, J.M., Enriquez, C.M., Wilson, C.L., Behnke, E.J., Fried, I., Engel, J. (2002). Regional analyses of CNS microdialysate glucose and lactate in seizure patients. *Epilepsia, 43,* 1360–1371.

[85] During, M.J., Fried, I., Leone, P., Katz, A., Spencer, D.D. (1994). Direct measurement of extracellular lactate in the human hippocampus during spontaneous seizures. *Journal of Neurochemistry, 62,* 2356–2361.

[86] Sierra-Paredes, G., Galan-Valiente, J., Vazquez-Illanes, M.D., Aguilar-Veiga, E., Soto-Otero, R., Mendez-Alvarez, E., Sierra-Marcuno, G. (1998). Extracellular amino acids in the rat hippocampus during picrotoxin threshold seizures in chronic microdialysis experiments. *Neuroscience Letters, 248,* 53–56.

[87] Ayala, G.X., Tapia, R. (2005). Late N-methyl-D-aspartate receptor blockade rescues hippocampal neurons from excitotoxic stress and death after 4-aminopyridine-induced epilepsy. *European Journal of Neuroscience, 22,* 3067–3076.

[88] Oreiro-Garcia, M.T., Vazquez-Illanes, M.D., Sierra-Paredes, G., Sierra-Marcuno, G. (2007). Changes in extracellular amino acid concentrations in the rat hippocampus after in vivo actin depolymerization with latrunculin A. *Neurochemistry International, 50,* 734–740.

[89] Sierra-Paredes, G., Oreiro-Garcia, T., Nunez-Rodriguez, A., Vazquez-Lopez, A., Sierra-Marcuno, G. (2006). Seizures induced by in vivo latrunculin a and jasplakinolide microperfusion in the rat hippocampus. *Journal of Molecular Neuroscience, 28,* 151–160.

[90] Nishida, N., Huang, Z.L., Mikuni, N., Miura, Y., Urade, Y., Hashimoto, N. (2007). Deep brain stimulation of the posterior hypothalamus activates the

histaminergic system to exert antiepileptic effect in rat pentylenetetrazol model. *Experimental Neurology, 205*, 132–144.

[91] Ziai, W.C., Sherman, D.L., Bhardwaj, A., Zhang, N., Keyl, P.M., Mirski, M.A. (2005). Target-specific catecholamine elevation induced by anticonvulsant thalamic deep brain stimulation. *Epilepsia, 46*, 878–888.

[92] Zhang, B., Chu, J., Zhang, J., Ma, Y. (2008). Change of extracellular glutamate and gamma-aminobutyric acid in substantia nigra and globus pallidus during electrical stimulation of subthalamic nucleus in epileptic rats. *Stereotactic and Functional Neurosurgery, 86*, 208–215.

[93] Nail-Boucherie, K., Le-Pham, B.T., Gobaille, S., Maitre, M., Aunis, D., Depaulis, A. (2005). Evidence for a role of the parafascicular nucleus of the thalamus in the control of epileptic seizures by the superior colliculus. *Epilepsia, 46*, 141–145.

[94] Terzioglu, B., Aypak, C., Onat, F.Y., Kucukibrahimoglu, E., Ozkaynakci, A.E., Goren, M.Z. (2006). The effects of ethosuximide on amino acids in genetic absence epilepsy rat model. *Journal of Pharmacological Sciences, 100*, 227–233.

[95] Khongsombat, O., Watanabe, H., Tantisira, B., Patarapanich, C., Tantisira, M.H. (2008). Acute effects of *N*-(2-propylpentanoyl)urea on hippocampal amino acid neurotransmitters in pilocarpine-induced seizure in rats. *Epilepsy Research, 79*, 151–157.

[96] Rowley, H.L., Marsden, C.A., Martin, K.F. (1995). Differential effects of phenytoin and sodium valproate on seizure-induced changes in gamma-aminobutyric acid and glutamate release in vivo. *European Journal of Pharmacology, 294*, 541–546.

[97] Sierra-Paredes, G., Oreiro-Garcia, M.T., Vazquez-Illanes, M.D., Sierra-Marcuno, G. (2007). Effect of eslicarbazepine acetate (BIA 2-093) on latrunculin A-induced seizures and extracellular amino acid concentrations in the rat hippocampus. *Epilepsy Research, 77*, 36–43.

[98] Dailey, J.W., Reith, M.E., Yan, Q.S., Li, M.Y., Jobe, P.C. (1997). Carbamazepine increases extracellular serotonin concentration: lack of antagonism by tetrodotoxin or zero Ca2$^+$. *European Journal of Pharmacology, 328*, 153–162.

[99] Okada, M., Zhu, G., Yoshida, S., Kanai, K., Hirose, S., Kaneko, S. (2002). Exocytosis mechanism as a new targeting site for mechanisms of action of antiepileptic drugs. *Life Sciences, 72*, 465–473.

[100] Yoshida, S., Okada, M., Zhu, G., Kaneko, S. (2007). Carbamazepine prevents breakdown of neurotransmitter release induced by hyperactivation of ryanodine receptor. *Neuropharmacology, 52*, 1538–1546.

[101] Ueda, Y., Doi, T., Tokumaru, J., Nakajima, A., Nagatomo, K. (2005). In vivo evaluation of the effect of zonisamide on the hippocampal redox state during kainic acid–induced seizure status in rats. *Neurochemical Research, 30*, 1117–1121.

[102] Clinckers, R., Smolders, I., Meurs, A., Ebinger, G., Michotte, Y. (2005). Hippocampal dopamine and serotonin elevations as pharmacodynamic markers for the anticonvulsant efficacy of oxcarbazepine and 10,11-dihydro-10-hydroxycarbamazepine. *Neuroscience Letters, 390*, 48–53.

[103] Smolders, I., Van Belle, K., Ebinger, G., Michotte, Y. (1997). Hippocampal and cerebellar extracellular amino acids during pilocarpine-induced seizures in freely moving rats. *European Journal of Pharmacology, 319*, 21–29.

[104] Meurs, A., Clinckers, R., Ebinger, G., Michotte, Y., Smolders, I. (2006). Substantia nigra is an anticonvulsant site of action of topiramate in the focal pilocarpine model of limbic seizures. *Epilepsia, 47,* 1519–1535.

[105] Kanda, T., Kurokawa, M., Tamura, S., Nakamura, J., Ishii, A., Kuwana, Y., Serikawa, T., Yamada, J., Ishihara, K., Sasa, M. (1996). Topiramate reduces abnormally high extracellular levels of glutamate and aspartate in the hippocampus of spontaneously epileptic rats (SER). *Life Sciences, 59,* 1607–1616.

[106] Dalby, N.O. (2000). GABA-level increasing and anticonvulsant effects of three different GABA uptake inhibitors. *Neuropharmacology, 39,* 2399–2407.

[107] Sayin, U., Timmerman, W., Westerink, B.H. (1995). The significance of extracellular GABA in the substantia nigra of the rat during seizures and anticonvulsant treatments. *Brain Research, 669,* 67–72.

[108] Smolders, I., Khan, G.M., Lindekens, H., Prikken, S., Marvin, C.A., Manil, J., Ebinger, G., Michotte, Y. (1997). Effectiveness of vigabatrin against focally evoked pilocarpine-induced seizures and concomitant changes in extracellular hippocampal and cerebellar glutamate, gamma-aminobutyric acid and dopamine levels, a microdialysis–electrocorticography study in freely moving rats. *Journal of Pharmacology and Experimental Therapeutics, 283,* 1239–1248.

[109] Richards, D.A., Manning, J.P., Barnes, D., Rombola, L., Bowery, N.G., Caccia, S., Leresche, N., Crunelli, V. (2003). Targeting thalamic nuclei is not sufficient for the full anti-absence action of ethosuximide in a rat model of absence epilepsy. *Epilepsy Research, 54,* 97–107.

[110] Ahmad, S., Fowler, L.J., Whitton, P.S. (2005). Effects of combined lamotrigine and valproate on basal and stimulated extracellular amino acids and monoamines in the hippocampus of freely moving rats. *Naunyn-Schmiedeberg's Archives of Pharmacology, 371,* 1–8.

[111] Wang, X., Ai, J., Hampson, D.R., Snead, O.C., 3rd (2005). Altered glutamate and GABA release within thalamocortical circuitry in metabotropic glutamate receptor 4 knockout mice. *Neuroscience, 134,* 1195–1203.

[112] Löscher, W., Dekundy, A., Nagel, J., Danysz, W., Parsons, C.G., Potschka, H. (2006). mGlu1 and mGlu5 receptor antagonists lack anticonvulsant efficacy in rodent models of difficult-to-treat partial epilepsy. *Neuropharmacology, 50,* 1006–1015.

[113] Wu, H.Q., Rassoulpour, A., Goodman, J.H., Scharfman, H.E., Bertram, E.H., Schwarcz, R. (2005). Kynurenate and 7-chlorokynurenate formation in chronically epileptic rats. *Epilepsia, 46,* 1010–1016.

[114] Zhang, D.X., Williamson, J.M., Wu, H.Q., Schwarcz, R., Bertram, E.H. (2005). In situ-produced 7-chlorokynurenate has different effects on evoked responses in rats with limbic epilepsy in comparison to naive controls. *Epilepsia, 46,* 1708–1715.

[115] Meurs, A., Clinckers, R., Ebinger, G., Michotte, Y., Smolders, I. (2007). Sigma 1 receptor-mediated increase in hippocampal extracellular dopamine contributes to the mechanism of the anticonvulsant action of neuropeptide Y. *European Journal of Neuroscience, 26,* 3079–3092.

[116] Renming, X., Ishihara, K., Sasa, M., Ujihara, H., Momiyama, T., Fujita, Y., Todo, N., Serikawa, T., Yamada, J., Takaori, S. (1992). Antiepileptic effects of CNK-602A, a novel thyrotropin-releasing hormone analog, on absence-like and tonic

seizures of spontaneously epileptic rats. *European Journal of Pharmacology, 223*, 185–192.

[117] Stragier, B., Clinckers, R., Meurs, A., De Bundel, D., Sarre, S., Ebinger, G., Michotte, Y., Smolders, I. (2006). Involvement of the somatostatin-2 receptor in the anti-convulsant effect of angiotensin IV against pilocarpine-induced limbic seizures in rats. *Journal of Neurochemistry, 98*, 1100–1113.

[118] Andrews, N., Davis, B., Gonzalez, M.I., Oles, R., Singh, L., McKnight, A.T. (2000). Effect of gastrin-releasing peptide on rat hippocampal extracellular GABA levels and seizures in the audiogenic seizure-prone DBA/2 mouse. *Brain Research, 859*, 386–389.

[119] Mazzuferi, M., Binaschi, A., Rodi, D., Mantovani, S., Simonato, M. (2005). Induction of B1 bradykinin receptors in the kindled hippocampus increases extracellular glutamate levels: a microdialysis study. *Neuroscience, 135*, 979–986.

[120] Vazquez-Lopez, A., Sierra-Paredes, G., Sierra-Marcuno, G. (2006). Anticonvulsant effect of the calcineurin inhibitor ascomycin on seizures induced by picrotoxin microperfusion in the rat hippocampus. *Pharmacology Biochemistry and Behavior, 84*, 511–516.

[121] Fujisaki, Y., Yamauchi, A., Dohgu, S., Sunada, K., Yamaguchi, C., Oishi, R., Kataoka, Y. (2002). Cyclosporine A–increased nitric oxide production in the rat dorsal hippocampus mediates convulsions. *Life Sciences, 72*, 549–556.

[122] Mark, G.P., Finn, D.A. (2002). The relationship between hippocampal acetylcholine release and cholinergic convulsant sensitivity in withdrawal seizure-prone and withdrawal seizure-resistant selected mouse lines. *Alcoholism: Clinical and Experimental Research, 26*, 1141–1152.

[123] Celik, T., Kayir, H., Ceyhan, M., Demirtas, S., Cosar, A., Uzbay, I.T. (2004). CPP and amlodipine alter the decrease in basal acetylcholine and choline release by audiogenic stimulus in hippocampus of ethanol-withdrawn rats in vivo. *Brain Research Bulletin, 64*, 243–249.

[124] Halladay, A.K., Wagner, G.C., Sekowski, A., Rothman, R.B., Baumann, M.H., Fisher, H. (2006). Alterations in alcohol consumption, withdrawal seizures, and monoamine transmission in rats treated with phentermine and 5-hydroxy-L-tryptophan. *Synapse, 59*, 277–289.

[125] Peng, X.Q., Li, X., Li, J., Ramachandran, P.V., Gagare, P.D., Pratihar, D., Ashby, C.R., Jr., Gardner, E.L., Xi, Z.X. (2008). Effects of gabapentin on cocaine self-administration, cocaine-triggered relapse and cocaine-enhanced nucleus accumbens dopamine in rats. *Drug and Alcohol Dependence, 97*, 207–215.

[126] Tor-Agbidye, J., Yamamoto, B., Bowyer, J.F. (2001). Seizure activity and hyperthermia potentiate the increases in dopamine and serotonin extracellular levels in the amygdala during exposure to *d*-amphetamine. *Toxicological Sciences, 60*, 103–111.

[127] Mannaioni, G., Carpenedo, R., Moroni, F. (2003). 5-Hydroxyindole causes convulsions and increases transmitter release in the CA1 region of the rat hippocampus. *British Journal of Pharmacology, 138*, 245–253.

[128] Miyamoto, O., Sumitani, K., Nakamura, T., Yamagami, S., Miyata, S., Itano, T., Negi, T., Okabe, A. (2000). *Clostridium perfringens* epsilon toxin causes excessive

release of glutamate in the mouse hippocampus. *FEMS Microbiology Letters,* *189,* 109–113.

[129] Gorji, A., Stemmer, N., Rambeck, B., Jurgens, U., May, T., Pannek, H.W., Behne, F., Ebner, A., Straub, H., Speckmann, E.J. (2006). Neocortical microenvironment in patients with intractable epilepsy: potassium and chloride concentrations. *Epilepsia, 47,* 297–310.

[130] Bourne, J.A., Fosbraey, P. (2000). Novel method of monitoring electroencephalography at the site of microdialysis during chemically evoked seizures in a freely moving animal. *Journal of Neuroscience Methods, 99,* 85–90.

[131] Crick, E.W., Osorio, I., Bhavaraju, N.C., Linz, T.H., Lunte, C.E. (2007). An investigation into the pharmacokinetics of 3-mercaptopropionic acid and development of a steady-state chemical seizure model using in vivo microdialysis and electrophysiological monitoring. *Epilepsy Research, 74,* 116–125.

[132] During, M.J., Ryder, K.M., Spencer, D.D. (1995). Hippocampal GABA transporter function in temporal-lobe epilepsy. *Nature, 376,* 174–177.

[133] Ghijsen, W.E., da Silva Aresta Belo, A.I., Zuiderwijk, M., Lopez da Silva, F.H. (1999). Compensatory change in EAAC1 glutamate transporter in rat hippocampus CA1 region during kindling epileptogenesis. *Neuroscience Letters, 276,* 157–160.

[134] Samuelsson, C., Kumlien, E., Elfving, A., Lindholm, D., Ronne-Engstrom, E. (2003). The effects of PBN (phenyl-butyl-nitrone) on GLT-1 levels and on the extracellular levels of amino acids and energy metabolites in a model of iron-induced posttraumatic epilepsy. *Epilepsy Research, 56,* 165–173.

[135] Ueda, Y., Doi, T., Nagatomo, K., Willmore, L.J., Nakajima, A. (2007). Functional role for redox in the epileptogenesis: molecular regulation of glutamate in the hippocampus of FeCl3-induced limbic epilepsy model. *Experimental Brain Research, 181,* 571–577.

[136] Touret, M., Parrot, S., Denoroy, L., Belin, M.F., Didier-Bazes, M. (2007). Glutamatergic alterations in the cortex of genetic absence epilepsy rats. *BMC Neuroscience, 8,* 69.

[137] Ayala, G.X., Tapia, R. (2003). Expression of heat shock protein 70 induced by 4-aminopyridine through glutamate-mediated excitotoxic stress in rat hippocampus in vivo. *Neuropharmacology, 45,* 649–660.

[138] Takeda, A., Hirate, M., Tamano, H., Oku, N. (2003). Zinc movement in the brain under kainate-induced seizures. *Epilepsy Research, 54,* 123–129.

[139] Pan, J.W., Williamson, A., Cavus, I., Hetherington, H.P., Zaveri, H., Petroff, O.A., Spencer, D.D. (2008). Neurometabolism in human epilepsy. *Epilepsia, 49*(Suppl. 3), 31–41.

[140] Petroff, O.A., Errante, L.D., Rothman, D.L., Kim, J.H., Spencer, D.D. (2002). Glutamate–glutamine cycling in the epileptic human hippocampus. *Epilepsia, 43,* 703–710.

[141] Cavus, I., Kasoff, W.S., Cassaday, M.P., Jacob, R., Gueorguieva, R., Sherwin, R.S., Krystal, J.H., Spencer, D.D., Abi-Saab, W.M. (2005). Extracellular metabolites in the cortex and hippocampus of epileptic patients. *Annals of Neurology, 57,* 226–235.

[142] Cavus, I., Pan, J.W., Hetherington, H.P., Abi-Saab, W., Zaveri, H.P., Vives, K.P., Krystal, J.H., Spencer, S.S., Spencer, D.D. (2008). Decreased hippocampal volume

on MRI is associated with increased extracellular glutamate in epilepsy patients. *Epilepsia*, *49*, 1358–1366.

[143] Pan, J.W., Cavus, I., Kim, J., Hetherington, H.P., Spencer, D.D. (2008). Hippocampal extracellular GABA correlates with metabolism in human epilepsy. *Metabolic Brain Disease*.

[144] Chaurasia, C.S. (1999). In vivo microdialysis sampling: theory and applications. *Biomedical Chromatography*, *13*, 317–332.

[145] Scott, D.O., Sorenson, L.R., Steele, K.L., Puckett, D.L., Lunte, C.E. (1991). In vivo microdialysis sampling for pharmacokinetic investigations. *Pharmaceutical Research*, *8*, 389–392.

[146] Höcht, C., Lazarowski, A., Gonzalez, N.N., Auzmendi, J., Opezzo, J.A., Bramuglia, G.F., Taira, C.A., Girardi, E. (2007). Nimodipine restores the altered hippocampal phenytoin pharmacokinetics in a refractory epileptic model. *Neuroscience Letters*, *413*, 168–172.

[147] Lin, L.C., Chen, Y.F., Lee, W.C., Wu, Y.T., Tsai, T.H. (2008). Pharmacokinetics of gastrodin and its metabolite *p*-hydroxybenzyl alcohol in rat blood, brain and bile by microdialysis coupled to LC–MS/MS. *Journal of Pharmaceutical and Biomedical Analysis*.

[148] Graumlich, J.F., McLaughlin, R.G., Birkhahn, D., Shah, N., Burk, A., Jobe, P.C., Dailey, J.W. (1999). Carbamazepine pharmacokinetics–pharmacodynamics in genetically epilepsy-prone rats. *European Journal of Pharmacology*, *369*, 305–311.

[149] Van Belle, K., Sarre, S., Ebinger, G., Michotte, Y. (1995). Brain, liver and blood distribution kinetics of carbamazepine and its metabolic interaction with clomipramine in rats: a quantitative microdialysis study. *Journal of Pharmacology and Experimental Therapeutics*, *272*, 1217–1222.

[150] Patsalos, P.N., O'Connell, M.T., Doheny, H.C., Sander, J.W., Shorvon, S.D. (1996). Antiepileptic drug pharmacokinetics in patients with epilepsy using a new microdialysis probe: preliminary observations. *Acta Neurochirurgica*, *67*, 59–62.

[151] Lindberger, M., Tomson, T., Stahle, L. (1998). Validation of microdialysis sampling for subcutaneous extracellular valproic acid in humans. *Therapeutic Drug Monitoring*, *20*, 358–362.

[152] Stahle, L., Alm, C., Ekquist, B., Lundquist, B., Tomson, T. (1996). Monitoring free extracellular valproic acid by microdialysis in epileptic patients. *Therapeutic Drug Monitoring*, *18*, 14–18.

[153] Lindberger, M., Tomson, T., Lars, S. (2002). Microdialysis sampling of carbamazepine, phenytoin and phenobarbital in subcutaneous extracellular fluid and subdural cerebrospinal fluid in humans: an in vitro and in vivo study of adsorption to the sampling device. *Pharmacology & Toxicology*, *91*, 158–165.

[154] Lindberger, M., Tomson, T., Wallstedt, L., Stahle, L. (2001). Distribution of valproate to subdural cerebrospinal fluid, subcutaneous extracellular fluid, and plasma in humans: a microdialysis study. *Epilepsia*, *42*, 256–261.

[155] Lindberger, M., Tomson, T., Ohman, I., Wallstedt, L., Stahle, L. (1999). Estimation of topiramate in subdural cerebrospinal fluid, subcutaneous extracellular fluid, and plasma: a single case microdialysis study. *Epilepsia*, *40*, 800–802.

[156] Hack, A., Busch, V., Pascher, B., Busch, R., Bieger, I., Gempel, K., Baumeister, F.A. (2006). Monitoring of ketogenic diet for carnitine metabolites by subcutaneous microdialysis. *Pediatric Research*, *60*, 93–96.

[157] de Lange, E.C., Ravenstijn, P.G., Groenendaal, D., van Steeg, T.J. (2005). Toward the prediction of CNS drug-effect profiles in physiological and pathological conditions using microdialysis and mechanism-based pharmacokinetic–pharmacodynamic modeling. *AAPS Journal, 7*, E532–E543.

[158] Derendorf, H., Hochhaus, G., Mollmann, H., Barth, J., Krieg, M., Tunn, S., Mollmann, C. (1993). Receptor-based pharmacokinetic–pharmacodynamic analysis of corticosteroids. *Journal of Clinical Pharmacology, 33*, 115–123.

[159] de Lange, E.C., Danhof, M. (2002). Considerations in the use of cerebrospinal fluid pharmacokinetics to predict brain target concentrations in the clinical setting: implications of the barriers between blood and brain. *Clinical Pharmacokinetics, 41*, 691–703.

[160] Hammarlund-Udenaes, M., Paalzow, L.K., de Lange, E.C. (1997). Drug equilibration across the blood–brain barrier: pharmacokinetic considerations based on the microdialysis method. *Pharmaceutical Research, 14*, 128–134.

[161] Wang, Y., Welty, D.F. (1996). The simultaneous estimation of the influx and efflux blood–brain barrier permeabilities of gabapentin using a microdialysis–pharmacokinetic approach. *Pharmaceutical Research, 13*, 398–403.

[162] Collins, J.M., Dedrick, R.L. (1983). Distributed model for drug delivery to CSF and brain tissue. *American Journal of Physiology, 245*, R303–R310.

[163] Sechi, G.P., Petruzzi, V., Rosati, G., Tanca, S., Monaco, F., Formato, M., Rubattu, L., De Riu, P. (1989). Brain interstitial fluid and intracellular distribution of phenytoin. *Epilepsia, 30*, 235–239.

[164] Gross, P.M., Sposito, N.M., Pettersen, S.E., Fenstermacher, J.D. (1986). Differences in function and structure of the capillary endothelium in gray matter, white matter and a circumventricular organ of rat brain. *Blood Vessels, 23*, 261–270.

[165] de Lange, E.C., Danhof, M., de Boer, A.G., Breimer, D.D. (1997). Methodological considerations of intracerebral microdialysis in pharmacokinetic studies on drug transport across the blood–brain barrier. *Brain Research, Brain Research Reviews, 25*, 27–49.

[166] Rada, P., Tucci, S., Perez, J., Teneud, L., Chuecos, S., Hernandez, L. (1998). In vivo monitoring of gabapentin in rats: a microdialysis study coupled to capillary electrophoresis and laser-induced fluorescence detection. *Electrophoresis, 19*, 2976–2980.

[167] Welty, D.F., Schielke, G.P., Vartanian, M.G., Taylor, C.P. (1993). Gabapentin anticonvulsant action in rats: disequilibrium with peak drug concentrations in plasma and brain microdialysate. *Epilepsy Research, 16*, 175–181.

[168] Luer, M.S., Hamani, C., Dujovny, M., Gidal, B., Cwik, M., Deyo, K., Fischer, J.H. (1999). Saturable transport of gabapentin at the blood–brain barrier. *Neurological Research, 21*, 559–562.

[169] Thomas, S.A., Segal, M.B. (1998). The transport of the anti-HIV drug, 2′,3′-didehydro-3′-deoxythymidine (D4T), across the blood–brain and blood-cerebrospinal fluid barriers. *British Journal of Pharmacology, 125*, 49–54.

[170] Walker, M.C., Tong, X., Perry, H., Alavijeh, M.S., Patsalos, P.N. (2000). Comparison of serum, cerebrospinal fluid and brain extracellular fluid pharmacokinetics of lamotrigine. *British Journal of Pharmacology, 130*, 242–248.

[171] Urien, S., Pinquier, J.L., Paquette, B., Chaumet-Riffaud, P., Kiechel, J.R., Tillement, J.P. (1987). Effect of the binding of isradipine and darodipine to different plasma

proteins on their transfer through the rat blood–brain barrier. Drug binding to lipoproteins does not limit the transfer of drug. *Journal of Pharmacology and Experimental Therapeutics*, *242*, 349–353.

[172] Pardridge, W.M., Sakiyama, R., Fierer, G. (1983). Transport of propranolol and lidocaine through the rat blood–brain barrier: primary role of globulin-bound drug. *Journal of Clinical Investigation*, *71*, 900–908.

[173] Cornford, E.M., Young, D., Paxton, J.W., Sofia, R.D. (1992). Blood–brain barrier penetration of felbamate. *Epilepsia*, *33*, 944–954.

[174] Walker, M.C., Alavijeh, M.S., Shorvon, S.D., Patsalos, P.N. (1996). Microdialysis study of the neuropharmacokinetics of phenytoin in rat hippocampus and frontal cortex. *Epilepsia*, *37*, 421–427.

[175] Tong, X., Ratnaraj, N., Patsalos, P.N. (2008). Vigabatrin extracellular pharmaco-kinetics and concurrent gamma-aminobutyric acid neurotransmitter effects in rat frontal cortex and hippocampus using microdialysis. *Epilepsia*.

[176] Wang, X., Ratnaraj, N., Patsalos, P.N. (2004). The pharmacokinetic inter-relationship of tiagabine in blood, cerebrospinal fluid and brain extracellular fluid (frontal cortex and hippocampus). *Seizure*, *13*, 574–581.

[177] Abbott, N.J., Revest, P.A. (1991). Control of brain endothelial permeability. *Cerebrovascular and Brain Metabolism Reviews*, *3*, 39–72.

[178] Petito, C.K., Schaefer, J.A., Plum, F. (1977). Ultrastructural characteristics of the brain and blood–brain barrier in experimental seizures. *Brain Research*, *127*, 251–267.

[179] Nitsch, C., Klatzo, I. (1983). Regional patterns of blood–brain barrier breakdown during epileptiform seizures induced by various convulsive agents. *Journal of the Neurological Sciences*, *59*, 305–322.

[180] Seifert, G., Huttmann, K., Schramm, J., Steinhauser, C. (2004). Enhanced relative expression of glutamate receptor 1 flip AMPA receptor subunits in hippocampal astrocytes of epilepsy patients with Ammon's horn sclerosis. *Journal of Neuroscience*, *24*, 1996–2003.

[181] Janigro, D. (1999). Blood–brain barrier, ion homeostatis and epilepsy: possible implications towards the understanding of ketogenic diet mechanisms. *Epilepsy Research*, *37*, 223–232.

[182] van Vliet, E., van Schaik, R., Edelbroek, P., Voskuyl, R., Redeker, S., Aronica, E., Wadman, W., Gorter, J. (2007). Region-specific overexpression of P-glycoprotein at the blood–brain barrier affects brain uptake of phenytoin in epileptic rats. *Journal of Pharmacology and Experimental Therapeutics*, *322*, 141–147.

[183] Oby, E., Janigro, D. (2006). The blood–brain barrier and epilepsy. *Epilepsia*, *47*, 1761–1774.

[184] Sisodiya, S.M. (2003). Mechanisms of antiepileptic drug resistance. *Current Opinion in Neurology*, *16*, 197–201.

[185] Potschka, H., Löscher, W. (2001). Multidrug resistance-associated protein is involved in the regulation of extracellular levels of phenytoin in the brain. *Neuroreport*, *12*, 2387–2389.

[186] Potschka, H., Löscher, W. (2001). In vivo evidence for P-glycoprotein–mediated transport of phenytoin at the blood–brain barrier of rats. *Epilepsia*, *42*, 1231–1240.

[187] Potschka, H., Fedrowitz, M., Löscher, W. (2002). P-glycoprotein–mediated efflux of phenobarbital, lamotrigine, and felbamate at the blood–brain barrier: evidence from microdialysis experiments in rats. *Neuroscience Letters, 327,* 173–176.

[188] Clinckers, R., Smolders, I., Meurs, A., Ebinger, G., Michotte, Y. (2005). Quantitative in vivo microdialysis study on the influence of multidrug transporters on the blood–brain barrier passage of oxcarbazepine: concomitant use of hippocampal monoamines as pharmacodynamic markers for the anticonvulsant activity. *Journal of Pharmacology and Experimental Therapeutics, 314,* 725–731.

[189] Potschka, H., Fedrowitz, M., Löscher, W. (2003). Brain access and anticonvulsant efficacy of carbamazepine, lamotrigine, and felbamate in ABCC2/MRP2-deficient TR- rats. *Epilepsia, 44,* 1479–1486.

[190] Baltes, S., Fedrowitz, M., Tortos, C.L., Potschka, H., Löscher, W. (2007). Valproic acid is not a substrate for P-glycoprotein or multidrug resistance proteins 1 and 2 in a number of in vitro and in vivo transport assays. *Journal of Pharmacology and Experimental Therapeutics, 320,* 331–343.

[191] Potschka, H., Baltes, S., Löscher, W. (2004). Inhibition of multidrug transporters by verapamil or probenecid does not alter blood–brain barrier penetration of levetiracetam in rats. *Epilepsy Research, 58,* 85–91.

[192] Hesselink, M.B., Smolders, H., Eilbacher, B., De Boer, A.G., Breimer, D.D., Danysz, W. (1999). The role of probenecid-sensitive organic acid transport in the pharmacokinetics of *N*-methyl-D-aspartate receptor antagonists acting at the glycine(B)-site: microdialysis and maximum electroshock seizures studies. *Journal of Pharmacology and Experimental Therapeutics, 290,* 543–550.

[193] Clinckers, R., Smolders, I., Michotte, Y., Ebinger, G., Danhof, M., Voskuyl, R.A., Pasqua, O.D. (2008). Impact of efflux transporters and of seizures on the pharmacokinetics of oxcarbazepine metabolite in the rat brain. *British Journal of Pharmacology, 155*(7), 1127–1130.

[194] Rizzi, M., Caccia, S., Guiso, G., Richichi, C., Gorter, J.A., Aronica, E., Aliprandi, M., Bagnati, R., Fanelli, R., D'Incalci, M., Samanin, R., Vezzani, A. (2002). Limbic seizures induce P-glycoprotein in rodent brain: functional implications for pharmacoresistance. *Journal of Neuroscience, 22,* 5833–5839.

[195] Potschka, H., Fedrowitz, M., Löscher, W. (2001). P-glycoprotein and multidrug resistance-associated protein are involved in the regulation of extracellular levels of the major antiepileptic drug carbamazepine in the brain. *Neuroreport, 12,* 3557–3560.

[196] Honack, D., Löscher, W. (1995). Kindling increases the sensitivity of rats to adverse effects of certain antiepileptic drugs. *Epilepsia, 36,* 763–771.

[197] Löscher, W., Jackel, R., Czuczwar, S.J. (1986). Is amygdala kindling in rats a model for drug-resistant partial epilepsy? *Experimental Neurology, 93,* 211–226.

[198] Scheyer, R.D., During, M.J., Hochholzer, J.M., Spencer, D.D., Cramer, J.A., Mattson, R.H. (1994). Phenytoin concentrations in the human brain: an in vivo microdialysis study. *Epilepsy Research, 18,* 227–232.

[199] Tisdall, M., Russo, S., Sen, J., Belli, A., Ratnaraj, N., Patsalos, P., Petzold, A., Kitchen, N., Smith, M. (2006). Free phenytoin concentration measurement in brain extracellular fluid: a pilot study. *British Journal of Neurosurgery, 20,* 285–289.

[200] Rambeck, B., Jurgens, U.H., May, T.W., Pannek, H.W., Behne, F., Ebner, A., Gorji, A., Straub, H., Speckmann, E.J., Pohlmann-Eden, B., Löscher, W. (2006). Comparison of brain extracellular fluid, brain tissue, cerebrospinal fluid, and serum concentrations of antiepileptic drugs measured intraoperatively in patients with intractable epilepsy. *Epilepsia, 47,* 681–694.

[201] Pardridge, W.M. (2005). The blood–brain barrier and neurotherapeutics. *NeuroRx, 2,* 1–2.

[202] Graff, C.L., Pollack, G.M. (2005). Nasal drug administration: potential for targeted central nervous system delivery. *Journal of Pharmaceutical Sciences, 94,* 1187–1195.

[203] Battaglia, G., La Russa, M., Bruno, V., Arenare, L., Ippolito, R., Copani, A., Bonina, F., Nicoletti, F. (2000). Systemically administered D-glucose conjugates of 7-chlorokynurenic acid are centrally available and exert anticonvulsant activity in rodents. *Brain Research, 860,* 149–156.

[204] Fisher, R.S., Ho, J. (2002). Potential new methods for antiepileptic drug delivery. *CNS Drugs, 16,* 579–593.

[205] Wolfe, T.R., Macfarlane, T.C. (2006). Intranasal midazolam therapy for pediatric status epilepticus. *American Journal of Emergency Medicine, 24,* 343–346.

[206] Czapp, M., Bankstahl, J.P., Zibell, G., Potschka, H. (2008). Brain penetration and anticonvulsant efficacy of intranasal phenobarbital in rats. *Epilepsia, 49,* 1142–1150.

[207] Ahmad, S., Fowler, L.J., Whitton, P.S. (2004). Effects of acute and chronic lamotrigine treatment on basal and stimulated extracellular amino acids in the hippocampus of freely moving rats. *Brain Research, 1029,* 41–47.

[208] Chenel, M., Limosin, A., Marchand, S., Paquereau, J., Mimoz, O., Couet, W. (2003). Norfloxacin-induced electroencephalogram alteration and seizures in rats are not triggered by enhanced levels of intracerebral glutamate. *Antimicrobial Agents and Chemotherapy, 47,* 3660–3662.

[209] Feng, M.R., Turluck, D., Burleigh, J., Lister, R., Fan, C., Middlebrook, A., Taylor, C., Su, T. (2001). Brain microdialysis and PK/PD correlation of pregabalin in rats. *European Journal of Drug Metabolism and Pharmacokinetics, 26,* 123–128.

[210] Kanner, A.M. (2008). Mood disorder and epilepsy: a neurobiologic perspective of their relationship. *Dialogues in Clinical Neuroscience, 10,* 39–45.

[211] Smolders, I., Clinckers, R., Meurs, A., De Bundel, D., Portelli, J., Ebinger, G., Michotte, Y. (2008). Direct enhancement of hippocampal dopamine or serotonin levels as a pharmacodynamic measure of combined antidepressant–anticonvulsant action. *Neuropharmacology, 54,* 1017–1028.

[212] Follesa, P., Biggio, F., Gorini, G., Caria, S., Talani, G., Dazzi, L., Puligheddu, M., Marrosu, F., Biggio, G. (2007). Vagus nerve stimulation increases norepinephrine concentration and the gene expression of BDNF and bFGF in the rat brain. *Brain Research, 1179,* 28–34.

[213] Merali, Z., Kent, P., Michaud, D., McIntyre, D., Anisman, H. (2001). Differential impact of predator or immobilization stressors on central corticotropin-releasing hormone and bombesin-like peptides in fast and slow seizing rat. *Brain Research, 906,* 60–73.

[214] Maciejak, P., Lehner, M., Turzynska, D., Szyndler, J., Bidzinski, A., Taracha, E., Sobolewska, A., Walkowiak, J., Skorzewska, A., Wislowska, A., Hamed, A.,

Plaznik, A. (2008). The opposite role of hippocampal mGluR1 in fear conditioning in kindled and non-kindled rats. *Brain Research, 1187*, 184–193.

[215] Ando, N., Morimoto, K., Watanabe, T., Ninomiya, T., Suwaki, H. (2004). Enhancement of central dopaminergic activity in the kainate model of temporal lobe epilepsy: implication for the mechanism of epileptic psychosis. *Neuropsychopharmacology, 29*, 1251–1258.

[216] Goren, M.Z., Aker, R., Yananli, H.R., Onat, F.Y. (2003). Extracellular concentrations of catecholamines and amino acids in the dorsomedial hypothalamus of kindled rats. A microdialysis study. *Pharmacology, 68*, 190–197.

[217] Yananli, H.R., Terzioglu, B., Goren, M.Z., Aker, R.G., Aypak, C., Onat, F.Y. (2008). Extracellular hypothalamic gamma-aminobutyric acid (GABA) and L-glutamic acid concentrations in response to bicuculline in a genetic absence epilepsy rat model. *Journal of Pharmacological Sciences, 106*, 301–309.

7

MICRODIALYSIS IN LUNG TISSUE: MONITORING OF EXOGENOUS AND ENDOGENOUS COMPOUNDS

THOMAS FEURSTEIN AND MARKUS ZEITLINGER
Medical University of Vienna, Vienna, Austria

1. INTRODUCTION

Since microdialysis was established for measuring concentrations of substances within living organisms, the range of capable tissues has expended continuously [1]. The first microdialysis in lung tissue was performed in 1991 in rats, but it was 10 years before a microdialysis probe was inserted into a human lung. This might be explained by two major reasons: (1) ethical and safety concerns, and (2) alternative measurement techniques for lung. In anycase, today the value of microdialysis in lung is uncontested, as for interstitium of lung, no other technique both delivers close-meshed data in order to gain continuous concentration–time profiles and can discriminate between free and bound fractions of drug.

2. SPECIAL ASPECTS ASSOCIATED WITH LUNG MICRODIALYSIS COMPARED TO MICRODIALYSIS IN OTHER TISSUES

Although microdialysis is a semi-invasive technique in general, adverse events associated with the procedure are rare and minor. For example, when

Applications of Microdialysis in Pharmaceutical Science, First Edition. Edited by Tung-Hu Tsai.
© 2011 John Wiley & Sons, Inc. Published 2011 by John Wiley & Sons, Inc.

implanting a microdialysis probe in subcutaneous adipose tissue or peripheral muscle tissue, adverse events are restricted to moderate pain during insertion or to hematoma at the insertion site [2]. Microdialysis in interstitial lung tissue is more complicated because of the protected anatomical position and the high vulnerability of the organ; therefore, insertion of probes may, theoretically, cause severe adverse events. As observed in animal studies, insertion of microdialysis probes through the chest wall may lead to pneumothorax and consecutive collapse of lung [3]. To evade the risk of pneumothorax in human studies, thoracotomy is required, limiting microdialysis experiments in the human lung to patients with elective thoracic surgery. Thus, implementation of lung microdialysis on a broad base (in healthy volunteers) is not possible, due to safety concerns. Since the first human lung microdialysis experiments, a range of studies were performed in humans. For routine reasons, microdialysis probes were always implanted under direct vision at the end of conventionally performed lung surgery. No adverse events or clinical complications related to microdialysis have been reported in any of these studies [4–9].

The lung is a complex organ consisting of various tissues and anatomical structures [10,11]. Even though a probe is inserted under visual control, the probe is in touch not only with interstitial space but also with alveoli, alveolar air, and alveolar lining fluid tangent to the probe. However, membrane exchange will not take place with alveolar air, as rapid diffusion may happen only with the fluid phase [9]. Therefore, the concentrations in microdialysis samples will represent concentrations in the interstitial space and alveolar lining fluid exclusively. In case of visible bleeding at the insertion site, the probe should be withdrawn, as direct diffusion of substances from blood into the perfusion fluid could falsify results.

3. INSERTION OF MICRODIALYSIS PROBES INTO LUNG TISSUE

3.1. Technique in Humans

Lung microdialysis in clinical studies is performed in patients breathing spontaneously during and immediate after thoracic surgery. The insertion of a microdialysis probe in the lung tissue itself takes place during the time when the thorax is open and direct access to the lung is possible (Figure 1). Before closing the chest at the end of a conventionally performed surgical procedure (commonly, removal of tumor tissue or abscess formations), a venous cannula is inserted percutaneously through the chest wall near the thoracotomy. The tip of the flexible microdialysis probe is led into the thoracic cavity through this cannula and inserted into the reinflated lung under vision (often by use of a needle). After successful implantation of the probe, the insertion site is inspected for air leakage or bleeding. Afterward the chest is closed conventionally [4–9]. The probes are perfused continuously with the required solution at a flow rate of 1.5 to 4 µL/min; other flow rates might be chosen during probe

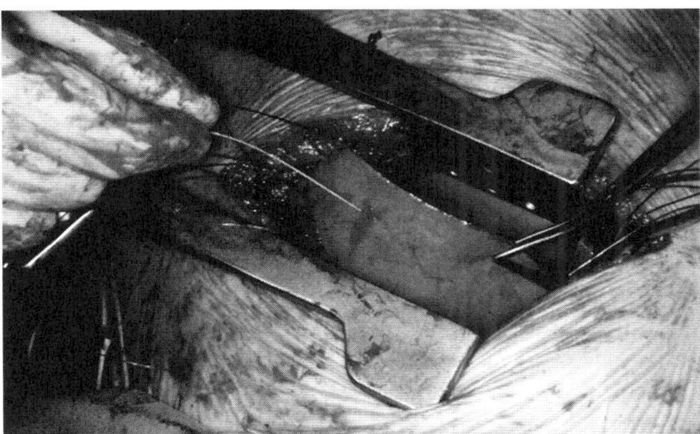

Figure 1   Insertion of a microdialysis probe into the lung under visual control during elective thoracic surgery. The microdialysis experiment started after closing of the chest and after the patient breathed spontaneously.

calibration or when appropriate. Substances present in the intracellular space and alveolar lining fluid will now diffuse out of the intracellular fluid through the semipermeable membrane into the dialysate (for details on the principles of microdialysis, see the general chapters). In most cases the equilibrium between intracellular tissue fluid and the perfusion medium is incomplete, and therefore concentration in tissue is higher than concentration in dialysate. The factor by which the concentrations are interrelated is called *recovery*. Recovery rates observed in human lung microdialysis so far have been high (varying from 43 to 78%), which allowed for accurate calculation of in vivo concentrations [9].

3.2. Technique in Animals

The most common technique for lung microdialysis in animals is to insert a probe into the intermediate lobe of the right lung through a small incision in the pleura. The probe is hold in place with ties. The lobe is put back in place carefully, and the chest is closed before the microdialysis procedure itself is begun.

4. INSERTION OF MICRODIALYSIS PROBES INTO THE BRONCHIAL SYSTEM

Due to its barrier function, the epithelial lining fluid (ELF) represents a compartment of special interest within the lung. It plays an important role in both inflammatory lung diseases and respiratory tract infection (RTI) and is thus a focus of many pharmacokinetic studies. The first step in pulmonary involvement of a respiratory tract infection is often bacterial adherence to the bronchial mucosa. High concentrations of antimicrobial drugs in the mucosa and ELF might inhibit germ adhesion and tissue invasion [10,12]. Therefore, assessing concentrations of substances, especial of antimicrobial agents, in the ELF is of particular interest. Until now, microdialysis studies measuring concentrations of agents in ELF have been performed exclusively in animals. For example, microdialysis probes have been inserted into the trachea of pigs, with the position of the probe confirmed by fiber-optic bronchoscopy [13]. However, most microdialysis studies in ELF take place in rats, in which case microdialysis probes are implanted into the bronchial system via tracheotomy [14] or tracheal puncture [15].

5. TYPES OF PROBES

In human studies, special custom-made flexible microdialysis probes (Metalant AB, Stockholm, Sweden) with a total length of 50 to 60 cm are commonly employed. The flexibility and length of the probe allow for pulmonal movement due to respiration. The tip of the probe usually consists of a 0.6×50 mm polyether sulfone semipermeable membrane with a molecular cutoff of 15,000 Da. Flow rates of 1.5 µL/min are most commonly used [4–9]. Characteristics of the probe as well as flow rates should, of course, be chosen on a case-by-case basis.

Shorter microdialysis probes are employed for small animals: for example, the CMA/20 microdialysis probe [16–23] (CMA, Stockholm, Sweden); a 70-mm microdialysis probe with a 2-mm regenerated-cellulose membrane tip (A-L-70-02, Eicom, Japan) [15]; and CMA/10 microdialysis probe [3,13,23–25]. Some microdialysis studies in animals are performed during cardiopulmonary bypass with cardiac arrest. In such settings, shorter microdialysis probes might also be sufficient for larger animals, due to the limited respiratory movement of the lung during cardiopulmonary bypass [22]. However, the use of shorter microdialysis probes leads to a substantial reduction in recovery rates, which might result in unreliable data [3].

6. ENDOGENOUS COMPOUNDS

Attempts to measure endogenous compounds (e.g., cytokines, hormones) in the lung by microdialysis are rare. Until now, assessing of the body's own substances and functions by lung microdialysis has only been carried out in animals. The majority of microdialysis studies in this field were performed in brain tissue, mostly in rats, rarely in mice or pigs [26–29]. In addition to indicating metabolism in brain tissue, available data demonstrate the power of microdialysis to assess the mucociliary function of bronchial endothelial cells and thus important pathophysiological processes in lung. Mucociliary transport and clearance represent the first line of defense against infections of the lower respiratory tract. The fast evacuation of particles and bacteria is considered a prerequisite for effective host defense. The remarkable role of the mucociliary device in maintaining the physiological condition in the lower respiratory tract can be observed most clearly by its absence in such genetic disorders as primary ciliary dyskinesia and cystic fibrosis, which predispose for RTIs [30,31]. Microdialysis in lung has been used to determine mucociliary transport in normal mice and those with cystic fibrosis [23]. A microdialysis probe was implanted into the lumen of the trachea, and then a small quantity of dye was placed on the surface of a distal bronchus. The distance from the site of the dye and the microdialysis probe was known, and the time until the first detection of the dye in the microdialysis samples was measured, enabling consecutive calculation of the rate of bronchial mucociliary transport.

In a similar setup, the influence of thyme extract, nigellone, and thymoquinone was investigated in the trachea of rats. Fluorescence dyes rhodamin B and rhodamin 123 served as markers. The rate of ciliary clearance was increased markedly by nigellone, while the other substances showed less influence on the mucociliary transport system [32,33]. These experiments demonstrate the power of microdialysis as a method for investigation of mucociliary transport rates influenced by pathophysiology or drugs.

As mentioned earlier, the ELF plays an important role in lung diseases, acute lung injury, and pharmacology. Commonly, samples of ELF are collected by bronchoalveolar lavage (BAL), but this method is invasive and may cause lung injury. Tyvold et al. evaluated bronchial microdialysis as a possible method for continuous monitoring of changes in the composition of the ELF. A microdialysis probe was implanted into the trachea near the carina, and another probe, serving as a reference, was implanted in the subcalvian artery in anesthetized pigs. Lactate, fluorescein–isocyanate–dextran, and urea were determined by microdialysis. Urea was used as an endogenous marker of dilution in samples collected from ELF. By correction of microdialysis results in bronchus by the urea gradient between arterial and bronchial microdialysis probes, total concentrations of endogenous substances in the lower respiratory tract could be measured [13].

In a murine study assessing the value of low-dose chemotherapy, microdialysis was employed to measure soluble biomarker by placing microdialysis probes in lung cancer tissue. By use of a novel technique allowing microdialysis in live freely moving animals for several days, a range of interleukins, growth factors, and other cytokines could be assessed during the course of low-dose chemotherapy [18].

7. EXOGENOUS DRUGS

Most lung microdialysis studies conducted so far investigated the pharmacokinetics of exogenous drugs in lung. The first lung microdialysis study, published in 1991, described the pharmacokinetics of theophylline in lung interstitium [34]. However, subsequent studies focused on the pharmacokinetics of antimicrobial agents in lung tissue and epithelial lining fluid. Lung concentrations of piperacillin, tazobactam, cefaclor, cefpirome, cefpodoxime, imipenem, meropenem, faropenem, levofloxacin, gatifloxacin, ulifloxacin, gentamycin, and tobramycin have been investigated by microdialysis in humans and animals [4–9,14–17,19–21,24,25].

The particular interest in the pharmacokinetics of antimicrobial compounds in the lung might be attributable to the clinical importance of infections in the lower respiratory tract. Over 60% of infections at intensive care units are infections of the lower respiratory tract [35]. In addition, the most common cause of death in the Western world due to infectious diseases is community-acquired pneumonia [36]. An impaired penetration of antimicrobial agents into the site of infection may cause malfunction of antibiotic therapy or lead to development of bacterial resistance [37–39]. For antibiotics, the importance of pharmacokinetic (PK)–pharmacodynamic (PD) relationships for microbiological and clinical efficacy has been documented extensively [40,41]. In the case of pneumonia, a close correlation between the clinical outcome and pharmacokinetic–pharmacodynamic surrogate parameters was demonstrated for humans [42]. Recently, regulatory authorities such as the U.S. Food and Drug Administration (FDA) and the European Agency for the Evaluation of Medicinal Products (EMEA) have therefore focused on tissue concentrations of anti-infective agents rather than plasma concentrations to predict the outcome of antimicrobial therapy [43,44].

8. ANIMAL DATA

Most lung microdialysis experiments were and still are performed in rats, with a few studies on hamsters and horses. Mostly, penetration of antimicrobial agents into interstitial tissue of lung was measured. Substances that have been investigated include β-lactam antibiotics (cefaclor, cefpodoxime, imipenem), aminoglycosides (tobramycin, gentamycin), and fluoroquinolones (gatifloxacin, ulifloxacin, levofloxacin) as well as a mucolytic agent (ambroxol) and a methylxanthine used in therapy for lung diseases (theophylline).

8.1. Cefaclor

A microdialysis study in rats measured and compared the penetration of cefaclor in plasma, lung, and skeletal muscle tissue. Concentration in muscle and lung were determined by microdialysis. The unbound concentrations of cefaclor in the interstitial space fluid of lung and skeletal muscle tissue were of similar quantity but lower than those in plasma. Thus, PK–PD considerations based on plasma concentrations could overvalue the clinical efficacy of cefaclor at its target site [25].

8.2. Cefpodoxime

Cefpodoxime, an oral cephalosporin, was investigated in one lung microdialysis study. Lung microdialysis probes were inserted into the intermediate lobe of the right lung through a small incision in pleura. Again, free concentrations of cefpodoxime in skeletal muscle and lung were similar, as indicated by plasma–tissue penetration ratios of 0.70 and 0.69 for muscle and lung tissue, respectively [20,24].

8.3. Tobramycin and Gentamycin in ELF

Microdialysis probes were installed in the bronchi of rats, and the pharmacokinetics of tobramycin and gentamycin in ELF was determined. Data obtained by microdialysis were comparable to results determined by BAL. This might be explained by the fact that aminoglycosides are known to distribute exclusively in the extracellular space fluid. Therefore, effusion of intracellular amounts of drug should not divert concentrations assessed by BAL as described previously for macrolides and fluoroquinolones (see below) [14].

8.4. Imipenem

Imipenem distribution in muscle and lung was investigated in noninfected and infected rats. Microdialysis probes were inserted into the jugular vein, skeletal muscle, and the right main pulmonary lobe. Free imipenem concentrations were virtually identical in blood, muscle, and lung for both noninfected rats and *Acinetobacter baumannii*–infected rats [19,21].

8.5. Ulifloxacin in ELF

A microdialysis probe was inserted into the bronchial system through tracheal puncture. Concentrations of ulifloxacin and urea (as a reference compound) were measured in ELF. The relative recovery of ulifloxacin and urea were not affected by their concentrations. The ratio of unbound ulifloxacin concentration in ELF to that in plasma was 2.90, confirming that fluoroquinolones accumulate in alveolar macrophages and subsequently in ELF [15].

8.6. Gatifloxacin

Another quinolone for which properties of target-site penetration were investigated by lung microdialysis is gatifloxacin. In peripheral muscle and lung tissue, microdialysis probes were inserted to correlate free tissue and plasma levels. In vitro and in vivo recovery of the compound were independent of its concentration. The area under the curve (AUC) was almost identical in muscle, lung, and plasma, suggesting that free plasma levels are a good surrogate for gatifloxacin levels at the infection site [17]. Comparing these data with the accumulation of ulifloxacin in ELF, it can be concluded that the accumulation of fluoroquinolones observed for ELF does not translate to the interstitium of lung.

8.7. Levofloxacin

Lung levels of levofloxacin following intravenous infusion at steady state were determined by microdialysis in lung, muscle, and blood of rats. Again, free concentrations of levofloxacin in plasma, muscle, and lung were equitable. However, experiments in humans could not fully verify these results (see below) [8,16].

8.8. Ambroxol in ELF

Ambroxol is one of the few compounds not belonging to the antibiotics that has been investigated by microdialysis in lung. The concentrations of ambroxol in plasma and ELF after tracheal administration were compared to those after intravenous administration in 12 rats. The concentration of ambroxol in ELF was measured by microdialysis. The AUCELF/AUC plasma ratio after tracheal administration (1.05 to 2.25) differed considerably from the ratio observed after intravenous administration (0.029 to 0.039). The results indicate that ambroxol administered as an inhalative achieves high concentrations at the targeted site, whereas systemic exposure is low, possibly transferring into good efficiency in the respiratory tract and concomitant low systemic side effects [45].

8.9. Theophylline

Theophylline was the first agent measured by microdialysis in lung. The oldest lung microdialysis study concerning theophylline investigated changes in the recovery of theophylline in lung and blood of rats during the experiment. The variation in recovery with time was significantly lower in lung tissue than in blood, suggesting that determination of the recovery rate at the beginning and end of the experiment might describe sufficiently well the recovery of theophylline in lung [34]. A second study assessed to what extent time–concentration profiles in plasma may reflect the concentration of theophylline in the lung of rats and horses. However, after intravenous application, the free concentration of theophylline in lung was considerably lower than the corresponding free concentration in blood in both rats and horses [3].

9. CLINICAL DATA

Six lung microdialysis studies in humans have been published to date. All were experiments aimed at describing the penetration of antimicrobial agents into lung tissue under normal or specifically defined abnormal conditions. Substances investigated belong to the classes of β-lactams (cefpirome, piperacillin, meropenem) and fluoroquinolones (levofloxacin).

9.1. Cefpirome

The first microdialysis in human lung was conducted in 2001. Cefpirome was chosen as a model β-lactam antibiotic. The study included continuous assessment of cefpirome concentrations in lung tissue after a single intravenous dose in five patients undergoing elective lung surgery. Unbound cefpirome concentrations in lung interstitial fluid were approximately. two-thirds of the corresponding total plasma values. These results confirmed the growing belief that target-site drug levels may be lower than corresponding plasma levels and rebut the former opinion that antibiotics display a complete tissue-to-plasma equilibrium [9].

9.2. Piperacillin and Meropenem

The concentrations of piperacillin in pneumonic lung were measured by inserting microdialysis probes into infected lung tissue. Concentrations of piperacillin in the lung, measured by two separate probes, varied substantially within an individual patient. This suggests that the pneumonic lung is not a homogeneous area. However, concentration obtained by microdialysis might also be influenced by other conditions, such as the distance of the sampling probe to a vessel. Simultaneously determined concentrations of the drug in healthy skeletal muscle tissue showed drug levels in muscle tissue comparable to those measured in infected lung tissue [4]. A similar approach investigated meropenem in inflamed lung tissue. Meropenem rapidly penetrated infected lung tissue but reached only two-fifths of the corresponding plasma concentration. In analogy to piperacillin, maximum concentration of meropenem obtained by two separate probes in one patient suggested a high degree of inhomogeneity with regard to antibiotic levels within pneumonic lung [5].

9.3. Levofloxacin

Three studies assessed the penetration of levofloxacin in lung tissue with microdialysis. The first study demonstrated the applicability of lung microdialysis in patients undergoing open-heart surgery [7]. The later study investigated the impact of atelectasis on the penetration of levofloxacin in lung tissue. For this study five patients undergoing coronary artery bypass grafting (CABG) with cardiopulmonary bypass serve as an "atelectasis model," while five other patients undergoing CABG by the off-pump coronary artery bypass technique represented the control group. The atelectasis model group showed substantially lower tissue concentrations in lung compared to the control group, indicating that atelectasis influences the penetration of antibiotics into tissue.

The authors concluded that clinical dosing schemas of antibiotic therapy should be reevaluated for diseases associated with atelectasis formation [6].

A third study compared the concentration of levofloxacin in lung with that of subcutaneous adipose and skeletal muscle tissue, each measured with microdialysis. The aim of the study was to verify if measurement of the concentration of levofloxacin in easily accessible peripheral soft tissue can predict penetration in lung tissue. Due to the low sample size, formal statistical comparison of lung and peripheral soft tissues was not possible. It should be noted that in this study different populations were used for the assessment of pharmacokinetics in lung and muscle tissue. However, although the pharmacokinetics in plasma was identical for both populations, the study found differences in levofloxacin concentrations between lung and peripheral soft tissues, suggesting that lung microdialysis cannot be replaced by peripheral microdialysis without comparing both techniques for each compound of interest [8].

10. COMPARISON OF PHARMACOKINETIC DATA IN LUNG OBTAINED BY MICRODIALYSIS AND OTHER TECHNIQUES

Early techniques for the determination of agents in the lower respiratory tract include sampling of sputum [46], respiratory secretions [47–49] and pleural fluid [50], and surgical collection of bronchial mucosa [46,51] and whole lung tissue [48,51]. At present, the BAL technique, developed in the late 1960s, is used most commonly. With BAL, both ELF [15,45,52–54] and alveolar macrophages [53,55] can be sampled and, subsequently, analyzed. In addition, imaging procedures such as planar scintigraphy and positron-emission tomography (PET) have been employed for pharmacokinetic research in lung [56–58]. However, the interpretation of data derived from all these methods is limited by several methodical problems:

1. Correlating pharmacokinetic data to a specific anatomical site. Pharmacokinetic data achieved by BAL represent an average concentration obtained from large parts of the lung, as instilled fluids may distribute freely in the bronchial system during the BAL procedure. However, considerable differences in concentrations of antibiotic compounds exist between bronchial secretions and acini in the lung [59].

2. Correlating pharmacokinetic data to a histological site. Imaging techniques such as scintigraphy and PET are not able to discriminate between different histological compartments. Analogically, biopsy with consecutive homogenization of tissue leads to a mixture of concentrations of intra- and extracellular content as well as blood.

3. Correlating the concentration in the sample and the concentration at the target site. During BAL, endogenous markers such as urea are commonly used to estimate the dilution rate due to instilled fluid. However, calculation of the true concentration may be imprecise, particularly if dilution is high (100-fold is common) [54]. Drug release from alveolar macrophages in the sample may

be another potential source of error when one would like to measure concentrations in the ELF [11,12,60].

4. Obtaining concentration–time profiles by the use of repeated measuring. In human tissue, for ethical reasons, biopsies cannot monitor dynamic changes in concentrations of substances in a single subject. Although BAL can be performed several times in one subject, frequent sampling at short intervals would falsify the results, due to alteration of the composition of ELF by saline solution that has been instilled into the lung [54].

5. Discrimination between free and bound fraction. Only the unbound fraction of a compound becomes active [61]. Imaging procedures or biopsies determine the sum of both intra- and extracellular concentrations and, in addition, include those factions of a drug that are bound to interstitial proteins or membrane structures [62].

In contrast, microdialysis enables continuous measurement of unbound drug concentrations at an exact histological site. However, to date there has been no study in which lung microdialysis was carried out simultaneously with a method for the determination of drug concentrations in the lung. Therefore, direct comparison of sampling techniques is difficult. Nevertheless, for some compounds, pharmacokinetic data assessed by different methods in lung are available from various studies. Table 1 presents data obtained in the human respiratory tract using microdialysis and other methods. As different dosing regimens have been employed in some of the studies and highly variable groups of patients have been included, interpretation of data in the Table 1 cannot be without restrictions.

Concentrations determined by microdialysis are commonly higher than corresponding drug levels assessed by other methods. This might be attributed to the fact that except for microdialysis, most maximal concentration values (C_{max} values) were obtained from very few time points. Therefore, the actual maximal concentration easily could have been missed by most techniques. In addition, data assessed by biopsy represent the average of concentrations of different compartments, which may falsify measured concentrations, as discussed previously. For β-lactam antibiotic agents, which are known for their poor cellular penetration, concentrations obtained by biopsy can be expected to be lower than those assessed exclusively in the interstitial space fluid by microdialysis [63]. In contrast, levofloxacin concentrations assessed by biopsy will be higher than those obtained by microdialysis, due to the property of fluoroquinolones to accumulate intracellularly [53].

11. PREDICTABILITY OF LUNG CONCENTRATIONS BY MEASUREMENTS IN OTHER TISSUES

Based on the relative challenge of lung microdialysis and its limitation to animals or restricted patient populations, researchers call for surrogate tissues,

TABLE 1 Maximal Concentration (C_{max}) of Levofloxacin, Meropenem, and Piperacillin in Lung Tissue and Plasma Determined by Different Techniques After Single Dose (SD) Administration or at Steady State (SS)

Antibiotic	Method	C_{max} Lung (mg/L)	Dose (mg)	C_{max} Plasma (mg/L)	No. of Sampling Time Points	No. of Subjects to Pool Data	Refs.
Levofloxacin	Microdialysis[a]	2.5 (2.0–2.9)/ 4.1 (3.7–11.8)	500 (SD)	7.2/ 13.5	12	5/ 5	[6]
	Microdialysis	6.8 (2.2–10.8)	500 (SD)	15.9	12	6	[7]
	Microdialysis	6.0 (4.3–7.7)	500 (SD)	9.4	12	5	[8]
	BAL (ELF)	9.9 ± 2.7	500 (SS)	5.3	3	12	[66]
	Bronchial mucosa biopsy	3.91 ± 2.33	200 (SD)	2.08	1	23	[67]
	Sputum	0.71 ± 0.63	200 (SD)	2.08	1	23	[67]
Meropenem	Microdialysis	11.4 ± 10.9	1000 (SS)	47.3	18	1	[5]
	Lung biopsy	3.9 (0.2–8.2)	1000 (SD)	approx. 20	3	3	[51]
	Bronchial muscosa biopsy	6.6 (3.0–13.3)	1000 (SD)	approx. 20	3	3	[51]
	BAL (ELF)	7.1 ± 2.9	1000 (SD)	26.0	6	6	[54]
	Bronchial secretion	0.53 ± 0.41	1000 (SD)	59.8	3	3	[47]
Piperacillin	Microdialysis	176 ± 105	4000 (SS)	326	12	1	[4]
	Bronchial muscosa biopsy	55.2 ± 12.8	4000 (SS)	196.3	1	1	[46]
	BAL (ELF)	13.6 ± 9.4	4000 (SS)	—	1	1	[68]
	Bronchial secretion	29.3 ± 25.1	4000 (SS)	184.8	6	1	[69]
	Bronchial secretion	31.4 ± 11.3	4000 (SS)	196.3	1	1	[46]
	Sputum	13.0 ± 6.7	4000 (SS)	347.9	3	1	[46]

Source: Adapted from [65].

[a]Two groups of patients were compared. The first values to the right refer to the atelectasis model. For more details, see Sections 7 and 9.

TABLE 2 Comparison of Area Under the Curve (AUC) and Maximal Concentration (C_{max}) Values of Different Antibiotic Agents in Lung and Skeletal Tissue Assessed by Microdialysis in Animal Studies

	AUC_{lung} (mg·h/L)	C_{max} Lung (mg/L)	AUC_{muscle} (mg·h/L)	C_{max} Muscle (mg/L)	AUC_{plasma} (mg·h/L)	C_{max} Plasma (mg/L)
Gatifloxacin (6 mg/kg, rat) [17]	4.1 ± 1.1	1.8 ± 0.8	3.9 ± 0.7	1.4 ± 0.3	3.8 ± 0.6	
Imipenem (120 mg/kg, rat) [21]	261.4 ± 81.8	285.0 ± 80	283.6 ± 33	292.0 ± 25	245.5 ± 30	253.0 ± 27
Cefaclor (50 mg/kg, rat) [25]	25.0 ± 12.9	29.2 ± 10.4	25.8 ± 9.6	28 ± 5.4	169.7 ± 97.9	270.2 ± 62.1
Cepidoxime (20 mg/kg, rat) [20]	175.4 ± 125.5	44.8 ± 28.8	221.2 ± 155.2	33.5 ± 6.6	589.5 ± 417.7	159.8 ± 27.5

TABLE 3 Comparison of Area Under the Curve (AUC) and Maximal Concentration (C_{max}) Values of Different Antibiotic Agents in Lung and Skeletal Tissue Assessed by Microdialysis in Human Clinical Studies

	AUC_{lung} (mg·h/L)	C_{max} Lung (mg/L)	AUC_{muscle} (mg·h/L)	C_{max} Muscle (mg/L)	AUC_{Plasma} (mg·h/L)	C_{max} Plasma (mg/L)
Levofloxacin (500 mg, single dose) [8]	37.8 (33.0–39.3)	6.0 (4.3–7.7)	73.0 (28.7–136.6)	6.3 (2.9–12.9)	56.3 (41.8–89.8)	9.4 (6.4–12.0)
Meropenem (1 g, at steady state, septic patients) [5]	36.2 ± 17.9	11.4 ± 10.9	44.6 ± 30	26.2 ± 25.2	95.4 ± 46.6	47.3 ± 21.0
Cefpirome (2 g)	174 ± 15 [9]	74 ± 24 [9]	163 ± 12 [9]	62 ± 4 [9]	261 ± 24 [9] 267 ± 19 [9]	175 ± 18 [9] 164 ± 14 [70]
Piperacillin (4 g, at steady state, septic patients) [4]	288 ± 167	176 ± 105	197 ± 122	76 ± 22	470 ± 142	326 ± 61

Source: Adapted from [65]; modified and supplemented.

which would allow the prediction of concentrations in the lower respiratory tract. Muscle tissue might be considered to represent the conditions in the lung, due to the similar content of water in muscle and lung tissue [64]. Furthermore, muscle tissue is easily accessible, and microdialysis in skeletal muscle is well implemented in both humans and animals. To date, several animal and human studies have been conducted to compare the concentrations of antimicrobial agents in skeletal muscle and lung tissue obtained by microdialysis.

In animal models, unbound concentrations in the interstitial space fluid of skeletal muscle and lung tissue as well as in plasma have been defined for different antimicrobial compounds. Substances investigated belong to the group of β-lactams (cefaclor, cefpodoxime, imipenem) and fluoroquinolones (gatifloxacin) (Table 2). For all four drugs, AUCs and maximal concentrations in the muscle and lung tissue were virtually identical and suggest that concentrations in the interstitial space fluid in skeletal muscle might be reasonable surrogates for concentrations in the lower respiratory tract [16,17,20,21,25].

Similarly, in humans the β-lactams (piperacillin, cefpirome) [4,9] and meropenem [5] investigated appeared to feature similar pharmacokinetics in peripheral muscle and lung tissue (Table 3). In contrast, levofloxacin showed somewhat different pharmacokinetics in skeletal muscle and the lower respiratory tract. Although similar peak concentrations between muscle and lung were observed, descriptively different AUC values were obtained for levofloxacin for muscle and lung. This was explained by shorter half-life time of levofloxacin in the lung tissue, which may in turn be ascribed to differences in perfusion in the investigated tissues and populations [8].

In summary, skeletal muscle tissue often seems to be a good surrogate to predict concentrations of β-lactams in lung. However, these findings cannot be transferred to all classes of drugs without further investigation. In addition, pathophysiological changes that have an impact on organ perfusion might affect the comparability of muscle and lung tissue.

REFERENCES

[1] Joukhadar, C., Müller, M. (2005). Microdialysis: current applications in clinical pharmacokinetic studies and its potential role in the future. *Clinical Pharmacokinetics, 44,* 895–913.

[2] Müller, M., Schmid, R., Georgopoulos, A., Buxbaum, A., Wasicek, C., Eichler, H.G. (1995). Application of microdialysis to clinical pharmacokinetics in humans. *Clinical Pharmacology & Therapeutics, 57,* 371–380.

[3] Ingvast-Larsson, C., Appelgren, L.E., Nyman, G. (1992). Distribution studies of theophylline: microdialysis in rat and horse and whole body autoradiography in rat. *Journal of Veterinary Pharmacology and Therapeutics, 15,* 386–394.

[4] Tomaselli, F., Dittrich, P., Maier, A., Woltsche, M., Matzi, V., Pinter, J., Nuhsbaumer, S., Pinter, H., Smolle, J., Smolle-Juttner, F.M. (2003). Penetration of piperacillin

and tazobactam into pneumonic human lung tissue measured by in vivo microdialysis. *British Journal of Clinical Pharmacology*, *55*, 620–624.

[5] Tomaselli, F., Maier, A., Matzi, V., Smolle-Juttner, F.M., Dittrich, P. (2004). Penetration of meropenem into pneumonic human lung tissue as measured by in vivo microdialysis. *Antimicrobial Agents and Chemotherapy*, *48*, 2228–2232.

[6] Hutschala, D., Kinstner, C., Skhirtladze, K., Mayer-Helm, B.X., Zeitlinger, M., Wisser, W., Müller, M., Tschernko, E. (2008). The impact of perioperative atelectasis on antibiotic penetration into lung tissue: an in vivo microdialysis study. *Intensive Care Medicine*, *34*, 1827–1834.

[7] Hutschala, D., Skhirtladze, K., Zuckermann, A., Wisser, W., Jaksch, P., Mayer-Helm, B.X., Burgmann, H., Wolner, E., Müller, M., Tschernko, E.M. (2005). In vivo measurement of levofloxacin penetration into lung tissue after cardiac surgery. *Antimicrobial Agents and Chemotherapy*, *49*, 5107–5111.

[8] Zeitlinger, M.A., Traunmuller, F., Abrahim, A., Müller, M.R., Erdogan, Z., Müller, M., Joukhadar, C. (2007). A pilot study testing whether concentrations of levofloxacin in interstitial space fluid of soft tissues may serve as a surrogate for predicting its pharmacokinetics in lung. *International Journal of Antimicrobial Agents*, *29*, 44–50.

[9] Herkner, H., Müller, M.R., Kreischitz, N., Mayer, B.X., Frossard, M., Joukhadar, C., Klein, N., Lackner, E., Müller, M. (2002). Closed-chest microdialysis to measure antibiotic penetration into human lung tissue. *American Journal of Respiratory and Critical Care Medicine*, *165*, 273–276.

[10] Bergogne-Berezin, E. (1995). New concepts in the pulmonary disposition of antibiotics. *Pulmonary Pharmacology & Therapeutics*, *8*, 65–81.

[11] Gail, D.B., Lenfant, C.J. (1983). Cells of the lung: biology and clinical implications. *American Review of Respiratory Disease*, *127*, 366–387.

[12] Baldwin, D.R., Honeybourne, D., Wise, R. (1992). Pulmonary disposition of antimicrobial agents: methodological considerations. *Antimicrobial Agents and Chemotherapy*, *36*, 1171–1175.

[13] Tyvold, S.S., Solligard, E., Lyng, O., Steinshamn, S.L., Gunnes, S., Aadahl, P. (2007). Continuous monitoring of the bronchial epithelial lining fluid by microdialysis. *Respiratory Research*, *8*, 78.

[14] Eisenberg, E.J., Conzentino, P., Eickhoff, W.M., Cundy, K.C. (1993). Pharmacokinetic measurement of drugs in lung epithelial lining fluid by microdialysis: aminoglycoside antibiotics in rat bronchi. *Journal of Pharmacological and Toxicological Methods*, *29*, 93–98.

[15] Aoki, M., Iguchi, M., Hayashi, H., Suzuki, H., Shibasaki, S., Kurosawa, T., Hayashi, M. (2008). Proposal of membrane transport mechanism of protein-unbound ulifloxacin into epithelial lining fluid determined by improved microdialysis. *Biological & Pharmaceutical Bulletin*, *31*, 1773–1777.

[16] Marchand, S., Frasca, D., Dahyot-Fizelier, C., Breheret, C., Mimoz, O., Couet, W. (2008). Lung microdialysis study of levofloxacin in rats following intravenous infusion at steady state. *Antimicrobial Agents and Chemotherapy*, *52*, 3074–3077.

[17] Tasso, L., Bettoni, C.C., Oliveira, L.K., Dalla Costa, T. (2008). Evaluation of gatifloxacin penetration into skeletal muscle and lung by microdialysis in rats. *International Journal of Pharmaceutics*, *358*, 96–101.

[18] Zhong, H., Han, B., Tourkova, I.L., Lokshin, A., Rosenbloom, A., Shurin, M.R., Shurin, G.V. (2007). Low-dose paclitaxel prior to intratumoral dendritic cell vaccine modulates intratumoral cytokine network and lung cancer growth. *Clinical Cancer Research, 13*, 5455–5462.

[19] Dahyot, C., Marchand, S., Pessini, G.L., Pariat, C., Debaene, B., Couet, W., Mimoz, O. (2006). Microdialysis study of imipenem distribution in skeletal muscle and lung extracellular fluids of *Acinetobacter baumannii*–infected rats. *Antimicrobial Agents and Chemotherapy, 50*, 2265–2267.

[20] Liu, P., Fuhrherr, R., Webb, A.I., Obermann, B., Derendorf, H. (2005). Tissue penetration of cefpodoxime into the skeletal muscle and lung in rats. *European Journal of Pharmaceutical Sciences, 25*, 439–444.

[21] Marchand, S., Dahyot, C., Lamarche, I., Mimoz, O., Couet, W. (2005). Microdialysis study of imipenem distribution in skeletal muscle and lung extracellular fluids of noninfected rats. *Antimicrobial Agents and Chemotherapy, 49*, 2356–2361.

[22] Eichler, W., Bechtel, M.J., Klaus, S., Heringlake, M., Hernandez, M., Toerber, K., Klotz, K.F., Bartels, C. (2004). Na^+/H^+ exchange inhibitor cariporide: effects on respiratory dysfunction after cardiopulmonary bypass. *Perfusion, 19*, 33–40.

[23] Grubb, B.R., Jones, J.H., Boucher, R.C. (2004). Mucociliary transport determined by in vivo microdialysis in the airways of normal and CF mice. *American Journal of Physiology: Lung Cellular and Molecular Physiology, 286*, L588–L595.

[24] Liu, P., Müller, M., Grant, M., Webb, A.I., Obermann, B., Derendorf, H. (2002). Interstitial tissue concentrations of cefpodoxime. *Journal of Antimicrobial Chemotherapy, 50*(Suppl.), 19–22.

[25] de la Peña, A., Dalla Costa, T., Talton, J.D., Rehak, E., Gross, J., Thyroff-Friesinger, U., Webb, A.I., Müller, M., Derendorf, H. (2001). Penetration of cefaclor into the interstitial space fluid of skeletal muscle and lung tissue in rats. *Pharmaceutical Research, 18*, 1310–1314.

[26] Pan, Y.F., Feng, J., Cheng, Q.Y., Li, F.Z. (2007). Intracerebral microdialysis technique and its application on brain pharmacokinetic–pharmacodynamic study. *Archives of Pharmacal Research, 30*, 1635–1645.

[27] Borjigin, J., Liu, T. (2008). Application of long-term microdialysis in circadian rhythm research. *Pharmacology Biochemistry and Behavior, 90*, 148–155.

[28] Linthorst, A.C., Reul, J.M. (2008). Stress and the brain: solving the puzzle using microdialysis. *Pharmacology Biochemistry and Behavior, 90*, 163–173.

[29] Neumann, I.D. (2007). Stimuli and consequences of dendritic release of oxytocin within the brain. *Biochemical Society Transactions, 35*, 1252–1257.

[30] Afzelius, B.A. (1976). A human syndrome caused by immotile cilia. *Science, 193*, 317–319.

[31] Robinson, M., Bye, P.T. (2002). Mucociliary clearance in cystic fibrosis. *Pediatric Pulmonology, 33*, 293–306.

[32] Wienkotter, N., Begrow, F., Kinzinger, U., Schierstedt, D., Verspohl, E.J. (2007). The effect of thyme extract on beta2-receptors and mucociliary clearance. *Planta Medica, 73*, 629–635.

[33] Wienkotter, N., Hopner, D., Schutte, U., Bauer, K., Begrow, F., El-Dakhakhny, M., Verspohl, E.J. (2008). The effect of nigellone and thymoquinone on inhibiting trachea contraction and mucociliary clearance. *Planta Medica, 74*, 105–108.

[34] Larsson, C.I. (1991). The use of an "internal standard" for control of the recovery in microdialysis. *Life Sciences*, *49*, PL73–PL78.

[35] Vincent, J.L., Bihari, D.J., Suter, P.M., Bruining, H.A., White, J., Nicolas-Chanoin, M.H., Wolff, M., Spencer, R.C., Hemmer, M. (1995). The prevalence of nosocomial infection in intensive care units in Europe. Results of the European Prevalence of Infection in Intensive Care (EPIC) Study. EPIC International Advisory Committee. *Journal of the American Medical Association*, *274*, 639–644.

[36] Bartlett, J.G., Mundy, L.M. (1995). Community-acquired pneumonia. *New England Journal of Medicine*, *333*, 1618–1624.

[37] Zeitlinger, M.A., Dehghanyar, P., Mayer, B.X., Schenk, B.S., Neckel, U., Heinz, G., Georgopoulos, A., Müller, M., Joukhadar, C. (2003). Relevance of soft-tissue penetration by levofloxacin for target site bacterial killing in patients with sepsis. *Antimicrobial Agents and Chemotherapy*, *47*, 3548–3553.

[38] Joukhadar, C., Frossard, M., Mayer, B.X., Brunner, M., Klein, N., Siostrzonek, P., Eichler, H.G., Müller, M. (2001). Impaired target site penetration of beta-lactams may account for therapeutic failure in patients with septic shock. *Critical Care Medicine*, *29*, 385–391.

[39] Lambert, H.P. (1978). Clinical significance of tissue penetration of antibiotics in the respiratory tract. *Scandinavian Journal of Infectious Diseases Supplementum*, 262–266.

[40] Craig, W.A. (1998). Pharmacokinetic/pharmacodynamic parameters: rationale for antibacterial dosing of mice and men. *Clinical Infectious Diseases*, *26*, 1–10; quiz 11–12.

[41] Drusano, G.L., Craig, W.A. (1997). Relevance of pharmacokinetics and pharmacodynamics in the selection of antibiotics for respiratory tract infections. *Journal of Chemotherapy*, *9*(Suppl. 3), 38–44.

[42] Schentag, J.J. (1990). Correlation of pharmacokinetic parameters to efficacy of antibiotics: relationships between serum concentrations, MIC values, and bacterial eradication in patients with gram-negative pneumonia. *Scandinavian Journal of Infectious Diseases Supplementum*, *74*, 218–234.

[43] European Agency for the Evaluation of Medicinal Products (2000). Points to consider on pharmacokinetics and pharmacodynamics in the development of antibacterial medicinal products. (http://www.emea.eu.int/pdfs/human/ewp/265599en.pdf).

[44] U.S. Food and Drug Administration (1998). Guidance for Industry: Developing antimicrobial drugs—general considerations for clinical trials. (http://www.fda.gov/cder/guidance/2580dft.pdf).

[45] Ren, Y.C., Wang, L., He, H.B., Tang, X. (2009). Pulmonary selectivity and local pharmacokinetics of ambroxol hydrochloride dry powder inhalation in rat. *Journal of Pharmaceutical Sciences*, *98*, 1797–1803.

[46] Marlin, G.E., Burgess, K.R., Burgoyne, J., Funnell, G.R., Guinness, M.D. (1981). Penetration of piperacillin into bronchial mucosa and sputum. *Thorax*, *36*, 774–780.

[47] Bergogne-Bérézin, E., Muller-Serieys, C., Aubier, M., Dombret, M.C. (1994). Concentration of meropenem in serum and in bronchial secretions in patients undergoing fibreoptic bronchoscopy. *European Journal of Clinical Pharmacology*, *46*, 87–88.

[48] Brun, Y., Forey, F., Gamondes, J.P., Tebib, A., Brune, J., Fleurette, J. (1981). Levels of erythromycin in pulmonary tissue and bronchial mucus compared to those of amoxycillin. *Journal of Antimicrobial Chemotherapy, 8,* 459–466.

[49] Klastersky, J., Thys, J.P., Mombelli, G. (1981). Comparative studies of intermittent and continuous administration of aminoglycosides in the treatment of broncho-pulmonary infections due to gram-negative bacteria. *Reviews of Infectious Disease, 3,* 74–83.

[50] Cole, D.R., Pung, J. (1977). Penetration of cefazolin into pleural fluid. *Antimicrobial Agents and Chemotherapy, 11,* 1033–1035.

[51] Byl, B., Jacobs, F., Roucloux, I., de Franquen, P., Cappello, M., Thys, J.P. (1999). Penetration of meropenem in lung, bronchial mucosa, and pleural tissues. *Antimicrobial Agents and Chemotherapy, 43,* 681–682.

[52] Valcke, Y., Pauwels, R., Van der Straeten, M. (1990). The penetration of aminoglycosides into the alveolar lining fluid of rats: the effect of airway inflammation. *American Review of Respiratory Disease, 142,* 1099–1103.

[53] Panteix, G., Harf, R., Desnottes, J.F., Gosselet, H., Leclercq, M., Diallo, N., Couprie, N., Desbos, A., Perrin Fayolle, M., Ballereau, M. (1994). Accumulation of pefloxacin in the lower respiratory tract demonstrated by bronchoalveolar lavage. *Journal of Antimicrobial Chemotherapy, 33,* 979–985.

[54] Allegranzi, B., Cazzadori, A., Di Perri, G., Bonora, S., Berti, M., Franchino, L., Biglino, A., Cipriani, A., Concia, E. (2000). Concentrations of single-dose meropenem (1 g iv) in bronchoalveolar lavage and epithelial lining fluid. *Journal of Antimicrobial Chemotherapy, 46,* 319–322.

[55] Harf, R., Panteix, G., Desnottes, J.F., Diallo, N., Leclercq, M. (1988). Spiramycin uptake by alveolar macrophages. *Journal of Antimicrobial Chemotherapy, 22*(Suppl. B), 135–140.

[56] Le Conte, P., Potel, G., Peltier, P., Horeau, D., Caillon, J., Juvin, M.E., Kergueris, M.F., Bugnon, D., Baron, D. (1993). Lung distribution and pharmacokinetics of aerosolized tobramycin. *American Review of Respiratory Disease, 147,* 1279–1282.

[57] Wollmer, P., Pride, N.B., Rhodes, C.G., Sanders, A., Pike, V.W., Palmer, A.J., Silvester, D.J., Liss, R.H. (1982). Measurement of pulmonary erythromycin concentration in patients with lobar pneumonia by means of positron tomography. *Lancet, 2,* 1361–1364.

[58] Brunner, M., Langer, O., Dobrozemsky, G., Müller, U., Zeitlinger, M., Mitterhauser, M., Wadsak, W., Dudczak, R., Kletter, K., Müller, M. (2004). [^{18}F]Ciprofloxacin, a new positron emission tomography tracer for noninvasive assessment of the tissue distribution and pharmacokinetics of ciprofloxacin in humans. *Antimicrobial Agents and Chemotherapy, 48,* 3850–3857.

[59] Braude, A.C., Hornstein, A., Klein, M., Vas, S., Rebuck, A.S. (1983). Pulmonary disposition of tobramycin. *American Review of Respiratory Disease, 127,* 563–565.

[60] Decre, D., Bergogne-Bérézin, E. (1993). Pharmacokinetics of quinolones with special reference to the respiratory tree. *Journal of Antimicrobial Chemotherapy, 31,* 331–343.

[61] Kunin, C.M. (1967). Clinical significance of protein binding of the penicillins. *Annals of the New York Academy of Sciences, 145,* 282–290.

[62] DeGuchi, Y., Terasaki, T., Yamada, H., Tsuji, A. (1992). An application of micro-dialysis to drug tissue distribution study: in vivo evidence for free-ligand hypothesis and tissue binding of beta-lactam antibiotics in interstitial fluids. *Journal of Pharmacobio-Dynamics*, *15*, 79–89.

[63] Nix, D.E., Goodwin, S.D., Peloquin, C.A., Rotella, D.L., Schentag, J.J. (1991). Antibiotic tissue penetration and its relevance: impact of tissue penetration on infection response. *Antimicrobial Agents and Chemotherapy*, *35*, 1953–1959.

[64] Poulin, P., Theil, F.P. (2000). A priori prediction of tissue: plasma partition coefficients of drugs to facilitate the use of physiologically-based pharmacokinetic models in drug discovery. *Journal of Pharmaceutical Sciences*, *89*, 16–35.

[65] Zeitlinger, M., Müller, M., Joukhadar, C. (2005). Lung microdialysis: a powerful tool for the determination of exogenous and endogenous compounds in the lower respiratory tract. *American Association of Pharmaceutical Scientists*, *7*, E600–8.

[66] Gotfried, M.H., Danziger, L.H., Rodvold, K.A. (2001). Steady-state plasma and intrapulmonary concentrations of levofloxacin and ciprofloxacin in healthy adult subjects. *Chest*, *119*, 1114–1122.

[67] Fujita, A., Miya, T., Tanaka, R., Hirayama, S., Isaka, H., Ono, Y., Koshiishi, Y., Goya, T. (1999). Levofloxacin concentrations in serum, sputum and lung tissue: evaluation of its efficacy according to breakpoint. *Japanese Journal of Antibiotics*, *52*, 661–666.

[68] Boselli, E., Breilh, D., Cannesson, M., Xuereb, F., Rimmele, T., Chassard, D., Saux, M.C., Allaouchiche, B. (2004). Steady-state plasma and intrapulmonary concentrations of piperacillin/tazobactam 4 g/0.5 g administered to critically ill patients with severe nosocomial pneumonia. *Intensive Care Medicine*, *30*, 976–979.

[69] Jehl, F., Muller-Serieys, C., de Larminat, V., Monteil, H., Bergogne-Bérézin, E. (1994). Penetration of piperacillin–tazobactam into bronchial secretions after multiple doses to intensive care patients. *Antimicrobial Agents and Chemotherapy*, *38*, 2780–2784.

[70] Joukhadar, C., Klein, N., Mayer, B.X., Kreischitz, N., Delle-Karth, G., Palkovits, P., Heinz, G., Müller, M. (2002). Plasma and tissue pharmacokinetics of cefpirome in patients with sepsis. *Critical Care Medicine*, *30*, 1478–1482.

8

MICRODIALYSIS IN THE HEPATOBILIARY SYSTEM: MONITORING DRUG METABOLISM, HEPATOBILIARY EXCRETION, AND ENTEROHEPATIC CIRCULATION

YU-TSE WU

Institute of Traditional Medicine, National Yang-Ming University, Taipei, Taiwan

TUNG-HU TSAI

Institute of Traditional Medicine, National Yang-Ming University, and Taipei City Hospital, Taipei, Taiwan

1. INTRODUCTION

1.1. Microdialysis Technique

Microdialysis, originally designed to study dopamine neurotransmission [1], has rapidly become a practical technique for in vivo sampling in pharmacokinetic studies since it allows sampling of exogenous and endogenous substances in the extracellular fluid surrounding the microdialysis probe. The principles of microdialysis are discussed in the literature [2,3]. At least three types of microdialysis applications have been proposed in previous studies. First, microdialysis can be used to introduce a substance into the extracellular space through a microdialysis probe by perfusion of a drug-containing perfusate at a higher concentration, allowing drug molecules to diffuse through the dialysis membrane and assess the target tissue. This technique is sometimes called *reverse microdialysis* [4]. Second, microdialysis can be used for drug molecule sampling by perfusion of an artificial biological solution (e.g., Ringer's solution or artificial cerebrospinal fluid) via a microdialysis probe. Here, drug molecules (or the endogenous substances) diffuse into the probe after drug administration to experimental subjects. This technique allows almost real-time monitoring of drug concentrations when coupled with suitable analytical systems, and high-temporal resolution pharmacokinetic data can be obtained [5]. Finally, microdialysis can be viewed as a sensor to monitor hydroxyl radical levels by in situ derivatization with salicylic acid as a trapping substance. Briefly, hydroxyl radicals react to the benzo ring of the salicylic acid contained in the perfusate, and this reaction results in the formation of 2,3-dihydroxyl benzoic acid and 2,5-dihydroxyl benzoic acid, which are measured using liquid chromatography coupled with electrochemical detection [6,7].

1.2. Pharmacokinetics

Pharmacokinetics describes the kinetics of drug absorption, distribution, metabolism, and excretion. Drug disposition often indicates the description of drug distribution and elimination. The development of biological sampling techniques, analytical methods for the measurement of drugs and metabolites, and procedures that facilitate data collection and manipulation are essential components of the experimental aspect of pharmacokinetics [8]. The major difference between conventional blood sampling and the microdialysis technique for pharmacokinetic studies is that the protein unbound (free-form) drug can be sampled by microdialysis, whereas drug concentrations determined by blood sampling contain bound and unbound drug molecules. In addition, discrete data from conventional blood sampling are likely to restrain the correct interpretation of pharmacokinetic results. More intensive sampling than conventional methods can be realized by microdialysis to acquire pharmacokinetic data with high temporal resolution. In addition, microdialysis can directly assess substance levels in the target tissue, which is very useful for comparing pharmacokinetic and pharmacodynamic responses. This feature adds to the pharmacological relevance, because the unbound drug level generally corresponds with the pharmacodynamic response in tissues. Furthermore, samples are protein-free and can be analyzed readily without the need for further analytical purification, due to the exclusion of protein by the low-molecular-weight cutoff values of the dialysis membranes [5,9].

1.3. Hepatobiliary System

The liver is a vital organ, responsible for many metabolic and excretion processes. The metabolism of drugs and xenobiotics takes place mainly in the liver and can be divided into phase I and phase II reactions. Phase I reactions contain oxidations, reductions, and hydrolyses. For example, cytochrome P450–dependent oxidation reactions include aromatic hydroxylation, aliphatic hydroxylation, epoxidation, oxidative dealkylation, N-oxidation, S-oxidation, deamination, desulfuration, and dechlorination. Cytochrome P450–independent oxidations include flavin monooxygenases, amine oxidases, and dehydrogenations. Reduction reactions include azo reductions, nitro reduction, and carbonyl reductions. Hydrolysis reactions include ester or amide hydrolysis. Usually, phase I metabolism does not lead to a great variation in molecular weight or solubility for the substrate, but it is significant because oxidative reactions add or expose sites where phase II metabolism can subsequently occur [10]. On the contrary, phase II conjugation typically results in an appreciable increase in molecular weight and water solubility, and phase II conjugations include UDP glucuronosyl transferase for glucuronidation, N-acetyltransferase for acetylation, glutathione-S-transferase for glutathione conjugation, acyl-CoA glycine-transferase for glycine conjugation, sulfotransferase for sulfation, transmethylases for methylation, and epoxide hydrolase for water conjugation [11].

The liver also produces and secretes bile, which provides a route for the excretion of bile acids, bilirubin, phospholipid, and cholesterol [12]. The biliary system is a convergent arrangement of canals that begins in the canaliculi, is followed by bile ducts, and ends with the common bile duct (coledochus). Both the structural and functional entirety of the biliary tree and the action of membrane transporters in hepatocytes and cholangiocytes contribute to normal bile secretion [13]. Bile consists primarily of water, with organic and inorganic substances in suspension, dissolved, or in equilibrium between both states. The organic composition of bile includes bile acids, phospholipids, cholesterol, proteins and biliary pigments, while the inorganic substances include electrolytes such as sodium, potassium, calcium, bicarbonate, and chloride [13]. For humans, the molecular-weight level of drugs that are mainly expelled in the bile is generally estimated to be 500 to 600 Da and drugs with molecular weight between 300 and 500 Da are evacuated in both urine and bile. Compounds with molecular weight below 300 Da are almost always excreted via the kidneys into urine. A minimum molecular weight of between 200 and 325 Da is required for significant biliary excretion in the rat, and 400 and 475 Da are the approximate values in the guinea pig and rabbit, respectively [14]. Substances expelled into bile usually need a strong polar group, such as many drug metabolites, and this is supported by the formation of a glucuronide by phase II reactions, which not only increase the molecular weight of the compound by nearly 200 Da, but also increase the polarity. For these drugs, a decrease in one excretory route results in an increase in another excretory route to compensate for the variation, which might also alter the drug's pharmacokinetic behavior [8].

Recently, an extended phase concept for drug transporters in pharmacokinetics has been proposed. For many xenobiotics, excretion in bile, which has been termed a phase III elimination reaction, is well known as a major pathway for the elimination of amphipathic, hydrophobic, and high-molecular-weight xenobiotics, and thus complements the renal elimination of hydrophilic compounds of low molecular weight [15]. Phase IV pathways, comprising the final steps of excretion, have also been proposed. Examples include the bile canaliculus of the liver and secretory steps in the luminal membrane, such as counteracting absorption in the gut. Phase IV is maintained primarily by directly driven uphill transport of drugs across cell membranes, which is accomplished via ATP-binding cassette carriers [16]. The phase III and phase IV pathways involving transporters both indicate the importance of hepatobiliary excretion in pharmacokinetic studies, and transport competition on the carrier leads to potential drug–drug interactions. In rats coadministered with cyclosporin A, the increased plasma concentration of cerivastatin, a 3-hydroxy-3-methylglutaryl coenzyme A inhibitor for the treatment of hypercholesterolemia, has been proven as a result of inhibition of transporter-mediated uptake [17]. In addition, an in vitro report indicates that changes in the uptake transporter function (organic anion transporting polypeptide 1B1 and organic anion transporting polypeptide 1B3) by oral antidiabetic drugs such as repaglinide should be considered as potential mechanisms for drug–drug interactions [18]. Similarly, these macrolide antibiotics (e.g., erythromycin and clarithromycin) have also been found to inhibit the transporter-mediated uptake of pravastatin in a concentration-dependent manner in cells expressing organic anion transporting polypeptides [19]. Furthermore, the biliary excretion of intravenously administered valsartan, a highly selective angiotensin II AT1-receptor antagonist for the treatment of hypertension, is severely damaged in multidrug resistance–associated protein 2–deficient Eisai hyperbilirubinemic rats compared with normal Sprague–Dawley rats, which may be the result of stopping the efflux transporter responsible for the efficient hepatobiliary transport of valsartan [20].

1.4. Enterohepatic Circulation

Enterohepatic circulation is a phenomenon in which drugs entering the intestinal tract are first absorbed into the portal vein of intestinal cells, extracted from blood into liver cells, secreted into the bile, and then dropped back into the intestinal lumen, where they may be absorbed again by enterocytes and available for circulating [14]. This reabsorption process may alter the absorption and elimination processes, and the biliary secretion process may become saturated, thus shifting the plasma drug concentration–time curve. A characteristic of drugs undergoing enterohapatic circulation is that a small secondary peak appears from time to time in the plasma drug concentration–time curve [21–23]. The first peak appears when the drug is depleted in the gastrointestinal tract; the secondary peak appears when biliary-excreted drug is again absorbed from the intestinal tract [8]. The pairing of baicalin and its aglycone, baicalein, is a good example of enterohepatic recirculation by the linked-rat model [24]. The area under the concentration–time curve of the recipient rats was compared to that in the donor rats to assess the extent of enterohepatic circulation taking place in the paired rats. The percentages of recycling were approximately 4.9 and 13.3% for baicalin and its conjugated metabolites, respectively, after intravenous administration of baicalin. The percentages of exposure to baicalin and its conjugated metabolites were 18.7 and 19.3%, respectively, after oral administration of baicalin. Similarly, whether *trans*-resveratrol in its aglycone and glucuronide forms underwent enterohepatic recirculation has been studied [25]. Enterohepatic recirculation was evaluated in the linked-rat model because significant plasma concentrations of resveratrol and its glucuronide metabolites were found in the bile of the recipient rats at 4 to 8 h. According to the area under the concentration–time curve (from time 0 to the last measurable plasma concentration) of bile-donor and bile-recipient rats, the percentages of exposure because of enterohepatic circulation were calculated to be 24.7 ± 15.1 and 24.0 ± 8.5% for resveratrol and its glucuronide metabolites, respectively. However, it remains to be clarified if this enterohepatic recirculation contributes significantly to the overall pharmacological activity of resveratrol.

2. EXPERIMENTAL CONSIDERATIONS OF PHARMACOKINETIC STUDIES

Conventional methods of investigating biliary excretion of a drug collect bile juice periodically over a long time, so body fluid loss and disruption of the bile flow are inevitable problems. The major advantages of microdialysis sampling for pharmacokinetic study include offering clean dialysate without a further cleanup procedure for analysis, allowing continuous monitoring of drug kinetics, causing almost no biological fluid loss during continuous sampling and providing highly temporal and spatial resolution [5]. The temporal and spatial resolution accomplished using microdialysis sampling gives in vivo information that was not readily obtainable by other staining or sampling methods. Regional metabolic differences in phenol have been examined using three concurrent microdialysis probes located in the middle lobe of the rat liver [26]. Designs for microdialysis probe include concentric cannula, flexible, linear, and shunt probes [27]. Sampling in heterogeneous target tissues such as the brain usually requires the use of concentric microdialysis probes to achieve good spatial resolution within a specific region. In addition, linear microdialysis probes are used for homogeneous target tissues such as skin and muscle to enhance recovery [5]. Furthermore, a shunt microdialysis probe is designed for sampling from moving fluids such as bile [28].

Calibration methods for the recovery (e.g., extraction efficiency) of microdialysis sampling have been reviewed [2,29]. Retrodialysis is considered the most acceptable method in recent in vivo pharmacokinetic studies [30–32]. Other calibration methods for microdialysis sampling include in vitro procedures [33,34], no-net-flux prodecures [35–37], dynamic no-net-flux procedures [38,39], and the use of reference substances for recovery assessments [40,41]. Generally, to increase recovery, one can increase the dialysis membrane, decrease the perfusate flow rate, and add some enhancers to maximize the extraction efficiency. However, these changes have some drawbacks. First, the position one wishes to sample cannot be located accurately using a probe with a long dialysis membrane, which results in lower spatial resolution. Second, suitable analytical systems must be chosen because the slow flow rate will lead to a very small sample volume. Finally, the addition of recovery enhancers may alter the osmotic property of the perfusion solution (e.g., Ringer's solution or artificial cerebrospinal fluid). Therefore, the parameter setting should be purpose-driven to maximize the benefit of microdialysis in the particular work.

2.1. Liver Sampling

Incubation with both isolated microsomes [42–44] and cultured hepatocytes [45,46] has been used to study the metabolism of drugs and xenobiotics. Microdialysis provides an alternative way to monitor drug metabolism. Scott et al. [47] have applied a microdialysis probe to the intact, in-place liver of a killed rat to investigate phenol metabolism and its conjugation by glutathione. By serving as its own control for each subject, this method provides the experimenter with some advantages, including maintaining the integrity of the organism and the physiological conditions and reflecting more accurately the metabolic processes of xenobiotic compounds. Different results were found between the in situ microdialysis experiments and the data obtained in vitro by liver homogenate and liver-microsomal protein. The pharmacokinetics of naringin, a predominant bioflavonoid in grapefruit, has been investigated by our group by simultaneous microdialysis sampling in rat liver, bile, brain, and blood [48]. In addition, the liver distribution and hepatobiliary of berberine have been evaluated in our lab using concurrent blood, liver, and bile microdialysis sampling of high-performance liquid chromatography (HPLC) [49]. Clinically, Nowak et al. [50] have determined glucose, pyruvate, lactate, and glycerol by microdialysis coupled with an enzymatical analyzer to monitor the tissue metabolism in transplanted liver. The authors concluded that the microdialysis procedure is safe and valuable in monitoring such pathological changes in a liver graft as thrombosis or early rejection.

Two types of microdialysis probes are used regularly for liver sampling. One is a flexible concentric probe, which is identical to the probe used for sampling in blood vessels and can be located in different lobes to monitor metabolic differences among them [51]. The other is a simple linear probe, constructed with fused silica as both the inlet and outlet tubing, and containing dialysis membrane between the silica tubes [47]. For liver probe implantation, the upper abdomen of the anesthetized rat was trimmed and cleaned by 70% (v/v) ethanolic solution or other disinfectant, and a midline incision was made at the xiphoid, extending approximately 2 to 4 cm posterior to expose the rat liver. A concentric probe was implanted in the liver lobe toward the head longitudinally and fixed to the abdominal muscle by a thread [5]. An equilibrium period of approximately 1 h was needed to clear the extracellular fluid space of substances released from cellular damage caused by the implantation procedure after probe insertion [52]. Although continuous pharmacokinetic information can be obtained from the same animal, long-term implantation of the probe in the liver tissue inevitably causes some cell necrosis at the implantation site after 12 h [53].

2.2. Bile Sampling

Clinically, patients suffering from gallbladder disease requiring resection of the gallbladder have been enrolled in studies to examine the extent of the biliary excretion of drugs. Patients received the drug for investigation before surgery, and the amount of the drug found in the gallbladder after removal was measured [54–57]. The drawback to this approach is that data can be obtained at only a single time point. Another way to investigate biliary excretion of drugs is to utilize a temporary bile shunt (T-tube) that diverts bile from the liver to collect bile samples from patients [58–61]. In addition, multichannel tubes have been employed in healthy volunteers to aspirate pancreatic–biliary secretions from the duodenum. Oroenteric tubes were employed to examine the pharmacokinetics of ^{99m}technetium mebrofenin [62] and piperacillin metabolism in humans [63]. Double lumen tubes positioned in the stomach and intestine were used to assess the human biliary secretion of [^{14}C] felodipine metabolites [64]. A Loc-I-Gut tube placed in the distal duodenum/proximal jejunum was used to determine the biliary secretion of rosuvastatin in healthy volunteers [65].

The most common way to investigate biliary excretion in animal studies is to collect bile periodically. Sugie et al. [66] compared the biliary excretion of azithromycin in Sprague–Dawley rats and male Eisai hyperbilirubinemic rats, and they also used bile collection to investigate the effects of cyclosporine and probenecid on biliary clearance of azithromycin in Wistar rats. Takayanagi et al. [67] evaluated biliary excretion of olmesartan, an anigotensin II receptor antagonist, in the normal rat and the multidrug-resistant protein 2–deficient rat. Bile samples were collected via the common bile duct cannulation with a PE-10 tube, and infusion through the femoral vein with a solution containing 3% human serum albumin in a 5% glucose solution during the experiment was used to compensate for the loss of body fluid. The biliary excretion of pravastatin [68] and the influence of genipin on the biliary excretion of cholephilic compounds in rats [69] have been estimated by the same bile sampling protocols.

The design for a shunt probe consisting of a linear dialysis probe inside a piece of plastic tubing was first proposed by Scott and Lunte [28]. For the most part, this design corresponded to the needs of bile sampling, and continuous bile sampling was accomplished with minimal fluid loss compared with that of the periodical bile collection described previously. Bile duct canulation has to be conducted before shunt microdialysis probe implantation. Procedures for bile duct catheterization are described in the method reported [70]. Briefly, a midline incision from the xiphisternum to the umbilicus was made after cleaning the surgical sites. The duodenal loop covered with liver sections was identified and exteriorized through the incision region to expose the bile duct. A cannulation site of the bile duct from the end of the liver was selected to avoid major blood vessels, and the duct was isolated from the surrounding vessels and tissue. A small cut was made to insert PE-10 tubing toward the liver, and bile flowing from the cannula can be observed. Similar procedures were carried out to insert a second PE-10 tube into the bile duct toward the small intestine, both ligatures were tightened, and the two cannulas were then connected to each other by an adaptor to maintain the normal bile flow. The duodenum was returned to the abdominal cavity. Connection of the bile probe to the bile duct cannula was discussed in an earlier paper [5]. Briefly, rat bile streamed from the common bile duct cannula into the microdialysis shunt probe, which was perfused with Ringer's solution (composition: 147 mM Na^+, 2.2 mM Ca^{2+}, 4 mM K^+, pH 7.0), and then flowed into the duodenum through the PE-10 tubing.

Bile sampling using microdialysis provides the experimenter with the following advantages. First, sampling from bile provides firsthand information regarding the hepatobiliary excretion and metabolism of drugs. Second, more normal bile composition is maintained even in long-term tests than with a conventional periodical bile collection method, which brings severe loss of bile salts and may obtain incorrect pharmacokinetic data. Homeostatic bile composition is preserved even with long-term sampling because the bile flow returns to the intestinal tract when using the shunt microdialysis probe [27]. Sometimes, bile salts have been added to the Ringer's solution as a perfusing solution to diminish osmotic differences between perfusate and bile [71]. Finally, microdialysis can easily be automated for pharmacokinetic studies.

Online microdialysis and HPLC systems have been used to determine meropenem in rat bile [72] and diclofenac in rat bile [73]. In our lab we have utilized a shunt microdialysis probe to assess the hepatobiliary excretion of selected antibiotic and antifungal drugs in male Sprague–Dawley rats (Table 1).

2.3. Enterohepatic Circulation by the Paired-Rat Model

A common method used to investigate the enterohepatic circulation is a paired-rat model. To prepare to use a paired-rat model (also called a linked-rat model), the bile duct cannulation procedures described in Section 2.2 must be conducted. After bile cannulation of both rats, the distal free end of the PE-10 cannula of the first rat (donor rat) is placed into the duodenum of the second rat (recipient rat), and the distal free end of the PE-10 cannula of the second rat (recipient rat) is inserted into the duodenum of the first rat, which balances the fluid between the donor and recipient rats [5]. The drug is then administered intravenously to the donor rat in a femoral vein cannula after a 2-h

TABLE 1   Selected Antibiotic and Antifungal Drugs Undergoing Hepatobiliary Excretion Assessed by a Shunt Microdialysis Probe in Male Sprague–Dawley Rats

Antibiotics	Transport by P-Glycoprotein	P-Glycoprotein Inhibitor Used	Ref.
Cefepime	Not investigated	—	[84]
Cefoperazone	(−)[a]	Berberine	[85]
Chloramphenicol	Not investigated	—	[86]
Fluconazole	(−)[a]	Cyclosporin A	[87]
Levofloxacin	(−)[a]	Cyclosporin A	[88]
Meropenem	Not investigated	—	[72]
Metronidazole	(−)[a]	Cyclosporin A	[89]
Pefloxacin	(−)[a]	Cyclosporin A	[90]

[a]Might not be related to.

surgical stabilization, and blood samples are collected for analysis. Blood collection causes body fluid loss in the experimental animal, so a blood microdialysis sample is taken at the jugular vein of the donor and recipient rats to avoid further body fluid loss. Combining the paired-rat model and microdialysis sampling allows multiple-site sampling without undue stress and biological fluid consumption [5]. The area under the concentration–time curve (AUC) in the recipient rats compared with the AUC in the donor rats (i.e., $AUC_{recipient}/AUC_{donor}$) is used as a recycling index and to quantitatively describe the degree of enterohepatic circulation occurring in the paired rats [74].

The biliary recirculation of droloxifene, a nonsteroidal estrogen agonist/antagonist developed for the treatment of osteoporosis and estrogen receptor–positive breast cancer, has been characterized in rats [75]. That study developed a modified linked-rat model for enterohepatic recirculation assessment and found that about 5% of droloxifene is subject to enterohepatic recirculation in the rat. Similarly, the possible impact of enterohepatic recirculation in extension of the elimination half-life of doxorubicin, an anthracycline antibiotic with a wide spectrum of antitumor activity and a long elimination half-life, was evaluated using a linked-rat model [76]. The authors concluded that approximately 22% of the dose administered to the donor rat was absorbed from the intestine of the receiver rat, which indicated that the cumulative cardiac toxicity and/or the increased efficacy of doxorubicin may result from the contribution of enterohepatic recirculation of the drug and its metabolites. The paired-rat model is used further to describe the mechanism of gastrointestinal disorders caused by mycophenolate mofetil [77] and to investigate the factors influencing the pharmacokinetics of tanshinone IIA [78]. In our lab we employed the paired-rat model with blood microdialysis to estimate the degree of enterohepatic circulation of chloramphenicol [74]. We found that the extent of recycling represented by the recycling index ($AUC_{recipient}/AUC_{donor}$) is around 1.8 and 4.9% for chloramphenicol and chloramphenicol glucuronide, respectively, after administering chloramphenicol to the donor rat. These examples demonstrate that this model is useful for preclinical studies of the enterohepatic circulation of drugs.

3. PHARMACOKINETIC AND HEPATOBILIARY EXCRETION STUDIES EMPLOYING MICRODIALYSIS

3.1. Hepatobiliary Excretion

Chen et al. [79] have proposed a microdialysis sampling method to study the pharmacokinetics of protein-unbound salvianolic acid B, a water-soluble polyphenolic antioxidant isolated from the root of *Salvia miltiorrhiza* Bunge (*Dan-Shen* in Chinese). A blood microdialysis probe was positioned within the rat's jugular vein/right atrium and perfused with anticoagulant solution (consisting of 3.5 mM citric acid, 7.5 mM sodium citrate, and 13.6 mM dextrose). Rat bile streamed from the common bile duct cannula into the microdialysis shunt probe, which was perfused with Ringer's solution, and then flowed into the duodenum. Salvianolic acid B (100 mg/kg) was administered intravenously by a femoral vein cannula after a 2-h stabilization following probe implantation. Each microdialysis sample was analyzed immediately using an online HPLC system. The results demonstrate that the concentration of salvianolic acid B in bile reaches a maximum concentration at around 30 min, which suggests an active transport of salvianolic acid B from blood vessels through the liver into bile duct. The hepatobiliary excretion of salvianolic acid B was defined as the blood-to-bile distribution ratio (AUC_{bile}/AUC_{blood}, where AUC represents the area under the concentration–time curve), which was 1.55 ± 0.21. This work revealed the hepatobiliary excretion of salvianolic acid B.

Wu and Tsai [80] explored the effects of silibinin, a polyphenolic flavonoid from the seeds of milk thistle, on the biliary excretion of pyrazinamide, a drug with hepatotoxicity used for treating active tuberculosis, in rats pretreated orally with silibinin (100 mg/kg/daily) for three consecutive days. A bile duct shunt microdialysis probe was constructed and used as described earlier. For post–bile duct cannulation, the microdialysis probe was then perfused with Ringer's solution at a 2.4-mL/min flow rate. According to the results, the hepatobiliary elimination of pyrazinoic acid may be affected by silibinin in the groups of long-term silibinin exposure and concomitant short-term silibinin exposure. The blood-to-bile distribution ratio (AUC_{bile}/AUC_{blood}) of pyrazinoic acid was reduced significantly in pyrazinoic acid + long-term silibinin exposure and the pyrazinoic acid + concomitant short-term silibinin exposure groups. After pyrazinamide administration, the blood, but not bile, levels of pyrazinoic acid were markedly raised in the pyrazinamide + long-term silibinin exposure and pyrazinamide + concomitant short-term silibinin exposure groups, but the blood-to-bile ratio of pyrazinoic acid was decreased. These results suggest that the excretion route of pyrazinoic acid may be blocked by the silibinin through xanthine oxidase and hepatobiliary excretion. In our lab

TABLE 2 Selected Herbal Components Undergoing Hepatobiliary Excretion Assessed by a Shunt Microdialysis Probe in Male Sprague–Dawley Rats

Herbal Compound	Source	Transport by P-Glycoprotein	P-Glycoprotein Inhibitor Used	Ref.
Baicalein	*Scutellaria baicallensis*	(+)[a]	Cyclosporin A	[91]
Baicalin	*Scutellaria baicallensis*	(−)[b]	Cyclosporin A	[92]
Berberine	*Berberis aristata* or *Coptis chinensis*	(+)[a]	Cyclosporin A	[82]
Esculetin	*Artemisia scoparia*	Not investigated	—	[93]
Gastrodin and HBA[c]	*Gastrodia elata* Blume	Not investigated	—	[94]
Genistein	Soybeans	(−)[b]	Cyclosporin A	[95]
Hesperidin	Citrus species	(+)[a]	Cyclosporin A	[96]
Kadsurenone	*Piper kadsura*	(−)[b]	Cyclosporin A	[97]
Mangiferin	*Anemarrhena asphodeloides* Bung	Not investigated	—	[98]
Naringin	Grapefruit	(−)[b]	Cyclosporin A	[48]
Salvianolic acid B	*Salvia miltiorrhiza* Bunge	Not investigated	—	[79]
Sinomenine	*Sinomenium acutum*	(+)[a]	Cyclosporin A	[99]

[a]May be related to.
[b]Might not be related to.
[c]*p*-Hydroxybenzyl alcohol.

we utilized a shunt microdialysis probe to assess the hepatobiliary excretion of selected herbal components in male Sprague–Dawley rats (Table 2).

3.2. Drug Undergoing Enterohepatic Circulation

A paired-rat model including one drug-treated donor rat and another blank recipient rat permits a more accurate evaluation of the enterohepatic circulation of chloramphenicol with minimal physiological disturbance [74]. The bile duct of the donor was cannulated proximal to the liver with PE-10 polyethylene tubing, the other end of which was inserted through the bile duct toward the duodenum of the recipient rat. The bile duct cannula of the recipient rat guides bile back to the donor rat in order to balance the fluid losses and gains of the rats. It must be noted that the donor and recipient rats have to be prepared simultaneously by trained operators in order to maintain similar degrees of recovery from surgery.

Chen et al. [81] investigated the pharmacokinetics of unbound colchicine in the rat and the interaction of unbound colchicine with cyclosporin A and proadifen using a microdialysis and liquid chromatographic assay method. The use of colchicine, an alkaloid derived from the plant *Colchicum autumnale*, which possesses antimitotic and anti-inflammatory characteristics, in the treatment of liver cirrhosis and primary biliary cirrhosis has been reported. The paired-rat model for investigating the enterohepatic circulation was designed as follows. A donor rat (for drug administration) and a recipient rat (no drug administration), with age and weight (280 to 320 g) matched, were first anesthetized with urethane 1 g/mL and -chloralose 0.1 g/mL (1 mL/kg, intraperitoneally). The bile duct of the donor rat was cannulated proximal to the liver with a 20-cm section of PE-10 tubing (inside diameter 0.28 mm × outside diameter 0.61 mm). The other end of the tubing was inserted through the bile duct into the duodenum of the recipient rat. To maintain the balance between fluid losses and gains in the donor and recipient rats, another bile duct cannula of the recipient rat was used to channel bile back to the donor rat. Colchicine (10 mg/kg) and cyclosporin A (20 mg/kg) were coadminisered intravenously into the femoral vein of the donor rat after the surgical procedures. The dialysates were collected from the jugular vein of the donor and recipient rats for additional assay by liquid chromatography. The high hepatobiliary excretion rate of colchicine indicates the potential for the drug to undergo enterohepatic circulation. Colchicine concentration in the blood declined in the donor rat, although the colchicine level increased in the recipient rat after colchicine and cyclosporin A administration. The results show that the area under the concentration–time curve of colchicine was 847.7 ± 141.0 and 55.5 ± 29.2 min·mg/mL in the donor and recipient rats, respectively. These results indicate that the hepatobiliary excretion of colchicine was regulated by P-glycoprotein and that the related acute diarrhea could be reduced by cyclosporin A. The enterohepatic circulation of colchicine was observed successfully using the paired-rat model.

3.3. Transporter-Involved Biliary Excretion of Drugs

Extracts of *Berberis aristata* and *Coptis chinensis* have been used in traditional oriental medicine for the treatment of gastrointestinal disorders, and berberine is one of the major alkaloids derived from their roots and bark. To investigate detailed hepatobiliary excretion mechanisms of berberine, an in vivo microdialysis coupled with HPLC has been conducted [82]. The blood and liver microdialysis probes were located separately within the jugular vein/right atrium and the hepatic middle lobe, and perfused with the anticoagulant citrate dextrose solution. Then the shunt bile probe connected with a bile duct catheter was perfused with Ringer's solution. Cyclosporin A (20 mg/kg used as a P-glycoprotein inhibitor), quinidine (10 mg/kg, used as an inhibitor of P-glycoprotein and organic cation transporter), SKF-525A (10 mg/kg, used as a nonspecific P450 inhibitor), and probenecid (100 mg/kg, used as a glucuronidation inhibitor) were examined to explore the metabolic and hepatobiliary excretion mechanisms of berberine. Since the value of the area under the concentration–time curve of berberine in bile was reduced significantly after the coadministration of cyclosporin A, it can be concluded that the bile efflux transport system responsible for berberine was probably regulated by P-glycoprotein. Similarly, coadministration of quinidine resulted in a relative reduction in the berberine level of bile, which revealed that biliary excretion of berberine might be affected by either P-glycoprotein or an organic cation transport inhibitor. The authors concluded that cyclosporin A, quinidine, and SKF-525A all interact with berberine and, additionally, that P-glycoprotein and P450 may play important roles in the regulation of hepatobiliary excretion and liver metabolism of berberine in rats.

Exploration of the mechanism related to the hepatobiliary excretion of ranitidine, a H2-receptor antagonist used for peptic ulcer, was investigated by profiling the pharmacokinetics of ranitidine alone and with the concomitant administration of cyclosporin A and quinidine, respectively, in rats [83]. The groups studied were as follows: the control group received ranitidine (10 and 30 mg/kg) intravenously; the second group was coadministered with ranitidine and cyclosporin A (10 mg/kg, used for P-glycoprotein inhibition); the third group received ranitidine with concurrent administration of quinidine (10 mg/ kg, used for P-glycoprotein and organic cation transporter inhibition). Blood and bile fluid sampling were achieved by microdialysis probes implanted into the jugular vein and connected to the bile duct cannula, respectively. Ranitidine concentrations in the microdialysis samples were determined by HPLC with ultraviolet detection. Ranitidine reached a maximum concentration in the bile between 20 and 30 min after drug administration. The mean value of area under the concentration–time curve of ranitidine in bile was 2460 min·mg/mL, and it decreased to 321 and 655 min·mg/mL after treatment with cyclosporine and quinidine, respectively. According to these results, the hepatobiliary excretion of ranitidine was partially controlled by P-glycoprotein or an organic cation transporter.

4. CONCLUSIONS

In this chapter we have described the use of microdialysis in pharmacokinetic studies related to the hepatobiliary system. Microdialysis sampling is a very useful tool for pharmacokinetic studies, providing continuous drug concentration monitoring without excessive body fluid loss and yielding data with high temporal and spatial resolution. This technique is advantageous for investigating drug metabolism by maintaining the integrity of the organism and the physiological conditions and reflecting more accurately the metabolic processes of drugs compared with isolated liver microsomes and cultured hepatocytes. Introduction of the shunt probe promotes the use of microdialysis to determine the hepatobiliary excretion and enterohepatic circulation. Biliary excretion of different antibiotics has been performed, and the information may help to determine the feasibility of treatment for biliary tract infections. In addition, it is important to identify the contribution of biliary clearance within total systemic clearance and to understand the cause of drug-induced hepatotoxicity, especially that caused by medicinal herbs and food supplements. Furthermore, exploring transports involved in biliary excretion can provide warnings of potential drug–drug interactions. Overall, microdialysis has become a widely recognized technique for pharmacokinetic studies, and its importance for clinical medicine and new drug development is increasing.

REFERENCES

[1] Ungerstedt, U., Pycock, C. (1974). Functional correlates of dopamine neurotransmission. *Bulletin der Schweizerischen Akademie der Medizinischen Wissenschaften*, *30*, 44–55.

[2] Plock, N., Kloft, C. (2005). Microdialysis: theoretical background and recent implementation in applied life-sciences. *European Journal of Pharmaceutical Sciences*, *25*, 1–24.

[3] Joukhadar, C., Müller, M. (2005). Microdialysis: current applications in clinical pharmacokinetic studies and its potential role in the future. *Clinical Pharmacokinetics*, *44*, 895–913.

[4] Hocht, C., Opezzo, J.A., Taira, C.A. (2007). Applicability of reverse microdialysis in pharmacological and toxicological studies. *Journal of Pharmacological and Toxicological Methods*, *55*, 3–15.

[5] Tsai, T.H. (2003). Assaying protein unbound drugs using microdialysis techniques. *Journal of Chromatography B: Analytical Technologies in the Biomedical and Life Sciences*, *797*, 161–173.

[6] Tsai, T.H., Cheng, F.C., Hung, L.C., Chen, C.F. (1999). Measurement of hydroxyl radical in rat blood vessel by microbore liquid chromatography and electrochemical detection: an on-line microdialysis study. *Journal of Chromatography B: Biomedical Sciences and Applications*, *734*, 277–283.

[7] Cheng, F.C., Jen, J.F., Tsai, T.H. (2002). Hydroxyl radical in living systems and its separation methods. *Journal of Chromatography B: Analytical Technologies in the Biomedical and Life Sciences*, *781*, 481–496.

[8] Shargel, L., Yu, A.B.C. (1999). *Applied Biopharmaceutics and Pharmacokinetics*, McGraw-Hill, New York.

[9] Kreilgaard, M. (2002). Assessment of cutaneous drug delivery using microdialysis. *Advanced Drug Delivery Reviews*, *54*(Suppl. 1), S99–S121.

[10] Zamek-Gliszczynski, M.J., Hoffmaster, K.A., Nezasa, K., Tallman, M.N., Brouwer, K.L. (2006). Integration of hepatic drug transporters and phase II metabolizing enzymes: mechanisms of hepatic excretion of sulfate, glucuronide, and glutathione metabolites. *European Journal of Pharmaceutical Sciences*, *27*, 447–486.

[11] Katzung, B.G. (2007). *Basic and Clinical Pharmacology*, McGraw-Hill, New York.

[12] Erlinger, S. (1982). Bile flow. In: Arias, I.M., Popper, H., Jacoby, W.B., Schachter, D., Shafritz, D.A. (Eds.), *The Liver: Biology and Pathobiology*, Raven Press, New York, pp. 407–427.

[13] Esteller, A. (2008). Physiology of bile secretion. *World Journal of Gastroenterology*, *14*, 5641–5649.

[14] Roberts, M.S., Magnusson, B.M., Burczynski, F.J., Weiss, M. (2002). Enterohepatic circulation: physiological, pharmacokinetic and clinical implications. *Clinical Pharmacokinetics*, *41*, 751–790.

[15] Vore, M. (1994). Phase III elimination: another two-edge sword. *Environmental Health Perspectives, 102*, 422–423.

[16] Petzinger, E., Geyer, J. (2006). Drug transporters in pharmacokinetics. *Naunyn-Schmiedeberg's Archives of Pharmacology, 372*, 465–475.

[17] Shitara, Y., Horie, T., Sugiyama, Y. (2006). Transporters as a determinant of drug clearance and tissue distribution. *European Journal of Pharmaceutical Sciences, 27*, 425–446.

[18] Bachmakov, I., Glaeser, H., Fromm, M.F., Konig, J. (2008). Interaction of oral antidiabetic drugs with hepatic uptake transporters: focus on organic anion transporting polypeptides and organic cation transporter 1. *Diabetes, 57*, 1463–1469.

[19] Seithel, A., Eberl, S., Singer, K., Auge, D., Heinkele, G., Wolf, N.B., Dorje, F., Fromm, M.F., Konig, J. (2007). The influence of macrolide antibiotics on the uptake of organic anions and drugs mediated by OATP1B1 and OATP1B3. *Drug Metabolism and Disposition, 35*, 779–786.

[20] Yamashiro, W., Maeda, K., Hirouchi, M., Adachi, Y., Hu, Z., Sugiyama, Y. (2006). Involvement of transporters in the hepatic uptake and biliary excretion of valsartan, a selective antagonist of the angiotensin II AT1-receptor, in humans. *Drug Metabolism and Disposition, 34*, 1247–1254.

[21] Ogiso, T., Kasutani, M., Tanaka, H., Iwaki, M., Tanino, T. (2001). Pharmacokinetics of epinastine and a possible mechanism for double peaks in oral plasma concentration profiles. *Biological & Pharmaceutical Bulletin, 24*, 790–794.

[22] Shin, B.S., Kim, J.J., Kim, J., Hu, S.K., Kim, H.J., Hong, S.H., Kim, H.K., Lee, H.S., Yoo, S.D. (2008). Oral bioavailability and enterohepatic recirculation of otilonium bromide in rats. *Archives of Pharmacal Research, 31*, 117–124.

[23] Granero, G.E., Amidon, G.L. (2008). Possibility of enterohepatic recycling of ketoprofen in dogs. *International Journal of Pharmaceutics, 349*, 166–171.

[24] Xing, J., Chen, X., Zhong, D. (2005). Absorption and enterohepatic circulation of baicalin in rats. *Life Sciences, 78*, 140–146.

[25] Marier, J.F., Vachon, P., Gritsas, A., Zhang, J., Moreau, J.P., Ducharme, M.P. (2002). Metabolism and disposition of resveratrol in rats: extent of absorption, glucuronidation, and enterohepatic recirculation evidenced by a linked-rat model. *Journal of Pharmacology and Experimental Therapeutics, 302*, 369–373.

[26] Davies, M.I., Lunte, C.E. (1996). Simultaneous microdialysis sampling from multiple sites in the liver for the study of phenol metabolism. *Life Sciences, 59*, 1001–1013.

[27] Davies, M.I., Cooper, J.D., Desmond, S.S., Lunte, C.E., Lunte, S.M. (2000). Analytical considerations for microdialysis sampling. *Advanced Drug Delivery Reviews, 45*, 169–188.

[28] Scott, D.O., Lunte, C.E. (1993). In vivo microdialysis sampling in the bile, blood, and liver of rats to study the disposition of phenol. *Pharmaceutical Research, 10*, 335–342.

[29] Elmquist, W.F., Sawchuk, R.J. (1997). Application of microdialysis in pharmacokinetic studies. *Pharmaceutical Research, 14*, 267–288.

[30] Lefeuvre, S., Marchand, S., Lamarche, I., Mimoz, O., Couet, W. (2006). Microdialysis study of imipenem distribution in the intraperitoneal fluid of rats with or without experimental peritonitis. *Antimicrobial Agents and Chemotherapy, 50*, 34–37.

[31] Shinkai, N., Korenaga, K., Mizu, H., Yamauchi, H. (2008). Intra-articular penetration of ketoprofen and analgesic effects after topical patch application in rats. *Journal of Controlled Release, 131,* 107–112.

[32] Dahyot, C., Marchand, S., Bodin, M., Debeane, B., Mimoz, O., Couet, W. (2008). Application of basic pharmacokinetic concepts to analysis of microdialysis data: illustration with imipenem muscle distribution. *Clinical Pharmacokinetics, 47,* 181–189.

[33] Solligard, E., Juel, I.S., Bakkelund, K., Jynge, P., Tvedt, K.E., Johnsen, H., Aadahl, P., Gronbech, J.E. (2005). Gut luminal microdialysis of glycerol as a marker of intestinal ischemic injury and recovery. *Critical Care Medicine, 33,* 2278–2285.

[34] Araujo, B.V., Silva, C.F., Haas, S.E., Dalla Costa, T. (2008). Microdialysis as a tool to determine free kidney levels of voriconazole in rodents: a model to study the technique feasibility for a moderately lipophilic drug. *Journal of Pharmaceutical and Biomedical Analysis, 47,* 876–881.

[35] Chefer, V.I., Kieffer, B.L., Shippenberg, T.S. (2003). Basal and morphine-evoked dopaminergic neurotransmission in the nucleus accumbens of MOR- and DOR-knockout mice. *European Journal of Neuroscience, 18,* 1915–1922.

[36] Bungay, P.M., Newton-Vinson, P., Isele, W., Garris, P.A., Justice, J.B. (2003). Microdialysis of dopamine interpreted with quantitative model incorporating probe implantation trauma. *Journal of Neurochemistry, 86,* 932–946.

[37] Ettinger, S.N., Poellmann, C.C., Wisniewski, N.A., Gaskin, A.A., Shoemaker, J.S., Poulson, J.M., Dewhirst, M.W., Klitzman, B. (2001). Urea as a recovery marker for quantitative assessment of tumor interstitial solutes with microdialysis. *Cancer Research, 61,* 7964–7970.

[38] Schaddelee, M.P., Groenendaal, D., DeJongh, J., Cleypool, C.G., IJzerman, A.P., De Boer, A.G., Danhof, M. (2004). Population pharmacokinetic modeling of blood-brain barrier transport of synthetic adenosine A1 receptor agonists. *Journal of Pharmacology and Experimental Therapeutics, 311,* 1138–1146.

[39] Cremers, T., Ebert, B. (2007). Plasma and CNS concentrations of gaboxadol in rats following subcutaneous administration. *European Journal of Pharmacology, 562,* 47–52.

[40] Schwalbe, O., Buerger, C., Plock, N., Joukhadar, C., Kloft, C. (2006). Urea as an endogenous surrogate in human microdialysis to determine relative recovery of drugs: analytics and applications. *Journal of Pharmaceutical and Biomedical Analysis, 41,* 233–239.

[41] Marchand, S., Frasca, D., Dahyot-Fizelier, C., Breheret, C., Mimoz, O., Couet, W. (2008). Lung microdialysis study of levofloxacin in rats following intravenous infusion at steady state. *Antimicrobial Agents and Chemotherapy, 52,* 3074–3077.

[42] Volotinen, M., Turpeinen, M., Tolonen, A., Uusitalo, J., Maenpaa, J., Pelkonen, O. (2007). Timolol metabolism in human liver microsomes is mediated principally by CYP2D6. *Drug Metabolism and Disposition, 35,* 1135–1141.

[43] Yamanaka, H., Nakajima, M., Katoh, M., Yokoi, T. (2007). Glucuronidation of thyroxine in human liver, jejunum, and kidney microsomes. *Drug Metabolism and Disposition, 35,* 1642–1648.

[44] Chiba, M., Ishii, Y., Sugiyama, Y. (2009). Prediction of hepatic clearance in human from in vitro data for successful drug development. *AAPS Journal, 11,* 262–276.

[45] Gomez-Lechon, M.J., Castell, J.V., Donato, M.T. (2007). Hepatocytes: the choice to investigate drug metabolism and toxicity in man: in vitro variability as a reflection of in vivo. *Chemico-Biological Interactions, 168*, 30–50.

[46] Li, A.P. (2007). Human hepatocytes: isolation, cryopreservation and applications in drug development. *Chemico-Biological Interactions, 168*, 16–29.

[47] Scott, D.O., Bell, M.A., Lunte, C.E. (1989). Microdialysis–perfusion sampling for the investigation of phenol metabolism. *Journal of Pharmaceutical and Biomedical Analysis, 7*, 1249–1259.

[48] Tsai, T.H. (2002). Determination of naringin in rat blood, brain, liver, and bile using microdialysis and its interaction with cyclosporin A, a P-glycoprotein modulator. *Journal of Agricultural and Food Chemistry, 50*, 6669–6674.

[49] Tsai, P., Tsai, T.H. (2002). Simultaneous determination of berberine in rat blood, liver and bile using microdialysis coupled to high-performance liquid chromatography. *Journal of Chromatography A, 961*, 125–130.

[50] Nowak, G., Ungerstedt, J., Wernerman, J., Ungerstedt, U., Ericzon, B.G. (2002). Clinical experience in continuous graft monitoring with microdialysis early after liver transplantation. *British Journal of Surgery, 89*, 1169–1175.

[51] Kannerup, A.S., Funch-Jensen, P., Gronbaek, H., Jorgensen, R.L., Mortensen, F.V. (2008). Metabolic changes in the pig liver during warm ischemia and reperfusion measured by microdialysis. *Journal of Gastrointestinal Surgery, 12*, 319–326.

[52] Stenken, J.A., Lunte, C.E., Southard, M.Z., Stahle, L. (1997). Factors that influence microdialysis recovery: comparison of experimental and theoretical microdialysis recoveries in rat liver. *Journal of Pharmaceutical Sciences, 86*, 958–966.

[53] Davies, M.I., Lunte, C.E. (1995). Microdialysis sampling for hepatic metabolism studies: impact of microdialysis probe design and implantation technique on liver tissue. *Drug Metabolism and Disposition, 23*, 1072–1079.

[54] Edmiston, C.E., Jr., Suarez, E.C., Walker, A.P., Demeure, M.P., Frantzides, C.T., Schulte, W.J., Wilson, S.D. (1996). Penetration of ciprofloxacin and fleroxacin into biliary tract. *Antimicrobial Agents and Chemotherapy, 40*, 787–791.

[55] Westphal, J.F., Brogard, J.M., Caro-Sampara, F., Adloff, M., Blickle, J.F., Monteil, H., Jehl, F. (1997). Assessment of biliary excretion of piperacillin–tazobactam in humans. *Antimicrobial Agents and Chemotherapy, 41*, 1636–1640.

[56] Swoboda, S., Oberdorfer, K., Klee, F., Hoppe-Tichy, T., von Baum, H., Geiss, H.K. (2003). Tissue and serum concentrations of levofloxacin 500 mg administered intravenously or orally for antibiotic prophylaxis in biliary surgery. *Journal of Antimicrobial Chemotherapy, 51*, 459–462.

[57] Petrikkos, G., Kastanakis, M., Markogiannakis, A., Kastanakis, S., Bastounis, E., Antonios, P., Daikos, G.L., Katsilambros, N. (2006). Pharmacokinetics of cefepime in bile and gall bladder tissue after prophylactic administration in patients with extrahepatic biliary diseases. *International Journal of Antimicrobial Agents, 27*, 331–334.

[58] Terhaag, B., Hermann, U. (1986). Biliary elimination of indomethacin in man. *European Journal of Clinical Pharmacology, 29*, 691–695.

[59] Westphal, J.F., Jehl, F., Schloegel, M., Monteil, H., Brogard, J.M. (1993). Biliary excretion of cefixime: assessment in patients provided with T-tube drainage. *Antimicrobial Agents and Chemotherapy, 37*, 1488–1491.

[60] Serra, M.A., Caballero, A., Del Olmo, J.A., Aparisi, L., Gilabert, M.S., Rodriguez, F., Escudero, A., Wassel, A., Ferrando, J., Reig, G., Rodrigo, J.M. (1997). Maximal biliary transport of sulfobromophthalein in patients with a T-tube placed in the common bile duct. *European Journal of Drug Metabolism and Pharmacokinetics*, 22, 135–139.

[61] Cheng, H., Schwartz, M.S., Vickers, S., Gilbert, J.D., Amin, R.D., Depuy, B., Liu, L., Rogers, J.D., Pond, S.M., Duncan, C.A., et al. (1994). Metabolic disposition of simvastatin in patients with T-tube drainage. *Drug Metabolism and Disposition*, 22, 139–142.

[62] Ghibellini, G., Johnson, B.M., Kowalsky, R.J., Heizer, W.D., Brouwer, K.L. (2004). A novel method for the determination of biliary clearance in humans. *AAPS Journal*, 6, e33.

[63] Ghibellini, G., Bridges, A.S., Generaux, C.N., Brouwer, K.L. (2007). In vitro and in vivo determination of piperacillin metabolism in humans. *Drug Metabolism and Disposition*, 35, 345–349.

[64] Sutfin, T.A., Lind, T., Gabrielsson, M., Regardh, C.G. (1990). Biliary secretion of felodipine metabolites in man after intravenous [^{14}C]felodipine. *European Journal of Clinical Pharmacology*, 38, 421–424.

[65] Bergman, E., Forsell, P., Tevell, A., Persson, E.M., Hedeland, M., Bondesson, U., Knutson, L., Lennernas, H. (2006). Biliary secretion of rosuvastatin and bile acids in humans during the absorption phase. *European Journal of Pharmaceutical Sciences*, 29, 205–214.

[66] Sugie, M., Asakura, E., Zhao, Y.L., Torita, S., Nadai, M., Baba, K., Kitaichi, K., Takagi, K., Takagi, K., Hasegawa, T. (2004). Possible involvement of the drug transporters P-glycoprotein and multidrug resistance-associated protein Mrp2 in disposition of azithromycin. *Antimicrobial Agents and Chemotherapy*, 48, 809–814.

[67] Takayanagi, M., Sano, N., Takikawa, H. (2005). Biliary excretion of olmesartan, an anigotensin II receptor antagonist, in the rat. *Journal of Gastroenterology and Hepatology*, 20, 784–788.

[68] Morisawa, Y., Takikawa, H. (2009). Effect of bile acids on the biliary excretion of pravastatin in rats. *Hepatology Research*, 39, 595–600.

[69] Mikami, M., Takikawa, H. (2008). Effect of genipin on the biliary excretion of cholephilic compounds in rats. *Hepatology Research*, 38, 614–621.

[70] Rath, L., Hutchison, M. (1989). A new method of bile duct cannulation allowing bile collection and re-infusion in the conscious rat. *Laboratory Animals*, 23, 163–168.

[71] Heppert, K.E., Flora, W.H., Davies, M.I. (1998). The importance of balancing bile salt concentration to avoid fluid loss when using the microdialysis shunt probe. *Current Separations*, 17, 61–63.

[72] Chan, Y.L., Chou, M.H., Lin, M.F., Chen, C.F., Tsai, T.H. (2002). Determination and pharmacokinetic study of meropenem in rat bile using on-line microdialysis and liquid chromatography. *Journal of Chromatography A*, 961, 119–124.

[73] Liu, S.C., Tsai, T.H. (2002). Determination of diclofenac in rat bile and its interaction with cyclosporin A using on-line microdialysis coupled to liquid chromatography. *Journal of Chromatography B: Analytical Technologies in the Biomedical and Life Sciences*, 769, 351–356.

[74] Tsai, T.H., Shum, A.Y., Chen, C.F. (2000). Enterohepatic circulation of chloramphenicol and its glucuronide in the rat by microdialysis using a hepato-duodenal shunt. *Life Sciences, 66*, 363–370.

[75] Nickerson, D.F., Tess, D.A., Toler, S.M. (1997). First-pass metabolism and biliary recirculation of droloxifene in the female Sprague–Dawley rat. *Xenobiotica, 27*, 257–264.

[76] Behnia, K., Boroujerdi, M. (1998). Investigation of the enterohepatic recirculation of adriamycin and its metabolites by a linked-rat model. *Cancer Chemotherapy and Pharmacology, 41*, 370–376.

[77] Saitoh, H., Kobayashi, M., Oda, M., Nakasato, K., Kobayashi, M., Tadano, K. (2006). Characterization of intestinal absorption and enterohepatic circulation of mycophenolic acid and its 7-*O*-glucuronide in rats. *Drug Metabolism and Pharmacokinetics, 21*, 406–413.

[78] Bi, H.C., Zuo, Z., Chen, X., Xu, C.S., Wen, Y.Y., Sun, H.Y., Zhao, L.Z., Pan, Y., Deng, Y., Liu, P.Q., et al. (2008). Preclinical factors affecting the pharmacokinetic behaviour of tanshinone IIA, an investigational new drug isolated from *Salvia miltiorrhiza* for the treatment of ischaemic heart diseases. *Xenobiotica, 38*, 185–222.

[79] Chen, Y.F., Jaw, I., Shiao, M.S., Tsai, T.H. (2005). Determination and pharmacokinetic analysis of salvianolic acid B in rat blood and bile by microdialysis and liquid chromatography. *Journal of Chromatography A, 1088*, 140–145.

[80] Wu, J.W., Tsai, T.H. (2007). Effect of silibinin on the pharmacokinetics of pyrazinamide and pyrazinoic acid in rats. *Drug Metabolism and Disposition, 35*, 1603–1610.

[81] Chen, Y.J., Huang, S.M., Liu, C.Y., Yeh, P.H., Tsai, T.H. (2008). Hepatobiliary excretion and enterohepatic circulation of colchicine in rats. *International Journal of Pharmaceutics, 350*, 230–239.

[82] Tsai, P.L., Tsai, T.H. (2004). Hepatobiliary excretion of berberine. *Drug Metabolism and Disposition, 32*, 405–412.

[83] Huang, S.M., Tsai, T.R., Yeh, P.H., Tsai, T.H. (2005). Measurement of unbound ranitidine in blood and bile of anesthetized rats using microdialysis coupled to liquid chromatography and its pharmacokinetic application. *Journal of Chromatography A, 1073*, 297–302.

[84] Chang, Y.L., Chou, M.H., Lin, M.F., Chen, C.F., Tsai, T.H. (2001). Determination and pharmacokinetic study of unbound cefepime in rat bile by liquid chromatography with on-line microdialysis. *Journal of Chromatography A, 914*, 77–82.

[85] Chang, Y.L., Chiou, S.H., Chou, Y.C., Yen, C.J., Tsai, T.H. (2007). Quantitative determination of unbound cefoperazone in rat bile using microdialysis and liquid chromatography. *Journal of Pharmaceutical and Biomedical Analysis, 45*, 158–163.

[86] Tsai, T.H., Hung, L.C., Chen, C.F. (1999). Microdialysis study of biliary excretion of chloramphenicol and its glucuronide in the rat. *Journal of Pharmacy and Pharmacology, 51*, 911–915.

[87] Lee, C.H., Yeh, P.H., Tsai, T.H. (2002). Hepatobiliary excretion of fluconazole and its interaction with cyclosporin A in rat blood and bile using microdialysis. *International Journal of Pharmaceutics, 241*, 367–373.

[88] Cheng, F.C., Tsai, T.R., Chen, Y.F., Hung, L.C., Tsai, T.H. (2002). Pharmacokinetic study of levofloxacin in rat blood and bile by microdialysis and high-performance liquid chromatography. *Journal of Chromatography A, 961,* 131–136.

[89] Tsai, T.H., Chen, Y.F. (2003). Pharmacokinetics of metronidazole in rat blood, brain and bile studied by microdialysis coupled to microbore liquid chromatography. *Journal of Chromatography A, 987,* 277–282.

[90] Tsai, T.H. (2001). Pharmacokinetics of pefloxacin and its interaction with cyclosporin A, a P-glycoprotein modulator, in rat blood, brain and bile, using simultaneous microdialysis. *British Journal of Pharmacology, 132,* 1310–1316.

[91] Tsai, T.H., Liu, S.C., Tsai, P.L., Ho, L.K., Shum, A.Y., Chen, C.F. (2002) The effects of the cyclosporin A, a P-glycoprotein inhibitor, on the pharmacokinetics of baicalein in the rat: a microdialysis study. *British Journal of Pharmacology, 137,* 1314–1320.

[92] Tsai, P.L., Tsai, T.H. (2004). Pharmacokinetics of baicalin in rats and its interactions with cyclosporin A, quinidine and SKF-525A: a microdialysis study. *Planta Medica 70,* 1069–1074.

[93] Tsai, T.H., Huang, C.T., Shum, A.Y., Chen, C.F. (1999). Simultaneous blood and biliary sampling of esculetin by microdialysis in the rat. *Life Sciences, 65,* 1647–1655.

[94] Lin, L.C., Chen, Y.F., Lee, W.C., Wu, Y.T., Tsai, T.H. (2008). Pharmacokinetics of gastrodin and its metabolite p-hydroxybenzyl alcohol in rat blood, brain and bile by microdialysis coupled to LC-MS/MS. *Journal of Pharmaceutical and Biomedical Analysis, 48,* 909–917.

[95] Tsai, T.H. (2005). Concurrent measurement of unbound genistein in the blood, brain and bile of anesthetized rats using microdialysis and its pharmacokinetic application. *Journal of Chromatography A, 1073,* 317–322.

[96] Tsai, T.H., Liu, M.C. (2004). Determination of extracellular hesperidin in blood and bile of anaesthetized rats by microdialysis with high-performance liquid chromatography: a pharmacokinetic application. *Journal of Chromatography B 806,* 161–166.

[97] Huang, S.P., Lin, L.C., Wu, Y.T, Tsai, T.H. (2009). Pharmacokinetics of kadsurenone and its interaction with cyclosporin A in rats using a combined HPLC and microdialysis system. *Journal of Chromatography B, 877,* 247–252.

[98] Lai, L., Lin, L.C., Lin, J.H., Tsai, T.H. (2003). Pharmacokinetic study of free mangiferin in rats by microdialysis coupled with microbore high-performance liquid chromatography and tandem mass spectrometry. *Journal of Chromatography A, 987,* 367–374.

[99] Tsai, T.H., Wu, J.W. (2003). Regulation of hepatobiliary excretion of sinomenine by P-glycoprotein in Sprague-Dawley rats. *Life Sciences, 72,* 2413–2426.

9

MICRODIALYSIS USED TO MEASURE THE METABOLISM OF GLUCOSE, LACTATE, AND GLYCEROL

GREG NOWAK

Karolinska Institute, Karolinska University Hospital Huddinge, Stockholm, Sweden

1. INTRODUCTION

Among a number of metabolites that can be measured with microdialysis, glucose, lactate, and glycerol have been used in particular. Concentrations of these three small molecules are easy to measure using simple enzymatic methods available commercially, and results can be obtained within minutes. The metabolism of glucose, lactate, and glycerol is strongly influenced by changes in tissue oxygenation and blood flow. Glucose and lactate together with pyruvate are markers for glycolysis; and glycerol is an indicator of lipolysis or cell membrane injury due to ischemia or trauma [1–10]. Therefore, these parameters are most often used to monitor pathophysiological processes and pharmacological intervention related to disturbances in glucose metabolism, lipolysis, and changes in microcirculation. Close monitoring with fast answers from measurements give us new possibilities for study and understanding, for example, ischemia and reperfusion-related injury. It also gives us new opportunities regarding timing for therapeutic interventions before irreversible injury occurs to monitored tissue or organs.

Glucose is used by cells to generate the energy needed for physiological processes. In humans, glucose is delivered to the tissue in the bloodstream after

Applications of Microdialysis in Pharmaceutical Science, First Edition. Edited by Tung-Hu Tsai.
© 2011 John Wiley & Sons, Inc. Published 2011 by John Wiley & Sons, Inc.

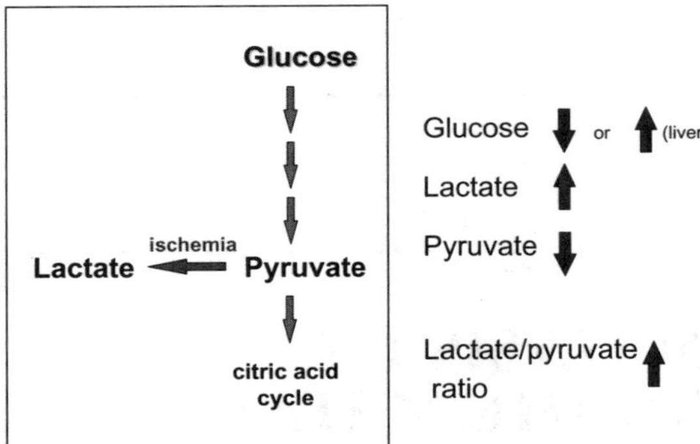

Figure 1   Glucose metabolism under anaerobic conditions/ischemia.

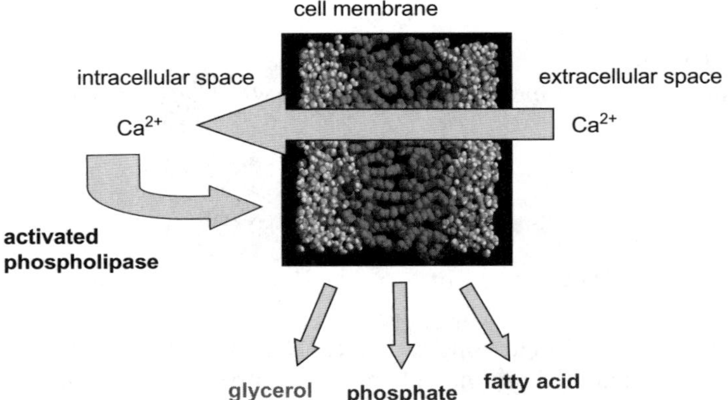

Figure 2   Glycerol release from the cell membrane due to ischemic event.

food intake. It can also be formed in the liver and skeletal muscle by the breakdown of glycogen (glycogenolysis) or glucose synthesis in the liver and kidneys (gluconeogenesis) [11]. In the cell, glucose is metabolized into pyruvate, which enters the citric cycle acid, and energy is produced in the form of ATP (Figure 1). In the case of insufficient oxygen delivery into the cell, pyruvate cannot be converted into acetyl-CoA, and lactate is produced [12–15]. Often, the lactate/pyruvate ratio is calculated, which reflects the cytoplasmatic redox state corresponding to the lactate dehydrogenase equilibrium. A typical pattern of changes with increasing lactate, decreasing pyruvate, and accelerated increase in lactate/pyruvate ratio corresponds to ischemia-induced metabolic changes in the tissue (Figure 1). In such a situation, glycerol can also indicate the level of cell injury due to ischemia event. During decreasing energy production, extracellular Ca^{2+} enters the cell and activate phospholipases, which degradate phospholipids in the cell membrane with the release of fatty acids and glycerol (Figure 2). In this situation, release of glycerol

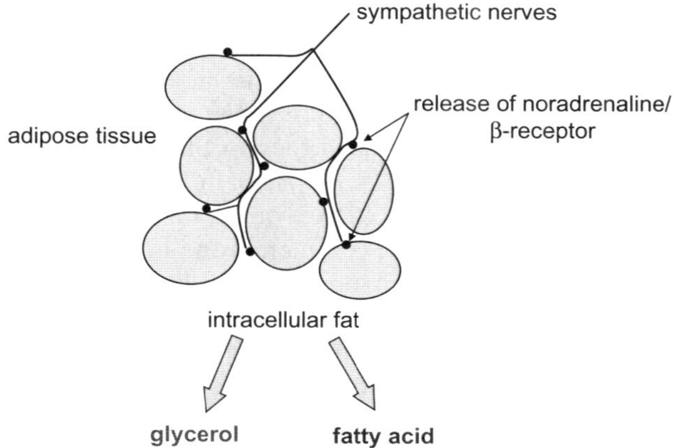

Figure 3 Glycerol release from adipose tissue related to stress.

indicates the severity of the cell membrane injury due to ischemia. This typical pattern of changes in markers of glycolysis and glycerol seems to be both specific and sensitive for detection of ischemia, due, for example, to vascular complications [16].

Glycerol can also be released due to lipolysis [2–5,8,10,17]. Under physiological conditions, lipolysis is controlled by the release of noradrenalin on the synapse of sympathetic nerves in adipose tissue. Noradrenalin in adipose tissue acts via the β_3-adrenergic receptor to split intracellular fat into glycerol and free fatty acids (Figure 3). Water-soluble glycerol is than easily wash out from the tissue and transported in the bloodstream.

Early microdialysis measurements of glucose, lactate, pyruvate, and glycerol are influenced by the extent of tissue injury that occurs after implantation of a catheter in monitored tissue. Implantation of the catheter results in tissue microinjury due to physical destruction of the cells as well as ischemia injury due to local disturbances in microcirculation. Histological studies show, however, that the implantation of small-diameter catheters does not induce major edema, severe inflammation, or bleeding in animals [18]. Also, circulation disturbances due to implantation of a catheter in the skin return to baseline within 40 min as measured by laser Doppler perfusion imaging [19]. In most studies it is estimated that normalization of the tissue metabolism around the implanted catheter takes around 30 to 60 min. After this "tissue equilibration" time, baseline levels are reached and the dialysate reflects the composition of the extracellular fluid [20–23]. An alternative for catheter implantation in the tissue is "free-floating" positioning of the catheter in anatomical cavities such as the abdominal cavity [24,25] or intraluminally such as in the gastrointestinal tract [26,27]. In comparison to the blood lactate, monitoring of intraperitoneal lactate was shown to be both a sensitive and a specific indicator of severe surgical complications in patients with acute abdominal status [28]. Intraperitoneal microdialysis samples can be obtained immediately after calibration of the system. However, when evaluating postoperative metabolism with intraperitoneal microdialysis, for example, results depend on the position of the microdialysis catheter and the tissue being monitored [29]. The relation of the catheter and the localization of tissue with a pathological event seems to be crucial and can follow each other, or there can be different results from differently placed catheters [25,30]. Also, in the case of intraperitoneal microdialysis, it would be difficult to figure out if, for example, increased glycerol levels are due to lipolysis or to cell membrane injury [31].

The validated measurement ranges for commercially available methods using enzymatic methods are glucose 0.1 to 25 mM, lactate 0.1 to 12 mM, pyruvate 10 to 1500 μM, and glycerol 10 to 1500 μM. The results obtained from microdialysis perfusate are easily coupled with those from other chemical analysis, such as high-performance liquid chromatography (HPLC) or mass spectrometry (MS). The results are often presented as absolute values but can also be presented as relative changes (percent of change over time), when the mean value of concentrations determined during the equilibration period is set as a reference for the baseline values (100%) [24].

The standard cutoff limit of the membrane is 20 kDa, which is large enough for the collection of such small molecules as glucose (180 Da), lactate (218 Da), pyruvate (110 Da), and glycerol (92 Da) [32]. When using membranes with cutoffs of 100 or 300 kDa, it is still possible to measure small molecules. It is, however, recommended that more osmotic perfusion solutions be used, to get results that follow results obtained from membranes of cutoff 20 kDa. Normally, results obtained from 20- and 100-kDa membranes closely follow each other, with differences in absolute values of below 20%. It was believed that small molecules are stable in microdialysis perfusate due to protein/enzyme-free perfusate. Often, analyses were performed even after a long period of frozen storage (−70°C recommended). Nowadays, however, at least regarding pyruvate, it is strongly recommended that the analysis be carried out within three months after sampling, due to the fact that pyruvate disintegrates with time of storage (recommendations from companies producing commercial kits for pyruvate detection). Regarding glucose, lactate, and glycerol, there are no indications for their disintegration over storage time, but it is still recommended that analyses be completed at the earliest convenience.

Microdialysis in pharmacological studies is increasingly popular. The popularity of the method is due primarily to possibilities to determine extracellular concentrations of small molecules of glucose, lactate, and glycerol, as well as bigger molecules such as hormones, various neurotransmitters, and drugs and their metabolites [2,3,33–41]. More and more interest is put into local tissue monitoring of small and large molecules, due to the fact that plasma concentration often has a poor predictive value regarding tissue substance concentration. It seems to be easy and practical to monitor the effect of pharmacological treatment using tissue or organ endogenous compounds as biological markers (e.g., lactate as a marker for glycolysis) [2]. Changes in local lactate levels can, for example, reflect the effect of drugs on blood flow, such as decreased blood flow in the presence of norepinephrine but increased flow in response to urapidil [3].

2. GLUCOSE

In physiology, glucose enters the aerobic glycolysis process in the cytosol. One molecule of glucose is converted into two molecules of pyruvate, which enter the mitochondria. In the mitochondrium, pyruvate undergoes oxidative decarboxylization to form acetyl coenzyme A, which is further oxidized into CO_2 in the citric acid cycle (i.e., the Krebs cycle). When there is an insufficient oxygen supply, anaerobic metabolism takes over and pyruvate is converted into lactate, with the accumulation of NADH. Pyruvic dehydrogenase converts pyruvate into lactate, allowing a regeneration of NAD^+ and continuation of glycolysis for a short time with little energy production. Accumulated NADH in the cell lactate is released from the cell and leads to acidosis and low pH. The redox equilibrium is shifted toward lactate, and the lactate/pyruvate ratio thus reflects the fact that the metabolism is aerobic or anaerobic. Lactate produced in this way is carried from peripheral tissue to the liver, where it is converted into pyruvate and enters gluconeogenesis, forming glucose, which again is returned to the tissue (Cori cycle).

In vivo monitoring with microdialysis has been used successfully in human subjects to monitor pathological processes and the effect of different substances on glucose uptake and its metabolism in various tissues [3–5,8,10,37,42–44]. Monitored with microdialysis, changes in extracellular glucose depend on glucose delivery into the tissue, which declines with decreasing blood flow [45]. It also depends on cell uptake and the consumption of available glucose, which results in a decrease of interstitial glucose as long as the uptake and metabolism take place [45]. However, when there are ischemia changes in the liver, glycogenolysis takes place and glucose levels increase as long as there is glycogen in hepatocytes (glycogenolysis) [46]. Such high levels of glucose can also be related to rearrangements of enzyme activity involved in carbohydrate metabolism. It is suggested that despite ischemia, gluconeogenesis can continue in the liver, thanks to the activity of the pentose–phosphate pathway, producing glucose for glycolysis [47]. At that time pyruvate levels are often undetectable, indicating decreased activity of glucokinase and phosphofructokinase [48]. After reperfusion, an increase in pyruvate indicates a recovery of the enzyme activity involved in glycolysis. After reperfusion there is also increased need for glucose in aerobic glycolysis, indicated by upregulation of glucose transporter 2 (GLUT2), which results in increased lactate and pyruvate levels monitored with microdialysis [46]. This situation is recognized as an example of hypermetabolism with increased lactate levels and a normal lactate/pyruvate ratio.

Hyperglycemia can also be a result of increased glucose production and decreased metabolic glucose clearance [49]. In surgical patients, gluconeogenesis accounts for more than 90% of total glucose production, due mainly to utilization of endogenous glycogen and stress-related release of gluconeogenic hormones such as catecholamines, glucagon, and cortisol. In patients under stress, gluconeogenesis uses amino acids released from muscle protein breakdown [49]. However, at the beginning of a surgical operation, anaesthesia may even induce opposite changes in the catabolic response. It has been shown that under general anaesthesia, within 2 h after the start of surgery, glucose clearance decreased by 30% while gluconeogenesis remained unchanged [50]. Sevoflurane anaesthesia results in a decrease in the intraoperative glucose production by 10%, whereas glucose concentration increases by 15% [51]. In the same study it was shown that gluconeogenic precursors such alanine, glutamine, and lactate, and counter-regulatory hormones such epinephrine and glucagons, increased. The kidney is the principal but possibly not the sole source of extrahepatic gluconeogenesis, primarily through the metabolism of alanine and glutamine [51].

Anaesthesia with pentobarbital and ketamine–xylazine in rats was shown to have an effect on glucose metabolism in the brain. The glucose levels of the intestinal fluid of the hippocampus were found to increase during anesthesia, which can be related to local neural activity [17]. It was also shown that intravenous infusion of low-dose prostacyclin, as a model of brain edema, results in a number of metabolic changes, such a decrease in glucose, an increase in lactate, and an increase in glycerol [4]. Microdialysis was also used to determine interstitial glucose concentrations in abdominal subcutaneous adipose tissue after the tissue was stimulated with the α_1-agonist norfenefrine, the $\alpha_{1,2}\beta$-agonist norepinephrine, and the α_1-antagonist urapidil in severely obese patients. Both norfenefrine and norepinephrine caused a concentration-dependent decrease in extracellular glucose concentration, with no relation to glucose uptake [3,42]. Preperfusion of adipose tissue with urapidil inhibited this decrease and enhanced extracellular glucose at high concentrations [3].

One of the most promising fields for the use of microdialysis is its clinical applications in monitoring glucose in diabetic patients. Several studies shown improved glycemic control in patients in intensive care units (ICUs), with a prevalence of hypoglycemia episodes well below 1% [52–54]. Despite the fact that subcutaneous microdialysis does not provide good enough control of glucose levels to be treated with insulin, new devices such as intravenous microdialysis catheters have promise. The simplicity of the method, use of user-friendly modules, and especially the possibility for continuous glucose monitoring can give us more correct information on postprandial glucose fluctuations than can single spot measurements, allowing improved insulin therapy [34,36,37,39]. This will result in delays or perhaps even in prevention of long-term complications related to hyperglycemia. Since the introduction of intensive insulin therapy in ICUs by van den Berghe et al. [55], there has been ongoing discussion on the concept of lowering morbidity and mortality by reduction of hyperglycemia under a variety of medical conditions [55–60]. The benefit seems to be more pronounced in patients treated in an ICU for a minimum of three days; short treatment periods did not elicit a significant effect [61]. In Freckmann et al.'s studies, continuously microdialysis monitored diabetic patients were given carbohydrate meals three times daily, which resulted in glucose concentrations (140 ± 13 mg/dL) close to the target value of 120 mg/dL, as well as stable levels during the night [34].

Combined glucose control and intensive insulin treatment should be consider as a strategy to prevent critical illness-associated complications triggered by persistent hyperglycemia and should not result in a higher incidence of hypoglycemia [61,62]. Further recommendations for glucose monitoring and insulin treatment will probably come from the outcome of ongoing large-scale multicenter randomized trials examining the issue of glycemic control in the ICU [e.g., the Normoglycaemia in Intensive Care Evaluation and Survival Using Glucose Algorithm Regulation (NICE-SUGAR) http://clinicaltrials.gov/ct/gui/show/NCT00220987].

Microdialysis is easy to use for studies on glucose levels in tissues such as in skeletal muscle and adipose tissue after treatment with a drug such as insulin or the insulin mimetic agent vanadate [3,42,63]. Reverse dialysis with vanadate dependently decreases interstitial glucose concentrations, probably secondary to an increased cellular glucose uptake [5]. Also, increasing lactate reflects the increased glucose uptake, which stimulates muscle glucose metabolism and indirectly confirms that vanadate mimics the effect of insulin in skeletal muscle in vivo [5].

3. LACTATE

Lactate is produced in the cell when there is an insufficient oxygen supply due to, for example, a systemic or local reduction in blood flow, known as the *Pasteur effect* [64]. However, depending on the hemodynamic state and liver function, systemic lactate concentration may vary widely [65–67]. In the case of vascular obstruction, the washout of lactate from ischemic tissue may go undetected because of reduced blood flow [68,69], and therefore microdialysis for local monitoring of lactate seems to be a good alternative for the systemic monitoring of lactate in the blood.

For years, separate lactate monitoring was recognized to be a good method for the detection and monitoring of anaerobic metabolism. There is a good correlation between arterial microdialysate lactate over 30 min and the mean arterial whole-blood lactate within the corresponding time period [70]. However, the whole-blood lactate concentrations are lower than the corresponding plasma concentrations when measured by the L-lactate oxidase–based method. The apparent higher concentration of lactate in the dialysate compared with the surrounding whole blood may be because microdialysate reflects the plasma concentration rather than the concentration in whole blood [71]. Increased lactate concentrations are often a result of hypoxia. However, there are studies showing that lactate can increase in brain, for example, despite adequate brain perfusion and oxygenation [72]. Experimental studies suggest that brain lactate increases because of partial impairment of enzymes involved in the tricarboxylic acid cycle by high ammonia concentrations in the brain, probably due to inhibition of pyruvate dehydrogenase and ketoglutarate dehydrogenase [73,74]. Functional impairment of enzymes in the metabolic process means that the brain cannot utilize oxygen, which results in anaerobic glycolysis with lactate accumulation within the brain.

In the liver after ischemia, one observes relatively high lactate levels despite the restoration of adequate perfusion and oxygenation. It was suggested that after reperfusion, due to the presence of H-ATPase damage in the mitochondrial inner membrane, despite the functional activity of glucokinase and phosphofructokinase, anaerobic glycolysis producing lactic acidosis persists [47]. Full recovery of oxidative phosphorylation results in cessation of anaerobic glycolysis and utilization of the lactate by hepatocytes with decreased lactate levels [47].

Microdialysis monitoring of lactate seems to have a special role in monitoring of splanchnic circulation. Concerning biochemical and metabolic changes in the digestive system, serum lactate may be one of the best serum markers of intestinal ischemia [75–78]. However, increasing lactate levels in the portal vein may not always result in systemic hyperlactatemia [79]. Hepatic uptake and metabolism of lactate might hide systemic changes in intestinal metabolism when monitored using systemic parameters, especially that the hepatic lactate uptake increases in response to hepatic lactate influx [80]. It was shown that local intestinal ischemia caused a substantial increase in lactate and glycerol and a decrease in glucose, as measured in the intestinal wall by microdialysis. In contrast, systemic markers did not change, and pCO_2, pO_2, and pH in the blood gas, as well as serum lactate and serum glucose, were completely unaffected [80]. In patients with normal liver function, systemic arterial hyperlactatemia is a relatively late sign of local splanchnic ischemia or is an indicator of a major ischemia event in the splanchnic region. Monitoring of circulation disturbances in splanchnic region is of special interest in patients in shock.

In animal models of endotoxin shock it has been shown that large-dose vasopressin induced redistribution of blood flow with reduction in the flow via the mesenteric artery and the maintenance of flow via the celiac trunk. This was also associated with systemic hyperlactatemia, despite increased hepatic lactate uptake [81]. In contrast, norepinephrine did not compromise either systemic or regional blood flow or tissue oxygenation [81]. This indicates that a large dose of vasopressin, but not norepinephrine, may decrease systemic and selectively low intestinal blood flow [81]. Epinephrine induces intraperitoneal lactate and glycerol release with systemic hyperlactatemia, acidosis, and both systemic and regional venous lactate/pyruvate ratio increase [82].

Many authors stress that even more important than monitoring lactate alone is monitoring pyruvate also. By calculating the lactate/pyruvate ratio, one is getting a more specific and sensitive marker of tissue redox status. The ratio calculated is not influenced by a changed metabolism ratio such as that during post-reperfusion hypermetabolism, where both lactate and pyruvate increase [1].

4. LACTATE/PYRUVATE RATIO

An increase in lactate concentration may be due to hypoxia, ischemia, and hypermetabolism [46], and therefore pyruvate measurements are of help to distinguishing the type of process in the tissue. Lactate levels and the lactate/pyruvate ratio indicate the extent of glycolysis and anaerobic metabolism in the tissue and are well-accepted markers of ischemia [15,83,84]. Aerobic hypermetabolism (e.g., in pre-septic patients) causes an increase in lactate. However, there is also a proportional increase in pyruvate, and the lactate/pyruvate ratio is stable. Therefore, the lactate/pyruvate ratio rather than lactate alone is considered to be a better indicator of whether aerobic or anaerobic metabolism takes place [46]. Under ischemic conditions, large quantities of lactate accumulate in cells in relation to pyruvate. Therefore, greater degrees of anaerobic metabolism lead to increased lactate concentrations without a significant rise in pyruvate, increasing the ratio between the two [85]. This is in comparison to situations such as hypermetabolism, in which lactate rises along with pyruvate levels without altering the lactate/pyruvate ratio significantly [47]. Backstrom and colleagues studied endotoxin and hemorrhagic shock–induced metabolic changes in the splanchnic region by the use of intravasal microdialysis in a pig model [86]. The most prominent differences were detected in the lactate/pyruvate ratio and in the increasing glycerol levels.

Studies in various tissues showed that a lactate/pyruvate ratio of 10 to 15 is considered normal [78,87–89]. In patients without complications, lactate/pyruvate ratio was also shown to be a sensitive indirect marker of inflammatory response after abdominal surgery when measured with intraperitoneal microdialysis and compared to the intraperitoneal level of TNF-α [90]. In patients with postoperative complications, an increase in the peritoneal lactate/pyruvate ratio preceded clinical signs of splanchnic ischemia [90]. The lactate/pyruvate ratio reflects tissue reaction to a changing oxygen and glucose supply [22]. Since glucose is the main substance for the formation of pyruvate, the lactate/glucose ratio was also tried as a marker of ischemia, with no improvement in sensitivity or specificity of the method compared to the lactate/pyruvate ratio [17,26].

5. GLYCEROL

There are two major sources of glycerol: cell membrane, where glycerol is a matrix for phospholipid structure; and adipose tissue, where glycerol is a part of the intracellular triacylglycerol. Glycerol as an integral component of the cell membrane can be used as a marker of cell membrane damage

[91–93]. Loss of energy and a decrease in pH result in changes in enzyme and transmembrane pump activity, which leads to an influx of calcium ions and the activation of phospholipases. Activated phospholipases split glycerol from the cell membrane phospholipids and fatty acids, which accumulate in the intracellular fluid. Therefore, increased levels of glycerol indicate the level of structural cell damage due to ischemia [89]. It was observed in studies on ischemia that with the time of the presence of ischemia and after an increase in lactate, glycerol increases continuously over time. It was also observed in ischemic brain tissue, where the level of glycerol in the brain could act as a marker for ischemia and cell membrane damage [57,84,91,92]. In a study on a liver graft during preservation, it was shown that despite use of a special organ preservation solution, there was continuous cell membrane injury over time of preservation with stable levels of glycolysis markers [46]. An accelerated decrease in glycerol after the restoration of blood flow through a liver graft was explained by the "washout" effect, where interstitial glycerol is washed out from the tissue with circulating blood into systemic circulation [46].

The second source of glycerol, adipose tissue, releases glycerol due to lipolysis, which contributes to energy production through oxidation and gluconeogenesis from the complexes of free fatty acids and glycerol [23,52,92,94]. This release of glycerol is an indirect marker of sympathetic tone stress mediated by release of noradrenalin. Absolute concentration of glycerol varies among tissues. Normal glycerol values in various organ tissues are up to 10 to 80 μM, whereas glycerol were intraperitoneally and subcutaneously have higher levels (up to 200 to 450 μM). Changes in glycerol levels can often be related to use of anaesthetic drugs. It has been demonstrated that isoflurane anaesthesia causes peripheral insulin resistance and decreased lipolysis [95], whereas anaesthesia and surgery cause significant lipolysis [96]. Halothane is known for the induction of lipolysis and increase in levels of glycerol as well as lactate and glucose [88].

REFERENCES

[1] Casaburi, R., Oi, S. (1989). Effect of liver disease on the kinetics of lactate removal after heavy exercise. *European Journal of Applied Physiology and Occupational Physiology*, 59, 89–97.

[2] de la Peña, A., Liu, P., Derendorf, H. (2000). Microdialysis in peripheral tissues. *Advanced Drug Delivery Reviews*, 45, 189–216.

[3] Flechtner-Mors, M., Jenkinson, C.P., Alt, A., Biesalski, H.K., Adler, G., Ditschuneit, H.H. (2004). Sympathetic regulation of glucose uptake by the alpha1-adrenoceptor in human obesity. *Obesity Research*, 12, 612–620.

[4] Gardenfors, F., Nilsson, A., Ungerstedt, U., Nordstrom, C.H. (2004). Adverse biochemical and physiological effects of prostacyclin in experimental brain oedema. *Acta Anaesthesiologica Scandinavica*, 48, 1316–1321.

[5] Hamrin, K., Henriksson, J. (2005). Local effect of vanadate on interstitial glucose and lactate concentrations in human skeletal muscle. *Life Sciences, 76,* 2329–2338.

[6] Marklund, N., Salci, K., Lewen, A., Hillered, L. (1997). Glycerol as a marker for post-traumatic membrane phospholipid degradation in rat brain. *Neuroreport, 8,* 1457–1461.

[7] Paschen, W., van den Kerchhoff, W., Hossmann, K.A. (1986). Glycerol as an indicator of lipid degradation in bicuculline-induced seizures and experimental cerebral ischemia. *Metabolic Brain Disease, 1,* 37–44.

[8] Rosdahl, H., Hamrin, K., Ungerstedt, U., Henriksson, J. (2000). A microdialysis method for the in situ investigation of the action of large peptide molecules in human skeletal muscle: detection of local metabolic effects of insulin. *International Journal of Biological Macromolecules, 28,* 69–73.

[9] Ungerstedt, U. (1997). Microdialysis in normal and injured human brain. In: Tucson, A.Z. (Ed.), *Physiology, Stress, and Malnutrition: Functional Correlates, Nutritional Intervention,* Lippincott Williams & Wilkins, Philadelphia, pp. 361–374.

[10] Yang, W.P., Oshida, Y., Wu, W., Sato, J., Ohsawa, I., Sato, Y. (1995). Effect of daily voluntary running on in vivo insulin action in rat skeletal muscle and adipose tissue as determined by the microdialysis technique. *International Journal of Sports Medicine, 16,* 99–104.

[11] Li, Y., Peris, J., Zhong, L., Derendorf, H. (2006). Microdialysis as a tool in local pharmacodynamics. *AAPS Journal, 8,* E222–E235.

[12] Cicalese, L. (2001). Pyruvate in organ transplantation. *Journal of Parenteral and Enteral Nutrition, 25,* 216–218.

[13] Sileri, P., Schena, S., Morini, S., Rastellini, C., Pham, S., Benedetti, E., Cicalese, L. (2001). Pyruvate inhibits hepatic ischemia–reperfusion injury in rats. *Transplantation, 72,* 27–30.

[14] So, P.W., Fuller, B.J. (2003). Enhanced energy metabolism during cold hypoxic organ preservation: studies on rat liver after pyruvate supplementation. *Cryobiology, 46,* 295–300.

[15] Tsung, A., Kaizu, T., Nakao, A., Shao, L., Bucher, B., Fink, M.P., Murase, N., Geller, D.A. (2005). Ethyl pyruvate ameliorates liver ischemia–reperfusion injury by decreasing hepatic necrosis and apoptosis. *Transplantation, 79,* 196–204.

[16] Jansson, K., Ungerstedt, J., Jonsson, T., Redler, B., Andersson, M., Ungerstedt, U., Norgren, L. (2003). Human intraperitoneal microdialysis: increased lactate/pyruvate ratio suggests early visceral ischaemia: a pilot study. *Scandinavian Journal of Gastroenterology, 38,* 1007–1011.

[17] Canal, C.E., McNay, E.C., Gold, P.E. (2005). Increases in extracellular fluid glucose levels in the rat hippocampus following an anesthetic dose of pentobarbital or ketamine–xylazine: an in vivo microdialysis study. *Physiology & Behavior, 84,* 245–250.

[18] Hickner, R.C., Ekelund, U., Mellander, S., Ungerstedt, U., Henriksson, J. (1995). Muscle blood flow in cats: comparison of microdialysis ethanol technique with direct measurement. *Journal of Applied Physiology, 79,* 638–647.

[19] Anderson, C., Andersson, T., Wardell, K. (1994). Changes in skin circulation after insertion of a microdialysis probe visualized by laser Doppler perfusion imaging. *Journal of Investigative Dermatology, 102,* 807–811.

[20] Benveniste, H., Huttemeier, P.C. (1990). Microdialysis: theory and application. *Progress in Neurobiology*, *35*, 195–215.

[21] Fellander, G., Nordenstrom, J., Ungerstedt, U., Arner, P., Bolinder, J. (1994). Influence of operation on glucose metabolism and lipolysis in human adipose tissue: a microdialysis study. *European Journal of Surgery*, *160*, 87–95.

[22] Hillered, L., Persson, L., Ponten, U., Ungerstedt, U. (1990). Neurometabolic monitoring of the ischaemic human brain using microdialysis. *Acta Neurochirurgica (Wien)*, *102*, 91–97.

[23] Lonnroth, P., Smith, U. (1990). Microdialysis: a novel technique for clinical investigations. *Journal of Internal Medicine*, *227*, 295–300.

[24] Klaus, S., Heringlake, M., Gliemroth, J., Bruch, H.P., Bahlmann, L. (2002). Intraperitoneal microdialysis for detection of splanchnic metabolic disorders. *Langenbeck's Archives of Surgery*, *387*, 276–280.

[25] Ungerstedt, J., Nowak, G., Ericzon, B.G., Ungerstedt, U. (2003). Intraperitoneal microdialysis (IPM): a new technique for monitoring intestinal ischemia studied in a porcine model. *Shock*, *20*, 91–96.

[26] Sommer, T., Larsen, J.F. (2003). Detection of intestinal ischemia using a microdialysis technique in an animal model. *World Journal of Surgery*, *27*, 416–420.

[27] Tenhunen, J.J., Jakob, S.M., Takala, J.A. (2001). Gut luminal lactate release during gradual intestinal ischemia. *Intensive Care Medicine*, *27*, 1916–1922.

[28] DeLaurier, G.A., Ivey, R.K., Johnson, R.H. (1994). Peritoneal fluid lactic acid and diagnostic dilemmas in acute abdominal disease. *American Journal of Surgery*, *167*, 302–305.

[29] Jansson, K., Strand, I., Redler, B., Magnuson, A., Ungerstedt, U., Norgren, L. (2004). Results of intraperitoneal microdialysis depend on the location of the catheter. *Scandinavian Journal of Clinical and Laboratory Investigation*, *64*, 63–70.

[30] Sommer, T., Larsen, J.F. (2004). Intraperitoneal and intraluminal microdialysis in the detection of experimental regional intestinal ischaemia. *British Journal of Surgery*, *91*, 855–861.

[31] Flechtner-Mors, M., Jenkinson, C.P., Alt, A., Adler, G., Ditschuneit, H.H. (2002). In vivo alpha(1)-adrenergic lipolytic activity in subcutaneous adipose tissue of obese subjects. *Journal of Pharmacology and Experimental Therapeutics*, *301*, 229–233.

[32] Tenhunen, J.J. (2003). Lipopolysaccharide preconditioning attenuates metabolic alteration induced by endotoxin shock: tissue-specific monitoring by microdialysis. *Intensive Care Medicine*, *29*, 515–517.

[33] Bolinder, J., Ungerstedt, U., Arner, P. (1993). Long-term continuous glucose monitoring with microdialysis in ambulatory insulin-dependent diabetic patients. *Lancet*, *342*, 1080–1085.

[34] Freckmann, G., Kalatz, B., Pfeiffer, B., Hoss, U., Haug, C. (2001). Recent advances in continuous glucose monitoring. *Experimental and Clinical Endocrinology & Diabetes*, *109* (Suppl. 2), S347–S357.

[35] Humpel, C., Ebendal, T., Olson, L. (1996). Microdialysis: a way to study in vivo release of neurotrophic bioactivity: a critical summary. *Journal of Molecular Medicine*, *74*, 523–526.

[36] Keck, F.S., Kerner, W., Meyerhoff, C., Zier, H., Pfeiffer, E.F. (1991). Combination of microdialysis and Glucosensor permits continuous (on line) s.c. glucose monitoring in a patient operated device: I. In vitro evaluation. *Hormone and Metabolic Research, 23*, 617–618.

[37] Keck, F.S., Meyerhoff, C., Kerner, W., Siegmund, T., Zier, H., Pfeiffer, E.F. (1992). Combination of microdialysis and glucosensor permits continuous (on line) SC glucose monitoring in a patient operated device: II. Evaluation in animals. *Hormone and Metabolic Research, 24*, 492–493.

[38] Liu, P., Müller, M., Derendorf, H. (2002). Rational dosing of antibiotics: the use of plasma concentrations versus tissue concentrations. *International Journal of Antimicrobial Agents, 19*, 285–290.

[39] Müller, M. (2002). Science, medicine, and the future: microdialysis. *British Medical Journal, 324*, 588–591.

[40] Tsai, T.H. (2003). Assaying protein unbound drugs using microdialysis techniques. *Journal of Chromatography B: Analytical Technologies in the Biomedical and Life Sciences, 797*, 161–173.

[41] Yokoyama, M., Suzuki, E., Sato, T., Maruta, S., Watanabe, S., Miyaoka, H. (2005). Amygdalic levels of dopamine and serotonin rise upon exposure to conditioned fear stress without elevation of glutamate. *Neuroscience Letters, 379*, 37–41.

[42] Djurhuus, C.B., Gravholt, C.H., Nielsen, S., Pedersen, S.B., Moller, N., Schmitz, O. (2004). Additive effects of cortisol and growth hormone on regional and systemic lipolysis in humans. *American Journal of Physiology: Endocrinology and Metabolism, 286*, E488–E494.

[43] Müller, M. (2000). Microdialysis in clinical drug delivery studies. *Advanced Drug Delivery Reviews, 45*, 255–269.

[44] Richterova, B., Stich, V., Moro, C., Polak, J., Klimcakova, E., Majercik, M., Harant, I., Viguerie, N., Crampes, F., Langin, D., Lafontan, M., Berlan, M. (2004). Effect of endurance training on adrenergic control of lipolysis in adipose tissue of obese women. *Journal of Clinical Endocrinology & Metabolism, 89*, 1325–1331.

[45] Hovda, D.A., Lee, S.M., Smith, M.L., Von Stuck, S., Bergsneider, M., Kelly, D., Shalmon, E., Martin, N., Caron, M., Mazziotta, J., et al. (1995). The neurochemical and metabolic cascade following brain injury: moving from animal models to man. *Journal of Neurotrauma, 12*, 903–906.

[46] Nowak, G., Ungerstedt, J., Wernerman, J., Ungerstedt, U., Ericzon, B.G. (2002). Metabolic changes in the liver graft monitored continuously with microdialysis during liver transplantation in a pig model. *Liver Transplantation, 8*, 424–432.

[47] Silva, M.A., Murphy, N., Richards, D.A., Wigmore, S.J., Bramhall, S.R., Buckels, J.A., Adams, D.H., Mirza, D.F. (2006). Interstitial lactic acidosis in the graft during organ harvest, cold storage, and reperfusion of human liver allografts predicts subsequent ischemia reperfusion injury. *Transplantation, 82*, 227–233.

[48] Inomoto, T., Tanaka, A., Awane, M., Kanai, M., Shinohara, H., Hatano, S., Sato, S., Gomi, T., Masuda, K., Someya, Y., et al. (1996). Changes in glucose transporter 2 and carbohydrate-metabolizing enzymes in the liver during cold preservation and warm ischemia. *Transplantation, 61*, 869–874.

[49] Battezzati, A., Bertoli, S. (2004). Methods of measuring metabolism during surgery in humans: focus on the liver–brain relationship. *Current Opinion in Clinical Nutrition and Metabolic Care, 7*, 523–530.

[50] Lattermann, R., Carli, F., Wykes, L., Schricker, T. (2002). Epidural blockade modifies perioperative glucose production without affecting protein catabolism. *Anesthesiology*, *97*, 374–381.

[51] Schricker, T., Galeone, M., Wykes, L., Carli, F. (2004). Effect of desflurane/remifentanil anaesthesia on glucose metabolism during surgery: a comparison with desflurane/epidural anaesthesia. *Acta Anaesthesiologica Scandinavica*, *48*, 169–173.

[52] Pachler, C., Plank, J., Weinhandl, H., Chassin, L.J., Wilinska, M.E., Kulnik, R., Kaufmann, P., Smolle, K.H., Pilger, E., Pieber, T.R., Ellmerer, M., Hovorka, R. (2008). Tight glycaemic control by an automated algorithm with time-variant sampling in medical ICU patients. *Intensive Care Medicine*, *34*, 1224–1230.

[53] Vogelzang, M., Loef, B.G., Regtien, J.G., van der Horst, I.C., van Assen, H., Zijlstra, F., Nijsten, M.W. (2008). Computer-assisted glucose control in critically ill patients. *Intensive Care Medicine*, *34*, 1421–1427.

[54] Vogelzang, M., Zijlstra, F., Nijsten, M.W. (2005). Design and implementation of GRIP: a computerized glucose control system at a surgical intensive care unit. *BMC Medical Informatics and Decision Making*, *5*, 38.

[55] van den Berghe, G., Wouters, P., Weekers, F., Verwaest, C., Bruyninckx, F., Schetz, M., Vlasselaers, D., Ferdinande, P., Lauwers, P., Bouillon, R. (2001). Intensive insulin therapy in critically ill patients. *New England Journal of Medicine*, *345*, 1359–1367.

[56] Grey, N.J., Perdrizet, G.A. (2004). Reduction of nosocomial infections in the surgical intensive-care unit by strict glycemic control. *Endocrine Practice*, *10* (Suppl. 2), 46–52.

[57] Nasraway, S.A., Jr. (2006). Hyperglycemia during critical illness. *Journal of Parenteral and Enteral Nutrition*, *30*, 254–258.

[58] Pittas, A.G., Siegel, R.D., Lau, J. (2006). Insulin therapy and in-hospital mortality in critically ill patients: systematic review and meta-analysis of randomized controlled trials. *Journal of Parenteral and Enteral Nutrition*, *30*, 164–172.

[59] van den Berghe, G., Wilmer, A., Hermans, G., Meersseman, W., Wouters, P.J., Milants, I., Van Wijngaerden, E., Bobbaers, H., Bouillon, R. (2006). Intensive insulin therapy in the medical ICU. *New England Journal of Medicine*, *354*, 449–461.

[60] van den Berghe, G., Wilmer, A., Milants, I., Wouters, P.J., Bouckaert, B., Bruyninckx, F., Bouillon, R., Schetz, M. (2006). Intensive insulin therapy in mixed medical/surgical intensive care units: benefit versus harm. *Diabetes*, *55*, 3151–3159.

[61] Ellger, B., Debaveye, Y., Vanhorebeek, I., Langouche, L., Giulietti, A., Van Etten, E., Herijgers, P., Mathieu, C., van den Berghe, G. (2006). Survival benefits of intensive insulin therapy in critical illness: impact of maintaining normoglycemia versus glycemia-independent actions of insulin. *Diabetes*, *55*, 1096–1105.

[62] Arabi, Y.M., Dabbagh, O.C., Tamim, H.M., Al-Shimemeri, A.A., Memish, Z.A., Haddad, S.H., Syed, S.J., Giridhar, H.R., Rishu, A.H., Al-Daker, M.O., et al. (2008). Intensive versus conventional insulin therapy: a randomized controlled trial in medical and surgical critically ill patients. *Critical Care Medicine*, *36*, 3190–3197.

[63] Djurhuus, C.B., Gravholt, C.H., Nielsen, S., Mengel, A., Christiansen, J.S., Schmitz, O.E., Moller, N. (2002). Effects of cortisol on lipolysis and regional interstitial

glycerol levels in humans. *American Journal of Physiology: Endocrinology and Metabolism*, *283*, E172–E177.

[64] Barker, J., Khan, M.A., Solomos, T. (1966). Mechanism of the Pasteur effect. *Nature*, *211*, 547–548.

[65] Gutierrez, G., Clark, C., Brown, S.D., Price, K., Ortiz, L., Nelson, C. (1994). Effect of dobutamine on oxygen consumption and gastric mucosal pH in septic patients. *American Journal of Respiratory and Critical Care Medicine*, *150*, 324–329.

[66] Jonas, J., Schwarz, S., Alebrahim-Dehkordy, A. (1996). [Behavior of the lactate level in occlusion and reperfusion of the right superior mesenteric artery: an animal experiment study]. *Langenbeck's Archiv fur Chirurgie*, *381*, 1–6.

[67] Perret, C., Poli, S., Enrico, J.F. (1970). Lactic acidosis and liver damage. *Helvetica Medica Acta*, *35*, 377–405.

[68] Heino, A., Hartikainen, J., Merasto, M.E., Koski, E.M., Alhava, E., Takala, J. (1997). Systemic and regional effects of experimental gradual splanchnic ischemia. *Journal of Critical Care*, *12*, 92–98.

[69] Schlichting, E., Lyberg, T. (1995). Monitoring of tissue oxygenation in shock: an experimental study in pigs. *Critical Care Medicine*, *23*, 1703–1710.

[70] Bland, J.M., Altman, D.G. (1986). Statistical methods for assessing agreement between two methods of clinical measurement. *Lancet*, *1*, 307–310.

[71] Tenhunen, J.J., Kosunen, H., Alhava, E., Tuomisto, L., Takala, J.A. (1999). Intestinal luminal microdialysis: a new approach to assess gut mucosal ischemia. *Anesthesiology*, *91*, 1807–1815.

[72] Tofteng, F., Jorgensen, L., Hansen, B.A., Ott, P., Kondrup, J., Larsen, F.S. (2002). Cerebral microdialysis in patients with fulminant hepatic failure. *Hepatology*, *36*, 1333–1340.

[73] Chatauret, N., Rose, C., Therrien, G., Butterworth, R.F. (2001). Mild hypothermia prevents cerebral edema and CSF lactate accumulation in acute liver failure. *Metabolic Brain Disease*, *16*, 95–102.

[74] Therrien, G., Giguere, J.F., Butterworth, R.F. (1991). Increased cerebrospinal fluid lactate reflects deterioration of neurological status in experimental portal-systemic encephalopathy. *Metabolic Brain Disease*, *6*, 225–231.

[75] Lange, H., Jackel, R. (1994). Usefulness of plasma lactate concentration in the diagnosis of acute abdominal disease. *European Journal of Surgery*, *160*, 381–384.

[76] Lange, H., Toivola, A. (1997). [Warning signals in acute abdominal disorders: lactate is the best marker of mesenteric ischemia]. *Lakartidningen*, *94*, 1893–1896.

[77] Murray, M.J., Gonze, M.D., Nowak, L.R., Cobb, C.F. (1994). Serum D(−)-lactate levels as an aid to diagnosing acute intestinal ischemia. *American Journal of Surgery*, *167*, 575–578.

[78] Poeze, M., Froon, A.H., Greve, J.W., Ramsay, G. (1998). D-Lactate as an early marker of intestinal ischaemia after ruptured abdominal aortic aneurysm repair. *British Journal of Surgery*, *85*, 1221–1224.

[79] Hamilton-Davies, C., Mythen, M.G., Salmon, J.B., Jacobson, D., Shukla, A., Webb, A.R. (1997). Comparison of commonly used clinical indicators of hypovolaemia with gastrointestinal tonometry. *Intensive Care Medicine*, *23*, 276–281.

[80] Jakob, S.M., Merasto-Minkkinen, M., Tenhunen, J.J., Heino, A., Alhava, E., Takala, J. (2000). Prevention of systemic hyperlactatemia during splanchnic ischemia. *Shock*, *14*, 123–127.

[81] Martikainen, T.J., Tenhunen, J.J., Uusaro, A., Ruokonen, E. (2003). The effects of vasopressin on systemic and splanchnic hemodynamics and metabolism in endotoxin shock. *Anesthesia & Analgesia*, *97*, 1756–1763.

[82] Martikainen, T.J., Tenhunen, J.J., Giovannini, I., Uusaro, A., Ruokonen, E. (2005). Epinephrine induces tissue perfusion deficit in porcine endotoxin shock: evaluation by regional CO(2) content gradients and lactate-to-pyruvate ratios. *American Journal of Physiology: Gastrointestinal and Liver Physiology*, *288*, G586–G592.

[83] Harken, A.H. (1976). Lactic acidosis. *Surgery, Gynecology & Obstetrics*, *142*, 593–606.

[84] Hillered, L., Persson, L. (1999). Neurochemical monitoring of the acutely injured human brain. *Scandinavian Journal of Clinical and Laboratory Investigation*, *229*, 9–18.

[85] Wasserman, K., Beaver, W.L., Davis, J.A., Pu, J.Z., Heber, D., Whipp, B.J. (1985). Lactate, pyruvate, and lactate-to-pyruvate ratio during exercise and recovery. *Journal of Applied Physiology*, *59*, 935–940.

[86] Backstrom, T., Liska, J., Oldner, A., Lockowandt, U., Franco-Cereceda, A. (2004). Splanchnic metabolism during gut ischemia and short-term endotoxin and hemorrhagic shock as evaluated by intravasal microdialysis. *Shock*, *21*, 572–578.

[87] Jansson, K., Jansson, M., Andersson, M., Magnuson, A., Ungerstedt, U., Norgren, L. (2005). Normal values and differences between intraperitoneal and subcutaneous microdialysis in patients after non-complicated gastrointestinal surgery. *Scandinavian Journal of Clinical and Laboratory Investigation*, *65*, 273–281.

[88] Lacoumenta, S., Paterson, J.L., Burrin, J., Causon, R.C., Brown, M.J., Hall, G.M. (1986). Effects of two differing halothane concentrations on the metabolic and endocrine responses to surgery. *British Journal of Anaesthesia*, *58*, 844–850.

[89] Nowak, G., Ungerstedt, J., Wernerman, J., Ungerstedt, U., Ericzon, B.G. (2002). Clinical experience in continuous graft monitoring with microdialysis early after liver transplantation. *British Journal of Surgery*, *89*, 1169–1175.

[90] Jansson, K., Redler, B., Truedsson, L., Magnuson, A., Ungerstedt, U., Norgren, L. (2004). Postoperative on-line monitoring with intraperitoneal microdialysis is a sensitive clinical method for measuring increased anaerobic metabolism that correlates to the cytokine response. *Scandinavian Journal of Gastroenterology*, *39*, 434–439.

[91] Goodman, J.C., Gopinath, S.P., Valadka, A.B., Narayan, R.K., Grossman, R.G., Simpson, R.K., Jr., Robertson, C.S. (1996). Lactic acid and amino acid fluctuations measured using microdialysis reflect physiological derangements in head injury. *Acta Neurochirurgica*, *67*, 37–39.

[92] Hillered, L., Valtysson, J., Enblad, P., Persson, L. (1998). Interstitial glycerol as a marker for membrane phospholipid degradation in the acutely injured human brain. *Journal of Neurology, Neurosurgery, & Psychiatry*, *64*, 486–491.

[93] Nilsson, O.G., Brandt, L., Ungerstedt, U., Saveland, H. (1999). Bedside detection of brain ischemia using intracerebral microdialysis: subarachnoid hemorrhage and delayed ischemic deterioration. *Neurosurgery*, *45*, 1176–1184; discussion, 1184–1175.

[94] Bazan, N.G., Jr. (1970). Effects of ischemia and electroconvulsive shock on free fatty acid pool in the brain. *Biochimica et Biophysica Acta, 218,* 1–10.

[95] Horber, F.F., Krayer, S., Miles, J., Cryer, P., Rehder, K., Haymond, M.W. (1990). Isoflurane and whole body leucine, glucose, and fatty acid metabolism in dogs. *Anesthesiology, 73,* 82–92.

[96] Fellander, G., Nordenstrom, J., Tjader, I., Bolinder, J., Arner, P. (1994). Lipolysis during abdominal surgery. *Journal of Clinical Endocrinology & Metabolism, 78,* 150–155.

10

CLINICAL MICRODIALYSIS IN SKIN AND SOFT TISSUES

MARTINA SAHRE, RUNA NAIK, AND HARTMUT DERENDORF
Department of Pharmaceutics, University of Florida, Gainesville, Florida

1. INTRODUCTION

Microdialysis makes if possible to measure actual free drug concentrations in various tissues and organs and subsequently to relate these pharmacokinetic (PK) findings to pharmacodynamic (PD) observations to predict clinical efficacy. The ability to measure the free concentrations at the site of drug action over time makes microdialysis a very valuable tool for the assessment of bioavailability and bioequivalence, and has been recognized by industry and regulatory authorities such as the U.S. Food and Drug Administration (FDA) [1]. Earlier, PK research was restricted to the measurement of drug concentrations from biological matrixes that are relatively easy to obtain, such as tissue biopsies, urine, saliva, or skin blister fluids, but the emerging knowledge of microdialysis and its benefits have slowly shifted the focus off these methods, and with good reason [2]. In this chapter we discuss various examples that help us understand why microdialysis is not just an important technique in the assessment of drug distribution in skin and soft tissues but also in general is very crucial in the clinical drug development process.

Misconceptions About Tissue Drug Distribution   Until recent years, research on pharmacokinetics (PK) was limited for practical reasons to concentration measurements derived from matrices that were easy to obtain, such as blood and total tissue specimens. These approaches have caused considerable confusion, however, as their interpretation was flawed by a few misconceptions.

Applications of Microdialysis in Pharmaceutical Science, First Edition. Edited by Tung-Hu Tsai.
© 2011 John Wiley & Sons, Inc. Published 2011 by John Wiley & Sons, Inc.

1. Drug- or plasma-based models refer to the process of penetration into hypothetical compartments as "tissue penetration." This concept is misleading, as it does not take into account the complexity of separate organ systems and disease processes. Therefore, although plasma-based modeling may provide useful information in many cases, it must be kept in mind that it assumes rapid, unrestricted, and homogeneous diffusion processes in hypothetical spaces, assumptions that do not always hold true.

2. A second misconception was the assumption that tissue is a uniform matrix. Measuring total antibiotic concentration measurements from biopsy specimens may be misleading for several reasons. Most important, it must be considered that the actual target space for anti-infective agents, with few exceptions, is the interstitial space fluid [3]. If only overall tissue drug concentrations are measured, the effective site concentrations of drugs that equilibrate exclusively with the extracellular space, such as β-lactams, may be underestimated [3]. This situation, in turn, will lead to an overestimation of the effective site concentrations of intracellularly accumulating drugs, such as quinolones or macrolides [3]. Thus, homogenization of various tissue fluids and cells will lead to a hybrid tissue drug concentration that is difficult to interpret [3].

3. Another misconception was the notion that the entire drug fraction present in various tissue spaces is responsible for pharmacological activity. In fact, it has been shown that only the unbound drug concentrations at the infection site have the ability to exert anti-infective efficacy, both in vitro and in vivo [3]. Besides the fact that only the free fraction exerts activity, it is also only the free drug that has the ability to be distributed to the target site. This information was shown experimentally by various investigators, who found that differences in penetration were related directly to the free drug concentrations in serum. Although this concept is best described for antibacterial agents, one may safely assume that similar concepts hold true for many other classes of drugs.

It follows from these three considerations that an appropriate definition of tissue drug concentrations should in most cases imply the meaning of unbound interstitial drug concentrations and that a method which is to be considered the gold standard for the measurement of tissue drug concentrations should allow for the direct measurement of unbound antibiotic concentrations in the interstitial fluid of a given organ.

2. TISSUE BIOAVAILABILITY

2.1. Microdialysis in the Skin

Microdialysis is an excellent method to assess local drug exposure at any site where a probe can be placed. Obviously, this includes the skin, where many microdialysis studies have been performed. The influence of changes in skin-barrier properties on drug penetration was tested using the microdialysis method [4,5]. Tissue levels of salicylic acid after the application of 5% solution in ethanol to intact skin or skin that was treated with 1 or 2% sodium lauryl sulfate (SLS), repeated tapestripping, and drying by acetone wipe were measured [4]. Skin irritation was highest in the 2% SLS and tape-stripped regions, moderate in the 1% SLS–treated skin, and mild after treatment with acetone. Although the area under the curve (AUC) of salicylic acid in tissue was increased only slightly after acetone treatment compared to intact skin, it increased 46-fold in moderately irritated skin and >140-fold in highly irritated skin (Figure 1). The authors also assessed whether an application site had an influence on drug penetration and found slight differences which were not significant. The authors conclude that microdialysis is a valuable tool for the assessment of concentrations of drugs in skin of various states of irritation and inflammation and for the evaluation of bioavailability of different study drug formulations. Acyclovir and penciclovir penetration through skin that was either intact, tape-stripped, and/or vasoconstricted with noradrenalin was tested [5]. The results showed that after the application of drug creams to intact, non-vasoconstricted skin, there were no concentrations measurable above the lower limit of detection. When intact skin was vasoconstricted, concentrations became measurable, and the mean (±S.E.M.) AUC_{0-5} was 13.3 ± 2.9 and $27.6 \pm 10.6\,ng/mL{\cdot}h$ for penciclovir and acyclovir, respectively. After tape-stripping the skin to the extent of removal of the stratum corneum (45 tape strippings) the mean (±S.E.M.) AUC_{0-5} became 17.5 ± 3.8 and $12.2 \pm 5\,\mu g/mL{\cdot}h$ (Figure 2). The authors concluded that for the two drugs under investigation, which are small molecules, water soluble, and have $\log P$ values of −1.8 (acyclovir) and −2.12 (penciclovir), the stratum corneum thus

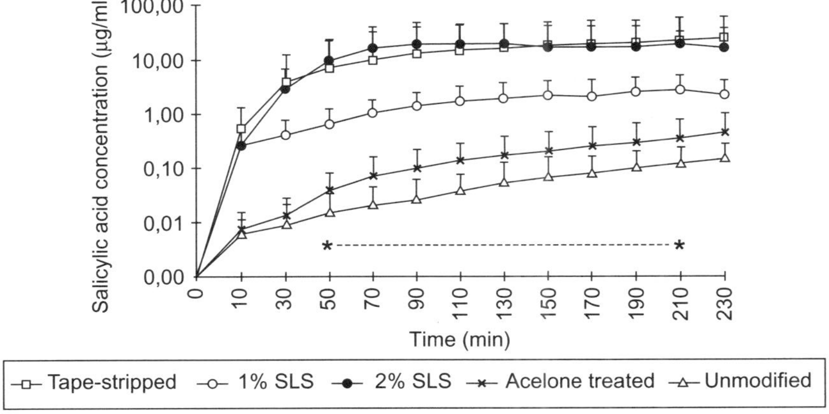

Figure 1 Influence of barrier perturbation on absorption of 5% salicylic acid solution through skin as measured by microdialysis. (From [4].)

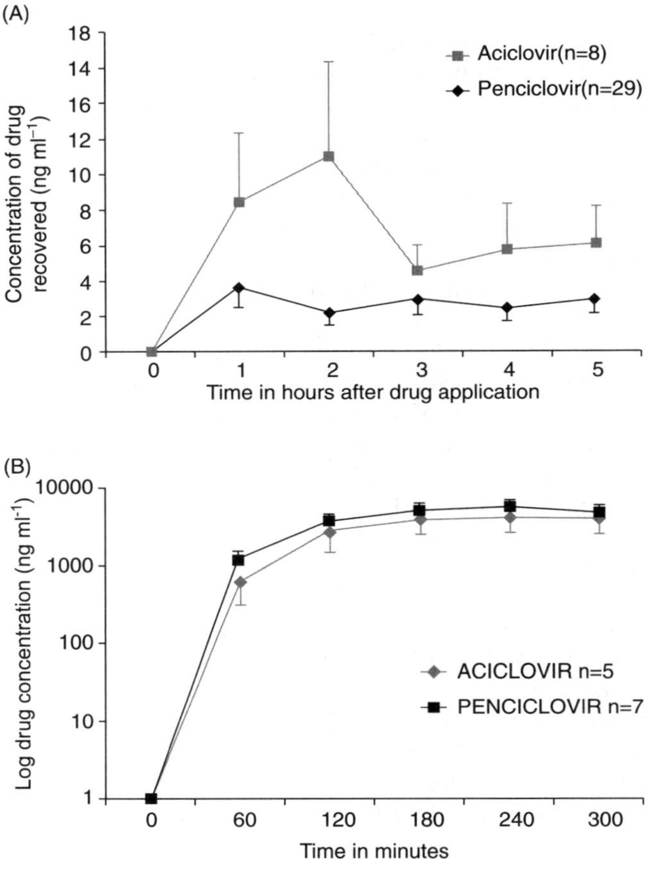

Figure 2 Acyclovir and penciclovir concentrations (mean ± S.E.M.) measured in microdialysate after topical application to (A) intact vasoconstricted skin; (B) skin from which stratum corneum was removed by repeated tape stripping. (From [5].)

presents the biggest barrier to drug penetration. Transepidermal water loss (TEWL) was used as a measure of the disruption of the stratum corneum barrier function. When TEWL was measured after every five tape strippings, water loss increased with increasing numbers of strippings until the stratum corneum had been removed completely after 40 to 45 strippings. Comparing the logarithm of the concentration of penciclovir per hour with the logarithm of the TEWL showed a strong positive correlation ($r^2 > 0.9$). Since no in vivo recovery for both drugs was measured, values given here are relative concentrations only, which serve to compare and show relative changes.

To assess the feasibility of using microdialysis for drugs that are either highly protein bound or highly lipophilic, tissue levels of fusidic acid (protein binding 97%) and β-methasone-17-valerate were evaluated [6]. In a human clinical trial with twice-daily application for 2 days, no measurable levels for either drug were found.

Microdialysis was used to assess skin penetration of acyclovir and salicylic acid into intact and tape-stripped skin (acyclovir only) [7]. For acyclovir, determinable dialysate levels were found only in tape-stripped skin, whereas salicylic acid levels were found in intact skin. At the concentrations used, salicylic acid has a keratolytic effect, so the authors conclude that a certain amount of skin penetration is to be expected.

The concentration–time course of penciclovir in skin-blister fluid and dialysate after an oral dose of its prodrug famciclovir was followed in a study by Borg et al. [8]. Comparing plasma, blister, and dialysate levels shows that plasma C_{max} levels are similar to recovery-adjusted dialysate C_{max} levels. C_{max} values found in blister fluid were slightly lower. The time to reach maximum concentrations in tissue (blister and dialysate) is about 60 min longer than in plasma, and the tissue half-life is also about 60 min longer than that of plasma. The authors point out that microdialysis measures free tissue concentrations only. For highly protein-bound drugs, the concentration in tissue could then be much smaller than total concentrations in plasma. It is pointed out that in the case of penciclovir, which has a protein binding of about 20%, plasma and tissue levels can be comparable. The study also tested for the influence of vasoconstriction on penciclovir concentrations in tissues. When adrenalin was added to the perfusate or the area above the probe site was cooled using a cooling pack, the AUC value for penciclovir in skin tissue was reduced to about one-third of the value achieved without interference.

The tissue concentrations of 8-methoxypsoralen after oral or topical application was also explored using microdialysis [9]. The compound is used for psoralen plus ultraviolet A (PUVA) therapy for a number of dermatoses. The study aimed to compare oral treatment and two topical treatments (bath and cream) to assess how to optimize treatment in regard to side effects such as the development of skin cancer after long-term PUVA therapy. The three treatments were compared in eight healthy volunteers in an open randomized three-way crossover study. Lower doses were given to three volunteers; the rest received higher doses to guarantee measurable concentrations, due to the assumption that 8-methoxypsoralen recovery may be low because of relatively high protein binding and lipophilicity. Maximum skin tissue concentrations after application of lower doses of 8-methoxypsoralen (i.e., 0.6 mg/kg oral dose, 0.001% cream, or 1 mg/L bath) were highest in volunteers treated by bath compared to cream and oral application. What is more, total plasma concentrations were up to 1000-fold higher after oral application compared to cream or bath application. Compared to T_{max} after oral application, the maximum concentrations in tissues after topical applications were reached faster (1 h compared to 1 to 4 h) and with less variability. Maximum plasma concentrations were reached at 1 to 4 h after oral application and after 1 to 3 h after cream and 1.5 h after bath application. Tissue concentrations of 8-methoxypsoralen were measurable after only 20 min and peaked within 1 h.

The authors conclude that topical treatment achieves high efficacy while having low systemic exposure, which could make topical treatment more desirable than oral application.

2.2. Antimicrobial Agents

Infections of skin and soft tissue can be caused by a variety of gram-positive and gram-negative pathogens and are treated routinely with antibiotics. Whereas penicillins and cephalosporins are drugs of first choice, agents of different classes (e.g., oxazolidinones, glycopeptides, macrolides, tetracyclines) have to be used in case of adverse events or the emergence of β-lactam resistance. To increase the chances of clinical success and to decrease the likelihood of toxic side effects as well as resistance development, selection of an appropriate antibiotic dosing regimen becomes extremely important [1]. The most rational approach is to link active drug concentrations to the respective pharmacodynamic outcome. However, efficacy predictions based on total plasma concentrations might be misleading, as most infections are not located in the bloodstream but, rather, in the interstitial fluid (ISF) of tissues, which is the usual target site for bacterial infections [10]. In fact, it is the free, unbound drug in the ISF that is responsible for antimicrobial efficacy. We shall discuss in detail different examples of how this can be achieved by use of microdialysis as a pharmacokinetic sampling technique and how it has the potential to streamline the decision process on proper drug dosing in drug development.

Moxifloxacin is used clinically in the treatment of uncomplicated skin and skin structure infections. A study by Burkhardt et al. [11] used microdialysis to compare the free protein-unbound moxifloxacin concentrations in normal as well as infected subcutaneous tissue. A single oral dose of 400 mg of moxifloxacin was administered to patients with spinal cord injury and decubitus ulcers, and drug concentrations were determined from serum, saliva, and subcutaneous tissue. The study found that moxifloxacin reaches adequate concentrations in normal subcutaneous (s.c.) tissue and decubitus ulcer tissue in patients with spinal injury. The concentrations measured in the ulcer tissue (C_{max}: 2.4 ± 1.6 mg/L) and healthy tissue (C_{max}: 2.7 ± 2.3 mg/L) were similar. The mean C_{max} value in serum was found to be 4.4 ± 2.7 mg/L and in saliva, 1.4 ± 0.4 mg/L. The AUC values for normal (9.2 ± 8.6 mg·h/L) and infected subcutaneous tissue (9.6 ± 6.8 mg·h/L) were approximately the same as that of the free AUC in serum [11]. Also, the tissue levels measured in the study correspond to the free protein-bound fraction of moxifloxacin in serum, which is about 50 to 60%, by which they concluded that poor blood flow does not affect tissue levels of moxifloxacin in patients with the injury.

Traditionally, plasma samples were taken to determine the pharmacokinetic properties of a compound and make predictions on the efficacy. However, these drug concentrations were sometimes presented as total drug concentrations, where as in reality only the free drug is pharmacologically active,

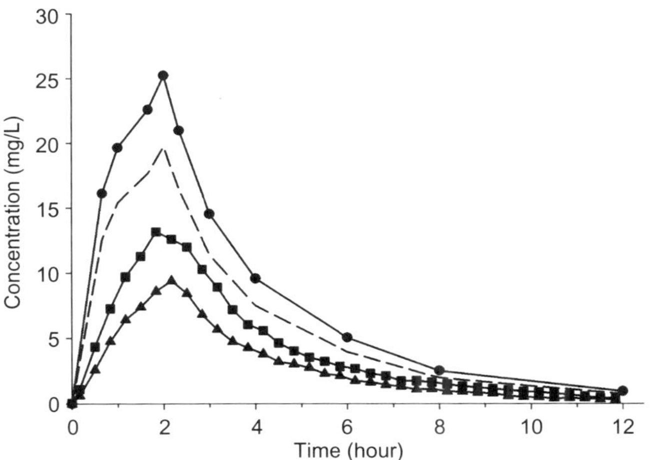

Figure 3 Mean ceftobiprole concentration in plasma (circles), skeletal muscle (squares), and s.c. adipose tissue (triangles) over 12 h in healthy volunteers ($n = 12$). The concentration of free drug in plasma (dashed line) was calculated based on the plasma protein binding of each patient. (From [15].)

and using total concentrations would overestimate the target-site concentrations and hence clinical efficacy. Microdialysis has proven useful in the measurement of free drug concentrations in muscle and s.c. adipose tissue [12–14]. A study by Barbour et al. [15] looked at the tissue penetration of ceftobiprole from plasma into skeletal muscle and adipose tissue after a single intravenous (i.v.) dose of 500 mg using microdialysis in healthy volunteers. The study measurements made it possible to show that ceftobiprole distributes into the ISF of skeletal muscle [free (t) $fAUC_{muscle}/fAUC_{plasma}$ of 0.69 ± 0.13] and s.c. adipose tissue ($fAUC_{s.c.}/fAUC_{plasma}$ of 0.49 ± 0.28). The $fAUC$ of plasma (76.0 ± 8.81 h·mg/L) and $fAUC$ of both tissues were significantly different, and a difference between the muscle (50.6 ± 10.9 h·mg/L) and s.c. tissue (34.3 ± 19.0 h·mg/L) was also noted (Figure 3). These findings confirmed that it is important to measure free active drug in each tissue and not make the assumption that free drug levels in plasma equal free drug levels in tissue, even in well-perfused tissues. The reason for the difference in penetration ratios of drug can be attributed to several factors, such as perfusion of particular tissue, local capillary density, degree of tissue binding, possibility of active transporters, loss of drug from peripheral compartments, and lipophilicity of the compound [15]. The MIC_{90} (minimum inhibitory concentration) value for ceftobiprole against methicillin-resistant *Staphylococcus aureus* [16] and penicillin-resistant *Streptococcus pneumoniae* [17] has been reported as 2 mg/L. Measurements from the study showed that concentrations in both skeletal muscle and adipose tissue met the efficacy breakpoint (i.e., remained above 2 mg/L for about 50% of the dosing interval). Hence, ceftobiprole qualifies as

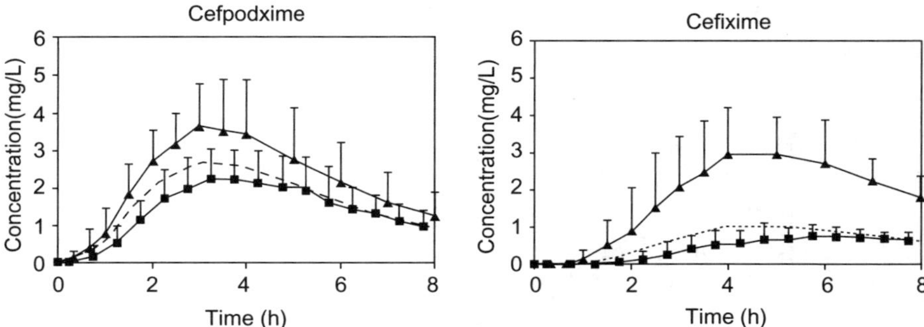

Figure 4 Plasma concentrations (m) and free, unbound concentrations in muscle (j) after a 400-mg oral dose of cefpodoxime (left) and cefixime (right) in six healthy male volunteers. The dashed lines are the calculated free plasma concentrations based on average protein-binding values (cefpodoxime, 25%; cefixime, 65%). Values are mean ± S.D. (From [14].)

a potential agent with penetration capabilities to treat complicated skin and skin structure infections, as it is the key determinant in clinical efficacy.

A comparison of the kinetic profiles of two cephalosporins (cefpodoxime and cefixime) in soft tissue given the same oral dose was conducted in six healthy volunteers [14]. In this study, the AUC value in the plasma of cefpodoxime was 22.4 mg/L·h ± 8.7 and for cefixime, 25.6 mg/L·h ± 8.5, which was similar. Interestingly, their target-site exposure was markedly different; AUC_{muscle} was 15.4 mg/L·h ± 5.1 and 7.3 mg/L·h ± 2.2 for cefpodoxime and cefixime, respectively. C_{max} in plasma for cefpodoxime and cefixime was found to be 3.9 mg/L ± 1.2 and 3.4 mg/L ± 1.1, and in muscle C_{max} was 2.4 mg/L ± 0.9 and 0.9 mg/L ± 0.3, respectively (Figure 4). The reason for the difference in target-site exposure of the two drugs despite the similar profile in plasma is the difference in the degree of protein binding. The study provided insight into the free interstitial levels of cefpodoxime and cefixime, which is more meaningful, and showed that it is not sufficient to consider only plasma concentration profiles when evaluating pharmacokinetic properties of anti-infective agents. Using total plasma concentrations may overestimate the therapeutic outcome because only the unbound fraction in plasma is able to cross the capillary membrane and reach the interstitial space where the infection is. This may be the reason that some antibiotic treatments fail despite good activity in vitro as well as resistance development [18,19]. The use of total plasma concentrations in a PK–PD approach to predicting the clinical efficacy of antibiotics is common where MIC values are compared with plasma concentrations of the antibiotic, usually without considering protein binding. It improves the accuracy of predicting the clinical efficacy of free antibiotic tissue levels by use of microdialysis. Cefpodoxime, with a low protein binding value of 25%, had a peak concentration in the muscle that was twice as high as that of cefixime (2.1 vs. 0.9 mg/L). The average tissue penetration ($AUC_{tissue,free}$/

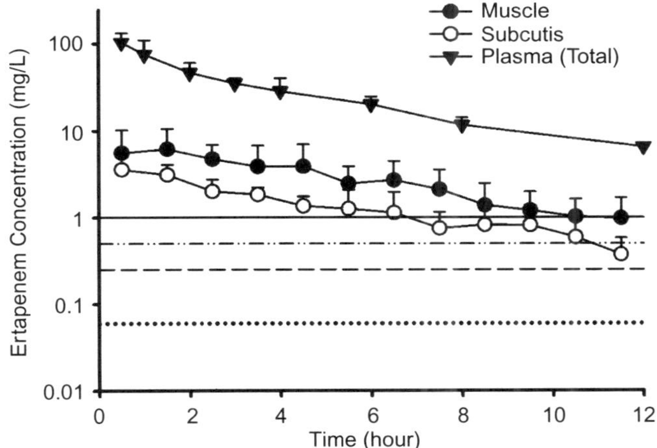

Figure 5 Ertapenem concentration (mean ± S.D.) time profiles in total plasma, skeletal muscle fluid, and interstitial adipose tissue (subcutis) following 1 g infusion for 30 min in healthy volunteers ($n = 6$). Horizontal lines indicate MIC_{90} values for methicillin-susceptible *Staphylococcus aureus* (dashed), *Streptococcus* spp. (dashed-dotted), extended-spectrum β-lactamase–producing Enterobacteriaceae (dotted), *Bacteroides fragilis*, and other anaerobic bacteria (solid). (From [11].)

$AUC_{plasma,total}$) of cefpodoxime (70%) was much higher than that of cefixime (29%), which is consistent with their protein binding values, suggesting that higher protein binding of the drug would lower the tissue penetration [18].

Burkhardt et al. [11] used microdialysis to measure and compare the free protein-unbound ertapenem concentrations in the interstitial space fluid of two soft tissues, skeletal muscle and subcutaneous adipose tissue, following the administration of 1 g infusion, and compared them with the respective plasma concentrations. Results from the study indicate that free, unbound ertapenem profiles in the ISF of both skeletal muscle and subcutaneous adipose tissue are lower than the corresponding total plasma concentrations (Figure 5). While free ISF concentrations of the skeletal muscle correlated well with free, unbound concentrations in plasma (4 to 16% of the total plasma concentration), they were comparably higher than free ISF concentrations in subcutaneous adipose tissue. This was observed in other studies as well, and differences in blood flow in these two tissues may be a possible explanation [10]. Free ertapenem concentrations of 1.13 ± 0.68 mg/L in the muscle, observed 12 h after single-dose i.v. infusion of ertapenem, exceeded the MIC_{90} value of methicillin-susceptible *S. aureus* (0.25 mg/L), *Streptococcus* spp. (0.5 mg/L), extended-spectrum β-lactamase–producing Enterobacteriaceae (0.03 to 0.06 mg/L), *Bacteroides fragilis*, and other anaerobic bacteria (≤1.0 mg/L) for at least 50% of the entire dosing interval. In comparison, free levels of 0.31 ± 0.16 mg/L in the subcutaneous adipose tissue (at 12 h) exceeded the MIC_{90} of the same skin and skin structure infection (SSSI) pathogens for at least 30% of the dosing interval [1]. The authors concluded that free, active ertapenem concentrations reached sufficient levels in noninfected ISF of muscle and subcutaneous adipose tissue.

When a drug is given orally, it gets absorbed from the intestine, following which it reaches the systemic circulation. Depending on the oral bioavailability of the drug, oral doses need to be adjusted to compare it with the respective i.v. dose to achieve similar therapeutic drug levels in the body. As a case in point, microdialysis was employed to compare free, active ciprofloxacin concentrations in the ISF of skeletal muscle and subcutaneous adipose tissue, after i.v. (400 mg) or oral ciprofloxacin (500 mg) administration, respectively [1,20]. Ciprofloxacin is a broad-spectrum anti-infective agent of the fluoroquinolone class used in the treatment of upper respiratory infections urinary tract infections, and skin infections. Free ciprofloxacin concentrations were measured in the ISF of skeletal muscle and subcutaneous adipose tissue, saliva, and cantharis-induced skin blister, as well as capillary plasma, and compared to total venous plasma concentrations. Mean fAUCs of both muscle and subcutaneous adipose tissue were significantly lower than the corresponding AUC for plasma after oral and i.v. administration. A $C_{\text{skin blister}}/C_{\text{plasma}}$ ratio > 4 is an indication that ciprofloxacin accumulates primarily in the inflamed lesions, while saliva and capillary blood concentrations were similar to total plasma [1,20].

In another study, Hollenstein et al. [21], using microdialysis, addressed the issue of tissue penetration of ciprofloxacin in obese subjects with a mean weight of 122 kg. They found significantly lower $AUC_{\text{tissue}}/AUC_{\text{plasma}}$ ratios in obese subjects (0.45 ± 0.27 vs. 0.82 ± 0.36) [14]. The results helped them conclude that the penetration of ISF is highly impaired in obese subjects, probably due to a reduced capillary permeability surface area in fat tissue.

An interesting application of microdialysis was extended to evaluate the effects of simulated microgravity (sμG) on the pharmacokinetics of ciprofloxacin [22]. Astronauts have been taking drugs during flights since the early days, but little information is available regarding the efficacy of drugs administered during space flights. Astronauts run the risk of infections due to long-term confinement of the spacecraft and impairment of the immune system [23,24]. Also, physiological changes induced by microgravity may affect the pharmacokinetics of antibiotics, resulting in altered concentrations at the infection site, thus affecting the way that antibiotics are given in space. The study determined and compared the unbound soft tissue concentrations of ciprofloxacin (250 mg orally) by microdialysis in normal gravity (1 g) and sμG and related them to plasma concentrations. Also, the disposition of ciprofloxacin in humans after 3 days of sμG in comparison to 1 g was examined. It was observed that the free interstitial ciprofloxacin concentrations measured by microdialysis in the medial vastus muscle ($AUC_{\text{free,tissue}}$) were slightly lower during sμG (1540 ± 879 ng/mL·h) than during 1 g (1861 ± 1091 ng/mL·h). The free plasma concentrations were not different in sμG (2547 ± 274 ng/mL·h) and 1 g (2518 ± 971 ng/mL·h). A slightly lower value of f was obtained in sμG

(0.61 ± 0.36 vs. 0.92 ± 0.63) suggesting that tissue penetration could be altered in microgravity [22]. However, the differences were not statistically significant, probably due to the small number of subjects that participated in our study.

Linezolid, an oxazolidinone, is approved for the treatment of nosocomial pneumonia and complicated SSSIs [8]. It shows good antimicrobial activity against various resistant gram-positive bacteria, including methicillin- and glycopeptide-resistant *S. aureus*. Although only free, unbound data are considered for antimicrobial efficacy, and most relevant pathogens are located in the ISF, most linezolid PK data available are based on total plasma concentrations [1,12]. Hence, a clinical microdialysis study was performed that evaluated the penetration of linezolid into soft tissues of healthy volunteers after single- and multiple-dose (MD) administration [12]. After calibration and baseline determination, 600 mg of linezolid was infused intravenously for 30 min and dialysate as well as blood samples were taken for up to 8 h. After withdrawal of the MD probes, volunteers were started on oral linezolid (600 mg) twice a day for 5 consecutive days [1]. The second set of MD experiments was started simultaneously with the last oral dose. Results show that after single i.v. administration of linezolid, $fAUC_{0\text{-}8}$ was 65.3 ± 18.2 and 75.8 ± 24.2 mg·h/L for skeletal muscle tissue and subcutaneous adipose tissue, respectively, which was significantly higher than the $fAUC_{0\text{-}8}$ of plasma (53.0 ± 11.6 mg·h/L) [12]. However, at steady state, no significant differences could be detected between concentrations in the ISF of skeletal muscle and subcutaneous adipose tissue. These findings indicated further that steady-state concentrations in both muscle ($fAUC_{24}$ muscle/MIC 58.9 ± 33.0 mg·h/L) and adipose subcutaneous tissue ($fAUC_{24}$ s.c./MIC 46.6 ± 15.9 mg·h/L) were sufficient to treat infections that are caused by pathogens with MICs of up to 4 mg/L.

3. PK–PD INDICES

The minimum inhibitory concentration is a well-established and routinely determined susceptibility breakpoint parameter for antibiotics. Combinations of this PD marker with free, unbound PK parameters to MIC-based PK–PD indices such as free time above the minimum inhibitory concentration ($fT_{>\text{MIC}}$), $fAUC/MIC$, and fC_{max}/MIC have led to a much better understanding of antibiotic dosing [1]. The first PK–PD index was developed for penicillins. It correlates in vivo efficacy with the amount of time that free drug levels stay above the MIC of the target organism ($fT_{>\text{MIC}}$) [1]. A common threshold of $fT_{>\text{MIC}} \geq 40\%$ seems to be sufficient for the clinical efficacy of β-lactam antibiotics, and fC_{max}/MIC index values of 10 to 12 seem to be a good predictor for aminoglycosides [1]. Target $fAUC_{24}/MIC$ values of 100 to 125 (gram negatives), 25 to 35 (gram positives), and fC_{max}/MIC index values of 10 have been identified for fluoroquinolones, and $fT_{>\text{MIC}} \geq 85\%$ is a good outcome predictor for oxazolidinones [1]. Once these MIC-based PK/PD indices are identified, they can support the identification of optimized dosing regimens and the prediction of treatment outcome. Knowledge of the free antibiotic concentration time course in the ISF is necessary to establish the respective $fT_{>MIC}$, fC_{max}/MIC, and $fAUC_{24}/MIC$ index values. Although several techniques are available for the determination of free, unbound concentrations, they are not all capable of characterizing dynamic changes in free ISF concentrations. Only microdialysis allows the combination of these two properties, which is essential in predicting clinical efficacy with accuracy. Therefore, it is a very valuable sampling tool and has become an inherent part of evaluation and establishment of the PK–PD approach.

3.1. NSAIDs

Nonsteroidal anti-inflammatory drugs (NSAIDs) are used widely to treat pain and rheumatoid and osteoarthritis, among others, but they have been shown to have side effects that can limit their application, at least for systemic dosing. Several studies have tried to assess and compare plasma and tissue concentrations of nonsteroidal anti-inflammatory drugs after systemic or topical application, to assess whether topical administration is a feasible way to utilize the drug effects while limiting or circumventing the side-effect potential.

Tissue and plasma exposure after an intramuscular or topical dose of 100 mg of ketoprofen was assessed in a crossover design study [25]. Most muscle and subcutaneous tissue dialysate samples after a topical application had ketoprofen concentrations below the limit of detection, but measurable tissue concentrations were 10-fold lower than these of intramuscular injections. Median ketoprofen concentrations in tissue after intramuscular (i.m.) injections were 0.05 and 0.04 µg/mL in muscle and skin, respectively. Plasma levels of the drug were around 4 to 10 µg/mL after i.m. injection versus a maximum 0.05 µg/mL after topical application. Also, tissue concentrations after topical application were much more variable than after intramuscular injection.

Exposure of skeletal muscle and subcutaneous adipose tissue after oral or topical dosing of ibuprofen was assessed by Tegeder et al. [26]. In eight male study participants, 800 mg of ibuprofen was given in tablet form or applied to the skin in a gel formulation, and R- and S-ibuprofen levels were measured. Plasma AUC for R- and S-ibuprofen after topical application was about 0.49 and 0.46% that of oral dosing, respectively. When ibuprofen dialysate concentrations were compared, ibuprofen AUCs after topical dosing reached 43% of the levels after oral dosing for muscle, but was about 10 times higher than oral-dosing AUCs in s.c. tissue. When comparing the dialysate concentrations after oral dosing, the authors found that they were concordant with unbound drug levels in plasma, assuming a protein binding of 99.75% after topical application; however, muscle dialysate concentrations were similar to unbound concentrations in plasma, while dialysate levels from subcutaneous adipose tissue were about 10 times higher than total plasma levels.

Diclofenac tissue levels at two different depths, one shallow (ca. 4 mm) and one deep (ca. 10 mm) were assessed after topical application of a gel formulation that is commercially available [27]. Dialysate concentrations of diclofenac were measurable in only 11 of the 20 healthy volunteers. The resulting concentrations in tissues after oral and topical applications were highly variable. In patients for whom diclofenac values were quantifiable after topical application, diclofenac AUC in shallow and deep tissue were 532 ± 197 and $438 \pm 249\,\mu g\cdot min/ml$, respectively.

Tissue levels of diclofenac after multiple- and single-dose administration were also compared [3]. In this study, volunteers received either 60 mg of diclofenac gel three times a day for 4 days or 300 mg of diclofenac as a gel preparation as a single dose. AUC values for tissue and plasma were higher after single-dose treatment. One factor in this study was also the high variability of tissue penetration and the resulting concentrations. For many microdialysis samples for multiple doses and slightly less for a single dose, the concentrations were below the lower limit of quantitation. The high variability is thought to stem from interindividual differences in skin barrier properties.

The tissue bioavailabilities of a spray gel and the oral application form of diclofenac were compared as well [12]. The authors found that the AUC in tissue was significantly higher after topical application than after oral application. Also, plasma AUCs were significantly lower after topical than after oral drug. When diclofenac was applied as a spray gel, the tissue/plasma AUC ratios were 78 and 52% for subcutaneous and muscle tissues, respectively. For oral application, tissue/plasma ratios were 0.54 and 0.56% for subcutaneous and muscle tissues, respectively. This was ascribed to the high plasma protein binding of diclofenac of >99%. In fact, after oral administration, unbound plasma concentrations were found to be similar to unbound concentrations in tissue. The authors conclude that effective concentrations are reached in the tissue after topical administration of the spray gel.

3.2. Anticancer Drugs

The availability of active drug at the site of action is of vital importance in anticancer therapy. Here, as for many antibiotic drugs, concentrations in the biophase are difficult to measure; thus, plasma concentrations are commonly used to extrapolate for dosing recommendations. Microdialysis has been used in some oncology trials to try to link drug available at the tumor site to response and possibly explain part of the tumor response variability by interindividual differences in bioavailability. A study done by Joukhadar et al. [19] measured the concentrations of dacarbazine and the active metabolite 5-aminoimidazole-4-carboxamide in the plasma, healthy subcutaneous tissue, and malignant tissue in metastatic malignant melanoma of seven patients. The authors found that there appear to be no limitations to the penetrability of dacarbazine and the active metabolite from plasma to malignant and healthy tissues. They observed a delay in T_{max} of about 30 to 60 min, which was attributed to the rate of diffusion being impaired due to increased tissue pressure, but not, as mentioned above, the extent of the diffusion. The mean $AUC_{ISF}/$

AUC_{plasma} for dacarbazine was one for tumor and subcutaneous tissue, respectively, at a dose of $200\,mg/m^2$ ($n = 4$). At a dose of $800\,mg/m^2$, the mean dacarbazine AUC_{ISF}/AUC_{plasma} was 0.76 ($n = 1$) in tumor tissue and 1 ($n = 2$) in healthy tissue. The mean AUC_{ISF}/AUC_{plasma} for the active metabolite after the $200\,mg/m^2$ was 0.93 in tumor tissue and 0.78 in healthy tissue ($n = 4$), for the $800\,mg/m^2$ dose, the ratio was 0.43 and 0.52 for tumor ($n = 1$) and healthy tissue ($n = 2$), respectively, and for the $100\,mg/m^2$ dose ($n = 1$), the ratio was 0.58 and 0.3 for tumor and unaffected subcutaneous tissue, respectively. The authors conclude, therefore, that limited response to dacarbazine in the clinical setting may not be explained by limited availability of the drug at the interstitial level.

The distribution of capecitabine, a 5-fluorouracil (5-FU) prodrug, was also tested employing the microdialysis method [28]. Tissue distribution was evaluated in 10 patients with breast cancer which had metastasized into the skin after a dose of $1250\,mg/m^2$ twice daily for two weeks. Microdialysis was performed on day 1 and again between days 12 and 14, which marked the end of the treatment period for one cycle. A CMA-10 microdialysis catheter (16 mm in length, 0.5 mm in diameter) was inserted into tumor tissue and into healthy subcutaneous tissue in the thigh. Capecitabine and its metabolites 5′-DFCR, 5′-DFUR, and 5-FU were assessed for penetrability into tissues. The authors found that capecitabine concentrations in tissues and plasma were quite variable, with C_{max} values spanning from below $1\,\mu g/mL$ to 16 for plasma and tumor and up to $28\,\mu g/mL$ for subcutaneous healthy tissue, respectively. Capecitabine AUC in malignant tissue was significantly higher than in plasma, with a median AUC_{tumor}/AUC_{plasma} ratio of 1.76. $AUC_{5\text{-}DFCR}$ in s.c. and tumor tissue was a median 67% compared to AUC_{plasma}, and $AUC_{5\text{-}DFUR}$ was 53% in tumor and 44% in s.c. tissue compared to AUC_{plasma}. Tissue distribution occurred quickly; within a median 15 to 45 min of plasma, the T_{max} value in tissue was reached. 5-FU concentrations were relatively low in plasma and tissue interstitium, reaching only a mean 0.25, 0.26, and 0.33 in plasma, tumor, and subcutaneous tissue, respectively. Mean $AUC_{tissue}/AUC_{plasma}$ was 0.25 and 0.36 for tumor and healthy tissue, respectively.

A study assessing the tissue distribution after i.v. administration of 5-FU was done by Müller et al. [27]. Microdialysis probes were implanted in the breast cancer and healthy tissue of 10 patients. Each patient received $600\,mg/m^2$ 5-FU as part of a three-drug regimen containing either cyclophosphamide and methotrexate or cyclophosphamide and epirubicin in addition to 5-FU as part of a preoperative chemotherapy. The authors also assessed response to the treatment based on World Health Organization criteria. It was found that 5-FU distributes quickly into healthy and malignant tissue, with a T_{max} value occurring within about 15 min in tissues. Distribution of drug into both tissues was not significantly different, yet absorption into tumor tissue had a longer half-life than that of healthy s.c. tissue. In addition to the pharmacokinetic analysis, patients were evaluated for tumor response. Two patients showed stable disease (tumor size $\pm$ 25% compared to pre-therapy conditions) and eight patients showed a partial response (≥50% tumor size reduction). When plasma and tumor tissue AUCs were compared, the two patients having stable disease had high plasma but low tumor levels. The conclusion of the authors was that plasma PK is, in this case, not predictive of tissue PK, and that response, even though observed only in a very small sample, correlated better with tissue exposure.

Plasma and malignant tissue distribution of cisplatin from two different treatment regimens was compared in patients with oral cancer [29]. The treatments were (1) an intraarterial infusion of $150\,mg/m^2$ in $500\,mL$ of saline and (2) an intraarterial embolization by giving $150\,mg/m^2$ in 45 to $60\,mL$ of saline, both via a microcatheter that was placed into a feeding artery close to the tumor. Both treatments were coupled with intravenous administration of $9\,g/m^2$ sodium thiosulfate (STS). The study found that after intraarterial administration, tumor tissue had a higher exposure to cisplatin than to plasma and also that the embolization technique yielded significantly higher tumor concentrations than the infusion technique with cisplatin AUCs after infusion of 19.8 and $8.6\,\mu mol/L{\cdot}h$, respectively, and AUCs after embolization being 89.6 and $0.15\,\mu mol/L{\cdot}h$ in tissue and plasma, respectively. C_{max} was reached later in the embolization treatment compared to the infusion technique in tissue. STS plasma levels were measured as well to allow for calculating the $AUC_{STS}/AUC_{cisplatin}$ ratio. The ratio after intraarterial infusion was 211 and 984 for the tumor tissue and plasma, respectively. After administration of the crystalline suspension, the ratios were 49 and 42,000 for tumor tissue and plasma, respectively. $AUC_{STS}/AUC_{cisplatin}$ in plasma should be higher than 500, whereas in malignant tissue it should be below100. Based on these limits, only the cisplatin embolization achieved these goals in all patients. Even though the tumor response was not assessed, it is speculated that, assuming linear dose–response characteristics, higher concentrations in tumor tissue could yield higher response rates.

Distribution of carboplatin in plasma, healthy subcutaneous tissue, and malignant tissue was assessed by Blöchl-Daum et al. [30] in six patients with metastatic malignant melanoma. Microdialysate was collected from tumor and healthy subcutaneous tissue. The patients received $400\,mg/m^2$ carboplatin as one dose. It was found that total serum levels exceeded tissue levels during the entire observation period of $240\,min$ after the start of the infusion. Tumor levels were slightly higher than subcutaneous levels. Mean ± S.E.M. C_{max} values in tumor and subcutaneous tissue were 7.5 ± 2.0 and 5.6 ± 1.2 compared to 14.6 ± 1.1 in serum, and mean AUC_{tissue}/AUC_{serum} ratios were 0.58 and 0.41 for tumor and healthy s.c. tissue, respectively, with $C_{max,\,tissue}/C_{max,\,serum}$ levels showing the same trend (Figure 6). There was no difference in AUC and C_{max} between the tissues, and tumor tissue pharmacokinetic parameters were not significantly correlated to serum parameters. The tissue exposure of carboplatin and the feasibility of using the microdialysis method for sampling for longer time periods from tissues were evaluated in a study by Konings et al. [31] The study assessed carboplatin concentrations in plasma and microdialysate from

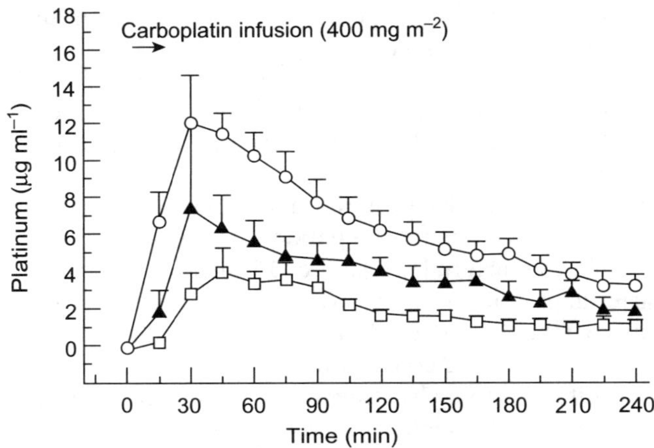

Figure 6   Time profile of carboplatin concentrations (mean ± S.E.M.) in serum (○), tumor (▲), and subcutaneous (□) tissue in cancer patients ($n = 6$). (From [30].)

catheters implanted into subcutaneous tumors and healthy subcutaneous adipose tissue after a 6 mg/mL·min dose over 1 h. The resulting doses ranged from 700 to 1150 mg of carboplatin. The results indicate that unbound platinum concentrations in plasma and tissues are very similar.

$AUC_{tissue}/AUC_{plasma}$ ratios (total plasma) range from 0.64 to 1.1 and adipose tissue ratios range from 0.71 to 1.46, even though carboplatin PK was measurable in only three of the six patients with assessable tumor concentrations. Probe recovery was stable in tumor tissue over the period of 48 h observed following drug infusion, thus indicating, as the authors point out, that microdialysis may be a useful tool for the evaluation of pharmacokinetics of antineoplastics in the future. Although the technique delivered stable recovery values for the tumor tissue, for healthy adipose tissue, recoveries fluctuated and in one case even declined, which accounts for only three of six patients dialysate concentrations being able to be used for pharmacokinetic analysis.

An exploratory study to assess the feasibility of the microdialysis method to assess δ-aminolevulinic acid concentrations after topical application was conducted by Wennberg et al. [32]. δ-Aminolevulinic acid concentrations were measured in basal cell carcinomas of 18 patients and healthy skin tissue. δ-Aminolevulinic acid accumulated rapidly in tumor tissue and to a much smaller degree in healthy skin tissue. Tumor T_{max} was reached within 15 to 30 min of application of an adhesive material containing the drug for 210 to 240 min. In two cases the cancerous regions were also exposed to 630-nm light. After light exposure, tumor levels dropped to one-third of their original level within 2 to 3 h, whereas the levels of δ-aminolevulinic acid in non-light-exposed areas remained unchanged. The authors also found a significantly increased perfusion in basal cell carcinoma (measured by scanning laser Doppler). In another study, δ-aminolevulinic acid and methyl aminolevulinic acid penetration into superficial and nodular BCCs with catheters at depths of less than 1 mm or up to 1.9 mm was elucidated [33]. For this study, 20 patients with 27 BCCs were treated with either δ-aminolevulinic acid or methyl aminolevulinic acid either with or without curettage. Both drug molecules can be found at a depth of less than 1 mm in almost all patients, whereas for more deeply situated microdialysis catheters, drug could be found in only six of 11 tumors where the catheter depth was higher than 1 mm. When peak concentrations in tissues were compared, however, there was no difference between deeper and shallower catheters. Also, no significant difference was found between tissue treated with curettage and untreated tissus. The authors concluded from this that the layers of the skin may already be compromised, giving curettage no added effect. Finally, there was no statistically significant advantage of using methyl aminolevulinate over non-methyl-substituted drug in terms of measurable tissue exposure. The authors conclude that lack of, or incomplete, tissue penetration, at least up to the concentrations observed in the study, may not be the only factors determining treatment failure.

4. TOPICAL BIOEQUIVALENCE

Topical bioequivalence has been assessed in a few clinical studies looking at the feasibility of using the microdialysis method for this purpose. The bioequivalence of two lidocaine topical formulations, cream and ointment, was determined using the microdialysis method and comparing its results to the tape-stripping method [34]. The authors found that when measured with microdialysis, the cream produced 4.8-fold-higher AUCs than these of the ointment formulation. The tape-stripping method found differences between the two formulations as well, even though the magnitude of the difference found with tape stripping was smaller than that found using microdialysis. The authors also assessed the number of subjects necessary for clinical trials to find differences between formulations based on the interindividual variability of 61% they found. To meet the requirements of the U.S. Food and Drug Administration (FDA) (90% confidence interval, 80 to 125% of AUC and $C_{\max}$), 18 subjects would be needed to compare two formulations if three probes were inserted per arm, and 27 subjects would be needed if two probes were inserted. The authors concluded that the high interindividual variability might result primarily from differences in skin-barrier properties between subjects. Kreilgaard et al. [35] evaluated bioequivalence between a lidocaine microemulsion and a commercially available lidocaine formulation. The authors also assessed the analgesic effect as a pharmacodynamic parameter. It was found that the microemulsion had a 4.3-fold-higher AUC than the ointment and a lagtime decreased by 20 min. The absorption coefficient estimate was 2.9-fold higher for the microemulsion than for the ointment. The pharmacodynamic assessment shows a significant pain reduction for both formulations compared to a drug-free microemulsion vehicle, but even though the microemulsion has a higher AUC (i.e., more free drug present at the site of action), there are no differences between the two active drug treatments. The authors attribute this to a maximum effect already achieved at the targeted nerve fibers. When dialysate concentrations were adjusted for each subject individually, the variability decreased. Here again, the authors conclude that a major contributing factor to interindividual variability in measurements is due to individual differences in skin properties. When assessing whether probe implantation depth was an influencing factor for variability, the authors found no significant contribution. The results in human skin were then compared to experiments done in rat skin. There it was found that whereas rat skin had shorter lagtimes, the absorption rates were comparable. A paper by McCleverty et al. [36] aimed at evaluating the degree and source of variability of using microdialysis as a sampling tool and, furthermore, how bioequivalence employing microdialysis needs to be designed to achieve FDA criteria. Variability estimates were generated from a study in eight healthy volunteers who received a topical application of methyl salicylate with three different probe perfusion fluids. Each subject had six probes implanted, two for each treatment, and the variability in AUC estimates was calculated. Between-subject variability was calculated for both the topically applied drug and its metabolite, salicylic acid, and was comparable for these two. Within-subject variability between the two probes was lower than between-subject variability, but higher in originally applied drug measurements. The authors calculated the necessary number of subjects for a bioequivalence study based on these estimates and found that, to be able to detect a difference of greater than 80 to 125% between two formulations with a power of 80%, the study needs to be designed as a crossover study with 20 subjects and two probes per subject.

5. ENDOGENOUS COMPOUNDS

Microdialysis has also been used for the study of growth factor and cytokine concentrations in healthy compared to tumor tissue. One study [37] in breast cancer patients compared vascular endothelial growth factor (VEGF) tissue levels in tumor compared to healthy breast tissue and found that VEGF was elevated in tumor tissue. Also, the study found that VEGF levels in plasma and breast tissue for the patients in the study was not related, and the authors pointed out that using plasma levels alone may not be indicative of VEGF levels measurable in the tissue. The same group also tried to elucidate the role of estradiol and its influence on IL-8 secretion in breast cancer patients [38].

Changes in levels of markers for inflammation and pain in tissue after irritation with heat, inflammation by UVB radiation, or both were evaluated using the microdialysis sampling method together with immunoassay quantification

[39]. In addition, the effect of two different doses of a COX-inhibitor, ibuprofen, on three cytokines, IL-1β, IL-6, and TNF-α, as well as heat and mechanical pressure tolerance, were measured. It was found that after inflammation and stimulation with heat a different cytokine profile is increased. Ibuprofen decreased IL-1β and IL-6 concentrations to a higher degree at an 800-mg dose compared to a 400-mg dose, but the two doses do not differ significantly in their heat and mechanical pain threshold reduction. The authors conclude, therefore, that ibuprofen has a discernible anti-inflammatory and antianalgesic component.

6. CONCLUSIONS

The examples presented in this chapter show that microdialysis is applicable for a variety of purposes, such as bioequivalence studies for topical formulations, pharmacokinetic studies with high temporal resolution, and tissue metabolism studies. Being that the method is comparatively young, several methodological issues remain to be addressed. Analytical methods for analyzing samples gathered via microdialysis frequently have to be very sensitive, due to the limited amount of sample volume available [40]. Probe calibration is another important part of any microdialysis experiment, especially when absolute interstitial space concentrations need to be reported. Recovery should be measured for each probe and individual experiment and is dependent on a variety of factors: for example, the tissue surrounding the probe, the flow rate, the length of the diffusible membrane, the composition of the perfusion fluid, and the molecule to be perfused [40,41]. In studies assessing topical bioavailability and bioequivalence (BA/BE), high interindividual variability in pharmacokinetic parameter measurements can be a problem [34–36]. This variability is considerably smaller after systemic application, suggesting that the skin properties have a vast influence on the inter- and intraindividual variabilities seen in pharmacokinetic parameters in bioavailability and bioequivalence studies [41]. In summary, microdialysis is an exciting new sampling technique for the evaluation of the pharmacokinetics and bioavailability of drugs in skin and soft tissues.

REFERENCES

[1] Schmidt, S., Banks, R., Kumar, V., Rand, K.H., Derendorf, H. (2008). Clinical microdialysis in skin and soft tissues: an update. *Journal of Clinical Pharmacology*, *48*, 351–364.

[2] Brunner, M., Derendorf, H. (2006). Clinical microdialysis: current applications and potential use in drug development. *Trends in Analytical Chemistry*, *25*, 674–680.

[3] Dehghanyar, P., Mayer, B.X., Namiranian, K., Mascher, H., Müller, M., Brunner, M. (2004). Topical skin penetration of diclofenac after single- and multiple-dose

application. *International Journal of Clinical Pharmacology and Therapeutics, 42,* 353–359.

[4] Benfeldt, E., Serup, J., Menne, T. (1999). Effect of barrier perturbation on cutaneous salicylic acid penetration in human skin: in vivo pharmacokinetics using microdialysis and non-invasive quantification of barrier function. *British Journal of Dermatology, 140,* 739–748.

[5] Morgan, C.J., Renwick, A.G., Friedmann, P.S. (2003). The role of stratum corneum and dermal microvascular perfusion in penetration and tissue levels of water-soluble drugs investigated by microdialysis. *British Journal of Dermatology, 148,* 434–443.

[6] Benfeldt, E., Groth, L. (1998). Feasibility of measuring lipophilic or protein-bound drugs in the dermis by in vivo microdialysis after topical or systemic drug administration. *Acta Dermatovenereologica, 78,* 274–278.

[7] Klimowicz, A., Farfal, S., Bielecka-Grzela, S. (2007). Evaluation of skin penetration of topically applied drugs in humans by cutaneous microdialysis: acyclovir vs. salicylic acid. *Journal of Clinical Pharmacy and Therapeutics, 32,* 143–148.

[8] Borg, N., Gotharson, E., Benfeldt, E., Groth, L., Stahle, L. (1999). Distribution to the skin of penciclovir after oral famciclovir administration in healthy volunteers: comparison of the suction blister technique and cutaneous microdialysis. *Acta Dermatovenereologica, 79,* 274–277.

[9] Tegeder, I., Brautigam, L., Podda, M., Meier, S., Kaufmann, R., Geisslinger, G., Grundmann-Kollmann, M. (2002). Time course of 8-methoxypsoralen concentrations in skin and plasma after topical (bath and cream) and oral administration of 8-methoxypsoralen. *Clinical Pharmacology & Therapeutics, 71,* 153–161.

[10] Brunner, M., Hollenstein, U., Delacher, S., Jager, D., Schmid, R., Lackner, E., Georgopoulos, A., Eichler, H.G., Müller, M. (1999). Distribution and antimicrobial activity of ciprofloxacin in human soft tissues. *Antimicrobial Agents and Chemotherapy, 43,* 1307–1309.

[11] Burkhardt, O., Brunner, M., Schmidt, S., Grant, M., Tang, Y., Derendorf, H. (2006). Penetration of ertapenem into skeletal muscle and subcutaneous adipose tissue in healthy volunteers measured by in vivo microdialysis. *Journal of Antimicrobial Chemotherapy, 58,* 632–636.

[12] Brunner, M., Dehghanyar, P., Seigfried, B., Martin, W., Menke, G., Müller, M. (2005). Favourable dermal penetration of diclofenac after administration to the skin using a novel spray gel formulation. *British Journal of Clinical Pharmacology, 60,* 573–577.

[13] Hollenstein, U., Brunner, M., Mayer, B.X., Delacher, S., Erovic, B., Eichler, H.G., Müller, M. (2000). Target site concentrations after continuous infusion and bolus injection of cefpirome to healthy volunteers. *Clinical Pharmacology & Therapeutics, 67,* 229–236.

[14] Liu, P., Müller, M., Grant, M., Obermann, B., Derendorf, H. (2005). Tissue penetration of cefpodoxime and cefixime in healthy subjects. *Journal of Clinical Pharmacology, 45,* 564–569.

[15] Barbour, A., Schmidt, S., Sabarinath, S.N., Grant, M., Seubert, C., Skee, D., Murthy, B., Derendorf, H. (2009). Soft-tissue penetration of ceftobiprole in healthy volunteers determined by in vivo microdialysis. *Antimicrobial Agents and Chemotherapy, 53,* 2773–2776.

[16] Jones, R.N., Deshpande, L.M., Mutnick, A.H., Biedenbach, D.J. (2002). In vitro evaluation of BAL9141, a novel parenteral cephalosporin active against oxacillin-resistant staphylococci. *Journal of Antimicrobial Chemotherapy, 50,* 915–932.

[17] Hebeisen, P., Heinze-Krauss, I., Angehrn, P., Hohl, P., Page, M.G., Then, R.L. (2001). In vitro and in vivo properties of Ro 63-9141, a novel broad-spectrum cephalosporin with activity against methicillin-resistant staphylococci. *Antimicrobial Agents and Chemotherapy, 45,* 825–836.

[18] Liu, P., Müller, M., Derendorf, H. (2002). Rational dosing of antibiotics: the use of plasma concentrations versus tissue concentrations. *International Journal of Antimicrobial Agents, 19,* 285–290.

[19] Joukhadar, C., Frossard, M., Mayer, B.X., Brunner, M., Klein, N., Siostrzonek, P., Eichler, H.G., Müller, M. (2001). Impaired target site penetration of beta-lactams may account for therapeutic failure in patients with septic shock. *Critical Care Medicine, 29,* 385–391.

[20] Brunner, M., Stabeta, H., Moller, J.G., Schrolnberger, C., Erovic, B., Hollenstein, U., Zeitlinger, M., Eichler, H.G., Müller, M. (2002). Target site concentrations of ciprofloxacin after single intravenous and oral doses. *Antimicrobial Agents and Chemotherapy, 46,* 3724–3730.

[21] Hollenstein, U.M., Brunner, M., Schmid, R., Müller, M. (2001). Soft tissue concentrations of ciprofloxacin in obese and lean subjects following weight-adjusted dosing. *International Journal of Obesity and Related Metabolic Disorders, 25,* 354–358.

[22] Schuck, E.L., Grant, M., Derendorf, H. (2005). Effect of simulated microgravity on the disposition and tissue penetration of ciprofloxacin in healthy volunteers. *Journal of Clinical Pharmacology, 45,* 822–831.

[23] Nicogossian, A.E., Sawin, C.F., Huntoon, C.L. (1989). Overall physiologic response to space flight. In: Nicogossian, A.E., Sawin, C.F., Huntoon C.L. (Eds.), *Space Physiology and Medicine,* Lea & Febiger, Philadelphia, pp. 139–154.

[24] Taylor, G.R., Konstantinova, I., Sonnenfeld, G., Jennings, R. (1997). Changes in the immune system during and after spaceflight. *Advances in Space Biology and Medicine, 6,* 1–32.

[25] Tegeder, I., Lotsch, J., Kinzig-Schippers, M., Sorgel, F., Kelm, G.R., Meller, S.T., Geisslinger, G. (2001). Comparison of tissue concentrations after intramuscular and topical administration of ketoprofen. *Pharmaceutical Research, 18,* 980–986.

[26] Tegeder, I., Muth-Selbach, U., Lotsch, J., Rusing, G., Oelkers, R., Brune, K., Meller, S., Kelm, G.R., Sorgel, F., Geisslinger, G. (1999). Application of microdialysis for the determination of muscle and subcutaneous tissue concentrations after oral and topical ibuprofen administration. *Clinical Pharmacology & Therapeutics, 65,* 357–368.

[27] Müller, M., Mader, R.M., Steiner, B., Steger, G.G., Jansen, B., Gnant, M., Helbich, T., Jakesz, R., Eichler, H.G., Blochl-Daum, B. (1997). 5-Fluorouracil kinetics in the interstitial tumor space: clinical response in breast cancer patients. *Cancer Research, 57,* 2598–2601.

[28] Mader, R.M., Schrolnberger, C., Rizovski, B., Brunner, M., Wenzel, C., Locker, G., Eichler, H.G., Müller, M., Steger, G.G. (2003). Penetration of capecitabine and its metabolites into malignant and healthy tissues of patients with advanced breast cancer. *British Journal of Cancer, 88,* 782–787.

[29] Tegeder, I., Brautigam, L., Seegel, M., Al-Dam, A., Turowski, B., Geisslinger, G., Kovacs, A.F. (2003). Cisplatin tumor concentrations after intra-arterial cisplatin infusion or embolization in patients with oral cancer. *Clinical Pharmacology & Therapeutics*, *73*, 417–426.

[30] Blöchl-Daum, B., Müller, M., Meisinger, V., Eichler, H.G., Fassolt, A., Pehamberger, H. (1996). Measurement of extracellular fluid carboplatin kinetics in melanoma metastases with microdialysis. *British Journal of Cancer*, *73*, 920–924.

[31] Konings, I.R., Engels, F.K., Sleijfer, S., Verweij, J., Wiemer, E.A., Loos, W.J. (2009). Application of prolonged microdialysis sampling in carboplatin-treated cancer patients. *Cancer Chemotherapy and Pharmacology*, *64*, 509–516.

[32] Wennberg, A.M., Larko, O., Lonnroth, P., Larson, G., Krogstad, A.L. (2000). Delta-aminolevulinic acid in superficial basal cell carcinomas and normal skin: a microdialysis and perfusion study. *Clinical and Experimental Dermatology*, *25*, 317–322.

[33] Sandberg, C., Halldin, C.B., Ericson, M.B., Larko, O., Krogstad, A.L., Wennberg, A.M. (2008). Bioavailability of aminolaevulinic acid and methylaminolaevulinate in basal cell carcinomas: a perfusion study using microdialysis in vivo. *British Journal of Dermatology*, *159*, 1170–1176.

[34] Benfeldt, E., Hansen, S.H., Volund, A., Menne, T., Shah, V.P. (2007). Bioequivalence of topical formulations in humans: evaluation by dermal microdialysis sampling and the dermatopharmacokinetic method. *Journal of Investigative Dermatology*, *127*, 170–178.

[35] Kreilgaard, M., Kemme, M.J., Burggraaf, J., Schoemaker, R.C., Cohen, A.F. (2001). Influence of a microemulsion vehicle on cutaneous bioequivalence of a lipophilic model drug assessed by microdialysis and pharmacodynamics. *Pharmaceutical Research*, *18*, 593–599.

[36] McCleverty, D., Lyons, R., Henry, B. (2006). Microdialysis sampling and the clinical determination of topical dermal bioequivalence. *International Journal of Pharmaceutics*, *308*, 1–7.

[37] Garvin, S., Dabrosin, C. (2008). In vivo measurement of tumor estradiol and vascular endothelial growth factor in breast cancer patients. *BMC Cancer*, *8*, 73.

[38] Bendrik, C., Dabrosin, C. (2009). Estradiol increases IL-8 secretion of normal human breast tissue and breast cancer in vivo. *Journal of Immunology*, *182*, 371–378.

[39] Angst, M.S., Clark, J.D., Carvalho, B., Tingle, M., Schmelz, M., Yeomans, D.C. (2008). Cytokine profile in human skin in response to experimental inflammation, noxious stimulation, and administration of a COX-inhibitor: a microdialysis study. *Pain*, *139*, 15–27.

[40] Chaurasia, C.S., Müller, M., Bashaw, E.D., Benfeldt, E., Bolinder, J., Bullock, R., Bungay, P.M., DeLange, E.C., Derendorf, H., Elmquist, W.F., et al. (2007). AAPS-FDA Workshop White Paper: Microdialysis Principles, Application, and Regulatory Perspectives. *Journal of Clinical Pharmacology*, *47*, 589–603.

[41] Kreilgaard, M. (2002). Assessment of cutaneous drug delivery using microdialysis. *Advanced Drug Delivery Reviews*, *54* (Suppl. 1), S99–S121.

11

MICRODIALYSIS ON ADIPOSE TISSUE: MONITORING TISSUE METABOLISM AND BLOOD FLOW IN HUMANS

Gijs H. Goossens, Wim H. M. Saris, and Ellen E. Blaak

NUTRIM School for Nutrition, Toxicology and Metabolism, Maastricht University Medical Centre, Maastricht, The Netherlands

1. INTRODUCTION

Adipose tissue is the main lipid storage depot in the human body. Importantly, adipose tissue is not just a passive organ for the storage of excess energy. It is a metabolically active organ that plays a central role in health and disease. Adipocytes express and secrete a variety of substances (adipokines), which may act at both the local (autocrine and/or paracrine) and systemic (endocrine) level. On the other hand, adipose tissue expresses numerous receptors that allow it to respond to several afferent signals. The acute metabolic responses in adipose tissue are regulated predominantly by insulin and catecholamines. Through this interactive network, adipose tissue is involved in the regulation of a variety of biological processes. Several pathological conditions, including obesity (adipose tissue excess) and type 2 diabetes mellitus, are characterized by disturbances in adipose tissue metabolism. Therefore, more insight into the regulation of adipose tissue metabolism in humans may contribute to more successful approaches to prevent and/or treat these conditions.

Applications of Microdialysis in Pharmaceutical Science, First Edition. Edited by Tung-Hu Tsai.
© 2011 John Wiley & Sons, Inc. Published 2011 by John Wiley & Sons, Inc.

Microdialysis is a relatively noninvasive technique used to study human metabolism in vivo. The technique makes it possible to measure concentrations of substances in the extracellular fluid and manipulation of the interstitial space of a certain tissue without affecting surrounding tissues and/or whole-body function. The technique was introduced more than 35 years ago by Delgado and colleagues [1] and was developed further by Ungerstedt and Pycock [2] to study brain function in rats. From this experimental field, the use of microdialysis gradually spread to other research areas, and it now plays an important role in the investigation of tissue concentrations of both endogenous and exogenous compounds. Microdialysis is also used to study adipose tissue metabolism in vivo in humans.

In the present chapter, the monitoring of human adipose tissue metabolism and blood flow using microdialysis is discussed. First, the principles and practical considerations for microdialysis in adipose tissue are explained. Thereafter, several approaches to the use of microdialysis on human adipose tissue are discussed in detail.

2. PRINCIPLES AND PRACTICAL CONSIDERATIONS IN THE USE OF MICRODIALYSIS ON ADIPOSE TISSUE

The measurement of concentrations of substances directly in the extracellular fluid using microdialysis has several advantages over the measurement of these concentrations in the circulation. First, it provides information about the concentration of the active form of a substance to which the cell is exposed. Second, it overcomes the fact that the distribution of a substance between the blood (circulation) and extracellular space (tissue) varies substantially depending on uptake, local production and/or secretion, and blood flow, as well as the physical characteristics of the substance, including molecular weight, molecular charge, and lipophilicity. It is, however, very important to understand the principles of microdialysis and to consider the practical issues that need to be taken into account for successful application of the technique.

2.1. Principles of Microdialysis

A microdialysis system consists of a precision pump, a microdialysis catheter (probe), and a microvial to collect the sample. The microdialysis probe, which is a hollow fiber that functions as an artificial blood vessel, is inserted into the tissue of interest. Commercially available microdialysis probes, which have been described in detail previously [3], consist of dialysis tubing glued to the end of a double-lumen polyurethane cannula (Figure 1). The dialysis solvent enters the probe through the inlet tube (inner cannula), passes down to the tip of the probe, streams upward in the space between the inner cannula and the outer dialysis membrane, and leaves the probe through the outer cannula via a sidearm, from which it is collected in a capped microvial. In general, the

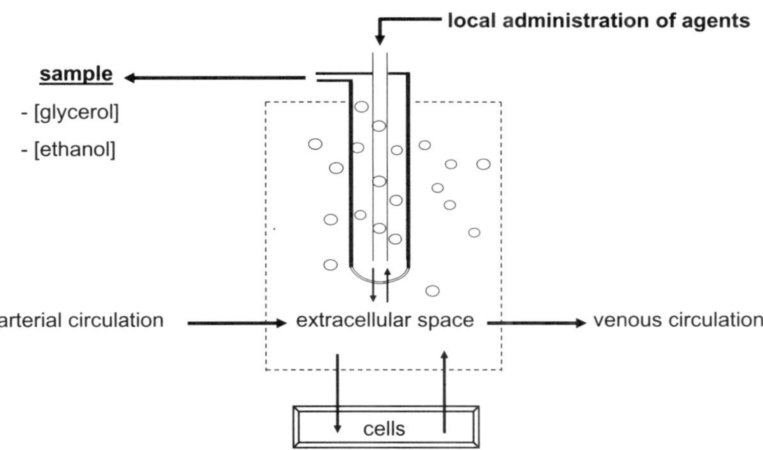

Figure 1 Principle of microdialysis with a double-lumen probe. After insertion of the catheter, the dialysis solvent enters the probe through the inlet tube (inner cannula), passes down to the tip of the probe, streams upward in the space between the inner cannula and the outer dialysis membrane, and leaves the probe through the outer cannula via a sidearm, from which it is collected in a capped microvial.

perfusate (perfusion fluid) is an aqueous solution (usually, saline or Ringer's solution) that mimics the composition of the surrounding medium in order to prevent excessive diffusion of molecules into or out of the periprobe fluid due to osmotic differences between the perfusate and extracellular fluid. Exchange of substances between the tube and the extracellular fluid occurs over the semipermeable membrane during the return through the outer tube. The diffusion direction is dependent on the concentration gradient between perfusate and the extracellular fluid. During perfusion of the microdialysis probe, molecules up to a certain molecular weight diffuse into (recovery) or out of (delivery) the perfusate. The dialysate (collected fluid in the microvial) is used for analysis of substances that are collected from the extracellular space, usually by a commercially analyzer that is capable of measuring the concentration of the substance of interest in small sample volumes. Therefore, microdialysis can be used to measure concentrations of substances in the extracellular fluid (e.g., metabolites) and offers the unique possibility of administering compounds locally (in situ) while avoiding systemic effects (e.g., unwanted side effects in vivo).

2.2. Practical Considerations in the Use of Microdialysis on Adipose Tissue

Although microdialysis is a very suitable technique for the investigation of adipose tissue metabolism and/or local drug delivery, a comprehensive understanding of the practical considerations for the application of this technique is needed. Several factors must be taken into account regarding the performance of the measurements as well as the interpretation of the results, as discussed below.

Insertion Trauma   Microdialysis probes are frequently placed in subcutaneous abdominal adipose tissue under sterile conditions 6 to 8 cm left and/or right of the umbilicus. One hour before insertion of the probes, the skin is anesthetized by means of a cream containing lidocaine (25 mg/g) and prilocaine (25 mg/g) (EMLA, AstraZeneca BV, Zoetermeer, The Netherlands). Insertion of the microdialysis catheter, although relatively noninvasive, will cause damage to a small number of cells. As a consequence, a local inflammatory reaction will occur. Obviously, this will lead to derangements in the interstitial fluid. It has been demonstrated that the initial trauma, measured as a rise in ATP concentration in the extracellular fluid, disappears approximately 30 min after implantation of the probe in human subcutaneous adipose tissue when using the commercially available double-lumen probe [4]. There is no evidence of edema or bleeding 12 h after insertion of the probe [5]. It is advised that an equilibration period of at least 45 min be allowed after probe insertion before starting microdialysis measurements on adipose tissue. The length of this recovery period depends on the microdialysis probe used and should thus be validated for every new approach.

Relative Recovery of the Microdialysis Probe   Under most circumstances, the concentration of a substance in the dialysate is only a fraction of the concentration of that substance in the extracellular fluid. The term *relative recovery* describes the ratio between the concentrations of a substance in the dialysate to that in the extracellular fluid. The relative recovery of the probe is generally below 100% (except when the perfusion rate is zero), because microdialysis is not an equilibrium process, due to continuous perfusion of the microdialysis probe, which does not allow complete diffusion equilibrium because of the fact that the fluid is "washed out" of the tissue. A number of factors determine the relative recovery of the probe in adipose tissue. The relative recovery is dependent primarily on the molecular weight cutoff (MWCO) of the membrane, the probe length, the perfusion rate, and the properties of the compound of interest (e.g., hydrophilicity).

Microdialysis probes with different MWCOs are available, ranging between 5 and 100 kDa. The MWCO is determined by the pore size of the probe membrane. Substances smaller than the membrane pores can diffuse into or out of the extracellular fluid. However, an acceptable relative recovery can only be obtained with substances having a molecular mass lower than approximately 25% of the MWCO determined under equilibrium conditions. Since measurements in vivo are performed under nonequilibrium conditions (due to perfusion), one can understand that the relative recovery becomes even smaller, depending on the molecular mass of the substance. Larger molecules cannot pass the semipermeable membrane or diffuse extremely slowly through the pores, causing their recovery to be negligible. Microdialysis of large molecules is discussed in more detail later in the chapter.

In addition to the molecular mass of the substance of interest, characteristics of the microdialysis membrane affect the relative recovery. Based on Fick's law of diffusion, the rate of diffusion across a membrane is proportional to its area. Therefore, increasing the length of the microdialysis membrane will result in increased relative recovery. The use of a very long dialysis membrane (60 to 70 mm) may thus increase the relative recovery. However, these long membranes are only available for use with single-lumen probes, and these probes cause more tissue trauma than do double-lumen probes. Another disadvantage of the single-lumen probes is that the MWCO is often much lower, and these probes are therefore less suitable for pharmacological experiments in which active compounds are administered to the extracellular space, as discussed in more detail later in the chapter.

The relative recovery of the microdialysis probe is also influenced by the perfusion rate. In general, low perfusion rates result in higher recovery rates, whereas higher perfusion rates will lead to lower recovery rates for microdialysis probe. With the use of a long probe (membrane length 30 mm) and a low perfusion rate (0.3 µL/min), a nearly complete exchange will be obtained for most metabolites. The dialysate concentration is therefore almost identical to that in the extracellular fluid. Thus, one could argue that the perfusion rate should be as low as possible. However, low perfusion rates result in small sample volumes. Therefore, use of the lowest perfusion rate as possible would increase the sample collection interval and thus lead to a poorer (less frequent) monitoring of changes over time. This hampers the use of a low perfusion rate in kinetic experiments. In contrast, a high perfusion rate ($\geq$10 µL/min) will result in dilution of the sample and might lead to increased pressure in the microdialysis probe, which may result in a net flow out of the probe into the extracellular fluid. Therefore, either a measurement technique should be available that requires only small sample volumes, or the detection limit should be low (lower quantification limit). In situations of incomplete recovery, calibration of the microdialysis probe is needed to obtain an estimate of the true extracellular concentration, as discussed in the next section.

In summary, the relative recovery of the microdialysis probe is associated positively with the MWCO and membrane length and inversely with the perfusion rate. Depending on the objective of an experiment and the analytical methods available, a balanced choice has to be made concerning probe characteristics (MWCO and membrane length) and the perfusion rate used in order to achieve acceptable probe recovery.

Calibration of the Microdialysis Probe   Obviously, it is of interest to determine the true concentrations of a substance in the extracellular fluid under steady-state conditions. When using a relatively high perfusion rate, a microdialysis probe has to be calibrated before any conclusions can be drawn about concentrations in the extracellular fluid. The true concentration of a substance in the extracellular fluid can be obtained when the relative recovery of that substance is known, since measured dialysate concentrations can be adjusted for the relative recovery of the probe ("calibration"). Many different in vitro and in vivo calibration methods have been used to assess probe recovery [6]. In earlier studies, the relative recovery was determined in vitro rather than in vivo. The in vitro relative recovery has been used to interpret in vivo results. However, the true concentration of a substance in the extracellular fluid is underestimated when the relative recovery is determined in vitro, as discussed previously [7,8]. Thus, it is not reliable to extrapolate the relative recovery obtained in vitro to the situation in vivo. Biological factors that complicate this extrapolation include the tissue tortuosity, the free fluid space between the cells, metabolism, uptake and release processes, intracellular and extracellular exchange, and possibly other factors [9,10]. In vitro probe recovery could be used in semiquantitative in vivo experiments in which the aim is to examine changes in the extracellular fluid concentration. However, several approaches are available to determine the relative recovery in vivo in humans [11–13]. Two calibration methods used frequently in microdialysis on adipose tissue are discussed below.

The first method that has been employed to determine the relative recovery in vivo is often referred to as the *equilibrium technique* [14]. Using this method, adipose tissue is perfused with increasing concentrations of the substance of interest while maintaining steady-state concentrations in the extracellular fluid. To accomplish this, the substance of interest is added to the perfusate in various known concentrations. The substance will diffuse out of the extracellular fluid into the periprobe fluid when the concentration of the substance in the perfusate is lower than that in the extracellular fluid. As a consequence, the dialysate concentration will be higher than the concentration in the perfusate. The opposite occurs when the concentration of the substance of interest is higher in the perfusate than in the extracellular fluid. Similarly, the concentrations in the perfusate and dialysate will be exactly the same when the concentration of the substance in the perfusate is identical to that in the extracellular fluid (equilibrium). Since there is a linear relationship between the ingoing perfusate concentration and the outgoing dialysate concentration, the extracellular fluid concentration of a particular substance can be estimated indirectly using simple regression analysis. The slope of this regression line corresponds to the relative recovery of the probe [15]. Since the perfusion rate is constant, this method is independent of mass transfer in the tissue, due to altered perfusion rate. A disadvantage of this method, however, is the time needed to calibrate with different perfusate concentrations and to reach steady-state measurements.

The second method that is often used to determine the relative recovery in vivo in adipose tissue is the *zero-flow method* [11,13]. In contrast to the equilibrium method described above, the microdialysis probes are perfused at increasing perfusion rates, but with the same solution. As mentioned earlier,

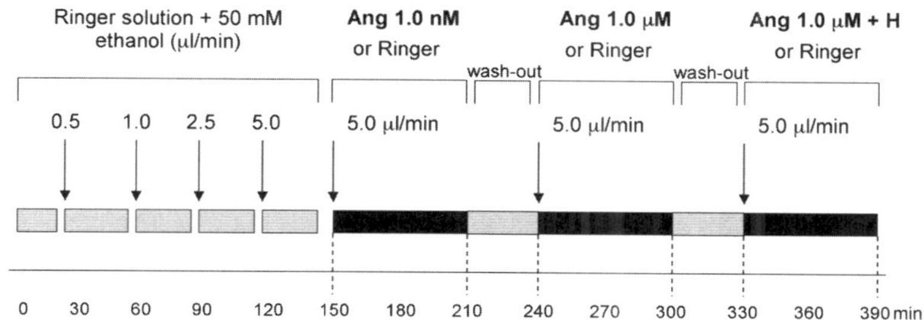

Figure 2 Microdialysis protocol to investigate the effects of angiotensin II on lipolysis and blood flow in human adipose tissue and skeletal muscle. Ang, angiotensin II; H, hydralazine. (© 2004 by The Endocrine Society.)

concentrations of a substance in the dialysate will decrease when the perfusion rate is higher. Equilibrium conditions would be attained at a rate of zero and would reflect the true concentration in the extracellular fluid. The values obtained at the different perfusion rates can be plotted against perfusion rates after log transformation of the concentrations measured in the dialysate, as has been described mathematically [16]. Linear regression analysis can be used to calculate the true concentration of the substance in the extracellular fluid (at a zero flow rate). The ratio between the dialysate glycerol concentration at a certain perfusion rate and the calculated true concentration in the extracellular fluid represents the in vivo recovery of the probe. Since the true value at a zero flow rate can only be estimated based on extrapolation (in contrast to interpolation when using the equilibrium method), it is recommended that multiple perfusion rates be used for the calibration, including a very low perfusion rate. More precisely, the probes can be perfused with Ringer's solution (147 mM sodium, 4 mM potassium, 2.25 mM calcium, and 156 mM chloride, Baxter BV, Utrecht, The Netherlands) at increasing perfusion rates, starting with a flow rate of 0.5 µL/min for 40 min (collect two 20-min fractions), followed by flow rates of 1.0, 2.5, and 5.0 µL/min for 30 min (collect three 10-min fractionsper/perfusion rate) (see also Figure 2).

The two calibration methods described above are precise but time consuming. An alternative calibration technique has been developed to shorten the calibration procedure. This technique, called the *internal reference technique*, uses the addition of a reference substrate, ideally a labeled equivalent to the substance being measured. The outflux fraction of the label is then the same as the relative recovery of the substance of interest. This calibration technique is convenient but should be validated carefully in each situation since the label may accumulate around the microdialysis probe and the extracellular fluid concentration many thus be overestimated [17]. This relatively short calibration procedure still needs steady-state conditions, which are not reached until the initial trauma after insertion of the probe has disappeared completely (see the section "Insertion Trauma" above).

Importantly, the in vivo recovery is not the same for different probes, even when the same substance is measured in the same tissue in the same person at the same time. Therefore, the in vivo recovery needs to be calculated for each probe separately, and dialysate concentrations measured in samples collected during the experiment can be adjusted for the in vivo recovery of the corresponding probe.

3. USE OF MICRODIALYSIS ON ADIPOSE TISSUE IN HUMANS

Microdialysis on adipose tissue can be used for kinetic experiments in humans, since this technique allows continuous sampling of dialysate and therefore continuous measurements of extracellular fluid concentrations. As explained earlier, the sensitivity of the kinetic experiment depends on the perfusion rate. For example, when the perfusion rate is low (e.g., $0.3\,\mu L/min$), it is difficult to detect rapid changes in the extracellular fluid since a certain volume is needed for sample analysis. On the other hand, when the perfusion rate is too high (e.g., $15\,\mu L/min$) it is very difficult to detect changes in the extracellular fluid concentration. Microdialysis is frequently used to measure adipose tissue metabolism and blood flow. In this section, the use of microdialysis to study adipose tissue lipolysis and blood flow in human adipose tissue is discussed.

3.1. Use of Microdialysis to Determine Adipose Tissue Lipolysis

When energy intake exceeds energy expenditure, the energy surplus is stored mainly in adipose tissue in the form of triacylglycerol (TAG). Whole-body energy balance therefore regulates the amount of TAG stored in adipocytes, which in turn reflects the net balance between fat deposition and fat mobilization. Fat mobilization (lipolysis) is described as the breakdown of circulating triacylglycerol (TAG)-rich lipoprotein particles (exogenous lipolysis) and stored TAG (endogenous lipolysis), and results in the release of fatty acids and glycerol. The regulation of adipose tissue lipolysis depends on the balance between lipolytic and antilipolytic hormones. Insulin and catecholamines are the dominant regulatory hormones in adipose tissue lipolysis. Interestingly, there is substantial evidence to suggest that impairments in lipolysis may contribute to the pathophysiology of obesity and type 2 diabetes mellitus, which highlights the importance of studying adipose tissue lipolysis.

Microdialysis has been used widely to determine the extracellular concentrations of several metabolites, including glycerol, glucose, and lactate, thereby providing insight into the regulation of lipolysis and carbohydrate metabolism in human adipose tissue. Fatty acids cannot diffuse from the tissue into the microdialysis probe, since their molecular weight is too high to pass the probe membrane. In contrast, glycerol is able to diffuse freely from the extracellular fluid into the probe. Therefore, the glycerol concentration in the dialysate can be used as an indicator of lipolysis.

Microdialysis can be applied to different adipose tissue depots to measure lipolysis. In general, subcutaneous adipose tissue depots are studied, although microdialysis on visceral adipose tissue has also been performed in humans during and/or after surgery [18]. Lipolytic activity is not the same in different adipose tissue depots [19–23]. It has also been shown using microdialysis in humans that lipolytic activity of abdominal subcutaneous adipose tissue differs from that in femoral subcutaneous adipose tissue [24–27]. Furthermore, even within the same adipose tissue depot, lipolysis may vary because of heterogeneity of the tissue. Since concentrations of a very limited area surrounding the microdialysis probe are measured (estimated to be equivalent to about 100 mg of tissue) [28], differences in lipolysis within the same adipose tissue depot can be assessed using microdialysis. Using this approach in humans, it has been demonstrated that the lipolytic rate is higher in superficial adipose tissue than in deep abdominal subcutaneous adipose tissue [29,30].

However, extreme care has to be taken with the interpretation of changes in glycerol concentration in the dialysate, since the extracellular fluid concentration of glycerol (and other metabolites) is influenced not only by adipocyte lipolysis but also by the local circulation. In other words, the concentration of a substance in the extracellular space depends on both local metabolism and tissue blood flow. In this respect it is important to realize that only the nutritive blood flow influences the extracellular concentration of a substance. The higher the nutritive blood flow, the lower the extracellular concentration of the substance of interest. Adipose tissue blood flow is of little or no importance for exchange between compartments. Therefore, it is necessary to determine local nutritive blood flow in order to adequately interpret (changes in) extracellular concentrations.

3.2. Use of Microdialysis to Determine Adipose Tissue Blood Flow

Local nutritive blood flow can be determined using the *ethanol dilution technique*. Ethanol is added to the perfusate, and its escape from the perfusion medium to the surrounding extracellular fluid is measured. Thus, the ethanol concentration in both the ingoing and outgoing perfusion solvent is measured to assess the ethanol outflow/inflow (out/in) ratio as an indicator of local nutritive blood flow [31,32]. Ethanol (at least up to 1055 mmol/L) does not alter metabolism, has no effect on local blood flow, and does not affect the properties of the microdialysis probe. The most effective ethanol concentration (50 mmol/L) has no effect on local adipose tissue lipolysis and blood flow [33]. Using the ethanol dilution technique, the direction of change in local nutritive blood flow to the tissue surrounding the microdialysis probe can be determined. A disadvantage of this method is that quantitative measurements of local nutritive blood flow are not possible. In other words, the ethanol dilution technique can at best be considered a semiquantitative method for the assessment of local nutritive blood flow. In addition, compared with the only quantitative method available for adipose tissue blood flow measurements—the

[133]Xe washout technique [34,35]—the ethanol dilution technique shows a delayed response to rapid physiological changes in adipose tissue blood flow [36]. Unlike other methods for the measurement of adipose tissue blood flow in humans, microdialysis can be used for long-term continuous blood flow monitoring (e.g., ambulatory for several days).

3.3. Pharmacological Studies

In addition to the use of microdialysis on adipose tissue to examine extracellular fluid concentrations of endogenous compounds, this technique can also be used to investigate the effect of certain exogenous compounds on tissue metabolism and blood flow. On the one hand, it can be used to deliver drugs or hormones locally and simultaneously to study their effects on tissue metabolism and blood flow. On the other hand, agents can be ingested orally or infused intravenously, and the combination of their effects at both the whole-body and tissue levels can be studied locally in the tissue of interest. In this section, several studies that have used the approaches described above are discussed to illustrate the applicability of microdialysis in kinetic experiments to assess drug effects on metabolism and blood flow in human adipose tissue.

Local Administration of Pharmacological Agents By the administration of pharmacological agents through the microdialysis catheter, the local, tissue-specific effects of metabolically active compounds can be investigated while avoiding systemic effects. This approach has been used by several investigators to examine the effects of the sympathetic activation of lipolysis in abdominal subcutaneous adipose tissue. For example, the role of the various β-adrenergic receptors in the regulation of lipolysis and local nutritive blood flow was investigated in abdominal subcutaneous adipose tissue in healthy lean volunteers [37]. The microdialysis probes were perfused with the nonselective β-adrenergic receptor agonist isoprenaline, the selective β_1-adrenergic receptor agonist dobutamine, the selective β_2-adrenergic receptor agonist terbutaline, or the selective β_3-adrenergic receptor agonist CGP 12177 (which also has β_1-/β_2-antagonistic properties). The findings of this study demonstrated that the β_1- and β_2-adrenergic receptor subtypes are involved primarily in the stimulation of both lipolysis and local blood flow, whereas the β_3-adrenergic receptors play a minor role in the regulation of these processes in abdominal subcutaneous adipose tissue [37].

Furthermore, our group has recently investigated whether local adipocyte β-adrenergic sensitivity and changes in nutritive blood flow in abdominal subcutaneous adipose tissue contribute to the impairment in β-adrenergically-mediated whole-body lipolysis observed in obese subjects [38]. Three microdialysis probes were placed in the abdominal subcutaneous adipose tissue depot of eight obese and nine lean men. Each probe was perfused with incremental doses of the nonselective β-adrenergic receptor agonist isoprenaline, the selective β_1-adrenergic receptor agonist dobutamine, or the selective β_2-adrenergic receptor agonist salbutamol, each dose for 45 min. During non-selective β-, β_1-, and β_2-adrenergic receptor stimulation, interstitial glycerol concentrations increased and ethanol out/in ratios decreased similarly in obese and lean men. Thus, these data indicate that the lipolytic and nutritive blood flow response to β_1-, β_2-, and nonselective β-adrenergic receptor stimulation in situ is comparable in lean and obese male subjects, suggesting that a blunted β-adrenergic receptor sensitivity of the abdominal subcutaneous adipocyte and an impaired local nutritive blood flow response do not contribute to the diminished whole-body β-adrenergically-mediated lipolytic response in obese males reported previously [38]. Rather, the impaired abdominal subcutaneous adipose tissue blood flow response in obese subjects [39–43] may affect the delivery of hormones to the tissue or the release and reuptake of fatty acids by the adipose tissue [38]. Alternatively, other tissues or intraabdominal adipose tissue might be at least partially responsible for the impaired β-adrenergically-mediated whole-body lipolytic response in these subjects. A follow-up microdialysis study by our group demonstrated that the capacity to increase skeletal muscle lipolysis and blood flow (gastrocnemius muscle) during direct β_2-adrenergic receptor stimulation is impaired in obese subjects compared with lean subjects [44].

In another study performed recently, we used microdialysis to examine the effects of angiotensin II (Ang II), the active component of the renin–angiotensin system, on adipose tissue and skeletal muscle blood flow and lipolysis in normal-weight and obese subjects [45]. Microdialysis probes were placed in the abdominal subcutaneous adipose tissue under sterile conditions 6 to 8 cm left and right from the umbilicus as well as in the medial portion of the gastrocnemius muscle of both legs after anesthesia (xylocaine 2% without adrenaline, AstraZeneca BV). Afters 90-min recovery, the true interstitial glycerol concentration was determined individually by means of the zero flow method (see Section 2.2). A calibration period with a flow rate of 5.0 µL/min was used as a baseline measurement for the second part of the experiment. During that part of the experiment, the experimental probes in adipose and skeletal muscle tissue were consecutively perfused with incremental doses of Ang II (perfusate supplemented with 50 mmol/L ethanol) at a flow rate of 5.0 µL/min, each dose for 60 min (with 30 min of washout between perfusion steps), to examine the effect of Ang II on local nutritive blood flow and lipolysis. The highest perfusion dose of Ang II was also administered in combination with the vasodilatator hydralazine to counteract the possible Ang II-induced vasoconstriction and therefore to distinguish between the possible direct and blood flow–mediated effects of Ang II on the interstitial glycerol concentration (Figure 2), as discussed earlier. The other probes in adipose and skeletal muscle tissue served as control probes and were perfused with Ringers solution (+50 mmol/L ethanol) at a similar perfusion rate during the entire experiment. All changes in ethanol out/in ratios and glycerol concentrations observed in the various experimental conditions were adjusted for corresponding

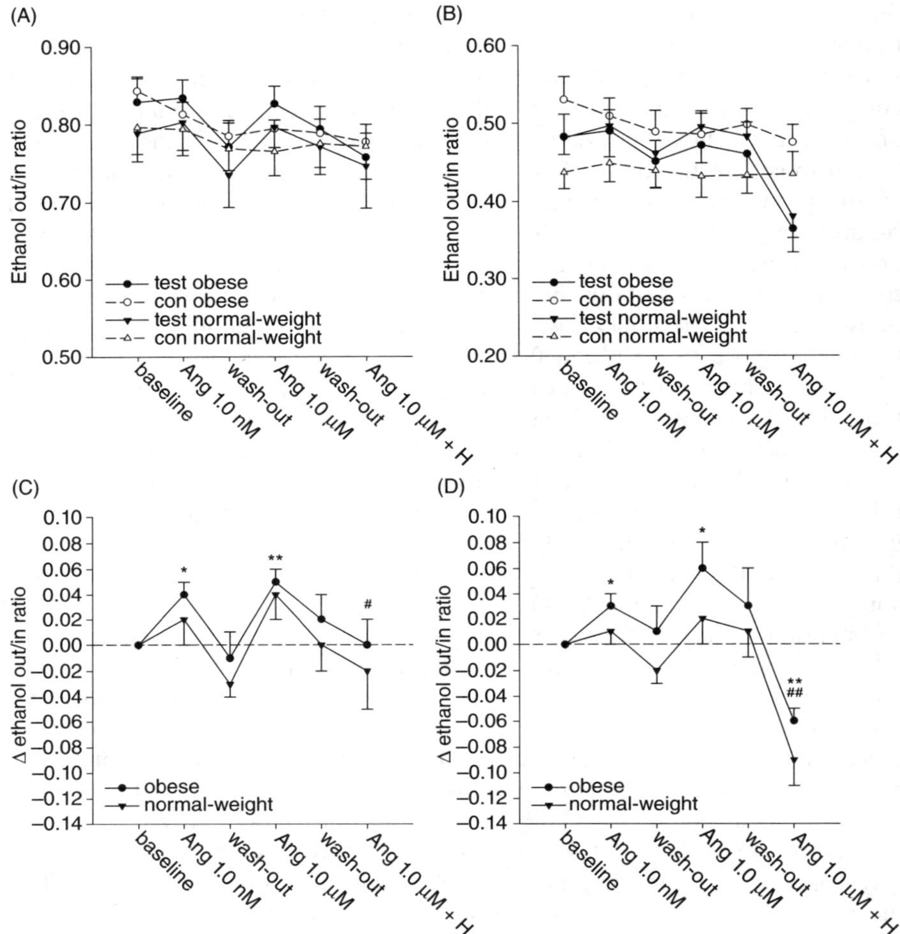

Figure 3 Effects of Ang II on the ethanol out/in ratio in abdominal subcutaneous adipose tissue (A) and skeletal muscle (B) in eight normal-weight and eight obese subjects. Data are adjusted for corresponding changes in the control probes and presented as changes from baseline values in adipose tissue (C) and skeletal muscle (D). $*p < 0.05$ vs. baseline; $**p < 0.01$ vs. baseline; $^{\#}p < 0.05$ vs. Ang 1.0 μM; $^{\#\#}p < 0.001$ vs. Ang 1.0 μM. Values are means ± S.E.M. (© 2004 by The Endocrine Society.)

changes in the control probes to take into account changes over time in tissue blood flow and lipolysis that were not due to the intervention. Because of the interindividual variations in baseline concentrations, these adjustments were made by subtracting changes in the experimental probes by changes in the control probes. Ang II caused a significant increase in ethanol outflow/inflow ratio compared with baseline values in both adipose tissue and muscle, indicating a decrease in local nutritive blood flow (Figure 3). The decrease in local

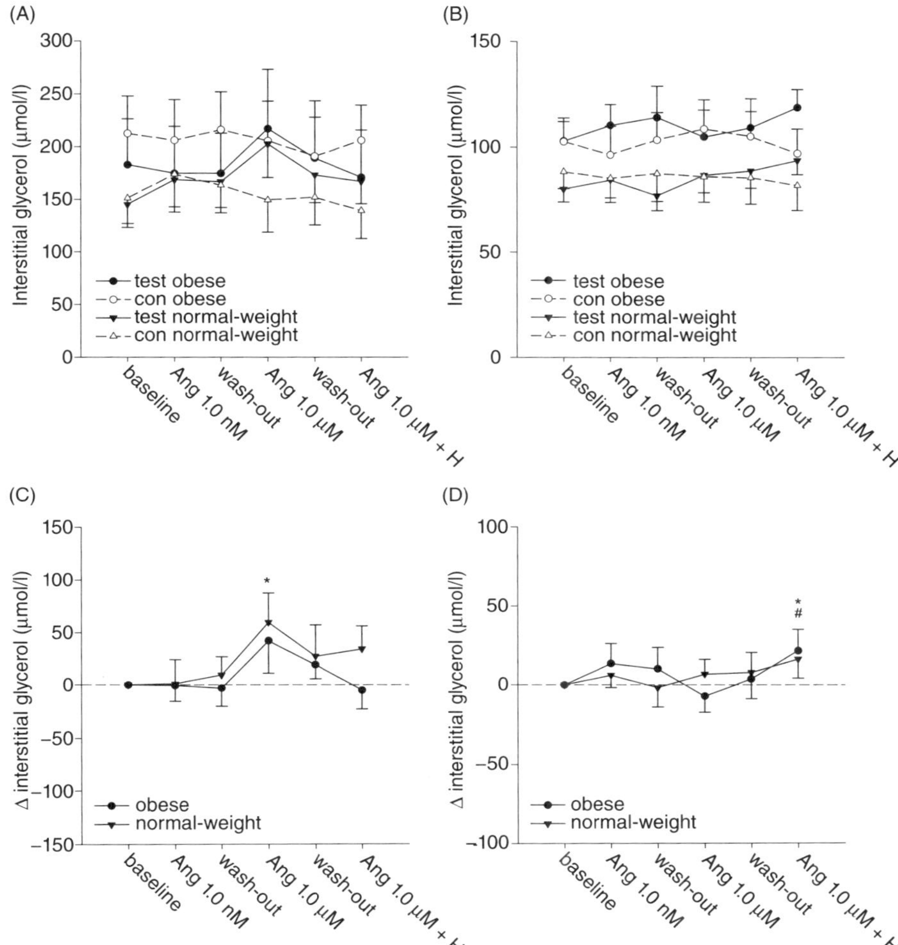

Figure 4 Effects of Ang II on the interstitial glycerol concentration in abdominal subcutaneous adipose tissue (A) and skeletal muscle (B) in eight normal-weight and eight obese subjects. Data are adjusted for corresponding changes in the control probes and presented as changes from baseline values in adipose tissue (C) and skeletal muscle (D). $*p < 0.05$ vs. baseline and $^{#}p < 0.05$ vs. Ang 1.0 μM. Values are means ± S.E.M. (© 2004 by The Endocrine Society.)

blood flow was accompanied by unchanged interstitial glycerol concentrations in adipose tissue (except during the supra-physiological dose) and skeletal muscle, suggesting that Ang II inhibits lipolysis in both tissues (Figure 4). These results indicate that Ang II decreases local nutritive blood flow in a dose-dependent manner and inhibits lipolysis both in adipose tissue and skeletal muscle of normal-weight and obese men [45].

Systemic Administration of Pharmacological Agents Beside the local administration of pharmacological compounds using microdialysis, many studies have focused on the clinical use of microdialysis in adipose tissue to investigate the kinetics of change of compounds that are orally ingested or intravenously infused. Boschmann and colleagues [46] performed a study to investigate the effects of short-term treatment with an angiotensin II type 1 receptor blocker on adipose tissue and skeletal muscle metabolism. Obese men were treated for 10 days with an angiotensin II type 1 receptor blocker or with placebo in a double-blind and crossover fashion. At the end of each treatment period, skeletal muscle and adipose tissue metabolism was assessed using microdialysis. Measurements were obtained at baseline and during the administration of incremental concentrations of a nonselective β-adrenergic receptor agonist via the microdialysis probe. As expected, nonselective β-adrenergic receptor stimulation evoked a dose-dependent increase in dialysate glycerol concentrations in both adipose and skeletal muscle. However, it was concluded that oral treatment with an angiotensin II type 1 receptor blocker for 10 days has no major effect on tissue lipolysis under resting conditions and during β-adrenergic stimulation in obese men.

As mentioned earlier in the chapter, insulin plays a major inhibiting role in adipose tissue lipolysis. In addition, skeletal muscle appears to be responsive to the antilipolytic effect of insulin. Stumvoll and colleagues [47] determined insulin dose–response characteristics of systemic, abdominal subcutaneous adipose tissue and skeletal muscle lipolysis in humans. A three-step hyperinsulinemic–euglycemic clamp was performed while systemic lipolysis and interstitial glycerol concentrations (microdialysis) were measured in 13 lean, healthy volunteers. The authors concluded that skeletal muscle is more sensitive than subcutaneous adipose tissue with respect to the net effect of circulating insulin (within the physiological insulin range) on the inhibition of lipolysis.

A similar microdialysis approach was used to evaluate the suppression of whole-body and regional adipose tissue lipolysis by insulin in nonobese (both endurance-trained and sedentary) and sedentary obese women [48]. The results of this study indicated that the sedentary obese women had whole-body resistance to the suppression of lipolysis by insulin. Since insulin resistance of the suppression of lipolysis was not evident in abdominal or femoral subcutaneous adipose tissue, it was suggested that intraabdominal adipose tissue may be the site of insulin resistance of lipolysis.

In addition to resistance of hormonal effects on tissue metabolism, desensitization of hormone responses can also be studied at the tissue level using microdialysis. For example, the dose dependency and desensitization of adipose tissue lipolysis to adrenaline was investigated [49]. To accomplish this, adrenaline was infused intravenously in healthy subjects for three periods of 35 min, with 30 min of washout between infusion periods. Whole-body and adipose tissue lipolysis was closely related to the arterial adrenaline concentration. Interestingly, both the increase in local blood flow and lipolysis in response to adrenaline were reduced by prior systemic infusion of adrenaline. Therefore, based on these microdialysis findings, the authors were able to conclude that in vivo adrenaline-mediated increments in adipose tissue lipolysis and blood flow are desensitized by prior adrenaline exposure.

These examples show that microdialysis is a very useful approach to localizing derangements of metabolism or effects on specific tissues or organs of hormones known to be present at the whole-body level. Unfortunately, it is difficult to distinguish between metabolic (e.g., lipolysis) and hemodynamic (blood flow) effects using microdialysis. The technique can be considered to be semiquantitative, since microdialysis alone is not suitable for quantitative measurements of uptake and/or release of metabolites in adipose tissue. However, microdialysis can be combined with arterial sampling and accurate measurements of adipose tissue blood flow, as well as with stable isotope methodology, to quantatively assess metabolic processes in adipose tissue.

3.4. Use of Microdialysis in Combination with Arterial Sampling and/or Stable Isotope Methodology

Microdialysis allows the measurement of sequential samples, and the changes observed over time can indicate changes in metabolic processes. As explained in the preceding section, changes in the extracellular glycerol concentration in adipose tissue may reflect alterations in lipolytic activity. Although this information can be of great interest, only semiquantitative results (increase/ decrease, dose-dependent effects, etc.) are obtained when microdialysis alone is used. To get more detailed information regarding tissue metabolism (net tissue uptake/release), it is necessary to compare the extracellular fluid concentration with the arterial concentration. Thus, when the extracellular fluid concentration exceeds the arterial concentration, a net release of the metabolite into the circulation is indicated. Vice versa, a lower extracellular fluid concentration than arterial concentration represents a net uptake of the metabolite from the circulation into the tissue.

Stallknecht and colleagues [50] used this approach to assess changes in adipose tissue and skeletal muscle lipolysis during exercise of different intensities. Glycerol concentrations in arterial blood and in the extracellular fluid of skeletal muscle and adipose tissue increased significantly during exercise [moderate intensity (65% of the maximal workload) > low intensity (25% of the maximal workload)]. In adipose tissue, the difference between the extracellular and arterial glycerol concentration increased significantly with increasing intensity, reflecting increased adipose tissue lipolysis. However, there was a slight net release of glycerol from skeletal muscle at rest and during low-intensity exercise, but not during moderate- or high-intensity exercise (85% of maximal workload). These findings were interpreted as skeletal muscle lipolysis peaking at low exercise intensities but could also indicate that glycerol is taken up in skeletal muscle at a rate that is increasing with exercise intensity [47].

For quantitative measurements of net uptake or release of a substance it is necessary to determine arteriovenous concentration differences in combination with the measurement of adipose tissue blood flow using ^{133}Xe washout [40,41,51–54]. Interestingly, quantitative estimates of net uptake or release of metabolites can also be obtained when microdialysis is combined with arterial sampling and quantitative blood flow measurements (^{133}Xe washout). Arteriovenous concentration differences can then be calculated by conversion of extracellular fluid concentrations to capillary venous concentrations. To accomplish this, knowledge of the permeability surface product area, which has been obtained empirically from the literature, is necessary. The capillary venous substrate concentration can be calculated as follows [55]:

$$V = (I - A)(1 - e^{-PS/Q}) + A$$

where V is the capillary venous substrate concentration, I the extracellular fluid concentration, A the arterial concentration, PS the permeability surface product area (for glycerol, ± 5 mL/100 g tissue·min [56–58]), and Q the adipose tissue plasma flow, calculated as adipose tissue blood flow (mL/100 g tissue·min) $\times$ (1 − hematocrit/100). Quantitative estimates of net uptake or release of metabolites can be made by multiplying the arteriovenous concentration difference by the adipose tissue plasma flow. Thus, microdialysis in combination with arterial sampling and quantitative measurement of adipose tissue blood flow can be used to obtain quantitative estimates of regional net tissue uptake or release. Inclusion of stable isotope methodology in addition to the above also allows examination of whole-body production or utilization rates.

We have used this approach to investigate abdominal subcutaneous adipose tissue lipolysis, adipose tissue blood flow, and whole-body lipolysis under baseline conditions and during nonselective β-adrenergic stimulation (intravenous infusion of isoprenaline) in upper-body obese subjects with type 2 diabetes and nonobese, nondiabetic controls [59]. Arterial glycerol concentrations were higher in subjects with type 2 diabetes than in controls, whereas the increases in arterial glycerol concentration in response to isoprenaline infusion were of a similar magnitude in both groups. Estimated abdominal subcutaneous adipose tissue glycerol release, expressed per unit fat mass, was not significantly different between type 2 diabetics and controls under baseline conditions and during β-adrenergic stimulation (comparable increase in both groups). Additionally, the β-adrenergic-induced increases in local abdominal subcutaneous adipose tissue lipolysis, systemic (whole-body) lipolysis, and abdominal subcutaneous adipose tissue blood flow responses were comparable between groups. These data are in contrast with previous findings in obese nondiabetic subjects [60–63] and therefore suggest that processes involved in lipid mobilization may differ in obesity and obesity-related type 2 diabetes [59].

Microdialysis in combination with arterial sampling, quantitative measurement of adipose tissue blood flow, and stable isotope methodology has also been used to investigate whole-body and regional (abdominal and femoral subcutaneous adipose tissue) lipolytic and adipose tissue blood flow sensitivity to epinephrine in lean and upper-body obese women [64]. Unfortunately, as opposed to the study described above, the authors did not directly measure the glycerol recovery of the microdialysis probes, but used an estimated glycerol recovery for all subjects. Therefore, their calculations on regional glycerol release should be considered qualitative (or semiquantitative) rather than absolute. Nevertheless, it was demonstrated that the basal whole-body rates of appearance of free fatty acids and glycerol were both significantly greater in obese woman than in lean women. Epinephrine infusion significantly increased the whole-body rate of appearance of free fatty acids and glycerol in lean but not obese subjects. In addition, lipolytic and ATBF sensitivity to epinephrine was blunted in abdominal but not femoral subcutaneous adipose tissue of obese compared with lean subjects. Therefore, it was concluded that whole-body lipolytic sensitivity to epinephrine is blunted in upper-body obese women because of decreased sensitivity in upper-body (abdominal subcutaneous) but not lower-body (femoral) subcutaneous adipose tissue [64].

An approach comparable to that described above was used in a more recent study that was performed to investigate the effects of systemic and local norepinephrine administration on lipolysis and blood flow rates in adipose tissue and skeletal muscle [65]. First, norepinephrine was systemically infused, and extracellular glycerol concentrations were measured in abdominal subcutaneous adipose tissue and the gastrocnemius muscle using microdialysis. Local blood flow was determined using the ^{133}Xe washout technique, and whole-body lipolysis rates were assessed using [^{2}H$_5$]glycerol. Second, incremental doses of norepinephrine were administered locally in adipose tissue and skeletal muscle via the microdialysis probe. Extracellular glycerol concentrations were measured and local nutritive blood flow was monitored with the ethanol dilution technique. The authors concluded that lipolysis and blood flow rates are regulated differently in adipose tissue and skeletal muscle. Adipose tissue displays a high but transient sensitivity to norepinephrine, resulting in increased lipolysis and blood flow rates. In skeletal muscle, physiological concentrations of norepinephrine decrease blood flow but have no stimulatory effect on lipolysis rates [65].

These studies demonstrate that the use of microdialysis in combination with arterial sampling (assessment of the arterioextracellular concentration differences) provides more information on the net uptake or release of metabolites across adipose tissue than does microdialysis alone. This approach is relatively straightforward from a practical point of view. Furthermore, quantitative estimates of net uptake or release of metabolites can be obtained when microdialysis is combined with arterial sampling and quantitative blood flow measurements (^{133}Xe washout), since capillary venous concentrations can then be estimated, allowing an estimation of net uptake or release. Finally, stable isotope methodology should be included to be able to investigate effects at both the whole-body (absolute production and/or utilization rates) and tissue levels.

3.5. Microdialysis of Large Molecules

Microdialysis has been used widely to recover low-molecular-mass (<20 kDa) endogenous substances, such as metabolites. More recently, attempts have been made to recover molecules of higher molecular mass, including cytokines, plasma proteins, and growth factors. The development of microdialysis probes suitable for the recovery of large molecules has been driven by the need to sample the extracellular fluid for bioactive proteins and regulatory peptides as markers of tissue homeostasis and tissue (dys)function. For microdialysis of large proteins, particularly cytokines, probes with 100-kDa MWCO or greater are necessary. Theoretically, the probes at the higher end of the MWCO range (MWCO > 100 kDa) may appear to be satisfactory for the microdialysis of large proteins. Practically, however, the number of pores within most membranes currently used that allow diffusion of these large proteins is small. As a consequence, several physical properties of the membrane are altered, introducing problems with diffusion, relative recovery, and kinetics. For example, the in vitro relative recovery of a 10-kDa protein across a commercially available 100-kDa MWCO probe at a perfusion rate of 1 µL/min is typically below 5% [66]. It should also be noted that continuous perfusion of high MWCO probes for long periods leads to a decrease in analyte recovery, with higher initial total protein concentrations compared with those recovered in samples collected later [67]. A possible explanation for this observation is that the amount of protein available for recovery decreases with time (removal > supply). Thus, it is highly unlikely that even at very low perfusion rates a diffusion equilibrium will be reached for high-molecular-mass proteins. An increase in tissue hydration due to fluid loss by ultrafiltration across the highly porous high-MWCO probe is another factor that may reduce the relative recovery and lead to an underestimation of the extracellular concentration of the substance of interest due to concentration dilution [67]. One way to increase the macromolecule recovery when using these high-MWCO probes is to add osmotic agents to the perfusate [68]. For example, some investigators have used bovine serum albumin to prevent fluid loss from the probe and to increase the relative recovery of the substance of interest [68]. Importantly, this benefit also has a big disadvantage: the collection of dialysate that is not free of proteins. Obviously, this may result in analytical problems, since a more complex sample preparation is necessary before analysis. In addition, this could lead to diffusion of previously bound tissue molecules out of the tissue in situations where the albumin concentration in the perfusate exceeds the tissue albumin concentration, resulting in an overestimation of the concentration of the active fraction of a substance. Another option to increase the relative recovery is to influence the solubility of a substance. This could be achieved by an alteration in pH [69]. However, it is important to realize that changes in the composition of the perfusate may induce physiological changes in the vicinity of the probe. Another challenge to microdialysis sampling of proteins as well as peptides is the low in vivo concentration of these substances in the extracellular fluid. The combination of low relative recovery and low analyte concentration requires either very sensitive methods [70] or methods to concentrate the sample prior to analysis [71].

In conclusion, the various restraints when using high-MWCO microdialysis probes, as discussed in this section, result in a very low relative recovery of high-molecular-mass proteins, even under optimal experimental conditions. In addition, the dialysate concentrations may not reflect actual tissue concentrations. Therefore, it may be better to aim for changes in large protein concentrations from baseline rather than absolute values.

4. SUMMARY AND CONCLUSIONS

The microdialysis technique is a very useful tool for the investigation of human adipose tissue. Its main advantages are that it can be used safely with low-grade invasiveness in humans, and thus allows continuous sampling over prolonged periods of time from specific adipose tissues without biopsies. Microdialysis can be used both to monitor tissue metabolism and to investigate the effect of certain compounds on tissue metabolism (e.g., local drug delivery). A relatively new development in the field of microdialysis on adipose tissue is the use of high-MWCO microdialysis probes. These probes allow the measurement of large proteins in the extracellular fluid, such as cytokines. However, further optimalization of high-MWCO probes is needed regarding the application of these probes in kinetic experiments on human adipose tissue. Microdialysis on adipose tissue currently is, and will certainly remain, a technique that contributes to an increased understanding of adipose tissue metabolism in health and disease and will provide better insight into the metabolic, hemodynamic, and inflammatory effects of various compounds on adipose tissue.

Acknowledgments

We are grateful to the colleagues who have worked with us in developing and applying microdialysis, particularly Peter Arner (Karolinska Institute, Stockholm, Sweden).

REFERENCES

[1] Delgado, J.M., DeFeudis, F.V., Roth, R.H., Ryugo, D.K., Mitruka, B.M. (1972). Dialytrode for long term intracerebral perfusion in awake monkeys. *Archives Internationales de Pharmacodynamie et de Therapie*, *198*, 9–21.

[2] Ungerstedt, U., Pycock, C. (1974). Functional correlates of dopamine neurotransmission. *Bulletin der Schweizerischen Akademie der Medizinischen Wissenschaften*, *30*, 44–55.

[3] Tossman, U., Ungerstedt, U. (1986). Microdialysis in the study of extracellular levels of amino acids in the rat brain. *Acta Physiologica Scandinavica, 128*, 9–14.

[4] Bolinder, J., Hagström, E., Ungerstedt, U., Arner, P. (1989). Microdialysis of subcutaneous adipose tissue in vivo for continuous glucose monitoring in man. *Scandinavian Journal of Clinical and Laboratory Investigation, 49*, 465–474.

[5] Lönnroth, P., Smith, U. (1990). Microdialysis: a novel technique for clinical investigations. *Journal of Internal Medicine, 227*, 295–300.

[6] Kehr, J. (1993). A survey on quantitative microdialysis: theoretical models and practical implications. *Journal of Neuroscience Methods, 48*, 251–261.

[7] Bungay, P.M., Morrison, P.F., Dedrick, R.L. (1990). Steady-state theory for quantitative microdialysis of solutes and water in vivo and in vitro. *Life Sciences, 46*, 105–119.

[8] Lafontan, M., Arner, P. (1996). Application of in situ microdialysis to measure metabolic and vascular responses in adipose tissue. *Trends in Pharmacological Sciences, 17*, 309–313.

[9] Chen, K.C., Hoistad, M., Kehr, J., Fuxe, K., Nicholson, C. (2002). Theory relating in vitro and in vivo microdialysis with one or two probes. *Journal of Neurochemistry, 81*, 108–121.

[10] Hsiao, J.K., Ball, B.A., Morrison, P.F., Mefford, I.N., Bungay, P.M. (1990). Effects of different semipermeable membranes on in vitro and in vivo performance of microdialysis probes. *Journal of Neurochemistry, 54*, 1449–1452.

[11] Boutelle, M.G., Fillenz, M. (1996). Clinical microdialysis: the role of on-line measurement and quantitative microdialysis. *Acta Neurochirurgica, 67*, 13–20.

[12] Plock, N., Kloft, C. (2005). Microdialysis: theoretical background and recent implementation in applied life-sciences. *European Journal of Pharmaceutical Sciences, 25*, 1–24.

[13] Stahle, L., Segersvard, S., Ungerstedt, U. (1991). A comparison between three methods for estimation of extracellular concentrations of exogenous and endogenous compounds by microdialysis. *Journal of Pharmacological Methods, 25*, 41–52.

[14] Lönnroth, P., Jansson, P.A., Smith, U. (1987). A microdialysis method allowing characterization of intercellular water space in humans. *American Journal of Physiology, 253*, E228–E231.

[15] Chaurasia, C.S. (1999). In vivo microdialysis sampling: theory and applications. *Biomedical Chromatography, 13*, 317–332.

[16] Jacobson, I., Sandberg, M., Hamberger, A. (1985). Mass transfer in brain dialysis devices: a new method for the estimation of extracellular amino acids concentration. *Journal of Neuroscience Methods, 15*, 263–268.

[17] Lönnroth, P., Strindberg, L. (1995). Validation of the "internal reference technique" for calibrating microdialysis catheters in situ. *Acta Physiologica Scandinavica, 153*, 375–380.

[18] Jansson, K., Redler, B., Truedsson, L., Magnuson, A., Ungerstedt, U., Norgren, L. (2004). Postoperative on-line monitoring with intraperitoneal microdialysis is a

sensitive clinical method for measuring increased anaerobic metabolism that correlates to the cytokine response. *Scandinavian Journal of Gastroenterology*, *39*, 434–439.

[19] Arner, P. (1995). Differences in lipolysis between human subcutaneous and omental adipose tissues. *Annals of Medicine*, *27*, 435–438.

[20] Bjorntorp, P. (1996). The regulation of adipose tissue distribution in humans. *International Journal of Obesity and Related Metabolic Disorders*, *20*, 291–302.

[21] Jensen, M.D., Johnson, C.M. (1996). Contribution of leg and splanchnic free fatty acid (FFA) kinetics to postabsorptive FFA flux in men and women. *Metabolism*, *45*, 662–666.

[22] Ostman, J., Arner, P., Engfeldt, P., Kager, L. (1979). Regional differences in the control of lipolysis in human adipose tissue. *Metabolism*, *28*, 1198–1205.

[23] Tan, G.D., Goossens, G.H., Humphreys, S.M., Vidal, H., Karpe, F. (2004). Upper and lower body adipose tissue function: a direct comparison of fat mobilization in humans. *Obesity Research*, *12*, 114–118.

[24] Boschmann, M., Rosenbaum, M., Leibel, R.L., Segal, K.R. (2002). Metabolic and hemodynamic responses to exercise in subcutaneous adipose tissue and skeletal muscle. *International Journal of Sports Medicine*, *23*, 537–543.

[25] Djurhuus, C.B., Gravholt, C.H., Nielsen, S., Mengel, A., Christiansen, J.S., Schmitz, O.E., Moller, N. (2002). Effects of cortisol on lipolysis and regional interstitial glycerol levels in humans. *American Journal of Physiology*, *283*, E172–E177.

[26] Hickner, R.C., Fisher, J.S., Kohrt, W.M. (1997). Regional differences in interstitial glycerol concentration in subcutaneous adipose tissue of women. *American Journal of Physiology*, *273*, E1033–E1038.

[27] Jansson, P.A., Smith, U., Lönnroth, P. (1990). Interstitial glycerol concentration measured by microdialysis in two subcutaneous regions in humans. *American Journal of Physiology*, *258*, E918–E922.

[28] Rosdahl, H., Ungerstedt, U., Jorfeldt, L., Henriksson, J. (1993). Interstitial glucose and lactate balance in human skeletal muscle and adipose tissue studied by microdialysis. *Journal of Physiology*, *471*, 637–657.

[29] Enevoldsen, L.H., Simonsen, L., Stallknecht, B., Galbo, H., Bulow, J. (2001). In vivo human lipolytic activity in preperitoneal and subdivisions of subcutaneous abdominal adipose tissue. *American Journal of Physiology*, *281*, E1110–E1114.

[30] Simonsen, L., Enevoldsen, L.H., Stallknecht, B., Bulow, J. (2008). Effects of local alpha2-adrenergic receptor blockade on adipose tissue lipolysis during prolonged systemic adrenaline infusion in normal man. *Clinical Physiology and Functional Imaging*, *28*, 125–131.

[31] Galitzky, J., Lafontan, M., Nordenstrom, J., Arner, P. (1993). Role of vascular alpha-2 adrenoceptors in regulating lipid mobilization from human adipose tissue. *Journal of Clinical Investigation*, *91*, 1997–2003.

[32] Hickner, R.C., Rosdahl, H., Borg, I., Ungerstedt, U., Jorfeldt, L., Henriksson, J. (1991). Ethanol may be used with the microdialysis technique to monitor blood flow changes in skeletal muscle: dialysate glucose concentration is blood-flow-dependent. *Acta Physiologica Scandinavica*, *143*, 355–356.

[33] Arner, P., Bulow, J. (1993). Assessment of adipose tissue metabolism in man: comparison of Fick and microdialysis techniques. *Clinical Science (London)*, *85*, 247–256.

[34] Goossens, G.H., Karpe, F. (2008). Human adipose tissue blood flow and micromanipulation of human subcutaneous blood flow. *Methods in Molecular Biology (Clifton, N.J.)*, *456*, 97–107.

[35] Larsen, O.A., Lassen, N.A., Quaade, F. (1966). Blood flow through human adipose tissue determined with radioactive xenon. *Acta Physiologica Scandinavica*, *66*, 337–345.

[36] Karpe, F., Fielding, B.A., Ilic, V., Humphreys, S.M., Frayn, K.N. (2002). Monitoring adipose tissue blood flow in man: a comparison between the (133)xenon washout method and microdialysis. *International Journal of Obesity and Related Metabolic Disorders*, *26*, 1–5.

[37] Barbe, P., Millet, L., Galitzky, J., Lafontan, M., Berlan, M. (1996). In situ assessment of the role of the beta 1-, beta 2- and beta 3-adrenoceptors in the control of lipolysis and nutritive blood flow in human subcutaneous adipose tissue. *British Journal of Pharmacology*, *117*, 907–913.

[38] Schiffelers, S.L., Akkermans, J.A., Saris, W.H., Blaak, E.E. (2003). Lipolytic and nutritive blood flow response to beta-adrenoceptor stimulation in situ in subcutaneous abdominal adipose tissue in obese men. *International Journal of Obesity and Related Metabolic Disorders*, *27*, 227–231.

[39] Blaak, E.E., van Baak, M.A., Kemerink, G.J., Pakbiers, M.T., Heidendal, G.A., Saris, W.H. (1995). Beta-adrenergic stimulation and abdominal subcutaneous fat blood flow in lean, obese, and reduced-obese subjects. *Metabolism*, *44*, 183–187.

[40] Goossens, G.H., Jocken, J.W., Blaak, E.E., Schiffers, P.M., Saris, W.H., van Baak, M.A. (2007). Endocrine role of the renin–angiotensin system in human adipose tissue and muscle: effect of beta-adrenergic stimulation. *Hypertension*, *49*, 542–547.

[41] Goossens, G.H., Jocken, J.W., van Baak, M.A., Jansen, E.H., Saris, W.H., Blaak, E.E. (2008). Short-term beta-adrenergic regulation of leptin, adiponectin and interleukin-6 secretion in vivo in lean and obese subjects. *Diabetes, Obesity & Metabolism*, *10*, 1029–1038.

[42] Jansson, P.A., Larsson, A., Lönnroth, P.N. (1998). Relationship between blood pressure, metabolic variables and blood flow in obese subjects with or without non-insulin-dependent diabetes mellitus. *European Journal of Clinical Investigation*, *28*, 813–818.

[43] Summers, L.K., Samra, J.S., Humphreys, S.M., Morris, R.J., Frayn, K.N. (1996). Subcutaneous abdominal adipose tissue blood flow: variation within and between subjects and relationship to obesity. *Clinical Science (London)*, *91*, 679–683.

[44] Blaak, E.E., Schiffelers, S.L., Saris, W.H., Mensink, M., Kooi, M.E. (2004). Impaired beta-adrenergically mediated lipolysis in skeletal muscle of obese subjects. *Diabetologia*, *47*, 1462–1468.

[45] Goossens, G.H., Blaak, E.E., Saris, W.H., van Baak, M.A. (2004). Angiotensin II–induced effects on adipose and skeletal muscle tissue blood flow and lipolysis in normal-weight and obese subjects. *Journal of Clinical Endocrinology and Metabolism*, *89*, 2690–2696.

[46] Boschmann, M., Engeli, S., Adams, F., Franke, G., Luft, F.C., Sharma, A.M., Jordan, J. (2006). Influences of AT1 receptor blockade on tissue metabolism in obese men. *American Journal of Physiology: Regulatory, Integrative, and Comparative Physiology, 290*, R219–223.

[47] Stumvoll, M., Jacob, S., Wahl, H.G., Hauer, B., Loblein, K., Grauer, P., Becker, R., Nielsen, M., Renn, W., Haring, H. (2000). Suppression of systemic, intramuscular, and subcutaneous adipose tissue lipolysis by insulin in humans. *Journal of Clinical Endocrinology & Metabolism, 85*, 3740–3745.

[48] Hickner, R.C., Racette, S.B., Binder, E.F., Fisher, J.S., Kohrt, W.M. (1999). Suppression of whole body and regional lipolysis by insulin: effects of obesity and exercise. *Journal of Clinical Endocrinology & Metabolism, 84*, 3886–3895.

[49] Stallknecht, B., Bulow, J., Frandsen, E., Galbo, H. (1997). Desensitization of human adipose tissue to adrenaline stimulation studied by microdialysis. *Journal of Physiology, 500*(Pt. 1), 271–282.

[50] Stallknecht, B., Kiens, B., Helge, J.W., Richter, E.A., Galbo, H. (2004). Interstitial glycerol concentrations in human skeletal muscle and adipose tissue during graded exercise. *Acta Physiologica Scandinavica, 180*, 367–377.

[51] Bickerton, A.S., Roberts, R., Fielding, B.A., Hodson, L., Blaak, E.E., Wagenmakers, A.J., Gilbert, M., Karpe, F., Frayn, K.N. (2007). Preferential uptake of dietary fatty acids in adipose tissue and muscle in the postprandial period. *Diabetes, 56*, 168–176.

[52] Frayn, K.N., Coppack, S.W., Humphreys, S.M. (1993). Subcutaneous adipose tissue metabolism studied by local catheterization. *International Journal of Obesity and Related Metabolic Disorders, 17*(Suppl. 3), S18–S21; discussion, S22.

[53] Jocken, J.W., Goossens, G.H., van Hees, A.M., Frayn, K.N., van Baak, M., Stegen, J., Pakbiers, M.T., Saris, W.H., Blaak, E.E. (2008). Effect of beta-adrenergic stimulation on whole-body and abdominal subcutaneous adipose tissue lipolysis in lean and obese men. *Diabetologia, 51*, 320–327.

[54] Karpe, F., Olivecrona, T., Olivecrona, G., Samra, J.S., Summers, L.K., Humphreys, S.M., Frayn, K.N. (1998). Lipoprotein lipase transport in plasma: role of muscle and adipose tissues in regulation of plasma lipoprotein lipase concentrations. *Journal of Lipid Research, 39*, 2387–2393.

[55] Lönnroth, P. (1997). Microdialysis in adipose tissue and skeletal muscle. *Hormone and Metabolic Research, 29*, 344–346.

[56] Jansson, P.A., Smith, U., Lönnroth, P. (1995). Microdialysis assessment of adipose tissue metabolism in post-absorptive obese NIDDM subjects. *European Journal of Clinical Investigation, 25*, 584–589.

[57] Lassen, N.A. (1967). Capillary diffusion capacity of sodium studied by the clearances of Na-24 and Xe-133 from hyperemic skeletal muscle in man. *Scandinavian Journal of Clinical and Laboratory Investigation, 99*, 24–26.

[58] Paaske, W.P. (1977). Absence of restricted diffusion in adipose tissue capillaries. *Acta Physiologica Scandinavica, 100*, 430–436.

[59] Blaak, E.E., Kemerink, G.J., Pakbiers, M.T., Wolffenbuttel, B.H., Heidendal, G.A., Saris, W.H. (1999). Microdialysis assessment of local adipose tissue lipolysis during beta-adrenergic stimulation in upper-body-obese subjects with type II diabetes. *Clinical Science (London), 97*, 421–428.

[60] Blaak, E.E., Van Baak, M.A., Kemerink, G.J., Pakbiers, M.T., Heidendal, G.A., Saris, W.H. (1994). Beta-adrenergic stimulation of energy expenditure and forearm skeletal muscle metabolism in lean and obese men. *American Journal of Physiology*, *267*, E306–E315.

[61] Blaak, E.E., Van Baak, M.A., Kemerink, G.J., Pakbiers, M.T., Heidendal, G.A., Saris, W.H. (1994). Beta-adrenergic stimulation of skeletal muscle metabolism in relation to weight reduction in obese men. *American Journal of Physiology*, *267*, E316–E322.

[62] Connacher, A.A., Bennet, W.M., Jung, R.T., Bier, D.M., Smith, C.C., Scrimgeour, C.M., Rennie, M.J. (1991). Effect of adrenaline infusion on fatty acid and glucose turnover in lean and obese human subjects in the post-absorptive and fed states. *Clinical Science (London)*, *81*, 635–644.

[63] Webber, J., Taylor, J., Greathead, H., Dawson, J., Buttery, P.J., Macdonald, I.A. (1994). A comparison of the thermogenic, metabolic and haemodynamic responses to infused adrenaline in lean and obese subjects. *International Journal of Obesity and Related Metabolic Disorders*, *18*, 717–724.

[64] Horowitz, J.F., Klein, S. (2000). Whole body and abdominal lipolytic sensitivity to epinephrine is suppressed in upper body obese women. *American Journal of Physiology*, *278*, E1144–E1152.

[65] Quisth, V., Enoksson, S., Blaak, E., Hagström-Toft, E., Arner, P., Bolinder, J. (2005). Major differences in noradrenaline action on lipolysis and blood flow rates in skeletal muscle and adipose tissue in vivo. *Diabetologia*, *48*, 946–953.

[66] Schutte, R.J., Oshodi, S.A., Reichert, W.M. (2004). In vitro characterization of microdialysis sampling of macromolecules. *Analytical Chemistry*, *76*, 6058–6063.

[67] Clough, G.F. (2005). Microdialysis of large molecules. *AAPS Journal*, *7*, E686–E692.

[68] Trickler, W.J., Miller, D.W. (2003). Use of osmotic agents in microdialysis studies to improve the recovery of macromolecules. *Journal of Pharmaceutical Sciences*, *92*, 1419–1427.

[69] Plock, N., Buerger, C., Kloft, C. (2005). Successful management of discovered pH dependence in vancomycin recovery studies: novel HPLC method for microdialysis and plasma samples. *Biomedical Chromatography*, *19*, 237–244.

[70] Freed, A.L., Cooper, J.D., Davies, M.I., Lunte, S.M. (2001). Investigation of the metabolism of substance P in rat striatum by microdialysis sampling and capillary electrophoresis with laser-induced fluorescence detection. *Journal of Neuroscience Methods*, *109*, 23–29.

[71] Haskins, W.E., Wang, Z., Watson, C.J., Rostand, R.R., Witowski, S.R., Powell, D.H., Kennedy, R.T. (2001). Capillary LC-MS2 at the attomole level for monitoring and discovering endogenous peptides in microdialysis samples collected in vivo. *Analytical Chemistry*, *73*, 5005–5014.

12

MICRODIALYSIS AS A MONITORING SYSTEM FOR HUMAN DIABETES

Anna Ciechanowska and Jan M. Wojcicki

Institute of Biocybernetics and Biomedical Engineering, Polish Academy of Sciences, Warsaw, Poland

Iwona Maruniak-Chudek

Medical University of Silesia, Katowice, Poland

Piotr Ladyzynski

Institute of Biocybernetics and Biomedical Engineering, Polish Academy of Sciences, Warsaw, Poland

Janusz Krzymien

Medical University of Warsaw, Warsaw, Poland

1. INTRODUCTION

Diabetes is a set of metabolic disorders resulted from a deficiency or lack of endogenous insulin secretion (type 1 diabetes), the occurrence of insulin resistance (type 2 diabetes), or the occurrence of both pathologies at the same time. The first description of a disease with the features of diabetes was found on an Egyptian papyrus scroll dating from 1500 B.C. Diabetes is characterized by continuous hyperglycemia, which creates very dangerous acute complications and early death. Diabetes is the major cause of kidney dysfunction, blindness in adults, nervous system damage, and low amputations. It is the main risk factor for heart disease, stroke, and malformation of newborns. Diabetes shortens life expectancy by about 15 years. Diabetes is one of the most costly human diseases from both an economic and a human point of view.

Applications of Microdialysis in Pharmaceutical Science, First Edition. Edited by Tung-Hu Tsai.
© 2011 John Wiley & Sons, Inc. Published 2011 by John Wiley & Sons, Inc.

Currently, it is estimated that the number of diabetes patients in the world exceeds 200 million peoples; however, it should be stressed that this number is growing very rapidly. In 1994, general estimations indicated 110.4 million diabetics. On the basis of the analysis performed by the World Health Organization, it is predicted that in the year 2010 the overall number of diabetes patients was 239.3 million, including 23.7 million type 1 and 215.6 million type 2 diabetic patients. These numbers indicate that diabetes is a social disease, one of the most critical health problems in the twenty-first century. The latest estimations performed in Poland indicated that there are about 2.0 million diabetics, close to 6% of population. The number of type 1 diabetes, patients without endogenous insulin secretion, in most countries is close to 10% of the entire diabetes population.

The diabetes research conducted for many years has proven the great practical value of the basic strategy of diabetes therapy: to maintain in diabetic patients levels of glycemia similar to those noted in normal subjects. It was proven in several long-term research programs performed in the United States [the Diabetes Control and Complications Trial (DCCT) study], Europe (the Steno project), and Japan (the Kumamoto study) that the use of this strategy in long-term intensive treatment delays the onset and slows the progression of the late micro- and macrovascular complications of diabetes. This strategy has not yet been fully implemented as a long-term therapy, because its realization is very difficult. It seems that only when new technical solutions are available for millions of diabetic patients can we expect significant improvement in the efficiency of diabetes treatment.

From a technical point of view, the process of treatment consists of two phases: (1) the monitoring phase, which includes blood glucose measurements with observation of other events important for treatment performed by the patient, and transfer of these data to the central registration system, and (2) the therapy phase, which consists of data analysis by application of simple mathematical indices and/or sophisticated algorithms allowing for an analysis of the patient's metabolic state and insulin dose adjustment. The last part of the structure described is associated with the selection of insulin preparations and method of administration (multiple injection technique, continuous infusion, or other methods of insulin delivery, which are at different stages of development). Figure 1 illustrates a system of intensive insulin treatment.

In such aspects of the therapy, a very crucial role is that of the patient's education and training in the initial phase of the therapy, frequent and routine control determined by the physician and his or her team, as well as corrections in the therapy and engagement of the patient in the implementation of these targets. Another very important aspect is the patient's comfort, which consists of the feeling of safety that is associated with an awareness of permanent medical care. However, implementation of the assumed therapy strategy would not be possible without effective techniques and technical solutions supporting the therapy. Today's technical solutions are far from perfect, but they undergo continuous improvement through introduction of new developments based on the newest achievements in technology.

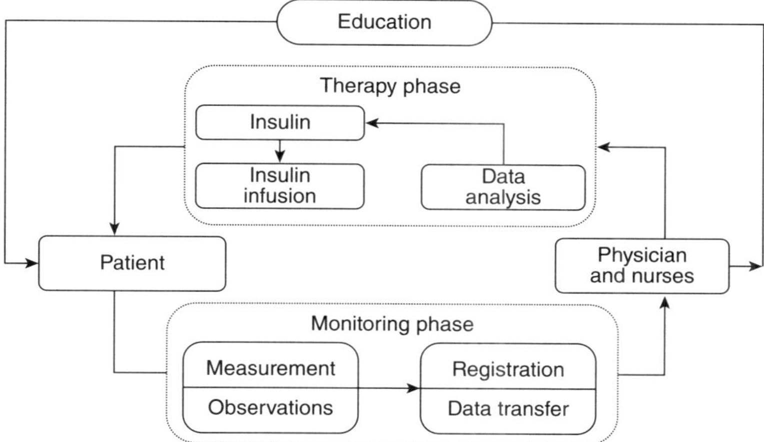

Figure 1   Scheme for intensive insulin therapy.

Microdialysis (MD) as a monitoring system for human diabetes aimed at continuous monitoring of glycemia was used for the first time in the early 1990s by a group of researchers from Ulm headed by Ernst Pfeiffer. After several years of research, the first prototype of such a system, called the Ulm Suger Watch, has been tested in vivo [1]. The development of glucose monitoring based on the MD technique is represented today by the AccuChek (Roche Diagnostics) [2], GlucoDay (Menarini) [3], and Gluc-Online (Roche/ Disetronic) [4] systems.

Another very important field of MD application in diabetes and diabetic complications treatments refers to monitoring of various substances, including glucose, in a variety of tissues and organs. Some of them could be considered as important markers of a diabetic patient's metabolic state.

Applications of MD systems for monitoring of acute and late diabetes complications, as well as assessment of insulin sensitivity as a marker of the diabetes treatment efficiency, are described below. Particular attention will be paid to:

- Monitoring the acute complications of diabetes to:
  - Monitor the ketoacidosis and hyperglycemic hyperosmolar states
  - Monitor hypoglycemia
- Monitoring the late complication of diabetes for:
  - Use of the microdialysis technique to ensure proper care of diabetic feet
- Monitoring infections in diabetes to:
  - Monitor antibiotic therapies in diabetic feet
- Monitoring insulin sensitivity in diabetic patients to:
  - Monitor insulin concentration in the interstitial fluid
  - Monitor muscle glucose uptake in diabetes

2. MONITORING THE ACUTE COMPLICATIONS OF DIABETES

2.1. Monitoring Ketoacidosis and the Hyperglycemic Hyperosmolar State

The most serious acute complications of diabetes are: diabetic ketoacidosis (DKA) and the hyperosmolar hyperglycemic state (HHS) [5]. Mortality rates are below 5% in DKA and about 15% in HHS [6–13]. The rate increase in HHS is up to 50% with aging and the presence of concomitant life-threatening illness [14,15]. Episodes of DKA are more frequent in younger than in older patients and are twice as frequent in females than in males [15]. In DKA and HHS, the fundamental metabolic abnormality arises from a combination of insulin deficiency and increased amounts of such counter-regulatory hormones as glucagon, catecholamines, cortisol, and growth hormone. Glucose production is increased, but its consumption in peripheral tissues decreases as a result of the insulin shortage and resistance induced by the raised levels of plasma catabolic hormones and free fatty acids (FFAs). The rise in blood glucose level (due to glycosuria) causes high blood osmolarity (higher in HHS—above 320 mOsm/L) and then osmotic diuresis, which results in fluid and electrolyte losses (particularly cations: sodium, potassium, and magnesium) [5,16]. The total body water deficit is usually about 5 to 7 L in DKA and 7 to 12 L in HHS (loss about 10 to 15% of total body weight) [14]. Tissue lipase is activated in DKA by significant insulin deficiency and elevated levels of counter-regulatory hormones, increasing lipolysis and releasing large amounts of glycerol and FFAs into the blood circulation. Glycerol is a precursor for gluconeogenesis in the liver, and the kidneys. FFAs are oxidized to the ketone bodies' aceto-acetate (strong organic acids) in the liver, leading to acidosis [16].

The current state of a patient is determined by blood examination (electrolytes, pH, blood glucose, etc.) in venous blood samples withdrawn every 2 to 8 h and monitoring of the capillary blood glucose every 1 to 2 h. However, this procedure does not provide continuous monitoring of the patient metabolic state in acute diabetes complication. Moreover, monitoring involves measurements from the blood, whereas most biochemical and pharmacological events take place in individual tissues, not in the blood.

Microdialysis: Clinical Studies   In one study [17], the results of an innovative application of microdialysis to monitor the metabolic parameters in the interstitial fluid (ISF) in the patients with acute diabetic complications are presented. The main objective of this study was to establish continuous courses of glucose, lactate, and glycerol in the ISF using microdialysis, to characterize the patient's metabolic state during the standard treatment of hyperglycemic crises. Clinical studies were carried out on 31 patients with acute diabetes complications, such as DKA (25 persons), HHS (4 patients), lactic acidosis (1 patient), and DKA with a lactic constituent (1 patient).

The patients' metabolic state was monitored using microdialysis during the first 48 h of the standard treatment (water and electrolyte supplementation and insulin therapy according to the current patient state). A microdialysis catheter (CMA-60; CMA Microdialysis, Solma, Sweden) was introduced in the patient's abdominal subcutaneous adipose tissue. First, in vitro tests on a solution of glucose, lactate, glycerol, and pyruvate in Ringer's solution were performed to choose the operating conditions for the microdialysis probe. The flow rate of the perfusion fluid was chosen as $0.3\,\mu L/min$, and for such flow conditions the in vitro recovery rate has been found to be 100%. The perfusion fluid (T1; CMA Microdialysis) was pumped to the microdialysis probe by a CMA-107 pump (CMA Microdialysis). Effluent from the catheter was collected in a quasi-continuous way (each drawing for 20 min) into microvials and then analyzed with the microsample analyzer CMA-600 (CMA Microdialysis). Biochemical parameters such as glucose, lactate, and glycerol in the interstitial fluid were estimated. As a reference, measurements in the blood circulation were made; venous glucose, venous lactate, and venous glycerol concentrations were measured every 8 h and the glucose concentration in finger capillary blood every 2 h.

To estimate to what degree glucose concentration in the ISF differs from the concentration in the capillary blood in patients with acute diabetes complications, the recovery coefficient was calculated as a ratio of the glucose concentration in the perfusion fluid (i.e., dialysate) of the microdialysis probe to that in the capillary blood.

Analysis of the data were done for the entire study group and two subgroups: obese [body mass index (BMI) $\geq 30\,kg/m^2$] and nonobese (lean and overweight with BMI $< 30\,kg/m^2$) patients. The data for glucose concentration were analyzed: for the 48-h study period, during the initial 12 h, and during the last 36 h of patient monitoring.

Evaluation of Microdialysis Effectiveness A comparison of the results obtained for the glucose, lactate, and glycerol monitored in the interstitial fluid and the capillary/venous blood for all the patients is presented in Figure 2. The values of the glucose concentration in the interstitial fluid (mean value equal to $8.4 \pm 2.6\,mmol/L$) during all monitoring were below the values of the glucose concentration in the capillary blood compartment (mean value equal to $10.9 \pm 3.9\,mmol/L$), and the mean value was 23% higher in the blood than in the ISF. The mean difference in the glucose concentration in the interstitial fluid and capillary blood was statistically significant (p = 0.000012). The time course of the lactate concentration in the ISF coincided with the time course of the lactate concentration in the blood compartment. The mean value of lactate concentration was $2.1 \pm 0.3\,mmol/L$ in the interstitial fluid and $2.0 \pm 0.5\,mmol/L$ in the blood, and the mean difference between these levels was not statistically significant (p = 0.35). The time courses of the glycerol concentration varied between the ISF (mean value $267 \pm 41\,\mu mol/L$) and blood (mean value $133 \pm 40\,\mu mol/L$), and the mean value in the interstitial fluid was 50% higher than that in the blood. The mean difference of the glycerol concentration in the interstitial fluid and capillary blood was statistically

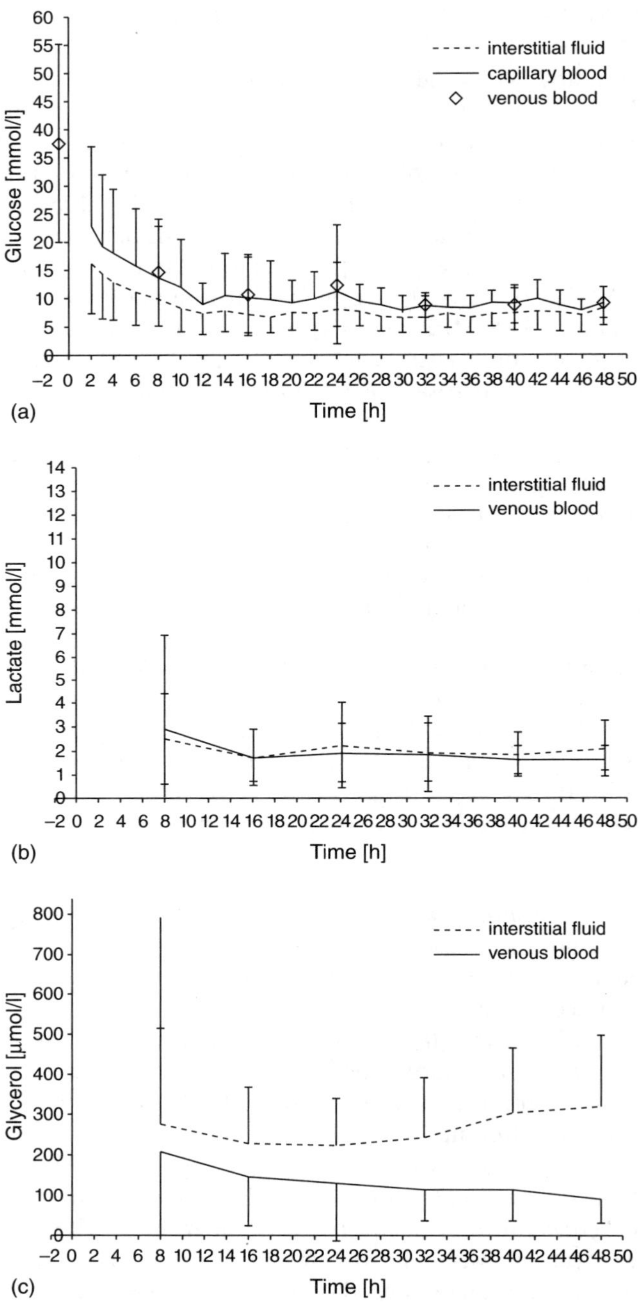

Figure 2 Comparison of the glucose (a), lactate (b), and glycerol (c) time courses during monitoring of all the patients in the interstitial fluid and in capillary and venous blood. The glucose concentrations were slightly lower, lactate concentrations were very similar, and glycerol concentrations were much higher in the interstitial fluid than in capillary/venous blood.

significant (p = 0.004). The greatest difference between glycerol concentrations in the interstitial fluid and blood compartments was observed at the end of the 48-h monitoring period. In [17] the most rapid changes of glucose concentration in patients with acute complications of diabetes occurred during the first 12 h of observation. This was a consequence of the treatment initiated. The long sample-collecting time induces differences between the glucose concentrations in the effluent from the microdialysis probe and in the plasma. The differences are more noticeable when there are rapid changes in glucose concentration in the patients' blood and interstitial fluid. On the other hand, by applying a low perfusion flow rate along the probe as was done by investigators in this study [17], it should almost be possible to reach the equilibrium glucose concentration at the end of the probe. In such a case, the effect of wound healing should not be observed [18]. This supposition was verified positively [17], confirming the applicability of the established procedure for monitoring a patient's state during hyperglycemic crises.

The average time course of the glucose concentration in the ISF of adipose tissue as well as the glucose concentration in blood plasma reflected the assumed goal of the treatment. This average time course was comprised of the initial 10 to 12 h long phase of fast decrease in glucose concentration, followed by a stabilization phase in which the glucose concentration was kept just below 10 mmol/L.

An average concentration of lactate in the effluent of perfusion fluid from a microdialysis probe [17] in nonobese patients (equal to 2.1 mmol/L) was 40% higher than the lactate concentration in the venous blood plasma. This difference was lower than that reported in studies carried out by other investigators, where venous blood arterialization was employed [19]. The average time course of the lactate concentration in the effluent of perfusion fluid from microdialysis probes was similar to the average time course of glucose concentration.

The ratio of glycerol concentrations in the effluent of perfusion fluid from microdialysis probes to those in venous blood plasma [17] was lower than that reported by other investigators [19,20] under a variety of experimental conditions, when blood was collected from arterial/arterialized veins or from the superficial abdominal vein. The absolute values measured [17] were higher than those in most previous studies. This was caused by the metabolic conditions in the study group. The time course of glycerol concentration in the interstitial fluid differed significantly from the time courses of the other metabolites [17]. After the initial decrease in glycerol concentrations, stabilization was reached, followed by an increase toward the initial values at the end of the second day of treatment. The glycerol time course observed seems to be justified by the insulin infusion profiles applied during the treatment and the presumable resulting insulinemia [21].

Another interesting feature of the time course of glycerol concentration was high variability, which was observed for some patients [17]. It was not clear [17] if the increased variability should be attributed to the rapid changes in patients metabolism or if it was simply a measurement problem. No distinct delays between measurements performed in the compartments studied were observed.

The time courses of the biochemical parameters monitored in the interstitial and blood compartments were found to be different in the two subgroups of patients: those with BMI < $30\,kg/m^2$ and those with BMI $\geq$ $30\,kg/m^2$. The glucose concentrations courses for patients with BMI < $30\,kg/m^2$ were slightly higher in the blood compartment (a mean value of $10.7 \pm 4.6\,mmol/L$) than in the ISF compartment (a mean value of $9.3 \pm 3.3\,mmol/L$) (Figure 3), and the mean value was by 13% higher in the blood than in the interstitial fluid. The time courses of the lactate and glycerol concentrations were significantly higher in the ISF compartment (mean values of 2.1 ± 0.3 and $245 \pm 41\,\mu mol/L$ for lactate and glycerol, respectively) than in the blood compartment (mean values of 1.5 ± 0.2 and $115 \pm 26\,\mu mol/L$ for lactate and glycerol, respectively). The mean values of lactate and glycerol concentrations were 29 and 53% higher in the interstitial fluid than in the blood, respectively.

The differences in the mean concentrations of all the metabolites [17] were statistically significant, with p values equal to 0.000008 for glucose, 0.0003 for lactate, and 0.003 for glycerol, respectively. The glucose and lactate concentrations time courses for obese patients with BMI $\geq$ $30\,kg/m^2$ were significantly higher in the blood compartment (a mean value of 14.1 ± 6.9 and $3.2 \pm 1.3\,mmol/L$ for glucose and lactate, respectively) than in the ISF compartment (a mean value of 6.0 ± 1.6 and $1.8 \pm 0.3\,mmol/L$ for glucose and lactate, respectively) throughout the entire monitoring period (Figure 4). The mean values of glucose and lactate concentrations were 57 and 44% higher in the blood than in the interstitial fluid, respectively. The difference in the mean concentrations of the glucose was statistically significant ($p < 0.000001$). This difference was 3.2 times higher than that in the entire study group and 5.8 times higher than that in the group of patients with BMI < $30\,kg/m^2$. The difference in the mean concentrations of the lactate was also statistically significant ($p = 0.03$). For the group of patients with BMI $\geq$ $30\,kg/m^2$, the lactate concentration in ISF was lower than that in blood, and this was opposite to the subgroup that consisted of nonobese patients. The glycerol concentration in the interstitial fluid was significantly higher than that in the blood compartment (mean value equal to 327 ± 69 and $185 \pm 85\,\mu mol/L$, respectively) and the mean value was 43% higher in the ISF than in the blood. Consequently, the mean difference between both compartments was statistically significant ($p = 0.02$).

The results of analysis of variance demonstrated that neither osmolarity nor interactions between BMI and osmolarity had any effect on glucose recovery. The results indicated that BMI was a factor that significantly influenced glucose recovery ($p = 0.000009$). The significant inverse correlation between a patient's BMI and glucose recovery ($r = 0.55$) was demonstrated by linear regression analysis.

A comparison of the results of glucose recovery during the initial 12 h, the final 36 h, and the entire monitoring period obtained for the entire study group,

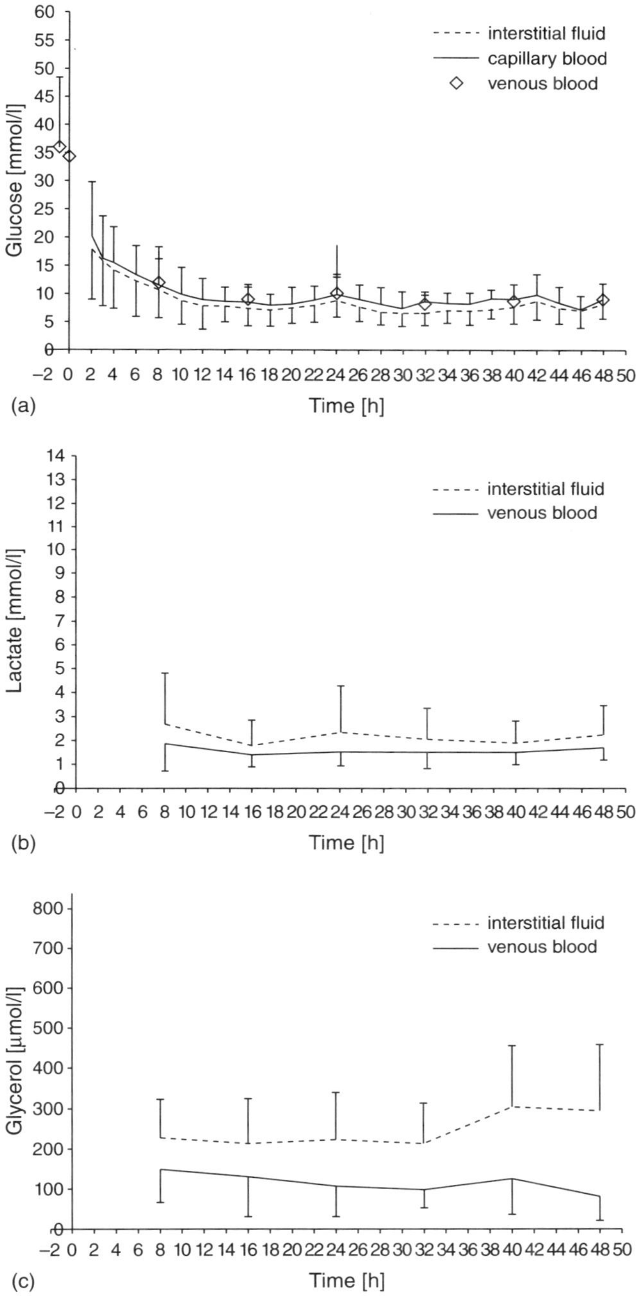

Figure 3 Comparison of the glucose (a), lactate (b), and glycerol (c) time courses during monitoring of patients with BMI < 30 kg/m^2 in the interstitial fluid and in capillary and venous blood. The glucose concentrations were very similar; lactate and glycerol concentrations were higher in the interstitial fluid than in capillary/venous blood.

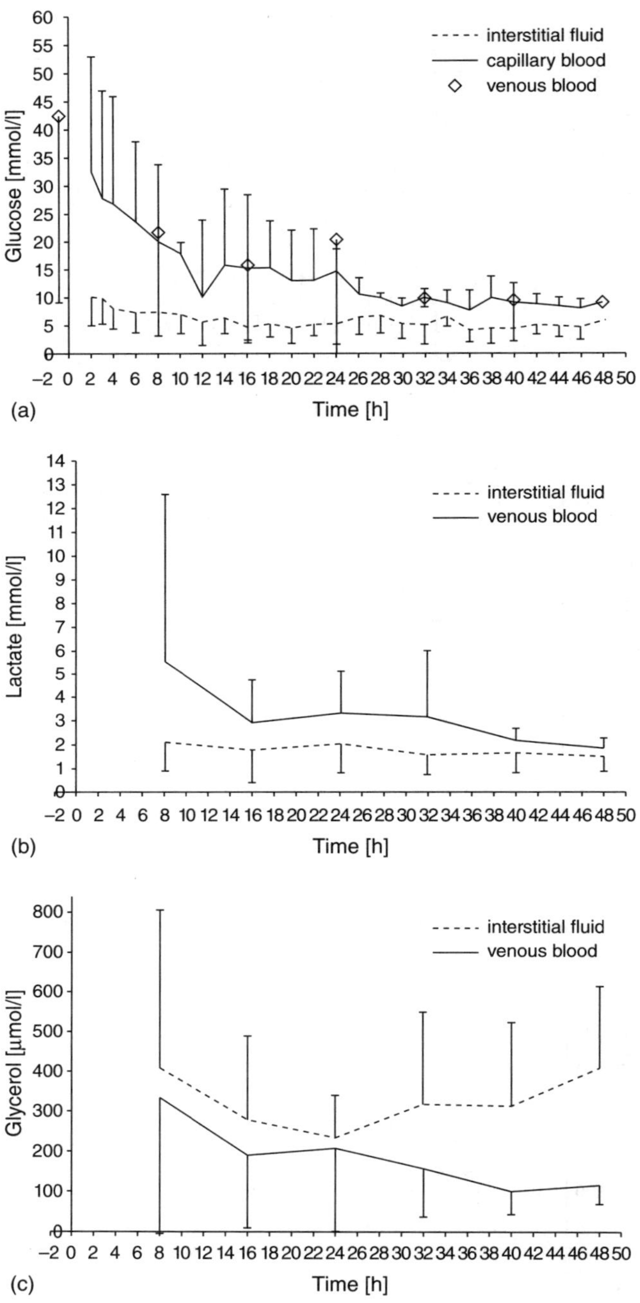

Figure 4 Comparison of the glucose (a), lactate (b), and glycerol (c) courses during monitoring of obese patients with BMI ≥ 30 kg/m², showing much lower glucose concentrations, much lower lactate concentrations, and much higher glycerol concentrations in the perfusion fluid of the microdialysis probe than in capillary and venous blood.

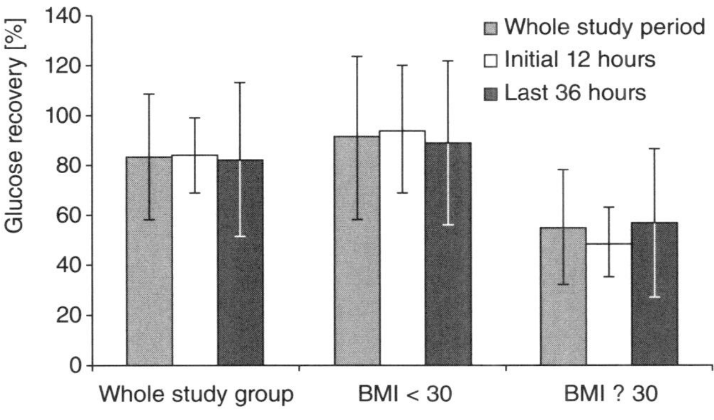

Figure 5   Glucose recovery.

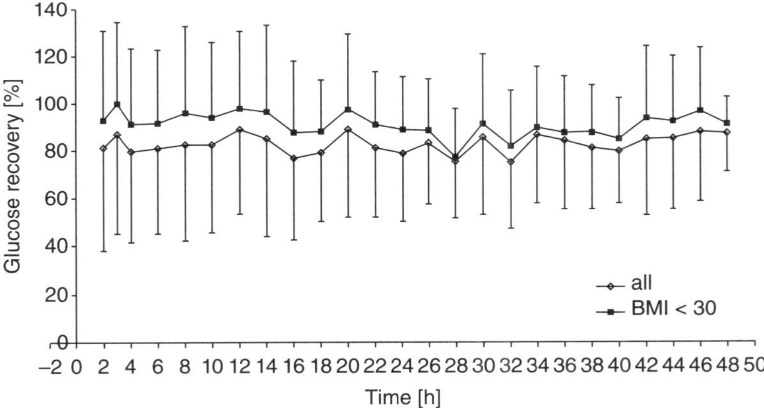

Figure 6   Comparison of the glucose recovery course during monitoring of the entire study group and patients with BMI < 30 showing higher values for patients with BMI < 30 than for the entire study group.

as well as for the subgroups (with and without obesity), is presented in Figure 5. Changes in glucose recovery with the passing of time were not statistically significant. It was found [17] that in a majority of the patients in the group, the glucose concentration in the perfusion fluid effluent from the microdialysis probe reflected the glucose concentration in the capillary blood. The glucose concentration was lower in the effluent than in the capillary blood of the entire study group. Consequently, the time course of the glucose recovery was less than 100% (Figure 6). The mean recovery value was equal to $83 \pm 25\%$. This value is similar to the result obtained during another study carried out on type 1 diabetic patients [22]. Analysis of the data [17] showed that the difference in glucose concentrations in interstitial fluid and blood compartments could

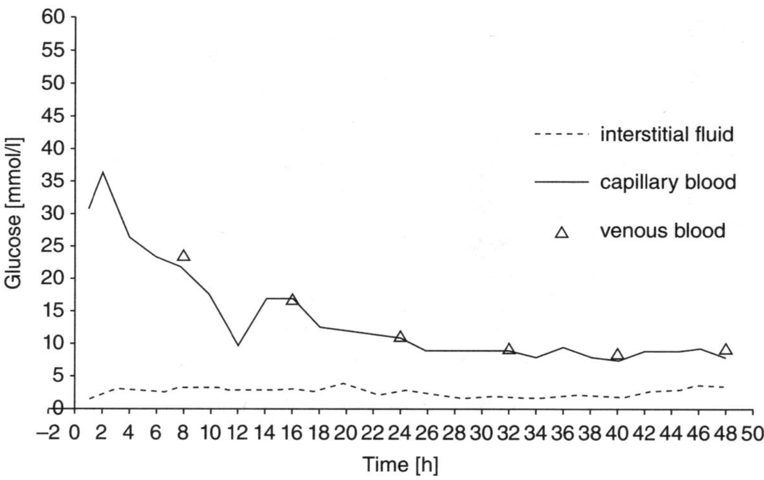

Figure 7   The glucose courses in the perfusion fluid and capillary and venous blood during monitoring of an exemplary patient with HHS and BMI $\geq 30\,\mathrm{kg/m^2}$. In this patient a lack of responsiveness of the microdialysis probe was observed.

be caused in part by the results of four obese female patients. Two of these patients with HHS were characterized as being very low and stable in time glucose concentration in the probe effluent despite a significant decrease in glucose concentration in the capillary blood caused by the treatment (Figure 7). In the remaining two obese patients with DKA and the concomitant hyperosmolarity, the time course of the glucose concentration in the probe effluent resembled the course registered from the capillary blood, but glucose recovery was below 50%.

The Decrease in glucose recovery in the obese female patients might have been caused by technical problems such as a decrease in the effective surface of the probe (e.g., membrane clogging or changes in the working conditions of the probe). It was also possible that obesity and/or some pathological conditions occurring in the course of DKA/HHS affected the glucose transport across the probes' capillary walls. Due to the limited number of obese female patients with acute complications of diabetes [17], further research is necessary to explain the problems described above in more detail.

Exclusion of the glucose concentrations data in the interstitial fluid and capillary venous blood of the obese patients from the entire study group led to a significant increase in the value of glucose recovery throughout the monitoring period (Figure 6). The average glucose recovery increased by 8%, up to $91 \pm 15\%$ (Figure 5). Similar results were announced by other researchers [23–25], who studied healthy volunteers and diabetic patients under diverse conditions using a variety of microdialysis systems and operating settings. The result [17] is in the upper part of a range of glucose recovery reported in the investigations mentioned above. Glucose concentration measured using microdialysis technique can be associated with inserting a microdialysis probe into subcutaneous adipose tissue [18,26]. Inserting the probe may cause a temporary decrease in the local equilibrium concentration around the probe and a rise in recovery thereafter. This is connected with a disruption in the tissue microstructure, causing a glucose-consuming inflammation reaction (the first stage) and then with a wound-healing process, leading to restoration of the microvasculature around the microdialysis probe (the second stage).

The first stage lasts about 1 h according to some investigators [26], or 12 to 18 h according to others [18], and the second stage lasts 4 to 6 days [18]. For the second stage, an inverse correlation was found between the recovery and skinfold thickness as a measure of obesity [27,28]. In the study [17] the 48-h monitoring period of biochemical parameters in the patients' interstitial fluid was preceded by a 1-h-long probe stabilization period. Thus, each of the effects described above (two stages of healing) could have affected results [17] and lowered the applicability of microdialysis for glucose monitoring. However, Ciechanowska et al. [17] did not observe any of there effects. They observed a lack of discrepancy between differences in glucose concentration in the effluent perfuse fluid from a microdialysis probe and in the plasma during the first 12 h of observation that attests to faster fading of the inflammation-induced increase in glucose consumption around the probe (minutes, not hours). It is also likely that the influence of this effect on glucose concentration is not strong enough to be observed in the presence of other phenomena, such as differences in glucose concentration in various tissues (the glucose concentration measured in abdominal adipose tissue is compared with the glucose concentration measured in plasma collected from a fingertip).

Summary A study of a group of patients with DKA and HHS using microdialysis [17] indicates that employing a 3-cm-long microdialysis probe with a polyamide membrane and a perfusion fluid rate of 0.3 μL/min may be an effective tool for continuous monitoring of the concentrations of different metabolites in the ISF of adipose tissue during the treatment of acute complications of diabetes. Application of quasicontinuous monitoring using microdialysis may create a completely new quality of data acquisition, leading to changes in the current standard treatment and making it possible to evaluate threats and prognoses in cases of acute diabetes complications. The applicability of this technique to monitor obese patients with acute complications of diabetes needs further research.

2.2. Monitoring Hypoglycemia

The American Diabetes Association Workgroup on Hypoglycemia (ADAWH) defines hypoglycemia (HG) as including "all episodes of an abnormally low plasma glucose concentration that expose the individual to potential harm" [29]. The Workgroup indicated the significance of the lowest glucose concentration level and the duration of a hypoglycemia episode, emphasizing the importance of the frequency of its occurrence in daily living [30–33]. The value of glucose concentration in HG varies in different age groups and clinical situations, but according to the ADA, HG is defined as ≤3.9 mmol/L [34]. According to the ADAWH [29], HG covers severe hypoglycemia, documented symptomatic hypoglycemia, asymptomatic hypoglycemia, probable symptomatic hypoglycemia, and relative hypoglycemia. HG episodes can be asymptomatic or symptomatic. Hypoglycemic symptoms can be mild and self-treated (autonomic symptoms: trembling, palpitations, sweating, anxiety, hunger, nausea, and several some unique idiosyncratic symptoms), moderate but still self-treated (autonomic and neuroglycopenia symptoms: difficulty in concentration and thinking, confusion, weakness, drowsiness, blurred vision, dizziness, difficulty speaking, and several unique psychological reactions), or severe and requiring the assistance of another person (confusion, seizures, unconsciousness, coma, and even death) [35]. Daytime and nocturnal HG episodes should be reported separately to be appropriate for both clinical interpretation and research [36,37]. Lack of awareness is believed to occur in 25 to 50% of patients with type 1 diabetes (suffering mainly from neuropathy or on intensive diabetes therapy) [38,39], primarily during sleep [30,31,40–42]. In children, the rate of asymptomatic HG is estimated to be very high, up to 50% of young patients [43–48].

According to various authors, nocturnal HG occurs in 30 to 40% of patients with type 1 diabetes, and the episodes may last as long as 2 to 4 h [44,49,50]. Nocturnal asymptomatic HG is regarded as being especially severe due to the mismatch between the action of insulin administered overnight and the nightly increase in hepatic sensitivity to insulin, the impaired catecholamine responses to HG during sleep [42,51], and other factors. Severe HG predisposes to further episodes (downgraded regulatory responses to the repeated HG events) [31].

In the Diabetes Control and Complications Trial (DCCT), patients undergoing intensive therapy had a threefold increase in hypoglycemic events compared with those on conventional therapy [37]. The rate of severe HG in intensively treated diabetic children and adolescents ranged from 15.6 to 26.9 per 100 patient-years [36,52–55], and in adults from 5 to 50 episodes per patient-year [37,53]. HG is more frequent in intensively treated type 1 diabetic patients with HbA1c of about 7% than in those with HbA1c of about 9% in conventional treatment [36]. Several pediatric studies have revealed that children below six year of age are at greater risk of severe hypoglycemia events [53,56] and that neurocognitive difficulties observed are associated with diabetes mellitus [57–61].

Hypoglycemia may have serious consequences for the health and well-being of diabetes patients, causing recurrent morbidity and serious problems for most people with type 1 diabetes and in many with type 2 diabetes [30]. The performance of critical tasks (e.g., driving) [62] during a period of symptomatic hypoglycemia with behavioral changes and cognitive-motor dysfunction may place both the patient and others at risk. Repeated episodes of HG, particularly in the youngest children and older people, can lead to cognitive impairment [61,63,64]. Also, the fact that the fear of HG discourages patients from injecting the optimal insulin therapy cannot be overlooked [49].

As HG is a critical limiting factor in the glycemic management of diabetes in both the short and long terms [30] according to the ADA Workgroup, a significant (≥30%) reduction in the frequency of documented HG [plasma glucose < 3.9 mmol/L (70 mg/dL/)] would be one of the most important steps toward improvement in diabetes management [29,65].

Standard Management of Hypoglycemia   The fundamental method used to diagnose hypoglycemia is blood testing, which requires skin puncture and blood withdrawal. The hexokinase method for deproteinized whole blood is recognized as highly precise and specific for glucose assessment, and is the reference method for HG evaluation [66].

The introduction of self-monitoring made possible remarkable advances in diabetes mellitus management and according to Thorsell, self-monitoring of blood glucose (SMBG) has probably been the most important advance in the field of diabetes since the discovery of insulin [65]. It is believed that more than 50% of HG episodes can be predicted when analyzing self-monitored plasma glucose concentrations [67]. However, detection by SMBG of all HG episodes, particularly nocturnal asymptomatic episodes, even with many (e.g., six or seven) glucose readings, is not possible [65]. Despite the fact that SMBG is still the basic method of tracking glucose excursions, its limitations must be noted: intermittent periodic measurements, painful fingersticks, risk of infection, and considerable blood volume lost in infants [65]. Especially in newborns, paper strips have been criticized for imprecision and underestimation of glucose values, which make them irrelevant for the diagnosis of neonatal hypoglycemia [66]. Reagent strip screening detects only about 85% of true cases of HG and 75% of babies who are truly normoglycemic [68,69]. Also, the very promising HemoCue glucose meter, which requires only a very small volume of blood, has been found unsuitable, due to imprecision in the measurement of low values [70–72].

The implementation of new devices for continuous glucose monitoring (CGM) fulfilled at least some of the criteria of an ideal glucose measuring tool. Indisputable attributes were the relatively small size of the meter, rapid processing, increased frequency of sampling, small samples of interstitial fluid (no blood loss), no need for preparatory steps (except occasional calibration), and presentation of the data seconded.

CGM makes it possible to assess HG awareness in patients with type 1 diabetes by entering an event whenever a patient experienced symptoms resembling HG [73] and also to detect asymptomatic HG episodes, including nocturnal episodes [73]. Selection of an alert limit permits immediate notification of a low glucose concentration [65]. According to Heise et al., the system should diagnose the majority (>75%) of HG episodes [74]. As the occurrence and duration of nocturnal HG episodes are closely associated with bedtime glucose levels and the mean nocturnal glucose concentrations, CGM may help to predict their occurrence [49].

All the methods of measuring the ISF were described as having to compensate for the time lag (physiological and physical) between blood and interstitial fluid measurements (from several seconds up to 18 min for different clinical conditions and various meters) [65,73–78]. However, a smaller delay than postulated previously has recently been demonstrated for novel models of needle-type sensors [79].

The values of glucose in ISF are usually reported to be lower than those in blood [80]. These differences can be corrected by the glucose sensor algorithms [65], but it has to be emphasized that a glucose monitoring system is calibrated using a capillary blood glucose value [74]. Inconsistent findings regarding physiological discrepancies between blood and ISF were observed during a dynamic blood glucose change (>3 mg/dL·min) [74,81]. The rate of change in blood glucose influences the accuracy of the sensor. The variation in the time lag between blood and ISF measurements [76,77] and the lower values of glucose in subcutaneous tissue may produce an unpredictable drift in the output, which seems to be an important limitation of this method [82,83]. The benefit of CGM is to inform on the trends in glucose concentration rather than on its absolute values [84]. The other features of CGM evaluation are accuracy, precision, and duration of operation [65,66,85], and those differ according to various types of measuring systems.

Three types of meters that pioneered in the field of CGM were (1) a needle-shaped sensor invented by Shichiri et al. [86] and developed by Medtronic MiniMed [87], CGMS System Gold, capable of capturing HG episodes [the lower range of measured glucose concentration: 2.2 mmol/L (40 mg/dL)], also those unrecognized by SMBG, but without notification in real time [88–90]; (2) a minimally invasive sensor, measuring glucose concentration in a liquid extracted from the skin by either suction or iontophoresis, working under the commercial name GlucoWatch Biographer [91]; and (3) a semipermeable membrane probe implanted under the skin (the device known as GlucoDay), whose working method is based on microdialysis principles.

A paper on the CGMS Medtronic MiniMed (Northridge, CA) sensor published by Metzger et al. in 2002 has confirmed that its usefulness in recording hypoglycemic episodes is over 75%, but has emphasized the differences between the results of capillary blood and interstitial fluid sampling [89], and also indicated the need for longer usability. The previous results in respect to the ability of CGMS to recognize HG episodes, especially nocturnal episodes, undetectable by other means were confirmed two years later by Yogev et al. [88]. The device appeared particularly helpful in evaluation of gestational diabetes mellitus [92] and in preventing recurrent severe hypoglycemia (educational process of management of diabetes) [75]. Tavris and Shoaibi [93] estimated as good (r = 0.73 to 0.092) the correlation between glucose readings performed by the CGMS MiniMed sensor and standard blood analysis, but equally important was their belief that it could detect episodes of asymptomatic HG. Chico et al. were of the same opinion [94]. In contrast, observations made by Korf et al. [95] raised the issue of the limited utility of a needle sensor for detection of nocturnal HG.

According to Kubiak et al. [96], mean CGMS-based glucose levels were significantly lower in unaware patients, who, on the other hand, experienced significantly more HG episodes (2.1 ± 1.5 vs. $4.6 \pm 2.6/24\,h$, $p < 0.05$). Even the correlation data between CGMS and calibration values were excellent ($r = 0.893$), but the hypoglycemic range was characterized by lower agreement, and the mean absolute difference (MAD) was 20.6%. Other authors also noted a weaker correlation between capillary blood and CGMS glucose values in the HG period [80,97].

CGMS MiniMed was also checked for its usefulness in neonates [98]. In this very fragile group of patients the correlation between ISF and blood glucose values was very good ($r^2 = 0.87$), yet not enough readings < 2.6 mmol/L were registered to evaluate its usefulness in HG. The fact that the low ISF glucose value sometimes preceded by 2 h the low value in blood put patients at risk of overdiagnosed cases, enabling researchers to conclude that the decreased values from the sensor should always be checked by blood glucose measurement before beginning an intervention. This phenomenon should not necessarily be attributed only to shortcomings of the CGMS, and the concept that decreases in tissue glucose concentrations may often precede the decrease in blood glucose (or the symptoms of neuroglycopenia) and may act as an early warning to the patient has been studied since the mid-1990s [73,78,84,99]. Disregarding these signals (artifact, the error of the sensor) would be the easiest but not a very challenging interpretation. As suggested by Reach [84], it may be speculated that glucose concentration in ISF is more relevant than blood glucose in respect to the symptoms of HG (especially neuroglycopenia).

The newer electrochemical sensors—Guardian (by Medtronic MiniMed), the fully implantable glucose monitoring system DexCom and STS-3 or 7 days (by DexCom, San Diego, CA) or TheraSense FreeStyle Navigator (Abbott, Alameda, CA)—make real-time measurement and data display possible, and are equipped with programmable alarms for HG. The lower range of glucose is 20 mg/dL for FreeStyle Navigator and 40 mg/dL for others. The measured glucose values are usually within 15% of the blood glucose level [73].

Only three years after the first publication on microdialysis as the technique for evaluation of glycemia in vivo, the device being the combination of MD with amperometric measurements of glucose in a flow chamber was described [100, 101].

Microdialysis and Its Usefulness in Monitoring Hypoglycemia The currently available glucose monitoring system, which operates based on microdialysis, is the GlucoDay (Menarini, Florence, Italy). The method that forms the basis for GlucoDay operation originated from microdialysis principles with an added collector and a glucose oxidase–based biosensor [49,102,103]. It can operate for a period of 48 h, and 3-min mean values of every second measurement are stored in the memory. To validate its accuracy during rapid and dynamic changes in glycemia, Fayolle et al. [104] conducted microdialysis during exercise sessions in type 1 diabetic patients. They found that venous glucose tended to decrease more rapidly than sensor measurements but could not confirm sensor accuracy during HG due to the lack of such events.

The reported average MAD is 9 to 13.6% [105,106], which may not be optimal but is still the best (most accurate) among currently available subcutaneous sensors [49]. The values of average MAD during hypoglycemia are better for GlucoDay (17.3 vs. 24.1%) than for CGMS system Gold [106]. The HG episodes were detected by the microdialysis sensor GlucoDay in patients after pancreas transplantation [107], and in another study, Sparacino et al. suggested that using GlucoDay, HG episodes could be predicted 20 to 25 min ahead in time [108]. Stimulated by the hypothesis that high local insulin concentration interferes with sensor readings, Hermanides et al. checked that microdialysis CGM can be performed accurately in the proximity of the CSII system, which made it possible to be incorporated in a closed-loop insulin delivery system [109].

Despite the fact that the usefulness of microdialysis in glucose concentration monitoring has been proved, the reliability of the method in relation to HG has been evaluated separately [103,110,111]. In the majority of observations a 0.3-μL/min flow rate was used as the perfusion rate recommended for optimal recovery. A decade age, Rosdahl et al. [28] found a very good correlation between plasma and adipose tissue glucose concentration when the flushing system was operating with slow flow rates (0.33 and 0.16 μL/min) and indicated that a lower perfusion flow in adipose tissue may be needed for people with a thick skinfold. This is an important remark to consider if one does not want to record false HG readings (up to 50% of the lower ISF glucose concentration in comparison to the plasma glucose concentration).

Microcirculatory disturbances are the factor that may influence microdialysis measurement and its accuracy in respect to the lower glucose concentration. This problem was investigated in neonates after surgery [112]. Under conditions of rapid changes in blood glucose (caused by pain or invasive procedures) and variations in the microcirculation, significant differences between momentary blood glucose and mean glucose measurements over 1 h should be expected [113–115]. In a study by Vlassealaers and van den Berghe et al. [116] in children in intensive care, a strong correlation was found between ISF and blood glucose ($r^2 = 0.86$). The authors found the microdialysis method unsuitable for replacement of frequent blood sampling to monitor TGC safely in this patient group (the risk that some HG episodes are undiagnosed and over diagnosed). They raised the problems of disturbed microcirculation and treatment with vasoactive drugs associated with microdialysis. In contrast, a clinical observation performed in patients recovering from major cardiac surgery revealed that arterial blood glucose fluctuations were well described using ISF glucose measurements from subcutaneous adipose tissue (SAT) [117]. Ninety-six percent of the glucose readings as derived from SAT would make it possible to establish tight glycemic control in this patient group. The glucose concentration in dialysate was calibrated using a two-step approach, first using the ionic reference technique to calculate the SAT glucose concentration (SATg), which improved the median correlation from 0.80 to 0.90, and second, using a one-point calibration procedure to obtain a glucose profile comparable to the SAT-derived blood glucose (BgSAT). Evaluation of the results of subcutaneous glucose measurements (as SAT-derived blood glucose) using insulin titration error grid analysis showed only one reading of BgSAT around 40 mg/dL with paired BG around 60 mg/dL, a few BgSAT between 40 and 60 mg/dL with paired BG 60 to 110 mg/dL, and some numbers of BgSAT readings between 110 and 300 mg/dL with paired BG in the range 80 to 200 mg/dL. There was no reading of BG < 60 mg/dL with paired BgSAT > 60 mg/dL; however, the correlation regarding the minimal value was estimated quite low ($r = 0.26/0.49$). The methodological aspects of tissue perfusion in critically ill patients with sepsis (impaired microvascular perfusion), which might also have an effect on the relation between blood and ISF glucose concentrations, were raised in this observational study [117–119].

The possible fluctuation of ISF glucose values in reference to blood glycemia causes a problem in the interpretation of microdialysis data. The existence of a time lag between the blood glucose measurement and the corresponding interstitial value is commonly known and present in all analytical methods incorporating ISF sampling. Kamel et al. carried out an experiment to investigate the dynamics between plasma and dialysate glucose during HG in children [120]. Insulin injection caused a decrease in glucose in blood and IFS followed by hypoglycemic stress response and rapid rebound in plasma glucose concentration, with a mean delay in the dialysate of 16 min. Sampling was performed every 15 min and all calculations were corrected for the lag phase of 10 min. The parallel examination of ISF and plasma glycerol concentration showed an almost unchanged plasma glycerol profile, while the ISF glycerol rose rapidly in the hypoglycemic stress response, overcoming the antilipolytic effect of insulin. This observation indicated the estimation of glycerol concentration as the important issue related to the confirmation of HG. As also found by Hagström-Toft et al. [111], adipose tissue glycerol increased markedly, by 75%, in response to HG and remained increased for at least 4 h following the lowest level glucose ($F = 3.70$, $p = 0.003$). The circulating levels of free fatty acids increased about threefold in parallel to the in situ lipolytic response.

A few studies by Moberg et al. dealt with the HG problem and its recognition using microdialysis. During insulin-induced HG, glucose levels were found to be significantly lower in adipose tissue and skeletal muscle than in plasma [76]. The glucose nadir was significantly lower in adipose tissue than in plasma (2.0 ± 0.1 mmol/L vs. 2.4 ± 0.1 mmol/L), the decline in glucose from the basal values to the nadir was 2.8 ± 0.1 mmol/L in adipose tissue (2.7 ± 0.1 mmol/L in plasma), and the time from starting the insulin infusion to the glucose nadir was increased significantly in adipose tissue 57 ± 2 min compared to plasma 39 ± 3 min. In addition, not surprisingly, perfusion at a higher velocity resulted in lower glucose levels, reflecting decreased glucose recovery. Moberg et al.

also studied the influence of various fractional sampling times on the detection of HG [121]. This experiment was conducted in healthy subjects during a standardized hyperinsulinemic HG clamp with blood and ISF being measured in 15-min intervals. Glucose values in plasma and ISF were similar before and after HG, but lowering of blood glucose affected the readings from the adipose tissue and the glucose nadir was significantly lower than that in plasma, followed by a corresponding increase in adipose tissue glucose delayed by 20 min. The authors concluded that 30-min fractions seem sufficient to detect HG while monitoring glucose concentration in ISF.

The long-lasting (over 3 days up to three weeks) observations of the MD monitoring implied that after several days the readings might be more accurate than at the beginning of the observation. The possible explanation of this phenomenon is the fact that a certain amount of time is needed to repair the microstructure of tissue around the probe, and after that the sensor reaches the steady state [24,122]. However, the inflammation surrounding the sensor placement is the alternative explanation. Nevertheless, it does not limit the utilization length of the probe or sensor.

Confirmation of an acceptable correlation between ISF and blood glucose values was provided by some observations done in newborns. Baumeister et al. [123] monitored a neonate with diabetes mellitus for 7 days. The range of blood glucose concentration measured was from 1.7 to 23.8 mmol/L and the variations in adipose tissue closely parallel changes in the capillary blood glucose with an overall correlation r = 0.89 (however, with considerable variation in the time lag). In the other study [124], the correlation between microdialysis and blood for glucose was similar (r = 0.88) and allowed the detection of symptomatic HG during the observational period (4 to 16 days). The diagnostic sensitivity of dialysate glucose ≤ 2.9 mmol/L to predict a blood glucose level ≤ 2.8 mmol/L was 92.3% with 88.1% specificity, and the time lag was evaluated for 15 min as reported by Holzinger et al. [125]. Microdialysis was used in neonatal diabetes mellitus for two weeks and a good correlation of subcutaneous and blood glucose concentration over a wide range of values was found. There were no episodes of blood HG or serious lowering of glucose values without parallel lower ISF glucose readings: in fact, quite the opposite. Blood glucose monitoring was not able to reveal all episodes of low ISF glucose values and there significant fluctuations. It was suggested that microdialysis should definitely be considered for use in avoiding hypoglycemic episodes.

There have been several trials concerning the effects of CGMS or GlucoWatch on the occurrence of HG, and only Tanenberg et al. [34] showed an improvement (twice HG episodes) with the use of CGMS. A decrease in HG over time was even demonstrated in other studies, but there was no clear evidence of a decreased rate of HG or any positive impact on the unawareness syndrome [113]. Also, the evaluation of real-time CGMS devices in six trials conducted recently has not provided reliable information on the specificity and sensitivity of a hypoglycemia alarm, except in one study (FreeStyle Navigator), which proved an 84.4% specificity of the HG alarm [126–131]. Compared with microdialysis they do not seem to have any advantage in the presence of HG. What is more important, the microdialysis technique showed the better accuracy than the other devices in evaluating of glucose concentration [49,106].

Summary   As with any other method, microdialysis has both advantages and limitations in respect to HG monitoring. The limitations are related to the influence of technical issues related to the microdialysis probe (bending, blocking, rupture of the membrane, extravasations of the blood in the region of probe insertion) and additional devices of a microdialysis system (syringe, pump) and also the securing of the subcutaneous probe (adhesive tape, good care of the insertion region). The time lag in the ISF versus blood glucose measurements as well as the disturbances in microcirculation and the properties of the tissue measured (e.g., the subcutaneous adipose tissue skinfold) have to be taken into consideration when interpreting microdialysis data. These are the reasons that direct extrapolation from ISF glucose values to blood glucose values is not possible, and when there are symptoms of hypoglycemia, the blood glucose should be estimated. The same action has to be taken when treating HG episodes, to avoid overdosage of carbohydrate.

Microdialysis is ideal for intensively treated patients (ICU patients), however, as self-control is not expected and 24-h care and monitoring are available. HG episodes can be confirmed by the parallel measurement of other parameters, mainly glycerol, and verified by blood sampling. Additional data (e.g., lactate concentration) may also provide information regarding microcirculation and help to evaluate the clinical condition comprehensively. Microdialysis enables almost online measurement for a long (>3 days) period of time and detection of symptomatic and asymptomatic HG episodes. Subcutaneous location of a probe seems to be safer than a vessel and protects against considerable blood loss. Microdialysis is a promising clinical, educational, and mostly research tool, which may help to control HG.

2.3. Monitoring Late Complications of Diabetes: Use of Microdialysis to Ensure Proper Care of the Diabetic Foot

The treatment of chronic wounds, including ulcers that develop in diabetic patients in the course of diabetic foot syndrome has been a serious burden to health care systems in Western societies. It has been estimated that costs associated with the treatment of chronic wounds consume 1 to 2% of the total health care expenditures in these countries [132]. Diabetic foot syndrome (DFS) is one of the most common causes of chronic wounds. DSF-related problems are the most common causes of hospital admittance of diabetic patients. Moreover, DFS is a primary cause of nontraumatic lower limb amputations all over the world. It has been estimated that about 80% of these amputations could be prevented if appropriate treatment procedures were used. To be successful, such procedures should be based on a thorough understanding of the biochemical processes associated with wound healing [132,133], yet knowledge related to local metabolism during the wound-healing process seems to be lacking [134,135]. It is known that this is a complex process involving interactions between epidermal and dermal cells, the extracellular matrix, and the vasculature. Wound fluid containing plasma exudates, soluble tissue, and cell-derived molecules is an important part of the local environment of the wound, and thus analysis of this fluid should provide valuable information concerning the healing process [132]. The following techniques have been used for the aspiration of chronic wound fluid:

- Aspiration of wound fluid using a syringe with a sterile needle after collecting the fluid under vapor-permeable polyurethane film or an occlusive dressing that has covered the wound for several hours [136,137]
- Direct aspiration of the wound fluid from its surface using blunted glass microcapillaries after accumulating the fluid for several hours [138]
- Elution of the wound fluid from disks of filtered paper that have been left on the surface of the wound for several hours [138]
- Immersion of the fluid from bandages covering the wound for a few days [139]
- Collection of the fluid from absorbent foam dressing covering the wound for several days [140]

All these techniques have been or could possibly be used for DFS ulcers. However, all of them are variable and poorly validated [139]. Moreover, some, if not all, may influence the quantity and quality of the fluid obtained, leading to spurious conclusions regarding the local environment of the wound. Additionally, all these techniques are unable to follow temporal and/or spatial changes within the wound, and their use is limited to collecting fluid from the surface of the wound. In this regard the microdialysis technique, looks very attractive. Using this technique, it is possible to sample many substances that might be of interest while studying the wound-healing process, such as glucose, lactate, glycerol, and other metabolites, proteins, cytokines, and growth factors. Moreover, an appropriate probe can effectively separate sampled compounds from the enzymes that might decompose them, which is impossible using the techniques mentioned earlier. It is also important that the microdialysis technique can be used to sample tissue inside or close to the wound but underneath the wound surface, and that a microdialysis probe does not influence significantly the tissue being sampled. It may be of interest in studying the healing process that microdialysis could monitor spatial changes with a resolution of less than 2 mm [141] and temporal changes with a time resolution of minutes [142,143]. Taking into account the aforementioned advantages of the microdialysis technique and the scale of medical, social, and economic problems associated with treatment of the diabetic foot syndrome, the number of reports on studies related to DFS and microdialysis is surprisingly limited.

Metabolism in Chronic Diabetic Foot Ulcers   Simonsen et al. investigated glucose and lactate metabolism in chronic diabetic foot ulcers in vivo [134]. The study group consisted of five IDDM and five NIDDM patients (nine men, one woman, median age 62 years, median duration of diabetes 10 years). Local tissue concentrations of glucose and lactate were measured at a distance of 0.5 to 1.0 cm from the edge of the ulcer and in normal skin in the contralateral foot using custom-made dialysis probes manufactured from a 3-cm-long fiber from an artificial dialysis kidney. The experiments lasted for 4 h. These experiments demonstrated for the first time that in diabetic patients with chronic foot ulcers the subdermal concentration of glucose is lower and that the concentration of lactate is higher on the edge of the ulcer than in a foot without a chronic wound. This finding is in line with data previously reported showing increased glucose uptake and lactate production in vitro in injured skeletal muscle [144,145] as well as in vivo in burn injuries [146]. The glucose uptake and lactate output in the ulcer region calculated by Simonsen et al. was found to be 80% higher than in the nonulcerated foot, the latter being at the same level as in abdominal subcutaneous adipose tissue (i.e., 1.7 µmol per 100 g per minute) [147]. This difference indicates that glucose metabolism of the ulcer is higher than the metabolism of the intact skin. However, the molar ratio between the lactate output and the glucose uptake was found to be similar in both regions investigated, indicating that the carbohydrate metabolism was qualitatively similar in the ulcer and the intact skin. Thus, glycolysis is the dominant metabolic process of the glucose metabolism in two of the regions studied. In the case of wounds within the ischemic legs, the lactate output was higher than from purely neuropathic ulcers, indicating an increased importance for the anaerobic metabolism. From a clinical point of view, these results meant that the disturbance in the glucose metabolism was probably not the major reason for the slow healing in chronic diabetic foot ulcers.

In 2004, Stolle and Riegels-Nielsen reported results of in vivo investigation of the metabolism of the diabetic foot in which glycerol concentration was monitored, in addition to glucose and lactate concentrations, for over 1 h [135]. Concentrations of these three metabolites on the edge of the diabetic ulcer were compared with corresponding concentrations in abdominal subcutaneous adipose tissue. The study group consisted of two IDDM and three NIDDM diabetic patients (three men, two women, median age of 53 years, median duration of diabetes 20 years). CMA-60 microdialysis catheters were used. This study showed significantly higher glycerol concentration in the ulcers than in the reference abdominal tissue. The quantitative results regarding glucose and lactate concentrations obtained in this study were not in conflict with Simonsen et al.'s data reported previously [134]. However, differences in mean levels of glucose and lactate concentrations between both body regions being compared were not appreciable enough to be statistically significant, which could be related to the small study group. It is noteworthy that two diabetic patients with high local interstitial concentrations of glycerol had an unfavorable outcome (i.e., necessity for surgical intervention). This result suggests that glycerol concentration in the vicinity of the wound might become a valuable predictor of the clinical outcome of the diabetic foot ulceration if it were confirmed in a clinical study that included a representative number of cases.

Summary   The microdialysis technique has been used to date to study the metabolism of inflamed skin following exposure to certain allergens [143], skin exposured to ultraviolet B [148] and psoriasis [149]. However, a very limited number of studies have been conducted that applied microdialysis to characterize the metabolism of chronic diabetic wounds. The studies described above demonstrated that it is possible using microdialysis to detect differences in concentrations of glucose, lactate, and glycerol in diabetic ulcers and reference tissues. Further investigations are needed to characterize the metabolism of chronic diabetic foot ulcers in more detail and to develop methods capable of predicting inflammation and necrosis.

2.4. Monitoring Infections in Diabetes: Antibiotic Therapies in the Diabetic Foot

Bacterial foot infections are frequent and serious complications in diabetic patients [150], which necessitate hospitalization more often than any other long-term complication of diabetes. Antimicrobial therapy is very important for patients with inflamed foot ulcers since the need for amputation is in many cases determined by the extent of inflammation [151]. Immediate empirical antibiotic therapy is introduced long before information is available regarding causal bacteria and their in vitro susceptibility to antimicrobial agents [152]. The successful treatment of bacterial infections depends on a antimicrobial agent being properly chosen for the microorganisms causing the infection and on appropriate concentration of this agent in the ISF of the infected tissue [152–154]. If a sufficient therapeutic level of the antimicrobial agent being used is not reached, the immediate empirical therapy fails to be effective despite the documented in vitro sensitivity of the causal bacteria against the agent applied. This view is supported by several studies showing that subinhibitory concentrations of antibiotics might be achieved in soft tissues despite adequate plasma concentrations [155]. The tissue concentration of antimicrobial agents has been evaluated and reported for the past few decades using, for example, tissue homogenates [156–158] or skin blisters [159,160]. However, the results obtained using these methods may be misleading, since the unbound interstitial drug concentration that determines the antibacterial activity may be substantially lower than the total tissue concentration [154,161,162]. Moreover, impaired target-site penetration may gain particular importance if the pathogens are located in poorly accessible peripheral sites or if drug penetration is hampered by interstitial diffusion barriers developed in the course of the inflammatory processes [154,163,164]. In general, tissue penetration of antimicrobial agents has been reported to be substantially affected in inflammatory states by changes in blood flow, loss of capillary integrity, altered capillary permeability, edema, changed pH, and altered protein content in the interstitial space fluid of the infected tissue (e.g., accumulation of fibrin) [165–167]. All of these factors should be considered in infections of diabetic foot ulcers. Additionally, in patients with long-lasting diabetes, the microenvironment in peripheral tissue is altered by peripheral neuropathy, macroangiopathy, and microangiopathy [168]. Furthermore, Joukhadar et al. demonstrated that peripheral arterial occlusive disease in patients with diabetes mellitus might reduce the tissue penetration of antibacterial agents [169].

Microdialysis has been used for the past decade to study the penetration of several antimicrobial agents into the ISF of the inflamed soft tissue of diabetic patients with diabetic foot infections. Results concerning ciprofloxacin [154], fosfomycin [152], and moxifloxacin [170] treatments, as well as, combination piperacillin (β-lactam antibiotic) and tazobactam (β-lactamase inhibitor) treatment [153], have been reported so far (Table 1). In all the studies, CMA-10 microdialysis probes were used. Müller et al. compared ciprofloxacin concentrations in six NIDDM patients (five male, one female, mean ± S.E.M.: age 72 ± 6 years, weight 70 ± 4 kg, height 166 ± 3 cm), who presented with a diabetic foot infection severe enough to require hospitalization and parenteral antibiotic therapy, in the ISF of the inflamed lesion close to the border of the primary necrosis, in serum, and in the unaffected subcutaneous adipose tissue of the ipsilateral extremity after administration of a single intravenous (i.v.) dose of 200 mg over 20 min [154]. The main finding of this study was that interstitial ciprofloxacin concentration at the inflammation site was not significantly different from the concentration in noninflamed interstitial tissue. This result was in contrast with previous reports showing that ciprofloxacin may preferentially distribute via phagocytes to inflamed sites [158,171] and reach a few-fold higher concentration than that in serum [172]. Interstitial ciprofloxacin concentrations in the ISF of both tissue compartments examined (i.e., inflamed and noninflamed) were significantly lower than the corresponding serum concentrations, which was in line with previous reports on this [161] and other antimicrobial agents [162,173].

In 2003, Legat et al. investigated the penetration of fosfomycin into inflammatory lesions in patients with diabetic foot syndrome [152]. The study group consisted of six diabetic patients (three IDDM, three NIDDM, three males, three females, mean ± S.D.: age 62.5 ± 7.1 years, duration of diabetes 12.9 ± 9.6 years, weight 76.3 ± 20.8 kg, height 168 ± 11.8 cm). Five of six patients had peripheral arterial occlusive disease (PAOD). The patients were treated with i.v. infusion of fosfomycin (200 mg/day per kilogram of body weight) combined with either clindamycin (1800 mg/day) or ceftrixone (2 g/day). The daily dose of fosfomycin was divided into three equal i.v. doses administered over 30 min every 8 h. The fosfomycin concentrations were measured on the third day of the treatment in the inflamed tissue, the upper subcutaneous layer of the tight (noninflamed area), and in the plasma. Distributions of fosfomycin in inflamed

TABLE 1 Studies Evaluating Antimicrobial Agents in Action in the Lower Limb Ulcerated Tissue of the Diabetic Patients

Ref.	Antimicrobial Agent	Dosage	Study Group	T (h)	Sites Examined	C_{max} (mg/L)	T_{max} (h)	$AUC_{0\text{-}T}$ (mg·h/L)
[154]	Ciprofloxacin	i.v. 200 mg over 20 min	6 NIDDM	5	Serum	2.83 ± 1.21	0.38 ± 0.13	4.2 ± 1.7
					Diabetic ulcer	2.18 ± 1.13	0.50 ± 0.18	3.5 ± 2.0
					Ipsilateral extremity	2.12 ± 0.55	0.38 ± 0.13	3.3 ± 1.3
[152]	Fosfomycin[a]	i.v. 200 mg/day·kg of body weight in three equal doses over 30 min every 8 h in 3 days	3 IDDM, 3 NIDDM	8	Plasma	320 ± 67.4	—	1331 ± 429
					Diabetic ulcer	136 ± 106.6	0.90 ± 0.22	782 ± 524
					Thigh	139 ± 76.7	1.15 ± 0.47	937 ± 848
[170]	Moxifloxacin	i.v. 400 mg over 60 min	6 diabetic patients[b]	8	Plasma	3.8 ± 1.0	1.0 ± 0.0	17.5 ± 5.4
					Diabetic ulcer	0.8 ± 0.5	2.8 ± 2.6	3.7 ± 1.9
					Contralateral thigh	1.4 ± 1.1	1.8 ± 0.7	8.1 ± 7.1
[155]	Levofloxacin	i.v. 500 mg over 30 min	7 patients including 3 with diabetes[c]	10	Plasma	8.37 ± 1.93	0.52 ± 0.26	32.3 ± 7.3
					Inflamed tissue	5.45 ± 3.74	1.06 ± 0.14	25.5 ± 16.8
					Contralateral thigh	4.42 ± 2.09	1.39 ± 1.29	23.5 ± 10.0
[153]	Piperacillin[d]	i.v. 200 mg/day·kg of body weight in three equal doses over 30 min every 8 h in 3 days	4 IDDM, 2 NIDDM	8	Plasma	341.0 ± 52.2	0.50 ± 0.18	672 ± 335
					Diabetic ulcer	102.2 ± 63.5	1.11 ± 0.25	283 ± 173
					Thigh	91.7 ± 83.0	1.11 ± 0.33	234 ± 169
	Tazobactam[d]	i.v. 200 mg/day·kg of body weight in three equal doses over 30 min every 8 h in 3 days	4 IDDM, 2 NIDDM	8	Plasma	15.8 ± 5.4	0.47 ± 0.18	25.1 ± 26.8
					Diabetic ulcer	10.4 ± 7.5	0.83 ± 0.00	18.6 ± 18.3
					Thigh	5.2 ± 3.7	1.11 ± 0.39	8.1 ± 8.5

[a]Fosfomycin was combined with ceftriaxone or clindomycin.
[b]Microdialysis data were collected in five of six diabetic patients only.
[c]Microdialysis data were collected in two of three diabetic patients only.
[d]Piperacillin and tazobactam were combined in this study.

and noninflammed subcutaneous tissue were similar, with no significant difference. The quantitative data obtained by Legat et al. showed that soft tissue penetration for fosfomycin is similar to that observed in healthy volunteers [174]. The fosfomycin concentration in the plasma was significantly higher than the concentration in the interstitial fluid, which, according to Legat et al., was probably due to diffusion barriers and not to protein binding of fosfomycin in the plasma [152]. Distribution of the drug from the plasma to the interstitial fluid could also be affected by concomitant medication taken by the patients and possible release of a variety of mediators during the inflammatory process.

Summing up, the main finding of Legat et al. in [152] was that the i.v. infusion of fosfomycin in the aforementioned dose was sufficient to reach and maintain a concentration of this drug in the ISF of the inflamed tissue high enough to inhibit the growth of relevant bacteria such as Staphylococcus aureus.

In 2005, Legat et al. reported results of the study evaluating the penetration of piperacillin and tazubactam into soft tissue of patients with diabetic foot syndrome [153]. Four subjects with IDDM and two with NIDDM (two male, four female, mean ± S.D.: age 72.3 ± 13.1 years, duration of diabetes 12.2 ± 6 years, weight 80.5 ± 17.2 kg, and height 168.3 ± 16.2 cm) were enrolled in the study. Four patients had PAOD. The study protocol was similar to that used by Legat et al. in a study described previously [152] related to fosfomycin distribution in the inflamed foot ulcers of diabetic patients. In particular, both antimicrobial agents were delivered intravenously in three equal doses over 30 min every 8 h for 3 days. Total daily doses of piperacillin and tazobactam were equal 12 g (154.8 ± 32.7 mg/day per kilogram of body weight) and 1.5 g (19.4 ± 4.1 mg/day per kilogram of body weight), respectively. On the third day of i.v. antibacterial treatment the maximum concentration of piperacillin was significantly higher in the plasma than in the interstitial fluid of inflamed and noninflamed soft tissue. However, concentrations of piperacillin in both interstitial compartments were similar to each other. The ability of piperacillin to penetrate inflamed and noninflamed interstitial space equally was not shared by tazobactam. Concentrations of tazobactam in plasma and in both interstitial compartments were considerably lower than the respective piperacillin concentrations and fell below 1 mg/L (i.e., the lowest detectable level) within 4 h after i.v. infusion began. This result was in agreement with the results of a previously reported animal study [175]. It should be mentioned that despite undetectable concentration of tazobactam, the treatment used in this study could still be effective since in vitro tazobactam concentrations of 0.05 to 0.5 mg/L were reported to be capable of inhibiting β-lactamases by 50% [176]. From concentrations of piperacillin–tazobactam in inflamed and healthy tissues, Legat et al. concluded that organisms for which the drug minimal inhibitory concentrations (MICs) were 16 mg/L and below should be eradicated successfully, whereas organisms for which the drug MICs are higher may not be sufficiently affected [153].

Bellmann et al. investigated the influence of local inflammation on target-site pharmacokinetics of levofloxacin in seven patients suffering soft tissue infections, including three diabetic subjects of whom two were monitored successfully using microdialysis [155]. The authors found high interindividual differences between plasma and tissue penetration by levofloxacin, although the mean time–concentration profiles were almost identical. Based on the MIC calculations performed by Bellmann et al. regarding Streptococcus pyogenes strains, it can be stated that on average, effective killing of these bacteria can be expected after administration of a single 500-mg i.v. dose of levofloxacin [155]. However, due to high interindividual variability in tissue penetration, some cases of clinical and microbial failure may not be ruled out.

The only results showing evidence that penetration of an antibiotic agent into inflamed subcutaneous lesions might be restricted in comparison with noninflamed tissue in diabetic patients concerned moxifloxacin and were reported by Joukhadar et al. [170]. Six diabetic patients (median: age 65 years, duration of diabetes 13.5 years, weight 74 kg, BMI 24.8 kg/m^2) with documented PAOD together with six nondiabetic subjects (median: age 57 years, weight 82 kg, BMI 25.7 kg/m^2), of whom two suffered from PAOD, took part in this study. As for the antimicrobial agents described previously, the free moxifloxacin concentrations in the ISF of healthy subcutaneous adipose tissues and inflamed lesions resembled the time–concentration course of free moxifloxacin in plasma. However, as stated above, the concentrations of moxifloxacin were descriptively lower in inflamed tissue than in healthy adipose tissue. The differences observed were not statistically significant, due to the limited number of patients and a high degree of interindividual variability. Joukhadar et al. hypothesized that a poor vascular supply in the lower limbs might be the reason for the lower concentration of moxifloxacin in inflamed tissue in diabetic patients [170]. According to these authors, local edema combined with impaired arterial flow at the inflammatory tissue of diabetic patients may balance the effect of the increased capillary permeability caused by the local inflammation itself, resulting in concentrations of moxifloxacin in the inflamed tissue of diabetic patients that are equal to the concentrations in unaffected tissue of nondiabetic patients. Despite differences in the moxifloxacin distribution in the soft tissue of diabetic and nondiabetic patients, Joukhadar et al. were able to demonstrate that in either case the concentrations of moxifloxacin were high enough to be effective against clinically relevant bacterial strains, including S. aureus and Streptococcus species [170].

Summary   Summing up, six agents used in antibiotic treatment of infected diabetic foot ulcers have been studied using microdialysis: ciprofloxacin, fosfomycin, moxifloxacin, levofloxacin, piperacillin, and tazobactam. In all cases CMA-10 microdialysis probes were used. The number of subjects participating in all studies was limited and in no case exceeded 10 patients. In all cases, clinically important findings have been reported that confirmed the applicability and usefulness of microdialysis in studies on the penetration of antimicrobial agents into the interstitial space fluid of the inflamed soft tissue of DFS patients.

2.5. Monitoring Insulin Sensitivity in Diabetic Patients

In nondiabetic persons, a hyperinsulinemic response to an oral glucose challenge is closely associated with decreases in insulin-mediated glucose uptake [177] as well as with a number of clinical syndromes associated with insulin resistance. High fasting insulin concentrations appear to be an independent predictor of ischemic heart disease in men [178]. Insulin resistance has been assigned a central place in the metabolic disturbances associated with obesity and type 2 diabetes. Type 2 diabetes is often associated with basal hyperinsulinemia, reduced sensitivity to insulin, and disturbances in insulin release [179]. Rizza et al. [180] found that continuous (40 h) hyperinsulinemia in humans significantly reduced glucose utilization and overall glucose metabolism at submaximally and maximally effective plasma insulin concentrations. Many tissues, including liver, skeletal muscle, and adipocytes, manifest resistance to insulin [181,182].

Skeletal muscle tissue represents the major site of glucose uptake after oral glucose load. Resistance to insulin-stimulated glucose uptake is present in the majority of patients with impaired glucose tolerance (IGT) or non-insulin-dependent diabetes mellitus (NIDDM) and in approximately 25% of non-obese individuals with normal oral glucose tolerance. In these conditions, deterioration of glucose tolerance can be prevented only if the β-cell is able to increase its insulin secretory response and maintain a state of chronic hyperinsulinemia [183].

Monitoring Insulin Concentration in the Interstitial Fluid and Muscle Glucose Uptake in Diabetes Frossard et al. [184] used microdialysis to measure glucose concentration in the serum and interstitial space fluid of skeletal muscle during an oral glucose tolerance test (OGGT) in patients with type 1 and type 2 diabetes and in young and middle-aged healthy volunteers. Regional blood flow in skeletal muscle was measured by the laser Doppler flow technique (LDF, Moor Instruments, Devon, UK). They concluded that the "arterial–interstitial gradient for glucose is intact in type 1 and absent in type 2 diabetes. These obtained data support the hypothesis that the capillary wall is not a rate limiting factor for glucose uptake in type 2 diabetic patients. Because no changes in muscular blood flow were detected in both groups, the might be considered as further consequence of insulin resistance with impaired cellular glucose uptake." Cline et al. [185] measured insulin concentrations in interstitial fluid by an open-flow microperfusion technique and used microdialysis to determine the gradient between plasma and interstitial glucose concentration during hyperglycemic–hyperinsulinemic clamp. They concluded that impaired insulin-stimulated glucose transport is responsible for the reduced rate of insulin-stimulated muscle glycogen synthesis in patients with type 2 diabetes.

The aims of the study performed in Goteborg, Sweden [186] were to measure interstitial muscle insulin and glucose in patients with type 2 diabetes to evaluate whether transcapillary transport is part of peripheral insulin resistance. The microdialysis measurements enable correct estimation of the rate of glucose, and lactate production under basal conditions. Plasma and interstitial insulin, glucose, and lactate (measured by the intramuscular in situ–calibrated microdialysis) in the medial quadriceps femoris were analyzed during a hyperinsulinemic euglycemic clamp. Blood flow in the contralateral calf was measured by vein plethysmography. The authors concluded that during hyperinsulinemia, muscle interstitial insulin and glucose concentrations did not differ between patients with type 2 diabetes and healthy controls, despite a significantly lower leg blood flow in diabetic patients. It is suggested that decreased glucose uptake in type 2 diabetes is caused by insulin resistance at the cellular level rather than by deficient access of insulin and glucose surrounding the muscle cell.

The forearm microdialysis technique was used to assess glucose uptake under ischemic conditions in the bronchioradialis muscle and in the subcutaneous adipose tissue. As estimated from the glucose elimination rate, results demonstrate that in the muscle, but not in the adipose tissue, glucose uptake is activated by ischemia in both type 2 diabetes and healthy subjects [186].

Summary   Based on the results presented, it could be concluded that the use of microdialysis to monitor insulin sensitivity in diabetic patients led to important results. Analysis of the insulin and glucose uptake during clamp studies in the interstitial compartment by the microdialysis technique allows for better understanding of the physiological processes occurring in the course of diabetes treatment. However, it should be stressed that microdialysis and analytical methods do not enable measurement of interstitial insulin at the lower physiological plasma insulin levels. Insulin transport remains unclear under these circumstances. In this situation in many studies, muscle interstitial insulin concentrations were obtained during clamping conditions with high plasma insulin levels.

REFERENCES

[1] Pfeiffer, E.F. (1995). Germany: The Sugar Watch. *Biosensors & Bioelectronics*, *10*(3), vi–vii(1).

[2] Schoemaker, M., Andreis, E., Röper, J., Kotulla, R., Lodwig, V., Obermaier, K., Stephan, P., Reuschling, W., Rutschmann, M., Schwaninger, R., et al. (2003). The SCGM1 system: subcutaneous continuous glucose monitoring based on microdialysis technique. *Diabetes Technology & Therapeutics*, *5*(4), 599–608.

[3] Maran, A., Poscia, A. (2002). Continuous subcutaneous glucose monitoring: the GlucoDay system. *Diabetes, Nutrition & Metabolism, 15*, 429–43.

[4] Jeckelmann, J., Seibold, A. (2002). GlucOnlineTM: a new approach to continuous glucose monitoring. *Diabetesprofile*, 2/02.

[5] Kitabchi, A.E., Umpierrez, G.E., Murphy, M.B., Barrett, E.J., Kreisberg, R.A., Malone, J.I., Wall, B.M. (2001). Management of hyperglycemic crises in patients with diabetes. *Diabetes Care, 24*, 131–153.

[6] Fishbein, H.A., Palumbo, P.J. (1995). Acute metabolic complications in diabetes. In: *Diabetes in America*, NIH Publ. 95-1468. National Diabetes Data Group, National Institutes of Health, Bethesda, MD, pp. 283–291.

[7] Umpierrez, G.E., Kelly, J.P., Navarrete, J.E., Casals, M.M.C., Kitabchi, A.E. (1997). Hyperglycemic crises in urban blacks. *Archives of Internal Medicine, 157*, 669–675.

[8] Kitabchi, A.E., Fisher, J.N., Murphy, M.B., Rumbak, M.J. (1994). Diabetic keto-acidosis and the hyperglycemic hyperosmolar nonketotic state. In: Kahn, C.R., Weir, G.C. (Eds.), *Joslin's Diabetes Mellitus*. 13th ed., Lea & Febiger, Philadelphia, pp. 738–770.

[9] Wachtel, T.J., Tetu-Mouradjain, L.M., Goldman, D.L., Ellis, S.E., O'Sullivan, P.S. (1991). Hyperosmolality and acidosis in diabetes mellitus: a three-year experience in Rhode Island. *Journal of General Internal Medicine, 6*, 495–502.

[10] Carroll, P., Matz, R. (1983). Uncontrolled diabetes mellitus in adults: experience in treating diabetic ketoacidosis and hyperosmolar coma with low-dose insulin and uniform treatment regimen. *Diabetes Care, 6*, 579–585.

[11] Hamblin, P.S., Topliss, D.J., Chosich, N., Lording, D.W., Stockigt, J.R. (1989). Deaths associated with diabetic ketoacidosis and hyperosmolar coma, 1973–1988. *Medical Journal of Australia, 151*, 439–444.

[12] Basu, A., Close, C.F., Jenkins, D., Krentz, A.J., Nattrass, M., Wright, A.D. (1992). Persisting mortality in diabetic ketoacidosis. *Diabetic Medicine, 10*, 282–289.

[13] Ellemann, K., Soerensen, J.N., Pedersen, L., Edsberg, B., Andersen, O. (1984). Epidemiology and treatment of diabetic ketoacidosis in a community population. *Diabetes Care, 7*, 528–532.

[14] Chiasson, J.L., Aris-Jilwan, N., Bélanger, R., Bertrand, S., Beauregard, H., Ékoé, J.M., Fournier, H., Havrankova, J. (2003). Diagnosis and treatment of diabetic ketoacidosis and the hyperglycemic hyperosmolar state. *Canadian Medical Association Journal, 7*, 168.

[15] Krentz, A., Nattrass, M. (2003). Acute metabolic complications of diabetes: diabetic ketoacidosis, hyperosmolar non-ketotic hyperglycaemia and lactic acidosis. In: Pickup, J.C., Williams, G. (Eds.), *Textbook of Diabetes 1*, 3rd ed. Blackwell, Oxford, U.K.

[16] English, P., Williams, G. (2004). Hyperglycaemic crises and lactic acidosis in diabetes mellitus. *Nigerian Postgraduate Medical Journal, 80*, 253–261.

[17] Ciechanowska, A., Ladyzynski, P., Wojcicki, J.M., Sabalinska, S., Krzymien, J., Pulawska, E., Karnafel, W., Foltynski, P., Kawiak, J. (2008). Microdialysis technique as a monitoring system for acute complications of diabetes. *Artificial Organs, 32*(1), 45–51.

[18] Schoonen, A.J., Wientjes, K.J. (2003). A model for transport of glucose in adipose tissue to a microdialysis probe. *Diabetes Technology & Therapeutics*, *5*(4), 589–598.

[19] Hagström-Toft, E., Enoksson, S., Moberg, E., Bolinder, J., Arner, P. (1997). Absolute concentrations of glycerol and lactate in human skeletal muscle, adipose tissue, and blood. *American Journal of Physiology: Cell Physiology*, *273*, 584–592.

[20] Coppack, S.W., Chinkes, D.L., Miles, J.M., Patterson, B.W., Klein, S. (2005). A multicompartmental model of in vivo adipose tissue glycerol kinetics and capillary permeability in lean and obese humans. *Diabetes*, *54*(7), 1934–1941.

[21] Kamel, A., Norgren, S., Persson B., Marcus, C. (1999). Insulin induced hypoglycaemia: comparison of glucose and glycerol concentrations in plasma and microdialysate from subcutaneous adipose tissue. *Archives of Disease in Childhood*, *80*, 42–45.

[22] Bolinder, J., Ungerstedt, U., Arner, P. (1992). Microdialysis measurement of the absolute glucose concentration in subcutaneous adipose tissue allowing glucose monitoring in diabetic patients. *Diabetologia*, *35*(12), 1177–1180.

[23] Sternberg, F., Meyerhoff, C., Mennel, F.J., Bischof, F., Pfeiffer, E.F. (1995). Subcutaneous glucose concentration in humans: real estimation and continuous monitoring. *Diabetes Care*, *18*(9), 1266–1269.

[24] Wientjes, K.J., Vonk, P., Vonk-van Klei, Y., Schoonen, A.J., Kossen, N.W. (1998). Microdialysis of glucose in subcutaneous adipose tissue up to 3 weeks in healthy volunteers. *Diabetes Care*, *21*(9), 1481–1488.

[25] Monsod, T.P., Flanagan, D.E., Rife, F., Saenz, R., Caprio, S., Sherwin, R.S., Tamborlane, W.V. (2002). Do sensor glucose levels accurately predict plasma glucose concentrations during hypoglycemia and hyperinsulinemia? *Diabetes Care*, *25*, 889–893.

[26] Stahl, M., Bouw, R., Jackson, A., Pay, V. (2002). Human microdialysis. *Current Pharmaceutical Biotechnology*, *3*, 165–178.

[27] Lutgers, H.L., Hullegie, L.M., Hoogenberg, K., Sluiter, W.J., Dullaart, R.P., Wientjes, K.J., Schoonen, A.J. (2000). Microdialysis measurement of glucose in subcutaneous adipose tissue up to three weeks in type 1 diabetic patients. *Netherlands Journal of Medicine*, *57*(1), 7–12.

[28] Rosdahl, H., Hamrin, K., Ungerstedt, U., Henriksson, J. (1998). Metabolite levels in human skeletal muscle and adipose tissue studied with microdialysis at low perfusion flow. *American Journal of Physiology: Endocrinology and Metabolism*, *274*(5), E936–E945.

[29] Workgroup on Hypoglycemia, American Diabetes Association (2005). Defining and reporting hypoglycemia in diabetes: a report from the American Diabetes Association Workgroup on Hypoglycemia. *Diabetes Care*, *28*, 1245–1249.

[30] Cryer, P.E., Davis, S.N., Shamoon, H. (2003). Hypoglycemia in diabetes. *Diabetes Care*, *26*, 1902–1912.

[31] Cryer, P.E. (2004). Diverse causes of hypoglycemia-associated autonomic failure in diabetes. *New England Journal of Medicine*, *350*, 2272–2279.

[32] Dagogo-Jack, S.E., Craft, S., Cryer, P.E. (1993). Hypoglycemia-associated autonomic failure in insulin-dependent diabetes mellitus: recent antecedent hypogly-

cemia reduces autonomic responses to, symptoms of, and defense against subsequent hypoglycemia. *Journal of Clinical Investigation, 91*, 819–828.

[33] Segel, S.A., Paramore, D.S., Cryer, P.E. (2002). Hypoglycemia-associated autonomic failure in advanced type 2 diabetes. *Diabetes, 51*, 724–733.

[34] Tanenberg, R., Bode, B., Lane, W., Levetan, C., Mestman, J., Harmel, A.P., Tobian, J., Gross, T., Mastrototaro, J. (2004). Use of the continuous glucose monitoring system to guide therapy in patients with insulin-treated diabetes: a randomized controlled trial. *Mayo Clinic Proceedings, 79*, 1521–1526.

[35] Daneman, D. (2006). Type 1 diabetes. *Lancet, 367*, 847–858.

[36] The Diabetes Control and Complications Trial Research Group (1993). The effect of intensive treatment of diabetes on the development and progression of long-term complications in insulin-dependent diabetes mellitus. *New England Journal of Medicine, 329*, 977–986.

[37] The Diabetes Control and Complications Trial Research Group (1997). Hypoglycemia in the Diabetes Control and Complications Trial. *Diabetes, 46*, 271–286.

[38] Clarke, W.L., Cox, D.J., Gonder-Frederick, L.A., Julian, D., Schlundt, D., Polonsky, W. (1995). Reduced awareness of hypoglycemia in adults with IDDM. *Diabetes Care, 18*, 517–522.

[39] Mastrototaro, J. (1999). The MiniMed Continuous Glucose Monitoring System (CGMS). *Journal of Pediatric Endocrinology & Metabolism, 12*(3), 751–758.

[40] Veneman, T., Mitrakou, A., Mokan, M., Cryer, P., Gerich, J. (1993). Induction of hypoglycemia unawareness by asymptomatic nocturnal hypoglycemia. *Diabetes, 42*, 1233–1237.

[41] Fanelli, C.G., Paramore, D.S., Hershey, T., Terkamp, C., Ovalle, F., Craft, S., Cryer, P.E. (1998). Impact of nocturnal hypoglycemia on hypoglycemic cognitive dysfunction in type 1 diabetes. *Diabetes, 47*, 1920–1927.

[42] Banarer, S., Cryer, P.E. (2003). Sleep-related hypoglycemia-associated autonomic failure in type 1 diabetes: reduced awakening from sleep during hypoglycemia. *Diabetes, 52*, 1195–1203.

[43] Gale, E., Tattersall, R. (1979). Unrecognized nocturnal hypoglycaemia in insulin treated diabetics. *Lancet, 1*, 1049–1052.

[44] Bendtson, I., Kverneland, A., Pramming, S., Binder, C. (1988). Incidence of nocturnal hypoglycaemia in insulin-dependent diabetic patients on intensive therapy. *Acta Medica Scandinavica, 223*, 543–548.

[45] Whincup, G., Milner, R.D. (1987). Prediction and management of nocturnal hypoglycemia in diabetes. *Archives of Disease in Childhood, 62*, 333–337.

[46] Shalwitz, R.A., Farkas-Hirsch, R., White, N.H., Santiago, J.V. (1990). Prevalence and consequences of nocturnal hypoglycemia in conventionally treated children with diabetes mellitus. *Journal of Pediatrics, 116*, 685–689.

[47] Matyka, K.A., Crawford, C., Wiggs, L., Dunger, D., Stores, G. (2000). Alterations in sleep physiology in young children with insulin dependent diabetes mellitus: relationship to nocturnal hypoglycaemia. *Journal of Pediatrics, 137*, 233–238.

[48] Porter, P., Keating, B., Byrne, G.C., Jones, T.W. (1997). Incidence and predictive criteria of nocturnal hypoglycaemia in young children with insulin dependent diabetes mellitus. *Journal of Pediatrics, 130*, 366–372.

[49] Wentholt, I.M., Maran, A., Masurel, N., Heine, R.J., Hoekstra, J.B., DeVries, J.H. (2007). Nocturnal hypoglycaemia in type 1 diabetic patients, assessed with continuous glucose monitoring: frequency, duration and associations. *Diabetic Medicine*, *24*, 527–532.

[50] Vervoort, G., Goldschmidt, H.M., van Doorn, L.G. (1996). Nocturnal blood glucose profiles in patients with type 1 diabetes mellitus on multiple (> or = 4) daily insulin injection regimens. *Diabetic Medicine*, *13*, 794–799.

[51] Jones, T.W., Davis, E.A. (2003). Hypoglycemia in children with type 1 diabetes: current issues and controversies. *Pediatric Diabetes*, *4*, 143–150.

[52] Davis, E.A., Keating, B., Byrne, G.C., Russell, M., Jones, T.W. (1998). The impact of improved glycaemic control on rates of hypoglycaemia in patients with IDDM. *Archives of Disease in Childhood*, *78*, 111–115.

[53] Danne, T., Mortensen, H.B., Hougaard, P., Lynggaard, H., Aanstoot, H.J., Chiarelli, F., Daneman, D., Dorchy, H., Garandeau, P., Greene, S.A., Hoey, H., et al. (2001). For the Hvidøre Study Group on Childhood Diabetes: persistent differences among centers over 3 years in glycemic control and hypoglycemia in a study of 3,805 children and adolescents with type 1 diabetes from the Hvidøre Study Group. *Diabetes Care*, *24*, 1342–1347.

[54] Nordfelt, S., Ludvigsson, J. (1999). Adverse events in intensively treated children and adolescents with type 1 diabetes. *Acta Paediatrica Scandinavica*, *88*, 1184–1193.

[55] Rewers, A., Chase, H.P., Mackenzie, T., Walravens, P., Roback, M., Rewers, M., Hamman, R.F., Klingensmith, G. (2002). Predictors of acute complications in children with type 1 diabetes. *Journal of the American Medical Association*, *287*, 2511–2518.

[56] Davis, E.A., Keating, B., Byrne, G.C., Russell, M., Jones, T.W. (1997). Hypoglycaemia: incidence and clinical predictors in a large population based sample of children and adolescents with IDDM. *Diabetes Care*, *20*, 22–25.

[57] Northam, E., Anderson, P., Jacobs, P., Hughes, M., Warne, G., Werther, G.A. (2001). Neuropsychological profiles of children with diabetes 6 years after disease onset. *Diabetes Care*, *24*, 1541–1546.

[58] Scoenle, E.J., Schoenle, D., Molinari, L., Largo, R.H. (2002). Impaired intellectual development in children with type 1 diabetes: association with HbA1c, age at diagnosis and sex. *Diabetologia*, *45*, 108–114.

[59] Rovet, J.F., Ehrlich, R.M., Hoppe, M. (1988). Specific intellectual deficits in children with early onset diabetes mellitus. *Child Development*, *59*, 226–234.

[60] Rovet, J. (1999). Neuropsychological sequelae of paediatric diabetes. In: Yeates, K., Ris, M., Taylor, H. (Eds.), *Paediatric Neuropsychologic Research, Theory and Practice*, Guilford Press, New York, pp. 336–365.

[61] Ryan, C., Vega, A., Drash, A. (1985). Cognitive deficits in adolescents who developed diabetes early in life. *Pediatrics*, *75*, 921–927.

[62] Cox, D., Clarke, W., Gonder-Frederick, L., Kovatchev, B. (2001). Driving mishaps and hypoglycaemia: risk and prevention. *International Journal of Clinical Practice Supplement*, *123*, 38–42.

[63] Ryan, C.M., Geckle, M.O., Orchard, T.J. (2003). Cognitive efficiency declines over time in adults with type 1 diabetes: effects of micro- and macrovascular complications. *Diabetologia*, *46*, 940–948.

[64] Brands, A.M., Biessels, G.J, de Haan, E.H., Kappelle, L.J., Kessels, R.P. (2005). The effects of type 1 diabetes on cognitive performance: a meta-analysis. *Diabetes Care*, *28*, 726–735.

[65] Thorsell, A., Gordon, M., Jovanovic, L. (2004). Continuous glucose monitoring: a stepping stone in the journey towards a cure for diabetes. *Journal of Maternal– Fetal & Neonatal Medicine*, *15*, 15–25.

[66] Williams, A.F. (1997). Hypoglycaemia of the newborn: a review. *Bulletin of the World Health Organization*, *75*, 261–290.

[67] Kovatchev, B.P., Cox, D.J., Kumar, A., Gonder-Frederick, L., Clarke, W.L. (2003). Algorithmic evaluation of metabolic control and risk of severe hypoglycemia in type 1 and type 2 diabetes using self-monitoring blood glucose data. *Diabetes Technology & Therapeutics*, *5*, 817–828.

[68] Ho, K.L., Loke, H.L, Tan, K.W. (1991). Accuracy and reliability of two methods of screening for hypoglycemia in neonates. *Journal of the Singapore Paediatric Society*, *33*, 156–158.

[69] Hall, D.M.B., Michael, J.M. (1995). Screening in infancy. *Archives of Disease in Childhood*, *72*, 93–96.

[70] Ellis, M., Manandhar, D.S., Manandhar, N., Land, J.M., Patel, N., del Costello, A.M. (1996). Comparison of two cotside methods for the detection of hypoglycaemia among neonates in Nepal. *Archives of Disease in Childhood Fetal Neonatal*, *75*, F122–F125.

[71] Dahlberg, M., Whitelaw, A. (1997). Evaluation of HemoCue Blood Glucose Analyzer for the instant diagnosis of hypoglycaemia in newborns. *Scandinavian Journal of Clinical and Laboratory Investigation*, *57*, 719–724.

[72] Leonard, M., Chessall, M., Manning, D. (1997). The use of a Hemocue blood glucose analyser in a neonatal unit. *Annals of Clinical Biochemistry*, *34*, 287–290.

[73] Buckingham, B., Caswell, K., Wilson, D.M. (2007). Real-time continuous glucose monitoring. *Current Opinion in Endocrinology, Diabetes, and Obesity*, *14*, 288–295.

[74] Heise, T., Koschinsky, T., Heinemann, L., Lodwig, V. (2003). Glucose Monitoring Study Group: Hypoglycemia warning signal and glucose sensors: requirements and concepts. *Diabetes Technology & Therapeutics*, *5*, 563–571.

[75] Bode, B.W., Sabbah, H., Davidson, P.C. (2001). What's ahead in glucose monitoring? New techniques hold promise for improved ease and accuracy. *Postgraduate Medicine*, *109*, 41–44, 47–49.

[76] Moberg, E., Hagström-Toft, E., Arner, P., Bolinder, J. (1997). Protracted glucose fall in subcutaneous adipose tissue and skeletal muscle compared with blood during insulin-induced hypoglycaemia. *Diabetologia*, *40*, 1320–1326.

[77] Aussedat, B., Dupire-Angel, M., Gifford, R., Klein, J.C., Wilson, G.S., Reach, G. (2000). Interstitial glucose concentration and glycemia: implications for continuous subcutaneous glucose monitoring. *American Journal of Physiology: Endocrinology and Metabolism*, *278*, E716–E728.

[78] Melki, V., Ayon, F., Fernandez, M., Hanaire-Broutin, H. (2006). Value and limitations of the Continuous Glucose Monitoring System in the management of type 1 diabetes. *Diabetes & Metabolism*, *32*, 123–129.

[79] Wentholt, I.M., Hart, A.A., Hoekstra, J.B., Devries, J.H. (2007). Relationship between interstitial and blood glucose in type 1 diabetes patients: delay and the push–pull phenomenon revisited. *Diabetes Technology & Therapeutics*, *9*, 169–175.

[80] McGowan, K., Thomas, W., Moran, A. (2002). Spurious reporting of nocturnal hypoglycemia by CGMS in patients with tightly controlled type 1 diabetes. *Diabetes Care*, *25*, 1499–1503. Erratum in: Diabetes Care, 2003, 26, 553.

[81] Diabetes Research in Children Network (Direcnet) Study Group, Buckingham, B.A., Kollman, C., Beck, R., Kalajian, A., Fiallo-Scharer, R., Tansey, M.J., Fox, L.A., Wilson, D.M., Weinzimer, S.A., Ruedy, K.J., Tamborlane, W.V. (2006). Evaluation of factors affecting CGMS calibration. *Diabetes Technology & Therapeutics*, *8*, 318–325.

[82] Moatti-Sirat, D., Capron, F., Poitout, V., Reach, G., Bindra, D.S., Zhang, Y., Wilson, G.S., Thévenot, D.R. (1992). Towards continuous glucose monitoring: in vivo evaluation of a miniaturized glucose sensor implanted for several days in rat subcutaneous tissue. *Diabetologia*, *35*, 224–230.

[83] Rebrin, K., Fischer, U., Hahn von Dorsche, H., von Woetke, T., Abel, P., Brunstein, E. (1992). Subcutaneous glucose monitoring by means of electrochemical sensors: fiction or reality? *Journal of Biomedical Engineering*, *14*, 33–40.

[84] Reach, G. (2008). Continuous glucose monitoring and diabetes health outcomes: a critical appraisal. *Diabetes Technology & Therapeutics*, *10*, 69–80.

[85] Reach, G. (2001). Which threshold to detect hypoglycemia? Value of receiver–operator curve analysis to find a compromise between sensitivity and specificity. *Diabetes Care*, *24*, 803–804.

[86] Shichiri, M., Kawamori, R., Yamasaki, Y., Hakui, N., Abe, H. (1982). Wearable artificial endocrine pancrease with needle-type glucose sensor. *Lancet*, *2*, 1129–1131.

[87] Mastrototaro, J. (1999). The MiniMed Continuous Glucose Monitoring System (CGMS). *Journal of Pediatric Endocrinology & Metabolism*, *12*(3), 751–758.

[88] Yogev, Y., Chen, R., Ben-Haroush, A., Phillip, M., Jovanovic, L., Hod, M. (2003). Continuous glucose monitoring for the evaluation of gravid women with type 1 diabetes mellitus. *Obstetrics and Gynecology*, *101*, 633–638.

[89] Metzger, M., Leibowitz, G., Wainstein, J., Glaser, B., Raz, I. (2002). Reproducibility of glucose measurements using the glucose sensor. *Diabetes Care*, *25*, 1185–1191.

[90] Jovanovic, L. (2000). The role of continuous glucose monitoring in gestational diabetes mellitus. *Diabetes Technology & Therapeutics*, *2*, 67–71.

[91] Rao, G., Glikfeld, P., Guy, R.H. (1993). Reverse iontophoresis: development of a noninvasive approach for glucose monitoring. *Pharmaceutical Research*, *10*, 1751–1755.

[92] Jovanovic, L., Yogev, Y., Hod, M. (2003). Optimal monitoring of glycemia and definitions of normoglycemia. In: Dunitz, M. (Ed.), *Textbook of Diabetes and Pregnancy*, Taylor & Francis, London, pp. 345–376.

[93] Tavris, D.R., Shoaibi, A. (2004). The public health impact of the MiniMed Continuous Glucose Monitoring System (CGMS): an assessment of the literature. *Diabetes Technology & Therapeutics*, *6*, 518–522.

[94] Chico, A., Vidal-Ríos, P., Subirà, M., Novials, A. (2003). The continuous glucose monitoring system is useful for detecting unrecognized hypoglycemias in patients with type 1 and type 2 diabetes but is not better than frequent capillary glucose measurements for improving metabolic control. *Diabetes Care*, *26*, 1153–1157.

[95] Korf, J., Tiessen, R.G., Venema, K., Rhemrev, M.M. (2003). Biosensors for continuous glucose and lactate monitoring. *Nederlands Tijdschrift voor Geneeskunde*, *147*, 1204–1208.

[96] Kubiak, T., Hermanns, N., Schreckling, H.J., Kulzer, B., Haak, T. (2004). Assessment of hypoglycaemia awareness using continuous glucose monitoring. *Diabetic Medicine*, *21*, 487–490.

[97] Gross, T.M., Mastrototaro, J.J. (2000). Efficacy and reliability of the continuous glucose monitoring system. *Diabetes Technology & Therapeutics*, *2*(1), S19–S26.

[98] Beardsall, K., Ogilvy-Stuart, A.L., Ahluwalia, J., Thompson, M., Dunger, D.B. (2005). The continuous glucose monitoring sensor in neonatal intensive care. *Archives of Disease in Childhood Fetal Neonatal*, *90*, F307–F310.

[99] Sternberg, F., Meyerhoff, C., Mennel, F.J., Mayer, H., Bischof, F., Pfeiffer, E.F. (1996). Does fall in tissue glucose precede fall in blood glucose? *Diabetologia*, *39*, 609–612.

[100] Keck, F.S., Kerner, W., Meyerhoff, C., Zier, H., Pfeiffer, E.F. (1991). Combination of microdialysis and GlucoSensor permits continuous (on line) s.c. glucose monitoring in a patient operated device: I. In vitro evaluation. *Hormone and Metabolic Research*, *23*, 617–618.

[101] Hashiguchi, Y., Sakakida, M., Nishida, K., Uemura, T., Kajiwara, K., Shichiri, M. (1994). Development of a miniaturized glucose monitoring system by combining a needle-type glucose sensor with microdialysis sampling method: long-term subcutaneous tissue glucose monitoring in ambulatory diabetic patients. *Diabetes Care*, *17*, 387–396.

[102] Meyerhoff, C., Mennel, F.J., Bischof, F., Sternberg, F., Pfeiffer, E.F. (1994). Combination of microdialysis and glucose sensor for continuous on line measurement of the subcutaneous glucose concentration: theory and practical application. *Hormone and Metabolic Research*, *26*, 538–543.

[103] Bolinder, J., Hagström-Toft, E., Ungerstedt, U., Arner, P. (1997). Self-monitoring of blood glucose in type I diabetic patients: comparison with continuous microdialysis measurements of glucose in subcutaneous adipose tissue during ordinary life conditions. *Diabetes Care*, *20*, 64–70.

[104] Fayolle, C., Brun, J.F., Bringer, J., Mercier, J., Renard, E. (2006). Accuracy of continuous subcutaneous glucose monitoring with the GlucoDay in type 1 diabetic patients treated by subcutaneous insulin infusion during exercise of low versus high intensity. *Diabetes & Metabolism*, *32*, 313–320.

[105] Maran, A., Crepaldi, C., Tiengo, A., Grassi, G., Vitali, E., Pagano Gbistoni, S., Calabrese, G., Santeusanio, F., Leonetti, F., Ribaudo, M., et al. (2002). Continuous subcutaneous glucose monitoring in diabetic patients: a multicenter analysis. *Diabetes Care*, *25*, 347–352.

[106] Wentholt, I.M., Vollebregt, M.A., Hart, A.A., Hoekstra, J.B., DeVries, J.H. (2005). Comparison of a needle-type and a microdialysis continuous glucose monitor in type 1 diabetic patients. *Diabetes Care*, *28*, 2871–2876.

[107] Esmatjes, E., Flores, L., Vidal, M., Rodriguez, L., Cortés, A., Almirall, L., Ricart, M.J., Gomis, R. (2003). Hypoglycaemia after pancreas transplantation: usefulness of a continuous glucose monitoring system. *Clinical Transplantation, 17,* 534–538.

[108] Sparacino, G., Zanderigo, F., Corazza, S., Maran, A., Facchinetti, A., Cobelli, C. (2007). Glucose concentration can be predicted ahead in time from continuous glucose monitoring sensor time-series. *IEEE Transactions on Bio-medical Engineering, 54,* 931–937.

[109] Hermanides, J., Wentholt, I.M., Hart, A.A., Hoekstra, J.B., DeVries, J.H. (2008). No apparent local effect of insulin on microdialysis continuous glucose-monitoring measurements. *Diabetes Care, 31,* 1120–1122.

[110] Bolinder, J., Ungerstedt, U., Arner, P. (1993). Long term continuous glucose monitoring with microdialysis in ambulatory insulin-dependent diabetic patients. *Lancet, 342,* 1080–1085.

[111] Hagström-Toft, E., Hellström, L., Moberg, E. (1998). Lipolytic response during spontaneous hypoglycaemia in insulin-dependent diabetic subjects. *Hormone and Metabolic Research, 30,* 586–593.

[112] Hildingsson, U., Sellden, H., Ungerstedt, U., Marcus, C. (1996). Microdialysis for metabolic monitoring in neonates after surgery. *Acta Paediatrica Scandinavica, 85,* 589–594.

[113] Anand, K.J.S., Brown, M.J., Causon, R.C., Christofides, N.D., Bloom, S.R., Aynsley-Green, A. (1985). Can the human neonate mount an endocrine and metabolic response to surgery? *Journal of Pediatric Surgery, 20,* 41–48.

[114] Anand, K.J.S., Sippel, W.G., Aynsley-Green, A. (1987). Randomised trial of fentanyl anaesthesia in preterm babies undergoing surgery: effects on the stress response. *Lancet, 1,* 243–248.

[115] Ward Platt, M.P., Tarbit, M.J., Aynsley-Green, A. (1990). The effects of anesthesia and surgery on metabolic homeostasis in infancy and childhood. *Journal of Pediatric Surgery, 25,* 472–478.

[116] Vlasselaers, D., Schaupp, L., van den Heuvel, I., Mader, J., Bodenlenz, M., Suppan, M., Wouters, P., Ellmerer, M., van den Berghe, G. (2007). Monitoring blood glucose with microdialysis of interstitial fluid in critically ill children. *Clinical Chemistry, 53,* 536–537.

[117] Ellmerer, M., Haluzik, M., Blaha, J., Kremen, J., Svacina, S., Toller, W., Mader, J., Schaupp, L., Plank, J., Pieber, T. (2006). Clinical evaluation of alternative-site glucose measurements in patients after major cardiac surgery. *Diabetes Care, 29,* 1275–1281.

[118] De Backer, D., Creteur, J., Preiser, J.C., Dubois, M.J., Vincent, J.L. (2002). Microvascular blood flow is altered in patients with sepsis. *American Journal of Respiratory and Critical Care Medicine, 166,* 98–104.

[119] Sair, M., Etherington, P.J., Peter Winlove, C., Evans, T.W. (2001). Tissue oxygenation and perfusion in patients with systemic sepsis. *Critical Care Medicine, 29,* 1343–1349.

[120] Kamel, A., Norgren, S., Persson, B., Marcus, C. (1999). Insulin induced hypoglycaemia: comparison of glucose and glycerol concentrations in plasma and microdialysate from subcutaneous adipose tissue. *Archives of Disease in Childhood, 80,* 42–45.

[121] Moberg, E., Hagström-Toft, E., Bolinder, J. (1997). Detection of hypoglycaemia by microdialysis measurements of glucose in subcutaneous adipose tissue. *Hormone and Metabolic Research*, *29*, 440–443.

[122] Rajamand, N., Ungerstedt, U., Brismar, K. (2005). Subcutaneous microdialysis before and after an oral glucose tolerance test: a method to determine insulin resistance in the subcutaneous adipose tissue in diabetes mellitus. *Diabetes, Obesity & Metabolism*, *7*, 525–535.

[123] Baumeister, F.A., Hack, A., Busch, R. (2006). Glucose-monitoring with continuous subcutaneous microdialysis in neonatal diabetes mellitus. *Klinische Padiatrie*, *218*, 230–232.

[124] Baumeister, F.A., Rolinski, B., Busch, R., Emmrich, P. (2001). Glucose monitoring with long-term subcutaneous microdialysis in neonates. *Pediatrics*, *108*, 1187–1192.

[125] Holzinger, A., Bonfig, W., Kusser, B., Eggermann, T., Muller, H., Munch, H.G. (2006). Use of long-term microdialysis for subcutaneous glucose monitoring in the management of neonatal diabetes: a first case report. *Biology of the Neonate*, *89*, 88–91.

[126] Bailey, T.S., Zisser, H.C., Garg, S.K. (2007). Reduction in hemoglobin A1C with real-time continuous glucose monitoring: results from a 12-week observational study. *Diabetes Technology & Therapeutics*, *9*, 203–210.

[127] Bugler, J.R., Bergenstal, R.M., Bode, B.W., Weinstein, R.L., Schwartz, S.L., Bernstein, R.M., El-Deiry, S.S. (2006). Home use evaluation of the Free-Style Navigator continuous glucose monitoring system. *Diabetologia*, *49*(Suppl), 0957.

[128] Deiss, D., Bolinder, J., Riveline, J.P., Battelino, T., Bosi, E., Tubiana-Rufi, N., Kerr, D., Phillip, M. (2006). Improved glycemic control in poorly controlled patients with type 1 diabetes using real-time continuous glucose monitoring. *Diabetes Care*, *29*, 2730–2732.

[129] Garg, S., Jovanovic, L. (2006). Relationship of fasting and hourly blood glucose levels to HbA1c values: safety, accuracy, and improvements in glucose profiles obtained using a 7-day continuous glucose sensor. *Diabetes Care*, *29*, 2644–2649.

[130] Garg, S.K., Schwartz, S., Edelman, S.V. (2004). Improved glucose excursions using an implantable real-time continuous glucose sensor in adults with type 1 diabetes. *Diabetes Care*, *27*, 734–738.

[131] Garg, S.K., Zisser, H., Schwartz, S., Bailey, T., Kaplan, R., Ellis, S., Jovanovic, L. (2006). Improvement in glycemic excursion with a transcutaneous, real-time continuous glucose sensor: a randomized control trial. *Diabetes Care*, *29*, 44–50.

[132] Clough, G., Noble, M. (2003). Microdialysis: a model for studying chronic wounds. *International Journal of Lower Extremity Wounds*, *2*(4), 233–239.

[133] Harding, K.G., Morris, H.L., Patel, G.K. (2002). Science, medicine and the future: healing chronic wounds. *British Medical Journal*, *324*(7330), 160–163.

[134] Simonsen, L., Holstein, P., Larsen, K., Bülow, J. (1998). Glucose metabolism in chronic diabetic foot ulcers measured in vivo using microdialysis. *Clinical Physiology*, *18*(4), 355–359.

[135] Stolle, L.B., Riegels-Nielsen, P. (2004). *Acta Orthopaedica Scandinavica*, *75*(1), 106–108.

[136] Latijnhouwers, M.A., Bergers, M., Veenhuis, R.T., Beekman, B., Ankersmit-Ter Horst, M.F., Schalkwijk, J. (1998). Tenascin-C degradation in chronic wounds is dependent on serine proteinase activity. *Archives of Dermatological Research*, *290*(9), 490–496.

[137] Weckroth, M., Vaheri, A., Myohanen, H., Tukiainen, E., Siren, V. (2001). Differential effects of acute and chronic wound fluids on urokinase-type plasminogen activator, urokinase-type plasminogen activator receptor, and tissue-type plasminogen activator in cultured human keratinocytes and fibroblasts. *Wound Repair and Regeneration*, *9*(4), 314–322.

[138] Weckroth, M., Vaheri, A., Lauharanta, J., Sorsa, T., Konttinen, Y.T. (1996). Matrix metalloproteinases, gelatinase and collagenase, in chronic leg ulcers. *Journal of Investigative Dermatology*, *106*(5), 1119–1124.

[139] Hoffman, R., Starkey, S., Coad, J. (1998). Wound fluid from venous leg ulcers degrades plasminogen and reduces plasmin generation by keratinocytes. *Journal of Investigative Dermatology*, *111*(6), 1140–1144.

[140] Fivenson, D.P., Faria, D.T., Nickoloff, B.J., Poverini, P.J., Kunkel, S., Burdick, M., Strieter, R.M. (1998). Chemokine and inflammatory cytokine changes during chronic wound healing. *Wound Repair and Regeneration*, *5*(4), 310–322.

[141] Petersen, L.J., Church, M.K., Skov, P.S. (1997). Histamine is released in the wheal but not the flare following challenge of human skin in vivo: a microdialysis study. *Clinical and Experimental Allergy*, *27*(3), 284–595.

[142] Clough, G.F. (1999). Role of nitric oxide in the regulation of microvascular perfusion in human skin in vivo. *Journal of Physiology*, *516*(2), 549–557.

[143] Church, M.K., Griffiths, T.J., Jeffery, S., Ravell, L.C., Cowburn, A.S., Sampson, A.P., Clough, G.F. (2002). Are cysteinyl leukotrienes involved in allergic responses in human skin? *Clinical and Experimental Allergy*, *32*(7), 1013–1019.

[144] Caldwell, M.D. (1988). Carbohydrate and energy metabolism in healing wounds. *Progress in Clinical and Biological Research*, *266*, 183–213.

[145] Daley, J.M., Shearer, J.D., Mastrofrancesco, B., Caldwell, M.D. (1990). Glucose metabolism in injured tissue: a longitudinal study. *Surgery*, *107*(2), 187–192.

[146] Wilmore, D.W., Aulick, L.H. (1978). Metabolic changes in burned patients. *Surgical Clinics of North America*, *58*(6), 1173–1187.

[147] Simonsen, L., Bülow, J., Madsen J. (1994). Adipose tissue metabolism in humans determined by vein catherization and microdialysis technique. *American Journal of Physiology*, *266*(3 Pt. 1), E357–E365.

[148] Rhodes, L.E., Belgi, G., Parslew, R., McLoughlin, L., Clough, G.F., Friedmann, P.S. (2001). Ultraviolet-B-induced erythema is mediated by nitric oxide and prostaglandin E2 in combination. *Journal of Investigative Dermatology*, *117*(4), 880–885.

[149] Krogstad, A.L., Lönnroth, P., Larson, G., Wallin, B.G. (1997). Increased interstitial histamine concentration in the psoriatic plaque. *Journal of Investigative Dermatology*, *109*(5), 632–635.

[150] Peterson, L.R., Lissack, L.M., Canter, K., Fasching, C.E., Clabots, C., Gerding, D.N. (1989). Therapy of lower extremity infections with ciprofloxacin in patients with diabetes mellitus, peripheral vascular disease, both. *American Journal of Medicine*, *86*(6 Pt. 2), 801–808.

[151] Lassen, N.A. (1973). General discussion on occlusive arterial disease in diabetes mellitus. *Scandinavian Journal of Clinical and Laboratory Investigation Supplement, 1128*, 235–237.

[152] Legat, F.J., Maier, A., Dittrich, P., Zenahlik, P., Kern, T., Nuhsbaumer, S., Frossard, M., Salmhofer, W., Kerl, H., Müller, M. (2003). Penetration of fosfomycin into inflammatory lesions in patients with cellulitis or diabetic foot syndrome. *Antimicrobial Agents and Chemotherapy, 47*(1), 371–374.

[153] Legat, F.J., Krause, R., Zenahlik, P., Hoffmann, C., Scholz, S., Salmhofer, W., Tscherpel, J., Tscherpel, T., Kerl, H., Dittrich, P. (2005). Penetration of piperacillin and tazobactam into inflamed soft tissue of patients with diabetic foot infection. *Antimicrobial Agents and Chemotherapy, 49*(10), 4368–4371.

[154] Müller, M., Brunner, M., Hollenstein, U., Joukhadar, C., Schmid, R., Minar, E., Ehringer, H., Eichler, H.G. (1999). Penetration of ciprofloxacin into the interstitial space of inflamed foot lesions in non-insulin-dependent diabetes mellitus patients. *Antimicrobial Agents and Chemotherapy, 43*(8), 2056–2058.

[155] Bellmann, R., Kuchling, G., Dehghanyar, P., Zeitlinger, M., Minar, E., Mayer, B.X., Müller, M., Joukhadar, C. (2004). Tissue pharmacokinetics of levofloxacin in human soft tissue infections. *British Journal of Clinical Pharmacology, 57*(5), 563–568.

[156] Burgmann, H., Georgopoulos, A., Graninger, W., Koppensteiner, R., Maca, T., Minar, E., Schneider, B., Stümpflen, A., Ehringer, H. (1996). Tissue concentration of clindamycin and gentamicin near ischaemic ulcers with transvenous injection in Bier's arterial arrest. *Lancet, 348*(9030), 781–783.

[157] Dan, M., Torossian, K., Weissberg, D., Kitzes, R. (1993). The penetration of ciprofloxacin into bronchial mucosa, lung parenchyma, and pleural tissue after intravenous administration. *European Journal of Clinical Pharmacology, 44*(1), 101–102.

[158] Fong, I.W., Ledbetter, W.H., Vandenbroucke, A.C., Simbul, M., Rahm, V. (1986). Ciprofloxacin concentrations in bone and muscle after oral dosing. *Antimicrobial Agents and Chemotherapy, 29*(3), 405–408.

[159] Müller, M., Brunner, M., Schmid, R., Putz, E.M., Schmiedberger, A., Wallner, I., Eichler, H.G. (1998). Comparison of three different experimental methods for the assessment of peripheral compartment pharmacokinetics in humans. *Life Sciences, 62*(15), PL227–PL234.

[160] Wise, R., Donovan, I.A. (1987). Tissue penetration and metabolism of ciprofloxacin. *American Journal of Medicine, 82*(4A), 103–107.

[161] Brunner, M., Hollenstein, U., Delacher, S., Jäger, D., Schmid, R., Lackner, E., Georgopoulos, A., Eichler, H.G., Müller, M. (1999). Distribution and antimicrobial activity of ciprofloxacin in human soft tissues. *Antimicrobial Agents and Chemotherapy, 43*(5), 1307–1309.

[162] Müller, M., Haag, O., Burgdorff, T., Georgopoulos, A., Weninger, W., Jansen, B., Stanek, G., Pehamberger, H., Agneter, E., Eichler, H.G. (1996). Characterization of peripheral-compartment kinetics of antibiotics by in vivo microdialysis in humans. *Antimicrobial Agents and Chemotherapy, 40*(12), 2703–2709.

[163] Ryan, D.M., Cars, O., Hoffstedt, B. (1986). The use of antibiotic serum levels to predict concentrations in tissues. *Scandinavian Journal of Infectious Diseases, 18*(5), 381–388.

[164] Seabrook, G.R., Edmiston, C.E., Schmitt, D.D., Krepel, C., Bandyk, D.F., Towne, J.B. (1991). Comparison of serum and tissue antibiotic levels in diabetes-related foot infections. *Surgery, 110*(4), 671–676.

[165] Joukhadar, C., Derendorf, H., Müller, M. (2001). Microdialysis. A novel tool for clinical studies of anti-infective agents. *European Journal of Clinical Pharmacology, 57*(3), 211–219.

[166] Bergogne-Bérézin, E. (2002). Clinical role of protein binding of quinolones. *Clinical Pharmacokinetics, 41*(10), 741–750.

[167] Maeda, H., Akaike, T., Wu, J., Noguchi, Y., Sakata, Y. (1996). Bradykinin and nitric oxide in infectious disease and cancer. *Immunopharmacology, 33*(1–3), 222–230.

[168] Chantelau, E. (1999). Pathogenesis of diabetic foot disease. *Internist (Berlis), 40*(10), 994–1001 (in German).

[169] Joukhadar, C., Klein, N., Frossard, M., Minar, E., Stass, H., Lackner, E., Herrmann, M., Riedmüller, E., Müller, M. (2001). Angioplasty increases target site concentrations of ciprofloxacin in patients with peripheral arterial occlusive disease. *Clinical Pharmacology & Therapeutics, 70*(6), 532–539.

[170] Joukhadar, C., Stass, H., Müller-Zellenberg, U., Lackner, E., Kovar, F., Minar, E., Müller, M. (2003). Penetration of moxifloxacin into healthy and inflamed subcutaneous adipose tissues in humans. *Antimicrobial Agents and Chemotherapy, 47*(10), 3099–3103.

[171] Capecchi, P.L., Blardi, P., De Lalla, A., Ceccatelli, L., Volpi, L., Pasini, L., Di Perri, T. (1995). Pharmacokinetics and pharmacodynamics of neutrophil-associated ciprofloxacin in humans. *Clinical Pharmacology & Therapeutics, 57*(4), 446–454.

[172] Licitra, C.M., Brooks, R.G., Sieger, B.E. (1987). Clinical efficacy and levels of ciprofloxacin in tissue in patients with soft tissue infection. *Antimicrobial Agents and Chemotherapy, 31*(5), 805–807.

[173] Müller, M., Schmid, R., Georgopoulos, A., Buxbaum, A., Wasicek, C., Eichler, H.G. (1995). Application of microdialysis to clinical pharmacokinetics in humans. *Clinical Pharmacology & Therapeutics, 57*(4), 371–380.

[174] Frossard, M., Joukhadar, C., Erovic, B.M., Dittrich, P., Mrass, P.E., Van Houte, M., Burgmann, H., Georgopoulos, A., Müller, M. (2000). Distribution and antimicrobial activity of fosfomycin in the interstitial fluid of human soft tissues. *Antimicrobial Agents and Chemotherapy, 44*(10), 2728–2732.

[175] Dalla Costa, T., Nolting, A., Kovar, A., Derendorf, H. (1998). Determination of free interstitial concentrations of piperacillin–tazobactam combinations by microdialysis. *Journal of Antimicrobial Chemotherapy, 42*(6), 769–778.

[176] Perry, C.M., Markham, A. (1999). Piperacillin/tazobactam: an updated review of its use in the treatment of bacterial infections. *Drugs, 57*(5), 805–843.

[177] Kim, S.H, Abbasi, F., Reaven, G.M. (2004). Impact of degree of obesity on surrogate estimates of insulin resistance. *Diabetes Care, 27*, 1998–2002.

[178] Despres, J.P., Lamarche, B., Mauriege, P., Cantin, B., Dagenais, G.R., Moorjani, S., Lupien, P.J. (1996). Hyperinsulinemia as an independent risk factor for ischemic heart disease. *New England Journal of Medicine, 334*(15), 952–957.

[179] Shanik, M.H., Xu, Y., Skrha, J., Dankner, R., Zick, Y., Roth, J. (2008). Insulin resistance and hyperinsulinemia: Is hyperinsulinemia the cart or the horse? *Diabetes Care, 31*(2), S262–S268.

[180] Rizza, R.A., Mandarino, L.J., Genest, J., Baker, B.A., Gerich, J.E. (1985). Production of insulin resistance by hyperinsulinemia in man. *Diabetologia*, *28*, 70–75.

[181] Groop, L.C., Bonadonna, R.C., Shank, M., Petrides, A.S., DeFronzo, R.A. (1991). Role of free fatty acids and insulin in determining free fatty acid and lipid oxidation in man. *Journal of Clinical Investigation*, *87*, 83–89.

[182] Abdul-Ghani, M.A., Matsuda, M., Balas, B., DeFronzo, R.A. (2007). Muscle and liver insulin resistance indexes derived from the oral glucose tolerance test. *Diabetes Care*, *30*, 89–94.

[183] DeFronzo, R.A. (1988). Lilly Lecture: The triumvariate: β-cell, muscle, liver, a collusion responsible for NIDDM. *Diabetes*, *37*, 667–687.

[184] Frossard, M., Blank, D., Joukhadar, Ch., Bayegan, K., Schmid, R., Luger, A., Müller, M. (2005). Interstitial glucose in skeletal muscle of diabetic patients during an oral glucose tolerance test. *Diabetic Medicine*, *22*(1), 56–60.

[185] Cline, G.W., Petersem, K.F., Kossak, M., Shen, J., Hundal, R.S., Trajanoski, Z., Inzucchi, S., Dreszer, A., Rothman, D.L., Hulman, G.I. (1999). Impaired glucose transport as a cause of decreased insulin stimulated muscle glycogen synthesis in type 2 diabetes. *New England Journal of Medicine*, *341*, 240–246.

[186] Niklasson, M., Holmäng, A., Sjöstrand, M., Strindberg, L., Lönnroth, P. (2000). Muscle glucose uptake is effectively activated by ischemia in type 2 diabetic subjects. *Diabetes*, *49*, 1178–1185.

13

MICRODIALYSIS USE IN TUMORS: DRUG DISPOSITION AND TUMOR RESPONSE

QINGYU ZHOU AND JAMES M. GALLO

Department of Pharmacology and Systems Therapeutics, Mount Sinai School of Medicine, New York, New York

1. INTRODUCTION

According to the World Cancer Report issued by the World Health Organization, there are over 10 million new cases of cancer worldwide each year, with approximately 6 million deaths annually [1]. In the United States, cancer is one of the five leading causes of death in all age groups among both males and females. In particular, cancer is the leading cause of death among women aged 40 to 79 years and among men aged 60 to 79 years. Overall, about one in four deaths is due to cancer, ranking it second only to heart disease. Analysis of the mortality data obtained from the National Center for Health Statistics (NCHS) has shown that despite a decrease in age-standardized death rates, 5424 more cancer deaths were reported in 2005 than in 2004. This is heavily weighted on solid tumors, which comprise the majority of human malignancies and account for about 85% of cancer deaths [2]. Hence, there is a pressing need for more effective therapeutic approaches in the management of cancer, particularly of malignant solid tumors.

Cancer chemotherapy is an integral component of cancer therapy. Unlike surgery and radiation therapy, which treat cancer in a specific area by physically removing a tumor or a part of it, chemotherapy is a systemic treatment

Applications of Microdialysis in Pharmaceutical Science, First Edition. Edited by Tung-Hu Tsai.

that is carried throughout the entire body by the bloodstream, using chemical agents to interact with cancer cells to eradicate or suppress the growth of cancer. Although extensive research has been carried out on the development of more specific and less toxic antineoplastic agents with tremendous progress in cancer chemotherapy being made over the past 50 years, the management of malignant solid tumors still represents a formidable therapeutic challenge. Various reasons are thought to account for the failure of cancer chemotherapy in the treatment of solid tumors, including the development of drug resistance and inadequate tumor drug concentrations [3,4].

Microdialysis, described elsewhere in this book as a diffusion-based sampling method that allows analytes from the interstitial space [i.e., extracellular fluid (ECF)] to diffuse across the semipermeable hollow fiber membrane, was developed originally for research on endogenous cerebral neurotransmitters and has since become an important tool in the investigation of the penetration of exogenous compounds such as small molecules and drugs across tissue barriers, and in monitoring their disposition under controlled conditions within well-defined tissue compartments [5,6]. A vital advantage of the use of microdialysis is the ability to measure the concentrations of unbound extracellular molecules in one living system over a period of time [7]. Moreover, this sampling technique causes less tissue damage than do other in vivo perfusion techniques, such as push–pull perfusion, because of the lack of direct contact of the perfusion fluid with the surrounding tissue [8].

The first study reporting the application of microdialysis to characterize tumor distribution of an antineoplastic agent, SR 4233, was published in 1994 [9]. Ever since, microdialysis has become an increasingly popular experimental technique for the characterization of in vivo drug disposition in solid tumors. Using the in vivo microdialysis sampling technique, tumor concentrations of several commonly used cytotoxic anticancer drugs, including cisplatin [10], epirubicin [11], gemcitabine [12], methotrexate [13–15], temozolomide [16–21], and topotecan [22–24], as well as some novel chemotherapeutic agents [25,26], have been investigated in humans or animals. This chapter focuses on considerations in the use of microdialysis methodology for pharmacokinetic (PK) and pharmacodynamic (PD) studies of anticancer drugs, and highlights the recent application of this sampling technique to understanding drug disposition and response in solid tumors.

2. MICRODIALYSIS AS A SAMPLING TECHNIQUE IN ONCOLOGY

2.1. Determining Tumor Exposure to Anticancer Drugs

Plasma PK has been related successfully to pharmacological effects for a variety of medications in different therapeutic classes, thus often being used as a surrogate for drug disposition at the site of action [27–29]. However, the heterogeneity of tumors and their aberrant microenvironment often hamper the penetration and delivery of drugs from plasma into the tumor and distribution within the tumor, leading to complex relationships between concentrations in plasma, interstitium, and neoplastic cells, thus precluding an accurate prediction of tumor drug delivery from blood data [30,31]. Given the complexities of drug accumulation in tumors, it is likely that measurement of tumor drug concentrations will be of greater value than plasma drug concentrations as a potential indicator of drug action and tumor response.

To better characterize drug penetration into a tumor and to understand the contribution it makes to treatment outcomes, various sampling methods and monitoring techniques that can obtain tumor-specific drug concentration measurements have been pursued. Those methods have generally been categorized as being invasive, semiinvasive, or noninvasive. The invasive approach, one in common preclinical use, is the serial sacrifice study design, which involves harvesting tumor at specific time points across a time course following the administration of a drug. Subsequent to tumor collection, drug concentration measurements are obtained from ex vivo analyses of tumor tissue extracts or biopsies by well-established and validated methods. Although easy to do and applicable to virtually any chemotherapeutic agent, such studies are resource intensive and limit measurement of drug concentrations to one point for any particular animal. The results derived from homogenized biopsy or whole tumor samples represent an average concentration from both vascularized and necrotic tumor tissue and of all tissue components extracted, including blood, intracellular fluid, interstitial fluid, and structural tissue components, and may thus cause confusion with regard to the actual concentration of a chemotherapeutic agent in a defined compartment. These data thus may not be optimal in characterizing the time course of drug concentrations at the target site: namely, tumor cells.

Noninvasive monitoring of drug distribution in tumors using imaging techniques, including positron-emission tomography (PET) [32,33], nuclear magnetic resonance spectroscopy (MRS) [34,35], and quantitative autoradiography (QAR) [36,37], has significant potential for both preclinical and clinical PK studies. With these techniques, it has been possible to obtain repetitively and/or continuously detailed quantitative information about drug PK and PD in various types of solid tumors. However, the wide use of those techniques is limited by many factors. For example, PET uses radiolabeled drugs, which must be prepared separately for individual subjects by a cyclotron/radiochemistry facility that has to be in close proximity to the imaging laboratory. Also, PET does not provide information on drug metabolism because only total radioactivity in tissue is measured. Although MRS can be used to determine different molecular species present in the tissue, the sensitivity of MRS is much lower than that of PET (10^{-5} to 10^{-3} mol/L [38] vs. 10^{-12} mol/L).

Microdialysis is known to be a semi-invasive sampling technique. Compared to technically demanding and expensive methods such as imaging techniques, in vivo microdialysis can readily be employed for experimental and clinical investigations in almost any research laboratory at a reasonable price. Of importance, microdialysis provides selective access to the unbound and thus pharmacologically active drug fraction in the extracellular space of the living tissue, which is the location nearest cellular biochemical events. This provides a considerable advantage over other methods traditionally used for tissue concentration measurements. However, microdialysis does have some drawbacks in PK studies of anticancer drugs as well as in general. First, microdialysis cannot readily be used to study hydrophobic compounds or compounds that bind strongly to carrier proteins. Second, the technique is not feasible for studying larger proteins such as the monoclonal antibodies targeting tumor growth factor receptors, although attempts have been made to measure soluble growth factor in the ECF of tumor tissues using membranes that have a high-molecular-weight cutoff (MWCO) point.

2.2. Evaluating Tumor Response to Anticancer Drugs

Since a rapid and accurate assessment of the antitumor efficacy of new therapeutic drugs could speed up drug discovery and improve clinical decision making, a method that can predict tumor response early in the course of a chemotherapeutic regime might find considerable use in animal models assessing drug potency or in the treatment of human cancer. *Biomarkers*, defined as molecular, cellular, or functional measurable parameters indicative of a particular genetic, epigenetic, or functional status of a biological system, are being investigated increasingly for their ability to predict therapy response and aid in the development of individualized treatment regimens [39]. Historically, immunohistochemical approaches were the most widely used method to evaluate the intratumoral protein expression levels despite the limited quantitative applicability [40]. Tumor homogenates or tumor cells isolated from fresh tumor tissues by enzymatic digestion may also provide an opportunity for the quantitative determination of cytokine production by ELISA or flow cytometry. Recently, laser capture microdissection has been used to extract from tumor sections a small number of cells that can be used in gene expression or proteomic assays [41]. However, none of these techniques enables dynamic measurement of tumor response to therapy with good temporal resolution in the same subject. Although the use of microdialysis to evaluate tumor response to chemotherapy has not yet been fully established, commercially available microdialysis probes with 100-kDa MWCO membranes have been reported for in vivo microdialysis sampling of a few selected cytokines, chemokines, and growth factors, with molecular weights ranging from 10 to 80 kDa in the intratumoral interstitial fluid (Table 1).

It is noteworthy that microdialysis sampling is a continuous process that results in substance concentration versus time data that are a reflection of the mean concentration at the sampling site over a discrete time interval. Thus, it may not identify rapid changes in the intratumoral levels of the molecules of interest.

TABLE 1 In Vivo Microdialysis Sampling of Biomarkers in the Intratumoral Interstitial Fluid

Biomarkers[a]	Molecular Weight (kDa)	Preclinical Study	Clinical Study	Calibration Method	Refs.
VEGF	42	MCF-7 human breast tumor xenografts	Breast cancer	In vitro	[60,61,63,64]
Cath D	52	MCF-7 human breast tumor xenografts	—	In vitro	[85]
Cat S	24	—	Astrocytoma	—	[86]
MMP-2 and MMP-9	72 (MMP-2) 92 (MMP-9)	MCF-7 human breast tumor xenografts	—	—	[87,88]
Multiple cytokines, chemokines, and growth factors (e.g., IL-2, IP-10, MCP-1, TNF-α, FGF-β, VEGF)	10–42	Murine 3LL lung carcinoma xenografts	—	—	[89]

[a]VEGF, vascular endothelial growth factor; Cath D, cathepsin D; Cat S, cathepsin S; MMP, matrix metalloproteinase; IL-2, interleukin-2; IP-10, IFN-γ inducible protein; MCP-1, monocyte chemoattractant protein-1; TNF-α, tumor necrosis factor-α; FGF-β, fibroblast growth factor-β.

3. EXPERIMENTAL CONSIDERATIONS

3.1. Recovery

In Vivo Calibration   As anatomical and experimental considerations do not allow 100% recovery of compounds from the ECF, the dialyzing properties of a microdialysis probe should be defined as the relative recovery (RR) of a particular substance, which relates the freely diffusing analyte concentration in the dialysate to the actual analyte concentration in the tissue [42–44]. For microdialysis sampling to be quantitative it is necessary to determine the RR of the microdialysis probe for the compound of interest. Several approaches to calibrating microdialysis probes in vivo have been reported and used in PK studies of established or new chemotherapeutic agents, including the extrapolation to the zero-flow method, zero-net-flux method, and retrodialysis [43,44].

The most commonly used in vivo microdialysis calibration approach is retrodialysis. This approach is based on the assumption that the diffusion process is quantitatively identical in both directions across a semipermeable membrane [45]. The RR is therefore calculated from the ratio of the concentration lost to the initial concentration in the perfusate:

$$RR\,(\%) = \frac{C_{\text{perfusate}} - C_{\text{dialysate}}}{C_{\text{perfusate}}} \times 100\%$$

Retrodialysis is usually conducted prior to the administration of a study drug. Following the calibration period, a sufficiently long washout period is allowed prior to studying drug administration so that subsequent concentration measurements are not confounded by drug originating from the retrodialysis experiment. By correcting with the in vivo RR value, drug concentration versus time profiles can be generated which are amenable to standard PK analyses. An alternative to the retrodialysis calibration method would be to use a calibrator with membrane permeability properties similar to those of the study drug, which could then be used to equate relative loss of the calibrator to the recovery of the study drug [46–49]. This method has been applied in a recent study of liposomal and nonliposomal CKD-602, a camptothecin analog, in mice bearing A375 human melanoma xenografts. Camptothecin was used as the calibrator in a retrodialysis manner to calibrate the microdialysis probe RR for liposomal or nonliposomal CKD-602 [25]. Similarly, in several earlier studies on drug uptake in rat glioma models reported by the Gallo group, determination of the in vivo recoveries of methotrexate and temozolomide were based on the retrodialysis calibration method using lomotrexol [13] and metronidazole [16], respectively, as the calibrator. In addition, in another study on the brain distribution of the novel antifolate pemetrexed, the unbound concentrations of pemetrexed in the brain and blood were calculated after correction for each probe recovery measured with the retrodialysis calibrator raltitrexed [50].

In an earlier series of preclinical studies on the tumor distribution of temozolomide, a cytotoxic alkylating agent used for the treatment of malignant brain tumors, the unbound temozolomide concentrations in the ECF of both subcutaneous and intracerebral tumors were determined using the extrapolation to zero flow microdialysis calibration method, which was carried out under steady-state temozolomide plasma concentrations in conjunction with variable microdialysis flow rates [17,18]. Based on the principle that drug recovery across the dialysis membrane and the associated dialysate drug concentrations are inversely related to perfusate flow through the microdialysis system, and that when the flow rate is zero the perfusate and interstitial fluid are in a state of complete equilibrium, the following equation was then fit to the measured temozolomide dialysate concentrations to obtain an estimate of the actual interstitial fluid concentration:

$$C_{\text{dial}} = C_0 (1 - e^{-rA/F})$$

where F is the perfusion flow rate, C_{dial} the analyte concentration in the dialysate at flow rate F, C_0 the actual analyte concentration in the interstitial fluid, r the mass transport coefficient, and A the surface area of the dialysis membrane [51]. This equation assumes that the other two parameters, dialysis membrane permeability and surface area, are constant during the experiment. Although the extrapolation to zero flow method provides an option to determine true unbound drug concentrations in tissue ECF, the main drawbacks are that it is inherently time consuming, as steady-state conditions are required for each flow rate, and it does not allow for the determination of the drug concentration versus time profile under non-steady-state conditions. The zero-net-flux method is another comprehensive method used to estimate ECF concentrations at steady state by varying the analyte concentration in the perfusate and then determining mass transport of analyte across the microdialysis membrane as a function of perfusate concentration. This method possesses about the same advantages and disadvantages as those of the extrapolation to zero flow method.

Recovery of Lipophilic and/or Highly Protein-Bound Drug Entities   The feasibility of using microdialysis to study the distribution profile of a therapeutic agent in tissue depends on the physical–chemical properties of the compound. Compounds that are lipophilic (log $P \approx 1$ to 4) and/or highly protein-bound drugs are less readily to diffuse through the probe membrane, thus causing a very low and variable probe recovery [52–54]. Moreover, because the analyte gained into the dialysate may not equal that lost to the surrounding ECF during retrodialysis due to the nonspecific binding of lipophilic compounds to the tubing and microdialysis probe components under the standard experimental conditions, the aforementioned retrodialysis method for in vivo recovery estimation is not applicable [55]. The fact that a number of chemotherapeutic agents are lipophilic and highly protein bound limits the use of microdialysis for an important class of drugs.

Several solutions have been proposed in the literature to overcome this limitation and allow the application of microdialysis sampling to lipophilic compounds. One approach is to include certain substances, such as albumin [11,56], lipid emulsion [57], cyclodextrin [58], and polysorbate 80 [55], in the perfusion solution to prevent the nonspecific drug binding after saturation of the dialysis probe and tubing is achieved. In a recent study by Loos et al. [55], an attempt was made to use microdialysis to assess the in vivo tissue distribution of docetaxel, a commonly used anticancer drug that is lipophilic and highly plasma protein bound (unbound fraction <2%) using microdialysis. Like many other lipophilic compounds, docetaxel has a very low recovery rate as a result of nonspecific binding to microdialysis probes. The use of Ringer's solution containing 1.0% nonionic surfactant polysorbate 80 significantly reduced the nonspecific binding of docetaxel to the microdialysis tubing and membrane, thus improving the recovery of docetaxel. Although this approach is considered efficient in terms of eliminating nonspecific binding, the additional substance in the perfusate may introduce a new compartment that has a composition which differs from that if the interstitial fluid, resulting in modified osmotic pressure and permitting convection of the analyte into the microdialysate. The distribution of the analyte in the tissue directly surrounding the microdialysis probes will thus be disturbed, and the analyte concentrations in the dialysate may not represent true tissue concentrations.

Another approach is to use a mathematical model to estimate the degree of drug binding to the dialysis equipment. Lindberger et al. [59] derived a set of equations to distinguish binding to the microdialysis equipment so that the actual recovery can be obtained from the apparent recovery determined experimentally. The modeling approach is regarded as a convenient approximation of the "binding" process, and further evaluation is needed.

Overall, the use of microdialysis for lipophilic compounds remains problematic and challenging. Other alternatives to improving the in vivo recovery of lipophilic compounds should be explored, such as alternative perfusion media, new tubing, and membranes of different materials.

Recovery of Macromolecules   The distinct advantages of microdialysis, including reasonable time resolution, minimal tissue damage, and continuous sampling within a single subject, have rendered it an attractive sampling technique. Efforts have thus been made to extend its application to obtain site-specific biochemical information in complex biological systems. Conventional microdialysis sampling can pose difficulties for in vivo collection of macromolecules because the large molecular sizes and small aqueous diffusion coefficients of macromolecules can restrain diffusion through the probe, resulting in low microdialysis recoveries. However, the development of microdialysis probes with 100-kDa MWCO membranes has made it possible to use this technique in research on macromolecules in extracellular space, including certain peptides and proteins, which can serve as the biomarkers of cancer progression and response to treatment [60–62]. A concern when using microdialysis probe with a 100-kDa MWCO membrane is that the imbalance between the hydrostatic and/or osmotic pressures over the membrane could lead to a substantial loss of perfusate into the interstitial space. In other words, a membrane of large pore size is more susceptible to fluid loss due to osmotic flux. To counteract this flux, a neutral osmotic agent can be added into the perfusate as a way to counterbalance the transmembrane hydrostatic driving pressure. One osmotic agent often used to prevent the fluid loss is dextran. In a series of studies by the Dabrosin group [60,61,63,64], a microdialysis system composed of a microdialysis catheter with a 100-kDa MWCO membrane and a perfusate containing 154 mM NaCl and 40 g/L dextrin 70 has been used successfully to assess the interstitial levels of VEGF, cathepsin D, and MMP-2 in breast tumors when coupled with enzyme-linked immunosorbent assay (ELISA). The mean in vitro recovery value at room temperature was 6% for VEGF at a flow rate of 0.6 µL/min and 6% for MMP-2 at a flow rate of 1 µL/min. As most endogenous macromolecules were usually studied by relating changes in their intratumoral levels to their baseline values or levels in the control tissue, determination of the probe recovery in vivo would not be a critical issue. A similar approach was used in a preclinical study to compare the extracellular levels of multiple (ca. 30) soluble cytokines, chemokines, and growth factors with a molecular weight of less than 100 kDa in the tumor mass to those in the normal subcutaneous tissue. In a clinical study by Flannery et al. [86], with the addition of dextran 60 to the perfusate (30 g of dextran to 60 L of Ringer's solution), the amount of capthesin S present in the microdialysate obtained from human brain tumors in vivo was detectable by ELISA in five of 10 cases.

Bovine serum albumin (BSA) can serve as an osmotic agent, and as demonstrated by Trickler and Miller [65], fluid loss due to convective flow out of the probe could be abolished completely by adjusting the BSA content of phosphate-buffered saline (PBS) as a function of the flow rate of the perfusate.

In addition to fluid loss from the microdialysis probe, nonspecific absorption of peptides or proteins onto the microdialysis apparatus can cause low analyte recovery during microdialysis sampling. To overcome this considerable low-recovery problem, various methods for enhancing mass transport across microdialysis probes and preventing nonspecific binding to the microdialysis membrane and polymeric tubing have been reported, including the addition of proteins (e.g., albumin), lipids, polymeric microspheres, antibody-immobilized microspheres, and different cyclodextrins to the perfusion media [66,67].

3.2. Placement of a Microdialysis Probe in a Tumor

Tumor mass is a heterogeneous three-dimensional composite of fibrous and connective tissues, stromal components, vasculature, and multiple clones of cancer cells. On top of that, tumor vasculature is spatially heterogeneous, with an avascular and seminecrotic region in the center, a relatively stable microcirculation at the periphery, and an advancing vascular front at the leading edge of the tumor [30,68]. Moreover, as the interstitial fluid pressure increases toward the deeper core layers of the tumor by a large gradient, the perfusion or flow in the tumor core is reduced considerably compared with the tumor rim [69]. Overall, these characteristics of the tumor microenvironment hinder the adequate and uniform distribution of drug molecules throughout tumor tissue, leading to significant differences in drug exposure within a tumor. The intratumoral differences in drug levels were demonstrated by Ekstrom et al. [70] in a study on the distribution of methotrexate in human osteosarcoma rat xenografts. In this study, two microdialysis probes were placed in the center and periphery of a tumor. It was shown that during the initial continuous intravenous infusion, the dialysate concentrations of methotrexate were higher at the periphery than those at the tumor center. However, upon termination of the methotrexate infusion, the situation reversed, with lower peripheral methotrexate concentrations than those at the center. The early high drug concentrations in the periphery of the tumor are probably a result of greater vascularity, enabling more rapid diffusion into the peripheral region than into the less well vascularized central regions, suggesting that the intratumoral differences in drug concentration–time profiles depend on the extent of angiogenesis and necrosis throughout the tumor. Similar results were observed by Zamboni et al. [10] in a study in which a dual-probe approach was used to evaluate the variability of topotecan disposition in the ECF of a single tumor. It was observed that the maximal difference in the topotecan dialysate concentrations between the two probes, which were placed parallel near the tumor surface, was about 2.4-fold, irrespective of the resistant or sensitive human neuroblastoma xenograft line used. However, in a study of the effect of an angiogenic inhibitor TNP-470 on the tumor uptake of temozolomide by Ma et al. [17], analysis of the TMZ tumor/plasma concentration ratios indicated that there were no significant differences between the central and peripheral regions for analogous treatments (i.e., comparing the vehicle control or TNP-470 groups) in either highly vascularized (i.e., VEGF-overexpressing or V+) or moderately vascularized (i.e., VEGF-deficient or V−) tumors (Figure 1). This may be attributed to a more uniform vasculature throughout the tumor and the lack of central necrotic areas because of the relatively small tumor sizes. Comparison of TMZ tumor/plasma concentration ratios revealed almost two-fold higher values in the V+ groups compared with the analogous V− groups, suggesting that the intratumoral drug levels are associated with the vascularity and membrane permeability in the tumors. Although it is possible to place two microdialysis probes in a single tumor, permitting collection of dialysate samples from the central and peripheral region, it could be technically burdensome and irreproducible. For subcutaneous tumors, microdialysis probes can be placed either in the tumor periphery, the area of active neovascularization, or the middle of the tumor, passing through both the peripheral and central regions using a probe with a membrane length of 10 mm or longer. Whatever the region selected, the location of a microdialysis probes in the tumor should be consistent among animals. In intracerebral tumors, however,

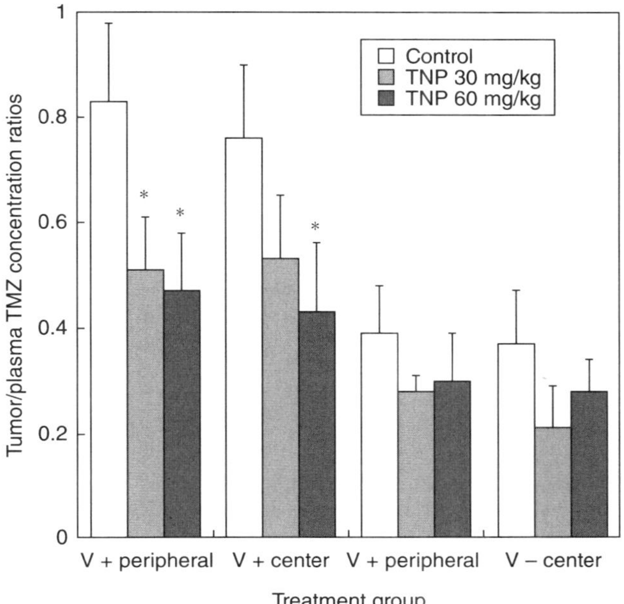

Figure 1 Steady-state TMZ tumor/plasma concentration ratios. Mean (±S.D.) ratios were obtained in vehicle control-, TNP-470 30 mg/kg-, and TNP-470 60 mg/kg-treated nude rats bearing subcutaneous human tumors obtained from either parental SF188 glioma cells (V–) or VEGF-overexpressing SF188 glioma cells (V+). In each animal, two microdialysis probes were inserted into the tumor with one in the central region and the other in the peripheral region. (From [17].)

stereotaxic implantation of the microdialysis guide cannulas at the time of tumor cell implantation does not permit selection of a microdialysis sampling site. This concern is critical in microdialysis experiments using mouse intracerebral tumor models, as the intracerebral tumor mass in mice is smaller than that in rats, and the physical constraints of the probe and tumor mass may prevent the probe from being encompassed by the tumor mass, causing sample collections outside the interstitial fluid of the brain tumor. Although the exact location of the dialysis probe can be verified by the standard postmortem histological staining following the completion of the experiment, verification of the implantation site of the dialysis probe in a tumor prior to the microdialysis experiment is difficult, if not impossible, thus imposing significant constraints on the use of this sampling technique to study drug penetration into the brain tumor in animal models.

3.3. Invasiveness of Microdialysis Probe Implantation

Although microdialysis preserves the integrity of the subject being examined, it is considered a semi-invasive technique. The small size of the probe does not preclude the potential of this technique to cause adverse tissue reactions, such as inflammation and hemorrhage, which may influence the measurements and interpretation of the data. The invasiveness of peripheral microdialysis has been examined in a variety of tissues, including tumor [9,71,72]. In an early study on in vivo dermal microdialysis sampling in rats [71], the extent of dermal tissue damage due to probe implantation was evaluated by histological examination and microdialysis delivery studies. The result showed that there was no immediate evidence of edema or tissue disruption in the skin, although lymphocyte infiltration started 6 h after the implantation procedure. The infiltration of lymphocytes was, however, considered not to affect the performance of the probe. In a study by Palsmeier and Lunte [9], histological examination of tumor tissue microdialyzed for 72 h has shown that the tumor tissue remains intact, with little or no inflammatory reaction adjacent to the microdialysis.

For intracerebral microdialysis, trauma induced by probe implantation has been characterized by both short- and long-term histological, physiological, and biochemical changes in surrounding neural tissue [73–76]. Generally, the BBB transcapillary transport function is inevitably disturbed in response to insertion of guide cannulas and/or microdialysis probes. The BBB dysfunction varies with time after insertion and occurs over a wide molecular range of solutes. However, the degree to which the trauma might compromise an interpretation of microdialysis measurements is generally unclear. From a pharmacokinetic viewpoint, the perturbation in normal blood–brain barrier (BBB) permeability by the implantation of guide cannulas and/or microdialysis probes may lead to overestimation of the rate of drug transfer into and out of the brain [73,75]. One way to attenuate the damage may be to leave a few days for experimental animals to reestablish the barrier properties before microdialysis is begun. Even in the event of tissue damage that could alter analyte recovery, a comparative study design should negate such effects, as each treatment group would be affected similarly.

4. EXAMPLES OF THE USE OF MICRODIALYSIS TO CHARACTERIZE DRUG DISPOSITION IN TUMOR

4.1. Preclinical Studies

There is a wealth of literature reporting on the application of microdialysis in preclinical studies to characterize tumor distribution of both conventional and novel antineoplastic agents, to evaluate novel formulations, and to scrutinize potential drug–drug interactions at the site of action (Table 2). As drug penetration into the interstitial space of solid tumors represents a rate-limiting step in tumor response to chemotherapy, determination of drug concentrations in the tumor interstitial fluid provides a key insight into drug disposition in tumors, extending an understanding not only of drug transport but also of pertinent dose–response relationships that could aid the drug development process. In

TABLE 2   Use of Microdialysis in Preclinical PK Studies of Anticancer Drugs

Anticancer Drug	Sampling Site and Reference	In Vivo Calibration Method
	Tumor ECF	
5-Fluororacil	DMBA-induced rat mammary tumors [90]	Retrodialysis
Carboplatin	Subcutaneous breast tumor xenografts [91]	Retrodialysis
Cisplatin	Subcutaneous B16 murine melanoma, H23 human NSCLC, and breast tumor xenografts [10,91]	Retrodialysis
Gemcitabine	Intracerebral C6 rat glioma xenografts [12]	Retrodialysis
Irinotecan	Subcutaneous NB1691 human neuroblastoma xenografts [92]	Retrodialysis
Methotrexate	Intracerebral B16 murine glioma [13], C6, and CNS1 rat glioma xenografts [14]	Retrodialysis
Temozolomide	Subcutaneous and intracerebral SF188V + human glioma xenografts [16–20]	Retrodialysis or zero-flow calibration
Topotecan	Subcutaneous NB1643 and NB1691 human neuroblastoma xenografts [22]	Retrodialysis
	Tissue ECF	
Pemetrexed	Brain [50]	Retrodialysis
Tirapazamine	Blood and muscle [93]	Retrodialysis
Topotecan	Brain and cerebrospinal fluid [24]	Retrodialysis

a study by Dukic et al. [14], microdialysis was used to compare the influence of two different brain tumors (C6 and CNS1 glioma) on local distribution in normal brain and brain tumors of methotrexate, an antimetabolite that acts as a folic acid antagonist to interfere with cell reproduction and is used widely in a variety of human cancers, including primary central nervous system lymphoma. In this study, two microdialysis probes were placed so that microdialysates from ECF in the normal brain and tumor tissue could be collected simultaneously. By using this dual-probe approach, tissue disposition of the drug in both tumor and normal regions can be examined in the same brain and comparative drug disposition data obtained from the same animal. The result of this study showed that concentration–time profiles of methotrexate in tumor tissue were different between the two tumor models. Although initially higher during the first 30 min than those measured in the C6 tumor tissue, ECF concentrations of methotrexate in CNS1 tumor tissue declined rapidly and became undetectable after the third hour. This difference could be related, at least in part, to the different blood vessel density found in the two tumor models. Despite the difference in elimination rates, methotrexate tumor penetration, expressed as the ratio $AUC_{tumorECF}/AUC_{plasma}$, was, however, in the same range in both tumor models. Moreover, the unbound methotrexate concentrations in the normal brain of CNS1 tumor-bearing rats were found to be two-fold lower than those of the C6 tumor-bearing rats. As a result, the uptake of unbound methotrexate in the CNS1 glioma was significantly higher than that in the normal brain, whereas there was no significant difference between the methotrexate distribution in brain tumor and in normal brain. Although it was not determined, it would be assumed for accurate comparison of the data from these two sites that implantation trauma would be equivalent in normal brain and brain tumor. The apparently more profound effect of C6-glioma on the integrity of the BBB compared to that of the CNS1 tumor was thought to be attributable to the high production of $VGEF_{121}$ in C6 tumor cells.

As the unbound drug concentration in the plasma is the driving force for unbound drug to distribute into the tissue, evaluation of tumor exposure in relation to systemic exposure would provide an unambiguous measure of the extent of drug distribution in the tumor. An elegant microdialysis study has been performed using simultaneous blood and brain microdialysis sampling to assess the brain distribution of pemetrexed, a novel antifolate compound structurally similar to methotrexate, in the absence and presence of indo-methacin, a known inhibitor of several active organic anion efflux transporters that exist in the BBB [50]. The results from this study indicate that pemetrexed has a limited central nervous septum distribution, which is indicated by the brain-to-plasma AUC ratio of pemetrexed and the brain-to-plasma ratio of steady-state concentrations. Additionally, it was noted that the brain level of pemetrexed rose to a maximum rapidly in the intravenous (i.v.) bolus study and also achieved a rapid steady state in the infusion study even though the clearance into the brain is low (about $2.9\,\mu L/min\cdot kg$) (Figure 2). This may be due to the small volume of distribution of pemetrexed in the brain, allowing a rapid achievement of distributional equilibrium in this tissue. Moreover, this kinetic behavior could be a result of an efficient efflux clearance by active efflux transporters at the BBB. Barring the technical difficulties of performing microdialysis at multiple sites in the body, the strategy is appealing.

Temozolomide is another anticancer drug suitable for performing microdi-alysis experiments because of its low molecular weight of 194.15 Da and low degree of protein binding (12 to 16% in human plasma, 20% in rat plasma). A series of preclinical studies conducted by the Gallo group have reported the use of microdialysis in characterizing temozolomide disposition in the brain, including cerebrospinal fluid (CSF), normal brain tissue, and brain tumor tissue [19], and evaluating the potential drug interactions between temozolomide and various angiogenic inhibitors, including TNP-470 [16,17], SU5416 [17] and sunitinib [21], at the target site. Based on the temozolomide tumor concentration–time profiles obtained in the preclinical tumor models,

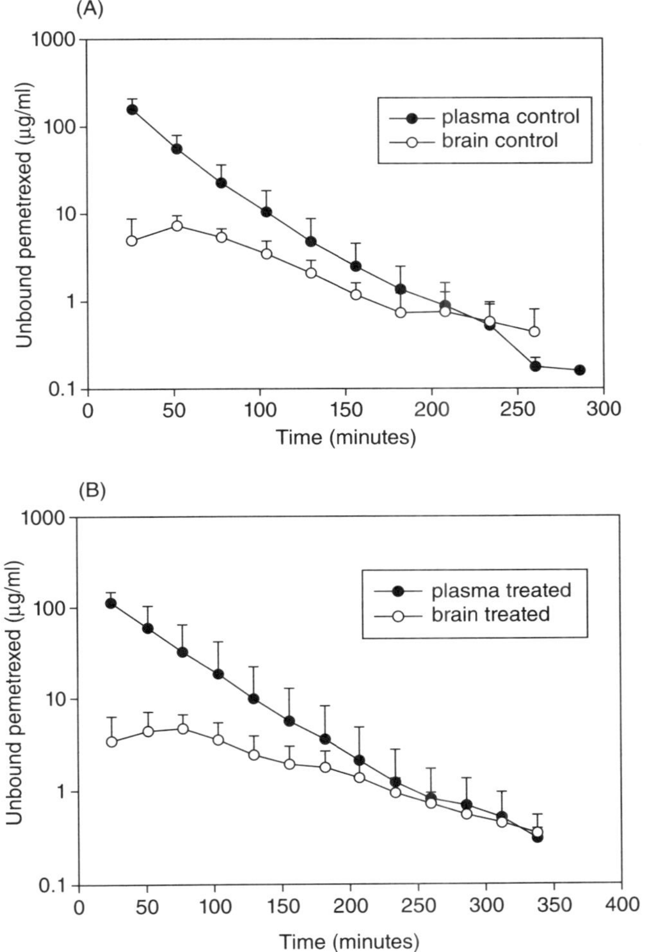

Figure 2 Plasma and brain concentration–time profile of unbound pemetrexed in rats. Pemetrexed (60 mg/kg) was administered by intravenous bolus without (control, A) or with (treated, B) the coadministration of indomethacin. Unbound pemetrexed concentrations in both the plasma and brain ECF were measured by simultaneous microdialysis. (From [50].)

physiologically based hybrid pharmacokinetic models could be developed and further scaled to predict human tumor drug concentrations [19].

One of the critical factors that contribute to the failure of a tumor to respond to chemotherapy is the presence of ATP-binding cassette (ABC) transporter proteins. In particular, the functional expression of these proteins at the brain barriers substantially restricts the capacity of antineoplastic agents to penetrate the brain in sufficient quantities to reach their therapeutic target and exert an intended effect. The traditional method of measuring drug concentrations in brain tissue homogenates poses limitations, as it reflects the combination of vascular, interstitial fluid, and intracellular compartments in the entire brain. To clarify the role of these transporters in the brain distribution of chemotherapeutic agents, it would be useful to distinguish among drug concentrations in the blood, in the CSF, and in the brain parenchyma. Topotecan, a topoisomerase I inhibitor with modest clinical efficacy demonstrated in patients with glioblastoma multiforme (GBM), is known to have high affinity for breast cancer resistance protein (BCRP) and moderate affinity for P-glycoprotein (P-gp), which are two important efflux drug transporters at the BBB. In a study by Zhuang et al. [24], modulation of BCRP and P-gp at the BBB and blood–cerebrospinal fluid barrier (BCSFB) was proposed to alter topotecan central neurons sqstom penetration. By using a microdialysis technique coupled with an online microbore high-performance liquid chromatography (HPLC) system, topotecan lactone concentrations were measured in either brain ECF or ventricular CSF in animals pretreated with gefitinib, an epidermal growth factor receptor (EGFR) tyrosine kinase inhibitor with demonstrated inhibitory effect on BCRP and P-gp at clinically relevant concentrations. The results from this study revealed that topotecan brain ECF penetration was lower than ventricular CSF penetration, whereas gefitinib treatment increased topotecan brain ECF penetration but decreased the ventricular CSF penetration, suggesting that the expression of Bcrp1 and P-glycoprotein at the apical side of the choroid plexus facilitates an influx transport mechanism across the BCSFB, resulting in high topotecan CSF penetration (Figure 3).

Another study was undertaken using the microdialysis technique to explore the role of MRP4 in conferring resistance to topotecan and protecting the brain from chemotherapy [23]. In that study, the topotecan disposition in ventricular CSF was characterized in both MRP4 knockout ($Mrp4^{-/-}$) and wild-type ($Mrp4^{+/+}$) mice. The microdialysis experiment was performed by stereotactically inserting guide cannulas into the lateral ventricles of mice and CSF microdialysate samples obtained at 15-min intervals over a 3-h period after injection of 2 mg/kg of topotecan via the tail vein. The results showed that mean CSF topotecan concentration in $Mrp4^{-/-}$ mice was almost 10-fold higher than that in $Mrp4^{+/+}$ mice at each time point, suggesting that Mrp4 in the choroid plexus and brain capillaries limits the transport of topotecan from the blood into the CSF.

4.2. Clinical Studies

The efficacy of cancer chemotherapy ultimately relies on the sensitivity of tumor cells to the action of the drug and the maintenance of adequate drug concentrations in the tumor. Although it is generally acknowledged that intratumoral drug concentrations are more predictive of clinical response than are plasma concentrations, conventional PK trials in oncology are usually restricted to drug concentration measurements from plasma and biological specimens

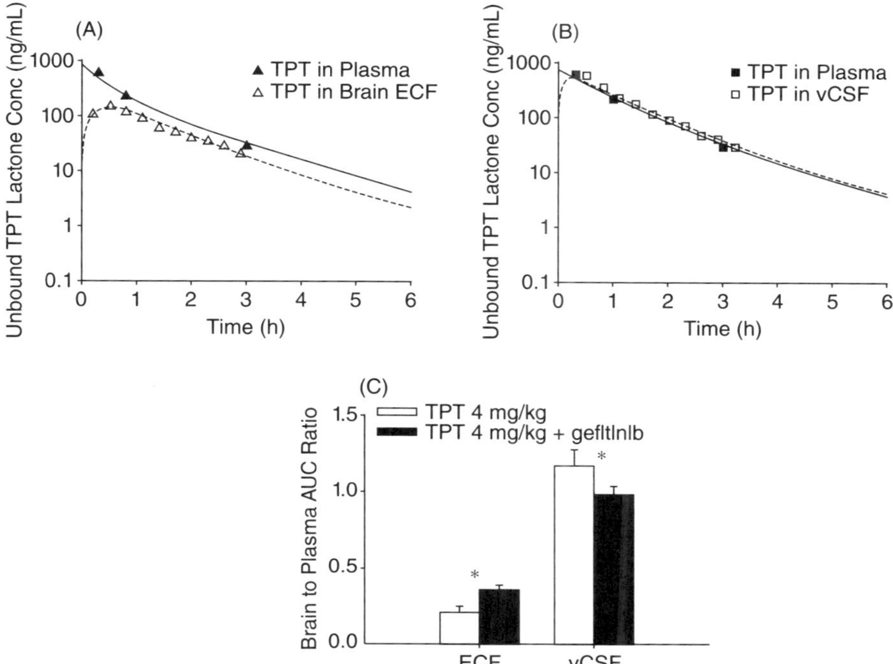

Figure 3 Representative unbound topotecan lactone concentration–time profiles in plasma, brain parenchymal ECF, and ventricular CSF in gefitinib-pretreated mice. Unbound topotecan lactone in plasma (▲, ■) and brain parenchymal ECF (A; Δ) or ventricular CSF (B; □) after 4 mg/kg topotecan i.v. injection. (C) Brain/plasma AUC ratio of unbound topotecan lactone after 4 mg/kg topotecan i.v. injection with and without gefitinib. *$p < 0.05$ (Student's t test). (From [24].)

that are relatively easier to obtain, such as biopsies and CSF, or to indirect modeling of tumor concentrations from plasma concentration–time profiles. In recent years, the development of new techniques and approaches has enabled continuous monitoring of drug distribution over an extended period of time in live tissues, including noninvasive imaging techniques and in vivo microdialysis. Results from studies using these techniques have emphasized the superiority of tumor concentrations to plasma concentrations as a potential predictor of clinical outcomes.

The first clinical oncological microdialysis study was performed to assess intratumoral distribution of the alkylating agent carboplatin in patients with metastatic malignant melanoma [77]. To date, clinical microdialysis has been employed for the characterization of intratumoral drug concentrations of 5-fluororacil [78], cisplatin [79], methotrexate [15], dacarbazine [80], capecitabine [81], and melphalan [82] and in various types of cancer, such as breast cancer, malignant melanoma, osteosacoma, Merkel cell tumor, and oral cancer.

One of the major advantages of microdialysis over noninvasive techniques such as PET and QAR is that microdialysis enables the characterization of both the parent drug and metabolites. One example is the study of capecitabine. Capecitabine is an orally available fluoropyrimidine carbonate that is preferentially metabolized to 5-fluororacil in tumors and liver via a three-step enzymatic process involving conversion into 5′-deoxy-S-fluorocytidine (5′-DFCR) by carboxylesterase, and then to 5′-deoxy-S-fluorouridine (5′-DFUR) by cytidine deaminase, and finally to 5-fluororacil by thymidine phosphorylase. The results from a PK study of capecitabine and its metabolites in blood, malignant tissue, and healthy tissue interstitium in breast cancer patients using microdialysis revealed that capecitabine and its metabolites 5′-DFCR and 5′-DFUR distribute extensively into the interstitium of both malignant and healthy tissues. Although direct assessment of the intracellular drug metabolism processes cannot be achieved by microdialysis, the efficient penetration of 5′-DFUR into the interstitium of malignant tissue suggested that the essential step of entry into cells followed by conversion to 5-fluororacil would be achieved [81].

Although it has been demonstrated that microdialysis can provide useful information for a better understanding of tumor PK and PD through direct and continuous sampling in the excellular fluid surrounding tumor cells, human studies involving microdialysis have been performed primarily in sites that are easily accessible due to the semi-invasive nature of this technique and hurdles to IRB approval. Nonetheless, the use of intracranial microdialysis as a tool to monitor intratumoral concentrations of anticancer drugs for the treatment of malignant brain tumors, including *p*-boronophenylalanine [83] and methotrexate [15], in glioblastoma patients has been reported. By the use of microdialysis, Bergenheim and co-workers [83] were able to assess concentrations of BPA continuously in the extracellular compartment of brain tumor tissue, brain adjacent to tumor, normal brain, and subcutaneous tissue in patients undergoing boron neutron capture therapy for glioblastoma multiforme, demonstrating variations in the temporal PK in different compartments in the very same patient. In a study by Blakeley et al. [15], the integrity of the BBB in the region of the tumor in which the microdialysis probe resided was evaluated by digitally fusing CT and contrast-enhanced MRI scans. Data from this study showed that the brain penetration of methotrexate, as indicated by the ratio of the area under the methotrexate concentration–time curves in tumor ECF and plasma, was considerably greater in contrast-enhancing tumor than in nonenhancing tissue, suggesting that the regional difference in BBB permeability in tumor is a crucial factor that must be taken into account in the design of microdialysis studies to assess drug penetration in brain tumors.

4.3. Pharmacokinetic Modeling Based on Microdialysis Data

Cancer chemotherapy is associated with a high failure rate and severe dose-limiting toxicities. Precise knowledge of the PK and PD of an anticancer drug can help select the optimal dosage and the best route of administration of the drug. However, the standard PK models employed are based on the plasma concentration–time profile data with limited capability to adequately address drug disposition in the tumor. Heterogeneous drug distribution in the tumor further diminishes the hope that the kinetics of a drug in plasma will parallel that in tumor. Therefore, a PK model including a tumor compartment that is relatively closely associated with the pharmacological effect may provide enhanced insight not only into drug accumulation in target tissues but also into pertinent PK–PD relationships, thus offering a quantitative basis to design and adjust therapies.

The microdialysis sampling technique allows determination of the free drug concentrations in tumor ECF as a function of time, thus providing detailed information for tumor drug disposition. The availability of both plasma and tumor drug concentrations enables compartmental modeling approaches to be applied to the data. Modeling approaches might vary, for example, between classical or physiologically based methods, yet models that can be extrapolated to patients, where tumor drug concentrations may be limited or unavailable, seem most pertinent.

Zamboni et al. [10] proposed a two-compartment model, which was fit to the plasma and tumor ECF concentration–time profiles of unbound platinum after cisplatin administration to tumor-bearing mice. In this model, the plasma concentration versus time profile and tumor ECF concentration versus time profiles were modeled separately. As the authors stated, it was impossible to obtain accurate estimates of the rate constant describing the movement of drug into the tumor ECF or the volume of the ECF by modeling the plasma and tumor distribution simultaneously due to the 100- to 1000-fold difference between the rate constants describing systemic and tumor disposition of unbound platinum. The difficulties encountered in this study seem unique. A physiologically based hybrid PK (PBPK) model developed by Gallo et al. [94] to characterize temozolomide disposition in subcutaneous tumor xenografts combined a classic compartment model to describe plasma drug concentrations with a physiological mass balance approach to describe drug disposition in tissues. The hybrid technique eliminates the burdensome task of determining whole-body drug distribution by focusing on selected tissues of particular interest, yet retains the physiological features of the organ. The great potential of this hybrid model approach lies in its ability to predict drug disposition in human tumors based on the preclinical (tumor disposition) and clinical (forcing function) data. The hybrid model derived from the preclinical study of temozolomide disposition in tumors was able to include a three-compartment tumor model that depicted vascular, interstitial fluid, and intracellular subcompartments due to the availability of tumor ECF drug concentrations. This structure introduces tumor blood flow and physiological volumes into the model, which provides not only a means to assess how these parameters affect drug disposition but that can also be replaced readily with human data when the hybrid model is applied to humans.

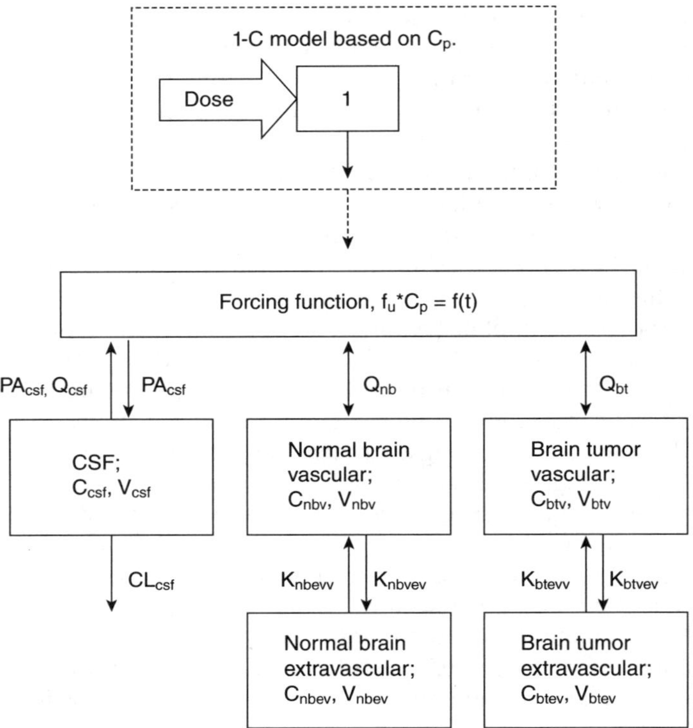

Figure 4 Schematic presentation of the hybrid PBPK model of temozolomide in brain. (From [19].)

This model was subsequently adapted and expanded to characterize the temozolomide brain disposition in an orthotopic brain tumor model (Figure 4) [19]. The design of this study was unique in the sense that it consisted of two groups of animals, with one group having microdialysis probes inserted in the lateral ventricle and brain tumor, and the other in the normal brain and brain tumor. A sequential approach was used to develop PBPK models of temozolomide in individual animals in both groups. In the first step, a compartmental model was fit to the plasma concentration–time data observed in each animal using maximum likelihood optimization. The resulting best-fit models were then cast as forcing functions that described the plasma concentration input into the microdialysis-sampled brain regions. Each brain region was defined by one or more compartments and fit to individual temozolomide concentration–time profiles. In the last step, the best-fit models for both groups were combined to form a single comprehensive brain model using a population approach. The PBPK model developed not only afforded a mechanistic and accurate prediction of temozolomide brain disposition in rats, but also enabled the prediction of temozolomide CSF concentrations in patients that was verified by analogous observed and model-predicted CSF/plasma AUC ratios of 0.2. Moreover, through a series of Monte Carlo model simulations it was shown that temozolomide accumulation in brain tumor was determined by BBB permeability and fractional tumor blood volume, but was minimally affected by clinical dosing regimens.

5. USE OF MICRODIALYSIS IN THE EVALUATION OF TUMOR RESPONSE TO THERAPY

An increasing number of studies have demonstrated the feasibility of microdialysis to assess chemotherapeutic responses or toxicities by monitoring changes in the concentration of one or more endogenous compounds in the blood or target tissue. In a clinical study by Castejon et al. [84], microdialysis was used to determine the levels of free 5-hydroxytryptamine (5-HT) and 5-hydroxyindoleacetic acid (5-HIAA) in blood after administration of cisplatin in cancer patients, as 5-HT, a monoamine neurotransmitter, is known to be involved in the production of emesis associated with cisplatin treatment. Although microdialysis is at its best for small molecules, the availability of microdialysis probes with higher-molecular-weight membrane cutoff has made it possible to sample biologically relevant macromolecules. Garvin and Dabrosin [63] examined the effect of tamoxifen on the secretion of VEGF in a mouse model of human MCF-7 breast cancer using microdialysis to sample $VEGF_{121}$ in the tumor interstitium, where VEGF is biologically active. The same approach has also been used successfully in a clinical study to explore the mechanism of breast cancer development by examining the association between steroidal sex hormones (i.e., estradiol and progesterone) and proangiogenic factors (i.e., VEGF and fibroblast growth factor-2) in normal breast tissue [61]. Given the fact that the extracellular space is the bioactive site for the many cytokines, chemokines, and growth factors associated with tumor growth, using microdialysis to sample those biomarkers at extracellular spaces opens up the possibility of monitoring tumor response to cancer chemotherapy. Moreover, if the time course of such responses could be quantified by the use of microdialysis and integrated with the PK profiles of the corresponding drug, it would help to define the PK–PD relationship, which is essential for the rational design of drug administration regimens in cancer patients.

6. CONCLUSIONS AND FUTURE PERSPECTIVES

Currently, most experience with the microdialysis sampling technique for oncological studies has been gained from the determination of tumor ECF drug concentration–time profiles. Limited attempts have also been made to use microdialysis to evaluate tumor response to chemotherapy. In this chapter we have highlighted the inherent strengths and shortcomings of microdialysis as well as the integrated use of microdialysis in both experimental and clinical studies of cancer chemotherapeutic agents. Overall, the investigation of drug disposition in tumors by microdialysis may help define not only drug transport but also pertinent PK–PD relationships that could aid in the selection of potential drug candidates and the design of optimal dosing regimens. Hopefully, with increasing acceptance of microdialysis as a well-validated sampling technique and continuous refinement of the microdialysis system to minimize the nonspecific binding of analytes, the microdialysis technique will be widely incorporated into the development of anticancer drugs.

Acknowledgment

Partial support was provided by National Institutes of Health grant CA72937.

REFERENCES

[1] Stewart, B.W., Kleihues, P. (2003). *World Cancer Report. World Health Organization*, Aencua, Switgudend.

[2] Jain, R.K. (2005). Normalization of tumor vasculature: an emerging concept in antiangiogenic therapy. *Science, 307*, 58–62.

[3] Mellor, H.R., Callaghan, R. (2008). Resistance to chemotherapy in cancer: a complex and integrated cellular response. *Pharmacology, 81*, 275–300.

[4] Redmond, K.M., Wilson, T.R., Johnston, P.G., Longley, D.B. (2008). Resistance mechanisms to cancer chemotherapy. *Frontiers in Bioscience, 13*, 5138–5154.

[5] Joukhadar, C., Müller, M. (2005). Microdialysis: current applications in clinical pharmacokinetic studies and its potential role in the future. *Clinical Pharmacokinetics, 44*, 895–913.

[6] Zhou, Q., Gallo, J.M. (2005). In vivo microdialysis for PK and PD studies of anticancer drugs. *AAPS Journal, 7*, E659–E667.

[7] Kitzen, J.J., Verweij, J., Wiemer, E.A., Loos, W.J. (2006). The relevance of microdialysis for clinical oncology. *Current Clinical Pharmacology, 1*, 255–263.

[8] Lisi, T.L., Westlund, K.N., Sluka, K.A. (2003). Comparison of microdialysis and push–pull perfusion for retrieval of serotonin and norepinephrine in the spinal cord dorsal horn. *Journal of Neuroscience Methods, 126*, 187–194.

[9] Palsmeier, R.K., Lunte, C.E. (1994). Microdialysis sampling in tumor and muscle: study of the disposition of 3-amino-1,2,4-benzotriazine-1,4-di-*N*-oxide (SR 4233). *Life Sciences, 55*, 815–825.

[10] Zamboni, W.C., Gervais, A.C., Egorin, M.J., Schellens, J.H., Hamburger, D.R., Delauter, B.J., Grim, A., Zuhowski, E.G., Joseph, E., Pluim, D., Potter, D.M., Eiseman, J.L. (2002). Inter- and intratumoral disposition of platinum in solid tumors after administration of cisplatin. *Clinical Cancer Research, 8*, 2992–2999.

[11] Hunz, M., Jetter, A., Warm, M., Pantke, E., Tuscher, M., Hempel, G., Jaehde, U., Untch, M., Kurbacher, C., Fuhr, U. (2007). Plasma and tissue pharmacokinetics of

epirubicin and paclitaxel in patients receiving neoadjuvant chemotherapy for locally advanced primary breast cancer. *Clinical Pharmacology & Therapeutics*, *81*, 659–668.

[12] Apparaju, S.K., Gudelsky, G.A., Desai, P.B. (2008). Pharmacokinetics of gemcitabine in tumor and non-tumor extracellular fluid of brain: an in vivo assessment in rats employing intracerebral microdialysis. *Cancer Chemotherapy and Pharmacology*, *61*, 223–229.

[13] Devineni, D., Klein-Szanto, A., Gallo, J.M. (1996). In vivo microdialysis to characterize drug transport in brain tumors: analysis of methotrexate uptake in rat glioma-2 (RG-2)-bearing rats. *Cancer Chemotherapy and Pharmacology*, *38*, 499–507.

[14] Dukic, S., Heurtaux, T., Kaltenbach, M.L., Hoizey, G., Lallemand, A., Gourdier, B., Vistelle, R. (1999). Pharmacokinetics of methotrexate in the extracellular fluid of brain C6-glioma after intravenous infusion in rats. *Pharmaceutical Research*, *16*, 1219–1225.

[15] Blakeley, J.O., Olson, J., Grossman, S.A., He, X., Weingart, J., Supko, J.G. (2009). Effect of blood brain barrier permeability in recurrent high grade gliomas on the intratumoral pharmacokinetics of methotrexate: a microdialysis study. *Journal of Neuro-Oncology*, *91*, 51–58.

[16] Devineni, D., Klein-Szanto, A., Gallo, J.M. (1996). Uptake of temozolomide in a rat glioma model in the presence and absence of the angiogenesis inhibitor TNP-470. *Cancer Research*, *56*, 1983–1987.

[17] Ma, J., Pulfer, S., Li, S., Chu, J., Reed, K., Gallo, J.M. (2001). Pharmacodynamic-mediated reduction of temozolomide tumor concentrations by the angiogenesis inhibitor TNP-470. *Cancer Research*, *61*, 5491–5498.

[18] Ma, J., Li, S., Reed, K., Guo, P., Gallo, J.M. (2003). Pharmacodynamic-mediated effects of the angiogenesis inhibitor SU5416 on the tumor disposition of temozolomide in subcutaneous and intracerebral glioma xenograft models. *Journal of Pharmacology and Experimental Therapeutics*, *305*, 833–839.

[19] Zhou, Q., Guo, P., Kruh, G.D., Vicini, P., Wang, X., Gallo, J.M. (2007). Predicting human tumor drug concentrations from a preclinical pharmacokinetic model of temozolomide brain disposition. *Clinical Cancer Research*, *13*, 4271–4279.

[20] Zhou, Q., Guo, P., Wang, X., Nuthalapati, S., Gallo, J.M. (2007). Preclinical pharmacokinetic and pharmacodynamic evaluation of metronomic and conventional temozolomide dosing regimens. *Journal of Pharmacology and Experimental Therapeutics*, *321*, 265–275.

[21] Zhou, Q., Guo, P., Gallo, J.M. (2008). Impact of angiogenesis inhibition by sunitinib on tumor distribution of temozolomide. *Clinical Cancer Research*, *14*, 1540–1549.

[22] Zamboni, W.C., Houghton, P.J., Hulstein, J.L., Kirstein, M., Walsh, J., Cheshire, P.J., Hanna, S.K., Danks, M.K., Stewart, C.F. (1999). Relationship between tumor extracellular fluid exposure to topotecan and tumor response in human neuroblastoma xenograft and cell lines. *Cancer Chemotherapy and Pharmacology*, *43*, 269–276.

[23] Leggas, M., Adachi, M., Scheffer, G.L., Sun, D., Wielinga, P., Du, G., Mercer, K.E., Zhuang, Y., Panetta, J.C., Johnston, B., et al. (2004). Mrp4 confers resistance to topotecan and protects the brain from chemotherapy. *Molecular and Cellular Biology*, *24*, 7612–7621.

[24] Zhuang, Y., Fraga, C.H., Hubbard, K.E., Hagedorn, N., Panetta, J.C., Waters, C.M., Stewart, C.F. (2006). Topotecan central nervous system penetration is altered by a tyrosine kinase inhibitor. *Cancer Research*, *66*, 11305–11313.

[25] Zamboni, W.C., Strychor, S., Joseph, E., Walsh, D.R., Zamboni, B.A., Parise, R.A., Tonda, M.E., Yu, N.Y., Engbers, C., Eiseman, J.L. (2007). Plasma, tumor, and tissue disposition of STEALTH liposomal CKD-602 (S-CKD602) and nonliposomal CKD-602 in mice bearing A375 human melanoma xenografts. *Clinical Cancer Research*, *13*, 7217–7223.

[26] Aluisio, L., Lord, B., Barbier, A.J., Fraser, I.C., Wilson, S.J., Boggs, J., Dvorak, L.K., Letavic, M.A., Maryanoff, B.E., Carruthers, N.I., Bonaventure, P., Lovenberg, T.W. (2008). In-vitro and in-vivo characterization of JNJ-7925476, a novel triple mono-amine uptake inhibitor. *European Journal of Pharmacology*, *587*, 141–146.

[27] Masson, E., Zamboni, W.C. (1997). Pharmacokinetic optimisation of cancer chemotherapy: effect on outcomes. *Clinical Pharmacokinetics*, *32*, 324–343.

[28] Simkin, P.A. (1988). Concentration–effect relationships of NSAID. *Journal of Rheumatology Supplement*, *17*, 40–43.

[29] Peck, C.C., Barr, W.H., Benet, L.Z., Collins, J., Desjardins, R.E., Furst, D.E., Harter, J.G., Levy, G., Ludden, T., Rodman, J.H., et al. (1992). Opportunities for integration of pharmacokinetics, pharmacodynamics, and toxicokinetics in rational drug development. *Clinical Pharmacology & Therapeutics*, *51*, 465–473.

[30] Jain, R.K. (1990). Physiological barriers to delivery of monoclonal antibodies and other macromolecules in tumors. *Cancer Research*, *50*, 814s–819s.

[31] Fukumura, D., Jain, R.K. (2007). Tumor microvasculature and microenvironment: targets for anti-angiogenesis and normalization. *Microvascular Research*, *74*, 72–84.

[32] Fischman, A.J., Alpert, N.M., Babich, J.W., Rubin, R.H. (1997). The role of positron emission tomography in pharmacokinetic analysis. *Drug Metabolism Reviews*, *29*, 923–956.

[33] Meikle, S.R., Matthews, J.C., Brock, C.S., Wells, P., Harte, R.J., Cunningham, V.J., Jones, T., Price, P. (1998). Pharmacokinetic assessment of novel anti-cancer drugs using spectral analysis and positron emission tomography: a feasibility study. *Cancer Chemotherapy and Pharmacology*, *42*, 183–193.

[34] Jynge, P., Skjetne, T., Gribbestad, I., Kleinbloesem, C.H., Hoogkamer, H.F., Antonsen, O., Krane, J., Bakoy, O.E., Furuheim, K.M., Nilsen, O.G. (1990). In vivo tissue pharmacokinetics by fluorine magnetic resonance spectroscopy: a study of liver and muscle disposition of fleroxacin in humans. *Clinical Pharmacology & Therapeutics*, *48*, 481–489.

[35] Artemov, D., Solaiyappan, M., Bhujwalla, Z.M. (2001). Magnetic resonance pharmacoangiography to detect and predict chemotherapy delivery to solid tumors. *Cancer Research*, *61*, 3039–3044.

[36] D'Souza, R.A., Partridge, E.A., Roberts, D.W., Ashton, S., Ryan, A., Patterson, A.B., Wilson, Z., Thurrell, C.C. (2007). Distribution of radioactivity and metabolite profiling in tumour and plasma following intravenous administration of a colchicine derivative (14C-ZD6126) to tumour-bearing mice. *Xenobiotica*, *37*, 328–340.

[37] Busk, M., Horsman, M.R., Jakobsen, S., Keiding, S., van der Kogel, A.J., Bussink, J., Overgaard, J. (2008). Imaging hypoxia in xenografted and murine tumors with

[18]F-fluoroazomycin arabinoside: a comparative study involving microPET, autoradiography, PO_2-polarography, and fluorescence microscopy. *International Journal of Radiation Oncology, Biology, Physics*, 70, 1202–1212.

[38] Christensen, J.D., Yurgelun-Todd, D.A., Babb, S.M., Gruber, S.A., Cohen, B.M., Renshaw, P.F. (1999). Measurement of human brain dexfenfluramine concentration by ^{19}F magnetic resonance spectroscopy. *Brain Research*, 834, 1–5.

[39] Sessa, C., Guibal, A., Del Conte, G., Ruegg, C. (2008). Biomarkers of angiogenesis for the development of antiangiogenic therapies in oncology: tools or decorations? *Nature Clinical Practice*, 5, 378–391.

[40] Bache, M., Kappler, M., Said, H.M., Staab, A., Vordermark, D. (2008). Detection and specific targeting of hypoxic regions within solid tumors: current preclinical and clinical strategies. *Current Medicinal Chemistry*, 15, 322–338.

[41] Silasi, D.A., Alvero, A.B., Mor, J., Chen, R., Fu, H.H., Montagna, M.K., Mor, G. (2008). Detection of cancer-related proteins in fresh-frozen ovarian cancer samples using laser capture microdissection. *Methods in Molecular Biology*, 414, 35–45.

[42] Ungerstedt, U. (1984). Measurement of transmitter release by intracerebral dialysis. In: Marsden, C.A. (Ed.), *Measurement of Neurotransmitter Release In Vivo*, Wiley, Chichester, U.K., pp. 81–105.

[43] Stahle, L. (2000). On mathematical models of microdialysis: geometry, steady-state models, recovery and probe radius. *Advanced Drug Delivery Reviews*, 45, 149–167.

[44] Lönnroth, P., Jansson, P.A., Smith, U. (1987). A microdialysis method allowing characterization of intercellular water space in humans. *American Journal of Physiology*, 253, E228–E231.

[45] Le Quellec, A., Dupin, S., Genissel, P., Saivin, S., Marchand, B., Houin, G. (1995). Microdialysis probes calibration: gradient and tissue dependent changes in no net flux and reverse dialysis methods. *Journal of Pharmacological and Toxicological Methods*, 33, 11–16.

[46] Larsson, C.I. (1991). The use of an "internal standard" for control of the recovery in microdialysis. *Life Sciences*, 49, PL73–PL78.

[47] Yokel, R.A., Allen, D.D., Burgio, D.E., McNamara, P.J. (1992). Antipyrine as a dialyzable reference to correct differences in efficiency among and within sampling devices during in vivo microdialysis. *Journal of Pharmacological and Toxicological Methods*, 27, 135–142.

[48] Van Belle, K., Dzeka, T., Sarre, S., Ebinger, G., Michotte, Y. (1993). In vitro and in vivo microdialysis calibration for the measurement of carbamazepine and its metabolites in rat brain tissue using the internal reference technique. *Journal of Neuroscience Methods*, 49, 167–173.

[49] Wang, Y., Wong, S.L., Sawchuk, R.J. (1993). Microdialysis calibration using retrodialysis and zero-net flux: application to a study of the distribution of zidovudine to rabbit cerebrospinal fluid and thalamus. *Pharmaceutical Research*, 10, 1411–1419.

[50] Dai, H., Chen, Y., Elmquist, W.F. (2005). Distribution of the novel antifolate pemetrexed to the brain. *Journal of Pharmacology and Experimental Therapeutics*, 315, 222–229.

[51] Jacobson, I., Sandberg, M., Hamberger, A. (1985). Mass transfer in brain dialysis devices: a new method for the estimation of extracellular amino acids concentration. *Journal of Neuroscience Methods, 15,* 263–268.

[52] Benveniste, H., Huttemeier, P.C. (1990). Microdialysis: theory and application. *Progress in Neurobiology, 35,* 195–215.

[53] Carneheim, C., Stahle, L. (1991). Microdialysis of lipophilic compounds: a methodological study. *Pharmacology & Toxicology, 69,* 378–380.

[54] Benfeldt, E., Groth, L. (1998). Feasibility of measuring lipophilic or protein-bound drugs in the dermis by in vivo microdialysis after topical or systemic drug administration. *Acta Dermatovenereologica, 78,* 274–278.

[55] Loos, W.J., Zamboni, W.C., Engels, F.K., de Bruijn, P., Lam, M.H., de Wit, R., Verweij, J., Wiemer, E.A. (2007). Pitfalls of the application of microdialysis in clinical oncology: controversial findings with docetaxel. *Journal of Pharmaceutical and Biomedical Analysis, 45,* 288–294.

[56] Traunmuller, F., Steiner, I., Zeitlinger, M., Joukhadar, C. (2006). Development of a high-performance liquid chromatography method for the determination of caspofungin with amperometric detection and its application to in vitro microdialysis experiments. *Journal of Chromatography B: Analytical Technologies in the Biomedical and Life Sciences, 843,* 142–146.

[57] Kurosaki, Y., Nakamura, S., Shiojiri, Y., Kawasaki, H. (1998). Lipo-microdialysis: a new microdialysis method for studying the pharmacokinetics of lipophilic substances. *Biological & Pharmaceutical Bulletin, 21,* 194–196.

[58] Ward, K.W., Medina, S.J., Portelli, S.T., Mahar Doan, K.M., Spengler, M.D., Ben, M.M., Lundberg, D., Levy, M.A., Chen, E.P. (2003). Enhancement of in vitro and in vivo microdialysis recovery of SB-265123 using Intralipid and Encapsin as perfusates. *Biopharmaceutics & Drug Disposition, 24,* 17–25.

[59] Lindberger, M., Tomson, T., Lars, S. (2002). Microdialysis sampling of carbamazepine, phenytoin and phenobarbital in subcutaneous extracellular fluid and subdural cerebrospinal fluid in humans: an in vitro and in vivo study of adsorption to the sampling device. *Pharmacology & Toxicology, 91,* 158–165.

[60] Dabrosin, C. (2005). Microdialysis: an in vivo technique for studies of growth factors in breast cancer. *Frontiers in Bioscience, 10,* 1329–1335.

[61] Dabrosin, C. (2005). Positive correlation between estradiol and vascular endothelial growth factor but not fibroblast growth factor-2 in normal human breast tissue in vivo. *Clinical Cancer Research, 11,* 8036–8041.

[62] Mellergard, P., Aneman, O., Sjogren, F., Pettersson, P., Hillman, J. (2008). Changes in extracellular concentrations of some cytokines, chemokines, and neurotrophic factors after insertion of intracerebral microdialysis catheters in neurosurgical patients. *Neurosurgery, 62,* 151–157; discussion, 157–158.

[63] Garvin, S., Dabrosin, C. (2003). Tamoxifen inhibits secretion of vascular endothelial growth factor in breast cancer in vivo. *Cancer Research, 63,* 8742–8748.

[64] Garvin, S., Dabrosin, C. (2008). In vivo measurement of tumor estradiol and vascular endothelial growth factor in breast cancer patients. *BMC Cancer, 8,* 73.

[65] Trickler, W.J., Miller, D.W. (2003). Use of osmotic agents in microdialysis studies to improve the recovery of macromolecules. *Journal of Pharmaceutical Sciences, 92,* 1419–1427.

[66] Ao, X., Stenken, J.A. (2006). Microdialysis sampling of cytokines. *Methods, 38*, 331–341.

[67] Duo, J., Fletcher, H., Stenken, J.A. (2006). Natural and synthetic affinity agents as microdialysis sampling mass transport enhancers: current progress and future perspectives. *Biosensors & Bioelectronics, 22*, 449–457.

[68] Endrich, B., Intaglietta, M., Reinhold, H.S., Gross, J.F. (1979). Hemodynamic characteristics in microcirculatory blood channels during early tumor growth. *Cancer Research, 39*, 17–23.

[69] Boucher, Y., Baxter, L.T., Jain, R.K. (1990). Interstitial pressure gradients in tissue-isolated and subcutaneous tumors: implications for therapy. *Cancer Research, 50*, 4478–4484.

[70] Ekstrom, P.O., Giercksky, K.E., Andersen, A., Bruland, O.S., Slordal, L. (1997). Intratumoral differences in methotrexate levels within human osteosarcoma xenografts studied by microdialysis. *Life Sciences, 61*, PL275–PL280.

[71] Ault, J.M., Riley, C.M., Meltzer, N.M., Lunte, C.E. (1994). Dermal microdialysis sampling in vivo. *Pharmaceutical Research, 11*, 1631–1639.

[72] Davies, M.I., Lunte, C.E. (1995). Microdialysis sampling for hepatic metabolism studies: impact of microdialysis probe design and implantation technique on liver tissue. *Drug Metabolism and Disposition, 23*, 1072–1079.

[73] Morgan, M.E., Singhal, D., Anderson, B.D. (1996). Quantitative assessment of blood–brain barrier damage during microdialysis. *Journal of Pharmacology and Experimental Therapeutics, 277*, 1167–1176.

[74] Grabb, M.C., Sciotti, V.M., Gidday, J.M., Cohen, S.A., van Wylen, D.G. (1998). Neurochemical and morphological responses to acutely and chronically implanted brain microdialysis probes. *Journal of Neuroscience Methods, 82*, 25–34.

[75] Groothuis, D.R., Ward, S., Schlageter, K.E., Itskovich, A.C., Schwerin, S.C., Allen, C.V., Dills, C., Levy, R.M. (1998). Changes in blood–brain barrier permeability associated with insertion of brain cannulas and microdialysis probes. *Brain Research, 803*, 218–230.

[76] Clapp-Lilly, K.L., Roberts, R.C., Duffy, L.K., Irons, K.P., Hu, Y., Drew, K.L. (1999). An ultrastructural analysis of tissue surrounding a microdialysis probe. *Journal of Neuroscience Methods, 90*, 129–142.

[77] Blochl-Daum, B., Müller, M., Meisinger, V., Eichler, H.G., Fassolt, A., Pehamberger, H. (1996). Measurement of extracellular fluid carboplatin kinetics in melanoma metastases with microdialysis. *British Journal of Cancer, 73*, 920–924.

[78] Müller, M., Mader, R.M., Steiner, B., Steger, G.G., Jansen, B., Gnant, M., Helbich, T., Jakesz, R., Eichler, H.G., Blochl-Daum, B. (1997). 5-Fluorouracil kinetics in the interstitial tumor space: clinical response in breast cancer patients. *Cancer Research, 57*, 2598–2601.

[79] Tegeder, I., Brautigam, L., Seegel, M., Al-Dam, A., Turowski, B., Geisslinger, G., Kovacs, A.F. (2003). Cisplatin tumor concentrations after intra-arterial cisplatin infusion or embolization in patients with oral cancer. *Clinical Pharmacology & Therapeutics, 73*, 417–426.

[80] Joukhadar, C., Klein, N., Mader, R.M., Schrolnberger, C., Rizovski, B., Heere-Ress, E., Pehamberger, H., Strauchmann, N., Jansen, B., Müller, M. (2001). Penetration of dacarbazine and its active metabolite 5-aminoimidazole-4-carboxamide into cutaneous metastases of human malignant melanoma. *Cancer, 92*, 2190–2196.

[81] Mader, R.M., Schrolnberger, C., Rizovski, B., Brunner, M., Wenzel, C., Locker, G., Eichler, H.G., Müller, M., Steger, G.G. (2003). Penetration of capecitabine and its metabolites into malignant and healthy tissues of patients with advanced breast cancer. *British Journal of Cancer*, 88, 782–787.

[82] Thompson, J.F., Siebert, G.A., Anissimov, Y.G., Smithers, B.M., Doubrovsky, A., Anderson, C.D., Roberts, M.S. (2001). Microdialysis and response during regional chemotherapy by isolated limb infusion of melphalan for limb malignancies. *British Journal of Cancer*, 85, 157–165.

[83] Bergenheim, A.T., Capala, J., Roslin, M., Henriksson, R. (2005). Distribution of BPA and metabolic assessment in glioblastoma patients during BNCT treatment: a microdialysis study. *Journal of Neuro-Oncology*, 71, 287–293.

[84] Castejon, A.M., Paez, X., Hernandez, L., Cubeddu, L.X. (1999). Use of intravenous microdialysis to monitor changes in serotonin release and metabolism induced by cisplatin in cancer patients: comparative effects of granisetron and ondansetron. *Journal of Pharmacology and Experimental Therapeutics*, 291, 960–966.

[85] Dabrosin, C., Johansson, A.C., Ollinger, K. (2004). Decreased secretion of cathepsin D in breast cancer in vivo by tamoxifen: mediated by the mannose-6-phosphate/IGF-II receptor? *Breast Cancer Research and Treatment*, 85, 229–238.

[86] Flannery, T., McConnell, R.S., McQuaid, S., McGregor, G., Mirakhur, M., Martin, L., Scott, C., Burden, R., Walker, B., McGoohan, C., Johnston, P.G. (2007). Detection of cathepsin S cysteine protease in human brain tumour microdialysates in vivo. *British Journal of Neurosurgery*, 21, 204–209.

[87] Nilsson, U.W., Dabrosin, C. (2006). Estradiol and tamoxifen regulate endostatin generation via matrix metalloproteinase activity in breast cancer in vivo. *Cancer Research*, 66, 4789–4794.

[88] Bendrik, C., Robertson, J., Gauldie, J., Dabrosin, C. (2008). Gene transfer of matrix metalloproteinase-9 induces tumor regression of breast cancer in vivo. *Cancer Research*, 68, 3405–3412.

[89] Zhong, H., Han, B., Tourkova, I.L., Lokshin, A., Rosenbloom, A., Shurin, M.R., Shurin, G.V. (2007). Low-dose paclitaxel prior to intratumoral dendritic cell vaccine modulates intratumoral cytokine network and lung cancer growth. *Clinical Cancer Research*, 13, 5455–5462.

[90] Stuhr, L.E., Salnikov, A.V., Iversen, V.V., Salvesen, G., Rubin, K., Reed, R.K. (2006). High-dose, short-term, anti-inflammatory treatment with dexamethasone reduces growth and augments the effects of 5-fluorouracil on dimethyl-alpha-benzanthracene-induced mammary tumors in rats. *Scandinavian Journal of Clinical and Laboratory Investigation*, 66, 477–486.

[91] Johansen, M.J., Thapar, N., Newman, R.A., Madden, T. (2002). Use of microdialysis to study platinum anticancer agent pharmacokinetics in preclinical models. *Journal of Experimental Therapeutics and Oncology*, 2, 163–173.

[92] Dodds, H.M., Tobin, P.J., Stewart, C.F., Cheshire, P., Hanna, S., Houghton, P., Rivory, L.P. (2002). The importance of tumor glucuronidase in the activation of irinotecan in a mouse xenograft model. *Journal of Pharmacology and Experimental Therapeutics*, 303, 649–655.

[93] McLaughlin, K.J., Faibushevich, A.A., Lunte, C.E. (2000). Microdialysis sampling with on-line microbore HPLC for the determination of tirapazamine and its reduced metabolites in rats. *Analyst*, 125, 105–110.

[94] Gallo, J.M., Vicini, P., Orlansky, A., Li, S., Zhou, F., Ma, J., Pulfer, S., Bookman, M.A., Guo, P. (2004). Pharmacokinetic model-predicted anticancer drug concentrations in human tumors. *Clinical Cancer Research*, 10, 8048–8058.

14

MICRODIALYSIS VERSUS IMAGING TECHNIQUES FOR IN VIVO DRUG DISTRIBUTION MEASUREMENTS

Martin Brunner

Medical University of Vienna, Vienna, Austria

1. INTRODUCTION

Variability in the response to drug therapy is a significant challenge for physicians, as it is often unpredictable how, or even if, patients receiving comparable doses of the same medication might respond to the agents administered. This notion is not new. Sir William Osler commented on this problem in 1903: "Variability is the law of life, and as no two faces are the same, so no two bodies are alike, and no two individuals react alike, and behave alike . . ." [1]. More than a century later, we are in the fortunate position of knowing more about the mechanisms and pathways that contribute to dose–response variability. Besides polymorphisms in genes encoding for enzymes, drug transporters, or receptor proteins [2,3], it has been recognized that most drugs, with few notable exceptions (e.g., heparin), exert their action in tissues rather than in plasma and do not distribute uniformly in the body but, rather, attain varying concentrations in different tissues. Assessing tissue chemistry and pharmacology has thus been viewed as a more rational way to provide clinically meaningful data on dose–response variability than gaining information from blood samples.

The in vivo assessment of drug distribution and target-site pharmacokinetics (PK), however, has long been treated as a "forgotten relative" by physicians, pharmacists, and clinical pharmacologists [4]. The main reason for this neglect was largely the lack of appropriate methodology providing in vivo access to

Applications of Microdialysis in Pharmaceutical Science, First Edition. Edited by Tung-Hu Tsai.
© 2011 John Wiley & Sons, Inc. Published 2011 by John Wiley & Sons, Inc.

the target sites in tissues and organs, and consequently, PK research was long restricted to drug concentration measurements from biological specimens that are relatively easy to obtain, such as tissue biopsies, urine, saliva, or skin blister fluid or to indirect modeling of tissue concentrations from plasma concentration curves, which in most cases only served as surrogates for true target-site concentrations. Data obtained by these approaches, however, caused considerable confusion about drug distribution and target-site delivery [5]. A current refined definition of target tissue drug concentrations should include in many cases the meaning of "unbound drug concentrations at anatomically distinct sites," such as the interstitial space fluid (ISF). Consequently, a suitable method for the measurement of tissue drug concentrations should allow for the direct measurement of unbound drug concentrations in a clearly defined space within the tissue of a given organ.

Recent years have seen the introduction of several new techniques and approaches to the assessment of drug distribution and target tissue PK in humans [6,7], including in vivo microdialysis (MD) and imaging techniques such as magnetic resonance spectroscopy (MRS) and positron-emission tomography (PET). Results from studies using these techniques have emphasized the importance of the previously neglected drug distribution process to the target site as a crucial determinant for clinical outcome. Furthermore, regulatory guidance documents issued by the FDA (Food and Drug Administration) in the United States and the CPMP (Committee for Proprietary Medicinal Products) in Europe have emphasized the value and importance of human tissue drug concentration data and conceptually support in particular the use of clinical MD to obtain this information [8,9]. In addition to other documents, and in light of the critical path initiative [10], the CDER report to the nation in 2003 indicated a need for tools that enable the measurement of tissue concentrations by stating that CDER continues to extend its "long-standing interest in the application of dose–response principles by viewing drugs and their actions directly at the level of the drug target, rather than indirectly via plasma concentrations" [11]. Currently, MD is the only tool available that explicitly provides data on the extracellular space. Although MD as a preclinical and clinical tool has been available for two decades, there is still uncertainty about the its use in drug research and development, from both a methodological and a regulatory point of view. Views on MD as a tool in drug research and development have been summarized recently [12].

In this chapter a short comparative description of the use of MD and imaging techniques for the assessment of in vivo tissue drug distribution, addressing advantages, limitations, and potential combinations is provided.

2. MICRODIALYSIS

MD is a semi-invasive focal sampling method, based on the use of probes with a semipermeable membrane at the probe tip. The MD probe, which is constantly perfused with a physiological solution at a low flow rate of 1 to 10 µL/min, is implanted into the tissue of interest, and substances in the interstitial space fluid pass the membrane by passive diffusion along their concentration gradient, resulting in a certain concentration in the perfusion medium. This dialysate is collected at timed intervals and is subjected ex vivo to various types of chemical analyses, which can be performed in either an off- or online fashion. Depending on the molecular cutoff of the membrane, large molecules such as proteins are usually excluded from the dialysate, which enables analysis without time-consuming sample preparation or sample storage without the immediate fear of enzymatic degradation. Due to small sample volumes, which are usually in the microliter range in human studies, there is no substantial biological fluid loss. Sample analysis, however, requires highly sensitive methods such as liquid chromatography–tandem mass spectrometry to deal with the low concentrations in a rather small sample volume. In most cases, MD is performed under nonequilibrium conditions and dialysate concentrations represent only a fraction of actual concentrations in the medium surrounding the MD probe. To obtain and quantify interstitial space fluid concentrations from dialysate concentrations, MD probes need to be calibrated. Given proper in vivo calibration procedures, intraindividual variation for interstitial space fluid measurements was shown to range between 10 and 20%, depending on the analyte. In contrast to other methods traditionally used for tissue concentration measurements, MD provides selective access to the unbound and thus pharmacologically active drug fraction in the interstitial space fluid of tissues, the true target site for drugs such as most antimicrobial agents, tumor chemotherapeutics, or substances that act by binding to cell surface receptors.

In clinical research, MD is currently employed to address various issues in different clinical fields, such as monitoring of secondary ischemia in neurointenisve care [13] or glucose monitoring for long-term metabolic control in patients with diabetes mellitus [14]. Further areas of research comprise studies on the local physiology of peripheral tissues [15] or the local administration of drugs by MD without inducing systemic side effects and simultaneous measurement of the corresponding tissue response [16]. In clinical pharmacology, research focuses on the use of MD to measure target-site concentrations of antibiotics [17] or anticancer drugs [18] in different tissues and organs and subsequently to relate target-site PK to pharmacodynamics [19]. Equally challenging is the characterization of skin penetration of analytes from transdermal therapeutic systems [20]. In contrast to other often technically demanding and expensive methods such as imaging techniques, MD can readily be employed for clinical studies in almost any research center at a reasonable price. Drawbacks of the technique stem from the semi-invasive nature of the technique. Consequently, most human studies have been performed in easily accessible tissues such as skeletal muscle, subcutaneous adipose tissue, skin, tendons, superficially located tumors, or blood. Combined with surgical procedures, however, almost every tissue within the human body is in reach for

MD probe implantation, as demonstrated by studies in brain, lung, bone, heart, liver, or the peritoneal cavity for metabolic monitoring following gastrointestinal surgery. For the latter two applications, special probes for use in humans have become available recently. For an in-depth view on methodological aspects and clinical applications of MD, we refer readers to recent review articles [7,12,21–26].

3. IMAGING TECHNIQUES

In the past 25 years, imaging techniques have evolved as powerful tools for the noninvasive study of drug distribution in vivo as well as for studying drug effects at their target sites. Imaging techniques that lend themselves to the study of drug distribution in humans are MRS and PET [27–32]. Although initially these techniques were introduced to clinical medicine for diagnostic purposes and for the study of tissue metabolism and blood flow, they also opened a unique opportunity for PK research by providing a means for non-invasive measurement of drug distribution from the plasma compartment to anatomically defined regions and for visualization of the entire pattern of drug distribution in given organs [33,34].

4. MAGNETIC RESONANCE IMAGING AND MAGNETIC RESONANCE SPECTROSCOPY

Magnetic resonance imaging (MRI) uses radio-frequency pulses and magnetic fields to obtain signals from changes in nuclear magnetic moments. A technique based on the same principle as MRI, but providing a greater degree of molecular characterization, is MRS, in which spectroscopic profiles of the chemical constituents within a sample are obtained. MRS measurements can be performed serially, thus making possible PK analysis with a temporal resolution on the order of minutes. Of importance, MRS is capable of resolving different chemical species, including metabolites, owing to different chemical shifts of the resonance signals. This poses a considerable advantage over nuclear imaging methods such as PET, which record nuclear decay events irrespective of the chemical surrounding of the decaying atom and therefore lump together all compounds labeled with the same radioactive atom. A main limitation of all nuclear MR-based methods is their inherent low sensitivity, which restricts the in vivo applicability primarily to molecules that are present in large concentrations in the human body. MRS has proven to be particularly feasible for fluorinated drugs, since ^{19}F is one of the lead isotopes for nuclear MRS, and several studies have been published describing brain PK of fluorinated psychiatric medications [35], tumor uptake of anticancer chemotherapeutics, and biodistribution and target tissue PK of fluorinated antibiotics [36,37]. Recently, the use of ^{19}F MRS was validated to quantify the experimental antihistamine tecastemizole in heart and liver [38]. Furthermore, in human liver studies, the heterogeneous metabolism of capecitabine was monitored in patients with metastatic colorectal cancer [39]. The authors of a recent review article state that MRS provides an attractive noninvasive way of studying drug distribution in vivo and that the widespread occurrence of fluorine in medicinal compounds, and its favorable MR properties, make it an effective probe for drug absorption, distribution, metabolism, and excretion studies. In the future, increasing clinical ^{19}F MRS use might be anticipated as high-field human scanners become widespread and requirements to demonstrate mechanisms underlying clinical effects become more pressing [40].

5. POSITRON EMISSION TOMOGRAPHY

In brief, PET is a nuclear imaging technique based on the use of molecules labeled with positron-emitting radioisotopes. The emitted positrons pass through tissue and are ultimately annihilated when combined with an electron, resulting in two 511-keV photons emitted in opposite directions. Detectors are arranged in a ring around the tissue of interest, and only triggering events that arrive nearly-simultaneously at diametrically opposite detectors are recorded (*coincidence detection*). The resulting PET images might yield three-dimensional information on tissue distribution of the positron-emitting molecules. The most commonly employed PET radionuclides are oxygen 15 (^{15}O), nitrogen 13 (^{13}N), carbon 11 (^{11}C), and fluorine 18 (^{18}F). Due to its comparably long half-life, ^{18}F is the most attractive PET radioisotope for drug distribution studies since it allows for imaging durations of up to 10 h. A considerable drawback, however, is that relatively few drug molecules contain fluorine in their native structure; consequently, despite the rather short half-life of ^{11}C (20.4 min), the majority of PET–PK experiments have relied on ^{11}C-labeled tracer molecules. So far, PET has been employed for studying the tissue distribution of radiolabeled antibiotics, antifungals, and inhaled drugs in patients and volunteers [33,34,41,42]. Furthermore, PET has proven to be a valuable noninvasive means for characterizing established and novel anticancer agents [43]. Using PET imaging, it was demonstrated that radioactivity uptake was correlated with response to chemotherapy [44] and the mechanisms postulated for an antineoplastic agent, temozolomide, were recently confirmed in vivo in glioma patients [45]. In the neurological field, PET was employed to compare intracerebral uptake of ^{11}C-labeled verapamil in volunteers differing in genotypes for the drug efflux transporter gene *ABCB1* [46]. To explain pharmacoresistance in epilepsy, a pilot study was conducted comparing [^{11}C]verapamil pharmacokinetics in epileptogenic and nonepileptogenic brain regions of patients with drug-resistant unilateral temporal lobe epilepsy [47]. In patients with Alzheimer's disease, distribution and local tissue pharmacokinetics of an investigational ^{11}C-labeled antiamyloid drug were assessed in its target organ, the human brain [48].

Important limitations stem from the fact that only drugs that lend themselves to radiolabeling may be studied. Second, the PET signal produced is not necessarily a measure of the intact drug concentration, and PET is not able to provide information about specific tissue compartments, such as the interstitial space fluid. Also, PET studies are bound to specialized centers, with on-site access to a cyclotron, radiochemistry, and a PET camera, which contributes substantially to the high costs of PET studies.

6. COMBINATION OF MICRODIALYSIS AND IMAGING TECHNIQUES

So far, most recently published combinations of MD and imaging techniques in humans have come from the neuroscience field. MD has been introduced as an intracerebral sampling method for clinical neurosurgery in 1990 [49,50] and since then has been embraced as a safe continuous monitoring technique to measure the neurochemistry of acute brain injury and epilepsy [51,52]. Although data from brain MD studies suggest strongly that changes in local markers of brain metabolism might precede the onset of secondary neurological deterioration, cerebral MD is still used primarily as a clinical research tool in neurosurgery, and its use to influence clinical therapeutic decision making has been restricted to only a few institutions worldwide [49]. Still, brain MD is one of a few methods for neurochemical measurements in the interstitial compartment of the human brain and has become a valuable translational research tool, providing new and important insights into the neurochemistry of acute human brain injury. Isolated interpretation of biomarkers derived from brain MD experiments, however, should be carried out cautiously and might require additional validation in particular in clinical studies, in which experimental conditions cannot easily be standardized. Therefore, the simultaneous use of complementary methods such as MRI (e.g., to get an additional estimate of intracellular changes or arterial–venous differences) or PET (to compare metabolic rates) might be crucial for biomarker interpretation [49]. Promising MD applications, as yet less explored in combination with imaging techniques, include local neurochemical provocations, drug penetration to the brain, and a technique to obtain surrogate endpoint(s) in neuropharmacological studies.

In a recent study in patients with refractory temporal lobe epilepsy, the relationship between the hippocampal volume and glutamate was investigated, combining quantitative MRI volumetrics to measure the hippocampal volume and MD to measure the interictal glutamate, glutamine, and γ-aminobutyric acid levels in the epileptogenic hippocampus [53]. From the finding that decreased hippocampal volume on MRI is associated with increased extracellular glutamate in epilepsy patients, the authors concluded that their work has implications for the understanding and treatment of epilepsy as well as other neurodegenerative disorders associated with hippocampal atrophy [53].

Few MD–PET combinations have been published in a clinical setting, most of them in the neurosurgical field to study alterations of brain metabolism as a consequence of brain trauma or surgery [54,55]. A recent study combined MD and PET with the aim to determine the occurrence of a metabolic crisis after brain injury by examining a representative region of brain tissue remote from the primary injury site [56]. In this region the authors measured the metabolic state of the tissue by PET and MD and, furthermore, validated the usefulness of MD as an indicator of ischemia in traumatic brain injury. The main findings of the study were that the injured brain had persistent impairments in metabolism that could be detected by brain MD. The lactate/pyruvate ratio, a MD marker that has been proposed to be a reliable marker of ischemia in previous MD studies [13], reflected impaired oxidative metabolism. However, the ratio was not specific for brain ischemia in the region investigated, as shown by simultaneous PET measurements, suggesting that energy perturbation unrelated to ischemia may contribute to secondary brain damage in traumatic brain injury.

A series of studies has employed the MD–PET combination for regional metabolic imaging [57–59] or to predict malignant edema progression in patients with brain infarctions [60]. Although all MD neuromonitoring parameters were significantly altered at the time of manifest malignant edema, causing a midline shift, only PET could predict this unfavorable clinical course by revealing larger volumes of ischemic core and irreversible neuronal damage during the first 24 h. The authors concluded that in contrast to PET, MD monitoring failed to predict a fatal outcome in time for successful therapeutic intervention. In a later study, however, the same group studied stroke patients, used other MD biomarkers, and came to the opposite conclusion [57]. The results from these two studies emphasize the importance of the combined use of MD and imaging to accurately interpret the results obtained and to translate the data into clinically useful information.

Apart from the neurological field, a recent study has combined MD and PET to assess intracellular drug pharmacokinetics in vivo [61]. PET yields a combined signal comprising the intracellular, the extracellular, and the intravascular fraction of a radiolabeled drug and its metabolites in a given volume of tissue, whereas MD describes unbound extracellular drug concentrations. As for several drugs, such as certain anti-infective and anticancer agents, the site of drug action is not the biophase surrounding the cells, but rather, an intracellular compartment, knowledge of intracellular rather than extracellular or total drug concentrations, is relevant in many cases. Fluorine-18-labeled ciprofloxacin ($[^{18}F]$ciprofloxacin) [62] was used as a model compound to perform simultaneous PET imaging and MD in healthy volunteers to describe intracellular drug pharmacokinetics in human muscle tissue for several hours. A three-compartment pharmacokinetic model was fitted to the tissue concentration–time profiles of ciprofloxacin measured by PET to estimate the rate constants of ciprofloxacin uptake and transport. The extracellular concentration–time profiles predicted by compartmental modeling were in good agreement with the measured MD data, and the results were in accordance with previous in vitro data describing cellular ciprofloxacin uptake and retention. The setting of this study is unique in the sense that it constitutes one of the rare occasions where the results of the compartmental modeling of PET data were validated directly by an independent measurement technique (i.e., HPLC quantification of samples collected by MD) in humans. The authors therefore concluded that the employed MD–PET combination might be useful during research and development of new drugs, for which knowledge of intracellular concentrations is of interest. Two further human studies have used the MD–PET combination to measure the lumped constant, an experimentally derived correction factor that accounts for the differences in transport and phosphorylation between 2-[^{18}F]fluoro-2-deoxy-D-glucose ([^{18}F]FDG), a widely used PET tracer for glucose utilization, and its endogenous counterpart, glucose [63,64]. Regional [^{18}F]FDG uptake was determined in skeletal muscle or adipose tissue by PET, whereas MD was used to monitor local interstitial glucose concentrations. Furthermore, regional tissue blood flow was determined by [^{15}O]H$_2$O/PET and regional glucose uptake was calculated.

One limitation of MD is derived from the fact that it provides focal information of neurochemical events. A combination of neuroimaging for neurochemical and neurophysiological monitoring, as used in clinical centers, and clinical MD might contribute to obtaining a broader picture of brain injury and metabolism. The combination of MD and imaging techniques has the potential to explore and describe exactly the fate and pharmacokinetics of a drug in the body. Exploiting the strengths of both approaches appears to be a straightforward way to predict drug action and therapeutic success and may be used for decision making in future drug research and development.

7. SUMMARY AND CONCLUSIONS

The use of MD and imaging techniques in human drug tissue distribution studies has highlighted inherent strengths and shortcomings of these techniques. MD is a comparably cheap method that is not bound to a research center with sophisticated technology and should be preferred over traditional techniques for the assessment of interstitial drug concentrations and tissue distribution. Assuming the availability of a suitable analytical assay to quantify the drug of interest, and an appropriate ethical setting, MD can be performed for virtually any drug molecule by physicians, who do not need extensive skills in MD probe placement, as the insertion process is similar to standard intramurcular or subcutaneous injections. However, experience and individual probe calibration are required for converting dialysate concentrations into absolute tissue concentrations. Besides measurement of the parent compound, metabolite monitoring is feasible. Issues such as radioactive waste handling, radiation exposure of patients/volunteers, and clinical staff are not relevant, as radiolabeled compounds are usually not employed. One major limitation stems from the fact that some tissues and organs, such as brain, lung, or liver, might be accessed only in combination with surgical procedures.

Nuclear imaging techniques and MRS, on the other hand, are fully noninvasive and are suitable for drug concentration measurements in virtually any organ. MRS imaging has a rather low spatial resolution. In contrast, PET offers excellent spatial resolution on the order of a few millimeters, which enables the assessment of regionally different drug concentrations in a given organ. MD is a focal sampling method and provides concentration measurements in a rather small volume of tissue, defined by the position of the MD probe, with a temporal resolution of 10 to 20 min. PET cannot provide chemical resolution; that is, bound and unbound drug or parent drug and metabolites give the same signal. Therefore, PET imaging should preferably be applied to metabolically stable analytes or substances without known tissue retention of metabolites. MRS can resolve metabolites and bound or unbound drug, due to chemical shift differences. Both PET and MRS, however, are not able to discern drug concentrations in different compartments in a given volume of tissue (e.g., intracellular, extracellular, intravascular). A limitation of PET, which is especially relevant for studies that aim at PK measurements over longer periods, stems from the short half-lives of the radioisotopes available. Whereas for MD, continuous sampling has been described for several days with appropriate probes, for PET studies with ^{11}C-labeled drugs the maximum possible imaging time ranges around 2 h. When labeled with ^{18}F, fluorinated compounds allow for extended imaging for up to 10 h with PET. Fluorine-containing compounds are also the compounds of choice for MRS, due to the high sensitivity and low natural background of the ^{19}F nucleus for NMR imaging. However, the percentage of fluorinated drug molecules is rather low, which restricts a broader use of both techniques in drug distribution studies. Because MRS does not involve patient radiation exposure, measurements can be repeated over prolonged periods of up to several weeks.

In conclusion, several techniques are currently available for assessing drug distribution and tissue pharmacokinetics in humans. Each of these techniques has proven its ability to provide new information on drug distribution for compounds, already marketed and potentially, each technique can provide valuable information during drug research and development. The choice of technique or the complementary combination should be based on the compound of interest, the region of the body, where distribution and tissue concentrations should be monitored, and the availability of technical and financial resources.

REFERENCES

[1] Osler, W. (1903). On the educational value of the medical society. *Boston Medical and Surgical Journal, 148,* 275–279.

[2] Wilkinson, G.R. (2005). Drug metabolism and variability among patients in drug response. *New England Journal of Medicine, 352,* 2211–2221.

[3] Evans, W.E., McLeod, H.L. (2003). Pharmacogenomics: drug disposition, drug targets, and side effects. *New England Journal of Medicine, 348*, 538–549.

[4] Eichler, H.G., Müller, M. (1998). Drug distribution: the forgotten relative in clinical pharmacokinetics. *Clinical Pharmacokinetics, 34*, 95–109.

[5] Müller, M., de la Peña, A., Derendorf, H. (2004). Issues in pharmacokinetics and pharmacodynamics of anti-infective agents: distribution in tissue. *Antimicrobial Agents and Chemotherapy, 48*, 1441–1453.

[6] Langer, O., Müller, M. (2004). Methods to assess tissue-specific distribution and metabolism of drugs. *Current Drug Metabolism, 5*, 463–481.

[7] Brunner, M., Langer, O. (2006). Microdialysis versus other techniques for the clinical assessment of in vivo tissue drug distribution. *American Association of Pharmaceutical Scientists Journal, 14*, E263–E271.

[8] http://www.fda.gov/cder/present/anti-infective798/073198.pdf. Accessed February 6, 2009.

[9] http://www.fda.gov/cder/guidance/2580dft.pdf. Accessed February 6, 2009.

[10] http://www.fda.gov/oc/initiatives/criticalpath/whitepaper.html. Accessed February 6, 2009.

[11] http://www.fda.gov/cder/reports/rtn/2003/rtn2003.PDF. Accessed February 6, 2009.

[12] Chaurasia, C.S., Müller, M., Bashaw, E.D., Benfeldt, E., Bolinder, J., Bullock, R., Bungay, P.M., DeLange, E.C., Derendorf, H., Elmquist, W.F., et al. (2007). AAPS–FDA Workshop White Paper: Microdialysis Principles, Application and Regulatory Perspectives. *Pharmaceutical Research, 24*, 1014–1025.

[13] Ungerstedt, U., Rostami, E. (2004). Microdialysis in neurointensive care. *Current Pharmaceutical Design, 10*, 2145–2152.

[14] Pickup, J.C., Hussain, F., Evans, N.D., Sachedina, N. (2005). In vivo glucose monitoring: the clinical reality and the promise. *Biosensors & Bioelectronics, 20*, 1897–1902.

[15] de la Peña, A., Liu, P., Derendorf, H. (2000). Microdialysis in peripheral tissues. *Advanced Drug Delivery Reviews, 45*, 189–216.

[16] Ekstrom, P.O., Andersen, A., Saeter, G., Giercksky, K.E., Slordal, L. (1997). Continuous intratumoral microdialysis during high-dose methotrexate therapy in a patient with malignant fibrous histiocytoma of the femur: a case report. *Cancer Chemotherapy and Pharmacology, 39*, 267–272.

[17] Joukhadar, C., Derendorf, H., Müller, M. (2001). Microdialysis: a novel tool for clinical studies of anti-infective agents. *European Journal of Clinical Pharmacology, 57*, 211–219.

[18] Brunner, M., Müller, M. (2002). Microdialysis: an in vivo approach for measuring drug delivery in oncology. *European Journal of Clinical Pharmacology, 58*, 227–234.

[19] Delacher, S., Derendorf, H., Hollenstein, U., Brunner, M., Joukhadar, C., Hofmann, S., Georgopoulos, A., Eichler, H.G., Müller, M. (2000). A combined in vivo pharmacokinetic–in vitro pharmacodynamic approach to simulate target site pharmacodynamics of antibiotics in humans. *Journal of Antimicrobial Chemotherapy, 46*, 733–739.

[20] Herkenne, C., Alberti, I., Naik, A., Kalia, Y.N., Mathy, F.X., Préat, V., Guy, R.H. (2008). In vivo methods for the assessment of topical drug bioavailability. *Pharmaceutical Research, 25*, 87–103.

[21] Joukhadar, C., Müller, M. (2006). Microdialysis: current applications in clinical pharmacokinetic studies and its potential role in the future. *Clinical Pharmacokinetics, 44,* 895–913.

[22] Schmidt, S., Banks, R., Kumar, V., Rand, K.H., Derendorf, H. (2008). Clinical microdialysis in skin and soft tissues: an update. *Journal of Clinical Pharmacology, 48,* 351–364.

[23] Lee, G.J., Park, J.H., Park, H.K. (2008). Microdialysis applications in neuroscience. *Neurological Research, 30,* 661–668.

[24] Li, Y., Peris, J., Zhong, L., Derendorf, H. (2006). Microdialysis as a tool in local pharmacodynamics. *American Association of Pharmaceutical Scientists Journal, 8,* E222–E235.

[25] Zeitlinger, M., Müller, M, Joukhadar, C. (2005). Lung microdialysis: a powerful tool for the determination of exogenous and endogenous compounds in the lower respiratory tract (mini-review). *American Association of Pharmaceutical Scientists Journal, 7,* E600–E608.

[26] Brunner, M., Derendorf, H., Müller, M. (2005). Microdialysis for in vivo pharmacokinetic/pharmacodynamic characterization of anti-infective drugs. *Current Opinion in Pharmacology, 5,* 495–499.

[27] Port, R.E., Wolf, W. (2003). Noninvasive methods to study drug distribution. *Investigational New Drugs, 21,* 157–168.

[28] Fischman, A.J., Alpert, N.M., Rubin, R.H. (2002). Pharmacokinetic imaging: a noninvasive method for determining drug distribution and action. *Clinical Pharmacokinetics, 41,* 581–602.

[29] Pien, H.H., Fischman, A.J., Thrall, J.H., Sorensen, A.G. (2005). Using imaging biomarkers to accelerate drug development and clinical trials. *Drug Discovery Today, 10,* 259–266.

[30] Phelps, M.E. (2000). PET: the merging of biology and imaging into molecular imaging. *Journal of Nuclear Medicine, 41,* 661–681.

[31] van der Veldt, A.A., Luurtsema, G., Lubberink, M., Lammertsma, A.A., Hendrikse, N.H. (2008). Individualized treatment planning in oncology: role of PET and radiolabelled anticancer drugs in predicting tumour resistance. *Current Pharmaceutical Design, 14,* 2914–2931.

[32] Bauer, M., Wagner, C.C., Langer, O. (2008). Microdosing studies in humans: the role of positron emission tomography. *Drugs R&D, 9,* 73–81.

[33] Fischman, A.J., Babich, J.W., Bonab, A.A., Alpert, N.M., Vincent, J., Callahan, R.J., Correia, J.A., Rubin, R.H. (1998). Pharmacokinetics of [18F]trovafloxacin in healthy human subjects studied with positron emission tomography. *Antimicrobial Agents and Chemotherapy, 42,* 2048–2054.

[34] Brunner, M., Langer, O., Dobrozemsky, G., Müller, U., Zeitlinger, M., Mitterhauser, M., Wadsak, W., Dudczak, R., Kletter, K., Müller, M. (2004). [18F]Ciprofloxacin, a new positron emission tomography tracer for noninvasive assessment of the tissue distribution and pharmacokinetics of ciprofloxacin in humans. *Antimicrobial Agents and Chemotherapy, 48,* 3850–3857.

[35] Lyoo, I.K., Renshaw, P.F. (2002). Magnetic resonance spectroscopy: current and future applications in psychiatric research. *Biological Psychiatry, 51,* 195–207.

[36] Wolf, W., Presant, C.A., Waluch, V. (2000). 19F-MRS studies of fluorinated drugs in humans. *Advanced Drug Delivery Reviews, 41,* 55–74.

[37] Griffiths, J.R., Glickson, J.D. (2000). Monitoring pharmacokinetics of anticancer drugs: non-invasive investigation using magnetic resonance spectroscopy. *Advanced Drug Delivery Reviews, 41,* 75–89.

[38] Schneider, E., Bolo, N.R., Frederick, B., Wilkinson, S., Hirashima, F., Nassar, L., Lyoo, I.K., Koch, P., Jones, S., Hwang, J., et al. (2006). Magnetic resonance spectroscopy for measuring the biodistribution and in situ in vivo pharmacokinetics of fluorinated compounds: validation using an investigation of liver and heart disposition of tecastemizole. *Journal of Clinical Pharmacology and Therapeutics, 31,* 261–273.

[39] Klomp, D., van Laarhoven, H., Scheenen, T., Kamm, Y., Heerschap, A. (2007). Quantitative ^{19}F MR spectroscopy at 3 T to detect heterogeneous capecitabine metabolism in human liver. *NMR in Biomedicine, 20,* 485–492.

[40] Reid, D.G., Murphy, P.S. (2008). Fluorine magnetic resonance in vivo: a powerful tool in the study of drug distribution and metabolism. *Drug Discovery Today, 13,* 473–480.

[41] Fischman, A.J., Alpert, N.M., Livni, E., Ray, S., Sinclair, I., Callahan, R.J., Correia, J.A., Webb, D., Strauss, H.W., Rubin, R.H. (1993). Pharmacokinetics of ^{18}F-labeled fluconazole in healthy human subjects by positron emission tomography. *Antimicrobial Agents and Chemotherapy, 37,* 1270–1277.

[42] Dolovich, M., Labiris, R. (2004). Imaging drug delivery and drug responses in the lung. *Proceedings of the American Thoracic Society, 1,* 329–337.

[43] Aboagye, E.O., Price, P.M. (2003). Use of positron emission tomography in anticancer drug development. *Investigational New Drugs, 21,* 169–181.

[44] Moehler, M., Dimitrakopoulou-Strauss, A., Gutzler, F., Raeth, U., Strauss, L.G., Stremmel, W. (1998). ^{18}F-labeled fluorouracil positron emission tomography and the prognoses of colorectal carcinoma patients with metastases to the liver treated with 5-fluorouracil. *Cancer, 83,* 245–253.

[45] Saleem, A., Brown, G.D., Brady, F., Aboagye, E.O., Osman, S., Luthra, S.K., Ranicar, A.S., Brock, C.S., Stevens, M.F., Newlands, E., Jones, T., Price, P. (2003). Metabolic activation of temozolomide measured in vivo using positron emission tomography. *Cancer Research, 63,* 2409–2415.

[46] Brunner, M., Langer, O., Sunder-Plassmann, R., Dobrozemsky, G., Müller, U., Wadsak, W., Krcal, A., Karch, R., Mannhalter, C., Dudczak, R., et al. (2005). Influence of functional haplotypes in the drug transporter gene *ABCB1* on central nervous system drug distribution in humans. *Clinical Pharmacology & Therapeutics, 78,* 182–190.

[47] Langer, O., Bauer, M., Hammers, A., Karch, R., Pataraia, E., Koepp, M.J., Abrahim, A., Luurtsema, G., Brunner, M., Sunder-Plassmann, R., et al. (2007). Pharmacoresistance in epilepsy: a pilot PET study with the P-glycoprotein substrate *R*-[(11)C]verapamil. *Epilepsia, 48,* 1774–1784.

[48] Bauer, M., Langer, O., Dal-Bianco, P., Karch, R., Brunner, M., Abrahim, A., Lanzenberger, R., Hofmann, A., Joukhadar, C., Carminati, P., et al. (2006). A positron emission tomography microdosing study with a potential antiamyloid drug in healthy volunteers and patients with Alzheimer's disease. *Clinical Pharmacology & Therapeutics, 80,* 216–227.

[49] Hillered, L., Vespa, P.M., Hovda, D.A. (2005). Translational neurochemical research in acute human brain injury: the current status and potential future for cerebral microdialysis. *Journal of Neurotrauma, 22,* 3–41.

[50] Helmy, A., Carpenter, K.L., Hutchinson, P.J. (2007). Microdialysis in the human brain and its potential role in the development and clinical assessment of drugs. *Current Medicinal Chemistry*, *14*, 1525–1537.

[51] Oddo, M., Schmidt, J.M., Carrera, E., Badjatia, N., Connolly, E.S., Presciutti, M., Ostapkovich, N.D., Levine, J.M., Le Roux, P., Mayer, S.A. (2008). Impact of tight glycemic control on cerebral glucose metabolism after severe brain injury: a microdialysis study. *Critical Care Medicine*, *36*, 3233–3238.

[52] Pan, J.W., Williamson, A., Cavus, I., Hetherington, H.P., Zaveri, H., Petroff, O.A., Spencer, D.D. (2008). Neurometabolism in human epilepsy. *Epilepsia*, *49* Suppl. 3, 31–41.

[53] Cavus, I., Pan, J.W., Hetherington, H.P., Abi-Saab, W., Zaveri, H.P., Vives, K.P., Krystal, J.H., Spencer, S.S., Spencer, D.D. (2008). Decreased hippocampal volume on MRI is associated with increased extracellular glutamate in epilepsy patients. *Epilepsia*, *49*, 1358–1366.

[54] Noske, D.P., Peerdeman, S.M., Comans, E.F., Dirven, C.M., Knol, D.L., Girbes, A.R., Vandertop, W.P. (2005). Cerebral microdialysis and positron emission tomography after surgery for aneurysmal subarachnoid hemorrhage in grade I patients. *Surgical Neurology*, *64*, 109–115.

[55] Hutchinson, P.J., Gupta, A.K., Fryer, T.F., Al-Rawi, P.G., Chatfield, D.A., Coles, J.P., O'Connell, M.T., Kett-White, R., Minhas, P.S., Aigbirhio, F.I., et al. (2002). Correlation between cerebral blood flow, substrate delivery, and metabolism in head injury: a combined microdialysis and triple oxygen positron emission tomography study. *Journal of Cerebral Blood Flow and Metabolism*, *22*, 735–745.

[56] Vespa, P., Bergsneider, M., Hattori, N., Wu, H.M., Huang, S.C., Martin, N.A., Glenn, T.C., McArthur, D.L., Hovda, D.A. (2005). Metabolic crisis without brain ischemia is common after traumatic brain injury: a combined microdialysis and positron emission tomography study. *Journal of Cerebral Blood Flow and Metabolism*, *25*, 763–774.

[57] Bosche, B., Dohmen, C., Graf, R., Neveling, M., Staub, F., Kracht, L., Sobesky, J., Lehnhardt, F.G., Heiss, W.D. (2003). Extracellular concentrations of non-transmitter amino acids in peri-infarct tissue of patients predict malignant middle cerebral artery infarction. *Stroke*, *34*, 2908–2913.

[58] Enblad, P., Valtysson, J., Andersson, J., Lilja, A., Valind, S., Antoni, G., Langstrom, B., Hillered, L., Persson, L. (1996). Simultaneous intracerebral microdialysis and positron emission tomography in the detection of ischemia in patients with subarachnoid hemorrhage. *Journal of Cerebral Blood Flow and Metabolism*, *16*, 637–644.

[59] Hutchinson, P.J., Gupta, A.K., Fryer, T.F., Al-Rawi, P.G., Chatfield, D.A., Coles, J.P., O'Connell, M.T., Kett-White, R., Minhas, P.S., Aigbirhio, F.I., et al. (2002). Correlation between cerebral blood flow, substrate delivery, and metabolism in head injury: a combined microdialysis and triple oxygen positron emission tomography study. *Journal of Cerebral Blood Flow and Metabolism*, *22*, 735–745.

[60] Dohmen, C., Bosche, B., Graf, R., Staub, F., Kracht, L., Sobesky, J., Neveling, M., Brinker, G., Heiss, W.D. (2003). Prediction of malignant course in MCA infarction by PET and microdialysis. *Stroke*, *34*, 2152–2158.

[61] Langer, O., Karch, R., Müller, U., Dobrozemsky, G., Abrahim, A., Zeitlinger, M., Lackner, E., Joukhadar, C., Dudczak, R., Kletter, K., Müller, M., Brunner, M.

(2005). Combined PET and microdialysis for in vivo assessment of intracellular drug pharmacokinetics in humans. *Journal of Nuclear Medicine*, *46*, 1835–1841.

[62] Langer, O., Mitterhauser, M., Brunner, M., Zeitlinger, M., Wadsak, W., Mayer, B.X., Kletter, K., Müller, M. (2003). Synthesis of fluorine-18-labeled ciprofloxacin for PET studies in humans. *Nuclear Medicine and Biology*, *30*, 285–291.

[63] Virtanen, K.A., Peltoniemi, P., Marjamaki, P., Asola, M., Strindberg, L., Parkkola, R., Huupponen, R., Knuuti, J., Lönnroth, P., Nuutila, P. (2001). Human adipose tissue glucose uptake determined using [(18)F]-fluoro-deoxy-glucose ([(18)F] FDG) and PET in combination with microdialysis. *Diabetologia*, *44*, 2171–2179.

[64] Peltoniemi, P., Lönnroth, P., Laine, H., Oikonen, V., Tolvanen, T., Gronroos, T., Strindberg, L., Knuuti, J., Nuutila, P. (2000). Lumped constant for [(18)F]fluoro-deoxyglucose in skeletal muscles of obese and nonobese humans. *American Journal of Physiology Endocrinology and Metabolism*, *279*, E1122–E1130.

15

IN VITRO APPLICATIONS OF MICRODIALYSIS

WEN-CHUAN LEE

Institute of Traditional Medicine, National Yang-Ming University, Taipei, Taiwan

TUNG-HU TSAI

Institute of Traditional Medicine, National Yang-Ming University, and Taipei City Hospital, Taipei, Taiwan

1. INTRODUCTION

Microdialysis is a technique originally developed for the neurological sciences that has been used widely in physiological, pharmacological, toxicological, and behavioral studies for the recovery of exogenous substances such as drugs and toxicants [1,2]. It is uniquely able to determine the free extracellular fluid levels in various tissues, such as muscle, brain, and lung, either in animals or humans, without directly extracting body fluids [3]. In addition, the microdialysis technique minimizes damage to the investigation site. It offers an alternative to traditional distribution studies that sample the tissue from animals by sacrificing at different times, providing a significant reduction in the number of animals that must be sacrificed. This technique allows real-time observation of living animals or humans. The basic principle of microdialysis sampling is that it is a diffusion-based membrane separation technique that has been used principally to obtain representative samples from the extracellular fluid of living tissue [4]. A dialysis probe is placed in the target tissue and connected with a precise micropump to delivery the perfusion fluid. During sampling, perfusion fluid is passed through the device due to the concentration gradient at low μL/min flow rates. Finally the dialysate is collected for further analysis. Because the typical molecular cutoff could exclude large molecules such as proteins, microdialysis samples can be injected directly into analysis systems such as chromatographic systems. Microdialysis is used to satisfy various issues in different clinical fields, such as monitoring for tumor assessment [5], early detection of secondary brain injuries [6], transplant rejection [7], and anastomosis leakage [8]. Microdialysis is also used in clinical pharmacology to measure concentrations of different drugs, such as antibiotics [9], anti-inflammatory [10], and psychoactive compounds [11], at target sites in different tissues and organs and, subsequently, to relate target-site pharmacokinetics to pharmacodynamics.

In addition to applications in pharmacokinetics studies in the field of living animals, or in vivo systems, microdialysis has also expanded considerably in in vitro fields. For in vitro systems, animal sacrifice is not necessary, although microdialysis applied in this field is based on the characteristics of continuously monitoring the target analyte in the specific environment. Through the continuous collecting sample technique, a clear kinetics model resulting from one individual subject could provide a series of time results over the entire experimental period. Furthermore, with microdialysis sampling, the samples collected are clean enough for direct analysis by detectors. Microdialysis for in vivo systems has been discussed extensively, and in vitro applications of microdialysis are described in this chapter. These applications cover microdialysis utilized for culture systems, including cell, tissue, and even organ, enzyme kinetics systems; fermentation systems; drug development, including plasma protein binding; environmental monitoring; etc and so on. These topics are covered in the following sections.

2. MICRODIALYSIS USED IN CULTURE SYSTEMS

The culture system was first demonstrated in 1885, when Wilhelm Roux [12] maintained embryonic chick cells in a warm saline solution outside the animal body for several days. In the following years, researchers developed different culture systems, cell systems, and cell lines for long-term culture. In these culture systems, mediua are widely used to support the growth of cells, tissues, or microorganisms. These mediua include basic components of amino acids, vitamins, salts, proteins, such as hormones or growth factors, and others such as glucose and penicillin. Analysis of the medium can provide information on the interactions between cultured targets and chemicals. Conventional medium sampling requires time-consuming pretreatments or complicated extraction prior to analysis, which may result in sample losses and increases in the amount of time needed to complete the experiments. In contrast, microdialysis sampling provides a relatively clean sample without the need for further pretreatment process, which makes real-time analysis possible. Microdialysis also speeds up the experimental procedure, provides high sensitivity, minimizes the sample volume required, enhances the detection limits, and decreases degradation of analyzed compounds [13]. Some examples that employ microdialysis sampling in culture systems are described next.

2.1. Tissue Culture

Tissue culture is often used generically to refer to both organ culture and cell culture. Here, the form is focused primarily on the culture of tissue collected from living creatures. Microdialysis was first used in tissue culture systems in the early 1990s by Einspanier et al. [14]. In their study, the microdialysis probe was inserted into bovine corpus luteum tissue collected from cows and incubated in an organ culture chamber. This in vitro culture system provided luteal cells to maintain cell-to-cell contact. The dialysate was collected to detect the variation of progesterone and oxytocin when stimulated by different doses of insulin-like growth factor-I (IGF-I). The authors suggested that IGF-I could be important in regulating the function of bovine corpus luteum and may act in an autocrine–paracrine way. Other investigators utilized the same or similar microdialysis sampling techniques to develop studies involving the alternation of hormones [15–20], growth factors [21–23], proteins [24], and peptides [25,26] stimulating the secretory function of the bovine corpus luteum. Microdialysis was also used by Maas et al. [27] in the investigation of paracrine actions of oxytocin, prostaglandin F2 α, and estradiol within the human corpus luteum. Einspanier and Hodges [28] employed microdialysis sampling to study the stimulation of human luteinizing hormone and human chorionic gonadotrophin (hCG) for the release of progesterone in intact luteal tissue of the marmoset monkey. In a recent investigation, Tanaka et al. [29] infused angiopoietin-2 to the bovine corpus luteum through a microdialysis probe to study the effect of angiopoietin-2 on progesterone release from the corpus luteum at different stages of the estrous cycle. Piotrowska et al. [30] demonstrated that phytoestrogens and their active metabolites may disrupt bovine corpus luteum function by inhibiting luteinizing hormone and prostaglandin-stimulated progesterone secretion. Beindorff et al. [31] delivered hCG and collected target chemical progesterone through a microdialysis probe to investigate the effect of hCG on progesterone secretion.

In addition to the corpus luteum tissue culture system, microdialysis sampling has been used to analyze hippocampal slice cultures. In 1991, Stoppini and co-workers [32] described a technique for preparing hippocampal organotypic cultures whereby brain slices are maintained on a porous and transparent membrane at the interface a between a culture medium and the atmosphere. These tissue cultures do not spread as monolayers but retain a three-dimensional organization that preserves the different cell types. Dendritic processes of pyramidal neurons and the time course of synaptic development resemble those observed in situ at comparable developmental stages. In 1993, the research group of Menéndez et al. [33] applied this culture technique, focusing on the biological function of rat hippocampal slices. They investigated the dependence of taurine release induced by N-methyl-d-aspartate (NMDA) on extracellular calcium and/or on calcium mobilization from intracellular stores. In this study, the microdialysis probe was used not only for collecting samples but also for administrating the drug. NMDA was administered through a microdialysis probe inserted into rat hippocampal slices, and the probe was also used to collect amino acids from extracellular space. In a later investigation report from Robert et al., [34], a combination of extracellular electrophysiological multirecordings from hippocampal organotypic slice cultures with application of drugs to and sampling of extracellular fluid from a restricted region of the slice using a microdialysis probe was investigated. This in vitro hippocampal organotypic culture system was used to mimic in vivo situation in a simpler way. The hippocampal slice was carried by a Millicell-CM membrane that was placed on the Anopore membrane of the lower part of the multielectrode array system. The microdialysis probe was place on the surface of the brain slice and connected to capillary electrophoresis equipped with laser-induced fluorescence detection (CE-LIFD), which provided a real-time analysis of samples. The investigation focused on the effect of K^+-induced depolarization, glutamate uptake blockade *trans*-pyrrolidine-2,4-dicarboxylic acid (PDC), and high-frequency electrical stimulation on the concentration variation of glutamate, which is one of the most important neurotransmitters in the central nervous system and is involved in many aspects of normal brain functioning. In their follow-up investigations, they provided the effect of PDC administration and ischemia on glutamate and *O*-phosphoethanolamine (PEA) concentrations [35].

In addition, microdialysis has been utilized to characterize the in vitro blood–brain–barrier (BBB) model [36]. In this investigation, the researchers not only used microdialysis for sampling but also combined microscopy to observe the morphology of the co-culture assembly with electrophysiology to recording the synaptic activity [excitatory postsynaptic potential (EPSP)] from the extracellular medium used to assess the permeability of the BBB to neuroactive solutes. This in vitro BBB model was set up by co-culture assembly. First, endothelial cells were seeded on the collagen-treated patches of a transparent and porous membrane, and then hippocampal slices were laid on top of endothelial cell monolayers. In this study, microdialysis probes were placed on the surface of the hippocampal slices to monitor the diffusion of exogenous molecules through the slices. Neurotransmitter dopamine, its precursor L-dopa, and glutamate with different BBB permeabilities were perfused through the culture medium chamber, and their concentrations were determined in brain slice microdialysates. Since the microdialysis probe was placed on the surface of tissues without breaking the BBB, this in vitro BBB model provides a useful tool for assessing drug permeability and revealing factors that influence the activity of the endothelial cells in their maintenance of a constant environment for the brain.

Recently, microdialysis has been used to detect local changes in cellular metabolism within tissue-engineered constructs. A chondrocyte–alginate construct system [37] and a bovine caudal intervertebral disk system [38] were investigated. Tissue engineering provides a potential method for the repair of damaged or failing tissues or organs by seeding appropriate cells into a three-dimensional scaffold and culturing the resulting construct in a bioreactor under conditions that promote cell and tissue growth. Since cellular activity or any biochemical parameters changes with time are important for successful tissue-engineered constructs, microdialysis is a useful tool that provides continuous sampling to monitor the local concentration of metabolites and tissue components in the extracellular fluid of the construct. In this culture system, local lactate and glucose concentrations were taken as markers of cellular activity. Such information during the growth of tissue-engineered constructs would allow either corrective action or an early termination of an unsuccessful test. Use of the system developed could be extended to the development of engineered tissue and to monitor tissue repair in vivo.

2.2. Cell Culture

Microdialysis has been used in cell culture systems such as nerve cells [39], neural stem cells [40,41], liver cells [42], mammalian cells [43], and even macrophages [44]. In these cell culture systems, the nerve cell culture system was the most extensively discussed, perhaps because the microdialysis technology was first developed for detection neurotransmitters. In these cell culture systems, the target cells are cultured in a petri dish, flask, or even a bioreactor. Then a microdialysis probe is generally immersed in the culture medium and the target analyte is exchanged out by eluting the perfusate. Usually, a shunt probe is used in the culture systems (Figure 1a). This type of probe is widely used for profiling analytes in the rat bile duct since it produces disruption of enterohepatic circulation of bile flow (Figure 1b). Sometimes, a needle-type microdialysis probe is also used for the cell culture system. A cell culture system coupled to the microdialysis sampling technique is relatively efficient,

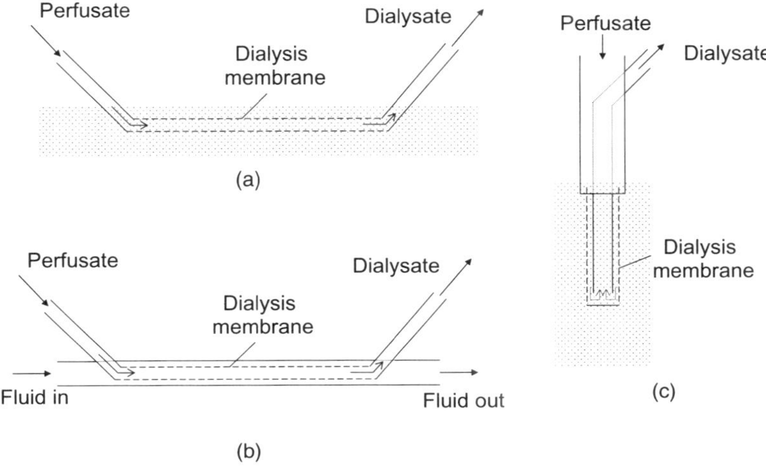

Figure 1   Microdialysis probes: (a) linear probe; (b) shunt probe; (c) needle probe.

cost-effective, and less vulnerable to human error then are conventional studies.

Many researchers utilize microdialysis sampling when culturing nerve cell systems to determine the correlations between every type of stimulation and cell life, including growth, proliferation, metabolism, apoptosis, and cell death. Since the microdialysis samples are clean enough for direct analysis without further pretreatment such as extraction, real-time analysis becomes possible for in situ observation biological behavior. This research includes investigation of extracellular L-glutamate concentration change by stimulating a single nerve cell with KCl or by stimulating a brain slice electrically [39]; the effect of KCN-induced hypoxia on the release of catecholamines in pheochromocytoma cell (PC-12 cells) [45]; the effect of electromagnetic impulse on neurotransmitter metabolism in nerve cells [46]; the mechanism of nitric oxide donor–induced changes in dopamine secretion [47] and ascorbic acid in 3-morpholinosydnonimine (SIN-1)-induced changes in dopamine secretion from PC-12 cells [48]; the role of endogenous melatonin in the oxidative homeostasis of the extracellular striatal compartment in PC-12 cells [49]; the biological effects, cell proliferation, neuroprotection effects, differentiation capability, and functional production of PC-12 cells treated with nerve growth factor [50]; and the effects of various drugs on dopamine secretion through direct comparison of extracellular dopamine variation in PC12 cells [51].

There are also investigations focused on neural stem cells. (NSCs), which are derived from the hippocampus and other germinal centers of the brain and are self-renewing multipotent cells that generate the main phenotypes of the nervous system [52]. In an investigation by Chen et al. using microdialysis sampling in the neural stem cell culture system, the authors found that serotonin secretion in neural stem cells has been regulated by the treatment of fluoxetine, which is a popular antidepressant drug in the treatment of depression as well as other psychiatric and neurodegenerative diseases, and the expression of the anti-apoptotic *Bcl-2* gene[40]. The same research group utilized the same technique to investigate the functional release of serotonin in the process of serotoninergic differentiation of imipramine, an antidepressant drug, on treated NSCs, concomitantly increasing and mediated by activation of the BDNF/MAPK/ERK pathway/*Bcl-2* cascades [41].

Microdialysis sampling has been used in the primary liver cell culture system [42]. The investigation provides variation of pyruvate and lactate under hypoxia culture conditions. Lactate and pyruvate have been used as important biochemical markers of ischemia [53–55]. The results showed that pyruvate concentrations decreased dramatically, to 73%, during hypoxia, while lactate increases to about 110%. It also revealed that the production of pyruvate is affected significantly by hypoxia, and the production rate recovers slowly after hypoxia. Meanwhile, the production of lactate is affected slightly by hypoxia, but the production rate of lactate continues to increase following hypoxia. These results provided information regarding ischemia in experimental animals and clinical studies.

Macrophages are a class of inflammatory cells believed to act in nonspecific defense and to initiate specific defense mechanisms. Macrophages are major phagocytic cells that release various molecules to regulate inflammation. Nitro oxide is an important signal transmitter related to different inter- and intracellular communication processes. Microdialysis sampling has been utilized to collect the nitro oxide produced from activated macrophages stimulated by lipopolysaccharide [44,56]. Microdialysis can be used to measure and monitor nitro oxide production in biological systems without interfering with cell viability. These investigations make it possible to measure various chemical mediators from activated macrophages with temporal resolution, and provide a greater understanding of different cell–cell interactions and signaling events during inflammatory responses. Microdialysis sampling is employed widely in various in vivo systems, and with a microdialysis probe implanted into the tissue, macrophages may activate the immune system to damage the semipermeability of membrane. Mou and Stenken investigated the differences in the sampling extraction efficiencies of glucose and 2-deoxyglucose in the dorsal subcutis of rats and in a macrophage culture system [57]. In this in vivo and in vitro comparison study, the authors found that macrophage activation in vivo at implant sites is much lower than that of highly activated macrophages in vitro.

2.3. Sampling Macromolecules Through Microdialysis

Besides collecting small molecules, microdialysis has been utilized to collect large molecules in culture systems, includeing peptides, proteins, and glycoproteins, which are responsible for cellular activities. Since the molecular weights of these molecules are high, the molecular weight cutoff of the membrane is a key factor that must be considered when microdialysis is used to collect samples. Different large molecules have been the target collected analytes in in vitro investigations. Schutte et al. [58] provided some information regarding in vitro sampling macromolecules through microdialysis technique. In their investigation, proteins and dextrans with molecular weight from 3000 to 150,000 Da were the target analytes. They found that the choice of a 100,000-Da molecular weight cutoff probe made possible the collection of proteins and dextrans. The collection efficiencies of proteins of interests with molecular weights from 5700 to 66,200 Da, ranged from 8.1 to 18.5%, respectively. Relative recoveries for 150,000-Da dextran and IgG, both of which exceeded the molecular weight cutoff of the membrane, were very small but not zero. They also tried to give a possible model to investigate the correlation between relative recovery and the molecular weight of large molecules. Wang and Stenken also focused on the differences in microdialysis sampling of macromolecules and small molecules [59]. In this study, the authors provided greater physical insight into the sampling process for larger molecules. These results gave later investigators important information for microdialysis collection of macromolecules in an in vivo system.

Cytokines, which are proteins, peptides, or glycoproteins, play a crucial role in many normal and pathological conditions. The understanding of cytokine networks and their interactions has become a priority. The goals of controlling inflammation, augmenting resistance to infection, or accelerating tissue repair have been elusive in large part because of the complexity of cytokine networks. Phillips et al. utilized microdialysis to investigate cell secretion of regulatory cytokines in a low-cell-density culture system [60] or a single-cell-level culture system [61,62]. Ao et al. [63] investigated the cytokine secretion in a lipopolysaccharide (LPS)-stimulated macrophage culture system. Rosenbloom et al. [64] established microdialysis as an alternative method of studyind the multiple cytokine and chemokine responses occurring in living tissues [64]. They focused on six proteins in a size range from 17 to 80 kDa. It is found that conventional microdialysis with a 100-kDa molecular weight cutoff probe allows efficient and sustained sampling of proteins up to at least 29 kDa. The use of dextran in the circulating buffer helps to counterbalance the high osmolarity of tissue interstitial fluid to prevent probe volume loss. These investigations provide the possibility of utilizing a microdialysis system to collect large molecules.

2.4. Bioprocess

Another application that does not require high temporal resolution but where continuous online monitoring is useful is the in vitro monitoring of products of bioreactors. Here, the time course of the experiment is usually several hours or days. In this case, sample collection and analysis every 15 min can provide adequate information regarding the progress of the reaction system [65,66]. Fermentation and mammalian cell culture are important bioprocesses since these culture systems provide desired products that we need, including acids, alcohols, lactic acids, viral vaccines, enzymes, hormones, antibodies, interleukins, lymphokines, and anticancer agents. Placed in a bioreactor, a microdialysis sampling probe subjected to the environment can disturb the dialysis process (e.g., by adsorption of compounds on the membrane surface). Buttler et al. [67] described how the dialysis performance of various membranes for alcohols and carbohydrates is affected in the presence of matrices. Marko-Varga et al. [68] established an online bioprocess monitoring system with the aid of microdialysis sampling. In this study, the authors gave two examples of this monitoring system; one is a well-established penicillin fermentation and the other a fermentation process presently under study for the optimization of ethanol production from industrial waste from the paper industry. They demonstrated the usability of a microdialysis probe as a sampling device in fermentation systems. The follow-up investigation applied this online monitoring system to monitoring the ethanol produced in 70-h fermentation of a lignocellulose hydrolysate [69], the production of inositol trisphosphate from phytic acid, the fermentation of lignocellulose hydrolysate to fuel ethanol [70], the enzymatic hydrolysis of mannan with endi-1,4-β-mannanase

[71], and the production of desalted enzymatic hydrolysates of lignocellulosic substrates [72].

Glucose monitoring becomes very important in bioprocesses because glucose is the major energy source for cells, and optimal glucose concentrations are crucial for healthy cell growth and efficient product yield. The microdialysis sampling technique is a good candidate for collecting glucose from bioprocesses since the semimembrane can allow target analyte penetrating through the membrane without affecting the cell cycles. Palmqvist et al. [73] provided online monitoring of glucose and ethanol in the continuous fermentation of enzymatic hydrolysates of spruce with *Saccharomyces cerevisiae*. They utilized microdialysis sampling to set the dilution rate and to study the influence of pH on fermentation. The results demonstrated that pH affected the cell growth rate and that in such a culture system, it is necessary to employ cell recycling in continuous fermentation of spruce hydrolysate to obtain high productivity. Through microdialysis, continuous monitoring of the cell cycle in the fermentation system is possible. According to the information collected, it is easy to regulate the cell growth at a sufficient level to compensate for cell death and to get the maximum biomass product. Ge et al. [43] employed a similar real-time technique to monitor the variation of glucose concentration in the mammalian cell culture system. In their investigation, the correlation of glucose concentration and mammalian cell cycle was clearly demonstrated. They also provided information on the effects on dialysis efficiency of bulk glucose in the medium, perfusion flow rates, and temperature.

3. MICRODIALYSIS USED IN ENZYME KINETICS

Enzymes are proteins with catalytic properties and are the reaction catalysts of biological systems. They have extraordinary catalytic power, often a high degree of substrate specificity, and can greatly accelerate specific chemical reactions [74]. Enzyme kinetics is the study of the chemical reactions that are catalyzed by enzymes, with a focus on their reaction rates. The study of an enzyme's kinetics reveals the catalytic mechanism of this enzyme, its role in metabolism, how its activity is controlled, and how a drug or poison might inhibit the enzyme. Monitoring enzyme kinetics is feasible not only for well-characterized enzymes under ideal conditions but also for enzymes under nonideal conditions (e.g., pH, temperature), with alternative substrates, or for new enzymes. Microdialysis sampling provides time-dependent sampling and rapid continuous sampling, and these characters avoid enzymatic degradation of the sample. This technique allows researchers to sample the reactant mixture continuously to obtain a complete kinetic profile [75].

In 1996, Zhou and colleagues [76] first utilized the microdialysis sampling technique to investigate the kinetics of enzyme catalysis. 2′,3′,5′-Triacetyl-6-azauridine (azaribine) with porcine liver esterase (PLE) and *N*-acetylphenylalanyl-3,5-diiodotyrosine (AcFY′) with pepsin were used as model compounds. Microdialysis sampling permitted the rapid separation of low-molecular-weight analytes from macromolecules, thus achieving simultaneous cleanup of the samples and quenching of the reaction. In their follow-up investigation, in vitro metabolism of azaribine by porcine liver esterase and in human plasma using the same technique coupled with a fast gradient-liquid chromatographic analysis method was studied [77]. The authors developed two simplified kinetic schemes to describe the time course of azaribine, the intermediates, and the final metabolites with five and four rate constants for the metabolism of azaribine by PLE and in human plasma, respectively. In an investigation by Zook and LaCourse [78], microdialysis sampling was used in a study of glucose oxidase reaction. In this investigation, a microdialysis cell rather than a probe was designed to accept a wide range of commercially available flat membranes and as an alternative to commercial probes designed for in vivo use. Other investigations utilized microdialysis to detect the activity of enzymes, such as proteolytic enzyme, porcine elastase [79], and the carbohydrate enzyme almond β-glucosidase [80].

Metabolism is a set of chemical reactions that occur in living organisms to maintain life. Enzymes are crucial to metabolism because they allow organisms to drive desirable biological reactions. Enzymes also allow the regulation of metabolic pathways in response to changes in a cell's environment or signals from other cells. Numerous enzyme activities are associated with the microsomal fraction. Through a microsomal incubation technique, in vitro investigation of the metabolism of chemicals or drugs is possible. Traditional sampling techniques for studying microsomal incubations or pharmacokinetics use discontinuous sampling at specified intervals. Microsomal studies are normally carried out by incubating the parent drug at 37°C with liver microsomes containing cytochrome P450 enzymes. The reaction is usually started by adding the enzyme cofactor NADPH or an NADPH-generating system. After a fixed interval the reaction is terminated by adding a quenching agent or an inhibitor to stop the enzymatic process. The samples drawn from the microsomal mixture need to be cleaned up to remove protein prior to the analytical system. Since samples collected by microdialysis are protein-free, further sample cleanup is not required before analysis. Gunaratna and Kissinger [81] investigated phase I metabolism of salicyclic acid, diazepam, and ibuprofen in rat liver microsomes by microdialysis. They applied the same technique to the stereoselective metabolism of amphetamine in rat liver microsomes [82]. Kolanczyk el al. [83] applied this technique to a metabolic scheme for metabolism of 4-methoxyphenol by rainbow trout microsomes. Wen et al. [84] provided not only the metabolism of calycosin and formononetin but also drug–drug interactions by dynamic microdialysis sampling. Sun et al. [85] studied 11β-hydroxysteroid dehydrogenase type 1 enzyme activity of conversion of stable-isotope-labeled cortisone to cortisol in human, dog, and monkey liver microsomes. Here, microdialysis was used to sample not only the cortisol from microsome incubation, but also infused cortisone to microsome incubation. In this way, interconversion from cortisone to cortisol and the cortisol production rate could both be investigated.

4. MICRODIALYSIS USED IN PROTEIN BINDING

The interactions of proteins with various ligands create the basis of an interlocked set of dynamic processes providing a communication and regulation pathway within and between different structures of a living organism. Most drugs bind to proteins or other biological materials, such as albumin; α_1-acid glycoprotein; lipoproteins; α-, β-, and γ-globulins; and erythrocytes. These drugs undergo some degree of reversible binding to plasma proteins, a process that may have significant effects on the overall activity profile of the compounds. Free, unbound drug concentrations in plasma decrease as the degree of binding to these compounds increases. It is a well-recognized fact that at least for small molecules, only free, unbound drug distributes into the extravascular space and is responsible for pharmacological activity and/or side effects. Meanwhile, unbound drug concentrations are believed to be more relevant than total drug to pharmacological and toxicological responses. To adjust the optimum therapeutic dose of a drug in humans it is necessary to know the ratio of bound to unbound drug. It is evident that a more profound understanding of the molecular basis of the interactions between drugs and proteins is of great practical and theoretical importance. Various methods have been developed to study drug–protein phenomena in vitro and in vivo. Because drug–protein binding is a reversible and kinetically rapid interaction, it should be analyzed without disturbing the binding equilibrium. Until now, equilibrium dialysis [86,87], ultrafiltration [88], and ion-selective electrodes [89,90] have been used widely to determine unbound drug concentrations in both in vivo and in vitro samples. However, these conventional methods suffer from drawbacks. Equilibrium dialysis experiments require large quantities of drug and are time consuming, due to the long periods required to reach equilibrium. The ultrafiltration method also requires large amounts of drug, and the drug–protein equilibrium may be altered by changes in sample concentration during filtration.

In recent years, microdialysis has been introduced to measure free concentrations of various drugs in vivo and in vitro for studying the binding properties between drugs and plasma protein [91–95]. This technology has some important benefits. It takes advantage of the large size difference between small molecules and macromolecules and does not require labeling, immobilization, mobility difference, or stopped flow. Wang et al. [96] provided a series of investigations of drug–protein binding in vitro. They utilized microdialysis to investigate the protein-binding kinetics of sulfamethoxazole. They discussed the protein-binding promise of carbamazepine, which has a narrow therapeutic plasma-concentration range (4 to 12 µg/mL), and human serum albumin [97]. They also demonstrated that the enhancement of microdialysis recovery by affinity trapping agents can be used to evaluate binding kinetics using ketoprofen and human serum albumin [98]. The results found fast binding of ketoprofen to human serum albumin and fast dissociation of the complex formed, which are consistent with the reversible binding property of drug and human serum albumin. In a Schmidt et al. investigation [99], microdialysis was used to collect the highly protein-binding antibiotics ceftriaxone and ertapenem. The researchers evaluated the effect of protein binding of these antibiotics on their respective antimicrobial activity against microorganisms. There are some other investigations focused on drug–protein binding, such as those on streptomycin sulfas [100], gefitinib [101], and even the traditional Chinese medicine *Flos Lonicerae Japonicae*, which is composed of many different active compounds [102]. In the latter investigation the authors provided not only drug–protein binding interaction but also evaluated the binding kinetics in a magnitude of seconds.

Besides drugs, protein binding of metal elements or ions has been discussed. Trace metal elements such as Cu(II) and Zn(II) play an important role in human growth, development, cell splitting, and the synthesis of proteins and DNA. Most exchangeable metal elements are bound to human serum albumin. The binding of metal ions to human serum albumin is a very critical theme for biochemistry. Guo et al. [103] investigated the binding of Cu(II) and Zn(II) with human serum albumin (HSA). They estimated the binding constants between Cu(II), Zn(II), and HSA. Two classes of binding sites on HSA have been observed, and the affinity of Cu(II) is much stronger than Zn(II) to this secondary binding site. Meanwhile, a similar technique has been used in the detection of metal elements found in industrial and environmental pollutants. Prozialeck et al. [104,105] discussed the interaction of cadmium (Cd^{2+}), which causes severe damage to a variety of organ systems and is teratogenic and carcinogenic, with cadherin. Cadherin is a Ca^{2+}-dependent cell adhesion molecule that is localized at the adhering junctions of epithelial cells. The results demonstrated that Cd^{2+} can interact with the Ca^{2+} binding site on the peptide B molecule and distort the secondary structure of the peptide, and that E-cadherin may be a direct molecular target for Cd^{2+} toxicity.

5. CONCLUSIONS

In this chapter, many in vitro applications of microdialysis sampling techniques have been summarized. These applications are based on the characteristics of microdialysis: continuous sampling, which provides real-time observation; sampling without further separation, which makes direct analysis possible; and sampling located at a specific site, which constructs a local variation detection. These in vitro investigations give the following in vivo studies more information, thus saving time and expense.

REFERENCES

[1] Elmquist, W.F., Sawchuk, R.J. (2000). Use of microdialysis in drug delivery studies. *Advanced Drug Delivery Reviews*, *45*, 123–124.

[2] de Lange, E.C., de Boer, A.G., Breimer, D.D. (2000). Methodological issues in microdialysis sampling for pharmacokinetic studies. *Advanced Drug Delivery Reviews*, *45*, 125–148.

[3] de la Peña, A., Liu, P., Derendorf, H. (2000). Microdialysis in peripheral tissues. *Advanced Drug Delivery Reviews, 45*, 189–216.

[4] Müller, M. (2002). Science, medicine, and the future: microdialysis. *BMJ, 324*, 588–591.

[5] Kitzen, J.J., Verweij, J., Wiemer, E.A., Loos, W.J. (2006). The relevance of microdialysis for clinical oncology. *Current Clinical Pharmacology, 1*, 255–263.

[6] Helmy, A., Carpenter, K.L., Hutchinson, P.J. (2007). Microdialysis in the human brain and its potential role in the development and clinical assessment of drugs. *Current Medicinal Chemistry, 14*, 1525–1537.

[7] Waelgaard, L., Thorgersen, E.B., Line, P.D., Foss, A., Mollnes, T.E., Tonnessen, T.I. (2008). Microdialysis monitoring of liver grafts by metabolic parameters, cytokine production, and complement activation. *Transplantation, 86*, 1096–1103.

[8] Matthiessen, P., Strand, I., Jansson, K., Tornquist, C., Andersson, M., Rutegard, J., Norgren, L. (2007). Is early detection of anastomotic leakage possible by intraperitoneal microdialysis and intraperitoneal cytokines after anterior resection of the rectum for cancer? *Diseases of the Colon and Rectum, 50*, 1918–1927.

[9] Hutschala, D., Skhirtladze, K., Kinstner, C., Mayer-Helm, B., Muller, M., Wolner, E., Tschernko, E.M. (2007). In vivo microdialysis to measure antibiotic penetration into soft tissue during cardiac surgery. *Annals of Thoracic Surgery, 84*, 1605–1610.

[10] Angst, M.S., Clark, J.D., Carvalho, B., Tingle, M., Schmelz, M., Yeomans, D.C. (2008). Cytokine profile in human skin in response to experimental inflammation, noxious stimulation, and administration of a COX-inhibitor: a microdialysis study. *Pain, 139*, 15–27.

[11] Fujiwara, M., Egashira, N. (2004). New perspectives in the studies on endocannabinoid and cannabis: abnormal behaviors associated with CB1 cannabinoid receptor and development of therapeutic application. *Journal of Pharmacological Sciences, 96*, 362–366.

[12] Alberts, B., Johnson, A., Lewis, J., Raff, M., Roberts, K., Walter, P. (2007). *Molecular Biology of the Cell*, Garland Science, New York, pp. 501–510.

[13] Tsai, T.H. (2003). Assaying protein unbound drugs using microdialysis techniques. *Journal of Chromatography B: Analytical Technologies in the Biomedical and Life Sciences, 797*, 161–173.

[14] Einspanier, R., Miyamoto, A., Schams, D., Muller, M., Brem, G. (1990). Tissue concentration, mRNA expression and stimulation of IGF-I in luteal tissue during the oestrous cycle and pregnancy of cows. *Journal of Reproduction and Fertility, 90*, 439–445.

[15] Okuda, K., Uenoyama, Y., Berisha, B., Lange, I.G., Taniguchi, H., Kobayashi, S., Kobayashi, S., Miyamoto, A., Schams, D. (2001). Estradiol-17beta is produced in bovine corpus luteum. *Biology of Reproduction, 65*, 1634–1639.

[16] Miyamoto, A., Schams, D. (1991). Oxytocin stimulates progesterone release from microdialyzed bovine corpus luteum in vitro. *Biology of Reproduction, 44*, 1163–1170.

[17] Miyamoto, A., von Lutzow, H., Schams, D. (1993). Acute actions of prostaglandin F2 alpha, E2, and I2 in microdialyzed bovine corpus luteum in vitro. *Biology of Reproduction, 49*, 423–430.

[18] Schams, D. (1992). Regulation of bovine intra-luteal function by peptide hormones. *Journal of Physiology and Pharmacology*, *43*, 117–129.

[19] Sauerwein, H., Miyamoto, A., Gunther, J., Meyer, H.H., Schams, D. (1992). Binding and action of insulin-like growth factors and insulin in bovine luteal tissue during the oestrous cycle. *Journal of Reproduction and Fertility*, *96*, 103–115.

[20] Kobayashi, S., Miyamoto, A., Berisha, B., Schams, D. (2001). Growth hormone, but not luteinizing hormone, acts with luteal peptides on prostaglandin F2alpha and progesterone secretion by bovine corpora lutea in vitro. *Prostaglandins*, *63*, 79–92.

[21] Schams, D., Berisha, B., Kosmann, M., Einspanier, R., Amselgruber, W.M. (1999). Possible role of growth hormone, IGFs, and IGF-binding proteins in the regulation of ovarian function in large farm animals. *Domestic Animal Endocrinology*, *17*, 279–285.

[22] Kobayashi, S., Berisha, B., Amselgruber, W.M., Schams, D., Miyamoto, A. (2001). Production and localisation of angiotensin II in the bovine early corpus luteum: a possible interaction with luteal angiogenic factors and prostaglandin F2 alpha. *Journal of Endocrinology*, *170*, 369–380.

[23] Miyamoto, A., Okuda, K., Schweigert, F.J., Schams, D. (1992). Effects of basic fibroblast growth factor, transforming growth factor-beta and nerve growth factor on the secretory function of the bovine corpus luteum in vitro. *Journal of Endocrinology*, *135*, 103–114.

[24] Acosta, T.J., Miyamoto, A., Ozawa, T., Wijayagunawardane, M.P., Sato, K. (1998). Local release of steroid hormones, prostaglandin E2, and endothelin-1 from bovine mature follicles in vitro: effects of luteinizing hormone, endothelin-1, and cytokines. *Biology of Reproduction*, *59*, 437–443.

[25] Miyamoto, A., Bruckmann, A., von Lutzow, H., Schams, D. (1993). Multiple effects of neuropeptide Y, substance P and vasoactive intestinal polypeptide on progesterone and oxytocin release from bovine corpus luteum in vitro. *Journal of Endocrinology*, *138*, 451–458.

[26] Acosta, T.J., Berisha, B., Ozawa, T., Sato, K., Schams, D., Miyamoto, A. (1999). Evidence for a local endothelin–angiotensin–atrial natriuretic peptide system in bovine mature follicles in vitro: effects on steroid hormones and prostaglandin secretion. *Biology of Reproduction*, *61*, 1419–1425.

[27] Maas, S., Jarry, H., Teichmann, A., Rath, W., Kuhn, W., Wuttke, W. (1992). Paracrine actions of oxytocin, prostaglandin F2 alpha, and estradiol within the human corpus luteum. *Journal of Clinical Endocrinology and Metabolism*, *74*, 306–312.

[28] Einspanier, A., Hodges, J.K. (1994). LH- and chorionic gonadotrophin-stimulated progesterone release in vitro by intact luteal tissue of the marmoset monkey (*Callithrix jacchus*). *Journal of Endocrinology*, *141*, 403–409.

[29] Tanaka, J., Acosta, T.J., Berisha, B., Tetsuka, M., Matsui, M., Kobayashi, S., Schams, D., Miyamoto, A. (2004). Relative changes in mRNA expression of angiopoietins and receptors tie in bovine corpus luteum during estrous cycle and prostaglandin F2alpha-induced luteolysis: a possible mechanism for the initiation of luteal regression. *Journal of Reproduction and Development*, *50*, 619–626.

[30] Piotrowska, K.K., Woclawek-Potocka, I., Bah, M.M., Piskula, M.K., Pilawski, W., Bober, A., Skarzynski, D.J. (2006). Phytoestrogens and their metabolites inhibit

the sensitivity of the bovine corpus luteum to luteotropic factors. *Journal of Reproduction and Development*, *52*, 33–41.

[31] Beindorff, N., Honnens, A., Penno, Y., Paul, V., Bollwein, H. (2009). Effects of human chorionic gonadotropin on luteal blood flow and progesterone secretion in cows and in vitro–microdialyzed corpora lutea. *Theriogenology*, doi:10.1016/j.theriogenology.2009.04.008.

[32] Stoppini, L., Buchs, P.A., Muller, D. (1991). A simple method for organotypic cultures of nervous tissue. *Journal of Neuroscience Methods*, *37*, 173–182.

[33] Menendez, N., Solis, J.M., Herreras, O., Galarreta, M., Conejero, C., Martin del Rio, R. (1993). Taurine release evoked by NMDA receptor activation is largely dependent on calcium mobilization from intracellular stores. *European Journal of Neuroscience*, *5*, 1273–1279.

[34] Robert, F., Parisi, L., Bert, L., Renaud, B., Stoppini, L. (1997). Microdialysis monitoring of extracellular glutamate combined with the simultaneous recording of evoked field potentials in hippocampal organotypic slice cultures. *Journal of Neuroscience Methods*, *74*, 65–76.

[35] Robert, F., Bert, L., Parrot, S., Denoroy, L., Stoppini, L., Renaud, B. (1998). Coupling on-line brain microdialysis, precolumn derivatization and capillary electrophoresis for routine minute sampling of *O*-phosphoethanolamine and excitatory amino acids. *Journal of Chromatography A*, *817*, 195–203.

[36] Duport, S., Robert, F., Muller, D., Grau, G., Parisi, L., Stoppini, L. (1998). An in vitro blood-brain barrier model: cocultures between endothelial cells and organotypic brain slice cultures. *Proceedings of the National Academy of Sciences of the United States of America*, *95*, 1840–1845.

[37] Boubriak, O.A., Urban, J.P., Cui, Z. (2006). Monitoring of metabolite gradients in tissue-engineered constructs. *Journal of the Royal Society, Interface*, *3*, 637–648.

[38] Li, Z., Boubriak, O.A., Urban, J.P., Cui, Z.F. (2006). Microdialysis for monitoring the process of functional tissue culture. *International Journal of Artificial Organs*, *29*, 858–865.

[39] Niwa, O., Torimitsu, K., Morita, M., Osborne, P., Yamamoto, K. (1996). Concentration of extracellular L-glutamate released from cultured nerve cells measured with a small-volume online sensor. *Analytical Chemistry*, *68*, 1865–1870.

[40] Chen, S.J., Kao, C.L., Chang, Y.L., Yen, C.J., Shui, J.W., Chien, C.S., Chen, I.L., Tsai, T.H., Ku, H.H., Chiou, S.H. (2007). Antidepressant administration modulates neural stem cell survival and serotoninergic differentiation through bcl-2. *Current Neurovascular Research*, *4*, 19–29.

[41] Peng, C.H., Chiou, S.H., Chen, S.J., Chou, Y.C., Ku, H.H., Cheng, C.K., Yen, C.J., Tsai, T.H., Chang, Y.L., Kao, C.L. (2008). Neuroprotection by Imipramine against lipopolysaccharide-induced apoptosis in hippocampus-derived neural stem cells mediated by activation of BDNF and the MAPK pathway. *European Neuropsychopharmacology*, *18*, 128–140.

[42] Wu, Y.S., Tsai, T.H., Wu, T.F., Cheng, F.C. (2001). Determination of pyruvate and lactate in primary liver cell culture medium during hypoxia by on-line microdialysis–liquid chromatography. *Journal of Chromatography A*, *913*, 341–347.

[43] Ge, X., Rao, G., Tolosa, L. (2008). On the possibility of real-time monitoring of glucose in cell culture by microdialysis using a fluorescent glucose binding protein sensor. *Biotechnology Progress*, *24*, 691–697.

[44] Sun, L., Stenken, J.A. (2003). Quantitation of nitric oxide–derived nitrite from activated macrophages using microdialysis sampling. *Journal of Chromatography B: Analytical Technologies in the Biomedical and Life Sciences*, *796*, 327–338.

[45] Cheng, F.C., Kuo, J.S., Huang, H.M., Yang, D.Y., Wu, T.F., Tsai, T.H. (2000). Determination of catecholamines in pheochromocytoma cell (PC-12) culture medium by microdialysis–microbore liquid chromatography. *Journal of Chromatography A*, *870*, 405–411.

[46] Xu, F., Gao, M., Wang, L., Jin, L. (2002). Study on the effect of electromagnetic impulse on neurotransmitter metabolism in nerve cells by high-performance liquid chromatography–electrochemical detection coupled with microdialysis. *Analytical Biochemistry*, *307*, 33–39.

[47] Serra, P.A., Rocchitta, G., Delogu, M.R., Migheli, R., Taras, M.G., Mura, M.P., Esposito, G., Miele, E., Desole, M.S., Miele, M. (2003). Role of the nitric oxide/cyclic GMP pathway and extracellular environment in the nitric oxide donor-induced increase in dopamine secretion from PC12 cells: a microdialysis in vitro study. *Journal of Neurochemistry*, *86*, 1403–1413.

[48] Serra, P.A., Migheli, R., Rocchitta, G., Taras, M.G., Mura, M.P., Delogu, M.R., Esposito, G., Desole, M.S., Miele, E., Miele, M. (2003). Role of the nitric oxide/cyclic GMP pathway and ascorbic acid in 3-morpholinosydnonimine (SIN-1)-induced increases in dopamine secretion from PC12 cells: a microdialysis in vitro study. *Neuroscience Letters*, *353*, 5–8.

[49] Rocchitta, G., Migheli, R., Mura, M.P., Esposito, G., Marchetti, B., Desole, M.S., Miele, E., Serra, P.A. (2005). Role of endogenous melatonin in the oxidative homeostasis of the extracellular striatal compartment: a microdialysis study in PC12 cells in vitro and in the striatum of freely moving rats. *Journal of Pineal Research*, *39*, 409–418.

[50] Chiou, S.H., Kao, C.L., Chang, Y.L., Ku, H.H., Tsai, Y.J., Lin, H.T., Yen, C.J., Peng, C.H., Chiu, J.H., Tsai, T.H. (2007). Evaluation of anti-Fas ligand-induced apoptosis and neural differentiation of PC12 cells treated with nerve growth factor using small interfering RNA method and sampling by microdialysis. *Analytical Biochemistry*, *363*, 46–57.

[51] Migheli, R., Puggioni, G., Dedola, S., Rocchitta, G., Calia, G., Bazzu, G., Esposito, G., Lowry, J.P., O'Neill, R.D., Desole, M.S., Miele, E., Serra, P.A. (2008). Novel integrated microdialysis–amperometric system for in vitro detection of dopamine secreted from PC12 cells: design, construction, and validation. *Analytical Biochemistry*, *380*, 323–330.

[52] Goldman, S. (2005). Stem and progenitor cell-based therapy of the human central nervous system. *Nature Biotechnology*, *23*, 862–871.

[53] Hillered, L., Hallstrom, A., Segersvard, S., Persson, L., Ungerstedt, U. (1989). Dynamics of extracellular metabolites in the striatum after middle cerebral artery occlusion in the rat monitored by intracerebral microdialysis. *Journal of Cerebral Blood Flow and Metabolism*, *9*, 607–616.

[54] Persson, L., Hillered, L. (1992). Chemical monitoring of neurosurgical intensive care patients using intracerebral microdialysis. *Journal of Neurosurgery*, *76*, 72–80.

[55] Mizock, B.A., Falk, J.L. (1992). Lactic acidosis in critical illness. *Critical Care Medicine, 20,* 80–93.

[56] Stefansson, B.V., Bjornson, A.L., Haraldsson, B., Nilsson, U.A. (2005). A new method for monitoring nitric oxide production using Teflon membrane microdialysis. *Free Radical Biology & Medicine, 39,* 249–256.

[57] Mou, X., Stenken, J.A. (2006). Microdialysis sampling extraction efficiency of 2-deoxyglucose: role of macrophages in vitro and in vivo. *Analytical Chemistry, 78,* 7778–7784.

[58] Schutte, R.J., Oshodi, S.A., Reichert, W.M. (2004). In vitro characterization of microdialysis sampling of macromolecules. *Analytical Chemistry, 76,* 6058–6063.

[59] Wang, X., Stenken, J.A. (2006). Microdialysis sampling membrane performance during in vitro macromolecule collection. *Analytical Chemistry, 78,* 6026–6034.

[60] Phillips, T.M., Kennedy, L.M., De Fabo, E.C. (1997). Microdialysis-immunoaffinity capillary electrophoresis studies on neuropeptide-induced lymphocyte secretion. *Journal of Chromatography B Biomedical Sciences and Applications, 697,* 101–109.

[61] Phillips, T.M. (2001). Analysis of single-cell cultures by immunoaffinity capillary electrophoresis with laser-induced fluorescence detection. *Luminescence, 16,* 145–152.

[62] Kalish, H., Phillips, T.M. (2009). Application of immunoaffinity capillary electrophoresis to the measurements of secreted cytokines by cultured astrocytes. *Journal of Separation Science, 32,* 1605–1612.

[63] Ao, X., Wang, X., Lennartz, M.R., Loegering, D.J., Stenken, J.A. (2006). Multiplexed cytokine detection in microliter microdialysis samples obtained from activated cultured macrophages. *Journal of Pharmaceutical and Biomedical Analysis, 40,* 915–921.

[64] Rosenbloom, A.J., Ferris, R.L., Sipe, D.M., Riddler, S.A., Connolly, N.C., Abe, K., Whiteside, T.L. (2006). In vitro and in vivo protein sampling by combined microdialysis and ultrafiltration. *Journal of Immunological Methods, 309,* 55–68.

[65] Torto, N., Laurell, T., Gorton, L., Marko-Varga, G. (1998). Recent trends in the application of microdialysis in bioprocesses. *Analytica Chimica Acta, 374,* 111–135.

[66] Torto, N., Laurell, T., Gorton, L., Marko-Varga, G. (1999). Recent trends in the application of microdialysis in bioprocesses [Reprinted from *Analytica Chimica Acta, 374*]. *Analytica Chimica Acta, 379,* 281–305.

[67] Buttler, T., Nilsson, C., Gorton, L., Marko-Varga, G., Laurell, T. (1996). Membrane characterisation and performance of microdialysis probes intended for use as bioprocess sampling units. *Journal of Chromatography A, 725,* 41–56.

[68] Marko-Varga, G., Buttler, T., Gorton, L., Gronsterwall, C. (1993). A study of the use of microdialysis probes as a sampling unit in online bioprocess monitoring in conjunction with column liquid-chromatography. *Chromatographia, 35,* 285–289.

[69] Buttler, T., Gorton, L., Jarskog, H., Marko-Varga, G., Hahn-Hagerdal, B., Meinander, N., Olsson, L. (1994). Monitoring of ethanol during fermentation of a lignocellulose hydrolysate by on-line microdialysis sampling, column liquid chromatography, and an alcohol biosensor. *Biotechnology and Bioengineering, 44,* 322–328.

[70] Buttler, T., Jarskog, H., Gorton, L., Marko-Varga, G., Ramnemark, L. (1994). The use of microdialysis for sampling in column liquid-chromatography. *American Laboratory*, *26*, I28–M28.

[71] Torto, N., Buttler, T., Gorton, L., Marko-Varga, G., Stalbrand, H., Tjerneld, F. (1995). Monitoring of enzymatic-hydrolysis of ivory nut mannan using online microdialysis sampling and anion-exchange chromatography with integrated pulsed electrochemical detection. *Analytica Chimica Acta*, *313*, 15–24.

[72] Rumbold, K., Okatch, H., Torto, N., Siika-Aho, M., Gubitz, G., Robra, K.H., Prior, B. (2002). Monitoring on-line desalted lignocellulosic hydrolysates by microdialysis sampling micro-high performance anion exchange chromatography with integrated pulsed electrochemical detection/mass spectrometry. *Biotechnology and Bioengineering*, *78*, 821–827.

[73] Palmqvist, E., Galbe, M., Hahn-Hagerdal, B. (1998). Evaluation of cell recycling in continuous fermentation of enzymatic hydrolysates of spruce with *Saccharomyces cerevisiae* and on-line monitoring of glucose and ethanol. *Applied Microbiology and Biotechnology*, *50*, 545–551.

[74] Marangoni, A.G. (2003). How do enzymes work? In: *Enzyme Kinetics: A Modern Approach*, Wiley Hoboken, NJ, pp. 41–43.

[75] Torto, N., Gorton, L., Laurell, T., Marko-Varga, G. (1999). Technical issues of in vitro microdialysis sampling in bioprocess monitoring. *Trends in Analytical Chemistry*, *18*, 252–260.

[76] Zhou, J., Shearer, E.C., Hong, J., Riley, C.M., Schowen, R.L. (1996). Automated analytical systems for drug development studies: V. A system for enzyme kinetic studies. *Journal of Pharmaceutical and Biomedical Analysis*, *14*, 1691–1698.

[77] Zhou, J., Riley, C.M., Schowen, R.L. (2001). In vitro metabolism studies of the prodrug, 2′,3′,5′-triacetyl-6-azauridine, utilizing an automated analytical system. *Journal of Pharmaceutical and Biomedical Analysis*, *26*, 701–716.

[78] Zook, C.M., LaCourse, W.R. (1998). Pulsed amperometric detection of microdialysates from the glucose oxidase reaction. *Analytical Chemistry*, *70*, 801–806.

[79] Steuerwald, A.J., Villeneuve, J.D., Sun, L., Stenken, J.A. (2006). In vitro characterization of an in situ microdialysis sampling assay for elastase activity detection. *Journal of Pharmaceutical and Biomedical Analysis*, *40*, 1041–1047.

[80] Modi, S.J., LaCourse, W.R. (2006). Monitoring carbohydrate enzymatic reactions by quantitative in vitro microdialysis. *Journal of Chromatography A*, *1118*, 125–133.

[81] Gunaratna, C., Kissinger, P.T. (1997). Application of microdialysis to study the in vitro metabolism of drugs in liver microsomes. *Journal of Pharmaceutical and Biomedical Analysis*, *16*, 239–248.

[82] Gunaratna, C., Kissinger, P.T. (1998). Investigation of stereoselective metabolism of amphetamine in rat liver microsomes by microdialysis and liquid chromatography with precolumn chiral derivatization. *Journal of Chromatography A*, *828*, 95–103.

[83] Kolanczyk, R., Schmieder, P., Bradbury, S., Spizzo, T. (1999). Biotransformation of 4-methoxyphenol in rainbow trout (*Oncorhynchus mykiss*) hepatic microsomes. *Aquatic Toxicology*, *45*, 47–61.

[84] Wen, X.D., Qi, L.W., Li, B., Li, P., Yi, L., Wang, Y.Q., Liu, E.H., Yang, X.L. (2009). Microsomal metabolism of calycosin, formononetin and drug–drug interactions

by dynamic microdialysis sampling and HPLC–DAD–MS analysis. *Journal of Pharmaceutical and Biomedical Analysis*, *50*, 100–105.

[85] Sun, L., Stenken, J.A., Yang, A.Y., Zhao, J.J., Musson, D.G. (2007). An in vitro microdialysis methodology to study 11beta-hydroxysteroid dehydrogenase type 1 enzyme activity in liver microsomes. *Analytical Biochemistry*, *370*, 26–37.

[86] Pacifici, G.M., Viani, A. (1992). Methods of determining plasma and tissue binding of drugs: pharmacokinetic consequences. *Clinical Pharmacokinetics*, *23*, 449–468.

[87] Cheng, Y., Ho, E., Subramanyam, B., Tseng, J.L. (2004). Measurements of drug–protein binding by using immobilized human serum albumin liquid chromatography–mass spectrometry. *Journal of Chromatography B Analytical Technologies in the Biomedical and Life Sciences*, *809*, 67–73.

[88] Li, B.X., Zhang, Z.J., Zhao, L.X. (2002). Flow-injection chemiluminescence detection for studying protein binding for drug with ultrafiltration sampling. *Analytica Chimica Acta*, *468*, 65–70.

[89] Pavey, K.D., Lyle, E.L., Olliff, C.J., Paul, F. (2001). A quartz crystal resonant sensor (QCRS) study of HSA–drug interactions. *Analyst*, *126*, 426–428.

[90] Huang, Y., Zhang, Z., Zhang, D., Lv, J. (2001). Flow-injection analysis chemiluminescence detection combined with microdialysis sampling for studying protein binding of drug. *Talanta*, *53*, 835–841.

[91] Ekblom, M., Hammarlund-Udenaes, M., Lundqvist, T., Sjoberg, P. (1992). Potential use of microdialysis in pharmacokinetics: a protein binding study. *Pharmaceutical Research*, *9*, 155–158.

[92] Oravcova, J., Bohs, B., Lindner, W. (1996). Drug–protein binding sites: new trends in analytical and experimental methodology. *Journal of Chromatography B: Biomedical Sciences and Applications*, *677*, 1–28.

[93] Juan, Y.P., Tsai, T.H. (2005). Measurement and pharmacokinetics of vincamine in rat blood and brain using microdialysis. *Journal of Chromatography A*, *1088*, 146–151.

[94] Su, X., Kong, L., Li, X., Chen, X., Guo, M., Zou, H. (2005). Screening and analysis of bioactive compounds with biofingerprinting chromatogram analysis of traditional Chinese medicines targeting DNA by microdialysis/HPLC. *Journal of Chromatography A*, *1076*, 118–126.

[95] Guo, M., Su, X., Kong, L., Li, X., Zou, H. (2006). Characterization of interaction property of multicomponents in Chinese herb with protein by microdialysis combined with HPLC. *Analytica Chimica Acta*, *556*, 183–188.

[96] Wang, H.I., Zou, H.F., Feng, A.S., Zhang, Y.K. (1997). Binding of sulfamethoxazole to human serum albumin studied by a combined technique of microdialysis with liquid chromatography. *Analytica Chimica Acta*, *342*, 159–165.

[97] Wang, H.L., Zou, H.F., Zhang, Y.K. (1997). Determination of drug–protein interactions by combined microdialysis and high-performance liquid chromatography. *Chromatographia*, *44*, 205–208.

[98] Wang, H., Wang, Z., Lu, M., Zou, H. (2008). Microdialysis sampling method for evaluation of binding kinetics of small molecules to macromolecules. *Analytical Chemistry*, *80*, 2993–2999.

[99] Schmidt, S., Rock, K., Sahre, M., Burkhardt, O., Brunner, M., Lobmeyer, M.T., Derendorf, H. (2008). Effect of protein binding on the pharmacological activity

of highly bound antibiotics. *Antimicrobial Agents and Chemotherapy*, *52*, 3994–4000.

[100] Shi, G.Y., Xu, F., Zhou, H.G., Mao, L.Q., Jin, L.T. (1999). Flow-injection analysis electrochemical detection for the determination of drug–protein interactions with microdialysis sampling. *Analytica Chimica Acta*, *386*, 123–127.

[101] Li, J., Brahmer, J., Messersmith, W., Hidalgo, M., Baker, S.D. (2006). Binding of gefitinib, an inhibitor of epidermal growth factor receptor-tyrosine kinase, to plasma proteins and blood cells: in vitro and in cancer patients. *Investigational New Drugs*, *24*, 291–297.

[102] Qian, Z.M., Wen, X.D., Li, H.J., Liu, Y., Qin, S.J., Li, P. (2008). Analysis of interaction property of bioactive components in Flos Lonicerae Japonicae with protein by microdialysis coupled with HPLC–DAD–MS. *Biological & Pharmaceutical Bulletin*, *31*, 126–130.

[103] Guo, M., Zou, H.F., Wang, H.L., Kong, L., Ni, J.Y. (2001). Binding of metal ions with protein studied by a combined technique of microdialysis with liquid chromatography. *Analytica Chimica Acta*, *443*, 91–99.

[104] Prozialeck, W.C., Lamar, P.C., Ikura, M. (1996). Binding of cadmium (Cd^{2+}) to E-CAD1, a calcium-binding polypeptide analog of E-cadherin. *Life Sciences*, *58*, PL325–PL330.

[105] Prozialeck, W.C., Lamar, P.C. (1999). Interaction of cadmium (Cd(2+)) with a 13-residue polypeptide analog of a putative calcium-binding motif of E-cadherin. *Biochimica et Biophysica Acta*, *1451*, 93–100.

16

MICRODIALYSIS IN DRUG–DRUG INTERACTION

MITSUHIRO WADA, RIE IKEDA, AND KENICHIRO NAKASHIMA
Graduate School of Biomedical Sciences, Nagasaki University, Nagasaki, Japan

1. INTRODUCTION

A *drug–drug interaction* is broadly defined as any reaction between one medicament and another in the human. This is caused by simultaneous coadministration of two or more medicaments. This interaction can either cause deleterious effects or lead to ineffective therapy. At present, multiple-medicament use is increasing in clinical practice, especially in the treatment of the elderly or patients having several concurrently existing ailments. Accumulation of knowledge of drug–drug interaction may be helpful for the rational use of medicaments in clinical practice. Drug–drug interaction is classified as either pharmacokinetic or pharmacodynamic, according to its mechanisms. The former is caused in the process of absorption, distribution, metabolism, and excretion (ADME) of a medicament coadministered with others. The latter is caused in the relation between coadministered medicaments and their receptors or mechanism of action. Many review articles have covered the entire array of drug–drug (or food) interactions for some medicaments [1–4].

1.1. Mechanism of Pharmacokinetic Drug–Drug Interaction

Most drugs are given orally and thus are absorbed primarily through the intestinal mucosa. Coadministration with medicaments such as an H_2 receptor

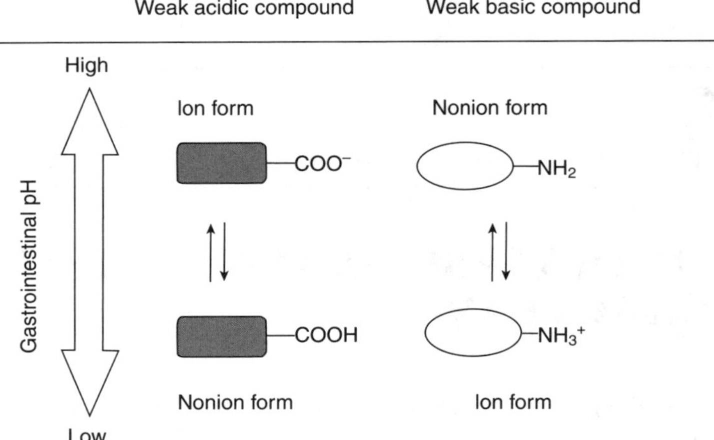

Figure 1   Weak ionic medicament form in various pH.

blocker, which changes the gastrointestinal pH, influences the ion-/non-ion-form ratio of weak ionic medicaments (Figure 1). As a result, the bioavailability (BA) of the medicament is influenced because the nonion form is easily absorbed from the gastrointestinal tract in general [5]. Other factors that cause drug–drug interaction in absorptional process are an intestinal permeability glycoprotein (P-gp) and an intestinal cytochrome P450 (CYP) enzyme. The former can actively pump medicaments back into the intestinal lumen, and the latter can metabolize the drug before it reaches the systemic circulation [6–9]. Induction and inhibition of P-gp and CYP3A4 by a coadministered medicament can thus influence the BA of other coadministered medicaments.

After being absorbed, medicaments generally bind to plasma proteins such as albumin. The level of protein binding plays a very important role in distribution and displacement interactions, which can affect the concentration of free-form medicament (Figure 2). However, it is believed that competition for plasma protein-binding sites does not carry weight with drug–drug interaction [10]. Moreover, a transporter such as P-gp influences the permeability of the blood–brain barrier (BBB). Therefore, induction and inhibition of a transporter by a coadministered medicament can influence the distribution of medicaments that display activity in the central nervous system (CNS) [6].

Drug–drug interaction mechanisms in a metabolic process are of two types: enzyme induction and enzyme inhibition. Induction of metabolic enzyme may decrease the plasma or tissue level of a medicament until it is noneffective. On the other hand, inhibition of enzyme increases the plasma or tissue level of a medicament to one with significant toxic consequences (Figure 3).

The metabolism of medicaments exhibits several phases. Phase I consists of hydrolysis, oxidation, and reduction, and is mediated primarily by CYP enzymes. Phase II consists of conjugation and is mediated by enzymes such as

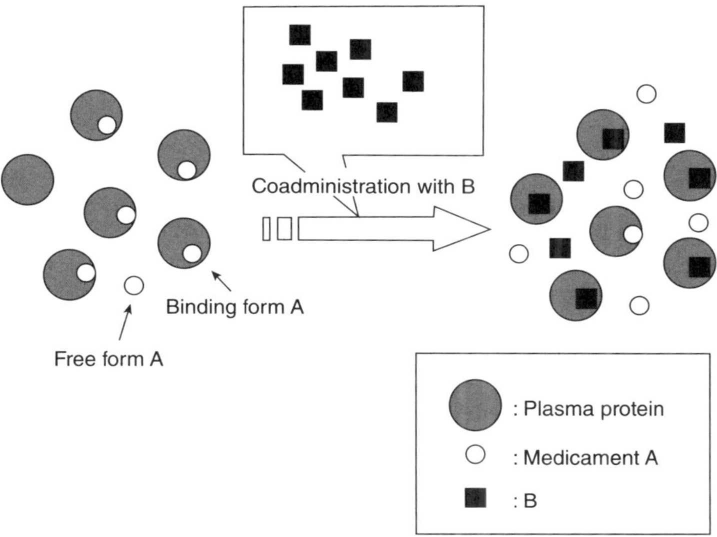

Figure 2 Displacement of medicaments on binding site of protein.

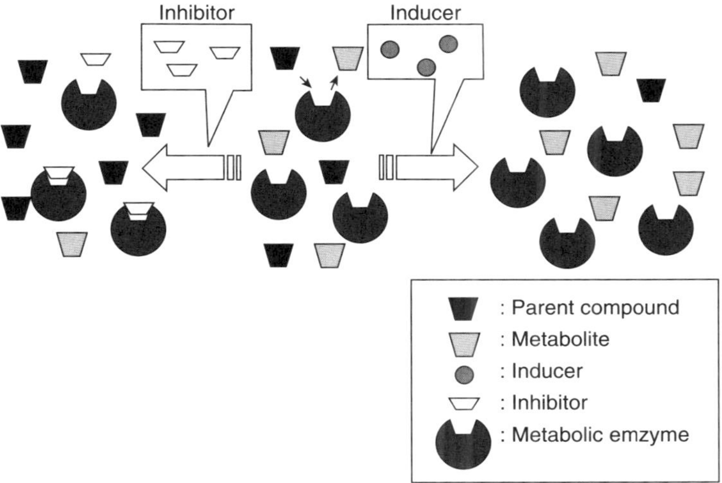

Figure 3 Inhibition and induction of metabolic enzyme.

uridine diphosphoglucuronosyl transferase, *N*-acetyl transferase, and glutathione *S*-transferase [11]. The elimination process through the primary active transport mechanism described below is designated as phase III in the detoxification process [22]. These phases play important roles in the detoxification and/or excretion rate of xenobiotics, so coadministration of a medicament with an inducer or inhibitor of these enzymes and transporters may cause drug–drug interaction.

Drugs are generally eliminated from the body by metabolism and/or excretion. The liver and kidney both play important roles in excretion of unchanged medicaments and their metabolites. For biliary excretion, a medicament must traverse the sinusoidal membrane of the hepatocytes by passive diffusion and/or hepatic uptake transporters. The sinusoidal membrane of hepatocyte contains a number of transporters responsible for the uptake of cations, anions, and endogenous substances into hepatocytes from the general circulation. Renal excretion of a medicament usually involves three processes: glomerular filtration, renal tubular secretion, and reabsorption from the renal tubular lumen. Glomerular filtration is a passive process by which only unbound medicament can be filtrated, whereas tubular secretion and reabsorption often involve active transporters. Nonfiltered medicaments must first cross the basolateral membrane and then the apical membrane of epithelial cells of the renal tubule, either by passive diffusion or by carrier-mediated processes [7]. Since various transporters play important roles in the excretory process, coadministration of an inducer and inhibitor of these may cause potential drug–drug interaction.

1.2. Mechanism of Pharmacodynamic Drug–Drug Interaction

With respect to pharmacodynamic drug–drug interaction, studies focus on the classification of synergetic and antagonistic drug interactions. This interaction may occur at the same active site or at a different site on a receptor. For example, medicaments that bind to the same site on the receptor and act as an agonist increase each pharmacological effect synergistically. However, coadministration of an agonist and antagonist that bind to the same receptor site cause each pharmacological effect to decrease or disappear competitively (Figure 4). Noncompetitive antagonists, also known as allosteric antagonists, bind to a distinctly separate binding site from the agonist, exerting their action on that receptor via the other binding sites [12].

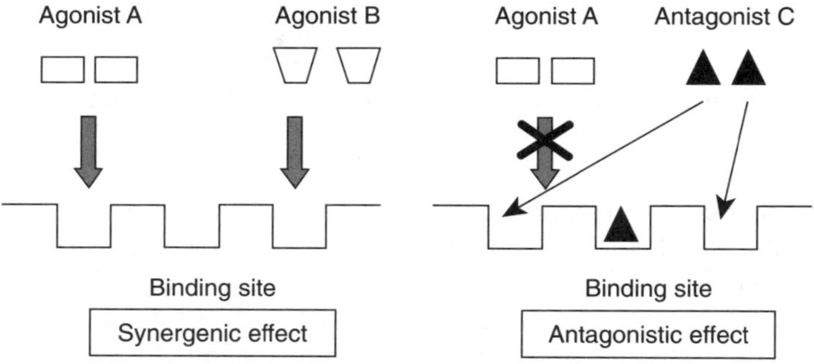

Figure 4   Mechanism of synergenic and antagonistic effect.

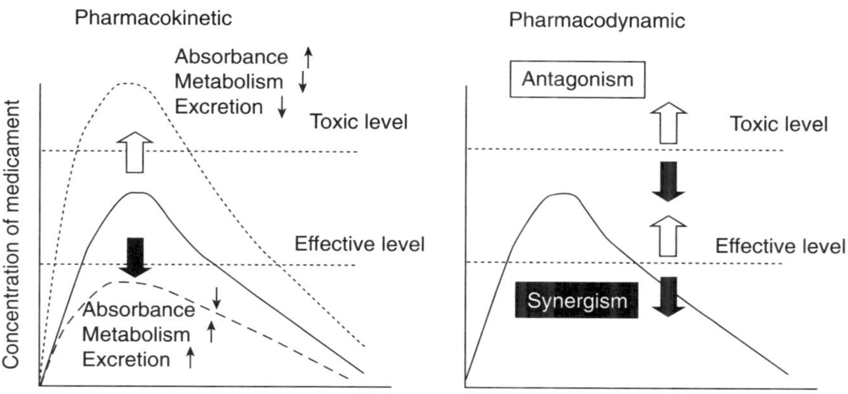

Figure 5 Difference between pharmacokinetic and pharmacodynamic drug–drug interactions.

1.3. Evaluation of Drug–Drug Interaction

To estimate pharmacokinetic drug–drug interaction, measurement of the concentration–time profile of medicaments or their metabolites in tissue or biological fluids is generally available. The change in the concentration–time profile of the target medicament is found when a pharmacokinetic drug–drug interaction occurrs. The appearance of an effect (or a side effect) of a medicament depends principally on the concentration of the free form of a medicament. On the other hand, no change could be found in the concentration–time profile of a medicament in pharmacodynamic drug–drug interaction, whereas change in the toxic or effective range will be apparent. This is a difference between pharmacokinetic and pharmacodynamic interactions with regard to the concentration–time profile (Figure 5). For an evaluation of the pharmacodynamic drug–drug interaction, measurement of the therapeutic effect and/or behavioral outcome is achieved. Especially for medicaments having activity in the central nervous system (CNS), quantification of the levels of neurotransmitters such as dopamine (DA) and serotonin (5-hydroxytryptamine, 5-HT) is available. To determine free-form medicaments in tissue or biological fluids, or neurotransmitters in the CNS, microdialysis is one of the most powerful techniques. As for the advantages of microdialysis, long-term sampling is possible with minimal tissue damage, and a cleanup procedure is not required in most cases. Therefore, this method is used to determine free-form medicaments (including drugs of abuse) for pharmacokinetic study [39,42,44,45]. However, sensitive analytical methods are required to determine very low concentration of analytes or quite a small size of sample [13,14]. For sensitive determination of medicament, a chromatographic method such as gas chromatography (GC) or high-performance liquid chromatography (HPLC) has proven to be most popular, due to their better sensitivity, selectivity, and applicability. HPLC, especially, shows excellent capability for the analysis of aqueous

TABLE 1 Studies on Pharmacokinetic Drug–Drug Interaction

Analyte	Concomitance	Route	Species	Target Organs	Notes	Ref.
Ciprofloxacin (20, 50 mg/kg)	Cyclosporin A (20 mg/kg)	i.v.	Rat	Jugular vein	$AUC_{Cip50} \uparrow$, $CL_{Cip50} \downarrow$	[15]
Pefloxacin (10 mg/kg)	Cyclosporin A (20 mg/kg)	i.v.	Rat	Jugular vein, striatum, bile duct	No significant impact of cyclosporin A on pharmacokinetics of pefloxacin	[16]
Norfloxacin (50 mg/kg)	Probenecid (20 mg/kg)	i.v.	Rat	Hippocampus	No effect on BBB transport of norfloxacin	[17]
Cefepime (20, 50, 100 mg/kg)	Cyclosporin A (20 mg/kg)	i.v.	Rat	Jugular vein, striatum	$MRT_{Cef50} \uparrow$ (blood) $AUC_{Cef50} \uparrow$ (blood, brain)	[18]
Amoxicillin (20 mg/kg)	Ivermectin (5 mg/kg)	i.v.	Rat	Jugular vein, striatum	$MRT_{Amox} \uparrow$ (blood) $AUC_{Amox} \uparrow$ (blood) $CL_{Amox} \downarrow$ (blood)	[19]
Cefoperazone (30 mg/kg)	Berberine (30 mg/kg)	i.v.	Rat	Bile duct	No effect on pharmacokinetics of cefoperazone	[20]
Fluconazole (10 mg/kg)	Cyclosporin A (20 mg/kg)	i.v.	Rat	Jugular vein, bile duct	Fluconazole undergoes hepatobinary excretion may be unrelated to the P-gp transported system	[21]
Berberine (10, 20 mg/kg)	Cyclosporin A (20 mg/kg), quinidine (10 mg/kg), SKF-525A (10 mg/kg), probenecid (100 mg/kg)	i.v.	Rat	Jugular vein, hepatic median lobe, bile duct	Berberine hepatobinary excretion $\downarrow$ (by cyclosporin A and quinidine); metabolism of berberine $\downarrow$ (by SKF-525A)	[22]
Diclofenac (1 mg/kg)	Cyclosporin A (20 mg/kg)	i.v.	Rat	Bile duct	$C_{max} \uparrow$ $AUC_{dic} \uparrow$	[23]
Baicalein (10, 30, 60 mg/kg)	Cyclosporin A (20 mg/kg)	i.v.	Rat	Jugular vein, hippocampus, bile duct	$AUC_{brain}/AUC_{blood} \uparrow$ $AUC_{bile}/AUC_{blood} \downarrow$	[24]

Baicalein (3, 10, 30 mg/kg)	Cyclosporin A (20 mg/kg), quinidine (10 mg/kg), SKF-525A (10 mg/kg)	i.v.	Rat	Jugular vein, striatum, bile duct	AUC_{bile}/AUC_{blood} ↑ (cyclosporin A, quinidine, SKF-525A)	[25]
Lamotrigine (10 mg/kg)	Valproate (300 mg/kg)	i.v.	Rat	Hippocampus	No effect on pharmacokinetics of lamotrigine	[30]
10, 11-Dihydro-10-hydroxy carbamazepine (MHD, 20 to 150 mg/kg)	Verapamil (5 mM)	i.p.	Rat	Hippocampus	MHD concentration in brain ↑	[31]
Triazolam (2.5 mg/kg)	Itraconazole (25 mg/kg, p.o.)	i.v.	Rat	Prefrontal cortex	C_{max} ↑ AUC ↑	[32]
Buprenorphine (30 mg/kg)	Flunitrazepam (40 mg/kg)	i.v.	Rat	Striatum	No effect on pharmacokinetics of buprenorphine	[33]
Morphine (20 mg/kg)	Probenecid (20 mg/kg)	i.v.	Rat	Jugular vein, striatum	Brain efflux clearance of Mor ↓	[34]
Morphine (10 mg/kg)	Diclofenac (5 mg/kg)	i.p.	Rat	Jugular vein, frontal cortex	No effect on pharmacokinetics of morphine	[37]
Hesperidol (30 mg/kg)	Cyclosporin A (20 mg/kg)	i.p.	Rat	Jugular vein, bile duct	AUC_{bile}/AUC_{blood} ↓	[40]
Phenylpropanolamine (PPA, 2.5 mg/kg)	Chlorpheniramine (0.4 mg/kg) Caffeine (5 mg/kg)	i.p.	Rat	Jugular vein, frontal cortex	C_{max} and AUC of PPA in brain ↑	[41]
Phentermine (1, 5 mg/kg)	Fenfluramine (5 mg/kg)	i.p.	Rat	Jugular vein, frontal cortex	C_{max} and AUC of Phen ↑ (blood and brain); CL of Phen ↓ (blood and brain)	[46]
MDMA (5 mg/kg)	Caffeine (20 mg/kg)	i.p.	Rat	Jugular vein, striatum	Brain efflux clearance of MDMA ↓	[49]

samples [53]. Most medicaments as well as endogenous components in biological samples are commonly nonvolatile polar compounds, and thus HPLC is more suitable than GC for their analysis. Another advantage of HPLC is a variety of detection methods, including ultraviolet (UV), mass spectrometry (MS), fluorescence (FL), and chemiluminescence (CL) detection, in addition to electrochemical detection (ECD). Additionally, a labeling reaction is often combined with these methods as demands. Labeling can be used to improve the sensitivity, selectivity, and chromatographic behavior in some cases.

In this section, special applications of microdialysis to evaluate drug–drug (or food) interaction involving drugs of abuse are described. Studies on both pharmacokinetic and pharmacodynamic drug–drug interactions in current publications are introduced.

2. PHARMACOKINETIC DRUG–DRUG INTERACTION

Pharmacokinetic interaction causes an extreme change of concentration of free-form medicament in blood and/or target tissue. Consequently, determination of free-form medicaments is helpful in assessing pharmacokinetic interaction. Many pharmacokinetic approaches to drug–drug interaction have been established in measurement of the total concentration of medicament in plasma. Some studies have also focused on free-form medicaments in target tissue (e.g., brain and bile). Determination of free-form medicaments in target tissue is expected to be very useful to predict more particular information about interactions [26–31,39]. Recently, metabolic enzymes and transporters have attracted attention as important factors, since most drug–drug interactions are believed to be mediated by these factors. The key enzymes responsible for drug metabolism are those of the CYP families. When medicaments pass through the liver, they are metabolized to active and/or nonactive metabolites. This metabolic process has a great influence on therapeutic efficacy. Moreover, transporters such as P-gp, multidrug resistance protein (MRP), and organic anion transport protein (OATP) are localized in the liver on the endothelium and epithelium membrane, as well as in the gastrointestinal track, kidney, and other organs for pumping out xenobiotics from the cell [52]. Since they are present in several types of cells and mediate cell exposure to medicaments, they can affect not only absorption, distribution, and excretion, but also concentration of medicaments in the target cell and change the therapeutic efficacy. Recent studies on pharmacokinetic drug–drug interaction are summarized in Table 1.

2.1. Antibiotics

Ciprofloxacin, a 4-quinolone antibiotic, is highly active on a broad spectrum of microbial pathogens. To investigate drug–drug interaction between ciprofloxacin and cyclosporin A (P-gp inhibitor) in rats, a blood microdialysis coupled to HPLC–UV was employed in this study [15]. A microdialysis probe

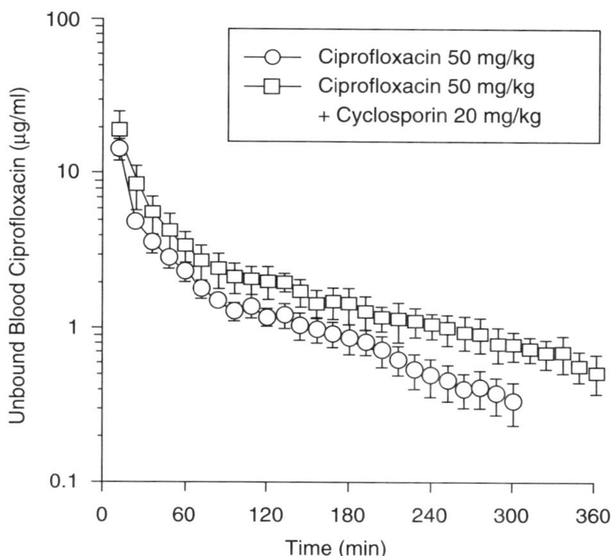

Figure 6 Mean free-form blood concentration–time profile of ciprofloxacin (50 mg/kg) with or without cyclosporin A treatment (20 mg/kg) ($n = 6$). (From [15], with permission from Elsevier Science.)

was inserted into the jugular vein of male Sprague–Dawley rat for sampling of free-form ciprofloxacin. The pharmacokinetic parameters of ciprofloxacin [intravenous (i.v.) doses of 20 or 50 mg/kg] were compared with those treated with a single i.v. dose of 20 mg/kg of cyclosporin A prior to ciprofloxacin administration. Figure 6 shows the concentration–time profile of ciprofloxacin (50 mg/kg). This study showed that a single-dose treatment with cyclosporin A resulted in a significant increase in the pharmacokinetic parameter of the area under the curve [AUC, min·μg/mL: 674 vs. 1074 ($p < 0.05$)] estimated for the higher dose of ciprofloxacin (50 mg/kg), whereas no significant difference was found at the lower dose of ciprofloxacin (20 mg/kg). These results suggested that cyclosporin A can alter the pharmacokinetics of ciprofloxacin in rats. A possible mechanism for this interaction includes the blocking by cyclosporin A of the protein binding of ciprofloxacin with P-gp in rat blood. Tsai et al. also examined the drug–drug interaction of cefoperazone coadministered with berberine (a P-gp enhancer) by the bile microdialysis method [20]. The probe for the bile microdialysis was made in-house and inserted in the bile duct (Figure 7). As a result, there was no significant difference in bile excretion of cefoperazone coadministered with or without berberine (30 mg/kg).

Moreover, using simultaneous sampling by a combination of bile or brain microdialysis with blood microdialysis, bile-to-blood or brain-to-blood distributions of medicament could be evaluated. The pharmacokinetics of pefloxacin (=norfloxacin, 10 mg/kg, i.v.) and its interaction with cyclosporin A (10 mg/kg,

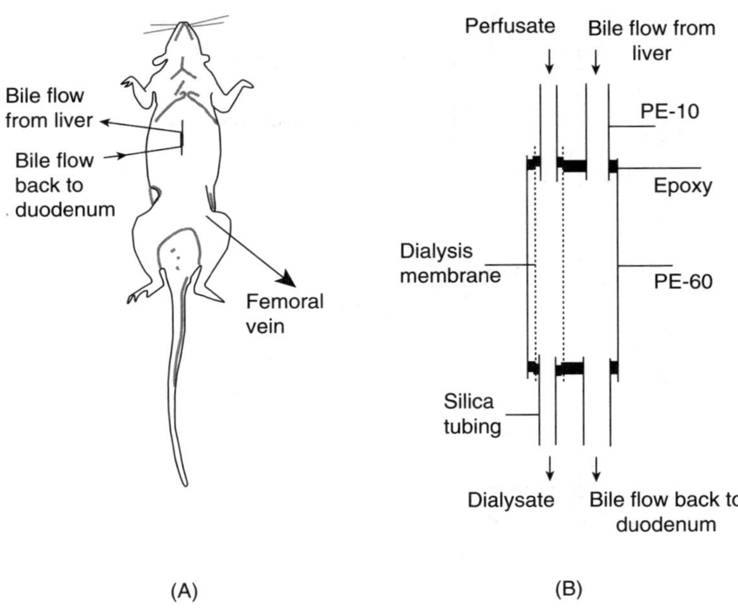

Figure 7 (A) Animal model for bile sampling, flow-through microdialysis probe used for rat bile sampling; (B) detailed description of a homemade bile microdialysis probe. (From [16], with permission from Nature Publishing Group.)

i.v.) were studied by the combination of blood and brain or bile microdialysis [16]. The probes for blood, brain, and bile microdialysis were inserted in the jugular vein, striatum, and bile duct of rat, respectively. The brain-to-blood coefficient of distribution (AUC_{brain}/AUC_{blood}) was 0.036, and the bile-to-blood coefficient of distribution (AUC_{bile}/AUC_{blood}) was 1.53. The results indicate that pefloxacin was able to penetrate the BBB and that the concentration in bile was greater than that in blood. These facts showed active biliary excretion of pefloxacin, and no significant impact of cyclosporin A on the pharmacokinetics of pefloxacin in rat blood and brain was observed. Marchand et al. studied the effect of probenecid, an inhibitor of anion transport proteins, on norfloxacin BBB transport [17]. In this case, the probe for brain microdialysis was implanted into hippocampus (hipp). However, probenecid had no effect on norfloxacin BBB transport. Furthermore, the drug–drug interactions for cefepime versus cyclosporin A [18] and amoxicillin versus ivermectin (a disruptor of P-gp) [19] were examined by blood and brain microdialysis.

2.2. Antifungal Agents

To investigate the mechanism of P-gp-related hepatobiliary excretion of fluconazole, the drug–drug interaction of fluconazole and cyclosporin A was examined [21]. Fluconazole was determined by HPLC with a 50-ng/mL detection limit, combined with simultaneous blood and bile microdialysis. The pharmacokinetic parameters, such as AUC, mean residence time (MRT), and AUC_{bile}/AUC_{blood} of fluconazole (10 mg/kg, i.v.), were comparable to those of fluconazole coadministered with cyclosporin A (20 mg/kg, i.v.). This means that fluconazole undergoes hepatobiliary excretion which may be unrelated to that of the P-gp transporter system.

Berberine, derived from the roots and bark of *Berberis aristata* and *Coptis chinensis*, has an antifungal effect. To investigate the detailed pharmacokinetics of berberine and its mechanism of hepatobilinary excretion, microdialysis coupled with HPLC–UV was used [22]. In the control group, rats were administered with berberine alone (10 or 20 mg/kg); in the drug-treated group, the rats were i.v. injected 10 min before berberine administration. Concomitant medications with berberine were cyclosporin A (a P-gp inhibitor, 20 mg/kg), quinidine [both organic cation transporter (OCT) and P-gp inhibitor, 10 mg/kg], SKF-525A (a CYP inhibitor, 10 mg/kg), and probenecid (100 mg/kg) to inhibit the glucuronidation. The results, shown in Figure 8, were as follows: (1) a proportional increase in AUC was observed; (2) berberine was processed though hepatobiliary excretion against a concentration gradient based on AUC_{bile}/AUC_{blood} (6.6 ± 1.6 for 10 mg/kg and 7.4 ± 0.9 for 20 mg/kg); (3) the active berberine efflux might be affected by P-gp and OCT since coadministration of berberine and cyclosporin A or quinidine at the same dosage of 10 mg/kg significantly decreased the berberine amount in bile; and (4) the metabolism of berberine was partially reduced by SKF-525A treatment, but glucuronidation of berberine was not obviously affected by probenecid under the present design.

2.3. Anti-inflammatory Agents

Diclofenac, a nonsteroidal anti-inflammatory drug, is a potent inhibitor of prostaglandin synthesis. To explore the mechanism of interaction with P-gp on hepatobiliary excretion of diclofenac, a single dose of cyclosporin A was coadministered with diclofenac [23]. To determine diclofenac, which binds to plasma protein in a high ratio, in rat bile and to investigate its hepatobiliary excretion, HPLC–UV coupled to a microdialysis sampling system was used. A microdialysis probe was inserted into the bile duct between the liver and the duodenum for continuous sampling of the medicament from bile fluids following i.v. administration of diclofenac (1 mg/kg). The diclofenac in rat bile appeared to have a slow elimination phase, with the peak concentration at 20 min following diclofenac administration (Figure 9). The pharmacokinetic parameters of diclofenac by administration of diclofenac with/without cyclosporin A (20 mg/kg) are shown in Table 2. The C_{max} and AUC values of diclofenac were increased by adding cyclosporin A, whereas $T_{1/2}$ and MRT were not significantly altered between the groups with or without cyclosporin A. The disposition of diclofenac in bile indicates that it may undergo biliary elimination and that the bile elimination of diclofenac was increased by coadministration of cyclosporin A.

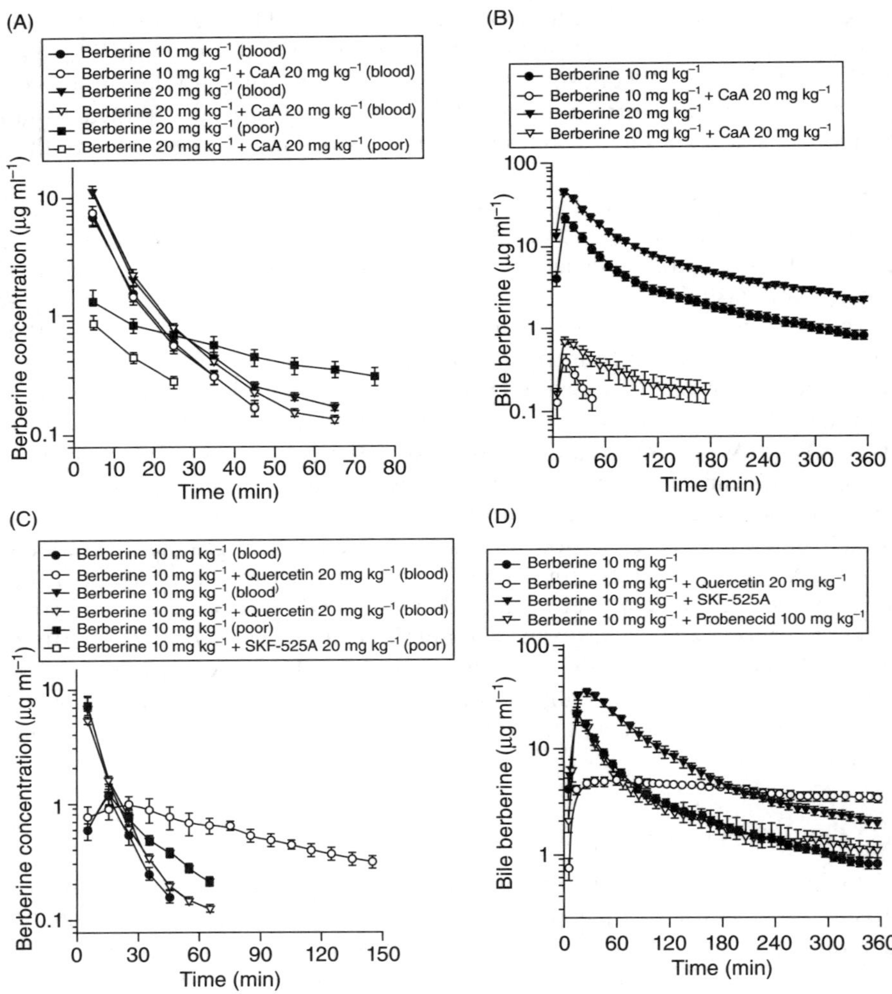

Figure 8 Concentration–time profiles for berberine in (A, B) blood and liver dialysates (C, D) bile dialysates. (A) and (C) obtained after berberine i.v. administration at dose of 10 and 20 mg/kg with or without 20 mg/kg cyclosporin A. (B) and (D) obtained after administration of 10 mg/kg berberine alone, and with 10 mg/kg quinidine, 10 mg/kg of SKF-525A, or 100 mg/kg probenecid. (From [22], with permission from The American Society for Pharmacology and Experimental Therapeutics.)

Baicalin, a flavone glucronide derived mainly from the root of *Scutellaria baicalensis*, has been used in traditional Chinese medicine as an anti-inflammatory and antiviral agent. Baicalin (3, 10, and 30 mg/kg, i.v.) was injected in rats, and microdialysis was coupled with HPLC–UV to study the effect of P-gp on the disposition of baicalin [25]. Probes for blood, bile, and

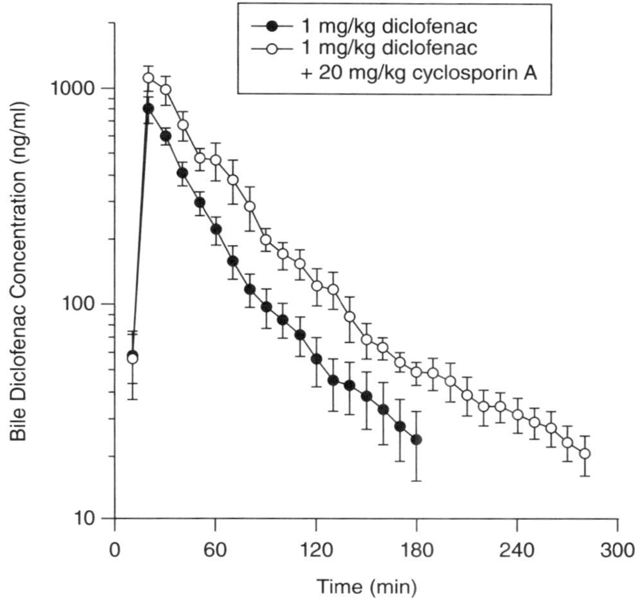

Figure 9 Pharmacokinetic curve of diclofenac in rat bile concentration versus time following i.v. administration of diclofenac (1 mg/kg, i.v.), both alone and pretreated with cyclosporin A (20 mg/kg, i.v., $n = 6$). Data are presented as mean ± S.E.M. (From [23], with permission from Elsevier Science.)

TABLE 2 Pharmacokinetics of Diclofenac in Rat Bile: i.v. Administration of 1 mg/kg[a] Diclofenac

	Diclofenac	
Drug Treatment	Without Cyclosporin A	With Cyclosporin A (20 mg/kg)
C_{max} (ng/mL)	798 ± 110	1187 ± 146[b]
$t_{1/2}$ (min)	31 ± 4	35 ± 4
AUC (min·ng/mL)	32,700 ± 2560	60,100 ± 6880[b]
MRT (min)	55 ± 7	65 ± 4

Source: From [23], with permission from Elsevier Science.

[a]Data are expressed as mean ± S.E.M. ($n = 6$).

[b]$p < 0.05$; significant different from the group without cyclosporin A.

brain microdialysis were inserted in the jugular vein, striatum, and bile duct, respectively. Coadministration of cyclosporin A (20 mg/kg) or quinidine (10 mg/kg), both P-gp inhibitors, was used to elucidate the role of P-gp on baicalin disposition, while SKF-525A (10 mg/kg, CYP inhibitor) could specifically inhibit the CYP enzyme catalysis of baicalin without being crossed with P-gp. Both cyclosporin A and quinidine significantly promoted the active transport of baicalin into bile and reduced its level in blood. The AUC_{bile}/AUC_{blood} for the control, cyclosporin A, and quinidine groups were 50.4 ± 4.4, 160 ± 13, and 154 ± 11, respectively. This result was the same as that obtained by treatment with SKF-525A (125 ± 13). Hence, the association of the involvement of P-gp in active baicalin efflux into bile seems to be excluded since cyclosporin A and quinidine are also CYP inhibitors. In addition, baicalin was not detected in the brain striatum after treatment with baicalin alone in the present study. Also, neither cyclosporin A nor quinidine coadministered with baicalin can induce measurable levels of baicalin in rat brain, which suggests that baicalin might not pass through the BBB. Moreover, baicalein, which is a aglycone of baicalin, showed same pharmacokinetic behavior, as shown by baicalin when it was coadministered with cyclosporin A [24].

2.4. Hypnotic Agents

Triazolam (TZ) is known as a short-acting hypnotic agent, and its clearance depends on hepatic metabolism of CYP3A4. Therefore, the inhibition of CYP3A4 caused by concomitant medication can result in pharmacokinetic drug–drug interaction. To study the effect of itraconazole (ITZ, CYP3A4 inhibitor) on the pharmacokinetics of TZ, a semimicro column HPLC method for TZ in rat plasma and brain microdialysates was used [32]. The probe for brain microdialysis was inserted in the prefrontal cortex. The pharmacokinetics was evaluated for TZ (2.5 mg/kg, i.v.) in a single simultaneous administration of TZ with ITZ [25 mg/kg, per os (p.o.)] and single administration of TZ after pretreatment with ITZ (25 mg/kg, p.o.) once a day for a week (Figure 10 and Table 3). In a single administration study, C_{max} and the AUC increased 3.4- and 2.9-fold, respectively, compared to the valus for TZ alone. On the other hand, there was no significant difference in TZ's C_{max} value in both ITZ

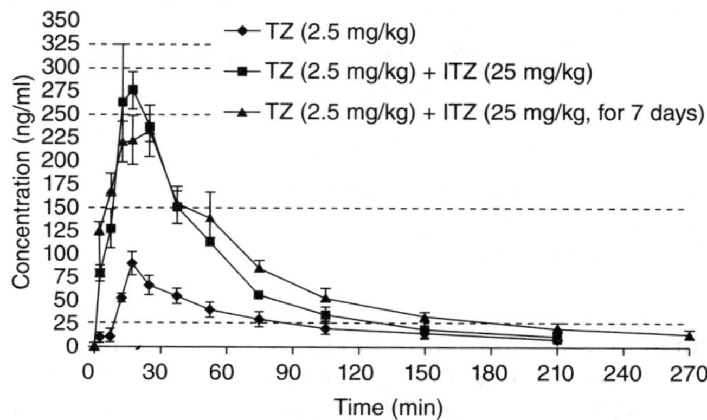

Figure 10   Mean concentrations of TZ in rat brain microdialysate after administration of TZ (2.5 mg/kg, i.v.) without and with ITZ (25 mg/kg, p.o.). TZ, triazolam; ITZ, itraconazole. (From [32], with permission from Elsevier Science.)

TABLE 3 Pharmacokinetic Parameters for 2.5 mg/kg i.v. TZ Administration[a]

	C_{max} (ng/mL)	T_{max} (min)	$t_{1/2}$ (min)	CL (ml/·min/kg)	AUC (ng·min/mL)	MRT (min)
TZ ($n = 5$)	91 ± 12	19.0 ± 2.0	37.0 ± 8.4	565.3 ± 138.1	5292 ± 937	63.1 ± 12.8
TZ + ITZ[b] ($n = 3$)	312 ± 40^{c}	15.8 ± 1.7	57.2 ± 8.9	163.6 ± 10.1	15404 ± 1012^{c}	65.2 ± 7.4
TZ + ITZ[d] (daily for 7 days) ($n = 4$)	268 ± 12^{c}	20.0 ± 3.1	78.6 ± 8.7^{e}	129.0 ± 13.3^{e}	19939 ± 1821^{c}	93.6 ± 5.9^{f}

Source: From [32], with permission from Elsevier Science.

[a]Data are expressed as mean ± S.E.M.

[b]25 mg/kg, p.o.

[c]Significant different ($p < 0.001$) from the TZ group.

[d]After 24 h of daily pretreatment of ITZ for 7 days in rat brain microdialysates.

[e]Significant different ($p < 0.05$) from the TZ group.

[f]Significant different ($p < 0.05$) from TZ + ITZ group.

treatments ($p > 0.2$); however, its $T_{1/2}$ value after daily pretreatment with ITZ increased significantly ($p < 0.05$). As a result, in single simultaneous administration of TZ with ITZ and single administration of TZ after daily pretreatment with ITZ to rats, ITZ seriously interfered with pharmacokinetic parameters of TZ in brain microdialysate.

The potential of lethal interaction has been reported among heroin addicts related to combinational use of buprenorphine with flunitrazepam. To determine the existence of a drug–drug interaction during the distribution phase passed through BBB of buprenorphine, the pharmacokinetics of buprenorphine was studied in plasma and in striatum using cerebral microdialysis after i.v. administration of buprenorphine alone (30 mg/kg), flunitrazepam alone (40 mg/kg), or both medications in rats [33]. Plasma buprenorphine kinetics were well described by a three-compartment linear model, with a distribution half-life of 7.4 ± 2.7 min and an elimination half-life of 463.9 ± 152.3 min. Neither plasma nor striatal buprenorphine kinetics were altered significantly by preadministration of flunitrazepam. Furthermore, partial pressure of carbon dioxide (PCO_2) of arterial blood was measured in both the administration of buprenorphine alone and that after rat pretreatment with flunitrazepam. In contrast to buprenorphine or flunitrazepam alone, buprenorphine in combination with flunitrazepam induced a significant, rapid, and sustained respiratory depression. Arterial PCO_2 was increased at 1.5 min (6.7 ± 0.2 vs. 5.4 ± 0.3 and 5.5 ± 0.3 kPa, respectively) (mean $\pm$ S.E.M., $p = 0.04$), and arterial pH decreased (7.37 ± 0.02 vs. 7.45 ± 0.02 and 7.45 ± 0.01, respectively, $p = 0.03$). The adverse interaction between flunitrazepam and buprenorphine could not be explained by a pharmacokinetic drug–drug interaction during the distribution phase of buprenorphine.

2.5. Narcotics

Morphine (Mor) is used most frequently as an analgesic drug for postoperative and cancer pain. Since there is a possibility that Mor will be coadministered with other medicaments, the elucidation of drug–drug interaction of Mor is important. Some studies on the pharmacokinetics of Mor using microdialysis sampling have been published [34–37]. Tunblad et al. reported the possible influence of probenecid on Mor transport across the BBB in rats [34]. The microdialysis probes were placed into the jugular vein and striatum of rats. Mor was administered as an exponential i.v. infusion over 4 h, with the goal of reaching 1800 ng/mL in plasma instantaneously. The experiment was repeated the next day with the addition of probenecid, administered as a bolus dose (20 mg/kg), followed by constant infusion (20 mg/kg/per hour). The steady-state ratio of 0.29 ± 0.07 of unbound Mor concentration in brain to that in blood indicates that Mor is actively effluxed at the BBB. Probenecid coadministration increased the ratio to 0.39 ± 0.04 ($p < 0.05$). The half-life in brain increased from 58 ± 9 min to 115 ± 25 min when probenecid was coadministered. Systemic clearance of Mor also decreased upon probenecid coadministration, and morphine-3-glucuronide (one of the metabolites of Mor) formation was decreased. This study indicates that Mor is a substrate for probenecid-sensitive transporters at the BBB. Moreover, the drug–drug interaction of Mor with diclofenac was examined by microdialysis with implanting of probes in the jugular vein and frontal cortex [37]. Rats were administered with a single intraperitoneal (i.p.) dose of Mor (10 mg/kg) with or without a single i.p. dose of diclofenac (5 mg/kg). Mor is a low-transitive compound for the BBB, since the AUC ratio (AUC_{brain}/AUC_{blood}), which indicates transitivity of Mor to brain, was 0.38. Following coadministration of Mor with diclofenac (5 mg/kg, i.p.), the AUC_{brain}/AUC_{blood} ratio did not change (0.31), and no pharmacokinetic parameters of Mor in either brain or blood were altered significantly.

The intake of grapefruit juice has been demonstrated to elevate plasma concentrations of several medicaments, such as calcium channel blockers [54]. Okura et al. studied the effect of grapefruit juice on the pharmacokinetics of Mor [36]. After repeated treatment with Mor (100 mg/kg, p.o.) for 5 days, the microdialysis probes were implanted into jugular veins and spinal intrathecal spaces in rats. The Mor concentrations in blood and microdialysate were gradually decreased by repeated treatment with Mor (Figure 11), and the grapefruit juice treatment (2 mL/rat, p.o.) significantly increased the blood concentration of Mor in Mor-tolerant rats. The antinociception of Mor (30 mg/kg, p.o.) was enhanced significantly by the oral administration of grapefruit juice. These results suggest that oral coadministration of grapefruit juice enhances the Mor antinociception by increasing the intestinal absorption of Mor.

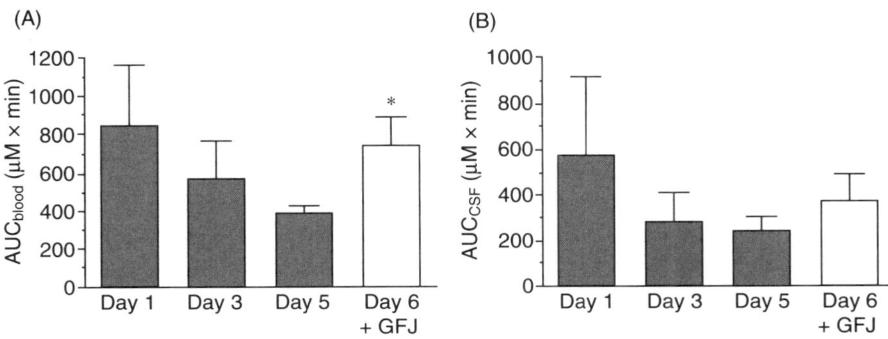

Figure 11 Area under the concentration curve of Mor in the blood (AUC_{blood}) (A) and spinal CSF (AUC_{CFS}) (B) of Mor after oral administration for 1 to 5 days in rats. Microdialysis probes were implanted into the jugular vein and intrathecal space. Rats received morphine orally (100 mg/kg) once a day. On the day after 5-day treatment with Mor (day 6), rats received grapefruit juice (2 mL/rat) 30 min before receiving Mor. The concentrations of Mor in the dialysates were measured on days 1, 3, 5, and 6. Each column represents the mean ± S.E. for three to four rats.* $p < 0.05$ vs. day 5. (From [36], with permission from the Pharmaceutical Society of Japan.)

2.6. Other Drugs

Camptothecin from *Camptotheca acuminate* plays an important role in cancer chemotherapy. The effect of P-gp modulators such as cyclosporin A, berberine, quercetin, naringin, and naringenin on camptothecin distribution into the bile and brain was examined. Pharmacokinetic parameters of camptothecin were assessed using an HPLC–FL method coupled with blood, bile, and brain microdialysis sampling [38]. The microdialysis probes for blood, bile, and brain were implanted in the jugular vein, bile duct, and striatum, respectively. Before 10 min of camptothecin (5 mg/kg, i.v.) injection, cyclosporin A (20 mg/kg) was injected via the femoral vein. For the cyclosporin A pretreatment group, 10 mg/kg of cyclosporin A was injected i.p. daily for 5 days; on the sixth day, cyclosporin A was injected 10 min prior to camptothecin injection. Berberine (20 mg/kg), quercetin (0.1 mg/kg), naringin (10 mg/kg), and naringenin (10 mg/kg) were also injected 10 min prior to camptothecin injection. As shown in Figure 12, the camptothecin crosses the BBB rapidly, within 20 min after camptothecin administration. The disposition of camptothecin in rat bile appeared to have a slow elimination phase and a peak concentration after 20 min of camptothecin administration. The AUC for camptothecin in bile significantly surpassed that in blood, suggesting active transport of hepatobiliary excretion ($AUC_{bile}/AUC_{blood} = 4.32 \pm 0.51$). In the presence of cyclosporin A, camptothecin AUC in the brain was elevated significantly (6.84 ± 0.36 vs. 11.48 ± 1.15, $p < 0.05$), but there was no significant change in the presence of berberine, quercetin, naringin, and naringenin. By treatment with smaller

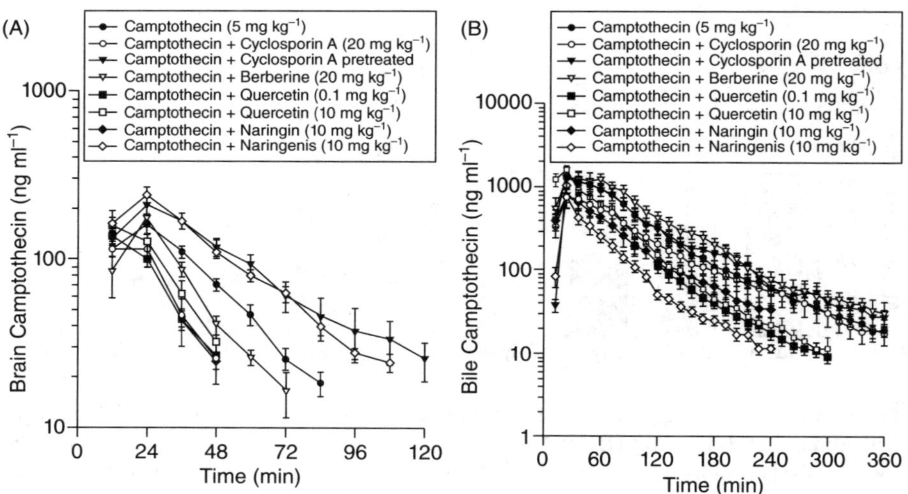

Figure 12   Mean camptothecin levels in rat (A) brain and (B) bile were divided into the groups above with six individual experimental animals used in each group. Data are presented as means ± S.E.M. (From [38], with permission from Nature Publishing Group.)

doses of quercetin (0.1 mg/kg), naringenin (10 mg/kg), and naringenin (10 mg/kg), they significantly diminished the camptothecin AUC in bile, but this was not altered by any treatment, such as berberine (20 mg/kg), a higher dose of quercetin (10 mg/kg), or cyclosporin A treated (20 mg/kg) and pretreated with cyclosporin A groups. The distribution ratio (AUC_{bile}/AUC_{blood}) of camptothecin in bile was decreased in the cyclosporin A-, quercetin-, naringin-, and naringenin-treated groups. However, the AUC_{brain}/AUC_{blood} in the brain was increased in the cyclosporin A groups (0.27 ± 0.0024 vs. $0.40 \pm 0.0035, p < 0.05$), while it was decreased in the groups treated with quercetin (vs. 0.16 ± 0.022), naringin (vs. 0.15 ± 0.023), and naringenin (vs. 0.14 ± 0.012). These results revealed that P-gp might modulate hepatobiliary excretion and BBB penetration of camptothecin.

Pyrazinamide (PZA) is widely used in combination with other drugs in chemotherapy for tuberculosis. However, the dose-related liver dysfunction is known as the main adverse effect of PZA and its active metabolite [pyrazinoic acid (PA)]. Silibinin is the main flavonoid extracted from milk thistle, and it displays hepatoprotective properties. The pharmacokinetics of PZA and PA and their interaction with silibinin in rats were evaluated [43]. The rats, divided into six groups, were administered PZA alone (50 mg/kg, i.v.), PZA after long-term silibinin (100 mg/kg, p.o. for three consecutive days), PZA with concomitant short-term silibinin (30 mg/kg, i.v.), PA alone (30 mg/kg, i.v.), PA after long-term silibinin (100 mg/kg, p.o. for three consecutive days), and PA with concomitant short-term silibinin (30 mg/kg, i.v.), respectively. The results indicated that the distribution ratio of PZA from bile to blood (AUC_{bile}/AUC_{blood}) in the PZA after long-term silibinin exposure and PZA with concomitant short-term silibinin exposure group was also not significantly different from that of the PZA-alone group. However, the AUC_{bile}/AUC_{blood} of PA (3.74 ± 0.41) decreased significantly in PA after long-term silibinin exposure (0.96 ± 0.16, $p < 0.05$) and in PA with concomitant short-term silibinin exposure groups (0.92 ± 0.08, $p < 0.05$). On PZA administration, the blood, but not bile, AUC of PA (890 ± 140) was markedly increased in the PZA after long-term silibinin exposure (3140 ± 300, $p < 0.05$) and PZA with concomitant short-term silibinin exposure groups (3730 ± 380, $p < 0.05$), but the AUC_{bile}/AUC_{blood} of PA was decreased. These results suggest that the excretion pathway of PA may be blocked by silibinin through hepatobiliary excretion.

Phenylpropanolamine (PPA) is a nonprescription sympathomimetic agent related structurally to amphetamine. PPA is now used as an appetite suppressant in the United States and has been used as a decongestant of over-the-counter (OTC) products, including caffeine and/or chlorphenylamine, in Japan. As the mechanism of drug–drug interaction of PPA has not yet been clarified, the possibility of pharmacokinetic interaction among PPA, caffeine, and chlorpheniramine was examined [41]. The pharmacokinetics of PPA in rat brain and blood was evaluated by the HPLC–FL method coupled with blood and brain microdialysis sampling. The blood probe was implanted within the jugular vein and the brain probe in the frontal cortex. Rats were administered with a single i.p. dose of PPA (2.5 mg/kg), by combination with caffeine (5 mg/ kg), with chlorpheniramine (0.4 mg/kg), with both caffeine and chlorpheniramine, and finally, as it existed in a particular OTC product (corresponding to 2.5 mg/kg of PPA, 5 mg/kg of caffeine, and 0.4 mg/kg of chlorpheniramine). The combinations of caffeine with PPA caused a 1.6-fold increase in the AUC in brain compared to that of a single administration of PPA, and the combination of chlorphenylamine with PPA caused a 1.8-fold increase ($p < 0.05$). The multiple combinations caused a 1.9-fold increase in the AUC ($p < 0.05$), which is similar to that in the AUC of PPA in brain compared to that of a single administration of the OTC product, including PPA, chlorphenylamine, caffeine, belladonna alkaloids, and lysozyme chloride ($p < 0.05$). The C_{max} value of PPA in brain was increased significantly by coadministration of caffeine, chlorphenylamine, or both (1.8-, 1.8- and 1.9-fold; $p < 0.05$, respectively), and the clearance of PPA in brain decreased significantly with coadministration (0.6-, 0.7-, and 0.6-fold; $p < 0.05$, respectively). Contrary to these results in brain pharmacokinetics, combined administration did not affect the pharmacokinetics of PPA in blood. These showed that the drug–drug interaction associated with PPA could be related to the significant increase in its levels in the brain by the concomitant medications.

2.7. Drugs of Abuse

In forensic cases, drugs of abuse might be interacted with each other by multidrug abuse, and this would lead to serious human toxic symptoms. Thus, drug–drug interactions including pharmacokinetics and/or pharmacodynamics should be clarified. Several studies on pharmacokinetic drug–drug interaction of drugs of abuse, such as stimulants [51], narcotics [49,50], cocaine [47], obesity drugs [46], and medicaments [41,48], have been reported.

3,4-Methylenedioxymethamphetamine (MDMA, called ecstasy) has become a popular recreational drug among young people. MDMA tablets often contain active components that have a hallucinogenic and/or stimulant effect, such as caffeine, ketamine, ephedrine, or methamphetamine. So the effect of caffeine on the pharmacokinetics of MDMA and MDA (an active metabolite of MDMA) was evaluated by simultaneous blood and brain microdialysis coupled with HPLC–FL detection [49]. For sensitive determination of MDMA, an FL labeling reagent, 4-(4,5-diphenyl-1H-imidazol-2-yl)benzoyl chloride (DIB-Cl), was used (Figure 13). The rat-inserted probes in jugular vein and striatum received MDMA (5 mg/kg, i.p.) with or without caffeine (20 mg/kg, i.p.). Figure 14 and Table 4 show the time–concentration profiles and pharmacokinetic parameters of MDMA and MDA in blood and brain microdialysates. After administration of MDMA with caffeine, the AUC_{0-300} of MDMA in blood increased significantly, to 1.7-fold higher ($91 \pm 5\,\mu g \cdot min/mL$) than that of MDMA alone, and the clearance (CL) of MDMA decreased by half, to $51 \pm 4\,mL/min \cdot kg$. These facts suggested that caffeine might interfere with the renal CL of MDMA. On the other hand, the C_{max} value of

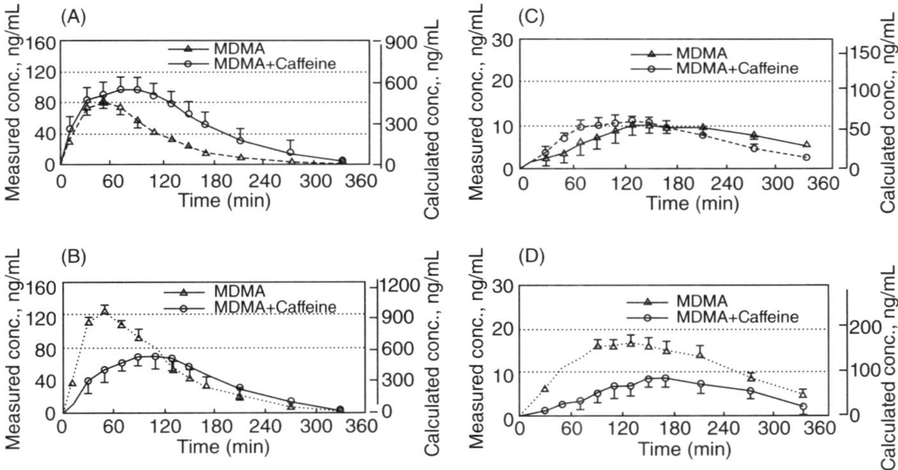

Figure 13 Labeling reaction of MDMA and MDA with DIB-Cl.

Figure 14 Concentration–time profiles of MDMA in rat (A) blood and (B) brain and MDA in (C) blood and (D) brain microdialysates after a single administration of MDMA (5 mg/kg) with or without caffeine (20 mg/kg). Data are expressed as mean ± S.E.M. (MDMA-alone group, $n = 5$; MDMA with caffeine group, $n = 3$). (From [49], with permission from John Wiley & Sons.)

MDMA and MDA in brain decreased by half (496 ± 80 and 93 ± 21 ng/mL, $p < 0.01$ and $p < 0.05$, respectively). Moreover, the T_{max} and MRT values of MDMA and MDA were extended (2.1- and 1.3-fold, respectively); however, the $T_{1/2}$ values of MDMA and MDA were unchanged. These showed that caffeine inhibits the transportation of MDMA and MDA to brain via the BBB, although the exclusions of MDMA and MDA from brain to blood were not

TABLE 4 Pharmacokinetic Parameters of MDMA and MDA in Rat Blood and Brain After Single Administration of MDMA (5 mg/kg) with or Without Caffeine (20 mg/kg)

Parameter	MDMA Alone ($n = 5$)		MDMA with Caffeine ($n = 3$)	
	MDMA	MDA	MDMA	MDA
In blood				
C_{max} (ng/mL)	477 ± 50	68 ± 9	550 ± 45	63 ± 10
T_{max} (min)	38 ± 5	114 ± 12	77 ± 24	183 ± 27^a
$t_{1/2}$ (min)	43 ± 8	118 ± 21	57 ± 6	164 ± 20
$AUC_{0\text{-}300}$ (μg·min/mL)	53 ± 5	13 ± 2	91 ± 5^b	12 ± 2
$MRT_{0\text{-}300}$ (min)	85 ± 4	140 ± 4	109 ± 11	164 ± 10^a
CL (mL/min·kg)	96 ± 10		51 ± 4^a	
In brain				
C_{max} (ng/mL)	1009 ± 28	168 ± 17	496 ± 80^b	93 ± 21^a
T_{max} (min)	46 ± 4	138 ± 23	97 ± 24	177 ± 18
$t_{1/2}$ (min)	44 ± 5	145 ± 26	54 ± 9	70 ± 15
$AUC_{0\text{-}300}$ (μg·min/mL)	118 ± 8	32 ± 3	93 ± 10	15 ± 4^a
$MRT_{0\text{-}300}$ (min)	92 ± 3	144 ± 4	122 ± 15	168 ± 15
CL (mL/min·kg)	42 ± 4		50 ± 7	

Source: From [49], with permission from John Wiley & Sons, Ltd.

Data are expressed as mean ± S.E.M.

[a] $p < 0.05$, significantly different from MDMA alone.

[b] $p < 0.05$, significantly different from MDMA alone.

influenced by caffeine. Therefore, it can be concluded that ingredients in MDMA tablets might influence the pharmacokinetics of MDMA, and thus abusers will take higher doses of MDMA, causing poisonously high levels of MDMA in the blood.

The combined use of phentermine (Phen) and fenfluramine (Fen) have been prescribed for obesity purpose. However, this combination was found to enhance the neurotoxic effect as well as the weight loss at lower doses. To clear the pharmacokinetic mechanism of this interaction, an HPLC method coupled with simultaneous blood and brain microdialysis was used [46]. The microdialysis probes were implanted in jugular vein and frontal cortex. Rats were administered individual doses of Phen (1 or 5 mg/kg) and Fen (1 or 5 mg/kg) or a combination of both drugs (1 or 5 mg/kg of Phen and 5 mg/kg of Fen). In addition, since the role of protein binding in drug interaction can be quite involved, the HPLC method was also used to determine the total and free-form Phen and Fen in rat plasma and ultrafiltrated blood sample, respectively. The profiles of pharmacokinetic parameters of Phen in rat blood and brain were altered significantly. In rat brain, Fen increased Phen levels significantly, with C_{max} and AUC values of Phen having increased 2.0- to 2.2-fold and 3.4-fold ($p < 0.05$ and $p < 0.01$, respectively). The T_{max} and MRT values of Phen were extended about two fold ($p < 0.05$ for both), and the CL values of Phen were notably decreased 0.2- to 0.3-fold ($p < 0.05$). In blood, similar results were obtained, in which the AUC of Phen was increased significantly ($p < 0.05$). Interestingly, the protein binding value of Phen (35%) was not altered by coadministration of Fen. On the other hand, neither Fen nor its active metabolite norfenfluramine (Norf) had a significant effect on the pharmacokinetic parameters. These results suggest that Fen and/or Norf may have a similar inhibiting effect on the transporters, which are responsible for Phen elimination, where efflux from brain to blood is inhibited. This would lead to higher brain distribution and hence alter the pharmacokinetic parameters. The consequence of such an alteration in the pharmacokinetics of a CNS drug may play a part in the enhancing effect as well as in the neurotoxicity of Phen and Fen.

The illicit use and abuse of 1,4-butanediol (1,4-BD) result from its presumed conversion to γ-hydroxybutyrate (GHB) and subsequent pharmacological effects on r-aminobutyric acid B (GABA-B) and GHB-specific receptors. Using brain microdialysis, the appearance of GHB in the striata of rats after i.p. administration of 1,4-BD (250, 500, and 700 mg/kg) was examined [48]. GHB appeared in the striatal microdialysates within 20 min following administration of 1,4-BD. GHB reached maximum concentration dose-dependently after 80 to 100 min of 1,4-BD administration, with peak values (mean $\pm$ S.E.M.) of 10.6 ± 2.9 (250 mg/kg), 25.3 ± 3.4 (500 mg/kg), and $48.1 \pm 7.1 \mu g/mL$ (750 mg/kg), respectively. The conversion of 1,4-BD to GHB was completely prevented by the administration of the alcohol dehydrogenase inhibitor 4-methylpyrazole (4MP) prior to 1,4-BD ($p < 0.05$). This phenomenon was evidenced by the failure of GHB detection in the striatal microdialysates.

3. PHARMACODYNAMIC DRUG–DRUG INTERACTION

Pharmacodynamic evaluation of drug interaction focuses on the biological and physiological effects of drugs and the mechanisms by which they produce such effects. In addition to pharmacokinetic evaluation, pharmacodynamic evaluation is important for the evaluation of effect, because it directly reflects drug action. The actions of drugs are the consequence of dynamic interactions between drug molecules and cellular components. The target of drug action can be classified as follows: (1) receptors or transporters, (2) signal transduction mediated by receptors, (3) target cell desensitization and hypersensitization, and (4) pharmacological effects not mediated by receptors (e.g. enzyme). These actions lead to changed amounts of physiological molecules such as neurotransmitters.

Drugs acting in the nervous system modulate neurotransmission in the CNS, which consists of the brain and the spinal cord, or in the peripheral nervous system, which includes the autonomic nervous system and the somatic

TABLE 5　Studies on Pharmacodynamic Evaluation for Drug–Drug Interaction

Target Compounds		Species	Target Organs	Analyte	Effect of [A or B]	Ref.
A	B					
Risperidone (0.05 mg/kg, p.o.)	Galantamine (0.05 mg/kg, p.o.)	Rat	Brain (PFC)	DA	DA levels on PCP-treated schizophrenia model mice, ↑ .	[55]
MPH (2.5 mg/kg, s.c.)	Cit (5 mg/kg, i.p.)	Rat	Brain (PFC, hipp, NAc, striatum)	5-HT, NA, DA	[B] MPH-induced increase in DA levels in PFC, NAc, and hipp, ↑ . [A] Cit-induced increase in 5-HT level in cortex, ↑ ; in hipp,↓.	[57]
Cit (10 μM, local infusion or 3.0 mg/kg, s.c.)	SB242084 (10 μM, inf. or 0.4 mg/kg s.c.)	Rat	Brain (hipp.)	5-HT, NA, DA	[B] locally infused SB242084 increased Cit-induced 5-HT release.	[58]
Cabergoline (0.25, 1, 2 mg/kg, s.c.)	Milnacipran (30 mg/kg, i.p.)	Rat	Brain (mPFC)	NA, DA, 5-HT	[B] NA level in high-dose cabergoline (1, 2 mg/kg) treated group, ↑ . [B] No change in DA and 5-HT.	[59]
VPA (300 mg/kg)	LTG (10 mg/kg)	Rat	Brain (hipp)	ASP, GLU, TAU, GABA, 5-HT, DA	[A] reversed the decreases in amino acid release seen after LTG, and increased 5-HT release was greatly prolonged by coadministration. No changes on pharmacokinetic parameters of LTG.	[30]
Isoflurane (2.5%)	Clozapine (10 mg/kg, i.p.) Risperidone (10 mg/kg, i.p.) Cit (10 mg/kg, i.p.) Fluoxetine (10 mg/kg, i.p.)	Rat	Brain (striatum)	DA, 3-MT, DOPAC, HVA	[A] clozapine- and risperidone-induced increase in DA, DOPAC, HVA,↓. [A] fluoxetine-induced decrease in DA level, ↑ .	[61]

Ketamine (100 mg/kg, i.p.)	Orexin A (1 nmol, i.c.v.)	Rat	Brain (mPFC)	NE	[B] ketamine-induced increase in extracellular NE release, ↓.	[62]
AP (1 mg/kg, i.p.) MDMA (5 mg/kg, i.p.)	Nandrolone (5, 20 mg/kg, i.m., subchronic)	Rat	Brain (NAc)	DA, DOPAC, HVA 5-HT, 5-HIAA	[B] AP or MDMA-induced increase in DA-level, ↑ . [B] MDMA-induced increase in 5-HT-level, ↓. No effect on pharmacokinetic parameters of MDMA.	[50]
MDMA (2.5, 5, 10 mg/kg, i.p.)	Rimonabant (10 mg/kg, i.p.) after chronically treatment with THC (20 mg/kg, i.p., twice daily, five consecutive days)	Mouse	Brain (PFC, NAc)	DA 5-HT	[A] MDMA increased 5-HT release but not DA release in THC-dependent mice.	[64]
d-AP (2 mg/kg, i.p.)	Kynurenic acid (100 nM, local perfusion)	Rat	Brain (mPFC)	Ach	[B] prevented AP-induced Ach release.	[65]
Cocaine (10 mg/kg, i.p.)	Ro 60-0175 (5 μg, intra-mPFC) SB 242084 and SB 243213, (0.5 or 1 μg, intra-mPFC)	Rat	Brain (NAc)	DA	[B] 5-HT$_{2C}$R agonist increased cocaine-induced DA outflow. [B] 5-HT$_{2C}$R antagonist reduced cocaine-induced DA outflow.	[66]

(*Continued*)

TABLE 5 (*Continued*)

Target Compounds		Species	Target Organs	Analyte	Effect of [A or B]	Ref.
A	B					
MDMA (30 mg/kg, i.p.)	Ethanol (3 g/kg)	Mouse	Brain (NAc shell)	DA	[B] DA-level in saline-treated mice but not in MDMA-treated mice, ↑.	[67]
MDMA (1.5 mg/kg, s.c.)	Music (65 to 75 dB) Noise (70 dB)	Rat	Brain (NAcc)	DA, 5-HT	[B] Music enhanced MDMA-induced DA and 5-HT release.	[68]
AP (0.5-5 mg/kg, s.c.)	Phencyclidine (15 mg/kg, chronic)	Rat	Brain (PFC, striatum)	DA	[B] AP-induced increase in DA-level in PFC, ↑. Glycine reversed PCP-induced stimulation of AP-induced DA release.	[51]
AP (0.25 mg/kg, i.p.) Nicotine (0.4 mg/kg, s.c.)	Prenatal stress	Rat	Brain (NAc)	DA	[B] AP-induced DA output, ↑. [B] nicotine-induced DA output in adolescent rat,↓. [B] AP- or nicotine-induced NA output only in adult, ↑.	[69]
Cocaine (5 mg/kg, i.v.)	AHN-1055 (10 mg/kg, i.v.)	Rat	Brain (NAc)	DA	The DA increase observed was lower than DA level calculated.	[47]
ʟ-Dopa (0.1 mg/kg, i.p.) + benzerazide (15 mg/kg, i.p.)	Trap-101 (10 mg/kg)	Rat	Brain (SNr, VMTh)	GLU, GABA	[A] neurochemical response to Trap-101 (shortening the latency on GLU, prolonging the duration on GABA), ↑.	[71]
Levodopa (6 mg/kg, s.c.)	WIN55,212-2 (1 mg/kg, i.p.)	Rat	Brain (striatum)	DA, GLU	[B] prevented levodopa-induced reduce in GLU level in 6-OH-DA-treated rat.	[72]

Mor (15 mg/kg, s.c.)	Nicotine (0.3 or 0.5 mg/kg, i.p.)	Mouse	Brain (VTA, SNr)	DA, DOPAC, HVA	[B] reversed Mor-induced decrease in GABA level in nipecotic acid-pretreated mice. No interaction effect on dopaminergic neurochemical response.	[73]
Mor (10 μg, i.t.)	Amitriptyrine (15 μg, i.t.)	Rat	i.t.	Excitory amino acids	[A and B] inhibited increase in excitory amino acids level in pertussis toxin-treated rats.	[74]
Mor (10 mg/kg and 20 mg/kg, s.c., chronic)	Agmatine (10 mg/kg, s.c.)	Rat	Brain (striatum)	DA, DOPAC, HVA	[B] Mor-enhanced DA release,↓.	[75]
Mor (100 μM, continuous i.t. infusion)	LPS (20 μg, i.t.)	Rat	i.t.	CCK-8S	[B] inhibited Mor-induced CCK release. Anti-NGF reversed inhibition of LPS on Mor-induced CCK release.	[76]
Alcohol (1 g/kg, intragastric)	Nicotine (0.4 mg/kg, s.c.) WIN 55,212-2 (1 mg/kg, s.c.).	Rat	Brain (NAc shell)	DA	[B] inhibited the DA release by previous (chronic) exposure to alcohol (14 days).	[77]
Ethanol (2 g/kg, i.p.)	Naloxonazine (15 mg/kg, i.p.)	Mouse	Brain (ventral striatum)	DA	[B] ethanol-stimulated DA release in female C57BL/6J-129SvEv genotype mice, ↓. No change on ethanol PK.	[78]
Alcohol (0.5 to 2.0 g/kg, i.p.)	Nicotine (0.25 to 1.0 mg/kg, i.p.)	Rat	Brain (NAc)	DA	Extracellular DA level ↑. Mecamylamine (0.5 to 2.0 mg/kg, i.p.) inhibited the effect of alcohol–nicotine combination.	[79]

system, which innervates the skeletal muscles. Synaptic transmission from neuron to neuron is intermediated by a neurotransmitter. Once neurotransmitters are released into the synaptic interval from a neuronal ending, they bind to a postsynaptic receptor and show physiological responses.

Microdialysis sampling has some advantages because it collects neurotransmitters released into a synaptic interval which directly reflect drug action on the nervous system. Monitoring of neurotransmitters using microdialysis sampling has been widely used for the evaluation of drug–drug interactions comprising drug–chemical, drug–herbal, or drug–food interaction. It is very useful to expect clinical assessment, which affects therapeutic effect by improving therapeutic response or an adverse drug response, and to reveal drug neuronal mechanisms by evaluating drug–drug interaction of a target drug to a well-known ligand of receptors. Recent studies of pharmacodynamic drug–drug interaction are summarized in Table 5.

3.1. Antipsychotic Agents

Schizophrenia is characterized by three broad types of symptoms—positive or psychotic, negative, and cognitive symptoms—most of which are related to dopaminergic aberration. Risperidone is an atypical antipsychotic drug with antagonistic properties on the dopamine D_2 receptor, 5-HT_{2A}, and α_1 receptors. Its effects are limited in practice by various dose-dependent side effects during clinical trials. Therefore, there is a need for adjuvant drugs or new drug-treating strategies. To investigate the effects of adjuvant drugs, Wang et al. studied the effect of coadministration of risperidone and galantamine on DA release in the medial prefrontal cortex (mPFC) in chronically phencyclidine (PCP)-treated schizophrenia model mice by microdialysis sampling coupled with HPLC–ECD determination [55]. Galantamine is an allosteric potentiating ligand at the nicotinic acetylcholine receptor (nAChR) and also displays the weakest acetylcholinesterase (AChE)-inhibiting effect. In PCP-treated mice, the basal extracellular DA level in mPFC was decreased compared to that of saline-treated mice (Figure 15a, $p < 0.05$). Treatment with risperidone (0.05 mg/kg) or galantamine (0.05 mg/kg) alone did not change the extracellular DA level in the mPFC of PCP-treated mice. In contrast to the individual treatments at their noneffective doses, coadministration of risperidone (0.05 mg/kg) and galantamine (0.05 mg/kg) increased the extracellular DA level significantly 60 min after the coadministration (Figure 15b, $p < 0.01$). Moreover, mecamylamine (3 mg/kg), an nAChR antagonist, abolished the synergistic effect on DA release in the mPFC ($p < 0.01$). At noneffective doses by themselves, coadministration of galantamine (0.05 mg/kg) and risperidone (0.05 mg/kg) also showed synergistic effects on PCP-induced impairment of social interaction. Galantamine has a synergistic effect with risperidone on the impairment of social interaction induced by repeated PCP treatment. These results showed that coadministration of risperidone and galantamine may be used as a new strategy for treating the negative symptoms of schizophrenia.

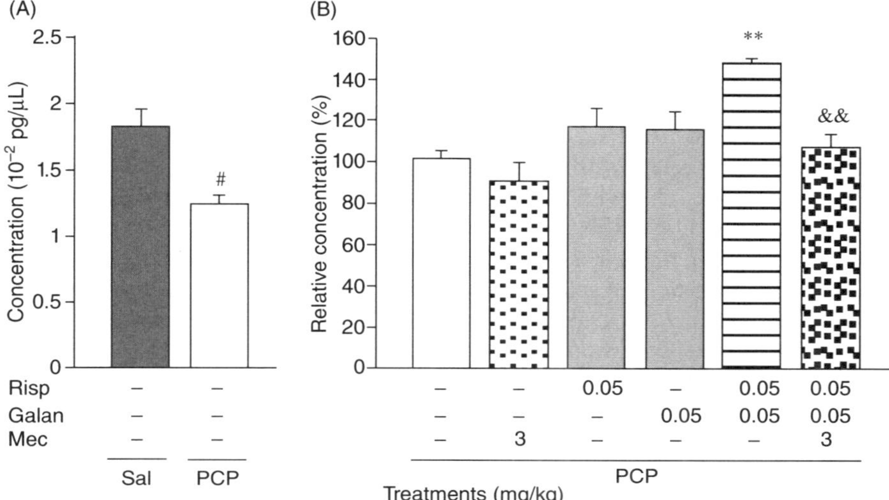

Figure 15 Synergistic effect of galantamine with risperidone on extracellular concentration of dopamine in the mPFC of PCP-treated mice and the involvement of nAChR in the effect. (a) Basel extracellular concentration of dopamine in the mPFC of Sal- and PCP-treated mice. [#] $p < 0.05$, compared to Sal-treated group. Results are expressed as means ± S.E.M., n 1/4 3e5, analyzed by Student's t-test. (b) Synergistic effect of galantamine with risperidone and the involvement of nAChR in the effect. Extracellular concentration of dopamine in the mPFC was tested 60 min after the coadministration. Results are expressed as means ± S.E.M., n 1/4 4e5, F5, 23 1/4 6.294 ($p < 0.01$), analyzed by one-way ANOVA, followed by the modified Tukey's test for multiple comparisons. [**]$p < 0.01$, compared to PCP/Vel-treated group, [&&]$p < 0.01$, compared to PCP/Risp/Galan-treated group. Sal, saline; PCP, phencyclidine; Risp, risperidone; Galan, galantamine; Mec, mecamylamine. (From [55], with permission from Elsevier Science.)

The synergistic effect may be mediated by D_1 receptors in the mPFC through nAChR activation–increased DA release.

3.2. Antidepressive Agents

The pharmacotherapies of depression rely on the use of drugs such as tricyclic antidepressants, selective serotonin reuptake inhibitors (SSRIs), serotonin–norepinephrine reuptake inhibitors (SNRIs), and monoamine oxidase inhibitors. But only 60 to 70% patients who are tolerant to antidepressants will respond to first-line monotherapy, and more than one-third of patients treated for depression will become treatment resistant. In managing treatment-resistant depression, combination therapy is one choice in a pharmacological strategy of concurrent optimization of dose and switching therapies [56]. Several studies of antidepressant drug interaction with another drug focusing on SSRI [57,58] and SNRI [59], have been reported.

Methylphenidate (MPH), a noradrenaline (NA)/DA inhibitor, is suggested to be useful in reducing such comorbid symptoms of depression as anhedonia, fatigue, motivation, and cognitive deficits by adjunctive use. Figure 16 shows time-extracellular 5-HT, NA and DA levels in the PFC of anesthetized or awake rats by HPLC–ECD determination of 5-HT and DA or HPLC–FL determination labeled with benzylamine for NA [57]. The combination of MPH (2.5 mg/kg, s.c.) with citalopram (Cit, 5 mg/kg, i.p.) showed synergic enhancement at DA levels to $425 \pm 68\%$ of basal levels compared to MPH alone ($230 \pm 17\%$, $p < 0.009$) and Cit alone ($365 \pm 39\%$, $p < 0.003$) in the PFC of anesthetized rats. In nucleus accumbens (NAc) and ventral hipp (but not in the striatum) similar results were obtained. Cit-induced increase in 5-HT levels ($463 \pm 76\%$) was attenuated by coadministration of MPH ($246 \pm 34\%$, $p < 0.001$) in the PFC, but strongly enhanced by adjunctive MPH in the hipp (MPH, no effect; Cit, $430 \pm 80\%$; combination, $550 \pm 82\%$). Combined treatment had no additive effect on the NA levels in PFC and hipp, but potentiated the increase of NA in NAc. However, Cit did not affect the extracellular levels of MPH in rat PFC following treatment with MPH alone and with a combination of MPH and Cit. These findings support the notion that combination therapy act as triple-acting reuptake inhibitors for 5-HT, DA, and NA, and suggest that proposed augmentation effects are probably associated with enhanced DA transmission in the corticolimbic areas, whereas 5-HT and NA levels show differential and region-specific responses [57].

3.3. Anticonvulsants

In treating all forms of epilepsy, monotherapy is preferred, due to improved patient compliance, a reduced likelihood of interactions, and fewer manifestations of the disorder. Rather than switching to an alternative monotherapy, combination therapy should be initiated if a trial of first-line monotherapy is ineffective [60].

Anticonvulsants such as sodium valproate (VPA) and lamotrigine (LTG) are used increasingly in combination in patients in whom monotherapy has failed to control seizures. Ahmad et al. reported the effect of a combined treat-

Figure 16   Effects of citalopram ($\bullet$, 5 mg/kg i.p., 0 min), methylphenidate (n, 2.5 mg/kg, s.c., –20 min) and methylphenidate given in combination with citalopram ($\square$) on extracellular 5-HT, NA, and DA levels in the PFC of anesthetized rats compared to the saline-treated ($\blacklozenge$) group. The insets in (A) and (C) show the corresponding effects of citalopram and methylphenidate treatment on the 5-HT and DA levels in awake rats. Extracellular concentrations of 5-HT, NA, and DA are expressed as a percentage of the basal levels of each respective monoamine in three fractions collected before the drug injection (mean $\pm$ S.E.M.; $\star\star\star p < 0.001$; $\star\star p < 0.01$; statistical significance between citalopram alone and citalopram + methylphenidate-treated groups, ANOVA, Fisher's PLSD-test). (From [57], with permission from Elsevier Science.)

(A)

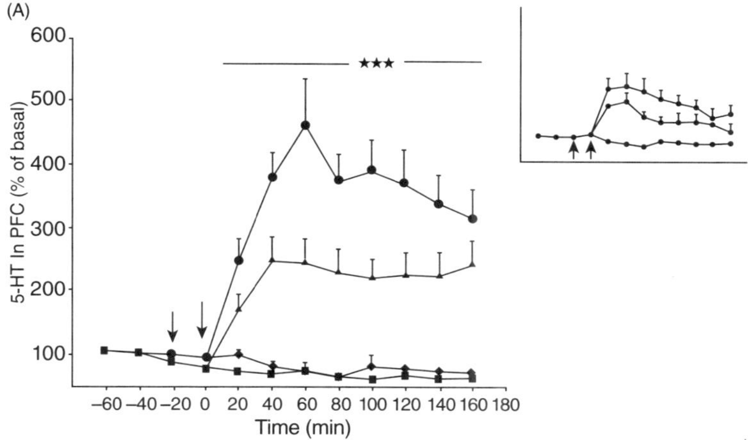

(B)

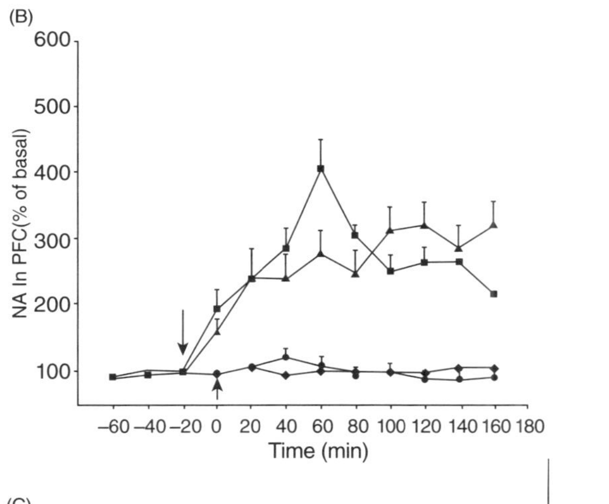

(C)

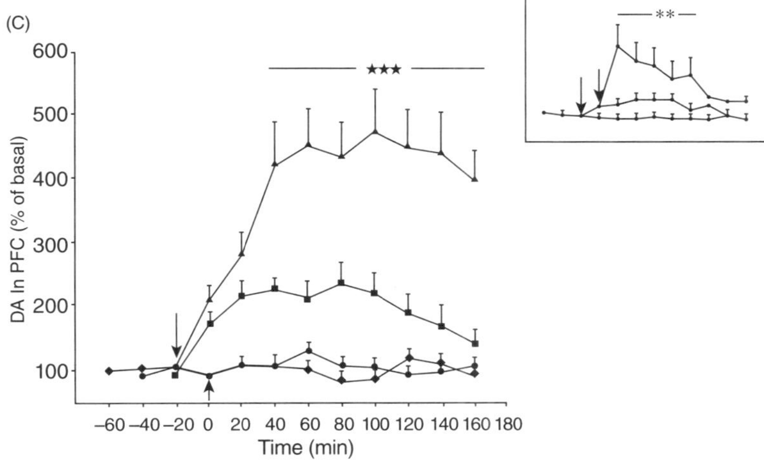

ment of anticonvulsants on extracellular amino acid [aspartate (ASP), gluta-mate (GLU), taurine (TAU), GABA] and catecholamine (5-HT and DA) release in hipp of freely moving rats using microdialysis [30]. Either LTG (10 mg/kg) or VPA (300 mg/kg) given alone significantly altered basal levels of ASP, GLU, or TAU. When given together, however, the two drugs reduced extracellular ASP [F (3, 11) = 10.43, $p < 0.001$] and GLU [F (3, 11) = 10.55, $p < 0.001$] significantly while increasing TAU levels [F (3, 11) = 10.76, $p < 0.001$; Figure 17]. In the case of GABA, LTG was free of the effect on basal levels of the transmitter, but these increased following VPA [F (3, 11) = 4.58, $p < 0.0086$], and this persisted with both drugs. VPA and LTG increased and decreased basal 5-HT [F (3, 11) = 8.154, $p < 0.003$] and DA [F (3, 11) = 2.903, $p < 0.05$], respectively. When given together, the increase in extracellular 5-HT was extended greatly, but no effect on DA release was obtained. However,

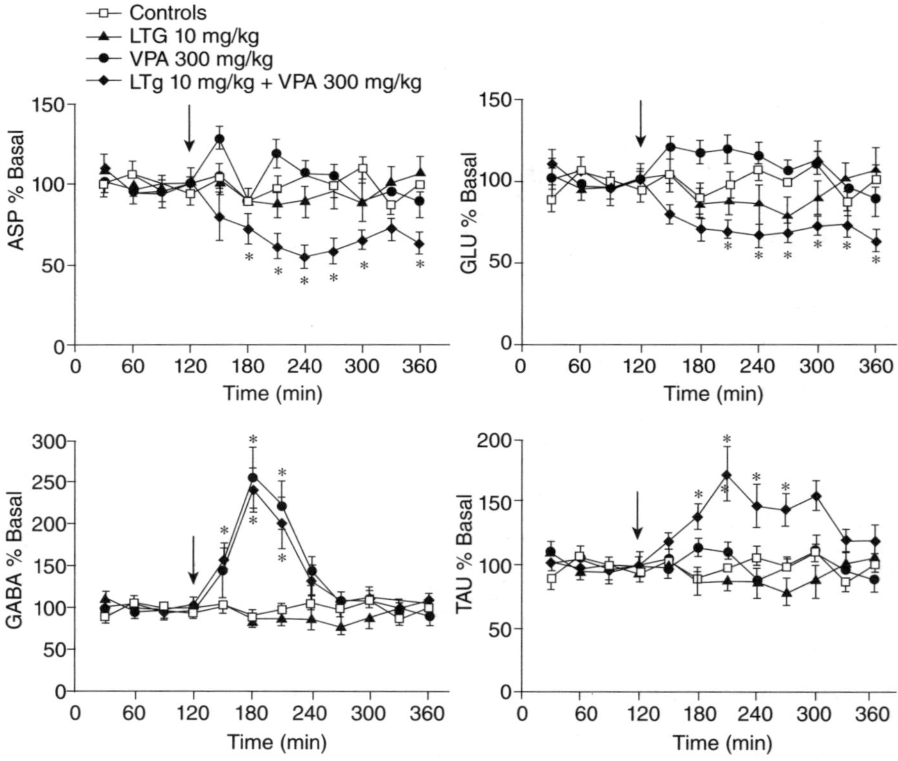

Figure 17   Effect of valproate (VPA), lamotrigine (LTG), or a combination of the two drugs on basal extracellular aspartate (ASP), glutamate (GLU), γ-aminobutyric acid (GABA), and taurine (TAU) in the hipp of freely moving rats. Data are the mean ± S.E.M. of six animals in each group. Administration of LTG or VPA is indicated by the arrow. *Denotes a significant difference from the first four basal values. (From [30], with permission from Springer-Verlag.)

coadministration of VPA with LTG showed no significant effect on LTG in any of the three compartments studied (plasma, whole brain, and brain dialysate), indicating that in this case a significant pharmacokinetic contribution is unlikely, which suggests that there should be a probable pharmacodynamic interaction of LTG and VPA.

3.4. Anesthetic Agents

Neurochemical evaluations of the anesthetic agents isoflurane on dopaminergic neuronal activity [61] and ketamine on noradorenergic neuronal activity [62] were investigated using a microdialysis sampling method. Isoflurane is a volatile anesthetic widely used in clinical anesthesia. Although both volatile anesthetics and many psychotropic drugs may modify DA release and metabolism, few studies have examined the interactions between volatile anesthetics and psychotropic drugs. Adachi et al. [61] assessed the effects of isoflurane anesthesia on changes in DA release and metabolism modulated by psychotropic drugs (e.g., clozapine, risperidone, and fluoxetine) by using in vivo microdialysis sampling connected to HPLC–ECD determination of DA, 3-methoxytyramine (3-MT), 3,4-dihydroxyphenylacetic acid (DOPAC), and homovanillic acid (HVA) in rat's brain dialysate. Rats were treated with psychotropic agents (10 mg/kg) intraperitoneally, and isoflurane (1.0 or 2.5%) was applied at a rate of 2 L/min, using air (23% oxygen) as a carrier to avoid hypoxia following initial introduction of 3% isoflurane at a rate of 3 L/min for about 5 min. A high concentration of isoflurane (2.5%, 2 L/min) anesthesia increased the extracellular concentration of DA during emergence from anesthesia and the level of DA metabolites increased in an isoflurane dose-dependent manner ($p < 0.05$, vs. the control value at each fraction). Treatment of the atypical antipsychotics clozapine (D_1, D_2, and 5-HT antagonist) and risperidone (D_1, D_2, and 5-HT antagonist) raised the DA and its metabolite levels, and inhalation of isoflurane attenuated the drug-induced increase in DA, DOPAC, and HVA during anesthesia. Another metabolite, 3-MT, was increased by clozapine and risperidone treatment, and the drug-induced increase was enhanced by isoflurane anesthesia. In the fluoxetine (an SSRI) treatment group, the extracellular DA level was decreased. During isoflurane anesthesia, the DA metabolite (DOPAC and HVA) levels were raised; however, a dose-dependent isoflurane-induced increase in DA metabolites was not observed with fluoxetine treatment. In addition the administration of citalopram showed no marked change in the concentration of DA and its metabolites, and isoflurane-induced changes were well reserved. These results suggest that isoflurane enhances DA metabolism, which might be influenced presynaptically by dopaminergic–serotonergic interaction during anesthesia.

3.5. Drugs of Abuse

Several studies on drug–drug interaction with other drugs of abuse have been reported for narcotics [50,63,64] and stimulants [50,51] to examine the recent trend toward multiple drug abuse. Drug–drug interactions for stimulants [65] and cocaine [47,66] have also been reported. Furthermore, drug–drug interaction to given receptor's ligands, drug–chemical interaction for narcotics [67] and drug–stress interaction (for narcotics [68] and stimulants [69]) have also been reported.

MDMA, a synthetic AP derivative, exhibits both psychostimulant and mild hallucinogenic properties and has unique psychopharmacological profiles for promoting the release of DA and 5-HT in multiple brain regions. Interactions of MDMA focused on dopaminergic, serotonergic, and cholinergic neurons were reviewed by Gudelskey and Yamamoto [63]. In this review, an inhibitor of the DA transporter (e.g., nomifensine, mazindol, GBR12909) attenuated MDMA-induced extracellular concentration of DA in striatum, and fluoxetine (a 5-HT uptake inhibitor) prevented MDMA-induced increase in 5-HT release in striatum.

Kurling et al. evaluated the acute neurochemical effects of coexposure of MDMA and nandrolone, anabolic–androgenic steroids (AASs), by HPLC-ECD determination of DA, 5-HT, and their metabolites, such as DOPAC, HVA, and 5-HIAA (Figure 18). MDMA (5 mg/kg) administration elevated extracellular DA and 5-HT levels ($p < 0.001$ for both) compared to the basal level (DA, 12.6 ± 1.1 fmol/40 µL; 5-HT, 7.4 ± 0.5 pmol/40 µL) [50]. Subchronic i.m. pretreatment with supraphysiological doses of nandrolone (20 and 5 mg/kg) attenuated dose-dependently the MDMA-induced extracellular DA concentration in NAc ($p < 0.001$, AUC). The lower dose of nandrolone (5 mg/kg) attenuated MDMA-induced increase in 5-HT levels. Nandrolone treatment also modified the effects of the metabolites DOPAC ($p = 0.001$) and HVA ($p < 0.047$), but not 5-HIAA levels. Moreover, behavioral changes focused on locomotor activity suggest that the effects of MDMA were attenuated dose-dependently by AAS treatment ($p = 0.008$), paralleling the DA results. But nandrolone did not change the pharmacokinetics of MDMA in brain or blood. These results showed that AAS pretreatment could moderate the reward-related neurochemical and behavioral effects of MDMA.

Cocaine is a naturally occurring substance found in the leaves of the *Erythroxylum coca* plant. Cocaine is the most frequent drug-related cause of emergency department visits in the United States [70]. Drug–drug interactions of cocaine and lidocaine (antiarrhythmics), β-adrenergic antagonists (succinylcholine), phenothiazines/butyrophenones (antidepressant drugs), and drugs of abuse (ethanol, marijuana, nicotine) focused on cocaine's safety were reviewed by Goldstein et al. [70].

To investigate drug interaction on the DA system which determines the treatment of cocaine abuse, the extracellular DA level in rat brain (NAc) was determined by HPLC–ECD after the administration of cocaine and AHN 1-055, a benztropine analog [47]. The DA level–time profile observed did not overlap the addictive effect calculated after the administration of cocaine (5 mg/kg, i.v.) and AHN 1-055 (5 mg/kg, i.v.). The DA levels observed at initial time periods for up to 5 h were significantly lower than the additive DA levels calculated. Otherwise, no significant effects were found in the pharmacokinetic

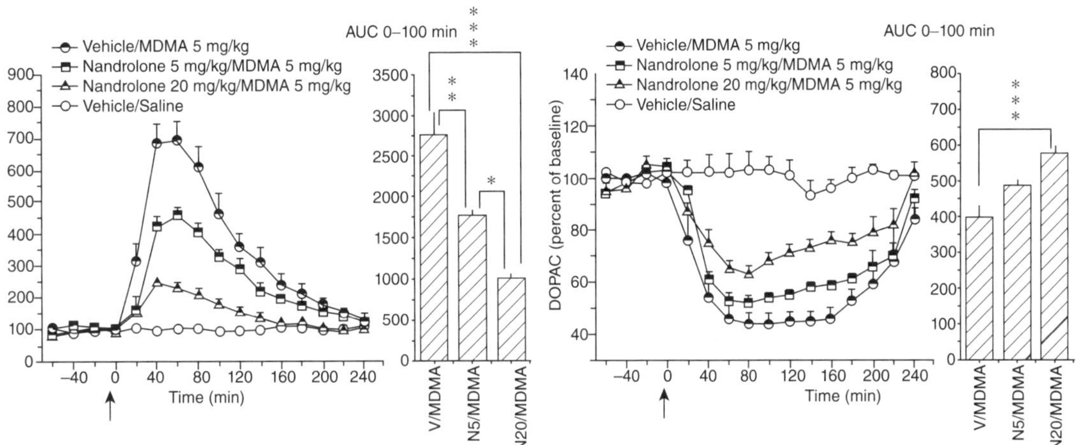

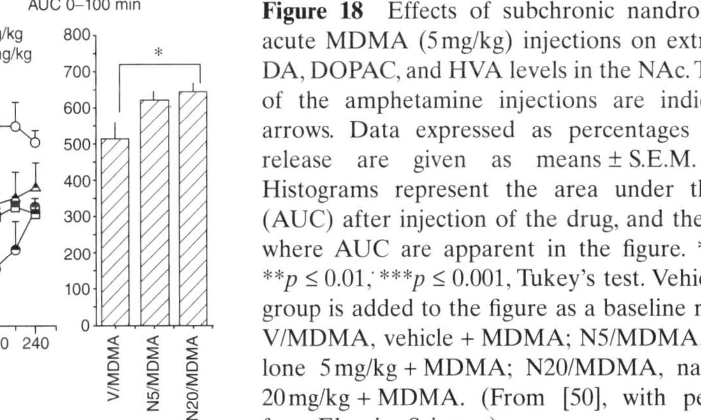

Figure 18 Effects of subchronic nandrolone and acute MDMA (5 mg/kg) injections on extracellular DA, DOPAC, and HVA levels in the NAc. The times of the amphetamine injections are indicated by arrows. Data expressed as percentages of basal release are given as means ± S.E.M. ($n = 6$). Histograms represent the area under the curve (AUC) after injection of the drug, and the minutes where AUC are apparent in the figure. *$p \leq 0.05$, **$p \leq 0.01$, ***$p \leq 0.001$, Tukey's test. Vehicle saline group is added to the figure as a baseline reference. V/MDMA, vehicle + MDMA; N5/MDMA, nandrolone 5 mg/kg + MDMA; N20/MDMA, nandrolone 20 mg/kg + MDMA. (From [50], with permission from Elsevier Science.)

TABLE 6   Pharmacodynamic Parameters for Cocaine and AHN 1-055 Alone and in Combination[a]

Compound	AHN 1-055 IC$_{50}$ (ng/mL)	Cocaine IC$_{50}$ (ng/mL)
AHN 1-055 alone	311.8 ($\pm$ 61.5)	
Cocaine alone	—	715.0 ($\pm$ 145.2)
AHN 1-055 + cocaine	277.3 ($\pm$ 61.4)	2224.8[b] ($\pm$ 926.1)

Source: From [47], with permission from John Wiley & Sons.

[a]Both competitive and noncompetitive types of interaction, according to the two-compartment pharmacokinetic model and the indirect response model (ADAPT II).
[b]Significant difference from drug alone.

parameters, including brain distribution of AHN 1-055 determined by HPLC–UV. It suggested that a pharmacodynamic interaction exists between cocaine and AHN 1-055. To quantify pharmacodynamic interaction, pharmacodynamic parameters were determined by an indirect response model using the ADAPT II software program. Modeling of pharmacodynamic data resulted in no significant changes in IC$_{50}$, the plasma concentration required to produce 50% of maximum inhibition, which was attributed to the drug for AHN 1-055 (Table 6). These results indicated that the binding of AHN 1-055 does not change in the presence of cocaine. However, the IC$_{50}$ value for cocaine was significantly reduced with AHN 1-055 (Table 6). It is suggested that AHN 1-055 may assume a possible desirable attribute in the development of a potential substitute therapeutic medication for the treatment of cocaine abuse.

3.6. Other Drugs

Several studies on other drugs acting on the CNS, such as antiparkinsonian agents, narcotics, and chemicals, focused on pharmacodynamic interaction using microdialysis. Levodopa, a DA precursor, is one of the most effective treatments at present for Parkinson's disease. Several investigations dealing with levodopa report on its additional therapeutic effects [71] and its preventive action on side effects, called dyskinesias [72]. In the former, extracellular GLU and GABA were investigated by HPLC-FL determination derivatized with *o*-phthalaldehyde connected to microdialysis sampling. In the latter, DA and HVA were determined by HPLC–ECD, and GLU and ASP were determined by an HPLC–FL determination labeled with *o*-phthalaldehyde connected to microdialysis sampling.

Morphine, used widely as an analgesic agent for cancer pain, increases DA release in the terminal areas of mesolimbic DA neurons. Drug–drug interaction on dopaminergic effect with other drugs was investigated [73–75]. To investigate in more detail, several studies focused on different neurotransmitters [73,76]. The neurochemical effects of interaction with chemicals can also be evaluated using microdialysis, and some researchers have investigated morphine's interaction with alcohol [77–79].

4. CONCLUSIONS

In this chapter, advanced applications of microdialysis for the evaluation of drug–drug interaction were introduced. Hereafter, the evaluation of drug–drug interaction in preclinical studies may carry more and more weight for the development of new medicaments in the diversification of medical treatment. For elucidation of the mechanism of drug–drug interaction, determination of the free form of medicaments or neurotransmitters is necessary. Microdialysis is one of the most powerful techniques that is available for this purpose. However, the lower recovery of analytes using probes and highly invasive surgery limit the applicable range of microdialysis. We hope that remarkable progress in the equipment needed to overcome these points will be made and that the wide applicability of microdialysis for the rational use of medicaments will be established in the near future.

REFERENCES

[1] Mullangi, R., Srinivas, N.R. (2008). Clopidogrel: review of bioanalytical methods, pharmacokinetics/pharmacodynamics, and update on recent trends in drug–drug interaction studies. *Biomedical Chromatography, 23*, 26–41.

[2] Buczko, W., Hermanowicz, J.M. (2008). Pharmacokinetics and pharmacodynamics of aliskiren, an oral direct renin inhibitor. *Pharmacological Reports, 60*, 623–631.

[3] Sousa, M., Pozniak, A., Boffito, M. (2008). Pharmacokinetics and pharmacodynamics of drug interactions involving rifampicin, rifabutin and antimalarial drugs. *Journal of Antimicrobial Chemotherapy, 62*, 872–878.

[4] Tomlinson, B., Hu, M., Lee, V.W.Y. (2008). In vivo assessment of herb–drug interactions: possibility utility of a pharmacogenetic approach? *Molecular Nutrition & Food Research, 52*, 799–809.

[5] Koneru, B., Cowart, D.T., Noorisa, M., Kisicki, J., Bramer, S.L. (1998). Effect of increasing gastric pH with famotidine on the absorption and oral pharmacokinetics of the inotropic agent vesnsrinone. *Journal of Clinical Pharmacology, 38*, 429–432.

[6] Balayssac, D., Authier, N., Cayre, A., Coudore, F. (2005). Does inhibition of P-glycoprotein lead to drug–drug interaction? *Toxicology Letters, 156*, 319–329.

[7] Lin, J.H. (2003). Drug–drug interaction mediated by inhibition and induction of P-glycoprotein. *Advanced Drug Delivery Reviews, 55*, 53–81.

[8] Masubuchi, Y., Horie, T. (2007). Toxicological significance of mechanism-based inactivation of cytochrome P450 enzymes by drugs. *Critical Reviews in Toxicology, 37*, 389–412.

[9] Babi, Z. (2008). Role of cytochrome P450 in drug interactions. *Nutrition and Metabolism, 5*, 1–10.

[10] Procyshyn, R.M., Ho, T., Wasan, K.M. (2005). The effects of competitive displacement on haloperidol's plasma distribution in normolipidemic and hyperlipidemic plasma. *Drug Development and Industrial Pharmacy, 31*, 901–905.

[11] Leucuta, S.E., Vlase, L. (2006). Pharmacokinetics and metabolic drug interactions. *Current Clinical Pharmacology, 1*, 5–20.

[12] Jonker, D.M., Visser, S.A.G., Graaf, P.H., Voskuyl, R.A., Danhof, M. (2005). Towards a mechanism-based analysis of pharmacodynamic drug–drug interactions in vivo. *Pharmacology & Therapeutics, 106*, 1–18.

[13] Lange, E.C.M., Boer, A.G., Breimer, D.D. (2000). Methodological issues in microdialysis sampling for pharmacokinetic studies. *Advanced Drug Delivery Reviews, 45*, 125–148.

[14] Nakashima, M.N., Wada, M., Nakashima, K. (2005). Microdialysis as an excellent sampling approach for biomedical analysis. *Current Pharmaceutical Analysis, 1*, 127–133.

[15] Tsai, T.H., Wu, J.W. (2001). Pharmacokinetics of ciprofloxacin in the rat and its interaction with cyclosporin A: a microdialysis study. *Analytica Chimica Acta, 448*, 195–199.

[16] Tsai, T.H. (2001). Pharmacokinetics of pefloxacin and its interaction with cyclosporin A, a P-glycoprotein modulator, in rat blood blood, brain and bile, using simultaneous microdialysis. *British Journal of Pharmacology, 132*, 1310–1316.

[17] Marchand, S., Forsell, A., Chenel, M., Comets, E., Lamarche, I., Couet, W. (2006). Norfloxacin blood–brain barrier transport in rat is not affected by probenecid coadministration. *Antimicrobial Agents and Chemotherapy, 50*, 371–373.

[18] Chang, Y.L., Chou, M.H., Lin, M.F., Chen, C.F., Tsai, T.H. (2001). Effect of cyclosporine, a P-glycoprotein inhibitor, on the pharmacokinetics of cefepime in rat blood and brain a microdialysis study. *Life Sciences, 69*, 191–199.

[19] Tsai, T.H. (2001). Effect of ivermectin on the disposition of amoxicillin in rat blood and brain using microdialysis sampling. *Analytica Chimica Acta, 431*, 279–285.

[20] Chang, Y.L., Chiou, S.H., Chou, Y.C., Yen, C.J., Tsai, T.H. (2007). Quantitative determination of unbound cefoperazone in rat bile using microdialysis and liquid chromatography. *Journal of Pharmaceutical and Biomedical Analysis, 45*, 158–163.

[21] Lee, C.H., Yeh, P.H., Tsai, T.H. (2002). Hepatobiliary excretion of fluconazole and its interaction with cyclosporin A in rat blood and bile using microdialysis. *International Journal of Pharmaceutics, 241*, 367–373.

[22] Tsai, P.L., Tsai, T.H. (2004). Hepatobiliary excretion of berberine. *Drug Metabolism and Disposition, 32*, 405–412.

[23] Liu, S.C., Tsai, T.H. (2002). Determination of diclofenac in rat bile and its interaction with cyclosporin A using on-line microdialysis coupled to liquid chromatography. *Journal of Chromatography B, 769*, 351–356.

[24] Tsai, T.H., Liu, S.C., Tsai, P.L., Ho, L.K., Shum, A.Y.C., Chen, C.F. (2002). The effects of the cyclosporine A, a P-glycoprotein inhibitor, on the pharmacokinetics of baicalein in the rat: a microdialysis study. *British Journal of Pharmacology, 137*, 1314–1320.

[25] Tsai, P.L., Tsai, T.H. (2004). Pharmacokinetics of baicalin in rats and its interactions with cyclosporin A, quinidine and SKF-525A: microdialysis study. *Planta Medica*, *70*, 1069–1074.

[26] Fang, J.Y., Tsai, T.H., Hung, C.F., Wong, W.W. (2004). Development and evaluation of the essential oil from *Magnolia fargesii* for enhancing the transdermal absorption of theophilline and cianidanol. *Journal of Pharmacy and Pharmacology*, *56*, 1493–1500.

[27] Jan, W.C., Lin, L.C., Chen, C.F., Tsai, T.H. (2005). Herb–drug interaction of *Evodia rutaecarpa* extract on the pharmacokinetics of theophilline in rats. *Journal of Ethnopharmacology*, *102*, 440–445.

[28] Ueng, Y.F., Tsai, T.H., Don, M.J., Chen, R.M., Chen, T.L. (2005). Alteration of the pharmacokinetics of theophilline by rutaecarpine, an alkaloid of the medicinal herb *Evodia rutaecarpa*, in rats. *Journal of Pharmacy and Pharmacology*, *57*, 227–232.

[29] Huang, S.P., Lin, L.C., Wu, Y.T., Tsai, T.H. (2009). Pharmacokinetics of kadsurenone and its interaction with cyclosporin A in rats using a combined HPLC and microdialysis system. *Journal of Chromatography B*, *877*, 247–252.

[30] Ahmad, S., Fowler, L.J., Whitton, P.S. (2005). Effects of combined lamotrigine and valproate on basal and stimulated extracellular amino acids and monoamines in the hippocampus of freely moving rats. *Naunyn-Schmiedeberg's Archives of Pharmacology*, *371*, 1–8.

[31] Clinckers, R., Smolders, I., Michotte, Y., Ebinger, G., Danhof, M., Voskuyl, R.A., Della Pasqua, O. (2008). Impact of efflux transporters and of seizures on the pharmacokinetics of oxcarbazepine metabolite in the rat brain. *British Journal of Pharmacology*, *155*, 1127–1138.

[32] Nakashima, K., Yamamoto, K., Al-Dirbashi, O.Y., Kaddoumi, A., Nakashima, M.N. (2003). Semi-micro column HPLC of triazolam in rat plasma and brain microdialysates and its application to drug interaction study with itraconazole. *Journal of Pharmaceutical and Biomedical Analysis*, *30*, 1809–1816.

[33] Megarbane, B., Pirnay, S., Borron, S.W., Trout, H., Monier, C., Risède P., Boschi, G., Baud, F.J. (2005). Flunitrazepam does not alter cerebral distribution of buprenorphine in the rat. *Toxicology Letters*, *157*, 211–219.

[34] Tunblad, K., Jonsson, E.N., Hammarlund-Udenaes, M. (2003). Morphine blood-brain barrier transport is influenced by probenecid co-administration. *Pharmaceutical Research*, *20*, 618–623.

[35] Groenendaal, D., Freijer, J., Mik, D., Bouw, M.R., Danhof, M., Lange, E.C.M. (2007). Influence of biophase distribution and P-glycoprotein interaction on pharmacokinetic–pharmacodynamic modeling of the effect of morphine on EEG. *British Journal of Pharmacology*, *151*, 713–720.

[36] Okura, T., Ozawa, T., Ito, Y., Kimura, M., Kagawa, Y., Yamada, S. (2008). Enhancement by grapefruit juice of morphine antinociception. *Biological & Pharmaceutical Bulletin*, *31*, 2338–2341.

[37] Wada, M., Yokota, C., Ogata, Y., Kuroda, N., Yamada, H., Nakashima, K. (2008). Sensitive HPLC–fluorescence detection of morphine labeled with DIB-Cl in rat brain and blood microdialysates and its application to the preliminary study of

the pharmacokinetic interaction between morphine and diclofenac. *Analytical Bioanalytical Chemistry*, *391*, 1057–1062.

[38] Tsai, T.H., Lee, C.H., Yeh, P.H. (2001). Effect of P-glycoprotein modulators on the pharmacokinetics of camptothecin using microdialysis. *British Journal of Pharmacology*, *134*, 1245–1252.

[39] Cheng, F.C., Ho, Y.F., Hung, L.C., Chen, C.F., Tsai, T.H. (2002). Determination of pharmacokinetic profile of omeprazole in rat blood, brain and bile by microdialysis and high-performance liquid chromatography. *Journal of Chromatography A*, *949*, 35–42.

[40] Tsai, T.H., Liu, M.C. (2004). Determination of extracellular hesperidin in blood and bile of anaesthetized rats by microdialysis with high-performance liquid chromatography: a pharmacokinetic application. *Journal of Chromatography B*, *806*, 161–166.

[41] Kaddoumi, A., Nakashima, M.N., Wada, M., Nakashima, K. (2004). Pharmakokinetic interactions between phenylpropanolamine, caffeine and chlorpheniramine in rats. *European Journal of Pharmaceutical Sciences*, *22*, 209–216.

[42] Chen, Y.F., Jaw, I., Shiao, M.S., Tsai, T.H. (2005). Determination an pharmacokinetic analysis of salvianolic acid B in rat blood and bile by microdialysis and liquid chromatography. *Journal of Chromatography A*, *1088*, 140–145.

[43] Wu, J.W., Tsai, T.H. (2007). Effects of silibinin on the pharmacokinetics of pyrazinamide and pyrazinoic acid in rats. *Drug Metabolism and Disposition*, *35*, 1603–1609.

[44] Fuh, M.R., Tai, Y.L., Pan, W.H.T. (2001). Determination of free-form of cocaine in rat brain by liquid chromatography-electrospray mass spectrometry with in vivo microdialysis. *Journal of Chromatography B*, *752*, 107–114.

[45] Fuh, M.R., Lin, H.T., Pan, W.H.T., Lin, F.R. (2002). Simultaneous determination of free-form amphetamine in rat's blood and brain by in-vivo microdialysis and liquid chromatography with fluorescence detection. *Talanta*, *58*, 1357–1363.

[46] Kaddoumi, A., Nakashima, M.N., Maki, T., Matsumura, Y., Nakamura, J., Nakashima, K. (2003). Liquid chromatography studies on the pharmacokinetics of phentermine and fenfluramine in brain and blood microdialysates after intraperitoneal administration to rats. *Journal of Chromatography B*, *791*, 291–303.

[47] Raje, S., Cornish, J., Newman, A.H., Cao, J., Katz, J.L., Eddington, N.D. (2006). Investigation of the potential pharmacokinetic and pharmacodynamic drug interaction between AHN 1-005, a potent benztropine analog used for cocaine abuse, and cocaine after dosing in rats using intracerebral microdialysis. *Biopharmaceutics & Drug Disposition*, *27*, 229–240.

[48] Kapadia, R., Böhlke, M., Maher, T.J. (2007). Detection of γ-hydroxybutyrate in striatal microdialysates following peripheral 1,4-butanediol administration in rats. *Life Sciences*, *80*, 1046–1050.

[49] Tomita, M., Nakashima, M.N., Wada, M., Nakashima, K. (2007). Sensitive determination of MDMA and its metabolite MDA in rat blood and brain microdialysates by HPLC with fluorescence detection. *Biomedical Chromatography*, *21*, 1016–1022.

[50] Kurling, S., Kankaanpää, A., Seppälä, T. (2008). Sub-chronic nandrolone treatment modifies neurochemical and behavioral effects of amphetamine and 3,4-methylenedioxymethamphetamine (MDMA) in rats. *Behavioural Brain Research, 189,* 191–201.

[51] Sershen, H., Balla, A., Aspromonte, J.M., Xie, S., Cooper, T.B., Javitt, D.C. (2008). Characterization of interactions between phencyclidine and amphetamine in rodent prefrontal cortex and striatum: Implications in NMDA/glycine-site-mediated dopaminergic dysregulation and dopamine transporter function. *Neurochemistry International, 52,* 119–129.

[52] Endres, C.J., Hsiao, P., Chung, F.S., Unadkat, J.D. (2006). The role of transporters in drug interactions. *European Journal of Pharmaceutical Sciences, 27,* 501–517.

[53] Aboul-Enein, H.Y., Hefnawy, M.M., Nakashima, K. (2004). *Drug Monitoring and Clinical Chemistry,* Elsevier Science, Amsterdam, pp. 15, 75.

[54] Uno, T., Yasui-Furukori, N. (2006). Effect of grapefruit juice in relation to human pharmacokinetic study. *Current Clinical Pharmacology, 1,* 157–161.

[55] Wang, D., Noda, Y., Zhou, Y., Nitta, A., Furukawa, H., Nabeshima, T. (2007). Synergistic effect of galantamine with risperidone on impairment of social interaction in phencyclidine-treated mice as a schizophrenic animal model. *Neuropharmacology, 52,* 1179–87.

[56] Souery, D., Papakostas, G.I., Trivedi, M.H. (2006). Treatment-resistant depression. *Journal of Clinical Psychiatry, 67,* 16–22.

[57] Weikop, P., Yoshitake, T., Kehr, J. (2007). Differential effects of adjunctive methylphenidate and citalopram on extracellular levels of serotonin, noradrenaline and dopamine in the rat brain. *European Neuropsychopharmacology, 17,* 658–671.

[58] Cremers, T.I.F.H., Rea, K., Bosker, F.J., Wikström, H.V., Hogg, S., Mørk, A., Westerink, B.H.C. (2007). Augmentation of SSRI effects on serotonin by 5-HT2C antagonists: mechanistic studies. *Neuropsychopharmacology, 32,* 1550–1557.

[59] Kitaichi, Y., Inoue, T., Izumi, T., Nakagawa, S., Tanaka, T., Masui, T., Koyama, T. (2008). Effect of co-administration of a serotonin–noradrenaline reuptake inhibitor and a dopamine agonist on extracellular monoamine concentrations in rats. *European Journal of Pharmacology, 584,* 285–290.

[60] Vajda, F.J. (2007). Pharmacotherapy of epilepsy: new armamentarium, new issues. *Journal of Clinical Neuroscience, 14,* 813–823.

[61] Adachi, Y.U., Yamada, S., Satomoto, M., Higuchi, H., Watanabe, K., Kazama, T., Mimuro, S., Sato, S. (2008). Isoflurane anesthesia inhibits clozapine- and risperidone-induced dopamine release and anesthesia-induced changes in dopamine metabolism was modified by fluoxetine in the rat striatum: an in vivo microdialysis study. *Neurochemistry International, 52,* 384–391.

[62] Tose, R., Kushikata, T., Yoshida, H., Kudo, M., Furukawa, K., Ueno, S., Hirota, K. (2009). Orexin A decreases ketamine-induced anesthesia time in the rat: the relevance to brain noradrenergic neuronal activity. *Anesthesia & Analgesia, 108,* 491–495.

[63] Gudelsky, G.A., Yamamoto, B.K. (2008). Actions of 3,4-methylenedioxymethamphetamine (MDMA) on cerebral dopaminergic, serotonergic and cholinergic neurons. *Pharmacology, Biochemistry and Behavior, 90,* 198–207.

[64] Touriño, C., Maldonado, R., Valverde, O. (2007). MDMA attenuates THC withdrawal syndrome in mice. *Psychopharmacology, 193,* 75–84.

[65] Zmarowski, A., Wu, H.Q., Brooks, J.M., Potter, M.C., Pellicciari, R., Schwarcz, R., Bruno, J.P. (2009). Astrocyte-derived kynurenic acid modulates basal and evoked cortical acetylcholine release. *European Journal of Neuroscience, 29,* 529–538.

[66] Leggio, G.M., Cathala, A., Moison, D., Cunningham, K.A., Piazza, P.V., Spampinato, U. (2009). Serotonin 2C receptors in the medial prefrontal cortex facilitate cocaine-induced dopamine release in the rat nucleus accumbens. *Neuropharmacology, 56,* 507–513.

[67] Izco, M., Marchant, I., Escobedo, I., Peraile, I., Delgado, M., Higuera-Matas, A., Olias, O., Ambrosio, E., O'Shea, E., Colado, M.I. (2007). Mice with decreased cerebral dopamine function following a neurotoxic dose of MDMA (3,4-methylenedioxymethamphetamine, "ecstasy") exhibit increased ethanol consumption and preference. *Journal of Pharmacology and Experimental Therapeutics, 322,* 1003–1012.

[68] Feduccia, A.A., Duvauchelle, C.L. (2008). Auditory stimuli enhance MDMA-conditioned reward and MDMA-induced nucleus accumbens dopamine, serotonin and locomotor responses. *Brain Research Bulletin, 77,* 189–196.

[69] Silvagni, A., Barros, V.G., Mura, C., Antonelli, M.C., Carboni, E. (2008). Prenatal restraint stress differentially modifies basal and stimulated dopamine and noradrenaline release in the nucleus accumbens shell: an "in vivo" microdialysis study in adolescent and young adult rats. *European Journal of Neuroscience, 28,* 744–758.

[70] Goldstein, R.A., DesLauriers, C., Burda, A.M. (2009). Cocaine: history, social implications, and toxicity—a review. *Disease-a-Month, 5,* 6–38.

[71] Marti, M., Trapella, C., Morari, M. (2008). The novel nociceptin/prphanin FQ receptor antagonist Trap-101 alleviates experimental parkinsonism through inhibition of the nigro-thalamic pathway: positive interaction with L-dopa. *Journal of Neurochemistry, 107,* 1683–1696.

[72] Morgese, M.G., Cassano, T., Gaetani, S., Macheda, T., Laconca, L., Dipasquale, P., Ferraro, L., Antonelli, T., Cuomo, V., Giuffrida, A. (2009). Neurochemical changes in the striatum of dyskinetic rats after administration of the cannabinoid agonist WIN55,212-2. *Neurochemistry International, 54,* 56–64.

[73] Vihavainen, T., Relander, T.R., Leiviskä, R., Airavaara, M., Tuominen, R.K., Ahtee, L., Piepponen, T.P. (2008). Chronic nicotine modifies the effects of morphine on extracellular striatal dopamine and ventral tegmental GABA. *Journal of Neurochemistry, 107,* 844–854.

[74] Lin, J.A., Tsai, R.Y., Lin, Y.T., Lee, M.S., Cherng, C.H., Wong, C.S., Tzeng, J.I. (2008). Amitriptyline pretreatment preserves the antinociceptive effect of morphine in pertussis toxin-treated rats by lowering CSF excitatory amino acid concentrations and reversing the down regulation of glutamate transporters. *Brain Research, 1232,* 61–69.

[75] Wei, X.L., Su, R.B., Wu, N., Lu, X.Q., Zheng, J.Q., Li, J. (2007). Agmatine inhibits morphine-induced locomotion sensitization and morphine-induced changes in striatal dopamine and metabolites in rats. *European Neuropsychopharmacology, 17,* 790–799.

[76] Xanthos, D.N., Kumar, N., Theodorsson, E., Coderre, T.J. (2009). The roles of nerve growth factor and cholecystokinin in the enhancement of morphine analgesia in

a rodent model of central nervous system inflammation. *Neuropharmacology*, *56*, 684–691.

[77] López-Moreno, J.A., Scherma, M., Rodríguez de Fonseca, F., González-Cuevas, G., Fratta, W., Navarro, M. (2008). Changed accumbal responsiveness to alcohol in rats pre-treated with nicotine or the cannabinoid receptor agonist WIN 55,212-2. *Neuroscience Letters*, *433*, 1–5.

[78] Job, M.O., Tang, A., Hall, F.S., Sora, I., Uhl, G.R., Bergeson, S.E., Gonzales, R.A. (2007). Mu (μ) opioid receptor regulation of ethanol-induced dopamine response in the ventral striatum: evidence of genotype specific sexual dimorphic epistasis. *Biological Psychiatry*, *62*, 627–634.

[79] Tizabi, Y., Bai, L., Copeland, R.I., Taylor, R.E. (2007). Combined effects of systemic alcohol and nicotine on dopamine release in the nucleus accumbens shell. *Alcohol and Alcoholism*, *42*, 413–416.

17

MICRODIALYSIS IN ENVIRONMENTAL MONITORING

MANUEL MIRÓ

University of the Balearic Islands, Palma de Mallorca, Illes Balears, Spain

WOLFGANG FRENZEL

Institut für Technischen Umweltschutz, Technische Universität Berlin, Berlin, Germany

1. INTRODUCTION

Environmental researchers have recognized that sampling and sample preparation are still the Achilles' heel of the overall analytical process for determination of environmentally relevant inorganic and organic species in terrestrial and aquatic media [1–3]. These steps are the sources of major bias and accidental errors that might have a decisive influence on the quality and reliability of analytical results. Sampling itself is a distortion of the natural environment, which can severely alter the composition of the sample and the analyte distribution originally present, as a consequence of transformations, losses, contamination, and change of equilibria from real scenarios [1,2]. Any further steps needed for sample preparation and preservation carried out in the laboratory (e.g., sample acidification, addition of complexing reagents, filtration, removal of microorganisms, cooling or deep freezing) enhance these risks and might call into question whether the analytical results obtained do actually reflect the actual concentrations and distribution of species in the environment.

In situ measurement and sample treatment techniques that are applied directly to the monitoring site would be ideal. In this context, the use of passive dosimetry [4–6] for sampling and in-field extraction of both inorganic and organic analytes has gained increasing popularity among analysts for simplification of the analytical procedure at the most critical step of sampling.

Applications of Microdialysis in Pharmaceutical Science, First Edition. Edited by Tung-Hu Tsai.
© 2011 John Wiley & Sons, Inc. Published 2011 by John Wiley & Sons, Inc.

Passive sampling is based on the free flow of analytes from the sample matrix to a collecting medium by either diffusion or permeation. Thus, as compared to grab sampling, conversion of analytes during transport and storage is minimized because of the isolation of target species from matrix ingredients. Moreover, the analytes in the perfusate are generally enriched relative to other matrix components, thereby reducing potential interferences in the subsequent determination step. Examples of passive dosimeters that have proven suitable for assessing the quality of environmental compartments are the semipermeable membrane devices (SPMDs) [7,8], ceramic dosimeters [9], solid-phase microextraction (SPME) devices [10], and diffusive gradients in thin-film (DGT) devices [11,12]. The latter were specifically designed for speciation of trace metals, monitoring of metal fluxes, and quantification under dynamic regime of dissolved concentrations of labile species [13], which are regarded as the most potentially available metal forms to biota.

The aim of this chapter is to illustrate that notwithstanding the fact that microdialysis is a mature technique for chemical sampling applications in neuroscience, pharmacokinetics, and drug metabolism studies [14–16], microdialyzers can be regarded as suitable passive dosimeters for environmental monitoring as well. To this end we describe the potential of microdialysis as a tool for automatic sampling, cleanup, and real-time monitoring of targeted analytes. Novel flow-through dialysis configurations are presented and compared critically with conventional concentric-type probes exploited in in vivo assays. The analytical performance of microdialysis-based procedures for environmental assays is discussed thoroughly, including novel strategies for improvement of relative recoveries of the analytes. A section is devoted to the critical evaluation of common calibration approaches in the neuroscience field (e.g., retrodialysis) for quantitative environmental analytical assays. Special attention is also paid to the positive attributes and limitations arising from the hyphenation of microdialysis probes with modern analytical detection instruments.

In addition, microdialysis units are presented as attractive miniaturized devices for implantation in solid samples of environmental and agricultural origin (e.g., plant tissue, soil, sediment, and sewage sludges) for in situ monitoring of interstitial or cellular water. A sketch of a potential microdialysis configuration for in situ sampling of soil pore water is shown in Figure 1. The unique features of microdialysis sampling in contrast to conventional active and passive samplers (e.g., suction cups and DGT, respectively) for investigation of the pools of available nutrients or hazardous metal ions for biota uptake are also pinpointed.

2. IN VIVO AND IN SITU SAMPLING: SIMILARITIES AND DIFFERENCES

As detailed in earlier chapters, microdialysis sampling involves the use of minimally invasive miniature dialyzers that can be implanted into a distinct

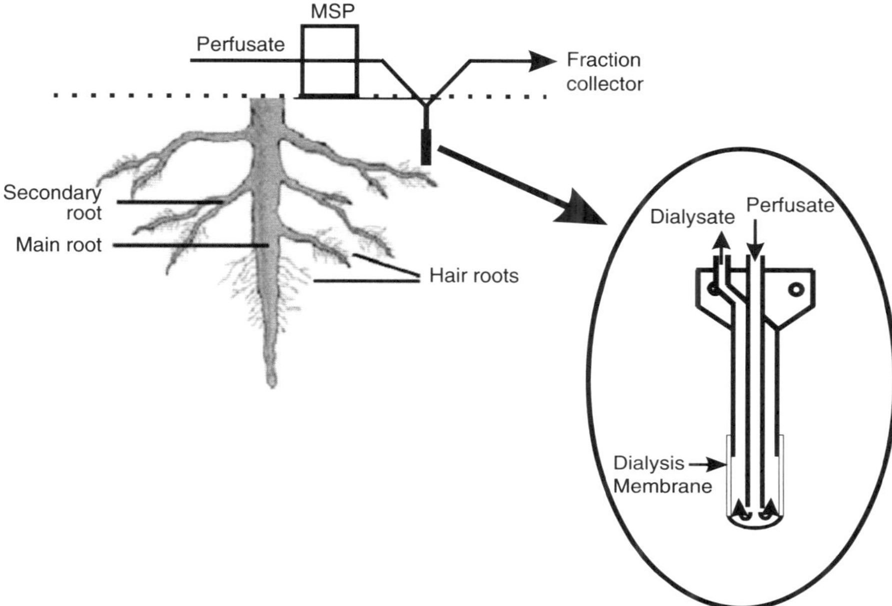

Figure 1 Implementation of microdialysis-based passive dosimeters at the soil–root interface for in situ monitoring of rhizosphere processes. The inset illustrates a magnified concentric-type microdialysis probe. MSP, microsyringe pump. (Inset from [18], with permission from Elsevier.)

tissue region to obtain an in vivo chemical snapshot of targeted low-molecular-weight species (e.g., neurotransmitters, neuropeptides, energy metabolites, and pharmaceuticals) over an integrated time period. Since no fluid is withdrawn or introduced into the tissue during microdialysis experiments, minimum changes in the fluid balance and metabolic processes result. In addition, only low-molecular-weight species in the extracellular fluid can cross the permselective membrane, thus eliminating any proteins from the sample matrix [16]. The immediate consequences of this sample cleanup inherent to microdialysis are the effective interruption of any further metabolism in the sample collected and the feasibility of injecting a preset volume of dialysate directly into the analytical instrument without any need for further sample pretreatment [17]. Analytical instruments applied to online determination of the target compound in the perfusate stream include chromatographic or electrophoretic separation systems, flow-through optical and electrochemical (bio)sensor systems, and less frequently, atomic spectrometric methods.

Considering in situ sampling, the ideal passive dosimeter should feature the following attributes [4]: It should be small enough to be implemented readily into the environmental compartment without significant intrusion, while offering an in situ snapshot of the concentration of target species or the ongoing chemical processes occurring at the implanted site over time. Very importantly, it must be insensitive to potential matrix ingredients, which could lead to interfering effects in the analytical detection step. In fact, the implantation of microdialyzers into aqueous or solid multiphase environmental compartments or plant tissues as passive dosimeters has fostered the simplification of environmental analytical assays [18] in terms of:

- Automation of the sampling step for low-molecular-weight species (e.g., metal ions, monosaccharides, carboxylic acids, nutrients, anions).
- Continuous monitoring in almost real time of natural events or anthropogenic occurrences by evaluation of trends in concentration of given analytes.
- Minimum disturbance of natural equilibria at the sampling site because the microdialyzer is only slightly invasive. The small recipient volume for microdialysate ensures minute uptake of target species from the medium probed, whereby the natural equilibria at the sampling site remain virtually unaltered.
- Sample cleanup with no requirement for further microdialysate processing because of low mass fluxes of high-molecular-weight interfering compounds (e.g., organic matter, including humic acids, humin compounds, and complexed species).
- Interruption of any further metabolism in the volume collected as microbial cells are excluded from microdialysates.
- In situ evaluation of pools of mobilizable forms of contaminants for biota uptake via the performance of retro-dialysis-based leaching tests in which the extracting reagent perfuses the implanted probe.
- Potential in field detection by online coupling of microdialyzers to downscaled flow through detectors (e.g., optical sensors).

As is the case in neurochemical and pharmacokinetic studies, a small probe bearing a permselective hydrophilic hollow-fiber membrane is utilized in environmental assays to attain a dialysate sample by perfusing the microdevice at μL/min flow rates. Thus, chemical species diffuse from the environmental compartment into the microdialysis probe and are delivered to the outlet for further chemical analysis [18,19]. However, while dialysis probes of minute size are mandatory in the biomedical field for in vivo probing of exogenous or endogenous species in a distinct tissue region, there is little limitation on probe size for in situ sampling of environmentally relevant species from natural waters, industrial effluents, or terrestrial environments. However, the inner volume of the microdialyzer should be kept far below that of the sample to avoid considerable mass depletion during sampling. The ideal microdialyzer for environmental sampling of micronutrients or trace-level contaminants would be that one working in a steady-state dialytic regime, where the analyte concentration in the dialysate (c_d) matches that of the outer microenvironment

(c_{ext}), to prevent dilution of the microdialysate and omitting calibration of the sampling step. Unfortunately, microdialysis units typically operate under dynamic sampling conditions, which in turn call for coupling the probing device to highly sensitive detection instruments and, most important, require the calibration of the device by calculation of the *relative recovery* (RR), also called the *extraction efficiency* or *extraction fraction*. When using an analyte-free perfusion solution, the relative recovery is defined as RR = c_d/c_{ext}. The diffusive mass-transfer model adopted for environmental monitoring [18,19] actually relies on the mathematical framework developed by Bungay et al. [20] for quantitative in vitro and in vivo tissue microdialysis. Accordingly, RR depends on the resistances of mass transfer through the membrane (R_m), the external media (R_e), and the dialysate (R_d) along with the perfusate flow rate (Q_p), detailed as follows:

$$ \text{RR} = 100 \times \left\{ 1 - \frac{1}{\exp\left[Q_p (R_m + R_e + R_d)^{-1} \right]} \right\} \tag{1} $$

where the term $1/R_m + R_e + R_d$ is called the *permeability factor* (P_f). It should be borne in mind that as opposed to in vivo microdialysis for pharmacokinetic investigations, where R_e frequently regulates RR due to the diffusive resistance exerted by the tissue region, the resistance to mass flux in environmental monitoring depends on the aggregate state of the sample. Hence, the RR of target species in freshwater systems or industrial streams is expected to be controlled largely by R_m, whereas R_e is the most significant contribution to P_f when probing interstitial waters in soil and sediments or plant tissues. As in clinical and pharmaceutical assays, the use of an appropriate calibration method would thus be a must in microdialysis-based environmental assays for accurate quantification of target species, as discussed below (see Section 5).

3. CRITICAL PARAMETERS INFLUENCING RELATIVE RECOVERIES

The mass transfer through a semipermeable thin membrane is influenced according to Fick's law by membrane properties, hydrodynamic variables in the two phases in contact with the membrane, and the diffusivity of the analytes of interest through the membrane and within both the outer and perfusate media. The RR in microdialysis-based environmental assays is thus affected by the material, the thickness, the inner tortuous structure, the aqueous-phase fraction and molecular weight cutoff of the membrane, the design and dimensions of the probe, the perfusate flow rate, and the conditions of both the perfusion liquid and the outer medium (e.g., chemical composition, physical barriers, viscosity, temperature), as detailed in the following sections.

3.1. Design and Dimensions of the Probe

Among the various configurations of microdialysis probes reported in the literature (i.e., concentric cannula, loop type, linear, side by side, and shunt type) [15], the concentric probe type has been the most widely utilized in environmental applications [19] (see the inset in Figure 1 for further details) as in both pharmacological and physiological studies. This preference denotes a characteristic advantage of using a well-designed robust configuration which is commercially available [21]. Yet, the main drawback of commercial probe systems intended for clinical and drug metabolism applications is the lack of flexibility for adaptation of the configuration to the needs of environmental assays. The costs and inherent rigidity of commercial microdialyzers limit the user's ability to change any part of the device without damaging it. In fact, microdialyzers with short effective membrane lengths (1 to 4 mm for tissue regions or ≤ 10 mm for subcutaneous implantation) glued to the tip of a hypodermic needle or flexible shaft yield low RRs that are suitable only for online dilution or cleanup of environmental samples. Recent trends to improve RRs have focused on the manipulation of the internal dimensions of custom-made arrangements. It has been demonstrated that the performance of concentric microdialyzers is enhanced by increasing the outer radius of the inner cannula [22,23], due to the minimization of R_d to mass transfer. Notwithstanding the fact that the effective dialysis membrane length might theoretically be increased at will to maximize RRs, maximum lengths of 10 mm have been reported so far in concentric microdialyzers [22,24]. This is probably due to the deterioration of the probe's performance for larger membrane dimensions.

To tackle this issue, flexible and cost-effective hollow fiber–based microdialyzers have been recently been designed in-house (see the Figure 2 inset for further details) and proved appropriate for sampling under quiescent conditions [25] and for in situ monitoring of the chemical variation of relevant parameters at the time scale of environmental events [26,27]. The most relevant asset of this configuration is the adaptation of the dynamic working range to the demands of the assays by the selection of suitable membrane surface areas. Further, the smaller volume fraction available for dialysates compared to common concentric-type designs assures low mass depletion and accurate measurements with minimal impact on the external microenvironment.

3.2. Membrane Characteristics

The properties of the dialysis membrane in environmental assays should be selected carefully in order to maximize RR by decreasing R_m. Common hydrophilic materials for fabrication of probes are polysulfone, polyether sulfone, polyamide, regenerated cellulose, and polycarbonate–polyether copolymer [19]. The choice of polymer should be done according to the surface reactivity,

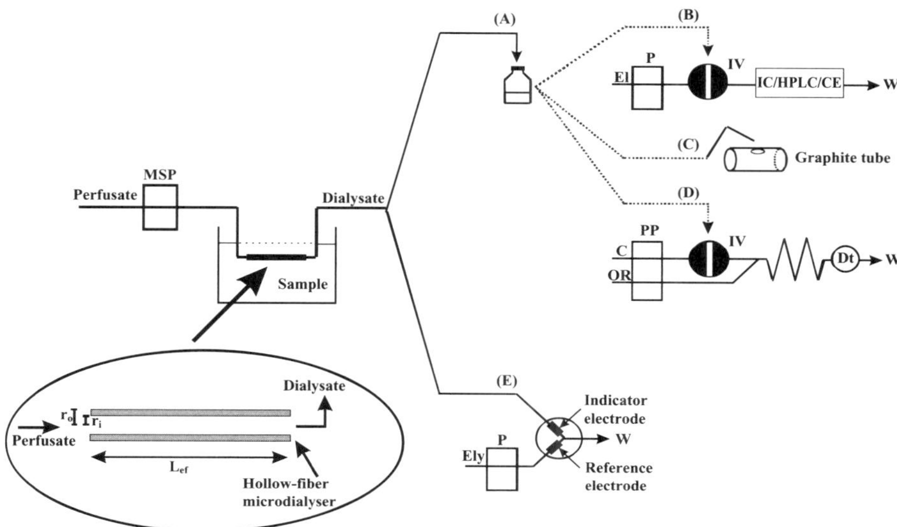

Figure 2 Analytical techniques utilized for analysis of dialysates. (A) Fraction collection and detection by: (B) column separation systems coupled to mass spectrometry or UV–visible spectroscopy; (C) electrothermal atomic absorption spectrometry; and (D) flow-based spectroscopy/fluorometry. (E) online coupling of microdialysis probes to miniaturized potentiometric systems or lab-on-chip microdevices. The inset shows a magnified view of a capillary-type microdialyzer. MSP, microsyringe pump; IV, injection valve; P, pump; CE, capillary electrophoresis; IC, ion chromatography; HPLC, high-performance liquid chromatography; El, eluent; C, carrier; OR, optional reagent; PP, peristaltic pump; Dt, detector; Ely, electrolyte; W, waste. (Adapted from [18], with permission from Elsevier.)

the polarity of the target species, and the physicochemical stability at the implantation site. Membrane–matrix interactions are undesirable for all microdialysis sampling applications because membrane fouling causes the deterioration of microdialyzer's performance in terms of RR and reusability. In environmental monitoring schemes, the eventual mineralization of the membrane by microbial activity should also be accounted for when selecting the chemical nature of membrane monomers.

Despite the fact that microdialysis is conventionally regarded as a molecular sievelike separation technique, analyte–membrane interactions have been reported to explain the dependence of the diffusive flux on membrane composition [28]. For example, in the determination of metal species, the higher the polarity of the membrane, the higher were the RRs observed, which was attributed to the electrostatic interactions between the analytes and nucleophilic groups at the barrier surface [24]. Yet the formation of strong ion-pair complexes between the analyte and ionic moieties of the membrane should be avoided to prevent decreasing mass transfer [29].

As to the thickness of the hollow fibers, membranes are usually $<50\,\mu m$ thick in either concentric or linear arrangements because the resulting P_f assures appropriate transfer rates under the common operational conditions of microdialysis experiments. To minimize R_m, the effective diffusion coefficient of the analyte through the membrane should be enhanced. To this end, membranes with a higher aqueous-phase fraction and minimal inner tortuous structure are preferred [19,30].

The molecular weight cutoff (MWCO) is another crucial parameter to be optimized whenever microdialysis is proposed for sample cleanup. From an environmental point of view, the isolation of low-molecular-weight compounds from particulate or colloidal matter, microbes, and high-molecular-weight organic interfering species such as humic acid and humin compounds is of particular interest. It is worth mentioning that for analytes even several orders of magnitude lower in molecular weight than the MWCO of the membrane, the higher the MWCO, the higher the RRs found [31]. This observation is attributed to the dynamic operation regime in microdialysis and the influence of the MWCO on the degree of solvation of the bulk membrane. By the use of membranes with low-to-moderate MWCO, microdialysis sampling might thus be exploited for speciation of trace metals by discrimination between free and complexed metal forms [32]. Further details on this subject are given in Section 6.1.

3.3. Perfusate Flow Rate

One of the main constraints for microdialysis sampling with concentric-type probes is the attainment of high RRs at ultraslow perfusing rates ($\leq 1.0\,\mu L/$ min). The resulting large response times for collection of an appropriate volume of microdialysate impose severe limitations for monitoring dynamic environmental processes such as root exudates, unless detection systems with low-volume requirements are employed (see Section 4). Better temporal resolution could be accomplished, however, by resorting to linear-type configurations. As reported in a recent paper [26], the major asset of the latter design is the wide interval of perfusate flow rates (1.0 to $60\,\mu L/min$ vs. 1.0 to $7.0\,\mu L/$ min for the commercial CMA-20 concentric probe [21]) which are applicable for prolonged operational periods. Besides, acceptable recoveries (RR ~ 20%) are even obtained for analytes (e.g., chloride) at flow rates above $20\,\mu L/min$. This configuration, in fact, solves in a simple manner the prevailing drawbacks for online interfacing of the perfusate stream (frequently delivered at 1 to $2\,\mu L/min$) with common analytical instrumentation.

When monitoring analytes (e.g., hazardous metals) which are expected to be encountered at low concentration levels, the application of stopped-flow approaches could be, in principle, applied for the improvement of RRs, as deduced from Bungay's formula [equation (1)]. However, dialysate enrichment is generally hindered as a result of back-diffusion of the entrapped analyte and dispersion of the minute dialysate plug during transportation toward the fraction collector or flow-through detector. Therefore, current efforts are directed to the modification of the chemical composition of the perfusate, as discussed in the following section.

3.4. Perfusate Composition

The chemical composition of the perfusing solution plays a decisive role in microdialysis based environmental assays involving the determination of charged species. The dependence of the permeation rate of a given ionic species on the electrolytes present in the sample medium causes different dialysate concentrations, and thus biased RRs, for the same concentration of analyte [33]. This is a consequence of differences in migration rates between the target ion and concomitant co-ions. In in vitro analysis, chemical effects due to the nature and concentration of matrix ingredients can be effectively suppressed by ionic strength adjustment of both samples and standard solutions. This strategy is, however, inappropriate for microdialyzers immersed in industrial effluents or aqueous/solid environmental sites because modification of the natural chemical conditions is, obviously, not feasible. An elegant solution to this drawback consists of using a reasonably high concentration of electrolyte as a perfusion solution for in situ sampling [26,32,34], in both clinical and pharmacological in vivo applications where Ringer's solution is utilized as a physiological fluid surrogate to prevent unbalanced osmotic pressure. The ionic strength within the diffusion layer of the sample solution is thus regulated with the ions permeated from the inner salt solution, which works as an effective electrolyte modifier of the outer microenvironment. For ionizable species, the use of a perfusing buffer solution for derivatization of the analyte into a noncharged compound would prevent unbalanced chemical potentials at both sides of the dialytic membrane and potential electrostatic interaction with membrane anchored groups. The use of ultrapure water as a perfusate carrier should thus be restricted to the sampling of noncharged organic compounds, since mass transfer is, in this case, unaffected by concomitant ions [29].

The nature of the constituents of the perfusion liquid is essential not merely for precluding matrix chemical interferences but also for attaining enhanced RRs and mass transfer via exploitation of exhaustive dialysis techniques. As in neurochemical applications of microdialysis where complexing reagents (e.g., cyclodextrins and ion exchangers) [35] and bioaffinity molecules, such as antibodies and antigens [36], have been added to the perfusion liquid to improve sensitivity, soluble chelating agents (e.g., 8-hydroxyquinoline [24], aminopolycarboxilic acids [37], humic acids [38,39] and small peptides [37,40] such as poly-L-aspartic acid and poly-L-histidine) have been exploited in analytical procedures for monitoring of trace-level concentrations of metal species in environmental matrices. For example, RR for Cu and Pb from aqueous samples increased 12- and 8-fold, respectively, when poly-L-aspartic acid, with a molecular weight of 35.7 kDa, was added to the perfusate [40]. Further, RRs of 100% for Cu and Ni are reported when using 0.05% (w/v) humic acid as a perfusate modifier [38]. Sorbent-assisted enhanced microdialysis based on the introduction of solid particles with affinity properties for the analytes into the perfusate channel can be regarded as a promising alternative [41] with potentials for environmental research.

The idea behind chemical derivatization in the perfusate is to yield a high-molecular-weight product that prevents the counter-current transfer of derivatized analyte. As a consequence, a steeper concentration gradient of analyte between sample and perfusate is ensured compared to conventional microdialysis sampling, with subsequent improvement in RRs.

4. DETECTION TECHNIQUES

Selection of the most appropriate analytical detection system for analysis of aqueous dialysates is not related solely to the chemical nature and expected concentration level of analytes in the environment, but, most importantly, to the minimum sample volume required by the instrument, which in turn determines the maximum temporal resolution for the environmental process under investigation. For the determination of trace concentrations of metal species in dialysates after in situ sampling, electrothermal atomic absorption spectrometry (ET-AAS) is a suitable choice, owing to its inherent high sensitivity and the minimum-volume requirements of the atomizer [24,37–40]. In fact, extremely short collection times are sufficient to accumulate the required volume of dialysate (typically, 5 to 10 µL) into the graphite tube prior to analysis [32]. Despite the fact that ET-AAS has conventionally been regarded as an element-specific detector, fast multielemental measurements are currently feasible by exploiting new instruments furnished with either hollow-cathode lamps or continuum sources [42,43]. For simultaneous determination of a large number of elements, inductively coupled plasma (ICP)-optical emission spectroscopy (ICP–OES) or ICP–mass spectrometry (ICP–MS) would be better options, yet a few hundred microliters of dialysates would be needed for reliable quantification of analytes.

Low-molecular-weight ionizable or noncharged organic compounds in dialysates are typically detected using column separation systems [44]. Capillary electrophoresis (CE) or micellar electrokinetic chromatography (MECK) in combination with UV–visible, laser-induced fluorescence or MS detection are attractive techniques for hyphenation to microdialysis sampling as a result of the minute sample volumes required, usually <30 nL, and the high resolution power of CE [45]. Unfortunately, the sensitivity of CE is far too low for monitoring of pollutants or environmentally relevant species, for which the use of reversed-phase high-performance liquid chromatography (HPLC) or ion chromatography (IC) coupled to electrochemical [46], UV–visible [27,29,34,47], or MS [48] detection has been boosted for sensitive quantification of analytes in dialysates. In this case, injection volumes typically range from 1.5 to 100 µL [27,29,34,46–48]. If dynamic processes are to be monitored, conventional HPLC could be replaced by downscaled systems, namely, micro-HPLC, where the volumes injected fall within the nanoliter level. Using miniaturized electrospray ionization mass spectrometers, no loss of sensitivity is expected, due to enhanced ionization efficiency [48].

The coupling of microdialysis sampling with several analytical instrumental devices is shown schematically in Figure 2. Detection of analytes in the dialysates is usually carried out off-line after fraction collection (see Figure 2A) using either column separation systems (Figure 2B) or atomic spectrometers (Figure 2C), as described above. It is also possible to exploit flow injection–based systems for appropriate analyte derivatization prior to detection, as illustrated in Figure 2D. For in situ real-time detection of analytes in the perfusate, miniaturized potentiometric detectors or electrochemical/optical sensors might be readily hyphenated to the microdialyzer (see Figure 2E). Even though current developments in micro-total analytical systems focus on coupling the microdialyzer with microfluidic devices [49], also termed *lab-on-a-chip*, for improving resolution in situ/in vivo, this interface has not yet reached a mature age with respect to real-life applications because of severe backpressure increase in downscaled channels under continuous-flow operation [50]. Also, its use in environmental assays remains a challenge, due to the lack of sensitivity and long-term reliability of the microchips.

5. CALIBRATION METHODS

As is the case in in vivo sampling, the accuracy of in situ microdialysis data is conditioned by the choice of a proper probe calibration method [51,52]. The in vivo method (called a *slow perfusion method*) [51,53] is the simplest way to assure reliable quantitative results without a need for probe calibration. It is based on the decrease in perfusion flow rate until the concentration of the analyte in the microdialysate is in equilibrium with the external concentration (i.e., RR ~ 100%). Hollow-fiber microdialyzers of appropriate lengths ($\geq$30 mm) have proved suitable to probe low-molecular-weight species such as chloride from aqueous samples under quiescent conditions at RR > 95% as a consequence of the decrease in R_m [26]. Unfortunately, RRs for many environmentally relevant analytes, even for perfusate flow rates below 1.0 µL/min, are typically << 100%, whereby P_f needs to be estimated. In the particular case of implantation of microdialyzers into surface waters or industrial effluents, in vitro calibration is deemed to be sound because the complexity of the sample matrix does not affect RR significantly [19]. Moreover, P_f as $1/(R_m + R_d)$ could be calculated using equation (1) from the linear plot between $\ln(1 - \text{RR})$ versus $1/Q$ using standard solutions in aqueous media.

On the contrary, in vitro calibration is not an adequate approach for the quantification of concentrations of analytes in terrestrial environments or plant tissues because the most critical variable governing RR is the resistance of the outer medium (R_e) to the diffusional transport. The in vivo retrodialysis method involving the addition of an internal standard (or analyte) to the perfusate and estimation of analyte RR from the delivery of the internal standard has been evaluated for probing interstitial soil solution [34]. Notwithstanding the fact that this in vivo method is a common practice when monitoring the extracellular space in medical and pharmacological applications, retrodialysis is not applicable to in-field measurements, inasmuch as isotropic diffusion exchange kinetics do not apply in solid matrices [34]. It should be borne in mind that the presence of solid particles in the outer medium increases the thickness of the diffusional layer for analyte recovery. Further, mass transfer in retrodialysis experiments occurs from a continuously flowing stream to a stagnant medium, whereby the concentration of analytes in the outer vicinity of the implanted microdialyzer is progressively increasing, with consequent stopping of the driving force for dialytic transport.

To calibrate a microdialyzer probe it is thus necessary to construct an exposure system in the laboratory that simulates environmental scenarios. The external resistance of environmental solids might be calculated as the direct RR of the analyte from inert soil particle surrogates (e.g., quartz sand moistened with a standard solution) [34]. As an alternative, R_e in solid substrates could theoretically be calculated if the effective diffusion coefficients of the analytes in the soil as well the volume fraction accessible for soluble species are known [54].

6. ENVIRONMENTAL APPLICATIONS OF MICRODIALYSIS

6.1. Sampling of Pollutants and Relevant Analytes in Natural Waters and Wastewaters

Direct implantation of microdialysis probes in fresh waters (e.g., river, lake, sea, stream) or industrial effluents for sample processing purposes is a simple task. Mechanical stress over the membrane surface due to particulate matter in the medium could be avoided by using protective sieves. The response time of microdialyzers is much shorter than that of conventional passive dosimeters, such as SPMD, SPME, and DGT devices, whereby fast fluctuations on the concentrations of pollutants or nutrients could be monitored continuously. It should be noted that SPMD-, SPME-, and DGT-based dosimeters are able to provide time-averaged concentrations only because of the need for removal of the implanted device prior to analyte quantification [4].

The first attempt to exploit microdialysis as an automatic in situ sampling and sample cleanup technique for probing harsh environmental matrices was reported by Torto et al. [46] in 2000. The authors assessed the capabilities of a custom-made concentric-type microdialyzer for sampling oligosaccharides from a septic tank of a brewery industry having a complexity equivalent to that of a domestic sewer. Yet no quantitative characterization of the industrial effluents was made. Further work was devoted to the use of microdialysis for fingerprinting wastewater effluents aimed at evaluating the content of nutrients (e.g., phosphate and nitrate), trace metals (e.g., Zn and Ni), and alkaline and alkaline earth elements [55]. Although microdialysis was employed in the above articles for monitoring concentration trends of target analytes, further articles demonstrated the potential of the technique for quantitative analysis of wastewaters. For example, Jen et al. [29] took advantage of the online coupling of microdialysis with HPLC for the determination of low-molecular-weight anthropogenic organic compounds (i.e., aromatic amines) in polymer industrial wastewater. The outstanding feature of the linear-type microdialyzer assembled was the capability of operating under RRs close to 100% at considerably high perfusing rates (ca. $10\,\mu L/min$). Unfortunately, the authors did not detail the effective membrane dialysis length, and thus the analytical performance of the flow system cannot be compared accurately with that of other reported configurations. The same research group has recently proposed a similar hollow-fiber arrangement interfaced to liquid chromatography for the determination of active components in plant extracts [27] and phthalate esters in plastic packages [47].

Dialysis sampling does not necessitate the implantation of a probe into the aqueous environmental compartment. Sample presentation to the separation membrane could actually be performed in a continuous-flow mode because of the large volumes available. Miniaturized sandwich-type units furnished with flat dialysis membranes rather than tubular hollow fibers have been constructed and implemented in fully automated flow-injection networks [31,56]. These flow-based dialyzers have been coupled to ion chromatography for the determination of soluble inorganic anions in olive-oil mill effluents containing a high organic load [57]. High recoveries were reported here for all of the anions assessed by resorting to a closed-loop design to allow for a halted receiver solution. It should be stressed, however, that the separation processes involved in processing the effluent are both dialysis and ultrafiltration, as a consequence of the morphological characteristics of the dialysis membrane selected ($0.2\,\mu m$ pore size, $100\,kDa$ MWCO), which differ from those of the homogeneous membranes ($<10\,nm$ pore size, $<30\,kDa$ MWCO) commonly used in dialysis and microdialysis assays. Readers are referred to the fundamental review [58] for further details concerning sandwich-type dialysis units, along with selected examples related to online monitoring of chemical parameters in effluent streams.

The potential of concentric-type microdialyzers for sampling and quantification of metal ions from beverage effluents was discussed by Torto and colleagues [24]. Particularly remarkable is the thorough survey of the influence of membrane variables (e.g., material and MWCO) as well as the nature of the perfusion liquid on the diffusive flux of metal ions. The evaluation of short-chain peptides (e.g., poly-L-aspartic acid and poly-L-histidine) [37,40] and humic acid [38,39] in the perfusate as selective and effective binding agents to enhance the RR of metal ions was also described. However, RRs > 90% for metal ions (e.g., Pb and Fe) are feasible in the absence of complexing reagents when using linear-type probes perfused at $2.0\,\mu L/min$ in lieu of concentric

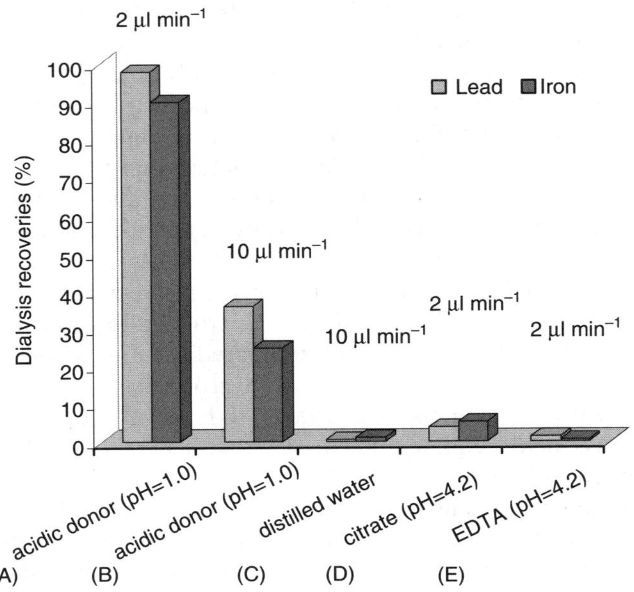

Figure 3 Investigation of the effect of perfusate flow rate and nature of complexing agents in the sample medium on relative recoveries of Fe and Pb for a 3-cm-long hollow-fiber-based microdialyzer. Dialysis recoveries are expressed as the ratio between the analyte dialysate concentration and that of the original sample. Metal ion concentrations: 25 ng/mL. (From [32], with permission from Springer-Verlag.)

probe designs [32], as shown in Figure 3A. This is attributed to the smaller dialysate volume per surface area unit of linear-type dialyzers compared to concentric probes. It is also particularly remarkable that RRs for both lead and iron with linear-type microdialyzers at considerably high flow rates (RR > 0.25 at 10 µL/min, see Figure 3B) are comparable to those reported with concentric arrangements at a dialysate flow rate as low as 1.0 µL/min [24].

As only free species (e.g., metal aquocomplexes) or low-molecular-weight chelates permeate the membrane, microdialysis could be exploited for fractionation analysis of metal ions [32]. According to the results presented in Figure 3C, hydrolyzable metal ions are not extracted appreciably (RR < 0.04%) under neutral conditions as a consequence of the formation of colloids and nondialyzable hydroxocomplexes. The influence of organic ligands with different metal binding strengths (e.g., EDTA and citrate) on the diffusive flux of metal ions is also shown in Figure 3D and E. The maximum RRs of 5.3 and 1.4% for citrate and EDTA, respectively, are in accordance with the differences in the metal stability constants of the two chelators selected and the lower diffusion coefficients of EDTA complexes [32]. These data demonstrated that microdialysis can efficiently discriminate between free and organically bound metal fractions.

6.2. Sampling of Target Analytes from Solid Substrates and Plant Tissues

The implantation of small microdialyzers into a solid or particulate matter–containing medium is not as straightforward as is discussed above for aqueous matrices, due to problems associated with mechanical strain over the semipermeable membrane, which remains a challenge to be faced by environmental and soil researchers. Soft plant tissues can be regarded as appropriate solid media for microdialysis sampling since no additional considerations on probe's stability have to be taken into account. Thus, in-house probes have been inserted directly into ripe tomato fruits grown from sewage sludge manure for assessment of metal uptake in plants [24]. Dialyzable concentrations were determined after sample harvest and pretreatment. From an environmental point of view, in vivo analyses based on the permanent attachment of the probe to living plant tissue are more attractive because concentration changes in analytes of interest may be monitored in near real-time during plant growth. Microdialysis sampling has also proven to be a suitable technique for fast screening of trace metal species (e.g., Cu and Ni) in leaf suspensions of metal hyper-accumulating plants [39].

New perspectives in microdialysis sampling have arisen from the exploitation of retrodialysis (stimulus–response) schemes in terrestrial environments [32,59]. The addition of a chemical agent (e.g., mild, acidic, chelating, reducing or oxidizing reagents from sequential extraction procedures) to the perfusion liquid will cause a change in the outer microenvironment of the probe that might be used for in-field prediction of trace metal pollution in soils or sediments as well as in situ metal availability for biota during natural events [59]. In fact, microdialysis probes could be used as biological surrogates to predict the actual (rather than potential) pool of pollutants available for plant uptake, due to similarities between the dialytic dosimeter and the capillary function of biological systems (e.g., plant roots).

Microdialysis sampling should also be regarded as a unique approach for elucidating metal–soil bonding mechanisms [32]. It has actually been feasible to differentiate the two mechanisms which contribute to the remobilization and dissolution of a given metal ion from soils in the presence of ethylenediamintetraacetic acid (EDTA) as a complexing agent [60,61]: (1) competition between the external reagent and the chelating sites of the soil phase for the target metal; and (2) dissolution of amorphous iron oxohydroxides by complexation of Fe with EDTA, with the subsequent mobilization of occluded trace metals. This is a consequence of the low RR values obtained for high-molecular-weight metal–EDTA chelates resulting from mechanism (1), whereby the free metal ion concentrations contribute primarily to the total metal content of the dialysate.

As a result of their miniature dimensions, dedicated microdialyzers are currently being used for in situ investigation of the influence of rhizosphere processes in the hyperaccumulation of trace elements by certain plants. It has been hypothesized that hyper-accumulator plants exudate large amounts of low-molecular-weight organic anions, which are accountable for the mobilization of metals from mineralogical phases [62]. Unfortunately, as stressed by Jones et al. [63], the spatial heterogeneity and actual role of carboxylate ions in the rhizosphere environment remain unknown so far because of the lack of chemical probes for organic acids. For this reason, in a recent communication of the authors' group, the potential of implantable linear-type microdialyzers for probing low-molecular-weight organic anions in rhizosphere soil solution has been addressed [34]. The beauty of the implanted microdialyzers is the inherent capability to follow the release of organic acids from hot spots such as fungal hyphae or the tip of a root hair, owing to the excellent spatial resolution (ca. 200 μm) of the assembled probes. Therefore, knowledge gaps concerning the fate of low-molecular-weight organic anions in soil samples can be replenished. Microdialysis probes, as opposed to suction cups, are actually the ideal tools for probing the soil solution since no liquid is withdrawn from the sampling site [54].

Further research in this field is being devoted to the accommodation of a series of dedicated microdialyzers at different depths and root-parallel layers in the rhizosphere for in vivo measurement of exudate release over short time periods according to the time scale of biological and environmental events, and for spatial resolution of root regions, whereby the exudation rates can be quantified accurately along the root length.

Table 1 summarizes the analytical performance of relevant methods involving microdialysis sampling for monitoring environmentally relevant analytes. The main variables affecting the dialysis yields (i.e., probe design, membrane composition and morphology, effective dialysis length, and internal hydrodynamics) are also given.

7. CONCLUSIONS AND FUTURE TRENDS

In this chapter we have shown that microdialysis, traditionally regarded as an in vivo sample processing technique, is equally applicable to in situ, real-time monitoring of environmental parameters in harsh aqueous and solid matrices. New perspectives in microdialysis sampling arise from the exploitation of stimulus–response schemes for in-field fractionation assays via introduction of a chemical extractant in the perfusate to assess the pools of mobile fractions of metals or other pollutants for plant uptake. Retrodialysis can also be regarded as a promising screening test for fast evaluation of the affinity (adsorption patterns) of soils, sediments, or sludges for uptake of trace metals. The introduction of anthropogenic metal ions into solid substrates by countercurrent diffusion from the perfusion liquid (using a metal-plug injection mode) provides a straightforward means to foresee their potential capabilities as remediation materials for metal-contaminated waste. Besides, as a consequence of the ability to be deployed at different depths with minor alteration of the natural site, microdialyzers are appropriate passive dosimeters to follow

TABLE 1 Analytical Performance of Microdialysis-Based Assays for Determination of Environmentally Relevant Analyte Parameters

Analytes	Detection Principle	Dialyzer Design	Effective Dialysis Length (mm)	Membrane Composition	Membrane Thickness (μm)	MWCO (kDa)	Perfusion Flow Rate (μL/min)	Sample Matrix	Ref.
Cd, Cr, Cu, Ni, and Pb	ETAAS	Concentric probe	10	Polysulfone and polyether sulfone	N.R.	3–30	1–9	Industrial effluent and plant tissues	[24]
Chloride	Dif-Pot	Linear type	30	Regenerated cellulose	8	5	1–60	Organic soils	[26]
Ginkgolide A,B and bilobalide	HPLC–UV	Linear type	NR	Regenerated cellulose	8	13	2–12	Plant extracts	[27]
Aniline and 2-chloroaniline	HPLC–UV	Linear type	NR	Cellulose acetate	45	5	10	Industrial (polymer) wastewater	[29]
Pb	ETAAS	Linear type	30	Regenerated cellulose	8	5	10	Sludge-amended soil	[32]
Oxalic acid and citric acid	HPLC–UV	Linear type	30	Regenerated cellulose	8	5	2	Soils	[34]
Cr, Ni, Cu, and Pb	ETAAS	Concentric probe	10	Polysulfone	N.R.	30	2–4	Natural and wastewaters	[37]
Cu and Ni	ETAAS	Concentric probe	10	Polysulfone	N.R.	3–100	1–8	Synthetic solutions	[38]
Cu and Ni	ETAAS	Concentric probe	10	Polysulfone	N.R.	10	3–4	Plant suspensions	[39]
Cu and Pb	ETAAS	Concentric probe	10	Polysulfone	N.R.	30	2	Wastewater	[40]
Saccharides (glucose, sucrose, fructose)	IC–ECD	Concentric probe	10	Polysulfone	N.R.	30 and 100	1.0	Industrial (brewery) wastewaters	[46]
Phthalate esters	HPLC–UV	Linear type	20	Polysulfone	160	10	0.1	Disposable plastic materials	[47]

IC, ion chromatography; HPLC, high-performance liquid chromatography; ECD, electrochemical detector; UV, ultraviolet; ETAAS, electrothermal atomic absorption spectrometry; Dif-Pot, differential potentiometry; N.R., not reported.

the mobility of compounds through soil horizons at the time scale of environmental events and to fingerprint plant exudates at the rhizosphere environment. In addition, the serial or parallel arrangement of several probes at a definite site might afford additional in situ information on the fate of pollutants after induction of a chemical variation in the outer boundary of one of the implanted probes.

Implantable microdialyzers are, however, not free of drawbacks for long-term assessment of pollutant mobility and availability. In this context we would draw the readers' attention to (1) the difficulty for accurate definition of the sample volume probed, (2) the need for membrane materials unalterable under microbial activity, (3) the lack of harmonization of in-field calibration methods and quality control tools for accurate determination of target species, and (4) the requirement for enrichment procedures whenever the concentration of contaminants in the perfusate is below the detection limits of the analytical instruments. In the years to come, it is expected that research interest in the environmental field will focus on the design of rugged lab-on-chip devices with integrated microdialysis probes and in situ optical or electrochemical sensing. Not less important, there is also the quest for appropriate on line preconcentration procedures for microdialysate enrichment in environmental assays.

REFERENCES

[1] Sliwka-Kaszynska, M., Kot-Wasik, A., Namieśnik, J. (2003). Preservation and storage of water samples. *Critical Reviews in Environmental Science and Technology*, *33*, 31–44.

[2] Huang, J.-H., Ilgen, G. (2004). Blank values, adsorption, pre-concentration, and sample preservation for arsenic speciation of environmental water samples. *Analytica Chimica Acta*, *512*, 1–10.

[3] Byron, K. (2003). Sampling and sample preservation for trace element analysis. *Comprehensive Analytical Chemistry*, Barceló, D. (Ed.), *41*, 1–21.

[4] Kot-Wasik, A., Zabiegała, B., Urbanowicz, M., Dominiak, E., Wasik, A., Namieśnik, J. (2007). Advances in passive sampling in environmental studies. *Analytica Chimica Acta*, *602*, 141–163.

[5] Vrana, B., Allan, I.J., Greenwood, R., Mills, G.A., Dominiak, E., Svensson, K., Knutsson, J., Morrison, G. (2005). Passive sampling techniques for monitoring pollutants in water. *Trends in Analytical Chemistry*, *24*, 845–868.

[6] Seethapathy, S., Górecki, T., Li, X.-J. (2008). Passive sampling in environmental analysis. *Journal of Chromatography A*, *1184*, 234–253.

[7] Huckins, J.N., Petty, J.D., Booij, K. (Eds.) (2006). *Monitors of Organic Chemicals in the Environment: Semipermeable Membrane Devices*, Springer Science+Business Media, New York.

[8] Esteve-Turrillas, F.A., Yusà, V., Pastor, A., de la Guardia, M. (2008). New perspectives in the use of semipermeable membrane devices as passive samplers. *Talanta*, *74*, 443–457.

[9] Bopp, S., Weiß, H., Schirmer, K. (2005). Time-integrated monitoring of polycyclic aromatic hydrocarbons (PAHs) in groundwater using the ceramic dosimeter passive sampling device. *Journal of Chromatography A, 1072,* 137–147.

[10] Ouyang, G.-F., Pawliszyn, J. (2006). Recent developments in SPME for on-site analysis and monitoring. *Trends in Analytical Chemistry, 25,* 692–703.

[11] Davison, W., Fones, G., Harper, M., Teasdale, P., Zhang, H. (2000). Dialysis, DET and DGT: In situ diffusive techniques for studying water, sediments and soils. In: Buffle, J., Horvai, G. (Eds.), *In Situ Monitoring of Aquatic Systems: Chemical Analysis and Speciation,* Wiley, New York, pp. 495–569.

[12] Garmo, Ø.A., Røyset, O., Steinnes, E., Flaten, T.P. (2003). Performance study of diffusive gradients in thin films for 55 elements. *Analytical Chemistry, 75,* 3573–3580.

[13] Zhang, H., Davison, W. (2000). Direct in situ measurements of labile inorganic and organically bound metal species in synthetic solutions and natural waters using diffusive gradients in thin films. *Analytical Chemistry, 72,* 4447–4457.

[14] Bourne, J.A. (2003). Intracerebral microdialysis: 30 years as a tool for the neuroscientist. *Clinical and Experimental Pharmacology and Physiology, 30,* 16–24.

[15] Weiss, D.J., Lunte, C.E., Lunte, S.M. (2000). In vivo microdialysis as a tool for monitoring pharmacokinetics. *Trends in Analytical Chemistry, 19,* 606–616.

[16] Stenken, J.A. (2006). Microdialysis sampling, In: Webster, J.G.(Ed.), *Encyclopedia of Medical Devices and Instrumentation,* 2nd ed., Vol. 4, Wiley, Hoboken, NJ, pp. 400–420.

[17] Davies, M.I., Cooper, J.D., Desmond, S.S., Lunte, C.E., Lunte, S.M. (2000). Analytical considerations for microdialysis sampling. *Advanced Drug Delivery Reviews, 45,* 169–188.

[18] Miró, M., Frenzel, W. (2005). The potential of microdialysis as an automatic sample processing technique for environmental research. *Trends in Analytical Chemistry, 24,* 324–333.

[19] Torto, N., Mwatseteza, J., Laurell, T. (2001). Microdialysis sampling challenges and new frontiers. *LC-GC Europe, 14,* 536–546.

[20] Bungay, P.M., Morrison, P.F., Dedrick, R.L. (1990). Steady-state theory for quantitative microdialysis of solutes and water in vivo and in vitro. *Life Sciences, 46,* 105–119.

[21] CMA/Microdialysis, Stockholm, Sweden. http://www.microdialysis.se.

[22] Torto, N., Mikeladze, E., Gorton, L., Csöregi, E., Laurell, T. (1999). Maximizing microdialysis sampling by optimising the internal probe geometry. *Analytical Communications, 36,* 171–174.

[23] Wisniewski, N., Torto, N. (2002). Optimization of microdialysis sampling recovery by varying inner cannula geometry. *Analyst, 127,* 1129–1134.

[24] Torto, N., Mwatseteza, J., Sawula, G. (2002). A study of microdialysis sampling of metal ions. *Analytica Chimica Acta, 456,* 253–261.

[25] Miró-Lladó, M., Frenzel, W. (2008). Microdialysis probe for automatic sampling and continuous monitoring of analytical parameters in solid samples. Patent ES 2272135, Application ES 200402825.

[26] Miró, M., Frenzel, W. (2004). Implantable flow-through capillary-type microdialyzers for continuous in-situ monitoring of environmentally relevant parameters. *Analytical Chemistry, 76,* 5974–5981.

[27] Chiu, H.-L., Lin, H.-Y., Yang, T.C.-C. (2004). Determination of ginkgolide A, B, and bilobalide in *Biloba* L. extracts by microdialysis–HPLC. *Analytical and Bioanalytical Chemistry*, *379*, 445–448.

[28] Torto, N., Ohlrogge, M., Gorton, L., Van Alstine, J.M., Laurell, T., Marko-Varga, G. (2004). In situ poly(ethylene imine) coating of hollow fiber membranes used for microdialysis sampling. *Pure and Applied Chemistry*, *76*, 879–888.

[29] Jen, J.-F., Chang, C.-T., Yang, T.C. (2001). On-line microdialysis–high-performance liquid chromatographic determination of aniline and 2-chloroaniline in polymer industrial wastewater. *Journal of Chromatography A*, *930*, 119–125.

[30] Torto, N., Mogopodi, D. (2004). Opportunities in microdialysis sampling of metal ions. *Trends in Analytical Chemistry*, *23*, 109–115.

[31] Miró, M., Frenzel, W. (2003). A novel flow-through microdialysis separation unit with integrated differential potentiometric detection for the determination of chloride in soil samples. *Analyst*, *128*, 1291–1297.

[32] Miró, M., Jimoh, M., Frenzel, W. (2005). A novel dynamic approach for automatic microsampling and continuous monitoring of metal ion release from soils exploiting a dedicated flow-through microdialyser. *Analytical and Bioanalytical Chemistry*, *382*, 396–404.

[33] Miró, M., Frenzel, W. (2004). Investigation of chemical effects on the performance of flow-through dialysis applied to the determination of ionic species. *Analytica Chimica Acta*, *512*, 311–317.

[34] Sulyok, M., Miró, M., Stingeder, G., Koellensperger, G. (2005). The potential of flow-through microdialysis for probing low-molecular weight organic anions in rhizosphere soil solution. *Analytica Chimica Acta*, *546*, 1–10.

[35] Ao, X.-P., Stenken, J.A. (2002). Water-soluble cyclodextrin polymers for enhanced relative recovery of hydrophobic analytes during microdialysis sampling. *Analyst*, *128*, 1143–1149.

[36] Fletcher, H.J., Stenken, J.A. (2008). An in vitro comparison of microdialysis relative recovery of Met- and Leu-enkephalin using cyclodextrins and antibodies as affinity agents. *Analytica Chimica Acta*, *620*, 170–175.

[37] Mogopodi, D., Torto, N. (2005). Maximizing metal ions flux across a microdialysis membrane by incorporating poly-L-aspartic acid, poly-L-histidine, 8-hydroxyquinoline and ethylenediaminetetraacetic acid in the perfusion liquid. *Analytica Chimica Acta*, *534*, 239–246.

[38] Mosetlha, K., Torto, N., Wibetoe, G. (2006). Enhancing the microdialysis recovery for sampling of Cu and Ni by incorporating humic acid in the perfusion liquid. *Analytica Chimica Acta*, *562*, 158–163.

[39] Mosetlha, K., Torto, N., Wibetoe, G. (2007). Determination of Cu and Ni in plants by microdialysis sampling: comparison of dialyzable metal fractions with total metal content. *Talanta*, *71*, 766–770.

[40] Mogopodi, D., Torto, N. (2003). Enhancing microdialysis recovery of metal ions by incorporating poly-L-aspartic acid and poly-L-histidine in the perfusion liquid. *Analytica Chimica Acta*, *482*, 91–97.

[41] Pettersson, A., Amirkhani, A., Arvidsson, B., Markides, K., Bergquist, J. (2004). A feasibility study of solid supported enhanced microdialysis. *Analytical Chemistry*, *76*, 1678–1682.

[42] Tseng, W.-C., Chen, P.-H., Tsay, T.-S., Chen, B.-H., Huang, Y.-L. (2006). Continuous multi-element (Cu, Mn, Ni, Se) monitoring in saline and cell suspension using on-line microdialysis coupled with simultaneous electrothermal atomic absorption spectrometry. *Analytica Chimica Acta, 576,* 2–8.

[43] Welz, B. (2005). High-resolution continuum source AAS: the better way to perform atomic absorption spectrometry. *Analytical and Bioanalytical Chemistry, 381,* 69–71.

[44] Davies, M.I., Lunte, C.E. (1997). Microdialysis sampling coupled online to microseparation techniques. *Chemical Society Reviews, 26,* 215–222.

[45] Ruiz-Jiménez, J., Luque de Castro, M.D. (2006). Coupling microdialysis to capillary electrophoresis. *Trends in Analytical Chemistry, 25,* 563–571.

[46] Torto, N., Lobelo, B., Gorton, L. (2000). Determination of saccharides in wastewater from the beverage industry by microdialysis sampling, microbore high performance anion exchange chromatography and integrated pulsed electrochemical detection. *Analyst, 125,* 1379–1381.

[47] Jen, J.-F., Liu, T.-C. (2006). Determination of phthalate esters from food-contacted materials by on-line microdialysis and liquid chromatography. *Journal of Chromatography A, 1130,* 28–33.

[48] Lanckmans, K., Sarre, S., Smolders, I., Michotte, Y. (2008). Quantitative liquid chromatography/mass spectrometry for the analysis of microdialysates. *Talanta, 74,* 458–469.

[49] Bergveld, P. (2000). Bedside clinical chemistry: from catheter tip sensor chips towards micro total analysis systems. *Biomedical Microdevices, 2,* 185–195.

[50] Chen, C.-F., Drew, K.L. (2008). Droplet-based microdialysis: concept, theory, and design considerations. *Journal of Chromatography A, 1209,* 29–36.

[51] Stenken, J.A. (2000). Methods and issues in microdialysis calibration. *Analytica Chimica Acta, 379,* 337–358.

[52] Peters, J.L., Yang, H., Michael, A.C. (2000). Quantitative aspects of brain microdialysis. *Analytica Chimica Acta, 412,* 1–12.

[53] Wages, S.A., Church, W.H., Justice, J.B., Jr., (1986). Sampling considerations for on-line microbore liquid chromatography of brain dialysate. *Analytical Chemistry, 58,* 1649–1656.

[54] Miró, M., Fitz, W.J., Swoboda, S., Wenzel, W.W. (2010). In-situ sampling of soil pore water: Evaluation of linear-type microdialysis probes and suction cups at varied moisture contents. *Environmental Chemistry, 7,* 123–131.

[55] Torto, N., Peloewetse, E., Lobelo, B., Mwatseteza, J. (2006). Profiling of wastewater effluent from a beverage industry by microdialysis sampling and a combination of analytical techniques. *Environmental Science, 1,* 1–7.

[56] Luque de Castro, M.D., Priego-Capote, F., Sánchez-Ávila, N. (2008). Is dialysis alive as a membrane-based separation technique? *Trends in Analytical Chemistry, 27,* 315–326.

[57] Buldini, P.L., Mevoli, A., Quirini, A. (2000). On-line microdialysis-ion chromatographic determination of inorganic anions in olive-oil mill wastewater. *Journal of Chromatography A, 882,* 321–328.

[58] van Staden, J.F. (1995). Membrane separation in flow injection systems: 1. Dialysis. *Fresenius' Journal of Analytical Chemistry, 352,* 271–302.

[59] Miró, M., Hansen, E.H. (2006). Recent advances and perspectives in analytical methodologies for monitoring the bioavailability of trace metals in environmental solid substrates, *Microchimica, Acta, 154*, 3–13.

[60] Sahuquillo, A., Rigol, A., Rauret, G. (2003). Overview of the use of leaching/extraction tests for risk assessment of trace metals in contaminated soils and sediments. *Trends in Analytical Chemistry, 22*, 152–159.

[61] Nowack, B., Kari, F.G., Krüger, H.G. (2001). The remobilization of metals from iron oxides and sediments by metal-EDTA complexes. *Water, Air, and Soil Pollution, 125*, 243–257.

[62] Wenzel, W.W., Bunkowski, M., Puschenreiter, M., Horak, O. (2003). Rhizosphere characteristics of indigenously growing nickel hyperaccumulator and excluder plants on serpentine soil. *Environmental Pollution, 123*, 131–138.

[63] Jones, D.L., Dennis, P.G., Owen, A.G., van Hees, P.A.W. (2003). Organic acid behaviour in soils: misconceptions and knowledge gaps. *Plant and Soil, 248*, 31–41.

INDEX